GROUNDS FOR GROUNDING

GROUNDS
FOR GROUNDING
A Circuit-to-System
Handbook

Elya B. Joffe
Kai-Sang Lock

IEEE Electromagnetic Compatibility Society, *Sponsor*

IEEE Press

A JOHN WILEY & SONS, INC., PUBLICATION

Published by John Wiley & Sons, Inc., Hoboken, New Jersey
Published simultaneously in Canada.

For general information on our other products and services please contact our Customer Care Department within the United States at (800) 762-2974, outside the United States at (317) 572-3993 or fax (317) 572-4002.

Wiley also publishes its books in a variety of electronic formats. Some content that appears in print, however, may not be available in electronic formats. For more information about Wiley products, visit our web site at www.wiley.com.

Library of Congress Cataloging-in-Publication Data is available.

ISBN 978-0471-66008-8

Printed in the United States of America.

10 9 8

To my beloved wife, Anat,
and the apple of my eye, my daughter Tami-Lee,
who are the center of my universe
—Elya B. Joffe

To my wife, Eyan,
and Angela, Andrew, and Anthony
for their love, support, and understanding
—Kai-Sang Lock

To the Beloved Memory of my Mother, Faiga Mary Joffe (Nee Bloom)

Faiga Mary Joffe (Nee Bloom)
1934–1984

This book is dedicated to the beloved memory of the first lady in my life—my mother, Faiga Mary Joffe (Nee Bloom), who was taken to the stay with the Lord in 1984, when she was only 50 years old. Her everlasting encouragement and attitude were the source of inspiration and motivation that gave me the strength to start (and complete) the writing of this book. Among her notes, we found after her passing, the following poem:

But once I pass this way.
And then . . . and then the Silent Door swings on its hinges;
Opens . . . Closes . . .
and no more I pass this way.
So, while I may, with all my might
I will essay sweet comfort and delight
to all I meet upon the pilgrim way.
For no man travels twice the Great Highway,
that winds through darkness up to light,
through night, to day.

—William Arthur Dunkerley (John Oxenham) 1852–1941

This was the story of her life. Thank you, Mom, for being there when I needed you so badly. I am so saddened that you do not see the results of your teachings.

Your Loving Son,
Elya B. Joffe

CONTENTS

FOREWORD

Israel procured several groups of F-16 Aircraft from my company, General Dynamics (now Lockheed Martin). Electromagnetic compatibility (EMC) engineering for these programs was accomplished by my group of thirty EMC engineers in Fort Worth, Texas. As the models of Israeli F-16 aircraft became more complex, direct interaction with Israeli engineers became appropriate. The first such meeting occurred in 1983 in Tel Aviv, and my attendance triggered the Israeli Air Force to find a suitable engineer to represent them.

They found Israeli Air Force Lieutenant Elya B. Joffe, who had a BSEE Degree from Ben Gurion University, and majored in wave propagation and electromagnetic theory. They told him he was now an EMC engineer, and sent him to my meeting! (Engineers usually wind up in EMC by accident, and Elya was no exception!)

Practical applications in engineering do not always follow the theory. In Israel, engineers were very knowledgeable of electromagnetic theory, but in Israel in 1983, the practice of the technology was limited. This limitation did not stop Elya. He readily grasped the differences between his technical background and the practical aspects of EMC, and progressed well. After 1983, he and I had many technical meetings both in Israel and in Fort Worth, and became good friends. One time, he and his wife, Anat, stayed at my home in Fort Worth while my wife and I were in San Francisco!

After completing his military service as a Major, he joined Israeli Aircraft Industries as an EMC engineer. He became very active in the Israeli Chapter of the IEEE EMC Society and was elected its chairman. Elya was nominated and elected to the EMC Society Board of Directors. In 2007, he was elected President-Elect of the IEEE EMC Society and served as president in the years 2008–2009; the first and only non-American ever to hold this office to-date!

Years ago, Elya recognized the need for a treatise on electrical grounding, the foundation for achieving system electromagnetic compatibility. He has spent many years researching the subject and preparing his manuscript. Now, he has provided the world with this valuable and long-needed reference.

Thank you, Elya!

JACK L. MOE

Fort Worth, Texas

FOREWORD

The topic of Grounding in EMC is one of the most controversial in this engineering discipline. A solution implemented in one system may not work in another, and "rules of thumb" and "good engineering practices" serve well mostly in presentations.

Elya B. Joffe and Dr. Kai Sang Lock have undertaken a monumental task and complied into this book, *Grounds for Grounding*, most of the data and information required by the designer. The treatment covers the electromagnetic basics, explanation of the reasoning behind the grounding solutions and up to their practical implementation. The book is perhaps the most comprehensive publication to date on the subject of grounding. It deals with every aspect of grounding, from component to system to facility. Because of its vast coverage and detailed discourse, it may take some time and effort to read through. However, each chapter can be approached for specific grounding solutions.

I applaud Elya B. Joffe, President of the IEEE EMC Society plus a full-time EMC Engineer, and Dr. Kai Sang Lock, the President of PQR Technologies Pte Ltd. in Singapore, for addressing the topic and spending the time and effort, despite their duties, to get this well written, important book, completed.

This book will not replace experience and experiment, but it will shorten the path to a successful design.

OREN HARTAL

Qiryat Tivon, Israel

PREFACE

The understanding of grounding as described in this book could not have been made possible without the achievements of Michael Faraday and the great philosopher who followed him, James Clerk Maxwell. The following excerpt was written by Maxwell after Faraday's death in 1867. It is from the introduction by T. F. Torrance to Maxwell's *A Dynamical Theory of the Electromagnetic Field* (Wipf and Stock Publishers, Eugene, OR, 1996).

> The high place which we assign to Faraday in electromagnetic science may appear to some inconsistent with the fact that electromagnetic science is an exact science, and that in some of its branches it had already assumed a mathematical form before the time of Faraday, whereas Faraday was not a professed mathematician, and in his writings we find none of these integrations of differential equations which are supposed to be the very essence of exact science. Open Poisson and Ampere, who went before him, and you will find their pages full of symbols, not one of which Faraday would have understood. It is admitted that Faraday made some great discoveries, but if we put these aside, how can we rank his scientific method so high without disparaging the mathematics of these eminent men?
>
> It is true that no man can essentially cultivate any exact science without understanding the mathematics of that science. But we are not to suppose that the calculations and equations which mathematicians find so useful constitute the whole of mathematics. The calculus is but a part of mathematics.
>
> The geometry of position is an example of mathematical science established without the aid of a single calculation. Now Faraday's lines of force occupy the same position in electromagnetic science that pencils of lines do in the geometry of position. They furnish a method of building up an exact mental image of the thing we are reasoning about. The way in which Faraday made use of his idea of lines of force in coordinating the phenomena of the magnetoelectric induction shows him to have been in reality a mathematician of a very high order—one from whom the mathematicians of the future may derive valuable and fertile methods. . . .
>
> [W]e are probably ignorant even of the name of science which will be developed out of the materials we are now collecting, when the great philosopher after Faraday makes his appearance.

ACKNOWLEDGMENTS

This book would have never come to be a reality had it not been for two special persons in my professional career. Without their guidance and mentorship, I would have never been able to attain the knowledge and experience in my profession. Both are not only good friends of mine but also of each other.

First, John (Jack) Moe, who initially introduced me to this magical discipline of EMC and with limitless patience, through "on-job training," revealed to me the very basics of EMC. The very first book I owned on EMC was presented to me by Jack. I have retained and value his hand-written sketches and notes explaining cable shield grounding.

Second, I would like to recognize Oren Hartal, under whose guidance (similar to most EMC engineers in Israel) I discovered the hidden mysteries of the science and art of EMC. The very first course on EMC I had taken was taught by Oren, and through his guidance I truly gained a deep understanding of this discipline. Many years of work with and encouragement by Oren in IEEE and URSI have, additionally, brought me to the international recognition I benefit from today.

I am especially grateful to my dear friend Mark M. Montrose, who encouraged me to undertake the tremendous challenge of writing this book. Through his support during the writing of the book, Mark redefined the term "friend." Mark's advice helped shape the outline of the book, and made significant contributions in many ways to the final manuscript. His technical review, professional scrutiny, and criticism of the material, made with a fine-tooth comb, helped ensure that concepts were clear, correct, and easy to follow, and that the language and style were appropriate. Without his guidance and support, the technical quality and clarity of this book would not have been as they are. I thank Mark wholeheartedly for his friendship, collegiality, and support.

A particular acknowledgment is given to Kai Sang Lock, the coauthor of this book. Kai Sang was the perfect choice for the task, with his expertise on high-power electrical systems. Without his contribution, a major aspect of grounding could not have been included in this book.

The authors are indebted and wish to gratefully acknowledge the contributions made by many individuals who helped to develop the material for this book, provided material, spent time, and exerted efforts in order to make this book of the high quality I believe it is. Acknowledgments of their contributions are included in the context of this book. Special gratitude is expressed to Bruce Archambeault (IBM), Douglas C. Smith, Keith Armstrong (Cherry Clough Consultants), Todd Hubing (Clemson University), Alexander Perez (Agilent), Edoardo Genovese, and David Johns (CST), as well as to Glenair, Inc. and MAJR Products, Inc., for their outstanding support and contribution of material used throughout this book.

Gratitude is extended to the editorial and technical reviewers of this book for their dedicated toil, and for spending long and frustrating hours of their personal time in tedious study and literal dissection the book. Their professionalism and experience have greatly enhanced the quality and value of this book.

My appreciation is also expressed to the many engineers and students whom I have educated and trained in EMC throughout the years, for their challenging and thought-provoking questions.

> I have learnt much from my teachers, but from my students I have learnt even more.
> —Shimon Ben Zoma, a second century Jewish Scholar, based on *Psalms,* 119:99

Above all, in a personal note, my most special gratitude is due to my wonderful, supporting, and understanding family: my mother, Faiga, who gave me the inspiration and taught me the power of words; my wife, Anat, who helped me appreciate the power of persistence and endured my endless frustrations and foibles without (or with little) protest while this book was being created. Most of all, my gratitude goes to my daughter, Tami Lee, whose beaming smile at the end of each day gave me the strength to go on with this effort.

Anat and Tami Lee must have often felt they had taken the back seat to this project and put up with the long hours during the years I spent at the computer keyboard. Combined with my extensive global traveling commitments, particularly as president of the EMC Society of the IEEE (2008–2009), this effort has no doubt taken its toll on their own lives. I shall forever treasure their sympathy to my passion and their faith in my ability to create this book.

ELYA B. JOFFE

Hod-Hasharon, Israel
October 2009

1

OVERVIEW

The term "ground" too often seems to be associated with a sort of cure-all concept, like snake oil, money or motherhood. Remember that, while you can always trust your mother, you should never trust your "ground." Examine and think about it. [1]

Grounding is probably the most important aspect of electrical or electronic system design, yet it is probably the least understood by most engineers. Often blended with misconceptions, in the final implementation it is typically necessary to strike a balance between electromagnetic interference (EMI) control, safety, and good engineering practices. Grounding theory is not intuitive. Achieving a functional grounding philosophy often results from battles of wits, perseverance, and the resolution of conflicts between instinct, intuition and engineering experience and judgment.

Electromagnetic interference and noise are generally pervasive in all electrical and electronic systems. Because this is true, it would be fair to say that every electrical and electronics design engineer will ultimately encounter grounding problems during the span of his career.

Ask two engineers for "their" solution to EMI or electrical noise problems and, if you are lucky, you will end up with only two different recommended approaches. It is for this reason that many design engineers are wont to say, "For every grounding problem there are many solutions, most of which are wrong. . . ."

When such casually provided approaches are attempted, it will often be revealed that what worked in one situation may not necessarily work well in another. Experi-

Grounds for Grounding. By Elya B. Joffe and Kai-Sang Lock
Copyright © 2010 The Institute of Electrical and Electronics Engineers, Inc.

ence *does* play a prime role in the details of the grounding design of a specific system. Unfortunately, that experience is generally not transferable to another system, even if they both utilize the same technology. Grounding design is so system-specific, that rarely is there a "generic" solution that is fit for all cases.

Many proposed solutions appear conflict with fundamental physical principle requirements imposed on the system. They may be based on myths and misconceptions regarding the very concept of grounding. Indeed, design of any grounding system, contrary to common belief, is founded on solid science. The watchful engineer may be generally aware of the correct principles but be guilty of their misapplication, either through inexperience, negligence, or even an intentional effort to avoid Maxwell's equations and their consequences.

Electrical and electronic design is usually taken to be "well founded." By that is meant it can be modeled, analyzed, and simulated using commonly known circuit analysis tools such as Spice, for instance. Electrical and electronic circuit design seems repeatable, the solution appears to be straightforward, and the parameters are typically simply outlined and implemented in a well-understood model.

For a successful analysis to be carried out, the engineer must be able to clearly state the issues of concern or the problem to be addressed down to the component level, including any related variables. This data is provided via measurement or simulation or both. However, when grounding problems are encountered, it is not so simple to identify the components involved, or even the path or paths of interest. This can be a challenging problem at DC or low frequencies. At higher frequencies in radio frequency (RF) circuitry as well as in the now commonly used high-speed digital switching circuits, components stubbornly obey the laws of physics, with capacitors acting like inductors, inductors exhibiting capacitive responses, and printed circuit boards introducing parasitic reactive circuit elements that cannot be found in the drawing.

As Don R. Bush, (1942–2001) said, "Anyone can construct a mathematical model and generate data. But if the predictions of your mathematical model do not match experimental data, either your model is worthless or your measurements are not done properly."*

Analysis of grounding systems, particularly in large installations, but even in small-scale circuits, may be a very complex issue and typically defies straightforward definition. The challenge increases when considering the risk of safety code violation, caused by misinterpretation and wrong implementation of grounding requirements, which may result in significant negative consequences.

The novice may be surprised to observe that electrical and electronics experts find the issue of grounding so complex. "After all," they may say, "what could ever be so difficult in connecting the 'zeros' all together?" The truth is that grounding problems, if not properly addressed in the early phases of design, are bound to surface at the end of the project, at which time a solution is likely to be both costly and complex. With budget and schedule virtually depleted in the panic of finding a solution, attempts may be made to modify the grounding system layout and design by disconnecting, reconnecting, removing, or adding grounding connections randomly in a trial-and-error ap-

*Paul, C., *IEEE EMC Society Newsletter,* Issue No. 192, Winter 2002.

proach. It is at this time that critical conflicts may be overlooked and safety-compromising situations may not be acknowledged.

When the attempts seem to yield good results, the solution is called "successful," leading all to believe that grounding system design is indeed founded in "black magic."

The typical electrical or electronics engineer will often avoid highly mathematical electromagnetic field theory. After all, "Physics is for the physicists," right? *Wrong!* All answers to electrical and electronic design questions, particularly those related to grounding questions, are well founded in electromagnetic field theory, more specifically, Maxwell's equations.

In any discipline, lack of knowledge and comprehension of the science behind the rules may bring about the use of "rules of thumb," resulting in either inadequate design, faulty or lacking (and possibly hazardous) results, or, worse still, overdesign and, thus, a costly solution.

"Rules of thumb," by their very essence, divest the engineer of his common sense and true engineering judgment. Such rules, which may have been appropriate years or decades before, are likely not applicable to new technologies. More importantly, they may not be compatible with current safety codes, which may have evolved independently through the years. Applying those rules of thumb may be inappropriate for the intended application (e.g., EMI control) or may introduce serious violation of current safety requirements. Many of these rules of thumb are like urban myths, evolved from misconceptions regarding grounding, and their use should be discouraged.

Most misconceptions and rules of thumb related to grounding theory were established in previous generations and were directly applicable to the technology available at the time. For example, how often do design engineers still recommend single-point grounding topologies for their high-speed digital circuits? Or still think that cable shields should be grounded at one end only? Or still use the 0.1 μF capacitors for decoupling of high-speed digital devices having rise/fall times on the order of tens of picoseconds?

Consider, for instance, the evolution of computing speeds and slew rates, coupled with the increased density on printed circuit boards. Today's digital circuits have gradually evolved into RF circuits, with their frequency content exceeding 10 GHz. The increase of switching speeds implies an increase in high-frequency spectral content, virtually from DC to daylight, leading to greater significance of parasitic reactive effects and heightened emissions and interference. Rules of thumb developed decades ago will not suffice for today's technology.

Traditional analog or digital design rules of thumb will not provide functional grounding system design or help control EMI either, and many, in fact that employ transmission line theory, for instance, are necessary today for explaining high-frequency effects encountered in modern circuits. Use of such techniques will yield appropriate grounding schemes, providing both performance and EMI control, while not compromising safety.

Grounding design for modern systems must cover many disciplines, including digital and analog design, power engineering, lightning protection, and many others, not to mention system safety concerns. Rather than an intuitive approach, the design of the grounding system must be founded on electromagnetic field theory, in particular Maxwell's equations.* A practical approach must be maintained, in consideration of safety and other codes and regulations, and conflicting situations, which may arise due to contradicting requirements in practical systems.

A key objective of this book is to dispel the mystery associated with grounding. This shall be accomplished by providing a methodical approach for the design of grounding systems, from circuits through systems and up to platforms and facilities.

The book attempts to meet the above challenge by putting grounding into the proper perspective. It outlines a physical foundation for explaining the concept of grounding, founded on electromagnetic field theory, while providing insight into practical aspects of grounding system implementation, particularly as related to its interdisciplinary nature, extending from circuits to facilities. It will be clearly demonstrated that grounding systems in facilities, systems, or circuits do, in fact, follow a consistent scheme.

From the topological perspective, there is no fundamental difference between a circuit, a rack, a platform, or a facility. The laws of electromagnetics revealed in Maxwell's equations remain unchanged regardless of the system dimensions. Only the manner and complexity of their application differs from one to the next.

In practice, though, circuit and equipment designers are electronic engineers, whereas the facility and platform designers are electrical engineers. The crossroad between the two, electronics and electrical engineering, also constitutes the borderline between power levels (milivolts and kilovolts, microamps and kiloamps) and, of greater significance, between frequency contents, particularly power frequency to signal frequencies ranging from DC to daylight.

Integrating equipment and systems in a large facility does require not only a new way of thinking and different comprehension of terms, but also a distinctive appreciation of numbers; the difference in units of measurement affect their actions because "they work on the opposite side of the decimal point" [1]. Can an electronic circuit designer truly internalize a 200 kA lightning return stroke current? And what concern are miliamps to the electrical power engineer?

This book is thus also targeted at developing a universal approach to the understanding of grounding at whichever tier is considered, delineating the distinctiveness of each while emphasizing the resemblance between them, hoping to remove the present fuzziness.

No doubt, the concepts presented herein may put several designers at unease. These concepts will conflict with the widespread notion that there is no scientific foundation for grounding, which is well known to amount to "black magic." The theories and practices discussed herein will diverge from the body of "common knowledge" related to the way grounding should really be accomplished. Without a doubt, many may choose to carry on utilizing former practices, finding that easier than attempting to understand these concepts.

---

*A detailed discussion of Maxwell's equations as they apply to grounding theory and practices is presented in Chapter 2.

We are confident that, eventually, this book will help to do away with old, outdated, and erroneous practices, which may have been acceptable where low frequencies were concerned, but constitute poor design practices for the high-frequency circuits and systems so widespread today.

Application of theory, observing physics principles working in practice, and proving Maxwell's equations' validity for grounding and EMC design practices have provided particular satisfaction and made this book, in the authors' opinion, of even greater consequence.

With this in mind, the book begins by introducing the reader to the fundamental concepts pertaining to grounding, starting with a discussion of Maxwell's equations, particularly as they apply to the topic of grounding. Essential terms and concepts relating to real-world electrical circuit behavior are also laid out in Chapter 2.

Chapter 3 presents the basics of grounding, beginning with a discussion of the term "ground" and the different objectives of grounding.

Chapter 4 provides an in-depth review of the fundamentals of grounding design. It discusses in detail the fundamental topologies of grounding systems and provides a novel yet practical systematic approach for planning grounding systems. The concept of "ground loops" is developed in Chapter 4 and solutions are presented. The implementation of the fundamental grounding architectures in large-scale systems and installation are further examined. Chapter 4 is completed with grounding-related case studies.

Chapter 5 explains the principles of bonding. The approaches of achieving low-impedance connections between metallic surfaces and structures as a fundamental objective for meeting the desired grounding objectives are portrayed.

Chapter 6 describes in detail safety-related grounding concerns. Rationale for electrical safety grounding requirements is provided and safety grounding design principles in power distribution and lightning protection systems are presented.

Chapter 7 covers grounding in wiring and cable systems. One of the most controversial and misunderstood aspects of system grounding design stems from the question of cable shield termination ("grounding"). This Chapter will clearly make a distinction between signal grounding and shield termination, putting this question to rest.

Chapter 8 provides the foundation for understanding the essential necessity of adequate grounding of EMI terminal protection devices (e.g., EMI filters and transient-suppressing devices) performance. The effect of acceptable versus objectionable grounding of such protective devices is clearly demonstrated.

In Chapter 9, the application of grounding in printed circuit boards (PCBs) is discussed in depth, particularly as related to power conditioning and signal return paths. The question of grounding in mixed analog/digital circuits is also addressed.

Chapter 10 leads the reader to the facility and platform levels. The design of integrated grounding systems in facilities is described. The complexity of and approaches to the integration of multiple subsystems into a larger system as related to grounding system design are discussed. It also expands the concept of grounding architecture design to the unique cases of mobile platforms, for example, tactical C^3I (command, control, communication, and intelligence) shelters, aircraft, space systems, and ships.

The several appendices in the book provide extensive supporting information and supplemental data, which will be of great use for the reader. Appendix A provides a glossary of grounding-related terms and definitions, with references to their sources, particularly when derived from official international standards and codes. When several definitions exist for a term, they are all included, with reference to the context of their applicability. Appendix B lists commonly used acronyms employed throughout the book for easy reference by the reader. Appendix C presents commonly used symbols associated with variables referred to throughout the book. Appendix D presents a list of many grounding-related standards, specifications, and codes and their scope. Appendix E demonstrates the equivalence between Ohm's Law and Fermat's "Least Time" Principle, which is useful for understanding the reason why current selects a particular return path. Finally, Appendix F provides an overview of S-parameters and their application for the evaluation of grounding performance, particularly on printed circuit boards, extensively used in Chapter 9.

With the emergence of new technologies—nanotechnology, in particular—the importance of proper grounding design is greater than ever. We are certain that this book, founded on fundamental physical principles on the one hand and on real-world, practical experience on the other, provides an excellent resource for achieving successful, cost-effective, and timely state-of-the-art designs of electronic and electrical equipment, systems, and networks.

BIBLIOGRAPHY

[1] Brokaw, P., "An I.C. Amplifier Users' Guide to Decoupling, Grounding, and Making Things Right for a Change," *Application Note, Analog Devices,* 1982.

[2] Perez, R. (Ed.), *Handbook of Electromagnetic Compatibility,* Academic Press, 1995.

[3] Morrison, R., Lewis, W. H., *Grounding and Shielding in Facilities,* Wiley, 1990.

[4] Hartal, O., *Electromagnetic Compatibility By Design,* R&B Enterprises, USA, 1994.

2

FUNDAMENTAL CONCEPTS

Things should be made as simple as possible, but not any simpler.—ALBERT EINSTEIN

Understanding grounding and, particularly, application of grounding design principles is close to impossible without a basic understanding of the fundamental concepts of electromagnetism. Dr. Howard Johnson, in his article "Why Digital Engineers Don't Believe in EMC," [1] points out some of the most fundamental errors of digital design engineers. Two of those deserve special attention:

1. Digital engineers don't believe current flows in loops
2. Digital engineers don't believe in H-fields

In this chapter, the concepts of the path of current flow and magnetic (H) fields constitute examples of such basic fundamentals, without which the concept of grounding literally can neither be rationalized nor properly applied.

2.1 MAXWELL'S EQUATIONS DEMYSTIFIED

Electromagnetic phenomena are characterized by Maxwell's equations, a set of four fundamental equations governing electromagnetism (i.e., the behavior of electric and magnetic fields). James Clerk Maxwell is generally regarded as the greatest theoretical physicist of the 19th century and is recognized as the "Father of Electromagnetic Theory." In Part III of his 1864 paper, "A Dynamical Theory of the Electromagnetic Field," entitled "General Equations of the Electromagnetic Field," Maxwell formulated eight equations, labeled (A) to (H). In this paper, Maxwell utilized the correction to Ampere's Circuital Law that he had made in Part III of his 1861 paper, "On Physical Lines of Force." In Part IV of the 1864 paper, Maxwell modified Ampere's Circuital Law to include displacement current with some of the other equations of electromagnetism and obtained a wave equation with a speed equal to the speed of light, commenting:

> The agreement of the results seems to show that light and magnetism are affections of the same substance, and that light is an electromagnetic disturbance propagated through the field according to electromagnetic laws.

These eight equations were to become known as "Maxwell's equations," until this term became applied instead to a set of four equations selected in 1884 by Oliver Heaviside. Heaviside's versions of Maxwell's equations are distinct by virtue of the fact that they are written in modern vector notation. They actually only contain one of the original eight, equation (G), "Gauss's Law." Another of Heaviside's four equations is an amalgamation of Maxwell's Law of Total Currents [equation (A)] with Ampere's Circuital Law [equation (C)], actually originally made by Maxwell himself in equation (112) in his 1861 paper, "On Physical Lines of Force."

Maxwell's derivation of the electromagnetic wave equation has been replaced in modern physics by a much less cumbersome method involving combining the corrected version of Ampere's Circuital Law with Faraday's law of electromagnetic induction, known as the "Heaviside form of Maxwell's equations."

Although not developed by him, his successful interpretation of Faraday's concept of the electromagnetic field resulted in four field equations bearing his name.* Formidable mathematical ability combined with great insight enabled Maxwell to lead the way in the study of the two most important areas of physics of his time [2].

> When I start describing the magnetic field moving through space, I speak of the *E*- and *B*-fields and wave my arms and you may imagine that I can see them. I'll tell you what I see. I see some kind of vague shadowy, wiggling lines—here and there is an E and B written on them somehow, and perhaps some of the lines have arrows on them—an arrow here or there which disappears when I look to closely at it. . . . I cannot really make a picture that is even nearly like the true waves. (Feynman, R., *The Feynman Lectures on Physics;* Vol. II; California Institute of Technology, 1989)

Considering also Maxwell's earlier mechanistic interpretations of the nature of a field, it becomes evident that even the best scientific minds are still unclear about many basic ontological questions regarding the nature of an electromagnetic field. That is, no clear mental picture can be drawn in which the field appears limpidly and

---

Physics World Magazine ranked Maxwell's Equations as the most important set of equations of all time.

uniquely. "Faraday's Lines of Force" were, as Maxwell puts it, "the key to the science of electricity." However, when firmly scrutinized, the exact nature of a field shifts and squirms like Proteus under Menelaus' grasp.

Maxwell's Equations are also quite challenging from the mathematical point of view and their solution is not simple. But this is not to diminish their practical implication, especially with respect to the theory of grounding, particularly since the intuitive understanding of their concepts is quite simple. These equations describe the distributed parameter nature of electromagnetic fields and sources, that is, how fields and their sources are distributed in space. However, quite often we may utilize approximations of these equations, such as applying lumped-circuit models in order to simplify analysis in certain cases, when dimensions of the problem are electrically small.

Maxwell's Equations form a set of partial differential equations, the variables of which are functions of the spatial parameters (x, y, z) in three-dimensional Cartesian space coordinates, as well as time (t). Accordingly, the $\vec{E}$ and $\vec{H}$ fields are time (t) and position dependant, but for the purpose of simplicity, the notation $\vec{E}$ rather than $\vec{E}(x, t)$ will be used.

In this section, only a brief overview of Maxwell's equations is presented, with practical implications of these equations to the understanding and design of grounding theory.

If you prefer not to be challenged by the concepts outlined in this section and, particularly, Maxwell's equations, you may skip Section 2.1. However, though the knowledge of Maxwell's equations is not required for the actual design of grounding systems, the understanding of the elementary concepts expressed by them, two in particular to be precise, is important. It is difficult to consider a scientific approach for grounding without their knowledge. The reader is therefore encouraged to pursue the discussion delineated in this section.

2.1.1 Fundamental Terms

Maxwell's equations are described in terms of vector and scalar values. The following are vector values:

$\vec{E}$ = Electric field strength (V/m)
$\vec{D}$ = Electric flux density (C/m^2)
$\vec{H}$ = Magnetic field strength (A/m)
$\vec{B}$ = Magnetic flux density (Tesla)
$\vec{J}$ = Conduction current density (A/m^2)

The following are scalar values:

Q = Charge (C)
ρ = Volume charge density (C/m^3)

Vector fields are italicized, with the vector notation above (for instance, if A is a vector field, it will be presented as $\vec{A}$). Other vectors (e.g., position, velocity, etc.) are italicized, such as v for velocity.

2.1.1.1 Electric and Magnetic Forces and Fields. Electric and magnetic fields are defined through the forces they exert on charged particles. An electric field $\vec{E}$ represents the force $\vec{F}$ applied on a charge q:

$$\vec{E} = \frac{\vec{F}}{q} \tag{2.1}$$

In the absence of an electric field, no force will be exerted on a charge when it is at rest. However, if a charge moves with velocity v, force is introduced, attributed to a *magnetic field*. The direction of this force is rather interesting as the force is perpendicular to the velocity and vanishes if the velocity is in some particular direction. For forces of this nature, the magnetic field is defined such that

$$v \times \vec{B} = \frac{\vec{F}}{q} \tag{2.2}$$

If magnetic charges were to exist, stationary magnetic charges q_m, would have experienced forces due to the magnetic field, $\vec{B}$, and if in motion, would have experienced a force $v \times \vec{E}$. However, magnetic charges, so painstakingly sought by physicists, have not been identified yet.

Both electric and magnetic fields are combined into the Lorentz Law of Force:

$$\vec{F} = q\left(\vec{E} + v \times \vec{B}\right) \tag{2.3}$$

The electric force is straightforward, being in the direction of the electric field if charge q is positive, however the "×" product in Equation (2.2) and Equation (2.3) denotes the "cross product" via which the unique directional properties of the magnetic field on the moving charge are demonstrated as given in Figure 2.1.

2.1.1.2 Biot–Savart Law. The Biot–Savart Law relates magnetic fields to current, which constitute their sources. From [17], taken as an arbitrary source, it can be observed that the magnetic field resulting from a current distribution along a conductor can be derived using the Biot–Savart Law. Figure 2.2 illustrates the manner in which magnetic fields surround a wire carrying a current I. Each infinitesimal current element

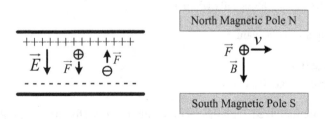

Figure 2.1. Lorentz Law of Force.

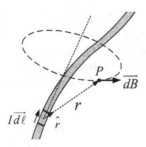

Figure 2.2. Biot–Savart Law, magnetic fields surrounding a current-carrying wire.

$I\overrightarrow{d\ell}$ contributes an infinitesimal magnetic field $\overrightarrow{dB}$, which is perpendicular to the current element and to the radius vector, $\hat{r}$ from the current element to the field point P. The relationship between the magnetic field contribution and its source current element is called the Biot–Savart Law:

$$\overrightarrow{dB} = \frac{\mu_0 I\overrightarrow{d\ell} \times \hat{r}}{4\pi r^2} \tag{2.4}$$

The direction of the magnetic field contribution follows the right-hand rule illustrated for a straight wire. This direction arises from the vector product nature of the dependence upon electric current.

2.1.1.3 Constitutive Relations. The following expressions, relating the response of matter to the applied field, are referred to as constitutive relations.

(a) The permittivity (or dielectric constant), ε, relates the electric displacement to electric field strength,

$$\vec{D} \triangleq \varepsilon \vec{E} \tag{2.5}$$

and

$$\varepsilon = \varepsilon_0 \cdot \varepsilon_r \tag{2.6}$$

where ε_r is the relative permittivity of the material containing the charge and ε_0 is the permittivity of free space, $\varepsilon_0 \cong (1/36\pi) \times 10^{-9} \approx 8.85 \times 10^{-12}$ [F/m].

The permittivity of free space ε_0 determines the intensity of the electric field occurring in free space resulting from electrical charges.

When an electric field is present in an environment other than free space, it has the effect of polarizing matter, causing charges to rearrange themselves in the direction of the field in a manner that tends to oppose the original applied field. This results in a reduction of the total net electric field in polarized material. This is taken into account by saying that the permittivity of the material, $\varepsilon = \varepsilon_0 \cdot \varepsilon_r$ is higher than ε_0 (or $\varepsilon_r > 1$).

Because permittivity changes discontinuously as the boundary enclosing the charges, the electric field strength will likewise be discontinuous. It is, therefore, advantageous to introduce the electric displacement, $\vec{D} \triangleq \varepsilon\vec{E}$. Unlike the E-field $\vec{E}$, electric displacement is continuous across the material boundary.

(b) The permeability μ relates the magnetic flux density to the magnetic field strength:

$$\vec{B} \triangleq \mu\vec{H} \tag{2.7}$$

and

$$\mu = \mu_0 \cdot \mu_r \tag{2.8}$$

where μ_r is the relative permeability and μ_0 is the permeability of free space, $\mu_0 = 4\pi \times 10^{-7}$ [H/m].

The permeability of free space μ_0 determines how easily magnetic fields are established in free space. When a magnetic field is present in matter, the characteristics of the material will augment the strength of the magnetic field. This effect is taken into account by saying that the permeability of the material, $\mu = \mu_0 \cdot \mu_r$, is different from μ_0 (or $\mu_r \neq 1$).

Because permeability changes discontinuously as the boundary enclosing the charge in the material, the magnetic field strength will, likewise, be discontinuous. It is, therefore, advantageous to introduce the concept of flux density $\vec{B} \triangleq \mu\vec{H}$. Unlike the magnetic field strength $\vec{H}$, magnetic flux density $\vec{B}$ is continuous across the material boundary.

(c) The conductivity σ relates current density to electric field:

$$\vec{J} \triangleq \sigma\vec{E} \tag{2.9}$$

Often, this relation is called Ohm's Law in Materials and

$$\sigma = \sigma_0 \cdot \sigma_r \tag{2.10}$$

where σ_r is the relative conductivity and σ_0 is the conductivity of copper (Cu), $\sigma_0 \approx 5.82 \times 10^7$ [S/m].

There are contexts in which the resistivity use of is more convenient than conductivity.

2.1.2 Maxwell's Equations

The differential forms of Maxwell's equations are:

$$\nabla \cdot \vec{D} = \rho \qquad \text{(Gauss's Law for Electric Field)} \tag{2.11}$$

$$\nabla \cdot \vec{B} = 0 \qquad \text{(Gauss's Law for Magnetic Field)} \tag{2.12}$$

$$\nabla \times \vec{E} = -\frac{\partial \vec{B}}{\partial t} \qquad \text{(Faraday's Law of Induction)} \qquad (2.13)$$

$$\nabla \times \vec{H} = \vec{J} + \frac{\partial \vec{D}}{\partial t} \qquad \text{(Ampere's Law)} \qquad (2.14)$$

In the above equations, the intrinsic vector ∇ (del) operator, for instance, in a Cartesian coordinates system $(\hat{x}, \hat{y}, \hat{z})$ denotes

$$\nabla \equiv \frac{\partial}{\partial x}\hat{x} + \frac{\partial}{\partial y}\hat{y} + \frac{\partial}{\partial z}\hat{z}$$

The integral forms of Maxwell's equations are:

$$\oint_S \vec{D} \cdot d\vec{s} = \int_V \rho dv \qquad \text{(Gauss's Law for Electric Field)} \qquad (2.15)$$

$$\oint_S \vec{B} \cdot d\vec{s} = 0 \qquad \text{(Gauss's Law for Magnetic Field)} \qquad (2.16)$$

$$\oint_C \vec{E} \cdot d\vec{l} = -\int_S \frac{\partial \vec{B}}{\partial t} \cdot d\vec{s} \qquad \text{(Faraday's Law of Induction)} \qquad (2.17)$$

$$\oint_C \vec{H} \cdot d\vec{l} = \int_S \left(\vec{J} + \frac{\partial \vec{D}}{\partial t} \right) \cdot d\vec{s} \qquad \text{(Ampere's Law)} \qquad (2.18)$$

The two equation sets, Equations (2.13) and (2.14), and (2.17) and (2.18), are sometimes referred to as the dynamic Maxwell's equations, since they have an explicit time dependence built into them.

The interactions described in the dynamic equations form the basis for understanding electromagnetic compatibility and in particular, grounding theory.

The fundamental sources depicted in Maxwell's Equations are charge and current density, defined through the relations,*

$$\rho = \lim_{V \to 0} \frac{Q}{V} \qquad \text{and} \qquad \vec{J} = \lim_{S \to 0} \frac{\vec{I}}{S} \qquad (2.19)$$

where Q is the total charge in the volume V in coulombs (C) and I is the total current flowing through the cross-section area S. With relation to these sources, the last relationship portrays the principle of Conservation of Charge, or the Law of Continuity:

$$\nabla \cdot \vec{J} = -\frac{d\rho_V}{dt} \qquad (2.20)$$

---

*Observe that currents and charges, not voltages and potentials, appear as sources in Maxwell's Equations.

Figure 2.3. Maxwell's "Field Equations." (Cartoon Courtesy of Tayfun Akgul.)

Equation (2.15), depicted above in differential form, can also be depicted in integral form:

$$\oint_C \vec{J} \cdot d\bar{s} = -\frac{d}{dt} \int_V \rho_V dv \tag{2.21}$$

The following is a short discussion of Maxwell's equations, unfolding their physical significance.

2.1.2.1 Gauss's Law for Electric Field. Maxwell's first equation, also known as Gauss's Law for Electric Fields or the Divergence Theorem, in its differential and integral forms [Equations (2.11) and (2.15)], is expressed as

$$\nabla \cdot \vec{D} = \rho \qquad \text{or} \qquad \oint_S \vec{D} \cdot d\bar{s} = \int_V \rho dv$$

This equation relates the (electrical) charge and electric flux density. It simply states that the net electric displacement emerging from a closed surface is equivalent to the net positive charge enclosed by the surface. Electric flux that begins at a positive charge must terminate at an equal negative charge (Figure 2.4).

It follows, therefore, that

$$\oint_S \vec{D} \cdot d\bar{s} = \begin{cases} 0, Q \text{ outside } S \\ Q, Q \text{ inside } S \end{cases} \tag{2.22}$$

A unique implication of this law, related to a charged piece of metal, is that, regardless of its shape and if current is zero, the electric field inside the piece of met-

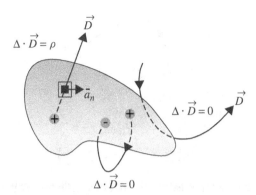

Figure 2.4. Gauss's Law for E-Fields. The net electric displacement, $\vec{D}$, through a surface of a closed volume is equal to the net positive charge enclosed in that volume.

al must be zero. Free charges in metal go to the surfaces of the metal and arrange themselves so that the electric field is zero everywhere inside the metal. Electrostatics is a linear theory, so equilibrium charge distribution is unique, that is, there is only one arrangement of charges on the surface, which produces zero electric field everywhere inside the metal. At equilibrium, there is no current in electrostatics so the electric field at the surface of a conductor is perpendicular to the surface of the metal everywhere. Furthermore, a closed metal surface, regardless of its shape, screens out external sources of electric field. Stated in general terms, if there are no charges inside a volume totally enclosed by a metal surface, the electric field is zero everywhere inside that volume.

A shielded enclosure, which is an application of this phenomenon, is often called a Faraday cage, after the physicist Michael Faraday, who built the first one in 1836.

2.1.2.2 Gauss's Law for Magnetic Field.
Applying a similar progression of ideas as those in Section 2.1.2.1 to magnetic fields leads to Maxwell's second equation, also known as Gauss's Law for Magnetic Field, in its differential and integral forms [Equations (2.12) and (2.16)]:

$$\nabla \cdot \vec{B} = 0 \qquad \text{or} \qquad \oint_S \vec{B} \cdot d\vec{s} = 0$$

Since there are no known isolated sources of magnetic fields ("magnetic charges"), no Gaussian surfaces can enclose any net magnetic charge, implying that all magnetic flux lines form closed paths (Figure 2.5).

2.1.2.3 Faraday's Law of Induction

Going to the south and circling to the north the wind goes round and round; and the wind returns on its circuit.—(ECCLESIASTES 1:6)

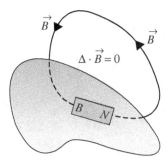

<u>Figure 2.5.</u> Gauss's Law for H-fields. The net magnetic flux $\vec{B}$ through the surface of any closed volume is zero.

Maxwell's third equation (Maxwell's Electrical Voltage Law), also known as Faraday's Law of Induction or simply Faraday's Law, in its differential and integral forms [Equations (2.13) and (2.17)] is

$$\oint_C \vec{E} \cdot d\vec{l} = -\int_S \frac{\partial \vec{B}}{\partial t} \cdot d\vec{s} \qquad \text{or} \qquad \nabla \times \vec{E} = -\frac{\partial \vec{B}}{\partial t}$$

This equation reveals that the electromotive force (emf) generated around a closed contour C is related to rate of change of the magnetic flux through an open surface S bounded by that contour (Figure 2.6). The notation $\nabla \times \vec{E}$ is referred to as the curl of $\vec{E}$, and can be thought of, loosely speaking, as quantifying the amount of "swirl" or "circulation" of the vector field $\vec{E}$. The curl itself is a vector. It points in the direction of the axis of the local swirl (Figure 2.6).

The name Farady's Law of Induction is attributed to Farady's Law by extension from Equation (2.17), whereby it can be expressed as

$$emf = \oint_C \vec{E} \cdot d\vec{l} = -\frac{\partial}{\partial t} \int_S \vec{B} \cdot d\vec{s} = -\frac{\partial \Phi}{\partial t} \qquad (2.23)$$

Equation (2.21) shows that in lumped circuits, electromotive force (emf) induced into a loop (closed contour) is a result of the change of the total magnetic flux through that loop. The induced emf can be considered as a voltage source U_{in} induced into the circuit. Note, however, that emf is a distributed parameter, whereas the lumped model of a voltage source is valid in electrically small circuits only.

When emf is generated by a change in magnetic flux according to Faraday's Law, the polarity of the induced emf is such that it produces a current whose magnetic field opposes the change of the magnetic field $\vec{B}$ producing it (see Figure 2.7). This is manifested in the negative sign in Faraday's Law, indicating that emf induced in a contour has a polarity that tends to generate an induced current, resulting in a magnetic flux opposing any change in the original magnetic flux that produced it.

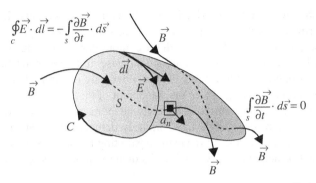

Figure 2.6. Faraday's Law: emf generated around a closed contour C due to varying magnetic flux through the open surface S.

This principle can easily be understood from the standpoint of conservation of energy. The energy volume density associated with a magnetic field $\vec{B}$ is expressed as

$$\eta_{\vec{B}} = \frac{Energy}{Volume} = \frac{1}{2}\frac{\vec{B}^2}{\mu} \tag{2.24}$$

whereas the rate of change of energy volume density, or the power volume density $P_{\vec{B}}$ is

$$P_{\vec{B}} = \frac{d}{dt}\left(\frac{1}{2}\frac{\vec{B}^2}{\mu}\right) = \frac{1}{\mu}\vec{B}\cdot\left(\frac{d\vec{B}(t)}{dt}\right) \xrightarrow{\frac{d\vec{B}(t)}{dt}\to\infty} \infty \tag{2.25}$$

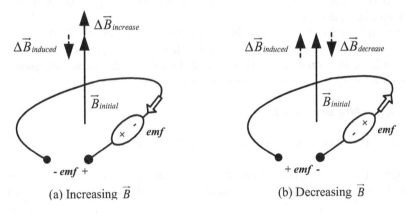

Figure 2.7. Lentz's Law: The emf induced in a contour will have a polarity that tends to produce an induced current resulting in magnetic flux opposing any change in the original magnetic flux that produced it.

The result obtained in Equation (2.25) should therefore be anticipated. If, contrary to the above principle, induced emf were not to produce current opposing the original magnetic flux, the magnetic flux through the contour due to the induced current would increase the induced emf, further aiding the increase in the magnetic field, eventually driving energy buildup η_B to infinity. This is obviously in contradiction to the principle of conservation of energy. From the standpoint of this principle, current must flow in the path such that energy stored in the consequent magnetic field is minimized.

The negative sign in Faraday's Law similarly demonstrates that time-varying currents will always seek to flow in the path constituting the smallest possible contour C. This important principle is recognized for its own merit and is known as Lenz's Law, expressed as

$$emf = -L\frac{dI}{dt} \tag{2.26}$$

where L is the inductance (see Section 2.3.1) of the contour C defined as

$$L \triangleq \frac{\int_S \vec{B}\cdot d\vec{s}}{I} = \frac{\vec{\Phi}}{I} \tag{2.27}$$

A special case of Faraday's Law occurs when no magnetic flux penetrates a surface bounded by a contour, that is, $\Phi = 0$. In this case, induced emf and, consequently, the sum of all voltages in the closed contour must be zero:

$$\oint_C \vec{E}\cdot d\vec{l} = \sum_i U_i = 0 \tag{2.28}$$

Related to lumped-circuit theory, this is the formulation of Kirchhoff's Voltage Law,* illustrated in Figure 2.8.

Faraday's Law is by far one of the two most important aspects of Maxwell's Equations associated with grounding theory (the other being Ampere's Law, discussed in Section 2.1.2.4), forming the basis for the definition of inductance, discussed in Section 2.3.1.

2.1.2.4 Ampere's Law

All the rivers flow to the sea, but the sea is not full.—(ECCLESIASTES 1:7)

Faraday's Law, discussed in Section 2.1.2.3, demonstrated that a time-varying magnetic field can produce an electric field. Similarly, Maxwell's fourth equation, also known

---

*Note that Kirchhoff's voltage and current laws (KVL and KCL, respectively) apply in lumped-circuit models only. Voltages and currents obtained from them are valid only so long as the largest physical dimension of the circuit is electrically small, that is, much less than a wavelength at the frequency of excitation, f, of that circuit.

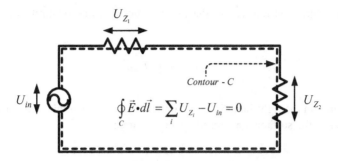

Figure 2.8. Equivalence between Faraday's Law and Kirchhoff's Voltage Law in the absence of magnetic flux.

as Maxwell's Electrical Current Law or Ampere's Law [Equations (2.14) and (2.18)] acknowledges that, in addition to net movement of electric charge, currents as well as time-varying electric flux crossing through an area, S, enclosed by a contour, C, can produce a magnetic field:

$$\oint_C \vec{H} \cdot d\vec{l} = \int_S \left(\vec{J} + \frac{\partial \vec{D}}{\partial t} \right) \cdot d\vec{s} \qquad \text{or} \qquad \nabla \times \vec{H} = \vec{J} + \frac{\partial \vec{D}}{\partial t}$$

Ampere's Law reveals, therefore, that magnetic fields emerge from two sources. The first source is the electric field associated with the current flowing in the form of transported charge (conduction current), $\vec{J} = \sigma \vec{E}$, and the second is the time varying electric fields crossing a surface of a closed-loop circuit (displacement current), $\partial \vec{D}/\partial t = \varepsilon \partial \vec{D}/\partial t$.

Similar to Faraday's Law, the notation $\nabla \times \vec{H}$ indicates that time-varying current $\vec{J}$ or time-varying electric displacement $\vec{D}$ introduce a "swirl" or "circulation" to the vector magnetic field, $\vec{H}$ (or $\vec{B}$). Therefore, for a current flowing in an upward direction, the magnetic flux flows as illustrated in Figure 2.9. The line integral of $\vec{H}$ along

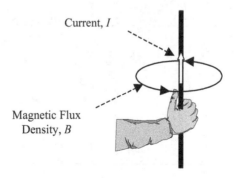

Figure 2.9. The right-hand rule: time varying current introduces a "swirl" to the magnetic field.

a closed contour C is referred to the magnetomotive force or mmf around the contour:

$$mmf = \oint_C \vec{H} \cdot d\vec{l} \qquad (2.29)$$

The first term on the right-hand side of Ampere's Law is the total conduction current I_c that penetrates the surface S enclosed by a contour C:

$$I_c = \int_S \vec{J} \cdot d\vec{s} \qquad (2.30)$$

The second term on the right hand side of Ampere's Law is the total displacement current I_d that penetrates a surface S enclosed by a contour C (Figure 2.10):

$$I_d = \frac{\partial}{\partial t} \int_S \vec{D} \cdot d\vec{s} \qquad (2.31)$$

The extension of Ampere's Law to include the displacement current can be justified using Figure 2.11, showing a current-carrying wire interrupted by a parallel plate capacitor.

According to the (original) Ampere's Law, the line integral $\oint_c H \cdot d\vec{l}$ is equal to the current I penetrating any surface enclosed by a contour C. If that surface were just a flat circle enclosed by a contour, however, the current is $I = 0$. On the other hand, the surface could be extended upward until it enclosed the capacitor created by the two plates, thus enclosing I. Maxwell settled this inconsistency by noting that current en-

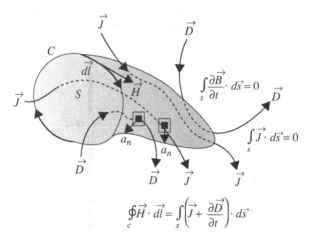

Figure 2.10. Ampere's Law: The mmf, generated along a closed contour C due to conduction and displacement currents penetrating a surface S bounded by a contour C.

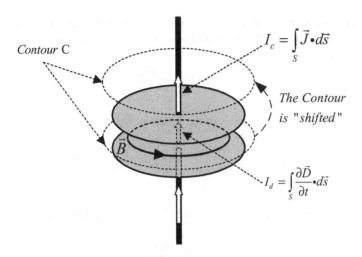

Figure 2.11. Ampere's Law was extended by Maxwell to include displacement current

tering the capacitor plate results in charging the capacitor, thus increasing the electric field, $\vec{E}$, within the capacitor:

$$Q(t) = C \cdot V(t) = C \cdot \vec{E}(t) \cdot d \qquad (2.32)$$

where C is the capacitance, $V(t)$ is the increasing voltage across the capacitor plates due to the increase in charge, and $\vec{E}(t)$ is the increasing electric field developing along the path, d, the separation between the capacitor plates. If the area of the plates is A, the capacitance is expressed as $c = \varepsilon_0 \cdot A/d$, yielding Equation (2.33):

$$\frac{Q(t)}{A} = \varepsilon_0 \cdot \vec{E}(t) \qquad (2.33)$$

But since, by definition, $I \triangleq dQ/dt$, Equation (2.33) reduces to

$$\frac{I}{A} = \varepsilon_0 \cdot \frac{d\vec{E}}{dt} = \frac{d\vec{D}}{dt} \qquad (2.34)$$

Therefore, time varying electric fields (or electric displacement) act in a manner similar to that of current density, the displacement current, I_d.

Interestingly, the concept of displacement current was the only original contribution by Maxwell to the four equations that bear his name. Maxwell established the foundation for comprehension of the concept of common-mode currents, essential for the awareness of grounding-related issues, for example, ground loops and balanced circuits. Common-mode currents are discussed in Section 2.7.2.1.

A direct outcome of Ampere's Law is the postulation that current always flows in closed loops. This is easily demonstrated now. We begin from Ampere's Law:

$$\nabla \times \vec{H} = \vec{J} + \frac{\partial \vec{D}}{\partial t} \qquad (2.35)$$

Taking the divergence of both sides of the equation, we obtain

$$\nabla \cdot \nabla \times \vec{H} = \nabla \cdot \left(\vec{J} + \frac{\partial \vec{D}}{\partial t} \right) = \nabla \cdot \vec{J} + \nabla \cdot \frac{\partial \vec{D}}{\partial t} \qquad (2.36)$$

However, from vector algebra we know that $\nabla \cdot \nabla \times \vec{H} \equiv 0$; therefore

$$\nabla \cdot \vec{J} + \nabla \cdot \frac{\partial \vec{D}}{\partial t} \equiv 0 \qquad (2.37)$$

can now be rewritten as

$$\nabla \cdot \vec{J} = -\frac{\partial}{\partial t} \left(\nabla \cdot \vec{D} \right) = -\frac{\partial \rho}{\partial t} \qquad (2.38)$$

Equation (2.38) can also be written, using Gauss's Theorem, as

$$\int_V \left(\nabla \cdot \vec{J} \right) dV = -\frac{\partial}{\partial t} \int_V \left(\nabla \cdot \vec{D} \right) dV = -\frac{\partial}{\partial t} \int_V \rho \, dV = -\frac{\partial Q}{\partial t} \qquad (2.39)$$

but

$$\int_V \left(\nabla \cdot \vec{J} \right) dV = \int_A \vec{J} \cdot dA = I = -\frac{\partial Q}{\partial t} \qquad (2.40)$$

This yields the equation presented above expressing the principle of Conservation of Charge, or the Law of Continuity [Equaiton (2.20)]. In general, this result demonstrates that:

- DC currents must always flow in closed galvanic loops, since for electrostatics, $\partial \vec{D} / \partial t = 0$, which mandates that $\nabla \cdot \vec{J} = 0$.
- AC and transient currents flow in closed circuits but the circuit may be comprised of conductive as well as displacement paths (which in circuits are known as capacitive paths), since for electrodynamics as a whole:

$$\nabla \cdot \vec{J} = -\nabla \cdot \left(\frac{\partial \vec{D}}{\partial t} \right) \qquad (2.41)$$

This outcome is of utmost importance for the understanding of electrical physical phenomena. For instance, lightning and electrostatic discharge (ESD) currents appear, at first glance, to violate the assertion that current always flows in closed loops, as they do not, seemingly, close any loops. However, the above derivation from Ampere's Law resolves this apparent paradox [Equations (2.41)]: When electrostatic and lightning discharges occur, a change (collapse) of the electrical field (change in flux or displacement) $-\partial \vec{D}/\partial t$ accompanies the conductive discharge current path $\vec{J}$, completing the loop, and maintaining continuity; thus, no contradiction exists.

A special case occurs when, in Ampere's Law, the length of the contour, C, on which integration of the magnetic field occurs, is reduced to zero. In that case, Ampere's Law states that the sum of all currents entering and leaving the volume must be zero. In this case, Ampere's Law transforms into Kirchhoff's Current Law, stating that the sum of all currents entering a lumped circuit node must be zero (Figure 2.12):

$$\oint_C \vec{H} \cdot d\vec{l} = \sum_i I_i = 0 \qquad (2.42)$$

Ampere's Law is the second of the two more important parts of Maxwell's Equations associated with grounding theory, demonstrating that current always flows in closed loops.

2.1.2.5 Law of Continuity. The Law of Continuity or Law of Conservation of Charge simply states that charge cannot be created or destroyed. The mathematical representation of the Law of Continuity in its differential and integral forms [Equations (2.19) and (2.20)] is

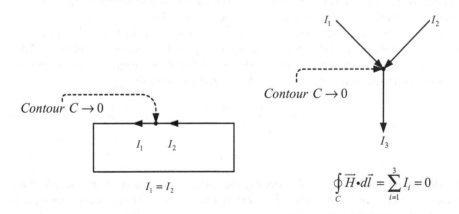

(a) Current always Flows in Closed Loops

(b) Total Net Current Entering any Junction in the Circuit is Zero

Figure 2.12. Equivalence between Ampere's Law and Kirchhoff's Current Law. Current does not accumulate anywhere in the circuit.

$$\oint_C \vec{J} \cdot d\vec{s} = -\frac{d}{dt}\int_V \rho_V dv \quad \text{or} \quad \nabla \cdot \vec{J} = -\frac{d\rho_V}{dt}$$

This equation illustrates the logical statement that any current leaving a closed surface S constitutes a decrease of the remaining charge in the volume enclosed within that surface.

Lessons Learned

- Maxwell's equations describe the basic interactions between electromagnetic sources and phenomena.
- Faraday's Law and Ampere's Law are fundamental for understanding grounding theory.
- All notions about inductance and its derivation stem from Faraday's Law of Induction, demonstrating that currents always flow in a loop area having the smallest possible contours.
- Ampere's Law demonstrates that current always flows in closed loops. Current must return to its source, whether in conductive paths or through displacement.
- Current, not voltage, constitutes the source of electromagnetic fields.

2.2 BOUNDARY CONDITIONS

Similar to all differential equations, Maxwell's equations have an infinite number of solutions. The uniqueness of a particular solution to Maxwell's equations that are partial differential equations is determined by specific boundary conditions. Boundary conditions are determined by characteristics of the medium and interfaces between different media with respect to their electromagnetic properties.

Assume a boundary between two physical media, a, b, characterized by their permittivity, permeability, and conductivity, ε_a, μ_a, σ_a, and ε_b, μ_b, σ_b, respectively. The tangential components of the electric field vector $\vec{E}$ and the magnetic field vector $\vec{H}$ must be continuous across the boundary between the two media:

$$\vec{E}_{t_a} = \vec{E}_{t_b} \tag{2.43}$$

$$\vec{H}_{t_a} = \vec{H}_{t_b} \tag{2.44}$$

In addition, from the Law of Continuity, the normal component of the electric flux density vector $\vec{D}$ and the magnetic flux density vector $\vec{B}$ must be continuous across the boundary between the two media:

$$\vec{D}_{n_a} = \vec{D}_{n_b} \tag{2.45}$$

$$\vec{B}_{n_a} = \vec{B}_{n_b} \tag{2.46}$$

Equation (2.45) for the electric flux density vector $\vec{D}$ holds a condition that no charges exist on the boundary between the two media. Figure 2.13 illustrates the boundary conditions and the field continuity across the boundary.

The effect of boundary conditions for electric and magnetic fields is fundamental to the concepts of image theory and "flux cancellation" provided by signal reference and current return planes, essential for EMI control in circuits and systems. In those cases, we are primarily concerned with boundary conditions, which apply when one of the media is a conductive surface, as would be the case with the use of reference (or return) planes in printed circuit boards. For simplicity, it can be assumed that medium b, for instance, can be approximated as a perfect electrical conductor, that is $\sigma_b \to \infty$. As a result, all fields within the perfectly conductive medium must vanish, that is, $\vec{E}_b = 0$, $\vec{D}_b = 0$, as well as $\vec{H}_b = 0$ and $\vec{B}_b = 0$. Equations (2.45) and (2.46) show that since all fields vanish in a perfectly conductive medium, the tangential electric field vector and normal magnetic flux density field vector components in medium a at the boundary must vanish too, resulting in

$$\vec{E}_{t_a} = 0; \qquad \sigma_b \to \infty \qquad\qquad (2.47)$$

$$\vec{B}_{n_a} = 0; \qquad \sigma_b \to \infty \qquad\qquad (2.48)$$

The tangential magnetic field vector and normal electric flux density vector components in medium a, do not vanish, however:

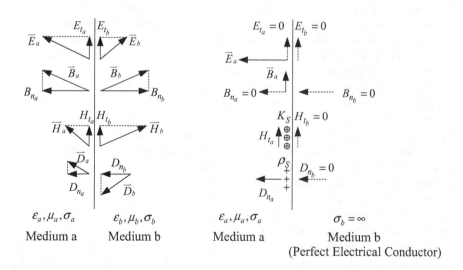

(a) Arbitrary Media   (b) Medium b is Perfectly Electrial Conductor

Figure 2.13. Electric and magnetic field vector continuity across a boundary between two media.

$$\vec{H}_{t_a} = K_S \text{ [A/m]}; \quad \sigma_b \to \infty \qquad (2.49)$$

$$\vec{D}_{n_a} = \rho_S \text{ [C/m}^2\text{]}; \quad \sigma_b \to \infty \qquad (2.50)$$

To satisfy the discontinuity, the tangential magnetic field component creates a surface current distribution K_S along the surface of the boundary that is orthogonal to $\vec{H}_t$. The units of K_S are in A/m, representing a surface current distribution along the surface of the intermedia interface. In a similar manner, the normal electric field flux density component deposits a surface charge distribution ρ_s along the surface of the intermedia interface. The units of ρ_s are in C/m^2.

Lessons Learned

- Boundary conditions across metallic surfaces preclude the existence of tangential electric field and normal magnetic flux density vectors.
- The effects of boundary conditions form the basis for image theory and "flux cancellation" concepts, essential for grounding and EMI control.

2.3 INTRINSIC INDUCTANCE OF CONDUCTORS AND INTERCONNECTS

The concepts of resistance, capacitance, and inductance are fundamental to the understanding of real-world grounding concepts. The meaning and calculation of resistance is clearly understood by all. When current I flows through a block of material and the resultant voltage drop V across it is measured, the resistance of the block of material is the ratio $R = V/I$. The units of resistance are in ohms (Ω).

Likewise, the concept and calculation of capacitance is easily understood. A potential difference V exists between any two bodies (which may be conductive or nonconductive), carrying equal and opposite charges, $+Q$ and $-Q$. The capacitance of this structure is defined as the ratio $C = Q/V$. The units of capacitance are in farads (F), named for Michael Faraday. Interestingly, Faraday had more to do with inductance than capacitance. Electric field $\vec{E}$ is usually visualized as lines of flux directed from the positively charged to the negatively charged body.

The magnitude of the electric field and, hence, the magnitude of the potential difference is directly proportional to the magnitude of the charges Q. Capacitance of this structure is thus dependent only on the shape and relative physical orientation of the two bodies, and the properties of the material they are immersed in. Alternatively, suppose we connect a voltage source, V, across the two bodies, resulting in a potential difference between them. Charge $Q = C \cdot V$ will subsequently be deposited from the source onto these bodies with the amount of charge Q that the bodies can hold, depending on the capacitance of the structure and the potential difference between them. Hence, capacitance represents the ability of a structure to maintain charge.

The concept of inductance appears to be less well comprehended, thereby bringing about misconceptions regarding its essence and mistakes in its calculation. Proper consideration of grounding design without a clear understanding of the concept of inductance is simply impossible. For instance, ever too often the performance of a grounding system is stated in terms of grounding resistance, totally neglecting the contribution of inductance to impedance. This section is dedicated to the explanation of the concept of inductance. Self and mutual and internal and external inductances are addressed. In particular, the concept of partial inductance, which allows the calculation of ground bounce and power rail collapse that are critically important in the design of electronic circuits for signal integrity, is presented in detail.

2.3.1 Concept of Inductance

The concept of inductance emerges directly from Faraday's Law of Induction:

$$\oint_C \vec{E} \cdot \vec{dl} = -\int_S \frac{\partial \vec{B}}{\partial t} \cdot \vec{ds} \qquad (2.51)$$

Whereas capacitance is associated with the separation of charge, inductance results from charges in motion, that is, electric current. All currents result in the creation of a magnetic field. Inductance is a phenomenon related to the storage of energy contained in the magnetic field surrounding current-carrying conductors. Conversely, if the magnetic field somehow vanishes, the inductance ceases to exist as well. Since inductance is conventionally defined for closed loops, how should the term "inductance of an isolated conductor" or "inductance of a segment of a conductor or portion of a loop" be understood?

A resolution to this confusing situation of determining self-inductance or, simply, the inductance of a current-carrying circuit is achievable by means of the concept of partial inductance. In a similar manner, mutual inductance between parallel conductors of a loop can be exclusively determined using the concept of partial mutual inductance.

2.3.2 Self-Inductance

For the definition of the concept of self-inductance we start from the Biot–Savart Law (from Section 2.1.1.2):

$$\overline{dB} = \frac{\mu_0 I \overline{d\ell} \times \hat{r}}{4\pi r^2} \qquad (2.52)$$

From the circuit depicted in Figure 2.14, we observe that a closed loop bounded by a contour C carrying a current I produces a magnetic flux density $\vec{B}$ emerging from the surface S of the circuit at a distance r from the current carrying element $\overline{d\ell}$ [4]. The contour C of the loop can be thought of as either a conducting material (as in the case of a wire) or an imaginary contour of nonconducting material.

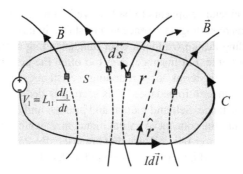

Figure 2.14. Model for formulation of the self-inductance concept.

Integrating Equation (2.52) with respect to the path length ℓ over the contour C we obtain the total flux density due to the loop's entire circumference:

$$\vec{B} = \frac{\mu_0 I}{4\pi} \oint_C \frac{\overrightarrow{d\ell'} \times \hat{r}}{r^2} \tag{2.53}$$

The total magnetic flux Φ penetrating a surface S bounded by the current carrying loop is therefore found by integrating the flux density $\vec{B}$ over the entire cross section of the loop:

$$\Phi \triangleq \int_S \vec{B} \cdot \vec{ds} \tag{2.54}$$

Substituting in Equation (2.53) $\vec{B}_1$, $d\ell_1$, and C_1 in Equation (2.54), and integrating over the surface the surface S_1 for a particular "loop 1," we end up with the self-inductance of the loop Φ_{11}:

$$\Phi_{11} = \int_{S_1} \vec{B}_1 \cdot \vec{ds} = \frac{\mu_0 I_1}{4\pi} \int_{S_1} \left(\oint_{C_1} \frac{\overrightarrow{d\ell_1'} \times \hat{r}}{r^2} \right) \cdot \vec{ds} \tag{2.55}$$

The self-inductance of a current-carrying loop is conventionally defined as the total magnetic flux penetrating the surface of the loop per unit current that produced it:

$$L_{11} \triangleq \frac{\Phi_{11}}{I_1} \tag{2.56}$$

Therefore,

$$L_{11} = \frac{\mu_0}{4\pi} \int_{S_1} \left(\oint_{C_1} \frac{\overrightarrow{d\ell_1'} \times \hat{r}}{r^2} \right) \cdot \vec{ds} \tag{2.57}$$

The subscripts "$_1$" and "$_{11}$" signify the self-inductance of a loop that is due to the flux produced by the current bounding it. The units of inductance are in henrys (H) honoring Joseph Henry of Albany, New York, who essentially discovered Faraday's Law at about the same time as Faraday.

Differentiating the left-hand part of Equation (2.56) with respect to time and the result with Equation (2.51) while considering total loop inductance results in one of the most commonly used expressions for the effect of magnetic induction, that of induced electromotive force or emf:

$$-E_L = -\oint_C \vec{E} \cdot d\vec{l} = \frac{d\Phi}{dt} \overset{\Phi=LI}{=} L \frac{dI}{dt} \tag{2.58}$$

According to Faraday's Law, therefore, the effect of the magnetic flux penetrating the surface of the contour C may be substituted by inserting an equivalent emf or voltage source V into the contour of the loop that encompasses the surface.*

Equation (2.58) depicts Lenz's Law (discussed in Section 2.1.2.3), demonstrating that a change in the current flow in the circuit will introduce an electromotive force (emf) E_L that opposes the rate of change of the magnetic flux through the loop (note the minus sign).

2.3.3 Mutual Inductance

Mutual inductance refers to magnetic flux penetrating a conducting loop C_2 produced by current I_1 flowing through another conducting loop C_1, causing induction of an electromotive force (emf) in the second circuit (Figure 2.15). Observe that the voltage source representing the induced emf in loop 2 has a polarity, according to Faraday's Law, such that it tends to induce a current in the second loop, which produces a magnetic flux that opposes the original magnetic flux passing through its surface, which is due to the current in loop 1.

Similar to Equation (2.56) for the definition of self-inductance, mutual inductance is conventionally defined as total magnetic flux penetrating one loop due to a unit current flowing in the contour of a second loop:

$$L_{12} \triangleq \frac{\Phi_{12}}{I_1} \tag{2.59}$$

Therefore,

$$L_{12} = \frac{\mu_0}{4\pi} \int_{S_2} \left(\oint_{C_1} \frac{\overline{d\ell_1}' \times \hat{r}}{r^2} \right) \cdot d\vec{s} \tag{2.60}$$

---

*Note that for the inductance of this loop to be of physical significance, the physical dimensions of this loop must be assumed to be electrically small. Furthermore, if the loop contour is assumed to be constructed of a conducting material (such as a metallic wire), these effects of the time-changing magnetic field can be represented as a single lumped voltage source placed anywhere in the loop contour.

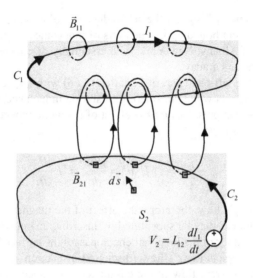

Figure 2.15. Model for formulation of the mutual inductance concept.

The subscripts "$_1$" and "$_{12}$" signify the mutual inductance between loops 1 and 2 by virtue of the flux penetrating loop 2 produced by current flowing in a closed current loop 1.

When most of the flux generated by loop 1 also couples into loop 2, the mutual inductance will approach the self-inductance of the circuit ($L_{12} \rightarrow L_{11}$). The loops are referred to as "tightly coupled" or "closely coupled." Hence, we may define the coupling factor k:

$$k \triangleq \frac{L_{12}}{L_{11}} \tag{2.61}$$

Note that k is upper bound by unity and will approach this value for tightly coupled loops.

2.3.4 Partial Inductance

It may be argued that in common expressions provided in the literature for conductors and traces, no indication of current return path and its association to inductance exists. It has been shown above that there is simply no such thing as a stand-alone conductor carrying current in a given direction; current *must* return to its source, thus current loops must subsequently exist. However, ample experimental evidence shows that when digital current passes through the return or "ground" conductor, a voltage drop will develop across two points on that return conductor when a change of logical state occurs at the digital driver, the current source (Figure 2.16) [5]. That voltage drop, commonly known as "ground bounce," appears to be proportional to the rate of change

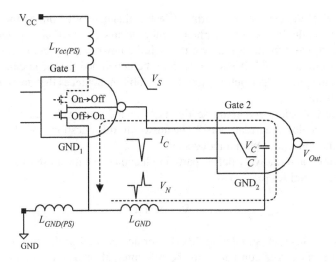

(a) Gate Switching from "Hi" ("1") to "Lo" ("0"),

Producing "Ground Bounce"

Figure 2.16. Ground bounce and "power rail collapse" generation.

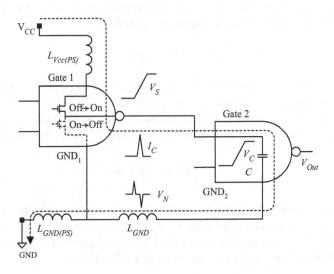

(b) Gate Switching from "Lo" ("0") to "Hi" ("1"),

Producing "Power Rail Collapse"

Figure 2.16. Ground bounce and "power rail collapse" generation.

(derivative) of the current waveform.* Clearly, this cannot be due to the resistance of that return conductor. A similar phenomenon occurs on the V_{CC} or power rail, when switching digital currents are drawn through the power supply conductor, referred to as "power rail collapse." It must, therefore, be concluded that segments of such conductors in the power distribution circuit do exhibit inductances that are uniquely attributable to them.

The question addressed in this section is, "Can inductances be uniquely attributed to segments of a closed current loop?" The answer to this question is yes! The key to doing so is the concept of partial inductance.

In order to quantitatively define partial inductance, we must use two important vector calculus identities [6]:

$$\nabla \cdot \nabla \times \vec{A} \equiv 0 \qquad (2.62)$$

And Stokes' Theorem, stipulating that the surface integral of the curl of a vector field through a surface S is equivalent to the line integral of that vector field around the closed contour C that encompasses this surface:

$$\int_S \nabla \times \vec{A} \cdot ds = \oint_C \vec{A} \cdot dl \qquad (2.63)$$

Gauss' Law for the magnetic field [Equation (2.12)] provides that all magnetic field lines must form closed loops, implying that the magnetic flux density vector $\vec{B}$ can be expressed in terms of the vector magnetic potential [6]:

$$\vec{B} = \nabla \times \vec{A} \qquad (2.64)$$

The vector magnetic potential $\vec{A}$ at all points around a current carrying wire can be shown to be parallel to the wire going to zero at infinity [5]. In addition, the vector magnetic potential is directly proportional to the current that produced it. Substituting Equation (2.64) into Stokes' Theorem [Equation (2.63)] yields the important result [8]

$$\int_S \vec{B} \cdot ds = \oint_C \vec{A} \cdot dl \qquad (2.65)$$

Hence, the magnetic flux through a surface S can be alternatively obtained as the line integral of the vector magnetic potential $\vec{A}$ around the closed loop contour C that encloses the surface. Substituting the result obtained in Equation (2.65) in the fundamental definition of inductance of a closed loop [Eq. (2.56)] results in [4]

$$L = \frac{\phi}{I} = \frac{\int_S \vec{B} \cdot ds}{I} = \frac{\oint_C \vec{A} \cdot dl}{I} \qquad (2.66)$$

*This phenomenon is known as "delta-I noise" and is discussed in detail in Chapter 9.

Therefore, the magnetic flux through the surface of a closed loop S, and subsequently the inductance of that closed loop, can be obtained by integrating the products of the components of the vector magnetic potential $\vec{A}$ tangent to the contour and the differential lengths of the contour around that closed contour $\vec{dl}$. It, therefore, follows that inductances may be uniquely attributed to segments of the loop contour C, which is represented as (Figure 2.17)

$$L = \sum_{i=1}^{n} \left(\frac{\oint_{C_i} \vec{A} \cdot dl}{I} \right) = \sum_{i=1}^{n} L_i \qquad (2.67)$$

Therefore, the closed loop contour C was divided into n segments such that $C = \sum_{i=1}^{n} c_i$, demonstrating that segments of a closed loop contour can indeed be uniquely attributed inductances, called self-partial inductances. As the vector magnetic potential $\vec{A}$ is produced by and proportional to the current I^* flowing through the ith segment of the loop contour c_i, it follows that the self-partial inductance of the segment that is associated with this segment is unique to that segment, regardless of the loop or loops that form part of it [9].

In a similar manner, the concept of mutual partial inductance between a segment of the contour of a closed loop c_i carrying current I_i and a second segment c_j of the same contour of the loop can be defined as

$$L_{ij} = \frac{\oint_{C_j} \vec{A} \cdot dl}{I_i} \qquad (2.68)$$

Figure 2.18 demonstrates the connection between the previously defined concept of the total closed loop C inductance with that of the sum of self- and mutual partial inductances associated with the loop's segments. Individual inductances L_{pii} are referred to as the self-partial inductances of the ith segment, whereas the inductances L_{pij} (where $i \neq j$) are referred to as the mutual partial inductance between the circuit's ith and jth segments [4].

Using this definition, the voltage developed across ith segment of a current-carrying loop can be exclusively and meaningfully obtained:

$$E_i = \sum_j \left[K_{ij} \left(L_{pij} \frac{dI_j}{dt} \right) \right]; \qquad K_{ij} = \pm 1 \qquad (2.69)$$

The K_{ij} factor is determined by the relative orientation of the currents assigned to segments i and ji, becoming negative ($K_{ij} = -1$) if the currents in segments i and j flow in opposite directions.

---

*This is clearly demonstrated by Ampere's Law.

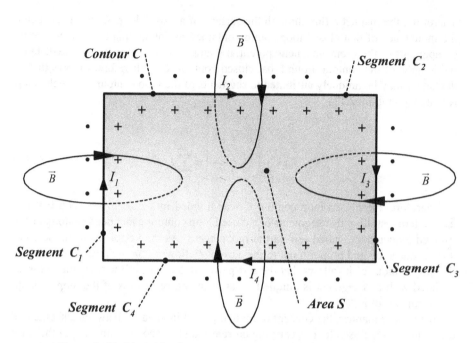

Figure 2.17. Model for formulation of the partial inductance concept.

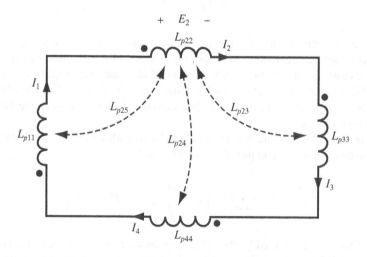

Figure 2.18. Equivalent circuit apportioning the loop inductance into individual inductances associated with segments of the loop.

Subsequently for segment 2, we obtain

$$E_2 = \sum_j L_{p2j} \frac{dI_j}{dt} = L_{p22}\frac{dI_2}{dt} + L_{p21}\frac{dI_1}{dt} + L_{p23}\frac{dI_3}{dt} - L_{p24}\frac{dI_4}{dt} \qquad (2.70)$$

As depicted in Figure 2.18, we identify different currents associated with the individual segments of the rectangular loop, I_i, where $i = 1, \ldots, 4$. However, $I_1 = I_2 = I_3 = I_4 = I$. Obviously, therefore, the total inductance of a segment of a loop constitutes the sum of the self- and mutual partial inductances associated with that segment:

$$L_i = \sum_j \left[K_{ij} L_{pij} \right]; \qquad K_{ij} = \pm 1 \qquad (2.71)$$

Consequently, the total inductance of the entire loop is obtained by

$$L_{Loop} = \sum_i \sum_j \left[K_{ij} L_{pij} \right]; \qquad K_{ij} = \pm 1 \qquad (2.72)$$

For a circuit with orthogonally arranged line segments, as depicted in Figure 2.18, the mutual partial inductances vanish when segments are mutually perpendicular; thus, $L_{p23} = L_{P21} = 0$ and $L_{p34} = L_{P14} = 0$. The total loop inductance [Equation (2.72)] reduces to

$$L_{Loop} = \sum_{i=1}^{n} L_i \qquad (2.73)$$

Here n is the total number of segments in the circuit.

Expressions for inductance can only be derived analytically for the simplest geometries; in most practical cases, some approximation is required. The partial inductances of a pair of parallel conductors depicted in Figure 2.19 from Equation (2.71) is (assuming $I_1 = I_2 = I$)

$$\begin{cases} L_{p1} = L_{p11} - L_{p12} \\ L_{p2} = L_{p22} - L_{p21} \end{cases} \qquad (2.74)$$

Equation (2.74) points to an interesting and extremely important conclusion: Because the partial mutual inductance between two conductors increases as the distance between the conductors shrinks, whereas the partial self-inductance remains unchanged, the total partial inductance decreases as the conductors are brought closer together, resulting in a reduction of inductive voltage drop developed across the conductors when brought closer together.

The low-frequency partial self-inductance of a round wire with a length l and radius r ($l \gg r$) is given by [4]

$$L_{p11\,(O)} = L_{p22\,(O)|_{Lo-F}} = \frac{\mu_0 l}{8\pi} + \frac{\mu_0 l}{2\pi}\left[\ln\frac{2l}{r} - 1 \right] \qquad (2.75)$$

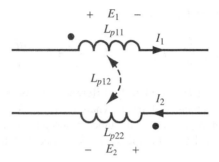

Figure 2.19. Total partial inductance of two parallel conductors in free space.

In this expression, the first term is denoted as the internal partial self-inductance and the last term the external partial self-inductance. At high frequency, the internal inductance vanishes, since the current is forced to flow on the surface only due to the skin effect (see Section 2.3.6), leaving

$$L_{p11\,(O)} = L_{p22\,(O)}\big|_{Hi-F} = \frac{\mu_0 l}{2\pi}\left[\ln\frac{2l}{r}-1\right] \qquad (2.76)$$

The low-frequency inductance of a pair of conductors with a rectangular cross section (such as traces on printed circuit boards) with a length l, thickness t, and width w ($l \gg w \gg t$) is similarly given by [4]

$$L_{p11\,(\Box)} = L_{p22\,(\Box)}\big|_{Lo-F} = \frac{\mu_0 l}{2\pi}\left[\ln\frac{8l}{w+t}-\frac{1}{2}\right] \qquad (2.77)$$

The high-frequency inductance cannot be straightforwardly extracted from Equation (2.71), but was derived separately [4]

$$L_{p11\,(\Box)} = L_{p22\,(\Box)}\big|_{Hi-F} = \frac{\mu_0 l}{2\pi}\left[\ln\frac{8l}{w}-1\right] \qquad (2.78)$$

If the separation between the pair of conductors is d, $l \gg d \gg w, r$, the mutual inductance between round or rectangular cross-sectioned conductors can be approximated as [4]

$$L_{p12\,(\Box,O)} = L_{p21\,(\Box,O)}\big|_{Hi-F} = \frac{\mu_0 l}{2\pi}\left[\ln\frac{2l}{d}+\frac{d}{l}\right] \qquad (2.79)$$

Using Equations (2-74) and (2-75) through Equation (2-79), the high-frequency total partial inductance for thin traces on printed circuit boards is therefore derived:

$$L_{p1\,(\square)} = L_{p2\,(\square)} = \frac{\mu_0 l}{2\pi}\left[\ln\frac{4d}{w} - \frac{d}{l}\right]^{d\ll l} \approx \frac{\mu_0 l}{2\pi}\left[\ln\frac{4d}{w}\right] \qquad (2.80)$$

The above expressions will be found to be useful in discussions found in other chapters of this book.

2.3.5 External and Internal Inductance

No current can flow within the cross section of perfect conductors and all current and charges exist on their surface only. Subsequently, magnetic flux occurs only outside the conductors, hence, inductance, as defined above, is called external inductance. The concept of loop inductance derived above implicitly considered external inductance only.

Real-world conductors, however, exhibit high but finite conductivity, and some nonzero current distribution can be observed within such conductors. Consequently, some magnetic flux exists within a conductor, associated with the portion of current flowing internal to it, resulting in inductance internal to the conductor, called internal inductance. The internal inductance of a wire of circular cross section with radius r, conductivity σ, permeability μ, and permittivity ε, is given by [4]

$$L_i = \begin{cases} L_{i,DC} = \dfrac{\mu_0}{8\pi} \approx 50\,nH/m = 1.27\,nH/\text{inch}, & \text{for } r \ll \delta \\[4mm] L_{i,HF} = \dfrac{2\delta}{r}\cdot L_{i,DC} = \dfrac{1}{4\pi\cdot r}\sqrt{\dfrac{\mu_0}{\pi\sigma}\cdot\dfrac{1}{f}}, & \text{for } r \gg \delta \end{cases} \qquad (2.81)$$

where δ is the skin depth, defined in Section 2.3.6 and particularly in Equations (2.86). The total inductance of a circuit or transmission line per unit length is the sum of both the internal and external inductance, L_i and L_e, respectively:

$$L = L_i + L_e \approx L_e \qquad (2.82)$$

Equation (2.81) demonstrates that internal inductance is inversely proportional to the square root of frequency, decreasing as $\sqrt{f}$. The value of the internal inductance of wires with circular cross section is upper bound by 50 nH/m. Therefore, when external inductance exceeds the internal inductance, total inductance is dominated by external inductance, hence, the approximation in Equation (2.82). In conclusion, at higher frequencies, internal inductance can be ignored.

Lessons Learned

- Inductance is everywhere!
- Inductance is an attribute of closed-circuit loops and is predominantly dominated by the area enclosed by a current loop.
- Partial inductance is a manner by which an overall loop can be broken into its sections in order to determine the total inductance of loop.

2.3.6 Skin Effect and Skin Depth

The terms skin effect and skin depth are frequently mentioned with respect to the propagating of high-frequency (HF) currents in conductive media. Unlike in DC applications, high-frequency current in a conductor does not flow throughout the entire cross section of the conductor. Alternating current flowing in the conductor tends to increasingly squeeze against the surface of the conductor, progressively decreasing the effective current-carrying cross section of the conductor, with the majority of the current concentrating in a thin annulus just below the surface of the conductor (Figure 2.20). Electromagnetic fields (and current) decay rapidly with depth inside a conductor. Resulting from this current redistribution is the rise of the conductor's apparent resistance with increasing frequency.

This tendency of alternating current to flow near the surface of a conductor, thereby restricting the total cross-sectional area and increasing the resistance to the flow of cur-

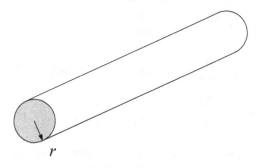

(a) DC current distribution

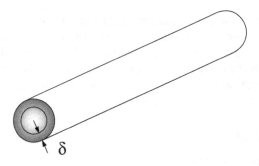

(b) High-frequency current distribution

Figure 2.20. Current distribution over a conductor's cross-section DC (a) and high frequencies (b).

rent, is called the skin effect.* The thickness of the conduction ring δ is referred to as the skin depth.

The skin effect constitutes an inductive phenomenon related to the rate of change of the magnetic field generated within the conductor due to the current flowing in the conductor. It can, thus, be explained using Ampere's Law [Equation (2.83)]:

$$\oint_C \vec{H} \cdot d\vec{l} = \int_S \left(\vec{J} + \frac{\partial \vec{D}}{\partial t} \right) \cdot d\vec{s} \qquad (2.83)$$

combined with Faraday's Law of Induction [Equation (2.84)]:

$$emf = \oint_C \vec{E} \cdot d\vec{l} = -\frac{\partial}{\partial t} \int_S \vec{B} \cdot d\vec{s} = -\frac{\partial \Phi}{\partial t} \qquad (2.84)$$

Time-varying magnetic fields $\vec{B}_{inc}$ crossing through a conductor create circulating eddy currents within the material. The eddy currents circulate about the penetrating flux (Ampere's Law), as illustrated in Figure 2.21, creating their own magnetic fields $\vec{B}_{eddy}$, opposing the incident field (Faraday's Law), resulting in a partial suppression of the penetrating field $\vec{B}_{pen}$ [10].

When time-varying current flows through a conductor's cross section, it creates a time-varying magnetic field around the current path, developing eddy currents, which flow along the current's path and oppose (cancel) the original current flowing through the core. As a result, the remaining net current is forced to flow on the circumference, resulting in increased current concentration on the surface, while decreasing its density within its center (Figure 2.22).

The largest current density is thus found at the surface of the conductor. The current density falls exponentially within the conductor. Denoting the value of the current density on the surface as $J(d = 0) = J_0$, the amplitude of the current density will decrease exponentially with increasing depth (d) in the material as

$$J(d) = J_0 \cdot e^{-\frac{d}{\delta}} \qquad (2.85)$$

The depth at which the current density amplitude is reduced to $J_\delta = J_0/e \approx 0.37 \cdot J_0$ and is called the skin depth or depth of penetration, δ, at the frequency f. For good conductors (e.g., metals), the displacement current is negligible compared to the conduction current ($\omega\varepsilon \ll \sigma$). Equation (2.86) provides the expression for the skin depth and Figure 2.23 illustrates the resultant current distribution within a conductor.[†] Typical values of skin depth in copper ($\sigma = 5.8 \times 10^7$ S/m, $\varepsilon_r = 1$, $\rho_r = 1$) versus frequency are provided in Table 2.1 [4].

---

*From FS-1037C, *Federal Standard 1037C, Telecommunication: Glossary of Telecommunication Terms*, [http://www.its.bldrdoc.gov/fs-1037/fs-1037c.htm], National Communications System Technology and Standards Division, August 7, 1996.
[†]This expression and the corresponding current distribution are valid as long as the skin depth is less than the radius of the conductor.

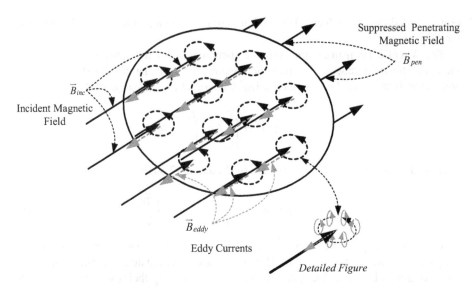

Figure 2.21. Eddy currents in a metal sheet due to time-varying penetrating magnetic flux generating an opposing magnetic field.

$$\delta = \frac{1}{\sqrt{\pi \cdot f \cdot \mu \cdot \sigma}}, \quad \text{meters} \qquad (2.86)$$

or

$$\delta = \frac{39.4 \cdot 10^3}{\sqrt{\pi \cdot f \cdot \mu \cdot \sigma}}, \quad \text{mils}$$

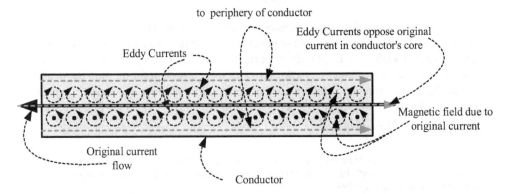

Figure 2.22. Generation of the skin effect in a current-carrying conductor.

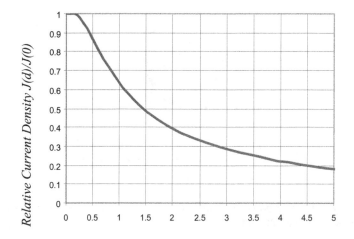

Depth in Metal, d (normalized to skin depths, δ)

Figure 2.23. Current density within a conductor falls exponentially with increased depth of penetration.

For frequencies sufficiently high, the curvature of a round conductor can be neglected and the current distribution can be approximated as uniformly distributed current across a flat conductor, having a width $W = 2\pi r$ and thickness δ (Figure 2.24).

Following on with the "flat" approximation of the current distribution on the round conductor's circumference, the resulting high frequency (HF) resistance can be approximated as [3]

$$R_{HF} \approx \frac{1}{2\pi r \sigma \delta}, \quad \Omega/m \qquad (2.87)$$

Table 2.1. Typical values of skin depth in copper

Frequency	Skin depth (δ)
60 Hz	8.5 mm
1 kHz	2.09 mm
10 kHz	0.66 mm
100 kHz	0.21 mm
1 MHz	2.6 mils*
10 MHz	0.82 mils
100 MHz	0.26 mils
1 GHz	0.0823 mils

*Note than 1 mil = 10^{-3} inch, hence 1 meter is approximately equivalent to 39,400 mils.

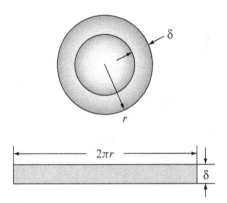

<u>Figure 2.24.</u> Approximation of the current distribution along the circumference of a round conductor due to skin effect.

Combining equations (2.86) and (2.87), we obtain

$$R_{HF} \approx \frac{1}{2\pi r \sigma \delta}^{\left(\times \frac{r}{r}\right)} = \frac{1}{\pi r^2 \sigma} \cdot \frac{r}{2\delta} = R_{DC} \cdot \frac{r}{2\delta}, \qquad \text{ohms} \qquad (2.88)$$

Reasonable approximations exist for resistance of rectangular cross-section conductors typical of traces on printed circuit boards. Assuming that at higher frequencies current concentrates predominantly in a rectangular cross section about the circumference of the conductor within a skin depth, the high-frequency resistance of such a conductor becomes [4]

$$R = \begin{cases} R_{DC} = \dfrac{1}{\sigma \cdot w \cdot t}, & \Omega/m \text{ for } t \ll \delta \\ R_{HF} = \dfrac{1}{2\sigma\delta \cdot (w+t)} \approx \dfrac{1}{2\sigma\delta \cdot w}, & \Omega/m \text{ for } t \gg \delta \end{cases} \qquad (2.89)$$

where t and w are the thickness and width of the rectangular conductor, respectively.

It follows, therefore, that conductors sufficiently thick with respect to the skin depth at the frequency of concern can be assumed to carry the current at a depth of several skin depths only from the surface, with the remainder of the conductor having virtually no effect on current propagation. The immediate result of this effect is the increase in the conductor's AC resistance with frequency. Following the skin depth's frequency dependence, the conductor's resistance increases with the square root of frequency. This current redistribution also results in the reduction of the conductor's inductance.

Lessons Learned

- Inductance is everywhere and is predominantly dominated by the area enclosed by the current path.

- Skin effect results from magnetic fields within the conductor, forcing high-frequency currents to flow only in a shallow band on the circumference of the conductor, utilizing only a very small fraction of the conductor's cross section.
- The effective depth at which the current density amplitude is reduced to $J_\delta = J_0/e \approx 0.37 \cdot J_0$ is called skin depth and is frequency dependent
- Resulting from the skin effect is the increase in the conductor's apparent resistance and decrease of its inductance with increase of frequency.
- Inductance of conductors comprises of external and internal inductance and is dominated by external inductance due to skin effect at higher frequencies

2.3.7 Proximity Effect

From the discussion on skin effect (Section 2.3.6), we would anticipate that when high-frequency current flows in a round conductor, the majority of the current would be distributed uniformly around the perimeter of the wire. That may be true for an infinitely long, "isolated" wire in free space. This effect is an extension of the skin effect and is called the proximity effect. The proximity effect tends to increase the AC resistance of a conductor to a value greater than that due to simple skin effect.

When two or more conductors are in close proximity, however, the current distribution is not uniform around the conductor. The current flowing in one conductor (e.g., a wire or plane) is redistributed because of the magnetic field produced by the current flowing in the adjacent conductor (e.g., a wire or PCB trace), introducing a slight nonuniformity to the current distribution around the perimeter of the conductor (Figure 2.25) or in the adjacent plane. This current redistribution, in turn, brings about an increase in the apparent resistance of the conductors beyond that expected due to the skin effect alone. In parallel conductors, this phenomenon is commonly entitled the proximity effect.

The proximity effect is not directly associated with the skin effect (related to high-frequency current distribution around the current-carrying conductor, as discussed in Section 2.3.6), and stands apart from the mechanical force effect discovered by Ampere, whereby adjacent conductors carrying equal but opposite DC currents repel each other [10]. Unlike the above, the proximity effect simply redistributes the AC current density around the perimeters of the adjacent wires.

Similar to the skin effect, the proximity effect results from an inductive mechanism occurring due to the time-varying magnetic fields. It perturbs the high-frequency currents but is insignificant for low-frequency currents. At frequencies beyond which the (frequency-dependent) proximity effect comes into force, the current distribution on the perimeter of the wires assumes a pattern of least inductance* and does not vary any further as frequency increases, as long as the transmission line formed by the conductors stays in the TEM propagation mode.

The proximity effect occurs as a result of a phenomenon similar to that taking place in the generation of the skin effect (see Figure 2.26). Magnetic flux formed by the cur-

---

*"Least inductance" high-frequency current propagation is discussed in Section 2.5.

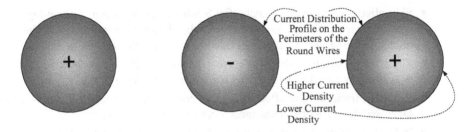

(a) Skin Effect in a Single and "Isolated"   (b) Proximity Effect in Two Adjacent

Current-Carrying Conductor   Current-Carrying Conductors

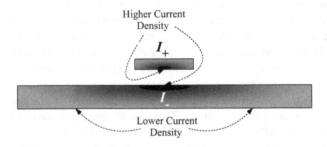

Figure 2.25. Approximation of the current redistribution along the circumference of a current-carrying conductor due to the proximity effect.

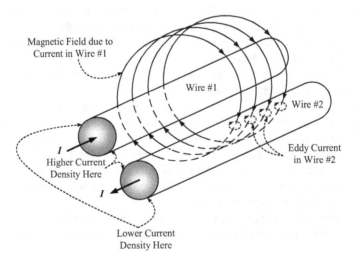

Figure 2.26. Generation of the proximity effect in parallel wires.

rent flowing in Wire #1 crosses through the surface of Wire #2, producing eddy currents at the entry and exit points on its surface, bringing about a magnetic field, which, on its part, induces a current opposing the original current flowing through it, on that surface. Consequently, the current flowing through Wire #2 is forced to concentrate on the surface facing Wire #1, bringing about the phenomenon known as the proximity effect. The proximity effect finds its greatest application in the understanding of the path of return current flow in transmission lines, for example, PCB traces and coaxial cables, in particular, "current at high frequencies assumes a distribution that minimizes the inductance of the loop formed by the signal and return path" [11]. This principle, discussed in detail in Section 2.5, constitutes a fundamental principle in shielding and grounding design of circuits and systems.

Lessons Learned

- The proximity effect is similar to the skin effect in that they are both associated with magnetic interactions, but is a result of interaction between two current-carrying conductors, rather than a self-induced phenomenon.
- Currents flowing in two adjacent conductors tend to somewhat distort the uniform current distribution of each other, forcing the current to flow in a manner that minimizes the size of the current loop.
- The proximity effect is consistent with, and is a manifestation of, the physical principle whereby "current at high frequencies, assumes a distribution that minimizes the inductance of the loop formed by the signal and return path" [11].

2.4 NONIDEAL PROPERTIES OF PASSIVE CIRCUIT COMPONENTS AND INTERCONNECTS

Behavior of circuit elements and interconnects in a manner not intended by the designer is one of the most important contributors to the generation and coupling of EMI, almost always resulting from the intrinsic characteristics of electrical conductors, often called parasitics.

Parasitics, often called the hidden schematic, constitute influential factors in the performance of circuit elements, deviating from their so-called ideal behavior as frequencies increase.

In DC and lower-frequency circuits, ideal conductors may not exhibit any resistance. However, practical conductors have finite resistivity and, hence, must be assumed to have some resistance. Inductance is associated with magnetic fields resulting from current distribution in conductors, whereas capacitance is related to the electrical fields, due to charge distribution in conductors. These "invisible circuit" attributes pop up and may even dominate at higher frequencies. It follows, therefore, that any current-carrying conductor will ultimately exhibit some resistance, inductance, and capacitance, which, if ignored, may demonstrate unexpected or bizarre system performance.

Hidden characteristics of conductors play an important role in determining the efficiency of grounding and bonding circuit implementation, particularly at higher frequencies. Each passive device will now be discussed separately with respect to its real world behavior.

2.4.1 Resistors

Resistors are among the most commonly used passive devices. Normally assumed to exhibit a frequency-independent resistance (at low frequencies), they will act at high frequencies like a series combination of inductance with resistance in parallel with a capacitor, as depicted in Figure 2.27.

Depending on the material and technology of the resistor (e.g., carbon composition, carbon film, mica, wire-wound, etc.) their high-frequency performance may be severely degraded: A wire-wound resistor is not suitable for high-frequency applications due to excessive inductance and inter-winding capacitance in the resistor wiring. Film resistors exhibit lower inductance and are often useful for high-frequency applications. Resistance itself will also be frequency dependent, increasing with frequency due to the skin effect.

Figure 2.28 depicts a typical frequency response for a real-world resistor. Values used for simulation of this resistor were $R = 10\ \Omega$, $L_S = 50$ nH and $C_P = 1$ nF.

$$Z_R = \frac{1}{\left(j\omega C_P + \dfrac{1}{R}\right)} + j\omega L_S \tag{2.90}$$

Clearly, at lower frequencies the resistor exhibits its ideal value, namely $R = 10\ \Omega$. At higher frequencies, however, the reactance dominates, driving the resistor to an inductive behavior due to its high internal series inductance. This could be the result of a large lead inductance or due to the resistor's technology, with wire-wound resistors exhibiting higher inductance.

A commonly overlooked aspect of resistors deals with package size and parasitic capacitance. Capacitance exists between the two terminals of the resistor. In wire-wound resistors, interwinding capacitance is present in the resistor's wiring. This parasitic capacitance can play havoc with extremely high-frequency designs, especially those in the GHz range. For most applications, however, parasitic capacitance between resistor leads is not of major concern.

(a) Low-Frequency Model   (b) High-Frequency Model

Figure 2.27. Nonideal properties of resistors.

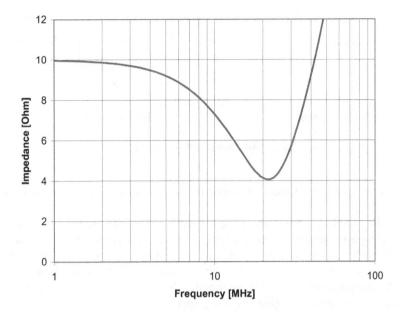

<u>Figure 2.28.</u> Frequency response of "real-world" (nonideal) resistors ($R = 10\ \Omega$, $L_S = 50$ nH and $C_P = 1$ nF).

In Figure 2.28, effect of the capacitance is observed in the intermediate frequency range, where a drop in the impedance is observed.

2.4.2 Capacitors

Capacitors are generally used for filtering, decoupling, bypassing, and certain grounding applications, as well as for many other high-frequency circuit applications. Capacitors exhibit a decreasing reactance with increasing frequency. This is shown in the expression $X_C = (2\pi f C)^{-1}$, where X_C is capacitive reactance (ohms). At higher frequencies, however, they perform as a series combination of inductance (equivalent series inductance or ESL) and resistance (equivalent series resistance or ESR), with a shunt resistance in parallel to the capacitor, representing the leakage resistance across the capacitor. A real-world capacitor C remains capacitive up to its self-resonant frequency, beyond which it exhibits inductive effects Figure 2.29):

$$Z_C = \frac{1}{\left(j\omega C + \dfrac{1}{R_P}\right)} + j\omega L_s + R_s \qquad (2.91)$$

Figure 2.30 depicts a typical frequency response of a real-world capacitor. Values used for simulation of this resistor were $C = 10$ nF, $R_S = 0.02\ \Omega$ (ESR), and $L_S = 5$

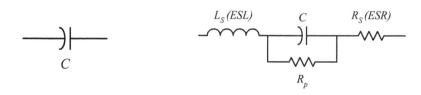

(a) Low-Frequency Model   (b) High-Frequency Model

Figure 2.29. Nonideal properties of capacitors.

nH (ESL). The parallel (shunt) resistance R_P is negligible for high-quality ceramic capacitors.

Depending on the composition of the dielectric material and technology of the capacitor, its high-frequency performance may be severely degraded. Electrical parameters of electrolytic capacitors with high values of equivalent series inductance (ESL) and equivalent series resistance (ESR) limit effectiveness of this particular type of capacitor to operation below 1 MHz, approximately. Ceramic capacitors, on the other hand, are more suitable for higher frequency applications.

In addition to intrinsic parasitic characteristics, the lead inductance of the capacitors must not be neglected when performing mathematical calculation of the self-res-

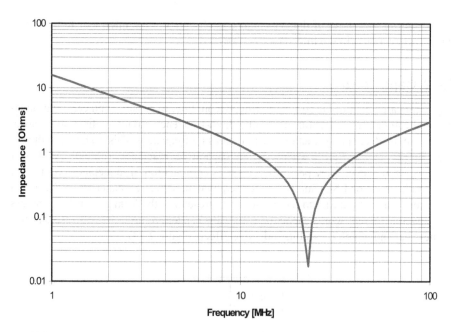

Figure 2.30. Frequency response of real-world (nonideal) capacitors (C = 10 nF, L_S = 5 nH, and R_S = 2 mΩ).

onant frequency and will contribute to additional inductance in series with the capacitor.

In summary, parasitic inductances and, to a lesser extent, the resistance in the capacitor's structure and lead bond wires, limit the performance of the capacitor to frequencies below its self-resonance frequency, beyond which the inductive behavior of the capacitor dominates.

2.4.3 Inductors

Inductors are used for EMI control, filtering, in certain grounding applications, and many other purposes. Inductors exhibit an increased reactance with an increase in frequency. This is described by $X_L = 2\pi f L$, where X_L is inductive reactance (ohms).

Similar to capacitors, inductors also exhibit parasitics: An inductor at high frequencies exhibits equivalent parallel interwinding capacitance. Resistance further limits, to a lesser extent, its performance (Figure 2.31).

Figure 2.32 illustrates a typical frequency response for a real-world inductor. Values used for simulation of this inductor were $L = 1$ μH, $R = 10$ mΩ, and $C_P = 10$ pF. The parallel (shunt) resistance R_P is typically negligible for high-performance inductors.

$$Z_L = \cfrac{1}{\cfrac{1}{R_P} + \cfrac{1}{j\omega L + R_S} + j\omega C_P}$$

(2.92)

2.4.4 Interconnects (Wires and PCB Traces)

Electrical conductors, for example, wires, traces, and interconnects, exhibit resistance as well as parasitic capacitance and inductance from the bond wires of the silicon die to the leads of passive devices. These (geometry-dependent) attributes influence the frequency-dependent impedance of the conductor. Depending on dimensions and geometry, a self-resonance will occur at a particular frequency, thus creating an efficient radiating antenna and an inefficient grounding conductor. Figure 2.33 illustrates

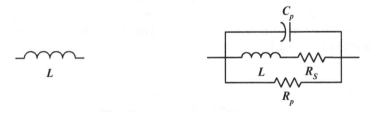

(a) Low-Frequency Model   (b) High-Frequency Model

Figure 2.31. Nonideal properties of inductors.

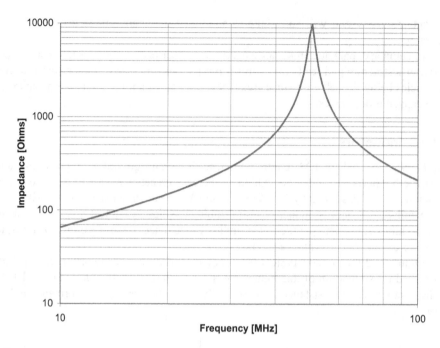

Figure 2.32. Frequency response of real-world (nonideal) inductors ($L = 1$ μH, $R = 10$ m Ω, and $C_P = 10$ pF).

the equivalent circuit of wiring. Note that this model is actually for an unbalanced line. The model for a balanced line is the same except that R and L are divided symmetrically on both conductors, with $R/2$ and $L/2$ both on the upper conductor and the lower conductor, as shown in Figure 2.33.

At lower frequencies, where the conductor is much shorter than $\lambda/10$, inductors can be modeled as a lumped circuit. The conductors are essentially resistive. However, at frequencies exceeding several kHz, shunt capacitance and inductive reactance of interconnects of any type of transmission line (e.g., twisted pair, coaxial cable, circuit board traces, etc.) cannot be ignored as their effect is greater than their resistance, even when considering the increase in resistance due to skin effect. At even higher frequencies, the conductor becomes reactive, and may be considered to be an efficient radiator of RF energy. In this situation, transmission line effects must be considered and are discussed in Section 2.5.

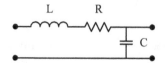

Figure 2.33. Equivalent circuit of wiring.

Lessons Learned

- Parasitics in passive circuit elements and interconnects limit their high performance.
- Consideration of the "hidden circuit elements" is crucial to the optimization of grounding design at high frequencies.
- With "electrically long" conductors, lumped circuit analysis can no longer be applied; transmission line (distributed) effects must then be employed.

2.5 RETURN CURRENT PATH IMPEDANCE

One Ring to rule them all, One Ring to find them, One Ring to bring them all and in the darkness bind them—J. R. R. TOLKIEN, *The Lord of the Rings*

There are few physical principles that have such broad implications as the one to be discussed in this section. Among the most well-known, wide-encompassing laws of nature are those of conservation of energy, conservation of momentum, and above all, the least action principle. This section addresses the question of the preferred path of current return in electrical and electronic systems.

No doubt, current flows in closed loops (hence the "ring"). However, can it be determined a-priori what constitutes that "loop," particularly when alternative paths exist? It will be shown in this section that an unambiguous and conclusive answer does indeed exist. Implications of the principles discussed in this section to grounding, in particular, and to electrical and electronic system design, in general, are far-fetched and are shown to be directly traceable to the most fundamental principles and postulates of modern science.

2.5.1 What Path Should Return Currents Follow?

The behavior of any conductor used for carrying current, particularly for grounding and return current paths, must be known from DC up to high frequencies, according to application. Electrical conductors exhibit resistive as well as reactive properties, which have critical effect on their high-frequency performance (see Section 2.3).

One of the most common misconceptions associated with circuit and system design is associated with the path of current return. When asked what path the return current follows in a circuit or system, the typical and intuitive replies vary from the shortest path to the path of least resistance. As a result, design of grounding conductors erroneously focuses on making their cross section as massive as reasonably achievable. This may be true for DC, but does not apply to higher frequencies.

Figure 2.34(a) illustrates the problem. Two arbitrary circuits, circuit #1 and circuit #2, are interconnected by signal and intended current-return wires. In addition, the two circuits also share a common signal reference (the different symbols used for the signal reference connection are indicative of the fact that some potential difference may exist between the two points).

Figure 2.34(b) demonstrates that one alternative for the current return is the shortest path, the path through the signal reference. This path will also normally exhibit the

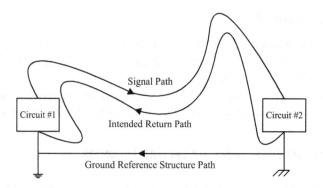

(a) What Path Will the Return Current Actually Take?

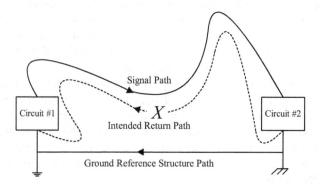

(b) Will It Take the Shortest Path?

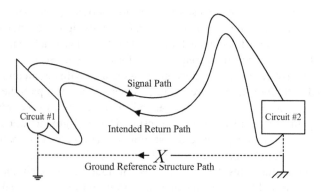

(c) Will It Take the (Longer) Intended Path?

Figure 2.34. Illustration of the problem, which path will the return current follow?

least resistance. If indeed this constitutes the return path, there would be no need for the intended return conductor. Could (and should) it be eliminated (indicated by the X on the return wire)?

Figure 2.34(c), on the other hand, demonstrates that the other alternative for the current return is via the intended return path. This path will definitely not exhibit the least resistance. Why would the return current take the path of higher resistance?

Of course, by disconnecting one of the circuits from the signal reference plane (indicated by the X in circuit 2), the current could be forced to flow through the return conductor, but is that desirable?

What then, is the actual current return path?

2.5.2 Equivalent Circuit Analysis

Figure 2.35 portrays a simplified equivalent circuit of the problem illustrated in Figure 2.34. Circuit #1 is depicted as a (differential) signal source, whereas Circuit #2 is depicted as a (differential) resistive load [12]. The L_i and L_s symbolize the partial self-inductance of the signal and intended return conductor, whereas M symbolizes the partial mutual inductance between the two. Signal current is symbolized by I_s, the current flowing through the intended path is I_i, and current flowing through the signal reference is I_g. Applying Kirchhoff's Voltage Law to the loop $A–R_i–L_i–B$, formed by the signal and intended return conductors, we obtain

$$\left(R_i + j\omega L_i\right) \cdot I_i - j\omega M \cdot I_s = 0 \tag{2.93}$$

Note that the current in the signal conductor is inductively coupled into this loop via the mutual inductance M between the signal and intended return conductor. Rearranging Equation (2.93) we obtain

$$\frac{I_i}{I_s} = \frac{j\omega M}{R_i + j\omega L_i} \approx \frac{j\omega L_i}{R_i + j\omega L_i} \tag{2.94}$$

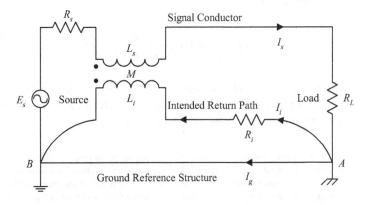

Figure 2.35. Simplified equivalent circuit of the system.

For tightly coupled signal and intended return conductors (i.e., when the spacing between the conductors d becomes infinitely small), the mutual inductance approaches the self-inductance of the conductors, that is,

$$\lim_{d \to 0} \{M_{Si}\} = L_{ii} \qquad (2.95)$$

where $M_{si} = M$ is the mutual inductance between the signal and the intended return current path, and $L_{ii} = L_i$ is the self-inductance of the intended return conductor. The tight coupling between the conductors justifies, therefore, the assumption $M \approx L_i$.* Further rearranging Equation (2.94),

$$\frac{I_i}{I_s}(\omega) = \frac{j\omega}{j\omega + R_i/L_i} = \frac{j\omega}{j\omega + \omega_c} \qquad (2.96)$$

ω_c constitutes the cutoff frequency of the current ratio in the intended return current conductor. The current ratio in Equation (2.96) is illustrated in Figure 2.36, which clearly demonstrates two distinct situations with respect to frequency (refer to Figure 2.36):

1. Low-frequency signal propagation
2. High-frequency signal propagation

The break frequency between "low" and "high" depends on the ratio between the resistance and inductive reactance of the circuit, clearly observed from Equation (2.96) to occur at

$$\omega_c = \frac{R_i}{L_i}, \qquad \text{or} \qquad f_c = \frac{R_i}{2\pi L_i} \qquad (2.97)$$

2.5.2.1 Return Current Path at Low Frequencies. At low frequencies ($\omega \to$ 0), Equation (2.96) reduces to

$$\lim_{\omega \to 0}\left(\frac{I_i}{I_s}\right) = \lim_{\omega \to 0}\left(\frac{j\omega}{j\omega + R_i/L_i}\right) = 0 \qquad (2.98)$$

Hence, $I_i \ll I_s$ and $I_i \ll I_g$. It follows therefore, that at very low frequencies, a large proportion of the return current flows through the signal reference, or the path of least resistance, rather than via the intended return conductor.

---

*Recall that when two conductors are closely spaced, the majority of the magnetic flux surrounding the signal conductor carrying a current, I, also surrounds the adjacent conductor, thus for a given source current the "same" flux surrounds the source conductor (contributing to its self partial inductance) and the adjacent conductor (contributing to its mutual partial inductance). If therefore follows that the self partial inductance and mutual partial inductance are "equal" in magnitude.

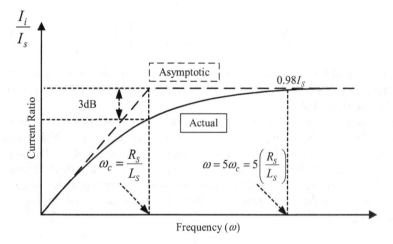

Figure 2.36. Ratio of the current in the intended return conductor versus signal current.

The impedance of the intended return path consists of both a resistive and a reactive (primarily inductive) component, $Z_i = R_i + j\omega L_i$. At low frequencies, $(\omega L_i \to 0, R_i \gg \omega L_i)$, impedance is dominated by the resistive component, R_i.

The solid line in Figure 2.37 depicts the preferred path for the return current at low frequencies, whereas the broken line (- - - - -) shows the (unused) intended return path. At low frequencies, the return current, therefore, follows the path of least resistance.

2.5.2.2 Return Current Path at High Frequencies. At higher frequencies, where $\omega \gg R_i/L_i$, Equation (2.96) approaches unity:

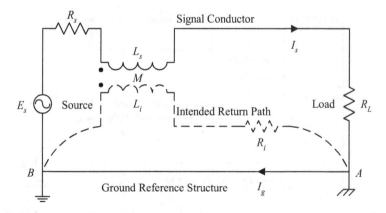

Figure 2.37. At low frequencies, the return current follows the path of least resistance.

$$\lim_{\omega \to \infty}\left(\frac{I_i}{I_s}\right) = \lim_{\omega \to \infty}\left(\frac{j\omega}{j\omega + R_i/L_i}\right) = 1 \qquad (2.99)$$

Hence, $I_i \approx I_s$ and $I_g \ll I_i$. This implies that at high frequencies the majority of the return current flows through the intended return path, as long as the return path is tightly coupled to the signal path.

Since the impedance of the intended return path consists of both resistive and reactive (primarily inductive) components, $Z_i = R_i + j\omega L_i$ and $\omega L_i \gg R_i$ at higher frequencies, impedance is dominated by the reactive (inductive) component, L_i. The solid line in Figure 2.38 depicts the preferred path for the return current at high frequencies, whereas the broken line (- - - - -) shows the ("abandoned") return path, the signal reference structure. At higher frequencies, therefore, the return current will follow the path of least inductance.

2.5.2.3 Return Current Flow in Inductive Return Paths.

Forcing return current at higher frequencies to follow a path other than the path of least inductance, for instance, by obstructing the intended current return path with a highly inductive return path connection, will cause current division. This will force a large proportion of the return current to follow a path of higher impedance. This ultimately results in an increase of EMI coupling (e.g., radiation to and from the circuit and crosstalk to adjacent conductors). In Figure 2.39, L_T represents the return path termination impedance.

Applying Kirchhoff's Voltage Law to the loop $A–L_T–L_i–R_i–B$, we now obtain instead of Equation (2.93),

$$\left[R_i + j\omega \cdot \left(L_i + 2L_T\right)\right] \cdot I_i - j\omega M \cdot I_s = 0 \qquad (2.100)$$

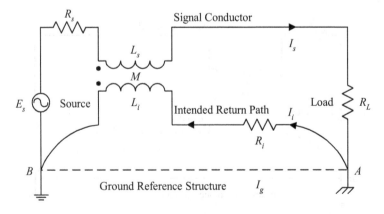

Figure 2.38. At high frequencies, the return current follows the path of least inductance.

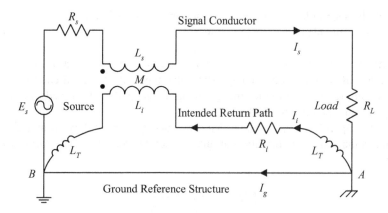

<u>Figure 2.39.</u> A highly inductive return path forces the return current to paths other than the path of least inductance.

Again, assuming $M \approx L_i$ and rearranging Equation (2.100), we arrive at

$$\frac{jw \cdot L_i}{R_i + jw \cdot (L_i + 2L_T)} = \frac{1 + jw \cdot \left(\dfrac{L_i}{R_i}\right)}{1 + jw \cdot \left(\dfrac{L_i + 2L_T}{R_i}\right)} \tag{2.101}$$

For higher frequencies, the right-hand part of this expression can be simplified to

$$\left.\frac{I_i}{I_s}\right|_{\text{High } w} \approx \frac{L_i}{L_i + 2L_T} \tag{2.102}$$

It follows, therefore, that for all return currents to follow the intended path at high frequencies ($w \to \infty$), L_T must be minimized ($L_T \to 0$).

2.5.2.4 Physical Generalization of the Path of Least Impedance Principle

The law that entropy always increases holds, I think, the supreme position among the laws of Nature. If someone points out to you that your pet theory of the universe is in disagreement with Maxwell's equations—then so much the worse for Maxwell's equations. If it is found to be contradicted by observation—well, these experimentalists do bungle things sometimes. But if your theory is found to be against the second law of thermodynamics I can give you no hope; there is nothing for it but to collapse in deepest humiliation.—SIR ARTHUR STANLEY EDDINGTON, *The Nature of the Physical World* (1927)

The principle presented above stems from wider-ranging physical premises, those of conservation of energy (or the first law of thermodynamics), further expanded through the second law of thermodynamics and the principle of least action or, more accurately, principle of stationary action.

PATH OF LEAST IMPEDANCE AND CONSERVATION OF ENERGY. In physics, the well-known law of conservation of energy states that the total amount of energy in any isolated system remains constant. Energy cannot be created or destroyed; it can only be changed from one form to another.

From Ampere's Law, we know that time-varying current flowing in a circuit will produce a magnetic flux density B around the conductors and, particularly, within the loop:

$$\oint_C \vec{H} \cdot d\vec{l} = \int_S \left(\vec{J} + \frac{\partial \vec{D}}{\partial t} \right) \cdot d\vec{s} \qquad (2.103)$$

When time-varying current flows in a circuit, it follows from Ampere's Law that magnetic flux will be present around the conductors, but will perish elsewhere (except in the very near vicinity of the loop), thanks to the cancellation of the flux outside the contour.

Faraday's Law of Induction illustrates that when time-varying magnetic flux crosses through a closed loop circuit, an electromotive force (emf) is induced into the circuit, having a polarity that tends to generate a current, which results in a magnetic flux that will opposes any change in the original magnetic flux that originally produced it (hence the negative sign):

$$emf = \oint_C \vec{E} \cdot d\vec{l} = -\frac{\partial}{\partial t} \int_S \vec{B} \cdot d\vec{s} = -\frac{\partial \vec{\Phi}}{\partial t} \qquad (2.104)$$

The energy volume density associated with the magnetic field $\vec{B}$ is expressed as

$$\eta_{\vec{B}} = \frac{Energy}{Volume} = \frac{1}{2} \frac{\vec{B}^2}{\mu} \qquad (2.105)$$

Clearly, the total energy stored in the magnetic field is directly proportional to the total volume, where the flux is nonzero, essentially within the contour of the current loop. Stated in terms of energy, current will flow in the path such that the total energy stored in the consequent magnetic field is minimized. Infinitely increasing the current loop area would ultimately drive the energy buildup in the current loop $\eta_{\vec{B}}$ to infinity, obviously in contradiction with conservation of energy

The negative sign in Faraday's Law therefore demonstrates that time-varying (i.e., high frequency) currents will always seek to flow in the path constituting the smallest possible contour C. Figure 2.40 illustrates this principle.

PATH OF LEAST IMPEDANCE AND THE SECOND LAW OF THERMODYNAMICS

Every star that burns, every planet whose orbit is slowly decaying, every breath you take and calorie you metabolize brings the universe closer and closer to the point when the en-

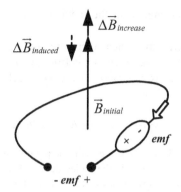

<u>Figure 2.40.</u> A rationalization for the current flow in the path that minimizes the total energy stored in the consequent magnetic field.

tropy is maximized, organized movement of any kind ceases, and nothing ever happens again. There is no escape. No matter how magnificent life in the universe becomes or how advanced, the slow increase in entropy cannot be stopped—the universe will eventually die.—WARREN FRIEDL, University of Windsor (2007)

Conservation of energy further stems from yet a broader physical law, the second law of thermodynamics. The second law of thermodynamics is an axiom of thermodynamics. It is an expression of the universal law of increasing entropy* and the direction in which thermodynamic processes can occur. Roughly speaking, the second law states that the entropy of an isolated system that is not in equilibrium will tend to increase over time, approaching a maximum value at equilibrium. In other words, the second law says that in an isolated system, concentrated energy disperses over time and, consequently, less concentrated energy is available to produce useful work. Energy dispersal also means that differences in temperature, pressure, and density even out. Hence, the second law is closely associated with the concept of entropy.

In simple terms, the second law is an expression of the fact that over time, ignoring the effects of self-gravity, differences in temperature, pressure, and density tend to even out in a physical system that is isolated from the outside world. Entropy is a measure of how far along this evening-out process has progressed.

Currents flowing in a circuit are driven by electric potential, which may be conceived of as "electric pressure." When this electrical pressure varies, an electric field exists, which creates a force on charged particles, resulting in the flow of electrical current. Generalizing to time-varying electrical potential fields, an effective potential

*Entropy is a measure of the unavailability of a system's energy to do work. It is a measure of the randomness (or "disorder") of molecules in a system and is central to the second law of thermodynamics and the combined laws of thermodynamics, which deal with physical processes and whether they occur spontaneously. Spontaneous changes in isolated systems occur with an increase in entropy. Spontaneous changes tend to smooth out differences in temperature, pressure, density, and chemical potential that may exist in a system, and entropy is, thus, a measure of how far this smoothing-out process has progressed.

difference is included, associated with the inductance of the circuit, also known as electromotive force (emf).

The second law of thermodynamics, therefore, implies that currents driven by "electrical pressure" will also seek maximum entropy, which is reached when the lowest energy state in the circuit is achieved. This state is achieved when current flows through the path of least impedance.

PATH OF LEAST IMPEDANCE AND THE LEAST ACTION PRINCIPLE

> It is impossible for a physicist to talk about the principle of least action without inadvertently imputing some kind of volition to the projectile. The ball seems to choose its path. It seems to know all the possibilities in advance.—Science historian J. GLEICK, 1993

The previous laws lead to the greatest generalization in all physical science, the principle of "least action".* Simply stated, it asserts that Nature always finds the most efficient course from one point to another (such as the path followed by water running downhill seeking the steepest descent, the fastest way downhill following the path of least action (defined as the path in which the total energy required to get from point A to point B is minimized). This is illustrated in Figure 2.41, which shows that a thrown ball follows the path of least action (a parabola in this case). The meandering dotted hypothetical course would require more action, defined in physics as the total difference between potential and kinetic energies (summated) for the entire path. In other words, Nature takes the most efficient path. The "least action" principle, therefore, represents the tendency of physical changes and processes to take the easiest or minimum path (also known in physics by the term "geodesy").

The "least action" principle is consistent with Fermat's premise of "least time" (consider the fact that light always finds the quickest trajectory through optical systems, in which it finds the path of least time, taking shorter paths in glass and water, in which it travels slower and longer paths through air). Ohm's Law, $\overset{=}{E} = (1/\sigma)\,\vec{J}$ [Equation (F) of Maxwell's eight original equations] as well as all Maxwell's equations satisfy conditions of least action. Current follows the path of least impedance, as that constitutes the path of least time and, thus, the path of least action (Appendix E presents the equivalence between Ohm's Law and the least time premise) [13,14].

2.5.2.5 Return Current Path, Conclusion. The above analysis has shown that at low frequencies, the return current will follow the path of least resistance, which would normally consist of the shortest path. At higher frequencies, however, above a cutoff frequency determined by the resistance of the conductors and the return current path inductance, the current will follow the path of least inductance, regardless of how low the resistance of the signal reference structure is.

Broadband (digital) signals comprise a spectral content extending over a broad frequency range (Section 2.7.1). For such signals, the lower frequency spectral compo-

---

*As Einstein discovered, all motion (speed) is relative and if all positions are relative, it appears that the concepts of "potential and kinetic energy" need to be, in some sense, qualified. Least action is curiously more powerful as a concept than the concepts of potential and kinetic energy it entails.

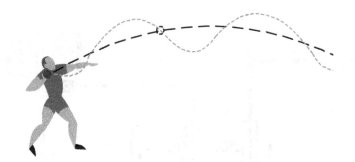

Figure 2.41. Trajectory of a thrown ball follows the path of least action.

nents will return through the shortest path, whereas the higher frequency spectral components will follow the path of least inductance, often significantly longer.

The following is an appropriate summary for the above discussion: "Current at high frequencies, if not altered by significant amount of impedance, always assumes a distribution that minimizes the inductance of the loop formed by the signal and return path" [11].

In terms of conservation of energy, this principle may be stated as: "Current flows in the path such that the energy stored, in the consequent magnetic field is minimized." This is consistent with the second law of thermodynamics, which states that "current driven by 'electrical pressure' flows in a path whereby the lowest energy state and thus maximum entropy in the circuit are reached."

Finally, in terms of the broader assertion of least action principle, "Current will flow in the 'most efficient' path requiring least time and action."

2.5.2.6 Experimental Validation for the Path of Least Inductance Principle.
For the purpose of demonstrating the validity of the principle of frequency dependence of the current path, particularly at higher frequencies, the setup depicted schematically in Figure 2.42(a) was assembled. A 50 Ω semirigid, U-shaped coaxial fixture was constructed. The length of each arm was approximately 30 cm. For the input and output ports of the fixture, TNC coaxial connectors were used. The input of the coaxial fixture was connected to a RF signal generator. A 50 Ω coaxial terminator was used to load the output port. In addition, a wide (1 inch or 2.5 cm) copper strap was used to interconnect the external conductor ("shield") of the semirigid conductor in between the input and output ports.

An RF current probe was placed on the copper strap. The probe's output was connected to a spectrum analyzer's input for the purpose of measuring the current flowing into the copper strap. The transfer impedance characteristics of the current probe were programmed into the spectrum analyzer; thus, its readings were actual current values in the copper strap. The output of the RF signal generator was swept between the frequencies 1 kHz to 10 MHz. The output power level of the signal generator was maintained fixed during the test. Figure 2.42(b) is a photograph of the setup.

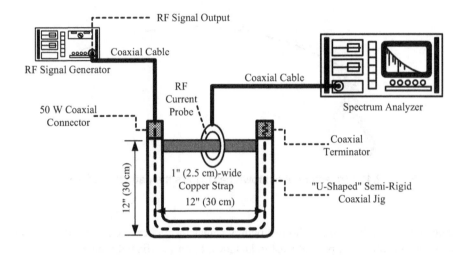

(a) Schematic Diagram

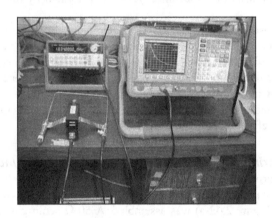

(b) Photograph of the setup

Figure 2.42. Setup for demonstrating the validity of the "path of least inductance" principle.

The measurement results of the current flow on the copper strap versus frequency are provided in Figure 2.43. It clearly demonstrates that as frequency increases from 1 kHz to 10 MHz, the return current redistributed itself so that at 10 MHz the current returning through the wide, low-resistance copper strap represents only 3.5% (–29 dB) approximately of the original return current at 1 kHz. It follows that almost 97% of the current returned through the outer conductor of the coaxial cable constituting the path of least impedance by virtue of the minimized inductance of this path.

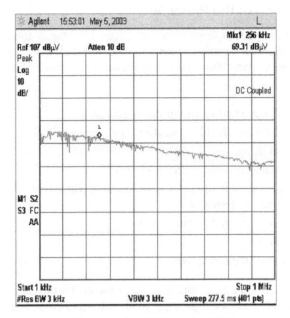

Low Frequencies ($1\text{kHz} \le f \le 1\text{MHz}$)

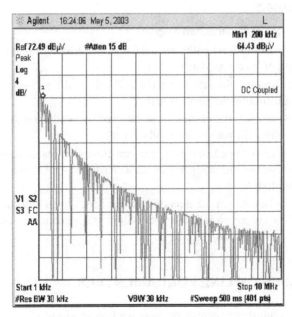

High Frequencies ($f \le 10\text{MHz}$)

(a) Measured Spectral Plots

Figure 2.43. Results of the "path of least inductance" principle experiment.

2.5.3 Implication of the Principle

The principle of the path of least impedance, describing the distribution of high-frequency return current propagation, is a principle of utmost importance for understanding EMI phenomena, particularly for the application of grounding design rules (in its broadest sense) related to:

- Selection of circuit and system grounding topology
- Application of cable shield termination
- Circuit signal balancing
- Filter and surge suppression circuit designs
- Printed circuit board designs, for ensuring compliance with EMC and signal integrity concerns, as related to:

    Transmission line and layout design

    Layer allocation

    Power decoupling and bypassing

Few principles in EMC are as important as this principle for the understanding and implementation of EMI control design, particularly related to grounding theory.

Lessons Learned

- Current always flows in closed loops.
- When alternative current paths are available, the return current will always follow the path of least impedance.
- At low frequencies, the return current will follow the path of least resistance, typically constituting the shortest path.
- At high frequencies, the current, if not altered by significant amount of impedance, will follow the path of least inductance.
- The path of least impedance principle constitutes an manifestation of the first and second laws of thermodynamics, as well as the "least action" premise as applied to electrodynamics.
- Excessive inductance in the intended current return path will force a large proportion of the current to follow a path of higher impedance, resulting in an EMI increase from the circuit.

2.6 TRANSMISSION LINE FUNDAMENTALS

Since standard circuit theory cannot be employed on an electrical network consisting of electrically long conductors,* an alternative analysis approach must be employed.

---

*A situation in grounding and bonding in which transmission line theory is particularly beneficial is in cases where long bonding conductors are used.

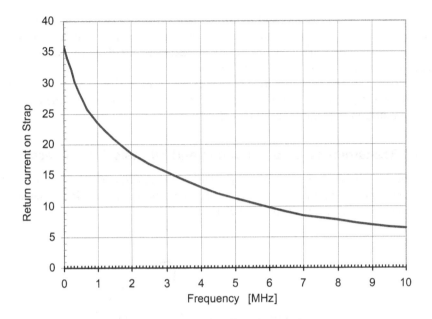

(b) Graphical Representation of Results (current in dBμA)

Figure 2.43. *Continued.*

Transmission line (T-line) theory bridges the gap between circuit theory and field analysis. The physical dimensions of electrical networks must be much smaller than the wavelength of the signal present. To facilitate common circuit analysis, we make use of lumped elements. Conversely, the length of a transmission line lies between a fraction of a wavelength and many wavelengths, thus a T-line must be considered a distributed parameter network. Consequently, the voltages and currents associated with a propagating wave in a T-line can vary in both phase and magnitude over the length of the line.

2.6.1 Transmission Line Definition

As the name implies, a transmission line constitutes a system of conductors used for transmitting electrical signals. The term transmission line in the field of electromagnetics is commonly reserved for structures of two or more closely spaced, parallel conductors that support transverse electromagnetic (TEM) wave propagation, characterized by the lack of longitudinal field components, and whose fields have a static distribution along the propagation direction, such that

$$\nabla^2 \vec{E} = \nabla^2 \vec{H} = 0 \qquad (2.106)$$

Implicit in most discussions of transmission-line theory is the assumption that the lines are uniform, that is to say, the conductor shape, size, and separation (transmission line geometry) are constant, and the electrical characteristics of the conductors and the medium between them (material) are uniform.

Coaxial lines and parallel-wire pairs are excellent examples of practical transmission lines. On printed circuit boards, the most common transmission line configurations are stripline and microstrip (Figure 2.44).

2.6.2 Transmission Line Equations and Intrinsic Parameters

Electrically long structures can be analyzed using circuit theory concepts, provided the problem is broken into infinitesimally small segments so that the circuit element dimensions are much smaller than a wavelength.

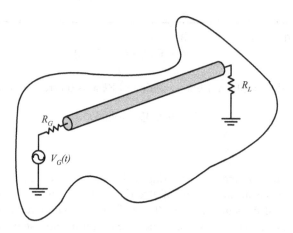

(a) Wire over Return Plane T-Line

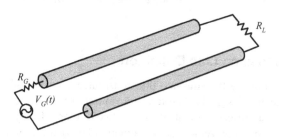

(b) Two-Wire T-Line

Figure 2.44. Common configurations of transmission lines.

Many excellent resources include the derivation of transmission line equations (see for instance [6] and [7]), which is beyond the scope of this book. Equation (22.107) presents the concluding expressions for the voltage and current distribution along the transmission line laid out along the x axis. The two terms in each equation denote traveling waves in the positive and negative x direction of the transmission line, respectively [6]:

$$\begin{cases} V(x) = V^+(x) + V^-(x) = V_0^+ e^{-\gamma x} + V_0^- e^{\gamma x} \\ I(x) = I^+(x) + I^-(x) = I_0^+ e^{-\gamma x} + I_0^- e^{\gamma x} \end{cases} \quad (2.107)$$

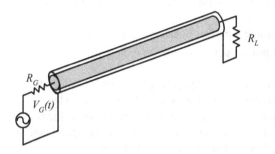

(c) Coaxial T-Line

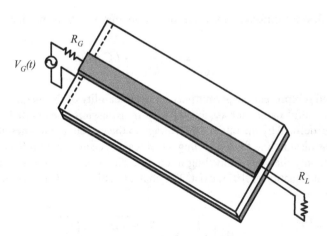

(d) Microstrip (Trace over Return Plane) on a Printed Circuit
Board T-Line

Figure 2.44. *Continued.*

The propagation constant γ of the T-line is defined as

$$\gamma = \alpha + j\beta = \sqrt{(R_x + j\omega L_x)(G_x + j\omega C_x)} \approx j\omega\sqrt{L_x C_x} \qquad (2.108)$$

where:

R_x = Resistance in both conductors per unit length, Ω/m
L_x = Inductance in both conductors per unit length, H/m
G_x = Conductance of the dielectric media per unit length, S/m
C_x = Capacitance between the conductors per unit length, F/m

In lossless transmission lines only C_x and L_x need to be considered, since

$R_x \to 0$, assuming a perfect (lossless) conductor
$G_x \to 0$, assuming a perfect (lossless) dielectric

The real part α of the propagation constant is called the attenuation constant in Np/m and the imaginary part β is called the phase velocity of the line in rad/m. In lossles transmission lines, γ reduces to $j\beta$, expressed in the right-hand part in Equation (2.108). TEM waves always propagate with velocity (, related to the phase velocity β, the wavelength λ in m, and frequency f in Hz by

$$\beta \approx \frac{2\pi}{\lambda}; \qquad v = \frac{2\pi f}{\beta} \qquad (2.109)$$

For lossless transmission lines and for any geometry of transmission line,

$$v = \frac{1}{\sqrt{\mu\varepsilon}} = \frac{1}{\sqrt{LC}} \to LC = \mu\varepsilon \qquad (2.110)$$

where ε and μ represent the permittivity and permeability of the medium, respectively.

A most useful parameter associated with transmission lines is their characteristic impedance, defined by the ratio of the voltage to the current amplitudes at any position on the transmission line for a traveling TEM wave. Assuming no reflections (see Section 2.6.3), only the waves traveling in the one direction [e.g., positive x, where only $V+$ and $I+$ in Equation (2.107)] exist. The characteristic impedance can then be written as

$$Z_0 [\Omega] = \frac{V_0^+}{I_0^+} = \sqrt{\frac{R_x + j\omega L_x}{G_x + j\omega C_x}} \approx \sqrt{\frac{L_x}{C_x}} \qquad (2.111)$$

The characteristic impedance constitutes an intrinsic property of the T-line and depends on the geometry of the line and the propagation medium. It does not represent "Ohmic" power loss (i.e., loss due to charge motion or resistance) but, rather, accounts for the power flow down the T-line. Note that the charactersitic impedance is independent of position along the line x.

2.6.3 Transmission Line Termination and Loading Conditions

Figure 2.45 depicts a diagram of a generally loaded T-line at the length l with characteristic impedance Z_0 and load impedance Z_L. Assuming a lossles T-line ($\alpha = 0$, $\gamma " j\beta$), the input impedance Z_i at point x' along the T-line can be expressed as [6]

$$Z_i(x') = \frac{V(x')}{I(x')}\bigg|_{x'=l} = Z_0 \frac{Z_L + Z_0 \tan \beta l}{Z_L + Z_L \tan \beta l} \qquad (2.112)$$

Note that when Z_L equals Z_0, the input impedance Z_i equals the characteristic impedance.

The impedance function can be applied to a generally loaded section of T-line using Equation (2.112) in order to determine the input impedance at a specific position along the line.

By investigating several different load alternatives, five special cases of particular interest for the sake of grounding design are identified, the most important of which the short circuit and quarter-wavelength T-line. Those are now discussed in brief.

a) Matched Load

In this situation, the impedance of the load at the end of the T-line is equal to the characteristic impedance of the line. As a result of this impedance match at the T-line load interface, there will be no reflection at the interface. This result can be mathematically noted in Equation (2.112) by substituting Z_L for Z_0 thus revealing that the line impedance equals the characteristic impedance:

$$Z_i(-l) = Z_0 \qquad (2.113)$$

b) Short Circuit ($Z_L = 0$)

A short-circuited load will produce full reflection of the incident wave, as can be observed by substituting $Z_L = 0$ into Equation (2.112), yielding a reflection coefficient of

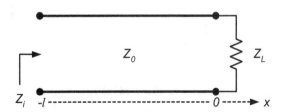

Figure 2.45. Schematic model of a generally loaded T-line.

$\Gamma_L = -1$. In terms of voltage and current, $V = 0$ at the load while the current I is at its maximum there. Since the impedance of a short is zero, the resulting input impedance due to the shorted load will have a periodic variation of (Figure 2.46):

$$Z_i(-l) = Z_0 \tanh \gamma l \approx jZ_0 \tan \beta l \qquad (2.114)$$

c) Open Line ($Z_L \to \infty$)

For an open load, the impedance is infinite ($Z_L \to \infty$) and the corresponding reflection coefficient at the load is $\Gamma_L = 1$. In this case, the voltage is a maximum at the load and $I = 0$. Again, using Equation (2.112), the resulting input impedance due to the open line will have a periodic variation of (Figure 2.47)

$$Z_i(-l) = -Z_0 \coth \gamma l \approx -jZ_0 \cot \beta l \qquad (2.115)$$

d) Quarter-Wavelength T-Line (Quarter-Wavelength Transformer)

For a quarter-wavelength T-line, the tangent function approached infinity and the resultant input impedance approaches

$$Z_i\left(l = \frac{\lambda}{4}\right) = \frac{Z_0^2}{Z_L} \qquad (2.116)$$

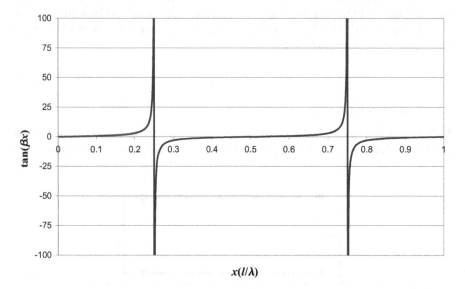

Figure 2.46. Periodic variation of the T-line input impedance (normalized to Z_0) for $Z_L = 0$, $\lambda = 1m$, and $\Gamma_L = -1$.

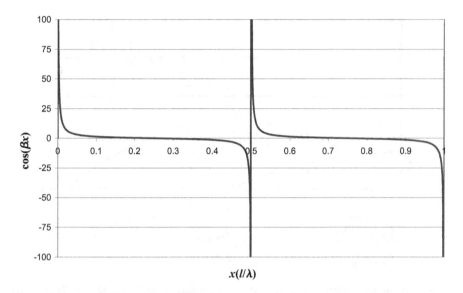

Figure 2.47. Periodic variation of the T-line input impedance (normalized to Z_0) for Z_L $\rightarrow \infty$, $\lambda = 1\text{m}$, and $\Gamma_L = +1$.

Note that for a short-circuit load ($Z_L = 0$), the input impedance of the T-line approaches infinity ($Z_i \rightarrow \infty$). This situation finds its equivalence in circuit theory as a tuned parallel resonant circuit, which exhibits infinitely high-input impedance.

Within transmission lines, this behavior can be understood intuitively in the following manner. At the "0V" point, the voltage is "strapped" to 0 V. The current would be at its peak at that point owing to the low impedance at that point. For a given frequency f, moving to the point $x = \lambda/4$ means that at this point the voltage reaches its peak and the current reduces zero, hence (Figure 2.48),

$$Z_{in} \triangleq \frac{V_{x=\lambda/4}}{I_{x=\lambda/4}} = \frac{V_{max}}{0} \rightarrow \infty \qquad (2.117)$$

e)   Half-Wavelength T-Line

In contrast to the quarter-wavelength T-line, in a half-wavelength T-line the tangent function vanishes and the resulting input impedance reduces to

$$Z_i\left(l = \frac{\lambda}{2}\right) = Z_L \qquad (2.118)$$

Note that for a short-circuit load ($Z_L = 0$), the input impedance of the T-line is also zero ($Z_i = 0$). This situation finds its equivalence in circuit theory as an ideal tuned series resonant circuit exhibiting zero impedance at resonance.

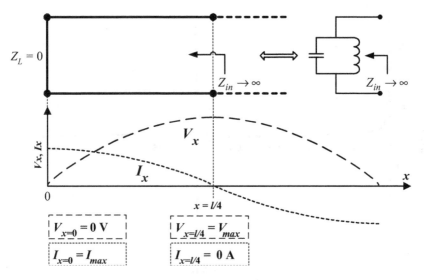

Figure 2.48. A short-circuited load (0 V) is converted to an "open circuit" (∞V) at the point $x = \lambda/4$.

An interesting relation between the short-circuit and open-circuit load becomes apparent by observing the periodic variation of Equations (2114) and (2.115) for the two different load impedances. For a T-line with a short-circuit load ($Z_L = 0$), for instance, Equation (2.114), the input impedance $|Z_{In}|$ will oscillate between ∞ and 0 at a quarter-wavelength and half-wavelength away from the load respectively.* This periodic behavior will continue to occur further away from a short-circuit load ($Z_L = 0$), with the input impedance alternating between $Z_i \to \infty$ and $Z_i = 0$ for parallel and series resonance of the T-line at odd and even multiples of $\lambda/4$, respectively, occurring at (Figure 2.49):

$$x = (2k+1)\cdot\frac{\lambda}{4}, \ k = 0, 1, 2, \ldots, \text{ parallel resonances, where } |Z_{in}| \to \infty$$

$$\text{for } Z_L = 0 \text{ (case d above)}$$

$$x = 2k\cdot\frac{\lambda}{4}, \ k = 1, 2, \ldots, \text{ series resonances, where } Z_{in} = 0$$

$$\text{for } Z_L = 0 \text{ (case e above)}$$

This feature is of great imporatance in understanding the frequency limitations associated with the application of different grounding topologies and techniques.

In the cases of unmatched T-lines, reflections will occur. The extent of which reflection occurs is determined by the difference between the characteristic and load im-

*The opposite will occur for an open-circuit load ($Z_L \to \infty$).

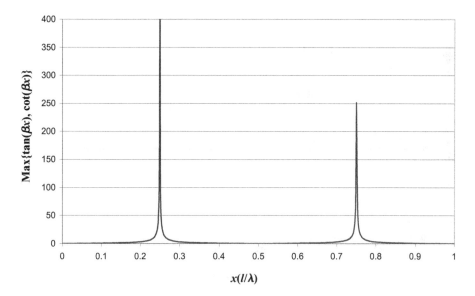

<u>Figure 2.49.</u> Input impedance (normalized to Z_0) of an "electrically long" T-line with a short-circuit load for $Z_L = 0$ and $\lambda = 1$ m.

pedances. For the T-line represented in Figure 2.45, the voltage reflection coefficient (at the load), Γ_L, is defined as the ratio of reflected to incident voltage amplitudes. This reflection coefficient is complex and carries both the phase and magnitude variations along the length of the line [6]:

$$\Gamma_L\left(x'\right)=\frac{V_0^-}{V_0^+}=\left|\Gamma_L\right|e^{j2\beta x'}=\frac{Z_L-Z_0}{Z_L+Z_0} \tag{2.119}$$

The combination of forward and reflected traveling waves produce a "standing wave," so named owing to the fact that the positions of the waveform "0 V nodes" are fixed in their location [6]:

$$\frac{\left|V\left(x\right)\right|^2}{\left|V_0^+\right|^2}=1+\left|\Gamma_L\right|^2+2\cdot\left|\Gamma_L\right|\cdot c\,os\left[2\beta x+\arg\left(\Gamma_L\right)\right] \tag{2.120}$$

(zeros are fixed in their x locations), where $\left|\Gamma_L\right|$ and $\arg(\Gamma_L)$ represent the magnitude and argument, or angle of Γ_L, respectively.

Note that the 0 V nodes of $|V(x)|$ occur at fixed locations along the T-line, for a given frequency (or wavelength). This expression is depicted in Figure 2.43 for various values of the reflection coefficients, Γ_L.

One immediate conclusion arising from the above discussion is that the length of grounding and bonding conductors should not exceed $\lambda/10$, corresponding to the par-

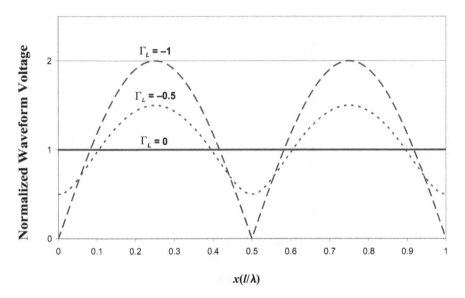

Figure 2.50. Normalized voltage waveform amplitude distribution along an electrically long T-line for various reflection coefficients, Γ_L, for $\lambda = 1$ m.

ticular frequency of interest. Considering this feature of conductors at high frequencies, resultant constraints in selection and implementation of grounding strategies will be addressed in subsequent chapters.

Lessons Learned

- A signal traveling along a transmission line has voltage and current waveforms related by the characteristic impedance of the line.
- Long transmission lines "transform" load impedances. A shorted T-line of length equal to an odd multiple of $\lambda/4$ exhibits infinite input impedance and, thus, appears as an open circuit.
- Interactions of electromagnetic fields and T-lines are enhanced at odd multiples of the $\lambda/4$ in short-circuit terminated T-lines.
- Reflections will occur at impedance boundaries, producing "standing waves."
- Length of T-lines should not exceed $\lambda/10$, corresponding to the highest frequency of interest.

2.7 CHARACTERISTICS OF SIGNALS AND CIRCUITS

Performance of grounding systems is frequency dependent. As will be shown in this section, commonly used signals encompass a broad spectrum, often by far exceeding the operating or switching frequency of the signal itself. It is a common error to consider the pulse repetition frequency (PRF) of a pulsed signal as the driving factor in the interference potential. The PRF determines the frequency (or number of times in a given period) that interference will occur due to the pulsed signal within a given range of frequencies. The transition time or edge rate, however, determines the spectral content and, thus, the potential for interference.

This section addresses the spectral content of signals, and certain features of signals, particularly with respect to their mode of propagation. In the following chapters, Chapter 4 in particular, the performance of grounding systems with respect to frequency and the mode of signal propagation are further elaborated.

2.7.1 Spectral Content of Signals

Electromagnetic compatibility is concerned with time-varying signals, with frequencies ranging from anything greater than DC to daylight. Such signals emerge both from the external environment, such as radio transmitters, and signals processed by the circuits and systems themselves.

Most of the current electronic technology, ranging from the wrist watch to large-scale C^4I (command, control, communications, computers, and intelligence) facilities, uses mainly digital signals. The time-domain waveform of digital waveforms is well defined, normally portrayed in amplitude versus time representation. Throughout the electronics industry, such signals are defined using interface control specifications such as TIA/EIA RS-232, RS-422, or RS-644, or IEEE 802.1 standards. However, such pulsed signals, while extremely efficient and versatile, are also associated with high levels of electromagnetic pollution or noise. Inasmuch as this may not be clearly observed in the time-domain representation of the signal waveform, an examination of their respective spectral content may reveal the source of that noise.

The EMC environment, as specified in most (though not all) EMI control standards, is provided in terms of amplitude versus frequency with limits pertaining to the permissible emission or required immunity level as a function of frequency. Therefore, conversion between time- and frequency-domain representations of signals is of great importance.

All signals can be represented either in the time domain (as a function of time) or in the frequency domain (as a function of frequency). The translation between the two is accomplished by the Fourier series or the Fourier transform for periodic (repetitive/harmonic) and stochastic processes, respectively. From the standpoint of EMC, particularly as associated with grounding systems design, knowledge of the spectral content of signals is vital since all characteristics of the grounding system are also frequency dependent. The performance of the system may significantly degrade as the frequency increases due to the non-ideal attributes of circuit elements presented in Section 2.4.

2.7.1.1 Periodic Pulsed Signals. Periodic or repetitive pulsed signals can be described using Fourier analysis. Such signals are characterized by a fixed duty cycle and a fixed spacing between pulses, $T = 1/f_0$, where f_0 is the pulse-repetition frequency (PRF) of the signal. The analysis performed yields the Fourier series, which provides a series of discrete harmonic signals. For zero rise/fall time pulses (square waves), the Fourier series is expressed as

$$e(t) = \frac{2A \cdot d}{T} \sum_{n=1}^{\infty} \frac{\sin(n\pi \cdot d / T)}{n\pi \cdot d / T} \qquad n = 1, 2, 3, \ldots \qquad (2.121)$$

See Figure 2.51 while considering a finite rise/fall time ($t_r, t_f > 0$). The spectral distribution envelope associated with Equation (2.121) is depicted in Figure 2.52.

The discrete frequencies at which the spectrum of this pulse is nonzero occur at $n/T = n \cdot f_0$. Since the spectrum is an assembly of discrete frequencies, it is considered narrowband noise.

Note: The term narrowband may be somewhat confusing: Although an infinitely large number of such harmonics exists, extending from DC to daylight, each harmonic can easily be resolved by a receiver with a properly set resolution bandwidth and, as such, each harmonic is considered "narrowband." If, however, the resolution bandwidth of the receiver is set wider than f_0 this may not be the case, as more than one harmonic may be captured within that bandwidth.

Practical pulsed waveforms, such as clock signals, exhibit a finite (nonzero) rise/fall time (trapezoidal pulses). The expression for the Fourier series for this waveform is given by:

$$e(t) = \frac{2A \cdot d}{T} \sum_{n=1}^{\infty} \frac{\sin(n\pi \cdot d / T)}{n\pi \cdot d / T} \cdot \frac{\sin(n\pi \cdot t_r / T)}{n\pi \cdot t_r / T} \qquad n = 1, 2, 3, \ldots \qquad (2.122)$$

The effect of the increased pulse transition (rise/fall) time has the effect of suppressing the amplitude of higher frequency spectral components.

When performing EMC analysis, worst-case situations are of interest, therefore, although Equation (2.122) provides the exact spectral distribution of the waveform. A worst-case notation can be used, the results of which are described as follows [15].

The low frequency spectrum amplitude is

$$e(f) = \frac{2A \cdot d}{T} \qquad (2.123)$$

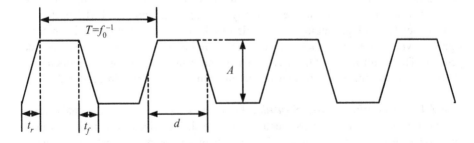

Figure 2.51. Repetitive trapezoidal waveform.

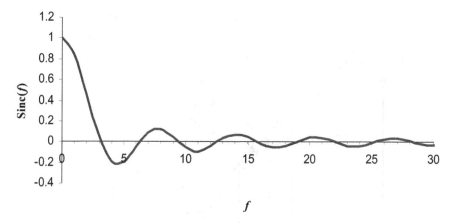

Figure 2.52. Generic spectrum of a rectangular-pulse waveform.

As no frequency terms exist in this expression, the amplitude observed results in a straight horizontal line, independent of frequency. The first "knee frequency"* occurs at

$$f_1 = \frac{1}{\pi \cdot d} \qquad (2.124)$$

The expression for the frequency-dependent amplitude function from this knee frequency is

$$e(f) = \frac{2A}{\pi \cdot Tf} \qquad (2.125)$$

This curve exhibits a linear decrease with increasing frequency, at a rate of 20 dB/decade of frequency. A second "knee frequency" occurs at:

$$f_2 = \frac{1}{\pi \cdot t_r} \qquad (2.126)$$

The amplitude curve beyond this knee frequency is expressed as

$$e(f) = \frac{2A}{f^2 \cdot \pi^2 t_r \cdot T} \qquad (2.127)$$

This corresponds to decrease at a rate of 40 dB/decade.

This simplified "spectral mask" is depicted in Figure 2.53. The first harmonic of the spectrum occurs at the repetition frequency, where $n = 1$, or at $1/T$. below this frequen-

*The first "knee" frequency is the breakpoint at which the slope of the spectral envelope changes (from 0dB/decade to –20 dB/decade or from –20 dB/decade to –40 dB/decade).

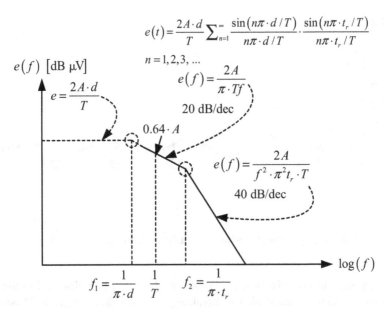

$$e(t) = \frac{2A \cdot d}{T} \sum_{n=1}^{\infty} \frac{\sin(n\pi \cdot d / T)}{n\pi \cdot d / T} \cdot \frac{\sin(n\pi \cdot t_r / T)}{n\pi \cdot t_r / T}$$

Figure 2.53. Fourier "spectral mask" of a repetitive trapezoidal signal.

cy, no signal exists. The separation between adjacent spectral components is also $1/T$ (i.e., the pulse repetition frequency or PRF). Note also that the scales used are logarithmic [$\log(f)$ and $\log(e(f))$].

Units used for description of narrowband signals are related to the amplitude of each spectral component, normally expressed in logarithmic units, for example, dB μA.

2.7.1.2 Random Pulsed Signals. Random (stochastic) pulsed signals (e.g., data signals) differ from the periodic signals analyzed in Section 2.1.1.2. Due to the stochastic nature of these signals, no duty cycle can be associated with them. The approach for evaluating the spectrum of such signals is to analyze the spectral content of the elementary pulse comprising the pulsed sequence.

Spectral analysis of random processes is performed using the Fourier transform, which, unlike the Fourier series, provides a continuous spectral density function spanning across the entire frequency spectrum, from DC to infinity.

Using the same notation found in Figure 2.51, the expression for the Fourier transform of a single pulse is [15]

$$e(f) = 2A \cdot d \cdot \frac{\sin(\pi \cdot f \cdot d)}{\pi \cdot f \cdot d} \cdot \frac{\sin(\pi \cdot f \cdot t_r)}{\pi \cdot f \cdot t_r} \qquad (2.128)$$

Using the same approach used for periodic signals, the expression for the low-frequency spectral distribution is

$$e(f) = 2A \cdot d \qquad (2.129)$$

Again, this expression corresponds to a frequency-independent value depicted as a straight horizontal line. Setting $\sin(\pi fd) = 1$ and considering that fd is quite small in this frequency range, the expression for the midrange spectrum density is obtained:

$$e(f) = \frac{2A}{\pi \cdot f} \qquad (2.130)$$

The knee frequencies, or breakpoints at which the slope of the spectral envelope changes (from 0 dB/decade to –20 dB/decade or from –20 dB/decade to –40 dB/decade) for random signals are identical to those derived for periodic pulsed signals, resulting in the simplified spectral envelope depicted in Figure 2.54.

With this, the analysis of the spectrum from a random pulsed signal is complete. As a random signal, the time domain characteristics, particularly the spacing between pulses, are random, resulting in a broadband spectrum called spectral density. Therefore, amplitude is bandwidth-related and units used for description of broadband signals are, therefore, related to bandwidth (such as MHz), normally expressed in logarithmic units, for example, dB μV/MHz.

2.7.1.3 Spectrum Conservation. From the standpoint of EMC, the designer's goal should be to limit the spectrum of unwanted signals and noise. Spectra extending

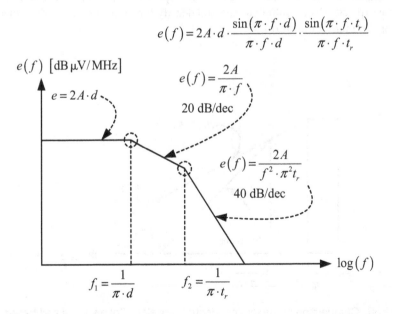

Figure 2.54. Fourier spectral "envelope" of a random trapezoidal pulse.

to hundreds of MHz have even been observed in switch-mode power supplies operating at 150 kHz, introducing a significant interference potential.

The above derivation clearly demonstrates that it is the transition (rise/fall) time of the pulse edges rather than the PRF that determines the effective interference bandwidth of a signal. Shorter transition times will drive the knee frequency f_2 higher.

Figure 2.55 presents a qualitative comparison between the expected spectra envelope of a square wave versus trapezoidal waveform. This figure is valid both for periodic and random waveforms. The benefit of the finite rise time in the trapezoidal waveform over the square wave (with zero rise time) with respect to spectrum conservation is self-evident.

The importance of spectral conservation and bandwidth of signals and the frequency effect on grounding and bonding performance and grounding topologies lies in the fact that at the higher frequency, components of a signal spectrum, even short grounding and bonding conductors, may become "electrically long," significantly degrading their effectiveness. The significance of frequency on grounding architecture and implementation is further discussed in Chapters 4 and 5.

Lessons Learned

- Any time-varying signal can be described in the frequency domain as a combination of an infinitely large number of sinusoidal signals.
- Repetitive signals are best represented as a discrete harmonic series of sinusoidal signals, commonly expressed using the Fourier series.
- Random, stochastic processes cannot be represented as a harmonic series; they are expressed by a continuous spectral density function, commonly described using the Fourier transform.

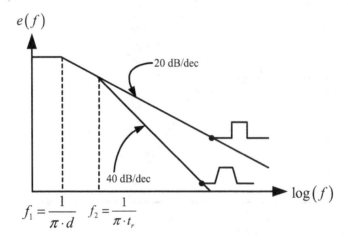

Figure 2.55. Comparison of the fourier spectral "envelope" of a square and trapezoidal pulse.

- The frequency content associated with a signal is commonly associated with the minimum rise/fall time of the signal, and is $f_{max} = 1/\pi t_{y/f}$. Spectrum conservation is, therefore achieved, primarily by controlling the edge rates of signals.
- Frequency considerations constitute a key factor in grounding and bonding performance.

2.7.2 Differential-Mode and Common-Mode Signals

Contradictions do not exist. Whenever you think you are facing a contradiction, check your premises. You will find that one of them is wrong.—AYN RAND, *Atlas Shrugged* [16]

A vital key to the understanding of the discipline of EMC, particularly as related to grounding theory, lies in (1) the recognition of the nature of currents to flow in closed loops (Ampere's Law) and (2) the realization that currents can flow in paths other than those intended.

From (1), as described by Kirchhoff's Current Law (KCL), the sum of all currents entering a circuit node must be zero. This implies that for a single conductor all currents flow in closed loops. Thus, a source within an electrical network will supply current at the positive terminal only if the same current is drained from its negative terminal!

The intended current in the circuit is called differential-mode* (DM) current, denoted I_{DM}. DM currents are present on both signal and return paths flowing in opposite directions (180° phase shift between outgoing and return signals) in the circuit conductors and are equal in magnitude.

Consider the circuit in Figure 2.56, where the DM current I_{DM} is propagating on conductors 1 and 2 to the load. However, another form of (undesired) current also exists in the circuit, appearing equally and in phase in all current-carrying signal and return conductors alike. This current, typically much smaller in magnitude than the DM current, is known as common-mode† (CM) current and is denoted I_{CM}.

The total current in the signal and return conductors, I_1 and I_2, respectively, results from the summation of the CM and DM currents in each conductor. By definition, the DM current is equal in amplitude but opposite in phase on both conductors, whereas the CM current propagates in equal amplitude and phase on both conductors, therefore,

$$\begin{cases} I_1 = I_{CM} + I_{DM} \\ I_2 = I_{CM} - I_{DM} \end{cases} \tag{2.131}$$

The CM and DM current components can, thus, be derived from the knowledge of the total currents in each conductor [Equation (2.132)]. Note the direction of both I_1 and I_2 is defined from source to load:

---

*Differential mode is also known as "symmetrical mode," "transverse mode," "metallic mode," or "normal mode."

†Common-mode is also known as "asymmetrical mode" or "longitudinal mode."

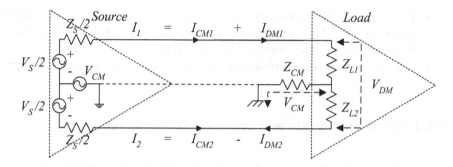

<u>Figure 2.56.</u> Generic model depicting the flow of common-mode and differential-mode currents in a circuit.

$$\begin{cases} I_{DM} = \dfrac{I_1 - I_2}{2} \\[2mm] I_{CM} = \dfrac{I_1 + I_2}{2} \end{cases} \tag{2.132}$$

The total CM current flowing in a circuit is obtained from the summation of the CM current flowing in all conductors constituting that circuit:

$$I_{CM\,(Total)} = \sum_{j=1}^{k} I_{CM\,j} \tag{2.133}$$

From Figure 2.56 it is obvious that the return conductor does not constitute the return path for the common-mode current. Since there seems to be no return path, a contradiction to Ampere's Law seems to be observed. How can this contradiction be settled? The key lies in the understanding of the mechanisms by which CM currents originate.

Figure 2.57 demonstrates the primary difference between the differential-mode (DM) and common-mode (CM) current paths. Whereas the DM current flows between the circuit conductors, the CM currents *must* follow alternative, unintentional paths in metallic structures, for example, the chassis of the enclosure serving as the reference structure. Unlike differential currents, the exact path of common-mode currents is difficult to predict.

In electrical circuits, the main concern regarding the presence of common-mode signals is that common-mode interference creates far more radiation and crosstalk than differential-mode current. Common-mode interference is an undesired by-product, causing problematic grounding and balancing situations.

2.7.2.1 Common-Mode Interference Generation Mechanism.
Common-mode interference can be generated from the intended signal by DM-to-CM conversion due to the imbalance between the transmitted signal and the return signal. When a

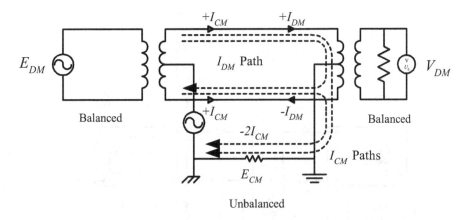

Figure 2.57. Illustration of the common- differential-mode current paths.

transmitted signal current and its associated return current are not exactly equal in amplitude and opposite in phase, the return current loses its pairing with the signal current path. The signal current from source to load is no longer completely balanced by the return current. The portion of the current not flowing in the intended return path constitutes the "common-mode" current.

The absence of a conductive path between the conductors and the signal reference structure does not preclude the generation of CM currents, as observed by Ampere's Law:

$$\oint_C \vec{H} \cdot d\vec{l} = \int_S \left(\vec{J} + \frac{\partial \vec{D}}{\partial t} \right) \cdot d\vec{s} \tag{2.134}$$

Ampere's Law states that time-varying electric fields can produce a current around contour C. However, contour C need not be conductive. The contour may be formed by displacement current flowing through stray parasitic capacitance and/or parasitic transfer impedance, within the network between the conductors to the signal reference structure (SRS),* commonly referred to "ground" or "0 V."

When imbalance occurs between the conductors of the circuit,[†] leakage current flows from the conductors to the SRS through the CM impedances. The current on both conductors will not be equal, resulting in generation of EMI.

The CM interference voltage V_{CM} in any path is equal to the sum of the common-mode sources E_{Cmi} in that path. For instance, in the source–receiver path in Figure 2.58,

$$V_{CM\,(Total)} = \sum_{j=1}^{3} E_{CM_j} \tag{2.135}$$

---

*Although the parasitic capacitance is illustrated as lumped capacitors, in practice, when the circuit is electrically large, distributed capacitance must also be considered; thus, inasmuch as the description may adequately illustrate the principles, it may not be scientifically accurate.
[†]Learn more about balanced and unbalanced circuits in Section 2.7.4.

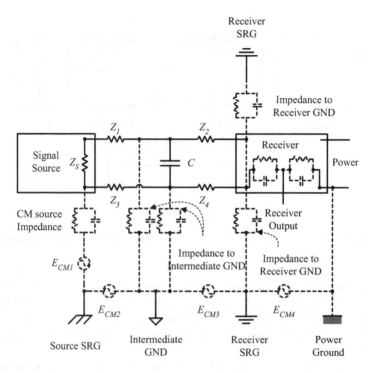

Figure 2.58. Model explaining common-mode interference-generation mechanism.

Figure 2.59 depicts the effects of DM-to-CM conversion in an unbalanced amplifier, which could result in the generation of CM EMI.

The following subsections provide further insight into some of the more common specific modes of CM noise generation.

2.7.2.1.1 CM GENERATION DUE TO "GROUND LOOPS." Figure 2.60 illustrates the mechanism of CM currents generation resulting from imbalance due to "ground loops,"* whereby the return conductor is connected to a signal reference structure (SRS) at two different locations, both at the source and the load (note the different symbols for the SRS connections).

The two reference point connections will have some potential difference between them due to the nonideal characteristics of the structure, for example, nonzero imped-ance (see Section 2.3), resulting in a voltage drop in the SRS. This voltage drop can re-sult, for instance, from some external interference source (e.g., between a motor and its driver) sharing the same reference structure as its current return path, subsequently forcing some current to flow through both the signal return conductors as common-mode current (Ohm's Law).

*Ground loops are discussed in Chapter 4.

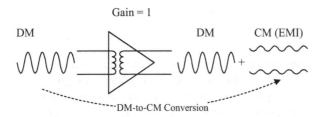

Figure 2.59. DM-to-CM conversion produces CM EMI.

As this interference mode is generated by the impedance of the SRS acting as a return path common to both circuit conductors, it is often called the common impedance interference coupling (CIIC).

Even in the absence of external interference source, common-mode noise can still be produced. Some current driven from the voltage source, V_{in}, can flow through the impedance of the reference structure, rather than the intended return conductor. The current flowing through the load will, therefore, split into two return paths, the intended return path through the signal return conductor and the unintentional return path through the reference structure. With respect to the latter case, we have from Kirchhoff's Current Law for node A in Figure 2.60,

$$\sum_{i=1}^{3} I_i = 0 \Leftrightarrow I_1 = I_2 + I_3 \tag{2.136}$$

Since I_1 is the transmitted signal and I_3 is the return current flowing in the intended return conductor, it follows from Equation (2.136) that $I_1 \neq I_3$. The CM current is therefore

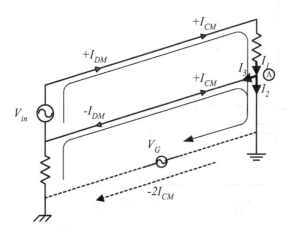

Figure 2.60. DM-to-CM conversion due to circuit imbalance formed by the alternative return path in the signal reference structure.

$$I_{CM} = \frac{I_1 + I_3}{2} = \frac{I_1 + (I_1 - I_2)}{2} = \frac{2I_1 - I_2}{2} \qquad (2.137)$$

This CM current splits equally between both signal and return conductors.

A simple example of this situation is illustrated in Figure 2.61. An oscilloscope probe is used to couple a signal from some point in a circuit to the oscilloscope terminals. The probe return ("ground") lead is connected to the circuit ground, which, in turn, is referenced to the facility ground system. Since there are generally stray currents flowing through the impedance of the facility ground (these are primarily at the power line frequency), the signal reference potential of the circuit is different from that of the oscilloscope. Common-mode noise voltages are impressed on both the circuit's signal reference and its signal point in the circuit with respect to the oscilloscope's signal reference, introducing possible measurement errors.

2.7.2.1.2 CM GENERATION DUE TO IMBALANCE IN DIFFERENTIAL DRIVERS. Observe the differential device in Figure 2.62. In this device, the differential input signal between ports A and B, as well as the output signals between ports C and D, should ideally be equal in amplitude and opposite in phase. The driver may also have a "center

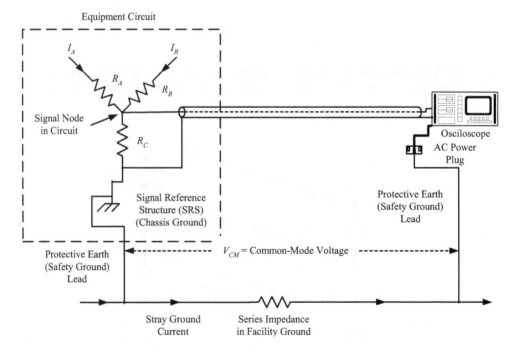

Figure 2.61. Illustration of common-mode generation due to common impedance ("ground loop") in the signal reference structure.

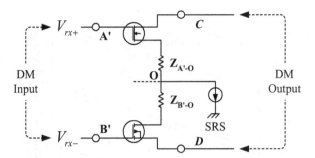

Figure 2.62. CM currents can be generated by imbalance in differential drivers.

tap" (O) constituting a "virtual ground,"* often grounded to the reference structure. Ideally, the output at C–D should be a result of the input differential signal only, rejecting CM currents between its inputs. In "real world" (nonideal) balanced devices (Section 2.7.4), the gain to the CM input ($V_{A'\text{-}O}$ and $V_{B'\text{-}O}$) is nonzero. Therefore, a DM input will result in some undesired CM signal at its output: DM to CM conversion takes place.

2.7.2.1.3 CM GENERATION DUE TO INDUCTION AND COUPLING. In addition to internal circuit imbalance, common-mode currents can also be produced due to external sources, for instance, by induction from externally incident electromagnetic fields, or due to capacitive crosstalk from an adjacent conductor Figure 2.63), or from an impinging radiated electromagnetic field. Electric flux emerging from a current-carrying conductor induces CM currents (I_{CM}) equal in magnitude and phase on both conductors simultaneously, AD and BC of the victim circuit. Induced CM currents will have no effect in perfectly balanced circuits, but may have adverse effects in nonbalanced circuits.

Figure 2.63 illustrates that perfectly balanced circuits do not exist in reality due to capacitive parasitics. Even with no conductive path between the circuit conductors and the signal reference plane, contours C_1 and C_2 are formed by displacement current flowing through the loops FADE and FBCE, respectively, via parasitic capacitance.[†]

Note: At high frequencies, the circuit can no longer be considered "electrically small." The CM capacitance will actually constitute distributed capacitance, expressed per unit length.

2.7.2.1.4 CM GENERATION DUE TO SIGNAL SKEW IN DIFFERENTIAL CONDUCTORS. Another example of CM current sources, which is of great importance in high-speed signal propagation, is signal skew. Signal skew (or phase difference) in a differential signal pair [$x^+(t)$, $x^-(t)$] may also generate CM interference due to a DM-to-CM conversion process [10, 17].

---

*This will occur if $Z_{A\text{-}O} = Z_{B\text{-}O}$, which is the definition of a balanced system.
[†]When capacitance between the circuits to a reference structure is infinitesimally small (for instance, when spacing is very large), induced currents will result in reradiation (or "scattering") from the wiring.

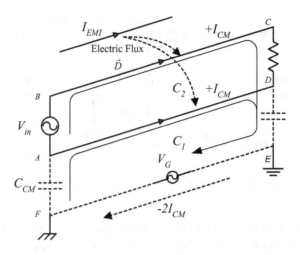

Figure 2.63. Generation of CM currents by electric field induction.

Assuming that the signal $x^+(t)$ is equal in amplitude to a negative signal $x^-(t)$, delayed by a small skew Δt, the CM signal $y(t)$ can be expressed as (Figure 2.64):

$$y(t) = \frac{x^+(t) + x^-(t)}{2} = \frac{x(t) - x(t - \Delta t)}{2} \tag{2.138}$$

For infinitesimal Δt (i.e., $\Delta t \to 0$ so that it may be considered small with respect to the rise time of the signal), Equation (2.138) can be rewritten as

$$y(t) = \frac{x^+(t) - x^-(t - \Delta t)}{2} \overset{\Delta t \to 0}{\approx} \frac{\Delta t \cdot \dfrac{dx(t)}{dt}}{2} \tag{2.139}$$

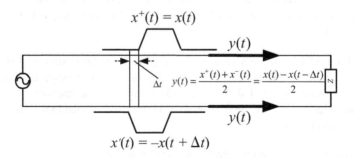

Figure 2.64. Signal skew (or phase difference) in a differential signal pair $[x^+(t), x^-(t)]$, may result in CM interference.

For a signal swing Δt during the rise time t_r, Equation (2.139) is transformed to

$$y(t) \approx \frac{\Delta t \cdot \dfrac{dx(t)}{dt}}{2} \approx \Delta t \cdot \frac{\Delta V / t_r}{2} = \frac{\Delta V}{2} \cdot \frac{\Delta t}{t_r} \tag{2.140}$$

When observing in Equation (2.140) the percentage of the CM signal $y(t)$ related to the DM signal swing, ΔV, as well as the percentage of the skew Δt to the rise time t_r, a skew of $S\%$ clearly produces $0.5\ S\%$ of CM signal:

$$S_{CM}\% = \left(\frac{y(t)}{\Delta V} \right)\% \approx \frac{1}{2} \cdot \left(\frac{\Delta t}{t_r} \right)\% \tag{2.141}$$

Figure 2.65 presents an illustration of the common-mode interference waveform, $y(t)$, resulting from signal skew in a differential signal pair carrying signals $x^+(t)$ and $x^-(t)$.

2.7.2.2 Mitigation of Common-Mode Interference Generation.

Generation of common-mode interference can be minimized by reducing the coupling between systems in such a manner that interaction between them does not produce interference in either one, by eliminating the source of the interference or by increasing the rejection of common-mode interference in the susceptible system.

Reducing the coupling into sensitive circuits on a printed circuit board is accomplished by minimizing the impedance of the reference plane, increasing the spatial separation between the traces, shielding, reducing the loop area of circuits, and balancing the signal lines.

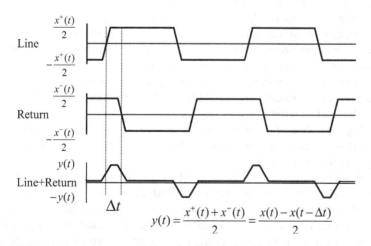

$$y(t) = \frac{x^+(t) + x^-(t)}{2} = \frac{x(t) - x(t - \Delta t)}{2}$$

Figure 2.65. Illustration of CM interference waveform resulting from signal skew (or phase difference) in a differential signal pair [x⁺(t), x⁻(t)]. (Courtesy of Dr. Todd Hubing, Clemson University.)

Minimizing Signal Reference Structure (SRS) Impedance. Minimizing the impedance of the SRS lowers the potential difference between any two points, thereby reducing the conductive coupling of interference. The impedance of the signal reference structure is reduced by minimizing both the resistance (R) and the series reactance (X) of the reference structure. The resistance is lowered with a decrease in either the length of the conductors or the signal frequency (resulting from the skin effect, Section 2.3.6) or with an increase in conductor cross-sectional area. The reactance is also lowered with a decrease in the signal frequency as well as with a decrease in the inductance of the conductors. The impedance of the signal reference can be reduced by making the signal reference conductors as short as possible and as wide as practical.

The overall impedance of the signal reference also depends upon the establishment of a low-impedance bond between ground conductors. (The various aspects of bonding and bond impedance are discussed in Chapter 5.)

Spatial Separation. As demonstrated in Section 2.7.2.1.3, common-mode noise may be developed due to induction and coupling of interference from an external source (e.g., a transmitter of other conductors in close proximity to the sensitive circuit).

Reduction of such coupling can be achieved by increasing the physical separation between the culprit (interference source) and victim circuits. Increasing the separation between the circuits will significantly decrease the interference level, particularly if the source is in near proximity (near-field coupling or crosstalk).

Reduction of Circuit Loop Area. Reducing the loop area of either or both the interference source or the susceptible circuit will decrease coupling via the mutual inductance. In Section 2.3.3, it was demonstrated that the mutual inductance between circuits can be minimized by reducing loop areas, for instance by running the signal-return conductor adjacent to the signal conductor.

In the case of wiring, a preferable approach to minimize coupling of EM interference is to twist the signal conductor with its return conductor. The use of twisted wire pairs (TWP), or twisted wire cables in the general sense, reduces the inductively coupled voltages into the circuit, since the voltage induced in each small twist area is approximately equal and opposite to the voltage induced in the adjacent twist area.

Note: Twisting of wires is an effective technique for reducing inductive coupling at low frequency, approximately up to 100 kHz. At higher frequency, common-mode interactions dominate interference coupling.

Shielding. Shielding of the conductors is yet another effective means for the reduction of interference coupling into circuits and interconnecting lines. Principles of cable shielding are presented in Chapter 7.

Balancing the Lines. In situations in which signal circuits must be grounded at both the source and the load, and, hence, establish conductive coupling paths, the use of balanced signal lines and circuits is an effective means of minimizing conductively coupled interference. In a balanced circuit, two signal conductors are routed symmetrically above the ground plane. At equivalent points on the two conductors, the desired signal is opposite in polarity and equal in amplitude relative to ground. A common-mode signal will appear with equal phase and amplitude on each of the conductors and will tend to cancel at the load. The amount of cancellation depends upon the degree to which the two signal lines are balanced relative to ground, as well as the circuit's CMRR.

If the signal lines are perfectly balanced, the cancellation would be complete and the coupled interference voltage at the load would be zero. If the source and load are not normally or cannot be operated in a balanced mode, transformers, common-mode chokes, optical isolators, and other coupling devices should be used at both the source and load ends of the signal line.*

Lessons Learned

- Common-mode (CM) currents do not violate Maxwell's equations. They flow in closed loops, whether conductive or capacitive.
- Common-mode currents constitute a dominant cause for grounding-related problems and EMI in circuits and systems.
- CM currents are generated owing to parasitics and imbalance between the transmitted signal and the return conductors.
- CM currents always involve the signal reference structure as the return path. By definition, they do not return via the circuit's intended conductors.
- Balanced circuits typically do not respond to CM signals; however, CM-to-DM conversion (due to imbalance in the circuit) will result in potential interference to circuit performance.
- "Ground loops" are a major source of CM interference. "Grounding-related interference" and "common-mode interference" problems are synonymous.
- Transformers, common-mode chokes, optical isolators, and other coupling devices may be used for improving balancing and reducing common-mode conversion and coupling in differential circuits.

2.7.3 Common-Mode (CM) to Differential-Mode (DM) Conversion

Common-mode (CM) currents were previously shown to be generated from external sources unrelated to the circuit's functional differential-mode (DM) currents. CM currents, however, could also be directly associated with signal currents. In balanced circuits, induced CM voltage will have little effect on performance. However, if CM current is converted into a DM signal, a differential interference across the circuit's load will be introduced.

Common-mode conversion is thus defined as the process by which DM interference is produced in a signal circuit by a CM signal. CM currents are converted to DM voltages through impedances Z_1, Z_2, Z_3, Z_4, Z_S, and Z_R, and capacitance C, shown in Figure 2.66 [18].

Common-mode currents existing in the circuit are converted into a differential-mode voltage developed across the receiver port that constitutes of a summation of

---

*These and other isolation devices and other techniques are discussed in detail in Chapter 4, within the context of preclusion of "ground loops."

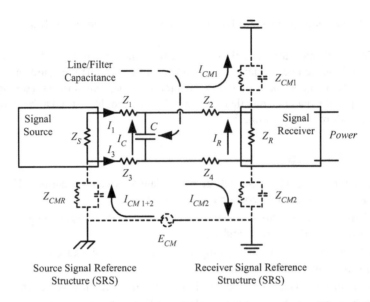

Figure 2.66. General mechanism for describing common-mode to differential-mode conversion.

voltage drops resulting from various currents flowing through those impedances. At various frequencies, some of those current components may be neglected, including those at DC and very low frequency, and at higher frequencies.

At DC and very low frequency,

$$V_{CM} = I_R \cdot Z_R \approx I_{CM1} \cdot (Z_S + Z_1 + Z_2) - I_{CM2} \cdot (Z_3 + Z_4) \qquad (2.142)$$

At higher frequency,

$$V_{CM} \approx I_{CM} X_C \cdot \left(\frac{Z_R}{Z_2 + Z_4 + Z_R} \right); \quad f > \frac{1}{C(Z_1 + Z_3 + Z_S)} \qquad (2.143)$$

From Equations (2.142) and (2.143) it can deduced that at DC and low frequencies the DM voltage developed on the load impedance Z_R, due to CM to DM conversion, is primarily dependant on the level of the circuit balance and the value of all CM currents. If the impedance to the signal reference structure (SRS) is capacitive only, whether intended or parasitic, that capacitance will have little effect on the DM noise developed on the load. At higher frequencies, on the other hand, particularly at frequencies at which the impedance to ground is very low due to the capacitance between the conductors to the SRS, the CM current flow through the SRS is increased. The circuit balance has little effect on the CM-to-DM conversion.*

*Refer to Section 2.7.4 for details.

By definition, $I_{CM1} = I_{CM2} = I_{CM}$. Thus, if $Z_{CM1} = Z_{CM2}$, no interference voltage is developed across Z_R by I_{CM}. However, if any imbalance exists in the network between the CM impedances ($Z_{CM1} \neq Z_{CM2}$), as could be expected due to the nature of the CM impedance, a voltage difference will develop proportional to the difference in CM impedance:

$$V_{DM} = I_{CM} \cdot Z_{CM\,1} - I_{CM} \cdot Z_{CM\,2} = I_{CM} \cdot \left(Z_{CM\,1} - Z_{CM\,2} \right) \tag{2.144}$$

Figure 2.67 depicts the possible results of conversion of CM to DM signals in an imperfectly balanced amplifier. The resultant DM interference signal is superimposed on the functional DM signal in the circuit, potentially adversely affecting its performance.

The mechanism of conversion from CM to DM in amplifiers can be enlightened with reference to Figure 2.68. In this figure, V_S represents some signal voltage from an unbalanced source; the output signal of a transducer or measuring amplifier with R_S represents the output impedance of this source. The source is connected to the input terminals of an electronic device, which is modeled as a two-terminal pair amplifier. Z_{S2} and Z_{S2} are the series impedances of the interconnecting conductors between the source and amplifier. The common-mode voltage source V_{CM} with output resistance R_{CM} represents a common-mode noise voltage, which causes the signal source to be at some voltage when measured with respect to the signal reference of the amplifier output. In Figure 2.68, the impedances Z_1 and Z_2 represent the input impedances of the two amplifier terminals. In a differential amplifier, these impedances are normally very high; however, in a single-ended amplifier, one is high and the other is very low, since it is tied directly to the signal reference terminal.

The ability of a differential amplifier to prevent the conversion of common-mode noise to differential-mode noise is described by its common-mode rejection ratio, or *CMRR*. The *CMRR* is equal to the common-mode noise voltage (V_{CM}) at the amplifier input multiplied by the voltage gain of the amplifier (K) and divided by the amplifier output V_O, due to V_{CM} [19, 20]. Expressed as a positive quantity, the *CMRR* is represented as

$$CMRR = \left| \frac{K V_{CM}}{V_O} \right|_{V_S = 0} \tag{2.145}$$

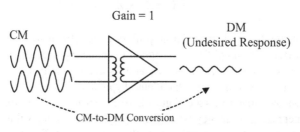

Figure 2.67. CM to DM conversion results in undesired response (susceptibility) to CM EMI.

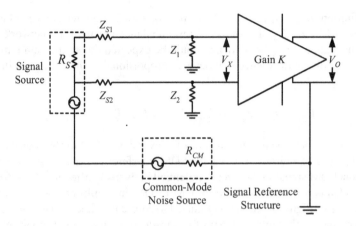

Figure 2.68. Common-mode generation in unbalanced systems.

The *CMRR* is often expressed in decibels (dB). Thus, from Equation (2.145),

$$CMRR\,[dB] = 20 \cdot \log\left(\left|\frac{KV_{CM}}{V_o}\right|_{V_S=0}\right) \qquad (2.146)$$

As defined, *CMRR* conveys a measure of how well the amplifier can reject a common-mode noise signal at its input. Ideally, the *CMRR* of an amplifier should be infinitely large. Under worst-case conditions, however, *CMRR* = 1 (or 0 dB). Typical values for a differential amplifier with balanced input impedances are *CMRR* = 1000 (or 60 dB), approximately.

The *CMRR* for the amplifier in Figure 2.69 can be shown to be estimated as [20]

$$CMRR = \left|\frac{1}{\dfrac{Z_1}{R_S + Z_{S1} + Z_1} - \dfrac{Z_2}{Z_{S2} + Z_2}}\right| \qquad (2.147)$$

It follows that the use of differential amplifiers is beneficial for common-mode suppression. A differential amplifier is designed to make Z_2 large compared to Z_{S2}, and Z_1 large compared to $Z_{S1} + R_S$. Since Z_1 and Z_2 are normally frequency dependent, the *CMRR* is likewise frequency dependent. Typically, Z_1 and Z_2 are modeled as resistors shunted by capacitors (see Figure 2.70). *CMRR* will therefore inevitably decrease with increasing frequency when the capacitive reactance value becomes smaller than the resistor. Consequently, a high *CMRR* is difficult to attain at high frequencies.

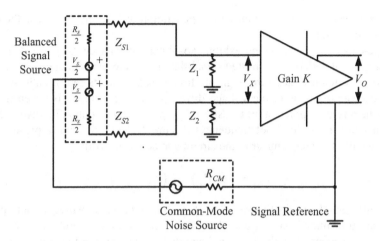

Figure 2.69. Common-mode noise in balanced systems.

Lessons Learned

- Balanced circuits typically do not respond to CM signals; however, CM-to-DM conversion (due to imperfect balance in the circuit) will result in potential interference to circuit performance.
- Balanced circuits are often a design objective. Perfect balance is rarely attainable in practice, though, owing to imbalance in drivers/receivers or due to imbalance in the interconnects.

2.7.4 Differential Signaling and Balanced Circuits

In Section 2.7.3, the effect of imbalance in a circuit on CM-to-DM conversion was presented. It was shown that the most important factor determining circuit perfor-

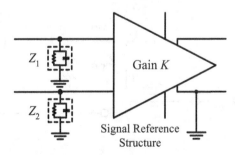

Figure 2.70. Practical model of the CM impedances Z_1 and Z_2 at the differential amplifier input.

mance with respect to the immunity to or generation of CM interference is the extent
of the circuits' balance.

Single-ended circuit signaling exhibits absolutely no balance. Single-ended signaling is performed with the transmitted signal flowing on the send conductor (e.g., wire), utilizing the reference structure (e.g., the frame/chassis) as its return path. A single-ended receiver responds to the voltage difference ΔV_{in} developed between its input (V_{rx}) to the reference potential of the SRS (V_{ref}), producing an output relative to ΔV_{in} (f represents "a function of," not frequency) [Figure 2.71(a)]. Any noise present in the reference will ultimately appear in the circuit and across its load:

$$V_{out} = f\left(\Delta V_{in}\right) = f\left(V_{rx} - V_{ref}\right) \qquad (2.148)$$

In a differential circuit [Figure 2.71(b)], on the other hand, two equal but opposite signals are transmitted on two complementary send lines, V_{rx+} and V_{rx-}. The differential receiver responds only to the voltage difference between the signals on its in-

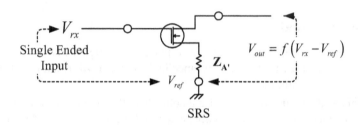

(a) Single-Ended Circuit

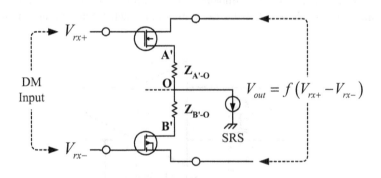

(b) Differential Circuit

Figure 2.71. Single-ended versus differential receivers.

puts, $\Delta V_{in} = V_{rx+} - V_{rx-}$. If we consider the voltage on each of the conductors to be relative to the reference potential on the SRS (V_{ref}), the receiver's output will be relative to ΔV_{in}:

$$V_{out} = f\left(\Delta V_{in}\right) = f\left[\left(V_{rx+} - V_{ref}\right) - \left(V_{rx-} - V_{ref}\right)\right] = f\left(V_{rx+} - V_{rx-}\right) \quad (2.149)$$

Clearly, the input voltage is independent of the reference potential, V_{ref}, present across the SRS.

The principles of the above discussion are now applied to (1) a typical single-ended digital-signaling circuit [Figure 2.72(a)] and to (2) a differential (e.g., TIA/EIA RS-422) signaling circuit [Figure 2.72(b)] at the presence of a noise voltage, V_N.

When Equation (2.149) is applied to the single-ended circuit shown in Figure 2.72(a), it is obvious that any noise voltage present on the signal reference will appear across the receiver input terminal (and subsequently at the driver output terminal), resulting in potential interference to the receiver,

$$\Delta V_{in}^{N} = V_{rx} - \left(V_{ref} + V_{N}\right) = \left(V_{rx} - V_{ref}\right) + V_{N} = \Delta V_{in} - V_{N} \quad (2.150)$$

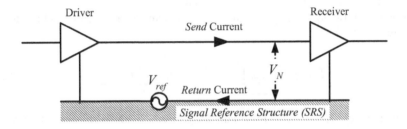

(a) Typical Single-Ended Signaling Circuit

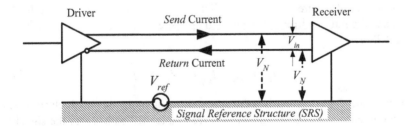

(b) Differential Signaling: EIA-RS-422

Figure 2.72. Single-ended versus differential signaling.

where the superscript "N" in ΔV_{in}^{N} indicates the voltage sensed at the receiver at the presence of noise voltage V_N. Conversely, in the case of differential signaling [Figure 2.72(b)], while using the same reasoning, the noise voltage in the signal reference structure (appearing as common-mode interference voltage) is equally introduced onto both the send and return conductors simultaneously. The receiver responds only to the difference, ΔV_{in}^{N}, between the voltages present on both lines. The noise present on both conductors has, therefore, no effect on the receiver's response:

$$\Delta V_{in}^{N} = \left[\left(V_{rx+} - \left(V_{ref} + V_N\right)\right) - \left(V_{rx-} - \left(V_{ref} + V_N\right)\right)\right] = \left(V_{rx+} - V_{rx-}\right) = \Delta V_{in} \quad (2.151)$$

In practice, the capability of differential receivers (e.g., the RS-422 line receiver) to reject common-mode noise (such as that present in the reference structure) is not unlimited. Typical maximum rating for common-mode voltage withstand capability of RS 422 receivers is approximately ±12V [21]. Common-mode voltages beyond those specified for the particular device could result in temporary upset or even permanent damage to the receiver.

Differential interfaces may exhibit degraded performance with respect to their common-mode rejection capability when unbalanced. The difference between the terms "differential" and "balanced" is now explained.

Ampere's Law, in its integral form,* states that an electric current yields a magnetomotive force:

$$\int_S \left(\nabla \times \vec{H}\right) \cdot ds = \int_S \left(\vec{J} + \frac{\partial \vec{D}}{\partial t}\right) \cdot ds = \sum_i I_{c_i} + \sum_j I_{d_j} \quad (2.152)$$

In Equation (2.152), the terms for conduction current, I_c, and displacement current, I_d, are

$$I_C = \int_S \vec{J} \cdot d\vec{s} \quad \text{and} \quad I_d = \frac{\partial}{\partial t}\int_S \vec{D} \cdot d\vec{s} \quad (2.153)$$

In an ideal circuit, where a single controlled conductive return path exists, the "send" current and the return current are equal and opposite, so Equation (2.152) reduces [see Figure 2.73(a)]

$$I_{C_S} + I_{C_R} = I_{C_S} + \left(-I_{C_S}\right) = 0 \quad (2.154)$$

When all return current comprises the current flowing in the intended return conductor only and no current flows through alternative unintended paths (e.g., the signal reference), the circuit is said to be balanced. In other words, all current flowing on the "send" conductors must return to the source on the intended return conductor.

*Note that surface integration is maintained for clarity as Stokes' Theorem was not applied in this expression.

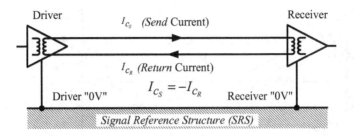

(a) Balanced Circuit

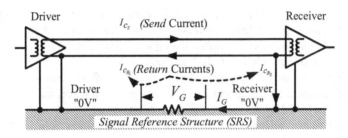

Figure 2.73. Balanced versus unbalanced differential circuits, receiver "0V" (Driver "0V."

When multiple conductive return paths exist, the return current splits between the various conductors; the current flowing in the ith conductor is then denoted I_{cRi}. In this case, we have [see Figure 2.73(b)]

$$I_{c_S} = -\sum_i I_{c_{Ri}} \qquad (2.155)$$

For the example shown in Figure 2.73(b), Equation (2.155) can be written as

$$I_{c_S} = -\left(I_{c_{R_1}} + I_{c_{R_2}}\right) \Rightarrow I_{c_S} \neq -I_{c_{R_1}} \Rightarrow \int_S \left(\nabla \times \vec{H}_{S+R_1+R_2}\right) \cdot d\vec{s} \neq 0 \qquad (2.156)$$

This result indicates that imbalance in the circuit results in degraded magnetic flux cancellation (and, subsequently, increased emissions).

Figure 2.74 demonstrates this situation in further detail. The source and load circuits are connected to the reference structure at two physically separate reference points, G_S and G_L, respectively, which are typically at different potentials (note the different symbols and names). A ground (common-mode) voltage due to this potential difference, $\Delta V_G = V_{CM}$, is thus present between the reference points across the imped-

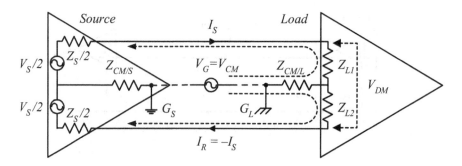

Figure 2.74. Generic model for demonstrating imbalance in a circuit.

ance $Z_G = Z_{CM/S} + Z_{CM/L}$. This noise voltage may result in erroneous response of the receiver. Due to the opposite direction of current flow, this ground voltage drop counteracts the signal voltage amplitude across the input terminals of the (differential) receiver. The circuit in this case is said to be "unbalanced."

Balancing of the circuit eliminates the development of common-mode voltage and the flow of common-mode current and subsequent interference, since any common-mode noise is rejected by the receiver (Figure 2.75).

Take note that a balanced circuit is not necessarily required to be floated! The term "floated circuit" is often confused with the term "balanced circuit." A floated circuit is a circuit not conductively connected anywhere to the signal reference structure. The circuit in Figure 2.76 is not floating since the load end is connected to the signal reference at one end. The circuit is balanced, however: no closed path exists for the return current to flow between the signal reference structure and the circuit conductors. Both the driver and the receiver are subsequently referenced to the same point (and "0V").

Inasmuch as the current must flow in closed loops, current loops may also be formed by a combination of conductive and displacement AC current paths, as shown in Figure 2.76(b). This implies that a network that may appear balanced at lower frequencies when no galvanic path to the signal reference is available may act as an unbalanced circuit at higher frequencies. This occurs when capacitance between the conductors and the reference structure allows sufficiently high displacement current to flow, splitting the return current between the intended return conductor and the reference plane.

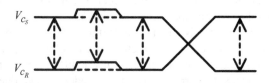

Figure 2.75. Common-mode noise rejected by the balanced receiver.

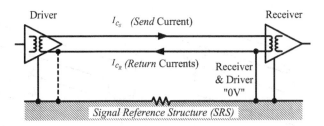

(a) Balanced (but Not Floated) Circuit: No Current Return Path Exists (Lower Frequencies)

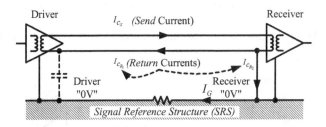

(b) Unbalanced Circuit: A Displacement Current Return Path Exists (Higher Frequencies)

Figure 2.76. Balanced circuit, Receiver "0V" = Driver "0V."

The concept of balanced circuits can also be extended to more complex networks, such as three-phase power systems. In Figure 2.77(a), a three-phase power circuit is depicted, each of the phase lines carrying current phase-shifted by $2/3\pi$ (120°) with respect to each other. In this case, we have a balanced three-phase system, with each of the conductors serving as an intended return path for the others. Adding a neutral line [Figure 2.77(b)] would still maintain the four-conductor circuit as a balanced (four-wire) system.

Figure 2.78 illustrates the response of ideal and real-world balanced-input receivers to an incoming common-mode (CM) signal. Clearly, in the ideal circuit [Figure 2.78(a)], the output of the receiver is independent of the input CM current, of whether regardless the circuit is fully balanced or has a balanced input only. In Figure 2.78(b), a real-world balanced receiver will also exhibit some response to a CM signal.

If a transmission line is not perfectly balanced, an undesired differential-mode signal will develop as a result of the common-mode signal, due to differences in amplitude and phase. A figure of merit for the capability of a balanced system to preclude CM to DM conversion is defined as the longitudinal conversion loss factor, or LCL [22], expressed in dB as (see Figure 2.79)

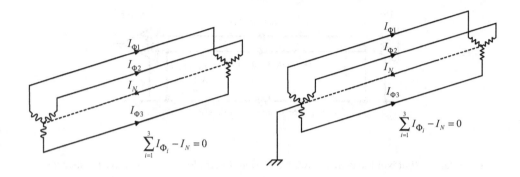

(a) Three-Wire Balanced and Floated Circuit   (b) Three-Wire Balanced but Not Floated Circuit

Figure 2.77. In a balanced circuit, the return current flows in the intended return path only.

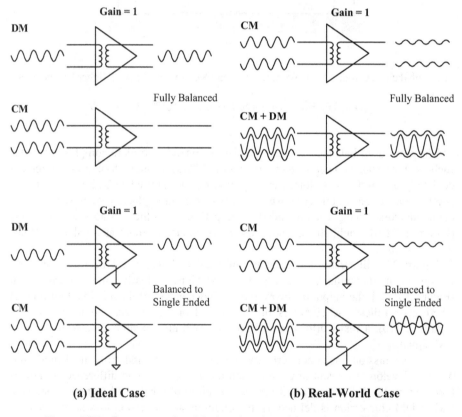

Figure 2.78. Response of an ideal and real-world receiver to CM signals.

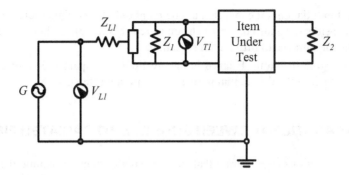

Figure 2.79. Definition of an LCL.

$$LCL\,[dB] = 20 \cdot \log\left(\frac{V_{CM}}{V_{DM}'}\right)\Bigg|_{V_O=\text{Constant}}$$

(2.157)

where:

V_{CM} = Common-mode voltage applied to the two balanced input ports of the system, when they are shorted together, required to generate the standard output voltage V_O

V_{DM}' = Differential-mode voltage applied between the two balanced input ports of the system, required to generate the standard output voltage V_O

The differential-mode voltage, V_{DM}' represents the portion of the CM noise converted to undesired DM interference and *not* the desired differential signal voltage.

LCL represents the ratio between the applied CM signal to the resulting DM signal, or the extent of common-to-differential-mode conversion. It serves as a measure of the degree of the unwanted signal produced at the terminals of the network due to the presence of a longitudinal signal on the connecting leads.

Lessons Learned

- Single-ended circuits respond to voltages appearing between the signal input terminal and signal reference. Differential circuits respond to the voltage appearing across their input terminals.
- Differential circuits are superior to single-ended circuits due to their enhanced CM noise immunity.
- The terms "differential" and "balanced" circuits should not be considered synonymous. A balanced circuit must be differential, whereas a differential circuit may not be balanced.
- Only balanced circuits can guarantee the full advantages of differential signaling.

- A differential circuit connected at one point only to the signal reference can still be considered balanced, but when connected at both points, this may severely violate the purpose of differential signaling.
- The degree of circuit balance is mostly limited by parasitic stray capacitance from each of the differential ports to the common reference

2.8 INTERACTION BETWEEN SOURCES TO RADIATED FIELDS

Maxwell's equations demonstrate that time-varying currents constitute the source of magnetic fields (ultimately generating time-varying electric fields). The relationship between time-varying currents and radiated electromagnetic fields will now be presented.

2.8.1 Radiation from Current-Carrying Conductors

The nature of the radiated fields is determined by the characteristics of the current sources. Time-varying currents materialize in two primary modes; conduction currents I_c and displacement currents I_d.

A typical electronic system consists of a variety of conductors such as wiring, traces on printed circuit boards, metallic planes and three-dimensional metallic structures. At higher frequencies, when the conductors become electrically long ($L > \lambda/10$, $L =$ cable length, and $\lambda =$ wavelength associated with the frequency under consideration), these conductors can become efficient radiators, emitting electromagnetic energy due to the time-varying currents flowing in them.

The fields emitted from the conductors originate from two elementary current sources (Figure 2.80):

1. Differential-mode sources (small magnetic loops)
2. Common-mode sources (short, loaded electric monopoles/dipoles)

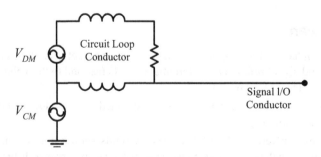

Figure 2.80. A circuit comprising both a DM (Loop) and CM (monopole) current sources of radiated EMI.

2.8.1.1 DM Current Sources (Small Magnetic Loops).

Consider a circuit containing a signal source (e.g., oscillator) and a load (Figure 2.81), which is derived from a part of the circuit of Figure 2.80. For a simplified analysis of emissions due to DM current flowing in a well-defined closed loop (comprising the signal and current return paths), such circuits are commonly represented as a loop antenna (or "magnetic dipole"), in which conduction current flows in closed loops (Section 2.7.4).

Using [6] as an arbitrary reference for antenna theory, it is shown that electric and magnetic time-varying fields emanating from a small elementary magnetic source ($\sqrt{A/\pi} \ll, r$) can be expressed as (see Figure 2.81):

$$H_\theta = \frac{IA\beta^3}{4\pi} \cdot \left[-\frac{1}{\beta \cdot r} - \frac{1}{j(\beta \cdot r)^2} + \frac{1}{(\beta \cdot r)^3} \right] \cdot \sin \theta \cdot e^{-j\beta r} \qquad (2.158)$$

$$H_r = \frac{IA\beta^3}{2\pi} \cdot \left[-\frac{1}{j(\beta \cdot r)^2} + \frac{1}{(\beta \cdot r)^3} \right] \cdot \cos \theta \cdot e^{-j\beta r} \qquad (2.159)$$

$$E_\phi = \frac{IA\beta^4}{4\pi\omega\varepsilon_0} \cdot \left[-\frac{1}{\beta \cdot r} - \frac{1}{j(\beta \cdot r)^2} \right] \cdot \sin \theta \cdot e^{-j\beta r} \qquad (2.160)$$

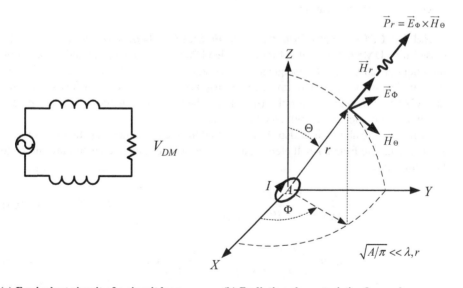

(a) Equivalent circuit of a circuit loop   (b) Radiation characteristics from a loop

Figure 2.81. A circuit carrying intended DM current is commonly represented as a loop antenna.

where:

$j = (-1)^{1/2}$
$\beta = 2\pi/\lambda$ = wave velocity, or phase velocity (radians/m)
λ = wavelength (m)
r = distance to point of observation (m)
A = current carrying loop area (m²)

When $r \gg \lambda/2\pi$ (in the "far field"), H_r vanishes, hence, only Equations (2.158) and (2.160) need to be considered. Therefore, when eliminating the phase components and after simplification, we end up with expressions for the radiation efficiency of an elementary magnetic loop:

$$\frac{|H_\theta|}{I_D} = \pi \cdot \frac{A}{\lambda^2} \cdot \left(\frac{1}{r}\right) \tag{2.161}$$

$$\frac{|E_\phi|}{I_D} = 120\pi^2 \cdot \frac{A}{\lambda^2} \cdot \left(\frac{1}{r}\right) \tag{2.162}$$

From observation, the emissions from the loop are proportional to (a) the amplitude of the DM current in the loop and (b) the electrical size of the loop (area of the loop, A, compared to wavelength). Also, the emissions are inversely proportional to (c) the distance to point of observation.

2.8.1.2 CM Current Sources (Small Electric Dipoles). In contrast to the closed-loop DM current source for unintended CM current, a well-defined current loop does not normally exist. Consequently, the conductor is best represented as an elementary dipole antenna in which time-varying displacement (CM) current flows along the dipole's length L between its poles (in practice, between the conductor and its "image" within the reference structure); (see Figure 2.82).

Similar to the above, it is also shown [6] that the electric and magnetic time-varying fields emanating from a small elementary electric source (dipole element) are shown to be expressed as

$$E_\theta = \frac{IL\beta^3}{4\pi\omega\varepsilon_0} \cdot \left[-\frac{1}{j\beta \cdot r} + \frac{1}{j(\beta \cdot r)^2} + \frac{1}{j(\beta \cdot r)^3}\right] \cdot \sin\theta \cdot e^{-j\beta r} \tag{2.163}$$

$$E_r = \frac{IL\beta^3}{2\pi\omega\varepsilon_0} \cdot \left[\frac{1}{(\beta \cdot r)^2} + \frac{1}{j(\beta \cdot r)^3}\right] \cdot \cos\theta \cdot e^{-j\beta r} \tag{2.164}$$

$$H_\phi = \frac{IL\beta^2}{4\pi} \cdot \left[-\frac{1}{j\beta \cdot r} + \frac{1}{(\beta \cdot r)^2}\right] \cdot \sin\theta \cdot e^{-j\beta r} \tag{2.165}$$

where $L \ll \lambda$, r, and $\beta = 2\pi/\lambda$.

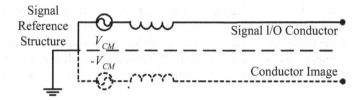

(a) Equivalent circuit of a CM current-carrying conductor

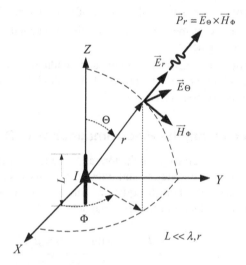

(b) Radiation characteristics from a dipole

Figure 2.82. A circuit carrying unintended CM current is commonly represented as a dipole antenna.

Similar to the magnetic loop, in the "far field," E_r vanishes and when removing the phase components. Equations (2.163) and (2.165) end up as the expression for the radiation efficiency of an elementary electric dipole:

$$\frac{|E_\theta|}{I_C} = 120\pi \cdot \frac{L}{\lambda} \cdot \left(\frac{1}{r}\right) \tag{2.166}$$

$$\frac{|H_\phi|}{I_C} = \frac{L}{\lambda} \cdot \left(\frac{1}{r}\right) \tag{2.167}$$

From observation, the emissions from the dipole are again determined by the following factors: (a) the amplitude of the CM current exciting the dipole and (b) the electrical size of the dipole (length of the loop, L, compared to wavelength). Also, the emissions are inversely proportional to (c) the distance to point of observation.

A conclusion of considerable significance from the derivation presented above is that an increase in circuit dimensions, in particular, the electrical length of conductors and the electrical loop areas of current loops, will lead to increased emissions from the circuit. This important conclusion will be further discussed in Chapter 9 when evaluating the effect of "ground planes" on the control of radiated emissions on printed circuit boards.

Lessons Learned

- Current, not voltage, constitutes the source of electromagnetic fields.
- CM currents are the prime contributors to electromagnetic emissions from circuits at higher frequencies.
- Electromagnetic emissions from circuits and conductors are proportional to the electrical length/loop area of the circuit, L/λ or A/λ^2, for dipole (CM) and loop (DM) circuit models, respectively.

2.8.2 Flux Cancellation, the Electromagnetics of Balancing

While considering current-carrying conductors as "radiators," a fundamental concept regarding EMI emissions from circuits and transmission lines is that of flux cancellation. Starting from Maxwell's Fourth Equation, Ampere's Law, we observe that current flowing through a conductor produces a magnetic field:

$$\oint_C \vec{B} \cdot d\vec{l} = \mu_0 I , \qquad \text{with} \qquad \vec{B} = \mu_0 \vec{H} \qquad (2.168)$$

For an infinitely long current-carrying conductor, this equation reduces to see [see Figure 2.83(a)]

$$\vec{B}_r = \frac{\mu I}{2\pi r} \hat{\Phi} \qquad (2.169)$$

where r is the radius of a circular integration contour C surrounding the conductor, and the direction of the magnetic flux vector is tangent to the contour in a cylindrical coordinate system, denoted as $\hat{\Phi}$.

Note: Although the concept of current flowing in a single conductor is introduced here, it is nonphysical in practice since current flows in closed loops only. This statement should be interpreted, therefore, as a flow of current in a considerably large loop with the return conductor infinitely far from the signal conductor. This situation may be encountered, for instance, when excessively large loops are formed by power-circuit return (neutral) and ground (protective earth) conductors.

When the return conductor is brought into near proximity of the signal conductor, such that they are closely spaced, a significant change in the magnetic flux pattern occurs. The flux produced by the signal conductor I_+ and the flux produced by the return conductor I_- will add or subtract depending on the point of observation in space. Assuming that the circuit carrying the current is balanced, that is, current flowing in the

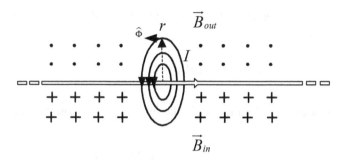

(a) Magnetic Field Flux Due to Current Flow in <u>One</u> Infinitely Long Single Conductor

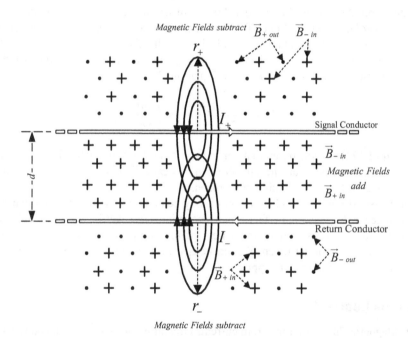

(b) Magnetic Field Flux Due to Current Flowing in <u>Two</u> Closely Spaced Infinitely Long Conductors

<u>Figure 2.83.</u> Magnetic field flux cancellation due to current flow in a pair of infinitely long conductors. Legend: • = flux exiting surface; + = flux entering surface.

signal and return conductors are equal in amplitude and reversed in phase, and assuming that the conductors are electrically short, opposing flux patterns will be generated by the conductors, resulting in the situation depicted in Figure 2.83(b).

In Figure 2.83, the fluxes entering and emerging from the circuit plane due to signal current I_+ are marked $\vec{B}_{+\text{in}}$ and $\vec{B}_{+\text{out}}$, respectively, whereas the fluxes entering and emerging from the circuit plane due to current I_- are marked $\vec{B}_{-\text{in}}$ and $\vec{B}_{-\text{out}}$, respectively.

With $I_- = I_+ = I$, it is evident that the magnetic fluxes due to both currents add between the conductors:

$$\vec{B}_{Total} = \vec{B}_{+in} + \vec{B}_{-in} = \frac{\mu_0 I}{2\pi}\left(\frac{1}{r_+} + \frac{1}{r_-}\right)\hat{\Phi} \qquad (2.170)$$

Outside the area enclosed by the circuit, the flux emerging from the conductors subtract, yielding

$$\vec{B}_{Total} = \vec{B}_{+in} - \vec{B}_{-in} = \frac{\mu_0 I}{2\pi}\left(\frac{1}{r_+} - \frac{1}{r_-}\right)\hat{\Phi} \qquad (2.171)$$

The direction $\hat{\Phi}$ or $-\hat{\Phi}$ of the residual flux vector will be determined by the ratio r_+/r_-. At the immediate vicinity of one of the conductors, for example, the return conductor, $r_+/r_- \gg 1$, so $\vec{B}_{-out} > \vec{B}_{+in}$ and a large residual flux will exist. However, when $r_+ \gg d$, $r_- \gg d$, and $r_+ \cong r_- = r$, which results in $\vec{B}_{-out} \cong \vec{B}_{+in}$, the flux efficiently subtracts and the total magnetic flux at a distance $r \gg d$ in Equation (2.171) [15]:

$$\vec{B}_{Pair} = \frac{\mu_0 I d}{2\pi r (r+d)} \approx \frac{\mu_0 I d}{2\pi r^2} \qquad (2.172)$$

Equation (2.172) demonstrates that for $r/d \to \infty$ the total magnetic field emanating from the conductor pair vanishes, resulting in flux cancellation. Thanks to flux cancellation, radiation from the circuit is minimized.

Of course, if the two conductors are exactly centered ($d \to 0$), total flux cancellation will occur. In practice, perfect flux cancellation can never be achieved, and some residual field will always remain in the near vicinity of the conductors.

Note: Although the term cancellation is used throughout this chapter, in practice, the term minimization is more appropriate.

Lessons Learned

- Magnetic flux emanating from currents circulating in closely spaced conductors, equal in amplitude and opposite in phase ("balanced circuits"), will vanish when $r \gg d$.
- Balancing of circuits is crucial for achieving effective magnetic flux cancellation.

BIBLIOGRAPHY

[1] Johnson, H., "One Reader's Opinion. . . . Why Digital Engineers Don't Believe in EMC," *IEEE EMC Society Newsletter, Spring 1998 Edition* (also published on line at http://www.sigcon.com/Pubs/news/noEMC.htm).

[2] Maxwell, J. C., *A Treatise on Electricity and Magnetism,* Vol. 1, New York: Dover Publications, 1954 (republication of the 3rd Ed., Published by the Clarendon Press, 1891).

[3]   Koehne, H., John, W., and Reichl, H., "Modeling and Extraction of Inductances on Chip Level," in *Proceedings of the 17th International Wroclaw Symposium and Exhibition on Electromagnetic Compatibility,* Wroclaw, Poland, June 2004.

[4]   Paul, C. R., *Introduction to Electromagnetic* Compatibility, New York: Wiley, 1992.

[5]   Paul, C. R. "Modeling Electromagnetic Interference Properties of Printed Circuit Boards," *IBM Journal of Research and Development,* Vol. 33, no. 1, pp. 33–50, January, 1989.

[6]   Jordan, E. C., and Balmain, K. G., *Electromagnetic Waves and Radiating Systems,* 2nd Ed., Englewood Cliffs, NJ: Prentice Hall, 1968.

[7]   Paul C. R., and Nasar, S. A., *Introduction to Electromagnetic Fields,* 2nd Ed., New York: McGraw-Hill, 1987.

[8]   Paul, C. R. "What Do We Mean by 'Inductance'? Part I: Loop Inductance," *IEEE EMC Society Newsletter,* Issue 215, Fall 2008.

[9]   Paul, C. R. "What Do We Mean by 'Inductance'? Part II: Partial Inductance," *IEEE EMC Society Newsletter,* Issue 216, Winter 2008.

[10]   Johnson, H., and Graham, M., *High Speed Signal Propagation, Advanced Black Magic,* Upper Saddle River, NJ: Prentice-Hall2003.

[11]   Johnson, H., "Minimum-Inductance Distribution of Current," *High Speed Digital Design Online Newsletter,* Vol. 6, Issue 7, August 7, 2003.

[12]   Ott, H. W., *Noise Reduction Techniques in Electronic Systems* (2nd Ed.), New York: Wiley-Interscience, 1988.

[13]   Briët, R., "Application of the LT-MP Principle to the Theory of Lightning Propagation," *Interference Technology Engineering Master (ITEM) 1997,* R&B Enterprises, West Conshohocken, PA, 1997.

[14]   Feynman, R., *The Feynman Lectures on Physics,* Vol. II, Reading, MA: Addison-Wesley, 1989.

[15]   Hartal, O., *Electromagnetic Compatibility by Design,* R&B Enterprises, West Conshohocken, PA, 1994.

[16]   Rand, A., *Atlas Shrugged,* Chapter VII, p. 98, 35th Anniversary Edition, New York: Penguin Group, 1992.

[17]   Johnson, H., "Common-mode Analysis of Skew," *EDN Magazine on Line* (Internet Edition, http://www.edn.com) January 22, 2004.

[18]   IEEE-100, *An Authoritative Dictionary of IEEE Standard Terms,* 7th Ed., New York: IEEE Press, 2000.

[19]   FS-1037C, Federal Standard 1037C, *Telecommunication: Glossary of Telecommunication Terms,* http://www.its.bldrdoc.gov/fs-1037/fs-1037c.htm, National Communications System Technology and Standards Division, August 7, 1996.

[20]   MIL-HDBK-419A, *Military Handbook, Grounding, Bonding, and Shielding for Electronic Equipments and facilities,* Washington, DC: U.S. Government Printing Office, 1987.

[21]   MIL-STD-188-114A, *Military Standard, Electrical Characteristics of Digital Interface Circuits,* Washington, DC: U.S. Government Printing Office, 1985.

[22]   ITU-T Telecommunication Standardization Sector of ITU Series O, *Specifications of Measuring Equipment—General Measuring Arrangements to Assess the Degree of Unbalance About Earth,* ITU-T Recommendation 0.9 (03/99), Geneva: ITU, 1999.

3

THE GROUNDS FOR GROUNDING

3.1 GROUNDING, AN INTRODUCTION

Grounding, though probably the most important aspect of system or circuit design, is one area of designing a product that is least understood by engineers. Lack of understanding of grounding brings about more confusion and misapprehension more than any other technical subject in electrical and electronic system design. It is not easy to understand the concept of grounding intuitively, since it is not given to straightforward definition, modeling, or analysis. Many uncontrolled factors affect grounding performance. Furthermore, regulations, standards, and codes do little to clear this fog and to demystify its concepts. Experience plays a prime role in the final choice of the system's grounding topology. This experience is not generally transferable to the grounding methodology of another system, even if they employ similar electronic technologies and applications.

3.1.1 "Grounding," One Term, Many Imports

"Ground is where potatoes and carrots thrive."—BRUCE ARCHAMBEAULT

Technical communications can take place only if the language used for transferring information is carefully and unambiguously defined. This refers to the need to use com-

Grounds for Grounding. By Elya B. Joffe and Kai-Sang Lock
Copyright © 2010 The Institute of Electrical and Electronics Engineers, Inc.

mon terminology understood by everyone. With the progress of science and electrotechnology, terms used to describe aspects of engineering take on different and specific meanings. Take, for instance, the terms "function" and "integration," which, although still utilized in their original and common forms of everyday use, acquire new and specific meanings when used by mathematicians. The terms are clear in usage within their context.

Several terms in electrical and electronic engineering are of a similar nature. In many cases, these terms do not fulfill the objective of being clear and unambiguous. Quite the opposite—several terms have been overused, let alone abused, by engineering disciplines to such an extent that only an insider to that discipline may decipher and comprehend their meaning. Furthermore, interpretation may be different between members of the same discipline.

To demonstrate the importance of correct definition and understanding on terms related to ground, the following true story was told by Yang, in his paper "The Myth about Ground and How to Correctly Use it in High-Speed Signal Integrity and Power Integrity Modeling and Simulation":

> One of the disciplinary tools we use with our nine-year-old daughter, Haley, is to remove privileges (TV, phone, visiting friends, etc.) I took Haley to work with me at Cisco last Thursday for "Take Our Children to Work Day." When it came time to take her on a tour of the Engineering Lab, I began to fit her with ESD heel straps (Heel straps are used to drain off electrostatic charges into the ground to prevent damage to sensitive electronic circuitry.) When she asked what they were for, I tried to explain, "This is so you'll be grounded. . . ."
>
> "She interrupted with "Grounded! GROUNDED? What did I do wrong?!?!"

Undoubtedly, terms such as "ground" and "earth" have been severely misused in the electronic engineering vocabulary. Excellent examples of ambiguous terms, having different imports for different people, follow.

IEEE-STD-1100 [1] provides the following definition of the term "Ground":

> A conducting connection, whether intentional or accidental, by which an electric circuit or equipment is connected to the earth, or to some conducting body of relatively large extent that serves in place of the earth.
>
> Note: Grounds are used for establishing and maintaining the potential of the earth (or of the conducting body), or approximately that potential, on conductors connected to it and for conducting ground currents to and from earth (or the conducting body).

The term "earth" is a fundamental part of this definition, and is associated primarily with the safety of equipment and personnel. With facilities and buildings constructed on the earth, with people present in those buildings, and with lightning striking buildings while discharging to earth, using the earth as part of the safety system for preclusion of electrical hazards was a natural consequence. Attention should be paid however, to the fact that grounding does not necessarily involve an electrical connection to the actual earth. For example, systems installed on aircraft operate quite well when the aircraft is airborne, even when the aircraft is struck by lightning and its potential raised to millions of volts! In that case, the aircraft structure constitutes the conducting body of relatively large extent that serves in place of the earth.

A power system engineer refers to the term "ground" with little ambiguity, implying the safety ground/protective earth connection as mandated by electrical safety and lightning protection codes, as the connection to "mother earth." When uttered by electronics and system engineers, the term is used to imply the common power supply connection or the connection to the metal enclosure that houses the system's components and circuits, serving as a common reference for those circuits; the earth connection is simply incidental and is a result of poor usage of terminology. To an antenna designer, "ground" implies the reference ("image") plane or the "other half" of the antenna, whereas an electronic circuit designer may interpret the term "ground" as a reference plane for analog and digital circuits and components or the signal current return path. In practice, three engineering disciplines are utilizing the same word "ground," but applying it in different ways, based on their respective field of interest (Figure 3.1).

A "ground" in a system is relative! Choose any point in a circuit or system to serve as the ground and reference all other voltages to that point. For instance, if you had a 48 V DC power supply (line to return terminals) and decided to make its "–" terminal the system's 0 V or ground, then the voltage at the other terminal of the power supply output would be at +48V DC with respect to ground [Figure 3.2(a)]; however, if you decided to make the "+" terminal the system's V or ground, then the voltage at the oth-

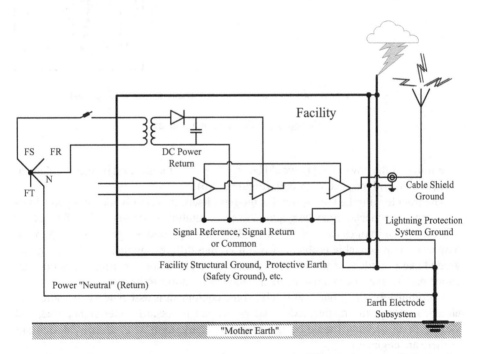

Figure 3.1. One ground serving many objectives with different (and conflicting) rules for each.

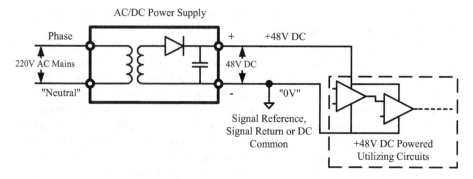

The 48V DC power supply output is at **+48V DC** with respect to "ground"

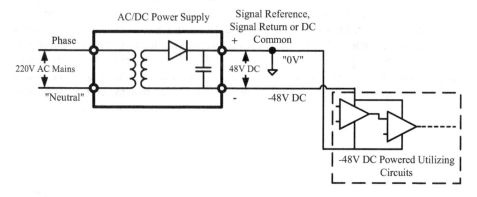

The 48V DC power supply output is at **−48V DC** with respect to "ground"

Figure 3.2. "Ground" is relative!

er terminal of the power supply would be at, −48V (commonly used in telecommunication circuits) with respect to ground. [Figure 3.2(b)].

The trouble with the term ground is the term itself; it is simply too ambiguous! Often, a single connection may be required to serve multiple purposes with different (and often conflicting) rules for each (Figure 3.3). Using a ground designed for one purpose may cause unpredictable problems when used in a different application. Furthermore, ground can provide "sneaky paths" for interference and noise: Current does not care which path it takes, even unintended, as along as it returns to its source.

Unfortunately, the grounding of electronic equipment is associated with more than electrical safety. In practice, safe systems are not necessarily interference-free, and vice versa. Grounding should be considered a generic term, with as many definitions as there are engineers.

In the design process, it is useful to develop a "ground map," supporting the need to identify all grounding applications while avoiding any unintended interactions. This

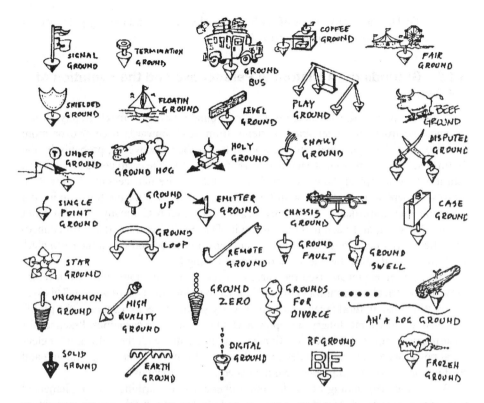

Figure 3.3. The problem with the term "ground" is that it is simply too ambiguous! (Cartoon courtesy of Dr. Bruce Archambeault.)

process involves considerations such as: What is the safety ground path? Where is the neutral connection? How are the power line filters connected to the EMI ground reference? Are there any unique requirements for signal ground isolation, such as between analog and digital circuits? Is there any unwanted mixing of signal return paths (50/60 Hz power frequency "hum" in analog systems or spikes in digital systems)? Are ground connections adequate for both current and frequency needs?

To prevent confusion, the following terms used throughout this book are defined as follows:

- Ground: A generic term to be avoided, unless used in conjunction with another term, such as "safety ground."
- Reference: A system of conductive paths among interconnected equipment that reduces noise-induced voltages to levels that minimize improper operation [1].
- Signal Return: A current-carrying path between a load and the signal source. It is the low side of the closed-loop energy-transfer circuit between a source–load pair [3].

- Earth: The conductive mass of earth, whose electric potential at any point is conventionally taken as equal to zero [4].

3.1.2 Grounding—A Historical Perspective and the Evolution of the Term

Historically, the first functional application of grounding evolved in telegraphy. Long-distance electromagnetic telegraph systems from 1820 onwards used two or more wires to carry the signal and return currents. It was then discovered, probably by the German scientist Carl August Steinheil in 1836–1837, that the earth could be used as a return path to complete the circuit, making the return wire unnecessary (Figure 3.4). However, there were problems with this system, exemplified by the transcontinental telegraph line constructed in 1861 by the Western Union Company between Saint Joseph, Missouri, and Sacramento, California. During dry weather, the earth connection often developed a high resistance, requiring water to be poured on the earth electrode to enable the telegraph to work or phones to ring [5].

Later, power engineers used the earth as a return path in single-wire earth-return (SWER) electrical distribution systems, in order to reduce the cost of wiring. This system is still in use in rural areas today. In the early years of the 20th century, the proliferation of electrical machinery and appliances caused fires and casualties. Practices for ensuring the preclusion of electrocution hazards associated with the utilization of electrical power were not totally understood. Resulting from this situation, grounding and earthing of electrical equipment were mandated for functional use.

With the understanding of the lightning phenomenon, earthing was implemented for addressing the risk of lightning strikes and electrostatic discharges, both being a potential source of damage to equipment and hazardous to people and domestic animals. Since the current path for both lightning and static electricity appear to involve the earth, it seems only reasonable to connect all electric equipment and structures to the earth, or in other words, ground them, in order to provide a controlled path for conducting the discharge currents to the earth.

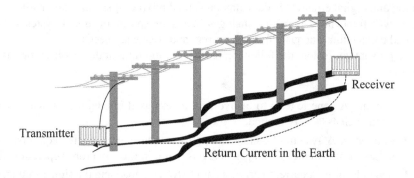

Figure 3.4. In early telegraph systems, the earth served as the current return path between the transmitter and the receiver.

Evolving from the reliance of power-fault control and lightning protection systems on the low-impedance earth connection, grounding was identified as the ultimate design practice. Without bearing in mind objectives, rationale, or consequence, a "good ground" was considered an ultimate need. With this, the legacy began. The terms grounding and earthing became therefore, synonymous.

Today, electrical product safety is still the primary and often *the* overriding consideration in system and facility grounding design. The need for earth connection of metal reinforcement members in modern facilities has become a source of confusion between the need to earth the facility and the need to ground the equipment.

Grounding of equipment and facilities does have, as a matter of fact, a much broader implication that just safety of equipment. The information age of the 20th century brought about the proliferation of large-scale distributed systems, particularly information technology equipment (ITE) networks comprising many circuits and subsystems of a diverse nature. Inspired by the utilization of the earth in the early telegraph systems, a large common signal reference structure between system components became indispensable for achieving reliable signaling between subsystems.

After World War II, with the proliferation of electromagnetic interference (EMI) sources, natural as well as man-made, grounding gradually began to be associated with the additional role of electromagnetic interference (EMI) control. Over the ensuing years, confusion about grounding on the part of electrical and electronic engineers further increased, leading to one of the greatest myths associated with grounding, namely, the belief that a "good ground" was synonymous with good EMI control measures. It was assumed that a good earth connection is mandatory for achieving reliable performance of electronic systems and equipment. Many engineers actually believe that they can "ground out" interference from a system. Furthermore, it has been asserted that the point of connection must be isolated from all other grounding connections. However, no support has been found to date to contradict the claim that well-designed equipment or systems actually do not require or benefit from such ground connections in the first place with respect to their functionality. Considerable evidence does exist, however, that such isolation may constitute a safety hazard to personnel and equipment, and that operational reliability and availability of the system can actually degrade when such grounding systems are implemented. In fact, the utilization of complex high-speed electronic systems on aircraft and space vehicles, where no earth connection is available or feasible, clearly demonstrates that such an earth connection is in fact, not required for functionality of the equipment.

In the early days of the integrated circuit (IC) and printed circuit board (PCB), the ground of the circuit board was directly connected to the metal enclosure and the protective earth (PE) lead or "green wire" of the AC power outlet. Since the signal speed was low and resistance of the ground structure was very low, ground parasitics could easily be ignored without compromising accuracy of the system. In parallel, with the evolution of circuit analysis and simulation tools, the ground was typically assumed to be ideal or a perfect electrical conductor (PEC) and could sink an infinitely large amount of current into the circuit schematics. As a result of those assumptions, circuit analysis became very simple for complex circuits, in particular for operational amplifier circuits.

The comprehension of the necessity for earth connection and the elective practice of using the earth connection for signaling purposes constitute therefore, the borderline between safety and functionality applications of grounding. This was the point at which the term "grounding" became puzzling.

Today, with lightning and power quality concerns being incorporated in the broader discipline of electromagnetic effects and EMI control, grounding and earthing of systems and facilities are often optimized for power faults and lightning protection but fail to properly address interference control. When considering equipment and circuits, where lightning and safety are of no immediate concern, confusion is further increased.

The release of SPICE (Simulation Program with Integrated Circuit Emphasis) from the University of California at Berkeley into public domain in 1972 marked the opening of an era of further confusion and misapprehension regarding the role of grounding in circuit and system design. SPICE has become a standard simulation tool throughout the electronics design industry. SPICE utilizes the concept of "global ground"* node. With a global ground node, other nodes in the circuit have well-defined absolute voltage values, so Kirchhoff's Current Law (KCL) and Kirchhoff's Voltage Law (KVL) can be easily implemented in circuit analysis solutions. In low-frequency circuits, the "global ground" concept worked well and provided a direct mapping between an ideal ground node and a physical point. Consequently, simulation results were consistent with measurements. One of the undesirable outcomes of the global ground concept is that some circuit designers, digital circuit designers in particular, believe that the ground connection is not important and the voltage signal could travel on a single wire without a return path. As a result of this misconception, designers often gave little or no attention to the ground and signal-return nets as opposed to the signal and power nets. Ignorance of return-current path is the root cause for most failures in EMC and high-speed signal/power integrity.

At high frequencies, electrical wavelength is greatly reduced, getting closer or even smaller than the physical dimensions of the circuit. When the signal propagation delay between two nodes is close or less than the signal cycle time, the voltage definition between these two nodes becomes ambiguous and has no physical meaning any longer. For this reason, the concept of "local ground" was introduced and additional voltage reference nodes were used to interoperate simulation results with physical and measurable meaning at high frequencies. This marked the closing days for the term "ground" as an "absolute" concept.

3.1.3 Grounding-Related Myths, Misconceptions, and Misapprehensions

The term "ground" is thought to be a sort of cure-all concept, like fish oil, money, or motherhood and apple pie. However, whereas you can always trust your mother, you should never trust your ground. Many grounding anomalies and interference problems result from expectations of the ground system to perform inappropriate functions substantially beyond its capability.

---

*Global ground, ideal ground, or virtual ground are important and powerful concepts in electronic circuit design and simulation. They have familiar meanings and are interchangeable in most cases.

The search for a perfect ground is very similar to the search for the Holy Grail in many respects. Tales about its existence are all around us. We all want and need it but cannot seem to find it. Observe any tier of system design at the circuit, equipment or facility level, and you will find designers continuously striving to create perfect grounding schemes.

It is no surprise, therefore, that myths and misconceptions have been associated with grounding [5, 7]. In addition to the associated belief that a good earth connection is necessary for reliable performance of electronic systems, other common misconception related to grounding exist. The conviction that a ground connection is unidirectional in nature and that it will sink unnoticeably any type and level of electrical noise will prevent a potential rise in the system. This implies that any signal designated as noise can be sent to a one-way connection point to be forever cast out in the grounding desert with no trace of it ever to be discovered again.

Reality shows that ground connections are bidirectional and as current flows into it, it is expected to emerge from it elsewhere; similarly, noise generated somewhere else in the system could emerge from this ground connection. AC currents of course, flow back and forth in their path; Ampére's Law states that current must flow in a closed loop, returning to its source. A practical ground system exhibits nonzero impedance of the ground connection and of the grounding conductors, implying that noise currents within the system can introduce substantial noise potentials in the grounding system (Figure 3.5).

With the nonzero impedance (ranging from milliohms to tens of ohms), currents associated with a lightning strike, for instance, will extend the potential of the external ground terminal to thousands of volts (Figure 3.6).

Related to the concept of the path for current flow is the myth related to the segregation of current flow according to the type of current permitted to flow in a particular path. Are the electrons actually provided with a particular symbol indicating which path they are to flow in?

Often, several current paths are given a name, such as signal ground, power ground, instrumentation ground, fault current ground, EMI ground, and, naturally, quiet ground. Yet, all are connected somewhere to the same structure. If the electrons were literate, they would know the road signs and find their way, but they cannot (Figure 3.7).

Misapprehension of the concept of grounding-path impedance leads to the consideration of resistance, rather than reactance, as the primary factor controlling the perfor-

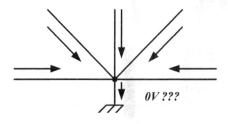

Figure 3.5. Grounding cannot provide an infinite current sink.

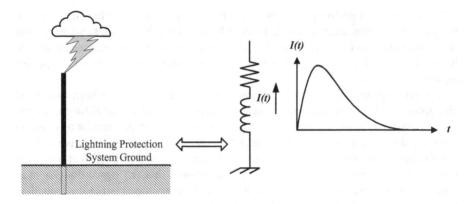

Figure 3.6. Grounding cannot prevent potential rise.

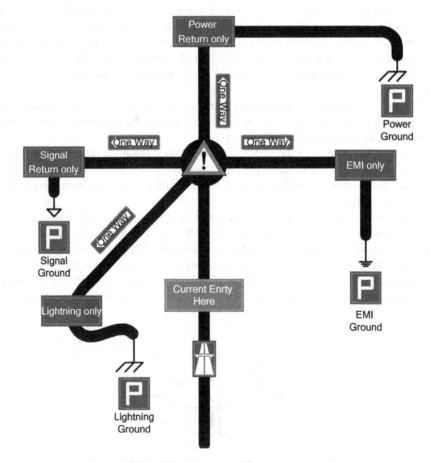

Figure 3.7. Do the electrons *really* know the correct road to follow to ground?

mance of the ground conductors and grounding system. At lower frequencies, resistance constitutes the most important term. On the other hand, at higher frequencies, such as those contained in the spectrum of a lightning discharge, impedance (primarily inductance) is most dominant. It sometimes comes as a surprise, therefore, that designers plan on using the power system ground for the purpose of lightning grounding, even though the lightning current spectrum ranges over a frequency band hundreds-fold greater than the power frequency.

The worst myth, however, is that, for the sake of best performance of the equipment, it is permissible to compromise safety regulations, such as electrical and lightning protection codes, by providing "isolated signal grounds." In practice, safety should be the primary consideration and must *never, ever* be compromised.

Lessons Learned

- There is nothing magical about "ground"; Ground is simply a path for flow of return current.
- The term "ground" is relative and is too ambiguous!
- To avoid confusion, avoid usage of the generic term "ground," unless utilized in conjunction with some descriptive term or adjective.
- Ground is a very convenient fantasy invented by engineers to simplify life but, like other fantasies, does not exist in reality.
- Ground is *not* a current sink. Current flowing in a circuit must return (somehow) to its source. This seemingly fundamental concept is often overlooked, bringing about EMI problems.
- Myths, misconceptions, and misapprehension of grounding theory, units, and quantities are inherent in the integration of sensitive electronic equipment into a facility or platform. A common terminology for the concepts of power and frequency is mandatory for achieving a successful and effective design.

3.2 OBJECTIVES OF GROUNDING

Electronic systems and equipment, in particular in large IT (information technology) and large C^3I (command, control, communication, and intelligence) facilities, often share common metallic structural members, which also serve as a path for power return, a path for lightning stroke discharge, a reference for electrical signaling, and a path for signal return. The same applies to vehicular installations such as aircraft, ships, automobiles, and even to satellites, where the structure of those vehicles fulfill the role of the ground, analogous to the functions satisfied by the earth for fixed terrestrial installations. Often, a single ground structure may fulfill multiple objectives with different (and conflicting) rules applied to each. The key to the success of equipment, system and installation design requires awareness of the characteristics associated with each of the grounding applications and the interrelationship between them in system implementation (Figure 3.8).

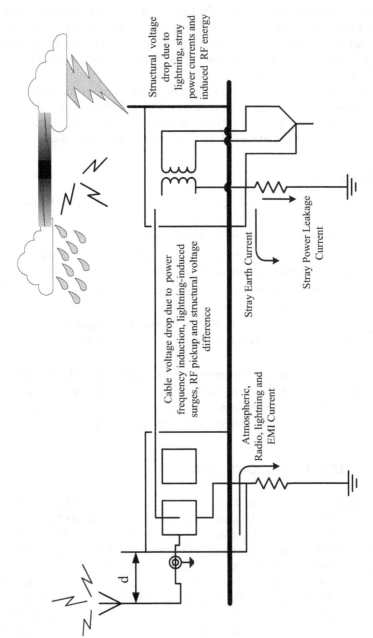

Figure 3.8. One common earth serving different "grounding" objectives.

Structural voltage drop due to lightning, stray power currents and induced RF energy

Cable voltage drop due to power frequency induction, lightning-induced surges, RF pickup and structural voltage difference

Stray Earth Current

Stray Power Leakage Current

Atmospheric, Radio, lightning and EMI Current

The term "ground" is commonly associated with the concept of a zero-impedance equipotential surface or structure, often mistakably characterized by its DC performance. As detailed below, this concept does not materialize in practice. Any conductor has some associated impedance, be it resistive or reactive. With current flowing through such conductors, a voltage drop will be introduced across it, defying the concept of an equipotential surface. Even superconductors, demonstrating approximately zero resistance, would still introduce some inductance,* again exhibiting nonzero impedance to non-DC signals.

The performance of a system, in particular, the impedance of the ground reference structure, is one of the most vital considerations related to the particular objectives of the system's performance. For applications associated with lower frequency operation (e.g., power frequencies) the concept of a zero-impedance/zero-potential surface may be considered valid. However, for higher frequency applications, significant impedance, primarily due to inductance of conductors, violates that concept, and a true zero-equipotential surface cannot be assumed.

This inconsistency highlights the primary distinction between the main objectives of grounding, which could be categorized into three key purposes, namely:

1. *Safety grounding,* which is related first and foremost to preclusion of electrical hazards such as avoidance of electric shock, risk of fire, or damage to the equipment. Safety concerns exist both under normal and faulty operating conditions within the equipment or the distribution network, or due to the impact of induced voltage and current, for example, by lightning. Safety grounding is principally associated with relatively low-frequency applications (power and lightning current surges).

2. *Signal grounding,* which is concerned predominantly with providing an acceptable reference for circuits and systems, for the purpose of ensuring acceptable equipment and system performance. When performing properly, signal grounding will prevent undue functional distortion or equipment malfunction and the risk of component failure. The signal ground also serves as a current return path for signals referenced to it. Signal grounding is concerned with frequencies ranging from DC to the highest signaling frequency anticipated.

3. *Electromagnetic interference (EMI) control grounding,* the objective of which is to gain satisfactory EMC of the system, primarily for the efficient performance of filters, transient and surge suppressors, and other terminal-protection devices and apparatus. Such devices may be incorporated in order to avoid undue emission of, or susceptibility to, electromagnetic energy under normal operating conditions, including the discharge of electrostatic energy.

As will be demonstrated, safety grounding mandates the presence of a low-impedance path from the system to earth. Signal and EMI control grounds, on the other hand, may or may not be connected to actual earth potential (visualize a satellite

---

*In Chapter 2, it was clearly shown that the amount of inductance associated with a current path is independent of the conductivity (σ) of the metallic surface.

functioning successfully with no earth connection). Difficulties may arise when both safety and signal grounds are interconnected, since a grounding system suitable for safety purposes might not be suitable for signal grounding and, in fact, could exacerbate EMI.

The following is a discussion of the main characteristics associated with systems applications.

3.2.1   Safety Grounding

The primary concern associated with safety grounding is the protection from hazards associated with high voltage/high current, which may be a threat to life or cause severe damage to equipment, installations, and facilities. In particular, safety grounding can be associated with three interrelated aspects [8]:

1. To limit voltages due to (1) lightning, (2) line surges, or (3) unintentional contact with higher voltage lines. These three sources of dangerous overvoltage conditions are provided with an alternative path around the electrical system at home, or at the workplace, by intentionally connecting the system to the earth.
2. To stabilize voltages. There are many sources of electricity. Every transformer can be considered a separate source. If there were not a common reference point for all voltage sources, it would be extremely difficult to calculate their relationships to each other. The earth, of course, is the most omnipresent conductive surface. It was adopted in the very beginnings as a nearly universal standard for all electric systems (there are a few exceptions where ungrounded systems are permitted).
3. To provide a path in order to facilitate the operation of overcurrent devices. From a day-to-day point of view, this last purpose of grounding is the most important one to understand with respect to safety grounding.

Safety grounding is intended, therefore, to protect from hazards associated with electric power faults and to preclude hazards associated with lightning strikes. It is noteworthy that, although under the same category of safety grounding, the requirements for protection against power fault hazards and against lightning strike hazards are governed by different safety codes, and different considerations and provisions apply to each.

3.2.1.1   Grounding for Preclusion of Power Fault Hazards. Figure 3.9 depicts a typical configuration for a commercial power distribution system. The intended power is provided to utilizing equipment between the phase (Φ) and neutral (N) conductors coming from the external distribution system. It enters the facility through the service entrance panel.

In addition to the above functional conductors, a third wire, the electrical safety ground conductor or ESGC, is provided, often designated "safety ground" and connected to the physical ground (earth) at the service entrance panel by means of an elec-

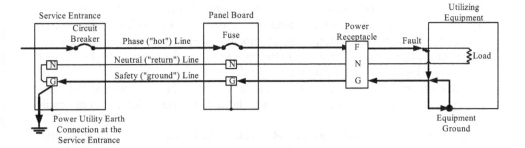

<u>Figure 3.9.</u> Power and safety grounding circuits (115/230 VAC single-phase configuration).

trode inserted deep into the earth. This protective earth connection is intended to provide electrical shock and fire hazard protection in the event of a power system fault. It guarantees that all exposed conductive surfaces are at the same electrical potential as the surface of the earth, in order to preclude hazards of electrical shock if a person comes into contact with a device in which an insulation fault has occurred. Likewise, it can avoid spark-over to any nearby structure, resulting in fire if flammable substances are present nearby. A grounded (earthed) neutral wire will ensure that the current due to the faults of the insulation are diverted to the earth. Bonding of all noncurrent-carrying metallic parts of equipment to earth ground also facilitates rapid detection of the fault by the branch circuit protection system, which will consequently interrupt the circuit.

For a 115/230 VAC configuration, a fuse or circuit breaker must be placed in series with the phase line connected to the load. The fuse precludes the fault current from flowing through the circuit, which, if not limited, could cause dangerous resistive power dissipation (I^2R losses), resulting in heating effects that could introduce a potential fire hazard.

A fault current of 100 A, for instance, flowing through a 0.1 Ω copper wire, will dissipate 1 kW in that conductor. This heat could melt the wire and set the equipment ablaze, igniting combustible substances, often with catastrophic outcome.

Consequently, at the utilizing equipment outlet, the phase and neutral lines provide power to the load, while a third conductor often called the protective earth (PE), connects the utilizing equipment metallic parts to earth at the service entrance. This earth conductor is intended to provide a path for unintended fault currents to flow back to the service entrance panel in order to open the circuit breaker or blow the fuse for that load, thus preventing an electric shock hazard to a person who may inadvertently come into contact with those metallic parts carrying unintentional hazardous currents.

When a power conductor is inadvertently short-circuited to the equipment's enclosure, the AC potential rises to that of the power conductor, becoming unsafe until the overcurrent protection device responds to the short-circuit condition. This could take from several cycles of the power frequency to minutes, depending on the magnitude of the short-circuit current. Anyone coming into contact with the enclosure and another grounded conductor can, therefore, be electrocuted when current passes through the human body. Current on the order of milliamps can produce a reaction. Higher current levels can bring about more harmful effects or be fatal.

If a low-impedance connection does not exist between the enclosure and the over-current protection device, the enclosure may become unsafe under any fault condition.

If the "hot" wire should come into electrical contact with an ungrounded metallic part of the system, and a person standing on a grounded surface (or on the earth itself) comes into contact with any metal part of the equipment, that person would most likely receive an electric shock. With no safety earth connection, the protective device will not trip. In this situation, assuming that the human body resistance is approximately 1 $k\Omega$ and the power line voltage is 230 V_{RMS}, the current flowing through the person's body would be approximately

$$I_L \approx \frac{V_\Phi}{R_B} \cong \frac{230 \text{ V}}{1000 \ \Omega} = 230 \text{ mA} \tag{3.1}$$

A current as high as 230 mA* is fatal when flowing through the human body!

If, on the other hand, all metallic parts of the equipment are grounded and connected back to the service entrance ground and earth via a low-impedance connection, no hazardous voltages will be present on the enclosure. These situations are illustrated in Figure 3.10(a), where the ESGC is not present, and Figure 3.10(b) for the safe situation. Voltages up to 60 V RMS or 42.4 V DC are not generally regarded as unsafe.[†] It is good practice, however, to provide an ESGC in all cases.

Under no circumstances should current be allowed to flow through the electrical safety ground conductor, as the intended return path for the power currents should be through the neutral conductor. Therefore, the safety ground conductor must be regarded as "equipotential" with respect to the power circuit.

The frequency of the current flowing on the conductors is low, either DC, 50 Hz, 60 Hz or 400 Hz[‡] The primary concern for the quality of safety grounding is the resistance in the grounding path.

The effect of unintentional stray impedance in the ground connection is shown in Figure 3.11, where the impedance between point a, being at potential V_1 with respect to the equipment enclosure, is depicted as Z_1. The impedance between equipment en-

---

*The current in short-circuit situations can be extremely high, reaching levels as high as 1000 A due to the low resistance of the mains distribution wiring.

[†]EN 60950, *Specification for Safety of Information Technology Equipment, Including Electrical Business Equipment.*

[‡]50 Hz and 60 Hz are the most common AC power frequencies used worldwide, whereas 400 Hz is commonly used in aviation and marine applications. Other power frequencies may also be found in special applications, such as 16 ⅔ Hz, used in trams and electric trains).

Power safety requirements do not apply only to AC power circuits, but also to any circuit with voltages exceeding 30 V_{RMS} or 42.4 V_{PK} AC or DC. Nevertheless, it is good practice to provide an ESGC in all cases.

closure and the earth is identified as Z_2. These impedances act a voltage divider. The equipment chassis potential, V_C, relative to the earth is

$$V_C = \left(\frac{Z_2}{Z_1 + Z_2} \right) V_1 \qquad (3.2)$$

The potential across Z_2 could reach hazardous levels, enough to cause a shock hazard.

An important point to consider is that floating the system does not eliminate potential safety hazards. At first glance, it would seem that floating is an ideal solution,

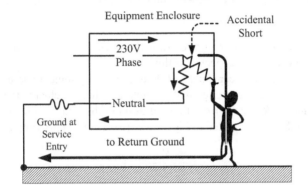

(a) No Electrical Safety Ground Conductor

Hazardous Condition

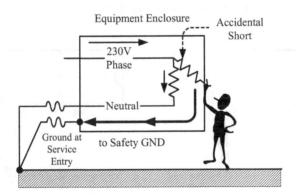

(b) Electrical Safety Ground Conductor Protection

Safe Condition

Figure 3.10. The electrical safety ground conductor (ESGC) shunts the fault currents to the earth.

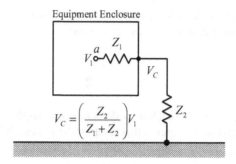

Figure 3.11. Stray impedance from PCB to chassis ground.

since theoretically, there can be no return path between the circuit and the earth and, hence, no current would be expected to flow.

Due to parasitic or stray capacitance C between the equipment and ground, leakage current could flow to earth through that capacitance, thereby closing the electrical circuit. When $1/\omega C \gg R_B$, the current flowing through the body of the person, having a resistance R_B in the case depicted in Figure 3.12(b), can be approximated by

$$I_L = \frac{V}{R_B + X_C} \cong V \cdot C\omega \tag{3.3}$$

where:

$X_C = 1/\omega C$, reactance of the capacitance between the power (or power return) line and the ground connection

$\omega = 2\pi f$, angular frequency, associated with the power frequency f

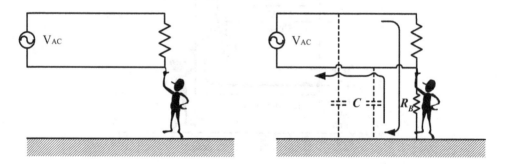

(a) Ideal Situation: No current flowing in an "open circuit"

(b) Practical situation: Leakage current flowing through stray capacitance

Figure 3.12. Floating power circuits could pose a serious safety hazard due to stray capacitance to earth.

For a 230 V_{RMS}/50 Hz (f = 50 Hz) supply, assuming a practical stray capacitance C = 3 nF, the leakage current to earth would be approximately 0.2 mA.

For AC mains power line EMI suppression, a power line filter is often incorporated at the mains input power port of electronic equipment Typical power line filters generally incorporate two groups of capacitors, namely "X" (differential-mode or line-to-line) capacitors connected between phase lines or between phase lines and neutral, and "Y" (common-mode or line-to-chassis) capacitors, connected between phase or neutral and the electrical safety ground. These "Y" capacitors are intended to shunt undesired RF currents to the SRS (Figure 3.13) but lead to a situation similar to that described above associated with stray capacitance.

Isolation, especially at higher frequencies, of the electrical safety ground conductor (ESGC) from interference sources within the system is desirable from the standpoint of interference control. This is of particular concern when chassis noise could easily couple into sensitive equipment due to such connections. Such isolation is best accomplished by inserting an inductor in series with the ESGC line. This choke limits the conducted emissions of interference currents from the system. This is also shown in Figure 3.13.

Figure 3.14 focuses on the possible effect of "Y" capacitors incorporated in power line filters on electrical functional safety. The "X" capacitors do not introduce a safety hazard, since any current flowing through them will not result in hazardous fault currents being made accessible to a metallic chassis. The "Y" capacitors, on the other hand, allow AC leakage current to flow through them to the chassis, thus making it "electrically hot," if not properly grounded. Leakage current can represent a serious shock hazard in case the chassis loses the electrical safety ground (protective earth) connection or if connected to earth ground through too high a resistance, R_T, compared to that of the human body, R_B. Any person coming into contact with the metallic equipment enclosure could experience an electrical shock, which can be hazardous [Figure 3.14(b)].

Note: Besides the leakage current due to the capacitors incorporated in the filter, leakage current associated with the electrical circuits themselves cannot be ignored. All these currents are cumulative at the protective earth conductor. This total amount of leakage current cannot exceed the specific value, mandated by electrical safety codes enforced by international and local regulatory bodies.

It must always be kept in mind, therefore, that particularly in the design of power line filters, these conflicting requirements must be addressed, and safety must always come first.

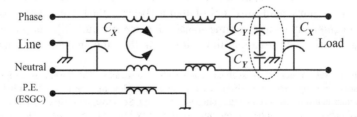

Figure 3.13. Line-to-chassis ("Y") capacitors incorporated in power line filters may introduce serious safety hazards if made too large.

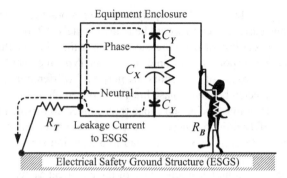

(a) Safe Situation: Equipment Enclosure Grounded through Low Impedance; Negligible Leakage Current through User's Body

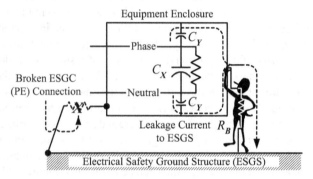

(b) Unsafe Situation: Equipment Enclosure Ungrounded due to Broken ESGC; Leakage Current Forced through User's Body

Figure 3.14. Impact of leakage current through power line filter capacitors on functional safety in power line filters.

   In addition to the potential hazardous situation arising from leakage current through excessive line-to-chassis capacitance, such currents may introduce power frequency interference in systems, platforms* or installations because the leakage current can couple into other equipment using the same signal reference structure [8]. For precluding this hazardous interference situation, limits for controlling the amount of leakage current

---

*Navy ships and submarines are an excellent example of such a situation. The power systems for Navy ships and submarines are ungrounded; however, leakage current can flow through power line filters' common-mode capacitance, ultimately reaching the ship's hull structure. Sensitive low-frequency radio and sonar receivers used onboard ships can be severely interfered with by low frequency (<100 kHz) current flowing through the hull, eventually penetrating sensitive equipment enclosures. The magnetic fields created by these currents can also couple into critical circuits and degrade their performance.

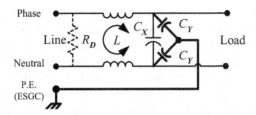

Figure 3.15. Single-phase power line filter with delta capacitive construction.

that may flow from the equipment to the reference structure are usually imposed by equipment specifications in compliance with national regulatory electrical safety codes.

The expression for an approximate calculation of the maximum leakage current of a single-phase filter with a delta capacitive construction (Figure 3.15) is defined by

$$I_L = 2\pi f \cdot 1.1 V \cdot 1.2 C_Y \qquad (3.4)$$

where:

I_L = maximum permissible leakage current through capacitors, amperes
V = rated line voltage, volts
F = rated line power frequency, Hz

Equation (3.4) allows for +20% tolerance on C_Y and +10% for a possible mains over-voltage condition. For three-phase filters, the maximum leakage current is measured with one phase connected and all other phases disconnected. Equipment safety standards (see Table 3.1) prescribe the discharge time limit of the total capacitance toward the supply mains. For this reason, many power line filter configurations incorporate a discharge or bleeder resistor (R_D in Figure 3.15).

Table 3.1. Most commonly used electrical safety requirements for household appliances (applies to nominal mains voltage 250 V, 50 Hz systems)

Equipment type	Insulation class	Maximum leakage current (mA)	Standard
Mobile	I	0.75	
Fixed	I	3.5	
All	0		IEC-335-1
	0I	0.5	EN-60335-1
	III		
All	II	0.25	

When applying electrical safety codes for a particular product, an applicable code should be considered. The data provided herein is for information and illustration purposes only. For other than household appliances refer to the relevant standards.

An example for standards-imposed requirements for EMI control can be found in MIL-STD-461E [8].* Line-to-structure capacitance for each power line is limited to no more than 100 nF for 60 Hz equipment, or 20 nF for 400 Hz equipment.

The maximum equivalent impedance for a given leakage current requirement, whether for meeting EMI control or preclusion of electrical safety hazard, is calculated by taking the power frequency AC voltage or the allowed voltage AC ripple present on DC power lines and dividing it by the maximum amount of allowed current leaking through the structure. Thus, a 1 V_{RMS} allowable voltage ripple and 5 mA maximum leakage current yields a maximum permissible equivalent resistance of 200 Ω:

$$Z_{Eq.} = \frac{1 \text{ V}_{RMS}}{5 \text{ mA}} = 200 \ \Omega \tag{3.5}$$

where Z_{Eq} = equivalent earthing impedance, ohms.

Next, the frequency at which the voltage ripple occurs must be known. If the voltage ripple is specified at 80 kHz, the maximum equipment capacitance to ground is calculated as

$$Z_{Eq.} = 200 \ \Omega = \frac{1}{2\pi (80 \text{ kHz}) \cdot C_Y} \tag{3.6}$$

Rearranging Equation (3.6) to solve for the maximum permissible capacitance C_Y yields

$$C_Y = \frac{1}{2\pi (80 \text{ kHz}) \cdot 200 \ \Omega} \cong 10 \text{ nF} \tag{3.7}$$

This capacitance represents the maximum combined capacitance to structure permitted for meeting interference-free objectives. The resultant value of the "Y" capacitors are often much lower than that necessary for providing sufficient common-mode EMI suppression. This deficiency is regularly compensated by complementary measures (e.g., common-mode chokes).

Electrical safety ground conductors must provide sufficiently high power-current conducting capability and low impedance according to the requirements of the relevant national electrical safety codes.

Lessons Learned

- EMI filters often rely on "Y" (line-to-chassis) capacitors for common-mode suppression.
- Conflicts may arise between contradictory requirements of superior suppression, requiring large capacitance values and limitation of leakage current, or, conversely, calling for reduction of line-to-chassis capacitance.

---

*This requirement applies for equipment installed on ships only.

- Power-frequency leakage currents to chassis can produce high-level structure-conducted EMI and even lead to electrical shock hazard situations.

- Safety must always be given due consideration when applying grounding and filtering practices in power systems, bearing in mind that safety must always come first.

3.2.1.2 Lightning Protection System (LPS) Grounding.

Lightning surges generally carry very large currents. The peak discharge current normally associated with a severe lightning return strike is 200 kA; its median (50%) is 30 kA and its maximum current gradient (di/dt) is 2×10^{11} A/sec [8]. Such high currents can destroy systems and damage structures, and constitute an electric shock hazard. Some form of lightning protection is therefore required.

A lightning protection subsystem must contain a controlled low impedance path for the energy associated with the lightning strike to dissipate to the earth electrode subsystem. Air terminals* are provided to intentionally attract the leader[†] of the lightning strike. Down conductors convey the high lightning discharge currents, diverting them away from susceptible elements in the facility and limiting voltage gradients developed by the high currents to safe levels.

Lightning discharge current need not flow into electrical circuits to cause damage. Ground potential differences in the vicinity of the strike can exceed 10 kV. If signal or power conductors are incorrectly designed or improperly grounded, transient energy can enter the facility through the ground conductors. Current surges flowing through the grounding grid[‡] will also increase the earth potential if the ground system impedance is not sufficiently low to preclude potential differences that might be present. Such earth potential differences may also constitute a hazard to personnel who may be standing on the ground in the vicinity of the striking point.**

Figure 3.17 [3] demonstrates the manner of current and potential distribution in a ground reference structure due to a lightning strike. This figure demonstrates that even with a well-grounded facility, ground currents and potentials can constitute a safety hazard.

Figure 3.18 illustrates a case in which an ungrounded cable arriving from a control center, penetrating a fuel tank, causes a voltage potential difference between the cable and the facility's structure. This strike triggered an explosion of the fuel tank. A proper protection device between the ungrounded cable and the facility's structure would have eliminated this catastrophe.

---

*Air terminals are the lightning rods or intended attachment conductors placed on or above a building, structure, or tower for the purpose of intercepting lightning.

[†]A leader (lightning) is the electric discharge that initiates each return strike in a cloud-to-ground lightning discharge. It is a channel of high ionization that propagates through the air by virtue of the electric breakdown at its front produced by the charge it lowers. The stepped leader initiates the first strike in a cloud-to-ground flash and establishes the channel for most subsequent strikes of a lightning discharge.

[‡]The ground grid is a system of interconnected bare conductors arranged in a pattern over a specified area and on or buried below the surface of the earth, intended to provide safety for workmen by limiting potential differences within its perimeter to safe levels in case of high currents that could flow if the circuit being worked became energized for any reason or if an adjacent energized circuit faulted.

**"Step voltage," as this phenomenon is called, is discussed in detail in Chapter 6.

<u>Figure 3.16.</u> Ben Franklin's big idea. (Illustration by Paddy Morrissey from *Code Check Electrical,* 4th Edition © 2005, Taunton Press, Inc. Reprinted with permission.)

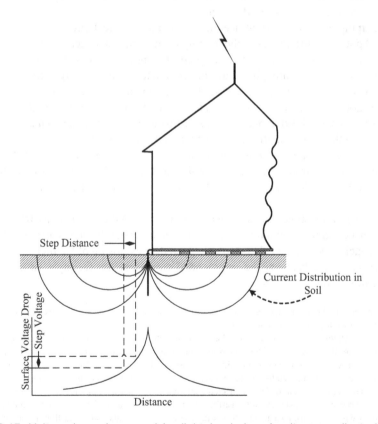

<u>Figure 3.17.</u> Voltage hazards caused by lightning-induced voltage gradients in the earth.

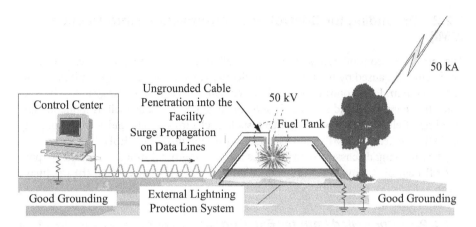

Figure 3.18. Improper grounding of a cable penetrating a facility caused a spark-over in a fuel tank.

If lightning surge current penetrates a facility's power or other ungrounded conductors, side flashes* through the air to adjacent grounded conductors may occur due to high potentials, which will subsequently develop on those conductors. Thus, high potentials may result in explosion of structural parts or cause fire.

The spectral content associated with the lightning strike current and subsequent field , due to the very short gradient or rate of change of the lightning discharge current waveform, reaches approximately 1 MHz. Grounding of the lightning protection system must be accomplished through a low-impedance path. Specifically, the high-frequency/high-gradient/high-current nature of the lightning waveform requires that resistance, R, as well as inductance, L, of the grounding system be minimized.

Note: Other aspects related to lightning-induced surges on equipment and its associated wiring, are discussed in Section 3.2.2.2.

Chapter 6 provides a detailed discussion of grounding systems for lightning protection.

Lessons Learned

- For lightning protection, the grounding system is intended to provide a controlled path for carrying and dissipating extremely high lightning-surge currents to earth.
- Impedance of the grounding and earthing system must be low at the frequencies associated with lightning (1 MHz, approximately) for ensuring effective protection.

---

*A lightning side flash is arcing occurring between the lightning-protection conductors to alternative (other than intended) discharge paths due to excessive impedance.

3.2.2 Grounding for Control of Electromagnetic Interference (EMI)

The term "EMI control" refers to the control of erroneous or sporadic behavior of electronic circuits caused by the influence of electromagnetic energy coupling into the circuit or system. EMI control covers all aspects of emission and susceptibility interactions. It encompasses radiated fields and conducted currents, as well as induction and coupling of electromagnetic phenomena such as electromagnetic pulse from lightning (L-EMP), nuclear (N-EMP) sources, and electrostatic discharge (ESD).

The following discussion demonstrates the role of ground structures for the purpose of EMI control. It covers (a) the controlled path for EMI current and (b) the image plane.

3.2.2.1 Controlled Path for EMI Current.

In order to understand the role of grounding in the control of EMI, it should be noted that a circuit totally enclosed within a perfectly constructed, six-sided metallic enclosure will not radiate, nor will it be susceptible to external incident radiated electromagnetic fields. However, apertures in the shielded enclosure, and particularly cable penetrations through the shield, will compromise shielding performance if not carefully controlled since interaction of conducted and radiated electromagnetic energy with circuits and systems commonly occurs by means of conductors (cables and transmission lines) connected to the circuit, penetrating the shielded enclosure. EMI control measures include interference suppression and cable shielding, the performance of which is depended on the proper implementation of grounding structure (Figure 3.19).

The majority of radiated and conducted EMI interactions are associated with common-mode currents. It follows, therefore, that effective suppression of such phenomena requires shunting of the EM energy to the signal reference structure (SRS). This is where the term "grounding" is applied and is to be understood as electrical connection

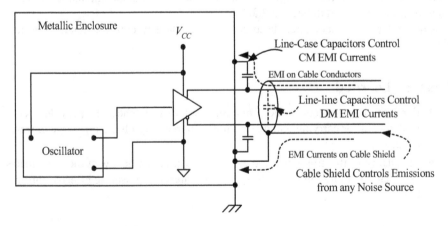

Figure 3.19. Grounding applications for EMI control.

of the protective device (e.g., the capacitor) between the protected lines to the signal reference structure. The same applies to filters and surge suppression devices, intended to provide CM suppression. In this respect, the "ground" provides a controlled low-impedance path for the flow of CM EMI, electrostatic discharge (ESD), and surge currents. High-quality, low-impedance "ground" paths thus provide an alternate path for EMI, ESD, and surge currents, diverting them away from sensitive circuits.

Incidentally, line-line capacitors or filters will have little if any effect on emissions due to common-mode currents but can significantly suppress differential-mode interference.*

An alternative or complementary manner for controlling radiated field and cable interactions is by means of shields placed on the cables. Cable shields suppress EM field coupling. Performance of cable shields and proper shield termination (i.e., grounding) provide a controlled path for shunting common-mode EMI currents induced on the cable shield into a reference system.

In both cases, that of protective devices and cable shields, the term "ground" is misleading, in the sense that no connection to ground (i.e., earth) is actually required. In fact, the term "ground" should be understood as "shield termination" or "connection to the signal reference structure or enclosure" and should not be used in this context.

Regardless of terminology, EMI ground is required to handle currents associated with EMI, ESD, and lightning surge currents, ranging in amplitude from microamperes to tens and even hundreds of amperes, respectively, and in frequency, virtually from DC to daylight.

In order to achieve an acceptable EMI ground, sufficiently low impedance to a ground reference is required for efficient connection of filters, cabinets, and cable shields. Due to the high-frequency nature of the EM phenomena, inductance constitutes the primary consideration in the grounding system design for EMI control.

The requirement to avoid undue emission of or susceptibility to electromagnetic energy under normal operating conditions may dominate the properties of the grounding system ahead of all the above requirements (i.e., safety and lightning protection).

Grounding for ESD is implemented on the basis of the same physical principles as those applied for "regular" EMI. Unique to grounding for ESD is the fact that often "soft grounds," in which a dissipative element (i.e., resistor) is incorporated in the grounding path, are used for limiting peak ESD currents from flowing through the protected sensitive circuit. As an example, electrically initiated explosive devices (EIEDs), when in the "safe" position, should be grounded through resistors having values of 10 to 100 kΩ [8].† In addition, such "soft grounds" serve for preclusion of hazardous situations that may occur if power currents are allowed to flow (typically under power fault conditions) into the human body.

Refer to Chapters 7 and 8 for detailed discussions of grounding of cable shields and EMI terminal protection devices. Also, Chapter 10 provides a discussion of grounding principles for ESD control in facilities and installations.

---

*Common-mode (CM) and differential-mode (DM) phenomena and their effects are discussed in Chapter 2.
†Special grounding considerations in EIEDs and ordnance circuits and systems are briefly discussed in Chapter 10.

3.2.2.2 Image Plane. Radiation from a printed circuit board is predominantly a result of common mode current with and even without attached cables. Common mode current drives the conductors in a manner similar to a monopole or dipole antenna.

For a monopole antenna, the ground plane constitutes a basic part of the antenna system and affects the antenna's characteristics. To obtain a first-order estimate of the effect of the ground plane on the antenna performance, image theory can be applied to the case of perfectly conductive surfaces [10].

It is a well-known principle in electromagnetic theory that the boundary conditions on the surface of a perfectly conductive plane mandate that tangential electric fields on its surface vanish; hence, the electric field must be normal to the conducting surface. In order to satisfy this requirement, a conducting charge above a conducting plane will cause charge redistribution on the surface equivalent to that which would be produced by an equal but opposite charge at the same distance below the plane's surface. Since current constitutes a movement of charge, the direction of the image of current in vertical, horizontal, and inclined conductors can henceforth be determined. For a vertically polarized monopole antenna, the vertical (normal to surface) field vector components will follow the same direction, whereas horizontal (tangential to the surface) components will be opposite, satisfying the boundary conditions on the surface of the plane. This is depicted in Figure 3.20.

The field in any direction above the image plane can be evaluated by substituting the plane by the image of the current-carrying conductor and calculating the result of the superposed field due to the actual conductor and its image. Below or within the ground plane, the field vanishes.

Figure 3.21 depicts the image plane current distribution due to a vertical monopole antenna above that plane. The current flowing through the monopole antenna creates an image within the image plane, thus acting as a center-fed dipole antenna with respect to the formation of the radiation pattern, antenna impedance characteristics, and other parameters of interest.

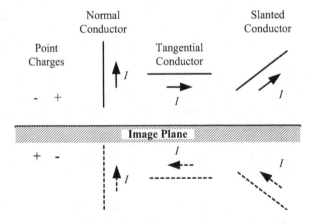

Figure 3.20. Images of current-carrying conductors above an image plane.

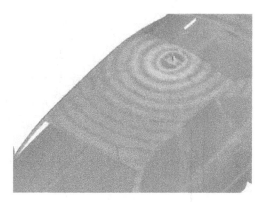

Figure 3.21. Implementation of reference structures for mobile monopole antenna structures on the roof of an automobile. (Images created using CST Microwaves Studio®, courtesy of CST.)

Although rigorously applied in antenna theory [Figure 3.22(a, b)], this phenomenon is equally valid for any arbitrary conductor above a conductive plane. The image plane concept finds an interesting and particular application for EMI control when extended to traces routed above solid return planes on printed circuit boards (PCBs)* as well as to cable conductors routed over the earth or a metallic surface. An image of the signal flowing in a conductor placed above the return plane is depicted in Figure 3.22(c). The equivalent image currents flowing within the plane effectively produce radiated fields that tend to cancel the fields generated by the current-carrying conductors above it (i.e., "flux cancellation"). Hence, the very presence of the ground plane beneath the circuit conductors serves to reduce radiated EMI emissions from the circuit or cable [2, 12, 13].

Lessons Learned

- An "EMI ground" is required to handle interference currents ranging in frequency across the entire spectrum, from DC to daylight, and in amplitude from microamperes to kiloamperes.
- In order to accomplish an acceptable ground performance, impedance should be minimized, requiring that inductance be the primary consideration in the grounding system design for EMI control.

3.2.3 Signal Grounding

3.2.3.1 Signal Reference Grounding. An electronic system is fundamentally an assembly of subsystems and circuits, typically interconnected with the objec-

---

*Chapter 9 provides a detailed discussion of application of image planes for emission control from printed circuit boards.

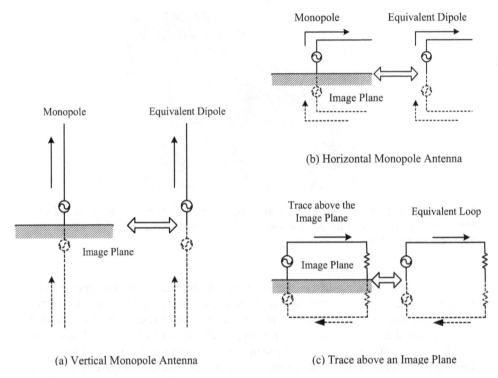

(b) Horizontal Monopole Antenna

(a) Vertical Monopole Antenna   (c) Trace above an Image Plane

Figure 3.22. Implementation of image theory for (a) vertical monopoles and (b, c) horizontal dipoles above a conductive (ground) plane.

tive of performing a common function. Essentially, all electrical equipment can be considered as a combination of sources and loads along with their respective interconnects, ranging from simple circuits to extremely complex subsystems and systems.

A common reference of high quality is a prerequisite for obtaining reliable, interference-free equipment performance. Satisfactory equipment operation should be achieved whether using balanced or unbalanced interfaces, or a combination of both. Distributed systems (such as local or wide-area networks), impose a special challenge: Each component of the system must perform its intended function as a stand-alone unit. Performance of a distributed system as a whole, even in the presence of interference, requires the presence of a path between the circuit grounds of the equipment at each end of the interconnection, serving as a common equipotential voltage reference or voltage reference ground between the different parts of the system. Grounding is required for the potential difference between the various parts of the system to be at sufficiently low levels.

Figure 3.23 depicts a system including source and load circuits. The output of the source circuit V_O (referenced to the local reference ground) arrives at the receiver at a

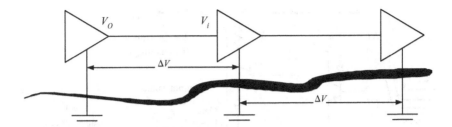

Figure 3.23. In single-ended signaling, the "ground" serves as a common signal reference structure for all system components.

level V_i (referenced to its local reference ground). V_O will typically not be equal to V_i due to the potential voltage drop on the nonperfect grounding system, ΔV.

In the case of single-ended signaling, a low-voltage drop across the ground system is mandatory, since any potential difference in the reference structure will appear across the load as well, causing degraded or disrupted operation.

An ideal ground structure would, consequently, exhibit a zero-potential voltage drop and, hence, a zero-impedance structure. Such a structure could serve as a reference for all signals in the associated circuitry and as a return path through which any undesirable return current could flow. An ideal ground plane would provide equipment with a common potential reference point, so that no potential difference could exist between any two locations.

A signal ground is usually defined, in this context, as an equipotential structure that serves as a common reference for components of the circuit or system.

In practice, regretfully, no ground reference structure is truly equipotential due to the physical properties and characteristics of the material. Some potential difference will always exist. The effect of this potential difference, if not accounted for, may actually exacerbate EMI rather than serving to control it. Observe the situation illustrated in Figure 3.24.

The shield of a coaxial signal cable (acting as a return conductor) is connected to an arbitrary point G' on the DGND (digital ground or signal return/reference) conductor of the logic gate, different from the actual circuit "0V" reference point (point G''), connected to the signal reference structure. Believing that the return conductor serves an equipotential conductor, the circuit designer paid no attention to the partial inductance between point G' and G'' of the return conductor, nor to the fact that during the gate's low-to-high ("0" $\rightarrow$ "1") transition, impulsive current due to the load capacitor's discharge through the gate's output flows through this inductance, developing impulsive voltage across this $L_{G'G''}$ inductance*:

$$I_C = C\frac{dV_L}{dt} \qquad (3.8)$$

*This phenomenon is well known as "ground bounce" or "simultaneous switching noise" (SSN), and is discussed in Chapter 9.

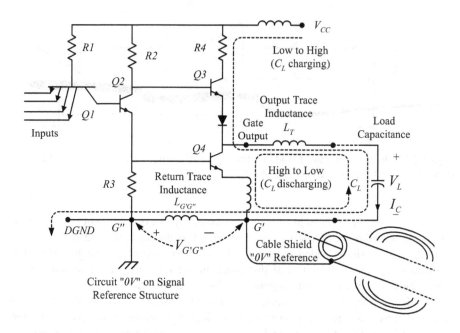

(a) Circuit Representation

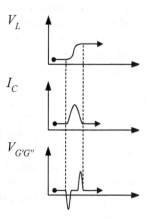

(b) Circuit Waveforms

Figure 3.24. Improper grounding of a cable shield to a circuit "0V" reference may increase emissions: A switching TTL gate example.

and

$$V_{G'G''} \approx -L_{G'G''} \cdot \frac{dI_C}{dt} = -L_{G'G''} C \cdot \frac{d^2 V_L}{dt^2} \qquad (3.9)$$

The shield is not at zero potential. On the contrary, the shield is excited by the voltage source as an effective monopole antenna, increasing emissions from the circuit via the driven shield.

The presence of an equipotential common reference is considered both from EMI and functionality standpoints: when not present, problems such as noise margin erosion may be the consequence, in addition to the development of ground noise voltage appearing between the circuits. This ground noise voltage will produce undesirable radiated and conducted emissions or ground loop situations.* Large potential differences between different parts of a system may even result in damage to circuits or components.†

Circuit designers often presume that a signal voltage reference structure is not meant to carry functional current and is solely intended to constitute a perfect and ideal voltage reference. Regretfully, the common voltage reference objective of the grounding system, normally defined as an equipotential structure, is often the only one considered by designers. As a result of this definition, mistakably founded on voltage, not current, considerations, designers often fail to concern themselves with what happens to the signal current once it has enters the ground system. As far as they are concerned, current is lost in the abyss of the grounding system, never to be seen again.‡ Hence, sufficient emphasis is not placed on the importance of the ground in providing the actual path taken by current in returning to its source.

In practice, however, the situation may be very different from that considered by designers. Figure 3.25 depicts an arbitrary yet common application of a digital circuit. Reference/return conductors are not drawn, and only voltage reference points are erroneously identified.** Designers often fail to consider the fact that any signal current flowing in the circuit actually returns to its source through a path or paths existing between such "voltage reference points."

Components U2, U3, and U4 will respond properly to the signals from U1, as they possess a common voltage reference. However, is the mere presence of a common voltage reference plane sufficient for signal propagation between the components?

3.2.3.2 Signal Current Return Path.
Digital and analog circuit designers generally assume that current return paths are irrelevant. "After all," they argue, "if

*See Chapter 4 for a discussion on ground loops.
†In RS-422 circuits, for instance, a potential difference in the order of 12 volts may result in damage to the receiver [7].
‡If a conductor never carries current, it is not required. Hence, the concept of noncurrent-carrying reference conductors constitutes a logical paradox, since in order to function as a reference, it must carry some amount of current in order to support the sensing or reference processes. It follows, therefore, that all reference conductors are rightfully assumed to carry current of some sort [1].
**Connections of the devices to the DGND is marked using dashed lines. Generally, these connections are not included in circuit diagrams.

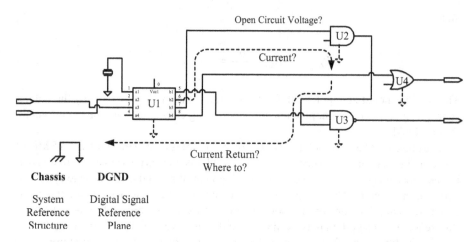

logic gates and drivers act as voltage generators, and the receivers act as voltage recep-
tors, why worry about current flow?" This constitutes the cornerstone of the misconception that "ground is not intended to carry any current but rather serve as a voltage
reference only."

In the above example, the circuit performs properly only when signal current
flows between devices and returns through a designated return path. This is an imperative function of the signal reference structure when single-ended signaling is
used. The reference ground structure functions also as a signal current return path.
Power as well as signal currents must flow between system components, requiring
the provision of current return paths, both for the power and signal current flowing
from the power source to the circuits and between the signal sources and loads, respectively.

Designers must always consider during the product design process the path that return current follows. Although it may be the designer's intent for the signals to return
through a designated return path, there is no guarantee that, in fact, the current will follow this route. Some frequency components of the signal spectral content may follow
the intended return path, while other frequency components of the same signal may
follow a different return path. Current always follows the path of least impedance. At
low frequencies, where $R \gg \omega L$, the return current will take the path of least resistance,
whereas at higher frequencies, where $R \ll \omega L$, the return current will follow the path of
least inductance.

Signal ground can thus better be defined as a low-impedance path for current to return to the source through a structure also intended to serve as a signal reference
throughout the system.

This definition focuses on and emphasizes the objective of the reference structure
as a current return path. This definition further implies that since current is flowing

through some low but finite impedance, there will be a potential difference between the physically separated points within the grounding system (Ohm's Law in circuits).

Lessons Learned

- The term "signal grounding" encompasses two primary and complementary objectives, namely signal voltage reference and signal current return path.
- In systems and facilities, a ground structure is typically intended to function as a common signal reference structure and to provide a near perfect voltage reference for the signal.
- The term "reference" implies voltage consideration rather than current (open-circuit voltage can exist, but no current can flow through an open circuit); designers thinking in terms of voltage may be tempted to ignore the need for adequate current return paths.
- For high-frequency signals, "ground" is a concept that does not exist in reality. Signal ground can better be defined as a low-impedance path for the signal's current to return to the source.
- The concept of equipotential reference defines an ideal objective of the grounding system, whereas the concept of current return path characterizes what the ground actually is.

3.2.3.3 *Summary of Grounding Objectives.* The trouble with the term "ground" is the term itself; it is too ambiguous! To avoid confusion, avoid the term "ground" unless used in conjunction with another descriptive term:

- *Safety Grounding* is intended for preclusion of hazards due to power faults or lightning strikes, which could set a facility ablaze and constitute a safety hazard to equipment and personnel.
- *EMI Grounding* is intended for controlling common mode EMI current drainage from cable shields and suppression devices as well to serve as an "image plane" for conductors routed adjacent to them.
- *Signal Grounding* essentially constitutes a "functional" or "technical" ground, intended to provide an equipotential signal voltage reference between components of the system and serve as a path for signal current return, particularly in unbalanced or single-ended interfaces. With the exception of electrical safety considerations and certain issues related to electrostatic shielding, connections of electronic circuits to the ground play no other role than to provide the signal and EMI current returns (whether desired or unintentional)

Real-world ground structures are nonideal in nature. Some potential difference always exists; thus, all reference conductors should be assumed to carry current, whether intended or unintended. The extent to which potentials in the ground system can be minimized and ground currents reduced will determine the effectiveness of the ground system. In high-speed signal circuits, in particular, "ground" is a meaningless concept. The important question is, "what path does return current follow?"

3.3 GROUNDING-RELATED CASE STUDIES

In this section, several grounding-related case studies are presented. The details of the cases, with respect to the design of the grounding system, are presented in the following chapters. These cases are intended, therefore, to illustrate the complexity and, particularly, the misconceptions and facts versus the fallacies associated with grounding.

3.3.1 Case #1: The Grounds for Electrostatic Discharge (ESD)

Case Description. In a textile factory, an RS-232 communications link between a new machine and the computer/controller was malfunctioning. All machines and controllers in the facility were bonded to each other by means of equipotential grounding structures; therefore, no potential difference between them was assumed to occur. Due to the nature of the facility and the known ESD (electrostatic discharge) phenomena associated with textile production (particularly where synthetic cloths are concerned), ESD was suspected as the primary cause for the interference.

Case Investigation. In order to investigate the source of interference, waveform measurements took place on the RS-232 lines, using a broadband digital oscilloscope and a current probe, revealing a waveform such as the one illustrated in Figure 3.26.

When the operation of the new machine was terminated, the interference on the RS-232 link ceased, and when communication from the new machine to the computer was activated, it was disrupted as well. Based on the above findings, it was deduced that this constituted an intrasystem (self-compatibility) problem. Probing the surface of the system using a small magnetic loop, the source of noise was found to be from the machine's electric motor, which was in electrical contact with both the machine's chassis and the signal return (GND) of the RS-232 link (Figure 3.27). In fact, it was found that the motor of the machine was the culprit and replacing it solved the problem.

Figure 3.26. Illustration of the waveform measured on the RS-232 link.

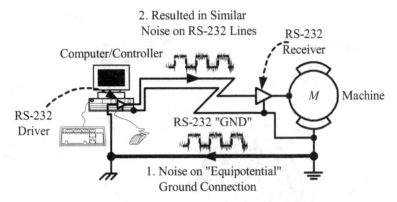

Figure 3.27. Illustration of the Machine Layout and Interference Mechanism

Case Resolution. The faulty (noisy) motor came into electrical contact with the machine's chassis and injected noise onto it. Since the RS-232 link is connected to the chassis at both ends of the link through the RS-232 reference "GND" line, a noisy ground loop was produced, allowing interference current to flow throughout the system by means of the equipotential structures. Any interference present on the "GND" line of the RS-232 link subsequently resulted in interference to the controller circuits, owing to the single-ended nature of the RS-232 interface and, therefore, the lack of immunity to common-mode-type interference. If a RS-422 or RS-485 were used instead, this interference would probably have not occurred.

Misconceptions, Fallacies, and Facts Demonstrated. This case study has illustrated that:

- "Equipotential ground structures" should not be assumed to be "quiet"; when noise is present on the structure, the noise will be propagate throughout the system.
- Single-ended, unbalanced digital communication links exhibit low immunity to noise and could fail when high-level interference is present across the reference terminals of the line transmitter and receiver.

3.3.2 Case #2: The Grounds for Lightning Protection

Case Description. A mobile remote control unit (denoted *Site R* herein) of a large communication system was placed approximately 1000 meters away from the main fixed facility (denoted Site F herein). The remote unit and the fixed installation were interconnected through several (differential) RS-422 links. Each of the sites was powered from an independent power source and grounded to earth at its near vicinity (Figure 3.28).

After lightning struck the fixed facility, the interfacing line driver/receiver circuits in the remote control unit were found to have burned out. At first glance, this outcome

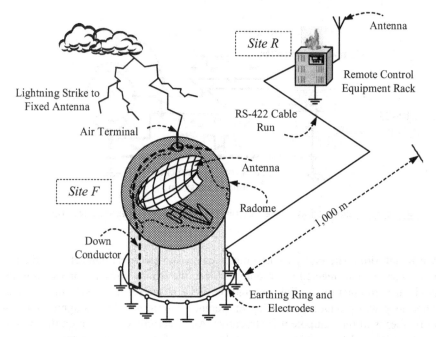

Figure 3.28. The communication facility and remote control layout and grounding scheme.

seemed anomalous since the remote control unit was grounded and, furthermore, differential signaling was used.

Case Resolution. In the event of a lightning strike, whether directly to earth or indirectly, through a grounded object, a voltage gradient in the soil will develop in the vicinity of the point where the lightning discharge enters the earth. In homogeneous soil of uniform resistivity, a significant voltage gradient will exist between two points that are at differing distances from the electrode. Figure 3.29 illustrates the nature of this voltage variation,* showing that a considerable voltage difference, ΔV, exists between the earth systems of Site F and Site R. As the local earth system serves as reference for all equipment racks and cabinets on each of the sites, it follows that the equipment racks and, consequently, the chassis of their respective equipment assemblies, will also be at different potential levels.

The situation was further worsened by a single earth electrode that was driven into the earth at the remote site (Site R). The resistance to earth of this electrode could exceed 25 Ω and still conform to binding safety codes. At the fixed facility (Site F), on the other hand, a well-designed earthing system consisting of horizontal ring electrodes and vertical rods was implemented, resulting in low resistance (1 Ω to 3 Ω) to

*The earth potential voltage gradient phenomenon is discussed in detail in Chapter 6.

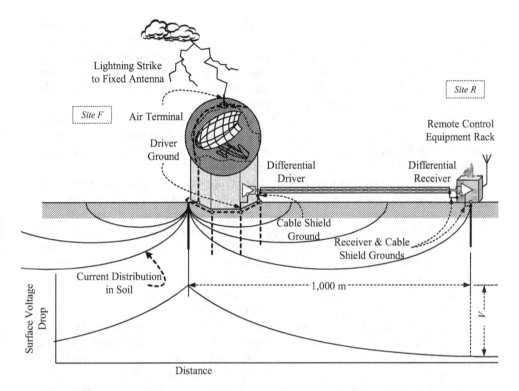

Figure 3.29. Earth voltage gradient due to a lightning strike to the air terminal of the fixed facility antenna radome.

earth. Assuming an earth resistance of 2 Ω, with only a median lightning discharge of 30 kA striking at Site F, the voltage potential of the complex would rise to 60 kV with respect to Site R and that portion of the earth not in the immediate vicinity of that structure.

At Site F, the communication cables' shield voltage would subsequently rise along with that of the building. This voltage surge would travel down the cable, successively raising the shield potential to as much as 60 kV with respect to the surrounding earth. Such high voltages cause insulation breakdown in the form of tiny pinholes where the lightning energy punches through. As the lightning surge travels down the cable, its amplitude diminishes due to cable resistance and dielectric losses. However, the amplitude of the surge was apparently sufficient to damage circuit components in terminating equipment in Site R.

Interconnecting the earth electrodes of the two sites to form one effective earth contact would not eliminate the lightning threat to the interconnecting cable between the two sites. As shown, the cable shield is connected to the equipment cabinets, that is, the facility grounds at each of the sites. In the event of a lightning strike as shown, the grounding infrastructure and power supply system in Site F would be elevated in potential relative to Site R. In particular, with the large distance between the two buildings, inductance, primarily, of the cable shield would prevent the cable from providing the low impedance necessary to keep the equipment racks in both sites at the same potential. Furthermore, if the shield of the cable is insulated from the earth, as is usually the case, the potential of the cable shield can become high enough with respect to the earth to exceed the insulation breakdown potential.

But weren't the interfacing circuits all differential RS-422 drivers and receivers? What about their common-mode rejection? Unfortunately, differential devices are attributed with higher immunity to common-mode interference than they actually have. A quick look at Figure 3.30 sheds light on the situation.

The RS-422 drivers and receivers in each of the sites were grounded (and referenced) to the local earth infrastructure at each site through the equipment assembly, cabinet, and local signal reference structure chain. Therefore, any potential difference between the earth infrastructures also appeared between the circuit grounds (denoted GND_F and GND_R for the fixed and remote sites, respectively). The RS-422 "totem-pole" circuits are not floated, therefore, from the chassis. Even when the driver is at an output "high" state (as illustrated in Figure 3.30) and transistors Q_{1L} and Q_{2L} are in "cutoff," the potential difference across these transistors will result in dielectric breakdown and permanent damage to the driver or receiver will occur.

So, why were the devices in the remote site damaged and not those at the fixed site (which was struck)? The answer is simple! At Site F, which was struck, the earth potential indeed rose to a much higher level; however, both the circuits (including their drivers and receivers and corresponding I/O interface lines) and the reference structure were at the same (high) potential. At the remote Site R, on the other hand, the local circuit reference was at a much lower potential; thus, a large potential difference was present between the signal reference, GND_R, and the I/O communication

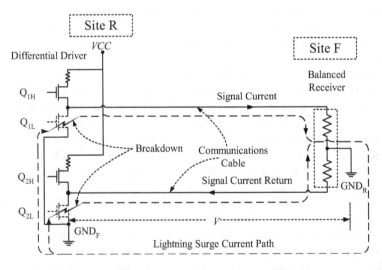

Figure 3.30. A Presumed differential "totem-pole" source driving a balanced load.

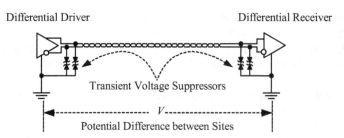

Figure 3.31. The solution: Installation of transient voltage suppressors (TVSs) on all interface circuits.

lines interconnecting to Site F. The stress on the I/O devices at Site R was, therefore, too high. As mentioned above, shielding the communications cable would provide no additional advantage, as it could not contribute to equalization of potential between the two sites.

As a solution to the problem, and in order to preclude future damage to the system, transient voltage suppressors (TVS),* compatible with the terminating components and hardware, were incorporated on all interfaces of the fixed and remote systems interconnected through the communication cables. This solution is illustrated in Figure 3.31.

Misconceptions, Fallacies, and Facts Demonstrated. This case study has illustrated that:

- Equipotential grounding in distributed systems is difficult if not impossible to accomplish.
- Differential line drivers/receivers are vulnerable to lightning surge voltages.
- Floating differential drivers/receivers will offer marginal advantage, as dielectric breakdown along communication cables will result in similar consequences to the I/O devices.
- Shielding communication cables does not preclude lightning-induced damage to I/O circuits. Potential differences between the earth and the cables will result in insulation breakdown in the cables and damage to the I/O circuits.
- Transient voltage suppressors (and terminal protection devices, in general) should be incorporated into interface circuits for preclusion of damage due to earth potential differences and subsequent voltage and current surges.

BIBLIOGRAPHY

[1] IEEE-1100, *IEEE Recommended Practice for Powering and Grounding Electronic Equipment,* IEEE Emerald Book, 1999.

*Grounding considerations in terminal protection devices (TPDs) are discussed in Chapter 8.

[2]   Montrose, M. I., *EMC and the Printed Circuit Board, Design, Theory and Layout Made Simple,* New York: IEEE Press/Wiley, 1999.

[3]   MIL-HDBK-419A, *Military Handbook, Grounding, Bonding, and Shielding for Electronic Equipments and Facilities,* Washington, DC: U.S. Government Printing Office, 1987.

[4]   ETS 300 253, *European Telecommunication Standard, Equipment Engineering (EE); Earthing and Bonding of Telecommunication Equipment in Telecommunication Centers,* January 1995.

[5]   Yang, Z., "The Myth about Ground and How to Correctly Use it in High-Speed Signal Integrity and Power Integrity Modeling and Simulation," in *Proceedings of the 2006 DesignCon Conference,* 2006.

[6]   Graf, W., and Nanevicz, J. E., "Topological Grounding Anomalies," in *Proceedings of the 1983 International Aerospace and Ground Conference on Lightning and Static Electricity, FAA Report DOT/FAA/CT-83/25,* pp. 9–1 to 9–13, June, 1983, Fort Worth, TX.

[7]   Perez, R. (Ed.), *Handbook of Electromagnetic Compatibility,* New York: Academic Press, 1995.

[8]   MIL-HDBK-461E, *Department of Defense Interface Standard, Requirements for the Control of Electromagnetic Interference Characteristics of Subsystems and Equipment,* Washington, DC: U.S. Government Printing Office, August, 1999.

[9]   Kardon R., *Code Check Electrical,* 4th Ed., Newtown, CT: Taunton Press, Inc., 2005.

[10] Johnson, R. C., and Jasik, H., *Antenna Engineering Handbook,* 2nd Ed., pp. 2–7 to 2–8, New York: McGraw-Hill, 1984.

[11] MIL-STD-188-114A, *Military Standard, Electrical Characteristics of Digital Interface Circuits,* Washington, DC: U.S. Government Printing Office, 1985.

[12] Paul, C. R., *Introduction to Electromagnetic Compatibility,* New York: Wiley, 1992.

[13] German, R. F., Ott, W. H., and Paul, C. R., "Effect of an Image Plane on Printed Circuit Board Radiation," in *Proceedings of the 1990 IEEE International Symposium on Electromagnetic Compatibility,* Washington, DC, August 1990.

4

FUNDAMENTALS OF GROUNDING DESIGN

4.1 GROUND-COUPLED INTERFERENCE AND ITS PRECLUSION

Grounding systems are interference distribution devices.—CARL E. BAUM

Grounding is considered a solution for many EMI problems. When improperly implemented, however, the grounding system chosen may in fact be part of the problem and could constitute a source of EMI coupling and undesired interactions. Discussion of ground-coupled interference modes is provided in this chapter. The objective of grounding system design could, therefore, be stated as follows: "Design the system such that in spite of the need for grounding, system performance will not be degraded due to ground-coupled interference."

The problem of ground-coupled interference is first addressed and illustrated through a simple example. Subsequently, a general model is described and a figure of merit for common-impedance coupling is introduced through the concept of transfer impedance, Z_T.

4.1.1 Grounding May Not Be the Solution; Rather, it Could be Part of the Problem

Per Ampere's Law, current flows in closed loops, returning to its source. The return current may split between numerous available return paths with the current amplitude in each path inversely proportional to the impedance of the individual path. In addition, unintended currents, created by external events (e.g., power faults and lightning strikes) can also travel through these same paths.

An aspect of great importance in circuit and system design is that the conductor or signal reference structure carrying the return current is often shared by multiple circuits. An ideal current-return path is assumed to exhibit zero impedance, thus producing zero potential difference across the entire frequency spectrum. In practice, all electrical conductors have some resistance and the current flowing through a conductor loop experiences some inductance. Whenever current flows across finite impedance, a nonzero voltage drop will result, producing common-mode current on cables attached to the circuits and resulting in interference. This interference coupling mechanism is known as common-mode or common-impedance coupling.

Figure 4.1 illustrates a possible outcome of nonzero return-path impedance, emphasizing the importance of the grounding system. Two single-ended circuits (source and load), are connected to the same current-return path (also known as common ground or common reference). The return current of each circuit travels through this common structure. A real-world structure is characterized by nonzero impedance, Z_G, which could be resistive and/or reactive (particularly, inductive).

Kirchhoff's voltage law* mandates that the sum of all voltage potentials across any closed circuit be zero. Since the current in Circuit #1 (denoted as "noisy") flow through the signal reference structure ("ground") also shared by Circuit #2, it introduces a potential difference between the ground reference connections of Circuit #2.

$$V_{NG} = E_{S1} \cdot \left(\frac{Z_G}{Z_{S1} + Z_{L1} + Z_G} \right) \cong E_{S1} \cdot \left(\frac{Z_G}{Z_{S1} + Z_{L1}} \right), \text{ volts; since } Z_G \ll Z_{S1} + Z_{L1} \quad (4.1)$$

This "ground noise voltage," V_{NG}, now appears as an second voltage source in Circuit #2, producing an additive interference voltage, V_{i2}, across the sensitive load of Circuit #2:

$$V_{i2} \cong V_{NG} \cdot \left(\frac{Z_{L2}}{Z_{S2} + Z_{L2}} \right), \text{ volts; since } Z_G \ll Z_{S2} + Z_{L2} \quad (4.2)$$

Note: In both Equations (4.1) and (4.2), the impedance of the ground structure is assumed to be negligible compared to the source and load impedances in Circuits #1 and #2, respectively.

Substituting Equation (4.1) into Equation (4.2) results in

$$V_{i2} \cong V_{NG} \cdot \left(\frac{Z_{L2}}{Z_{S2} + Z_{L2}} \right) = E_{S1} \cdot \left(\frac{Z_G \cdot Z_{L2}}{(Z_{S1} + Z_{L1}) \cdot (Z_{S2} + Z_{L2})} \right), \text{ volts} \quad (4.3)$$

Ultimately, the proper performance of Circuit #2 will be related to the signal to interference ratio (*SIR*) across Z_{L2}:

---

*Electrically small systems and interconnects are assumed for ascertaining validity of Kirchhoff's Law.

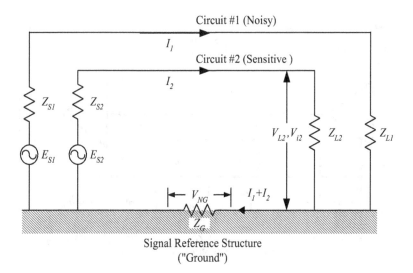

Figure 4.1. Common-impedance coupling in a signal reference (ground) structure.

$$SIR_2 = \frac{V_{S2}}{V_{i2}} \cong \frac{E_{S2} \cdot \left(\dfrac{Z_{L2}}{Z_{S2} + Z_{L2}} \right)}{E_{S1} \cdot \left(\dfrac{Z_G \cdot Z_{L2}}{(Z_{S1} + Z_{L1}) \cdot (Z_{S2} + Z_{L2})} \right)} = \frac{E_{S2}}{E_{S1}} \cdot \left(\frac{Z_{S1} + Z_{L1}}{Z_G} \right) \quad (4.4)$$

Clearly, the SIR is, first and foremost, a function of the amplitude of the sources E_{S1} and E_{S2}, and is inversely proportional to the impedance of the common return path, Z_G. The lower this impedance, the higher is the signal-to-interference ratio.

The following is an application of the above conclusions. We begin with Figure 4.2 where two single-ended circuits (such as TTL logic) are interconnected by a single conductor, with the signal reference and current-return path accomplished by means of a ground reference (in certain cases, such as RS-232, a return wire, denoted "GND" is also used and is shown in Figure 4.2 as a broken line (See also Figure 4.4.) Obviously, with no connection to the signal reference, such circuits cannot function properly.

The disadvantage of this configuration is obvious. Any switching noise due to the voltage difference between the connections, V_G, will appear as undesired noise in the system, since the reference constitutes part of the circuit.

Such noise need not result from an explicit noisy circuit sharing the same common return path. Consider, for instance, a vehicle in which the metallic frame typically serves as the return path for (noisy) power current to return back to the negative terminal of the vehicle's battery. Nominal +12 VDC power can exhibit transient voltages as high as 100 V (e.g., Test Pulse 1 of ISO 7673-1), due to switching events during starting of the vehicle's engine, resulting in large transient currents flowing through the return path. Any fraction of transient current present will cause a voltage drop across the

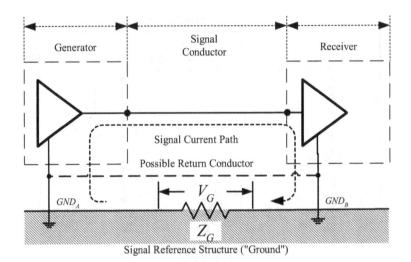

Figure 4.2. Single-ended circuits with a signal reference structure as a current-return path.

signal return (reference) between points GND_A and GND_B. Any potential difference will result in an undesired voltage at the input of the receiver. Figure 4.3 illustrates a similar situation with a PWM (pulse width modulation) motor acting as a source of interference to the sensitive sensing circuit through the common return-path impedance.

A slightly more complex situation is illustrated in Figure 4.4, typical of an unbalanced RS-232 digital communication system.

The RS-232 interface standard does not mandate a dedicated return line, requiring only the presence of a connection between the circuit grounds of the equipment at each

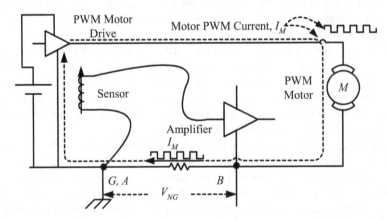

Figure 4.3. Coupling of interference from a PWM motor to a sensitive circuit through common return-path impedance.

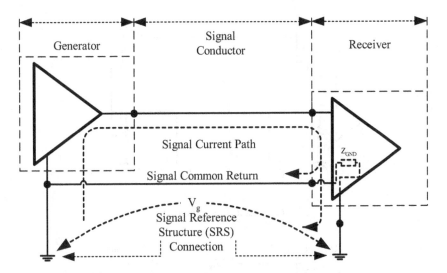

Figure 4.4. RS-232 unbalanced digital data interface utilizes the electrical safety ground conductor as a current-return path.

end of the circuit. One manner of accomplishing this is by utilizing a system ground connection, whereby both the generator and the receiver circuit grounds are connected to their respective circuit reference (chassis ground), which is also connected to earth ground (e.g., through the electrical safety ground conductor, the green wire included in the power cord; see Figure 4.5).

In this case, a noise source from an external system sharing the same system reference ground drives a fraction of the noise current I_N into the ESGC wire, which is common to both the generator and the receiver. Assuming that the load impedance Z_L is much greater than the impedance of the return wire Z_W and of the system reference ground Z_G (which applies to the case for RS-232), the noise voltage developed on the return wire, which adds to the signal voltage of the load, can be expressed as

$$V_N = I_N \cdot \left(\frac{Z_W}{Z_W + Z_G} \right) \cdot Z_G \qquad (4.5)$$

The signal reference ground, or chassis, does not exhibit a zero potential across its entire surface. Figure 4.6 [1] illustrates an artist's concept of a possible potential distribution. In this image, dark areas surround ground lugs or shields in high-voltage and/or high-current zones where the power is sufficiently high to create a signal on the physical structure. If an unbalanced system, such as that depicted in Figure 4.5, is referenced to a structure exhibiting such a potential distribution, interference will be introduced into the system. On certain platforms such as aircraft, RF potential differences can exceed 50 V under certain circumstances. If such an aircraft is equipped, for instance, with a short-wave (2 to 30 MHz) transmitter, RF currents exceeding 20 A have been observed along the aircraft fuselage during high-power

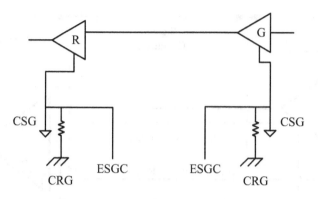

(a) Circuit Representation

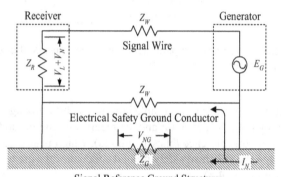

(b) Equivalent Circuit

Figure 4.5. Interconnection of an unbalanced digital interface circuits' signal return via a common signal reference return. CSG = circuit signal ground; CRG = circuit reference ground; ESGC = electrical safety ground.

transmissions! Surface impedance on the order of 1 to 10 mΩ would, hence, result in a voltage drop of 20 to 200 mV!

The arrangement depicted in Figure 4.7 appears to serve as a possible improvement compared to Figure 4.5. In this case, the SC (send common) wire of the generator is connected to the receiver circuit ground by means of a dedicated wire, which now serves as the common reference and return signal between the generator and the receiver. In order to preclude circulating ground currents, the signal reference is isolated from the ground structure at the receiver's end. The generator and the receiver-circuit signal grounds are now referenced to one single reference point within the system, the generator's circuit ground.

It is evident from Figure 4.7 that the noise current I_N cannot flow through the send common conductor, thus, ideally, it will have no effect on the receiver's response.

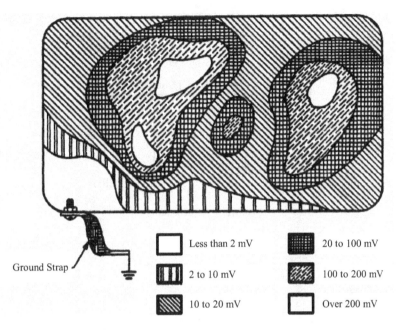

☐ Less than 2 mV	▦ 20 to 100 mV
⫼ 2 to 10 mV	▨ 100 to 200 mV
▧ 10 to 20 mV	☐ Over 200 mV

Ground Strap

Figure 4.6. Typical ground potential distribution at a given frequency.

Note: This model does not consider parasitic coupling such as mutual capacitance between the SC conductor and the circuit reference-ground structure, making this solution effective for relatively low frequencies only.

Common to the above examples is the nonzero impedance of the common ground-return path. If shared by more than one system, coupling will occur and is known as common-impedance coupling.

Three factors can be singled out as contributors to the resulting interference coupling into the load:

1. The nonzero impedance of the common return path
2. The actual sharing of a common return path by several systems
3. The low noise tolerance exhibited by the receiver circuit due to its unbalanced character

4.1.2 Controlling Common-Impedance Interference Coupling

Appreciating the fact that any physical conducting structure will exhibit finite impedance, as illustrated in the above examples, the problem of controlling common-impedance interference coupling reduces to the need to ensure that

$$Z_G \times I_N \leq S_V \qquad (4.6)$$

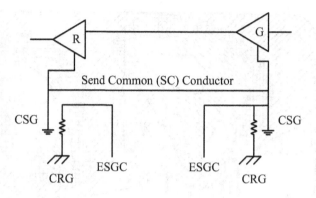

(a) Circuit Representation

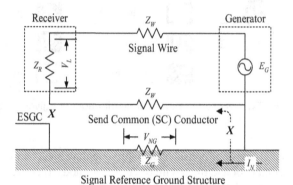

(b) Equivalent Circuit

Figure 4.7. Interconnection of an unbalanced digital interface circuits' signal return via the send common (SC) conductor. CSG = circuit signal ground; CRG = circuit reference ground; ESGC = electrical safety ground.

where Z_G and I_N represent the impedance of the signal reference ground structure and the noise current flowing through it, respectively, and S_V stands for the victim's sensitivity threshold level. In other words, the interference voltage level developed across the victim load due to the common return path impedance Z_G and the interference current I_N flowing across the common return path does not exceed the victim's sensitivity threshold level S_V. Several approaches for addressing the problem are:

1. **Lowering the impedance of the common return path,** Z_G, thereby reducing the voltage drop across the common return path so that the noise developed is below the sensitivity levels of victim circuits referenced to that path and does not constitute a threat to their performance. For the circuit configuration in Figure 4.5, this would constitute the only feasible solution.

2. **Precluding common current-return paths.** Provide each circuit with its own dedicated return path, or prevent interference current from flowing in the reference structure from entering the sensitive circuit's current-return path. Keep the noise voltage developed on the reference away from the victim load impedance. The circuit configuration depicted in Figure 4.7 puts this solution into practice.

3. **Designing noise-tolerant circuits.** Ensure that the sensitivity threshold S_V is above the expected common-impedance noise level. This can be achieved, for instance, by using balanced circuits exhibiting a high common-mode rejection. Unlike the previous two techniques, this approach does not eliminate the actual coupling between the circuits but instead ensures that the system is insensitive to the expected coupled interference levels.

Selection of the most appropriate technique (or a combination thereof) must be implemented during system design, depending on feasibility and cost, as well as frequency and safety considerations.

4.1.2.1 Reduction of Common Signal Reference Structure Impedance.

A complete grounding system contains metallic conductors that exhibit frequency-dependent impedance, associated with their material and geometry. Reduction of the impedance of grounding conductors lowers the potential difference between any two points, thereby reducing conductive coupling in susceptible circuits referenced to these points.

Intuitively, lower impedance is erroneously considered synonymous with lower resistance, leading to the belief that "the heavier the better" (i.e., the more massive the metal, the better). This notion holds for both direct current and very low-frequency alternating current only, where the resistance of the conductor is the controlling factor, with the exception of very unusual situations, such as when the signal to be processed is very low in amplitude or where interfacing equipment are physically far apart. An adequate ground can generally be realized for DC current in a relatively economical manner utilizing low-resistivity materials such as copper and aluminum.

Most systems, however, use AC signals so reactive characteristics of the conductors cannot be ignored.* Therefore, the frequency-dependent properties of conductors become important and both the conductors' resistance and reactance must be considered. Reduction of the conductors' impedance is reduced by minimizing both the resistance (R) and the (essentially inductive) series reactance ($X_L = \omega L$, $\omega = 2\pi f$) of the conductors forming the grounding system.

The alternating current (AC) impedance of a conductor is composed of two parts: the AC resistance and inductive reactance. Both the AC resistance and the reactance of a conductor vary with frequency due to skin effect[†]:

$$Z(\omega) = R(\omega) + jX_L(\omega) \tag{4.7}$$

---

*Nonideal behavior of conductors was discussed in Chapter 2.
[†]The skin effect phenomenon was discussed in Chapter 2.

where $Z(\omega)$, $R(\omega)$, and $X_L(\omega)$ symbolize the frequency-dependent impedance, resistance, and inductive reactance, respectively.

4.1.2.1.1 DC RESISTANCE. The DC resistance of a conductor of uniform cross section R (ohm) is proportional to the resistivity of the material ρ (ohm × mm²/m) and to the length in the direction of the current flow ℓ (m), and inversely proportional to the cross-sectional area perpendicular to the current flow A (mm²):

$$R = \rho \frac{\ell}{A}, \text{ ohm} \qquad (4.8)$$

Typical values of the resistivity, ρ, for two commonly used materials are:

$$\text{Copper (Cu)} = 1.7 \times 10^{-3} \ \Omega \times \text{mm}^2/\text{m}$$

$$\text{Aluminum (Al)} = 2.8 \times 10^{-3} \ \Omega \times \text{mm}^2/\text{m}$$

4.1.2.1.2 AC RESISTANCE. Whereas a DC current is uniformly distributed over the cross-sectional area of a conductor, AC currents tend to concentrate near the surface. The higher the frequency, the greater the concentration of current near the surface due to the skin effect. Skin depth is defined as the depth at which current density is attenuated to $e^{-1} \approx 1/2.7180$ or 37% of its value at the conductor surface. The effective cross section of the conductor available for current flow decreases, resulting in an increase of resistance. For a round conductor, skin effect is illustrated in Figure 4.8.

From Equation (4.8), the DC resistance of a conductor with a round cross can therefore be expressed as

$$R = \rho \frac{\ell}{A} = \rho \frac{\ell}{\pi r^2} \qquad (4.9)$$

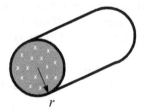

(a) DC Current Distribution   (b) High Frequency Current Distribution

Figure 4.8. Current distribution over a conductor's cross section at DC (a) and AC (b).

For conductors whose thickness is at least three times the skin depth, $-\delta$, this depth is given by

$$\delta = \frac{1}{\sqrt{\pi f \sigma \mu}}$$

(4.10)

where f is given in Hz.

$$\sigma = \sigma_0 \cdot \sigma_r$$
$$\mu = \mu_0 \cdot \mu_r$$

(4.11)

and

$$\sigma_0 = 5.82 \times 10^7 \, \text{Si/m}$$
$$\mu_0 = 4\pi \times 10^{-7} \, \text{H/m}$$

(4.12)

where:
μ = permeability of the material (H/m)
μ_0 = permeability of free space (H/m)
μ_r = material's relative permeability
σ = conductivity of the material (Si/m)
σ_0 = conductivity of copper (Cu) (Si/m)
σ_r = material's conductivity relative to that of copper

For both copper and aluminum, the value of the relative permeability μ_r is 1, whereas the relative conductivity σ_r is 1 for copper and 0.63 for aluminum. Note that copper exhibits a skin depth of 0.34 inch (8.63 mm) at 60 Hz but only 0.0026 inch (0.066 mm) at 1 MHz.

Approximating the effective cross section of the current flow as

$$A_{AC} \approx 2\pi r \cdot \delta = 2r \cdot \sqrt{\frac{\pi}{f \sigma_r \mu_r}}$$

(4.13)

the AC resistance can therefore be expressed as

$$R_{AC} \approx \rho \frac{\ell}{2r} \cdot \sqrt{\frac{f \sigma \mu}{\pi}}$$

(4.14)

Equation (4.14) could also be rewritten as

$$R_{AC} \approx R_{DC} \cdot \frac{r}{2\delta}$$

(4.15)

The AC resistance of a conductor with a cross section of an arbitrary shape can be determined from the skin depth if both the thickness and the radius of curvature of the conductor are much greater than the skin depth and if the radius of curvature does not vary too rapidly around the conductor's perimeter. For a conductor meeting these conditions with a circumference P, the AC resistance can be obtained by substituting in Equation (4.15) to get

$$r \approx \frac{P}{2\pi} \tag{4.16}$$

A round conductor exhibits the minimum perimeter for a given cross section and results, therefore, in a higher AC resistance compared to that of a flat grounding conductor with a rectangular cross section of an equal area.

4.1.2.1.3 REACTANCE. In addition to resistance, conductors also exhibit reactance, which at high frequencies constitutes the most important term in Equation (4.7). This reactance is generally inductive and is given by the product of the radian frequency and the self- (internal*) inductance L of the conductor. Inductance produces undesirable effects as the frequency of the AC current flowing through this path increases. The internal inductance of a conductor is a measure of the opposition to a change in the current flowing in the conductor. Because skin effect redistributes the current within a conductor with changes in frequency, the inductance of the conductor is also frequency dependent.

For a straight conductor with a circular cross section, with a length ℓ and diameter d (both in cm), and relative permeability μ_r (if surrounded by air), the internal partial† inductance at low frequencies, where current flow can be assumed to be uniform across the conductor cross section, is given by [2]

$$L[\mu H] = 0.002 \cdot \ell \cdot \left(2.303 \, \log\left(\frac{4\ell}{d} \right) - 1 + \frac{\mu_r}{4} \right) \tag{4.17}$$

As frequency increases, a limiting value of inductance, L_{HF}, is approached:

$$L_{HF}[\mu H] = 0.002 \cdot \ell \cdot \left(2.303 \, \log\left(\frac{4\ell}{d} \right) - 1 \right) \tag{4.18}$$

This equation demonstrates that inductance L increases linearly with length ℓ, whereas an increase in diameter d will reduce the total inductance logarithmically (a slowly varying function).

When a rectilinear grounding strap, for example, a flat conductor, is used in lieu of a circular conductor, the internal inductance of the conductor is expressed by (see Figure 4.9)

---

*The internal inductance of a conductor should be distinguished from the external inductance, which is a function of the loop area enclosed by the conductor.

†The concept of partial inductance was discussed in detail in Chapter 2, considering the fact that inductance is associated with current flow through a loop. Within the context of this section, unless stated otherwise, the term inductance refers to partial inductance.

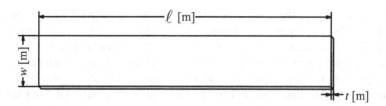

Figure 4.9. Dimensions of a rectilinear grounding strap.

$$L\ [\mu H] = 0.002 \cdot \ell \cdot \left[2.303 \cdot \log\left(\frac{2 \cdot \ell}{w+t}\right) + 0.5 + 0.2235 \cdot \left(\frac{w+t}{\ell}\right) \right] \quad (4.19)$$

The width w is assumed to be greater than the thickness t by a factor of 10 or more.

When comparing the expressions for the internal inductance for a circular [Equation (4.18)] and a rectilinear [Equation (4.19)] conductor, it is clearly observed that in the rectilinear conductor, the length to width (ℓ:w) aspect has a considerable effect on the total inductance presented along the conductor length, whereas altering the diameter of a circular wire will introduce insignificant reduction in the conductor's inductance. This makes straps more useful for obtaining low impedance ground paths at high frequencies. Figure 4.10 demonstrates the advantage of rectilinear over circular conductors with respect to their internal inductance [2].

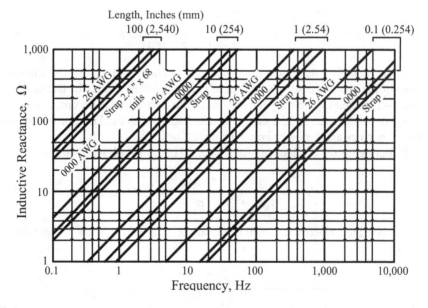

Figure 4.10. Advantage of rectilinear over circular conductors with respect to their internal partial inductance.

The impedance of the grounding conductors can be reduced by making them as short as possible and by using conductors with cross-sectional areas as large as practical. Paralleling conductors can also contribute to reduction of impedance.

Planes constitute an extension of a strap, where the $\ell{:}w$ aspect ratio is very small, resulting in very low inductance. This feature is applied in the design of printed circuit boards where solid planes are utilized for RF and high-speed digital applications to maintain low impedance for the return path at higher frequencies.

The overall impedance of the grounding system depends both on the impedance of the signal reference and the establishment of low-impedance bonds between ground system conductors. Various aspects of bonding are discussed in Chapter 5.

4.1.2.1.4 SURFACE IMPEDANCE OF PLANES. As previously mentioned, large planes do not offer zero impedance; however, their impedance is much lower than a single small wire or strap. Therefore, planes find much use in printed circuit boards where the power and return conductors are implemented as planes. Also, reference structures such as vehicle and equipment chassis can be considered as planes, similarly exhibiting very low impedance.

One of the most commonly used parameters associated with planes is surface resistance, expressed in terms of Ω/square (commonly written as $\Omega/\square$). Consider a conducting cube [exhibiting a resistivity ρ (Ω/m), or conductivity, σ (Si/m)] of which the length of each side is L (m). Shrinking the height of the cube to t (m), yields a box with the dimensions of $L \times L \times t$. Computing the resistance of a square with dimensions $L \times L \times t$ based on Equation (4.8), using length L and cross section, A (m^2) $= L \times t$, yields, therefore,

$$R_{DC} = \rho_{\Omega/m} \times \frac{L}{A} = \frac{L}{\sigma \cdot L \cdot t} = \frac{1}{\sigma \cdot t} \Omega / \text{sq} \qquad (4.20)$$

In Equation (4.20), the length of the square L is eliminated and the DC surface resistance of the plane is now independent of length. In other words, the DC surface resistance is obtained for arbitrary dimensions of the square, and is dependent on the resistivity (or conductivity) of the surface and its thickness alone. The dimensions of the surface resistance are, hence, "ohms per square," or $\Omega/\square$.

Substituting for the conductivity for copper 5.8005×10^7 Si, we obtain

$$R_{DC} = \frac{1}{\sigma_0 \sigma_r \cdot t} = \frac{1.72 \times 10^{-8}}{\sigma_r} \times \frac{1}{t}, \ \Omega/\text{sq} \qquad (4.21)$$

Planes also exhibit reactance as well as resistance. The RF reactance of a plane is expressed as

$$X_{RF} = \frac{369 \cdot \sqrt{\dfrac{\mu_r}{\sigma_r} \cdot F_{MHz}}}{1 - e^{-1/\delta}}, \ \mu\Omega/\text{sq} \qquad (4.22)$$

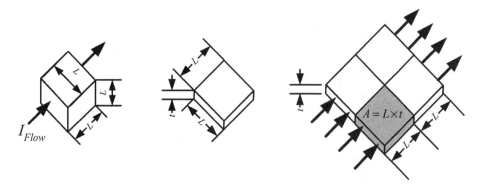

Figure 4.11. Demonstration of the concept of surface resistance per square.

The general expression for the impedance between two points on a plane is, therefore [1],

$$Z_{RF} = (R_{DC} + jX_{RF}) \cdot \left[1 + \tan\left(\frac{2\pi d}{\lambda}\right)\right] \approx R_{DC} + jX_{RF}, \ \Omega/\text{sq}; \ d \leq 0.005\lambda \quad (4.23)$$

where R_{DC} is the DC resistance of the plane, Z_{RF} is the impedance of the plane, both in $\Omega/\square$, and d is the distance between points across the plane. The wavelength λ is associated with the highest frequency of concern. Clearly, resistance is low from DC through lower frequencies, gradually increasing in value as the frequency rises. When the frequency increases to that at which dimensions of the plane approach a quarter of a wavelength, $\lambda/4$, where λ is associated with that particular frequency, the first resonant frequency is reached and the impedance swells asymptotically to infinity, sharply decreasing thereafter. At successive odd multiples of this wavelength, $(2k + 1) \times \lambda/4$, where $k = 1, 2, \ldots$, this behavior is repeated. At even multiples of this wavelength, $2k \times \lambda/4$, beginning from $\lambda/2$, antiresonances take place and the impedance of the plane is reduced to its RF resistance (considering the skin effect). It follows that for frequencies at which the dimensions of the plane are smaller than $\lambda/20$, optimal performance of the planes is achieved with respect to reactive impedance. This is depicted in Figure 4.12.*

  4.1.2.1.5   EXAMPLE: COUPLING VIA THE SIGNAL REFERENCE STRUCTURE. Consider the circuit in Figure 4.13, where a 20 kHz PWM (pulse width modulation) motor is referenced to a return current common to a sensor and associated amplifier. Assume that the motor current drive is 12 A and the resistance between the points A–B on the wire is 10 mΩ. Neglecting inductive effects at the 20 kHz frequency, the error voltage at the amplifier input due to the voltage drop across the return wire resistance is $V_{A\text{-}B} = 12$ A $\times$ 10 mΩ = 120 mV [Figure 4.13(a)], which may constitute a significant error at the input of the amplifier.

*See Chapter 2 for a discussion on transmission line effects in electrically long conductors.

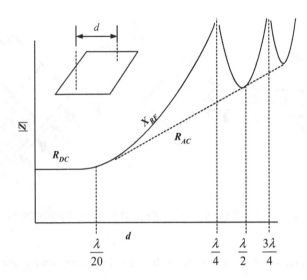

Figure 4.12. Surface impedance between two points across a plane versus separation.

If the wire is replaced by a solid plane, having a surface resistance of 0.5 m $\Omega/\square$, the voltage drop across the points A–B will now be $V_{A-B} = 12 \text{ A} \times 0.5 \text{ m}\Omega = 6 \text{ mV}^*$ [Figure 4.13(b)].

4.1.2.2 *Avoiding a Common-Impedance Return Path.* Another means to reduce common-impedance ground coupling is to avoid common-impedance coupling between circuits altogether. For achieving this goal, care must be taken to provide a separate and dedicated return path for each circuit through a dedicated conductor to the system single-point connection of the signal reference. This is illustrated in Figure 4.14.

In this case, the voltage developed on the load Z_{L2} is due to the source E_{S2} only (neglecting magnetic induction between the two circuits), regardless of the structure impedance, Z_G.

Returning to the example in Figure 4.13, it is evident that by connecting the sensor and amplifier to a common reference [Figure 4.15(b)] (serving as the return path of the motor current) at one point only (denoted as point A, B) no (or negligible) coupling can occur through the ground impedance. But for one common point there is no common current-return path between the two circuits [compare this situation to that in Figure 4.15(a), where, similar to Figure 4.13(a), a large voltage drop may occur between points

---

*Note that this conclusion applies for low frequencies only where significant spreading of the return current in the plane occurs, effectively reducing the resistance of the return path (this effect is known as "spreading resistance"). High-frequency current assumes a different distribution in the plane and tends to concentrate primarily under the signal conductor regardless of the width of the return plane and does not utilize its entire cross section. A detailed discussion of high-frequency current-return paths is provided in Chapter 2.

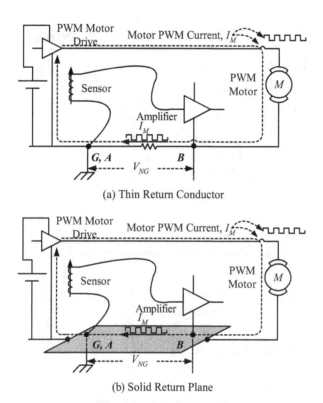

(a) Thin Return Conductor

(b) Solid Return Plane

Figure 4.13. Reduction of ground interference coupling by controlling the impedance of the return path.

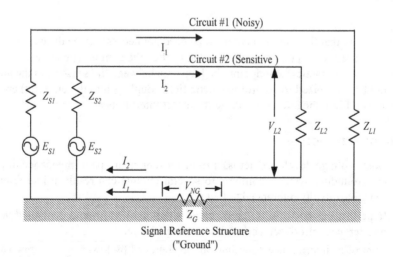

Figure 4.14. Avoiding common-impedance coupling by eliminating a common return path.

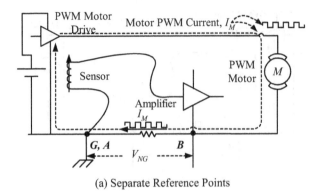

(a) Separate Reference Points

(b) Common Reference Point

<u>Figure 4.15.</u> Reduction of ground interference coupling by eliminating a common return path.

A and B]. Note that the common point (A, B) need not necessarily be the circuit's external common reference point G (although that is often the common practice).

Note: From a practical standpoint, the loop area between the sensor and the amplifier should be minimized to control magnetic field coupling into the circuit, thereby introducing emf into the circuit, resulting in similar interference.

Lessons Learned

- Noise voltage developed across a finite current-return path impedance of practical grounding systems common to several circuits can result in interference to sensitive circuits referenced to that path.
- Imperfect balance of differential circuits will convert ground-generated noise to interference introduced across the load impedance.
- Control of interference coupling can be achieved by lowering the impedance of the common return path, preclusion of common current-return paths, or by designing noise-tolerant circuits.

4.2 FUNDAMENTAL GROUNDING TOPOLOGIES

4.2.1 The Need for Different Topologies

The simple examples presented above have demonstrated that performance of circuits and systems can be significantly affected by the topology (or design) of the chosen grounding (or signal referencing and return path) methodology. Grounding within equipment is intended to realize multiple functions such as signal return (often divided to "noisy" and "sensitive" signals), DC/AC power return, and, of course, electrical safety ground connection.

From the standpoint of system performance and EMI control, the implementation of grounding should guarantee minimum interference coupling between circuits sharing the same grounding reference while providing a low-impedance return path for controlled current flow.

In order to meet this objective, proper signal and circuit identification, segregation of current-return paths, followed by careful location of ground nodes, must be considered during the design cycle.

Selection of an appropriate grounding topology should be considered first and foremost, prior to circuit realization or design. Primary features of common signal applications are:

1. **Low-Frequency Analog Signals.** Low-frequency analog signals are typically narrowband, low-level signals, often as low as μV or mV. On the other hand, analog circuits are gain devices, amplifying both signal and interference, with no signal regeneration (in other words, "noise in equals noise out"). Analog circuits must be provided with extremely noiseless dedicated return conductors (e.g., wires and traces). Return paths for such circuits should not be shared with any other circuit, in order to preclude any common-impedance coupling from high-level noisy signals (e.g., digital, PWN and other pulsed signals).

2. **High-Frequency Analog Signals.** High-frequency analog signals (RF, video, etc.) may cover a wide range of frequencies and signal levels. Similar to the previous category, RF and video circuits are highly sensitive to noise corruption, particularly at the receiver. For best performance, particularly when considering the high-frequency nature of certain signals, such circuits must be provided with low-impedance, noise-free return paths, generally implemented in the form of planes or their extensions (e.g., coaxial cables).

3. **Digital Signals.** Digital signals are typically broadband with moderate voltage levels, typically in the order of 5 V or less. Digital circuits exhibit no gain and provide moderate signal regeneration when a noisy digital signal is buffered or processed. Digital, especially high-speed digital circuit returns, require a low-impedance return path over the entire bandwidth of operation (determined by the "edge rate," not pulse repetition frequency), as power and signal returns share the same return paths.

4. **Powerful Load Signals.** Signals associated with powerful loads such as solenoids, motors, and relays exhibit a broadband signature, with extremely high

levels (possibly as high as several kV), with little or no control of signal parameters. Return paths of such signals should not be shared with any other signal as they have a high chance of severely interfering with each other.

Typically, each of the above circuit types, or variation thereof, must have separate return paths. Special cases such as electrically initiated explosive devices (EIEDs), firing circuits, or red/black* circuits may require isolated and dedicated return paths.

It is likely that some return conductors will share a common reference; however, to avoid circulating currents between the respective circuits (through some common impedance) they should have no more than one point in common. There are cases, however, in which trade-offs require that compromises be implemented due to practical constraints associated with the circuit design.

The following simple example will demonstrate the relationship between signal types and grounding topology. Figure 4.16(a) depicts the general circuit/system grounding nomenclature, whereas Figure 4.16(b) provides the interpretation of several commonly used schematic symbols for ground.

The symbol commonly identified as "earth ground" or "safety ground" is usually used only for chassis or safety ground connection. The uppermost two ground symbols are used interchangeably, although sometimes the left-hand one may be used for "analog return/ground" (AGND), whereas the other is for "digital return/ground" (DGND). When other ground symbols are used, the designation of the type of ground or reference connection the symbol refers to should be noted on the schematics.

Figure 4.17 illustrates a system comprising analog, digital, control, and display circuits, powered from a common power source [3]. In order to eliminate any interference from coupling via common impedance between the circuits, a "star" configuration is used for providing power and return paths to the various circuits. Application of this topology ultimately results in additional wiring and interconnects hardware.

Considering the similar nature of the digital, control, and display circuits, a compromise yielding simplification may be implemented by providing a common low-impedance path between circuits while still retaining a separate analog return path, resulting in the configuration depicted in Figure 4.18 [3]. This unification benefits from the reduction in the number of return conductors, while having no practical effect on the system performance. This configuration is often denoted as a "daisy chain."

In order to avoid cumbersome calculations, a simple yet effective approach may be implemented. In this method, separate power and signal-return paths are assigned to each subsystem identifiable as a distinct load according to its noise and signal characteristics (noisy, indifferent, sensitive, etc.). The following are examples of common signal categories' segregation applied in order of increased noise immunity:

---

*The concept of red/black engineering is associated with the separation of electrical and electronic equipment components, equipment, and systems processing classified or plain text information from those that process encrypted or unclassified information; "TEMPEST," an associated concept, is related to measures used to control compromising emanations [9].

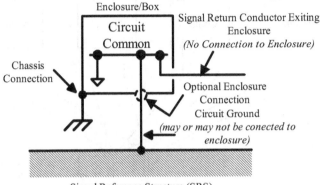

(a) Circuit/System Grounding Nomenclature

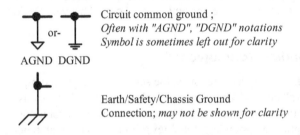

(b) Grounding Symbology

Figure 4.16. Grounding drawing nomenclature/key.

- Audio (low-frequency analog)
- Video (high-frequency analog)
- High-speed digital
- Control (high-level signals)
- Power supply (high-current DC and low-frequency AC)

Such categorization is best accomplished according to the noise-immunity margin of components. Once assigned, each of these categories will be allocated a separate return path, to be tied together at one common point only, typically at the system power supply (where all power leads branch out).

This example serves as a demonstration of the dependence of the system grounding topology on the circuit and signal characteristics. The following section discusses the properties of the different grounding topologies.

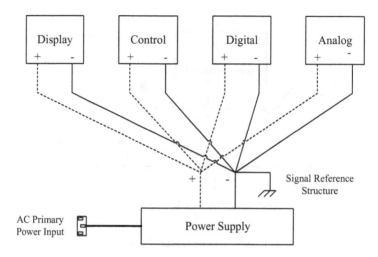

Figure 4.17. Separation of return-current paths for reduction of common-impedance coupling. (Source: Hartal, O., "Electromagnetic Compatibility by Design" [3].)

4.2.2 Grounding Topologies

In addition to the signal characteristics, the selection of a grounding topology is determined by several factors. These factors include system dimensions, system-specific separation/isolation requirements, and electrical safety considerations.

In implementing a grounding topology, three fundamental types are commonly used; a floating topology (which actually consists of no ground connection at all), sin-

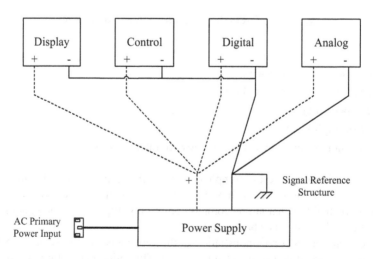

Figure 4.18. A system comprising separate and combined return-current paths. (Source: Hartal, O., "Electromagnetic Compatibility by Design" [3].)

gle-point grounding (SPG), and multipoint grounding (MPG). Most practical systems will consist of combinations of these, called a composite or hybrid grounding system.

Selection of a grounding topology best suited for a particular application requires careful consideration and is dependent on the product application. It is recommended, whenever possible, that grounding topology options be left open during the design and development phase. The optimal grounding method is an evolutionary procedure, and must be adaptable to the system being designed.

4.2.2.1 *Floating Topology.*

The floating system is a method to electrically isolate circuits or equipment from a common reference structure, or from any common wiring that might introduce circulating currents flowing through common impedance. In the floating topology, return currents do not flow through any reference structure.

A floating topology is shown in Figure 4.19. This system consists of three assemblies, System #1 through System #3, each enclosed within a metallic case (or chassis). For ensuring floatation, each is electrically isolated from its respective enclosure. For safety purposes, when applicable, the cases can be grounded using an electrical safety ground conductor (ESGC). This safety ground connection does not violate the floating characteristics of the system thanks to internal isolation between the circuits and their respective enclosure.

Systems #1 and #2 are powered from a common (isolated) power source, E_{S1}. As such, they are single-ended circuits, requiring that a common reference and current-return path be provided between the circuits. Systems #2 and #3 are transformer-coupled, allowing System #3 to be powered from a separately isolated power source (E_{S3}) while not requiring any common reference conductor.

In this topology, circuits are electrically isolated from the signal reference structure; hence, any noise currents present in the structure will not be conductively coupled to signal circuits. When employed in equipment design, signal returns are isolated from the equipment enclosure, thus preventing noise currents on the chassis from coupling directly into the signal circuits.

The effectiveness of floating systems depends on true isolation. To be effective, circuits in the floating topology must be truly "floating." In large systems and facilities, it

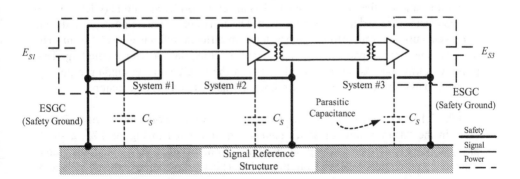

Figure 4.19. A floating topology.

is often difficult to achieve a completely floating system, and even if complete isolation is achieved it is difficult to maintain.

In addition, a floating system suffers from other limitations. For instance, static charge buildup on isolated signal circuits is likely to present a shock and spark hazard. In particular, if a floated system is located near high-voltage power lines, static buildup is probable. Furthermore, in most modern electronic facilities, external sources of energy such as commercial power are referenced to earth ground.

Thus, power faults to the signal system would cause a voltage increase to hazardous levels relative to other conductive objects in the facility or installation. Another danger is the threat of arc flashover between structures and the signal circuits in the event of a lightning strike to the facility. Not being conductively coupled, the system could be elevated to a voltage level sufficiently high relative to the signal ground so as to cause insulation breakdown, arcing, and electric shock hazard.

Finally, at higher frequencies, a floating system does not exist in practice, since parasitic capacitance from circuits to chassis, C_S, (Figure 4.19) will allow leakage due to displacement current through parasitic capacitances, limiting its effectiveness to low frequencies only. It follows that floating systems are typically used in small, battery-operated equipment or audio-frequency systems. Typical applications of floated systems are, therefore, associated with low-frequency AC (e.g., audio, power frequency, synchro, and control circuits) and balanced, transformer-isolated data buses (e.g., 10/100BaseT).

Examples of circuit isolation techniques are presented in Figure 4.20(a), which employs transformer isolation (commonly used in audio or synchro circuits), whereas Figure 4.20(b) obtains isolation by optical means (typically used for digital interfaces).

Lessons Learned

- Truly "floating" systems rarely exist, and they may be prone to shock and static-charge buildup.
- Floating systems are normally implemented at low frequencies only.

4.2.2.2 Single-Point Grounding. A single-point ground topology has a single physical location in the system designated as the ground-reference point (GRP). All return conductors are tied to that one location. The aim of using single-point grounding is to prevent currents from different subsystems (at different reference levels) from sharing a common return path, which could result in common-impedance coupling. Two primary configurations are commonly used in the single-point grounding topology, namely, "daisy chain" or series connection and "star" or parallel connection.

4.2.2.2.1 "DAISY CHAIN" (SERIES CONNECTION) SINGLE-POINT GROUNDING TOPOLOGY. In the "daisy chain" or series connection, illustrated in Figure 4.21, a single physical point is defined as the ground-reference point (GRP). A "ground bus" interconnecting all circuits extends from this point. This topology enjoys the advantage of simplicity: only one grounding conductor interconnects all circuits. All assemblies,

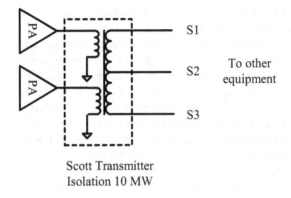

Scott Transmitter
Isolation 10 MW

(a) Typical Synchro Output-Isolation Circuitry

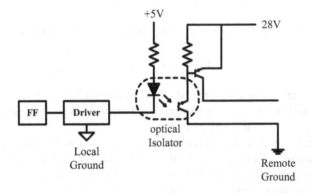

(b) Digital-to-Solid-State Relay-Isolation Circuitry

Figure 4.20. Two methods for achieving circuit isolation.

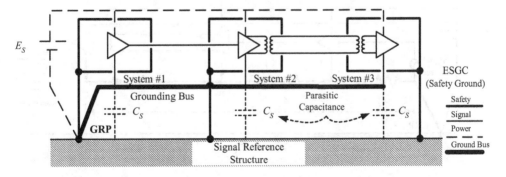

Figure 4.21. "Daisy chain" (series connection) single-point grounding topology.

Systems #1 through #3 are powered from a common source, E_S, which is possible as they also share a common reference.

The obvious disadvantage inherent to this topology is common-impedance coupling between circuits sharing the same return path. This is illustrated in Figure 4.22. The DC power source E_S as well as the signal return currents I_1, I_2, and I_3 are also shown.

In a large system, the major disadvantage of the single-point ground topology is the inescapable use of long conductors to connect to the GRP. The long conductors (>0.05 λ at the highest frequency of concern) prevent the realization of a satisfactory reference for higher frequencies because of large reactive self-impedance of the ground bus. This is due to the voltage drop developed along the ground bus due to return currents. For instance, at point "C" which serves as the reference point for the most remote circuit, the voltage drop from this point to the ground-reference point (GRP) can, by observation, be expressed as

$$V_C = (I_1 + I_2 + I_3) \cdot Z_1 + (I_2 + I_3) \cdot Z_2 + (I_3) \cdot Z_3 \tag{4.24}$$

At point "A," closest to the ground-reference point (GRP), the voltage drop is equal to:

$$V_C = (I_1 + I_2 + I_3) \cdot Z_1 \tag{4.25}$$

This configuration forces return signals from Systems #2 and #3 to be impressed on System #1 if the impedance along the line is not limited to very low values, undermining the objectives of single-point grounding.

"Daisy chain" single-point grounding is often used in equipment racks. It should not be used however, in systems comprising very diverse assemblies (with respect to their ground-noise emissions or susceptibility) such as high-gain amplifiers as this may introduce intrasystem interference. Since high-power systems produce large return currents that share a common ground bus, sensitive components may be affected due to common-impedance coupling. If this approach must be used, the most sensitive circuits must be located at the point closest to the ground-reference (point "A" in Figure

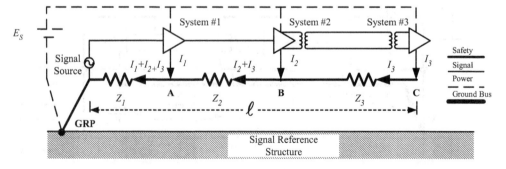

Figure 4.22. Common-impedance coupling is not eliminated in the "daisy chain" (series connection) single-point grounding topology.

4.22) and as far away as possible from high-level components and circuits (which, in turn, should be placed at the far end of the ground bus, at point "C" in Figure 4.22). A superior approach could be employed, whereby separate branches are employed for systems of different emission levels and sensitivity thresholds. In practice, this results in the "nested grounding" architecture discussed in Section 4.2.2.7.

Another concern associated with this topology is related to the high-frequency performance of this configuration. Owing to distributed stray capacitance along the ground bus to the signal reference structure C_S (Figure 4.21), single-point grounding topology essentially ceases to function optimally as the frequency is increased beyond several hundred kilohertz.

Up to this point, the length of the ground conductors when compared to wavelength (i.e. the "electrical length" of the conductors) was not considered. The conductors were assumed to be electrically short. As long as this condition prevails, the "daisy chain" topology can be used to an advantage.

At higher frequencies, at which this assumption does not hold, and the length of the ground bus ℓ exceeds 0.05 Ω, grounding wires running along the signal reference structure exhibit transmission-line (T-line) behavior.* Consequently, the RF impedance of the ground bus is dominated by (assuming a lossless conductor) a distributed (per-unit-length) inductance L_X and capacitance C_X [Figure 4.23(b)].

Note: The distributed (per-unit-length) resistance and conductance of the ground bus, R_X and G_X, respectively, are not presented, since the line is assumed to be lossless.

The point on the conductor connected to the GRP will be at zero volt potential ("0V"); however, other points along the line will exhibit a distance-dependent voltage and current distribution (V_X and I_X, respectively). Due to the varying distribution of the voltage and current along the transmission line, the impedance at any point along the line also varies. In particular, for a lossless T-line shorted at one end (to the "0V" point) the input impedance at any point x along the line, at a frequency of concern, f, exhibits a periodic variation expressed as [4]

$$Z_{in} \triangleq \frac{V_{In}}{I_{In}} = jZ_0 \cdot \tan\left(\beta \cdot x\right) = jZ_0 \cdot \tan\left(2\pi f \cdot x \cdot \sqrt{L_x C_x}\right) \qquad (4.26)$$

For lossless transmission lines, the phase velocity of the line, β, in rad/m, is equal to

$$\beta \approx \omega \cdot \sqrt{L_x C_x} = 2\pi f \cdot \sqrt{L_x C_x} \qquad (4.27)$$

and its characteristic impedance is

$$Z_0 \triangleq \frac{V_x}{I_x} = \sqrt{\frac{L_x}{C_x}} \qquad (4.28)$$

In the special case where $x = \lambda/4$ at a particular frequency, the phase difference between the voltage and current distribution is $\pi/2$ (or 90°). The short circuit at the load

---

*Transmission-line basics and phenomena are discussed in Chapter 2.

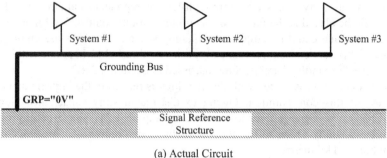

(a) Actual Circuit

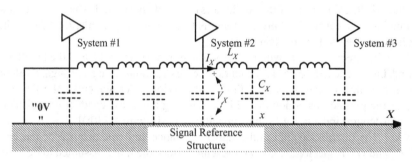

(b) Equivalent Transmission-Line Model of the Ground Bus

$[L_X$ = distributed (per-unit-length) inductance, C_X = distributed capacitance]

Figure 4.23. A grounding conductor as a transmission line with the signal reference structure.

is transformed to the input as an effective "open circuit" resulting in infinitely large input impedance at the frequency associated with this wavelength $-\lambda$:

$$Z_{in} = jZ_0 \cdot \tan\left(\frac{\pi}{2}\right)_{\Big|x=\frac{\lambda}{4}} \to \infty \qquad (4.29)$$

This situation finds its equivalence in circuit theory in a tuned parallel resonant circuit, which exhibits infinitely high input impedance when at resonance.

Moving away from the "0V" GRP along the grounding conductor, the input impedance will alternate between $Z_i \to \infty$ and $Z_i = 0$ for parallel and series resonance of the T-line at odd and even multiples of $\lambda/4$, respectively, occurring at (Figure 4.25):

a) $\qquad x = (2k+1)\cdot\dfrac{\lambda}{4}$, $k = 0,1,2...$, parallel resonances, where $|Z_{in}| \to \infty$

b) $\qquad\qquad x = 2k\cdot\dfrac{\lambda}{4}$, $k = 1,2...$, series resonances, where $|Z_{in}| = 0$

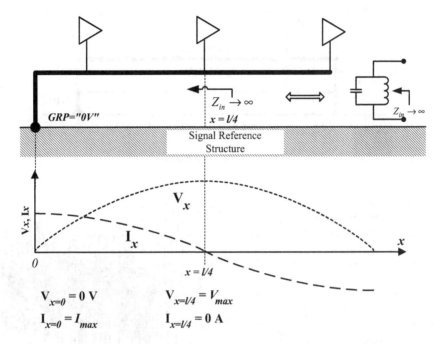

Figure 4.24. A 0 V ground potential is converted to "open circuit" (∞V) at the Point $x = \lambda/4$.

In practice, transmission-line resonant effects are superimposed on the AC resistance of the line, therefore, at series resonance frequencies, occurring in case (b), $|Z_{in}| \rightarrow R_{AC}$. The line input impedance is, therefore, limited by the line AC resistance, R_{AC} (Figure 4.25).

In practice, this results in the equipment located at all points $x = (2k + 1) \cdot \lambda/4$ along the grounding conductor, being practically isolated from the signal reference at the resonance frequency. In general, for a given frequency, the input impedance, Z_{In}, will vary at different points, x, along the ground bus; conversely, at a given point, x, along the ground bus, the input impedance, Z_{In}, will exhibit a frequency dependence, appearing as inductive, capacitive, open-circuit, or pure resistance, providing, therefore, an extremely poor and unreliable ground reference. Furthermore, in the presence of broadband signals such as high-speed digital, tuned circuits will ring at the specific frequencies at which they are resonant.

This situation is not an EMC problem per se but may evolve into one if a large RF voltage drop develops along the ground bus, resulting in enhanced radiation. This situation, however, could definitely result in functionality concerns.

4.2.2.2.2 "STAR" (PARALLEL CONNECTION) SINGLE-POINT GROUNDING TOPOLOGY. The ideal single-point signal ground network is the "star" or parallel connection. In this topology (Figure 4.26), a single physical location in the system is defined as the ground-reference point (GRP). From this point, separate and dedicated ground conduc-

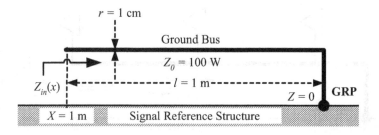

(a) Circuit Configuration

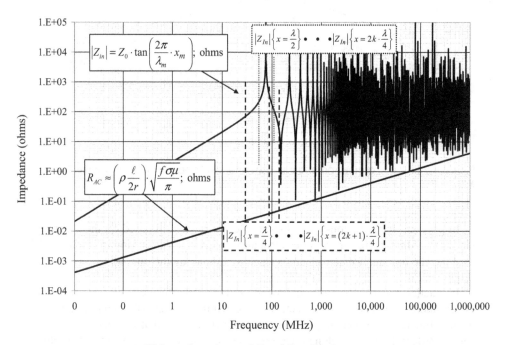

(b) Input Impedance of Grounding Conductor

<u>Figure 4.25.</u> Parallel and series resonances along a ground bus, $|Z_{in}|$ [Ω] versus frequency, for the geometry in (a).

tors extend to the return side of each circuit. This configuration requires a considerable number of conductors, which is generally not practical in large distributed systems.

The single-point grounding topology accomplishes the functions of signal return, while helping control common-impedance-coupled interference between systems. As illustrated in Figure 4.27, closed-loop paths for noise currents in the signal grounding network are avoided. The interference voltage V_{NG} in the signal reference is not coupled into the signal circuit paths. Hence, potential differences produced by lower fre-

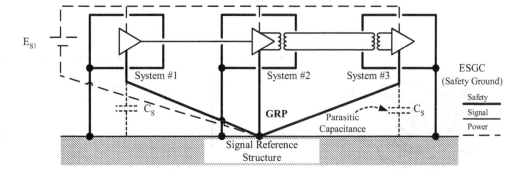

Figure 4.26. "Star" (parallel connection) single-point grounding topology.

quency noise currents have little or no effect on the system performance. Since each of the systems has a dedicated and independent signal-return path, common-impedance coupling between different branches is also precluded.

Similar to the "daisy chain" topology, the "star" configuration when used in large systems suffers from the same major problem associated with unavoidably long return conductors interconnecting circuits or systems and a GRP, in particular, the transmission-line resonant effects along the long return conductors, discussed in detail in Section 4.2.2.2.1. Furthermore, return conductors of different lengths may result in the circuits connected to them sensing different impedance to the GRP for a given frequency. This may result in functionality and EMI concerns.

The length of return conductors in a "star" topology may increase radiation from return conductors. The electrically long grounding conductors constitute efficient radiators, resulting in increased radiated emissions and crosstalk, depending on their respective physical spacing and layout. The extent of the radiation and crosstalk that may take place will also be a function of the spectral content of the return signal: higher fre-

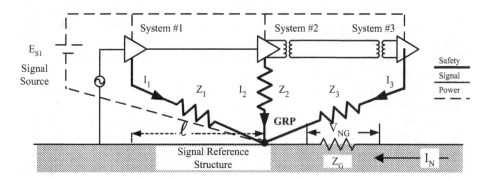

Figure 4.27. "Star" (parallel connection) single-point grounding topology minimizes the effects of common-impedance coupling.

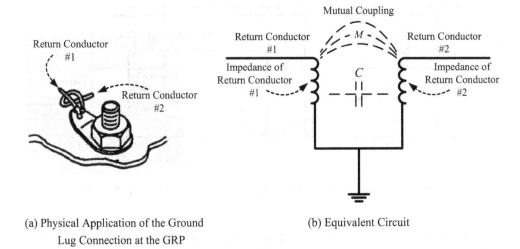

(a) Physical Application of the Ground   (b) Equivalent Circuit
    Lug Connection at the GRP

Figure 4.28. Mutual coupling (crosstalk) between grounding conductors in a "star" single-point grounding configuration.

quency components will radiate and couple more efficiently than lower frequency components. Figure 4.28 illustrates how crosstalk can be generated between ground conductors in a "star" topology.

Similar to the "daisy chain" topology, distributed stray capacitance, C_S, is present from each circuit to the signal reference (Figure 4.26); therefore, single-point grounding essentially ceases to exist as the signal frequency is increased.

This topology is particularly violated when coaxial cables (where the shield actually serves as the signal-return path) are used to interface between different assemblies, with the shield of the cable interconnecting the circuits' return and their cabinets or enclosures, in order to ensure functionality [Figure 4.29(a)]. In the presence of low-frequency magnetic fields, for instance, induced electromotive force (emf) due to the magnetic fields in the large loop shown in Figure 4.29(a) will occur. As performance of cable shields is limited at lower frequencies,* significant coupling of such interference into the circuits may occur. If, however, the shield of the coaxial cable is floated from the enclosure (i.e., not terminated or bonded to the enclosure), violation of the shielding integrity could result [Figure 4.29(b)].

It is concluded that use of the "star" grounding topology is limited to the lower frequencies, that is, DC up to 300 kHz, approximately. Typical applications of the "star" topology include audio circuits, analog instrumentation, DC and AC (50/60 Hz) power systems, and similar circuit applications. Nevertheless, even when used in lower frequencies, attention must be paid to the implementation of this topology, particularly where large loops are produced and high levels of low-frequency ambient magnetic fields are present. In certain cases, "star" grounding configurations may also be found

---

*Performance of cable shields is discussed in Chapter 7.

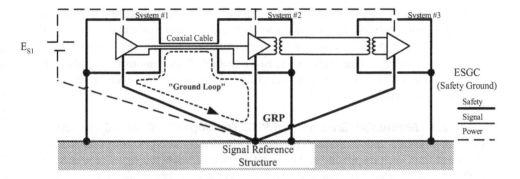

(a) Outer Conductor of Coaxial Cable Grounded to System #1 and System #2 Enclosures, Compromising the Single-Point Grounding

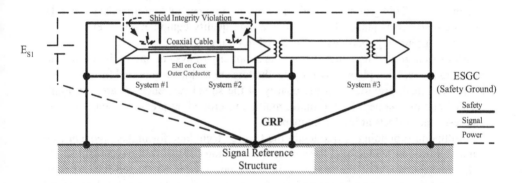

(b) Coaxial Cable's Outer Conductor Not Grounded to Enclosure, Violating the Enclosure's Shielding Integrity

Figure 4.29. Problems associated with coaxial interconnects in a "star" SPG configuration.

in higher frequency applications, where attention must be paid to proper implementation and functional use.

Lessons Learned

- three single-point grounding topologies—"daisy chain," "star," and "tree" topologies—are limited in their practical applications to lower frequency currents, primarily due to the electrical length of the grounding conductors at higher frequencies.
- Due to distributed stray capacitance along each circuit to the signal reference structure, single-point grounding essentially ceases to exist at higher signal frequencies, generally 300 kHz and above.

- Transmission-line resonant effects along the long return conductors used in single-point grounding topologies could result in high impedance to the ground reference, counteracting a desired low-impedance ground-reference connection.
- Because of the aforementioned reasons, use of all variants of the single-point grounding schemes is not practical at high frequencies, where length of the return conductor exceeds 0.05 λ.

4.2.2.3 Multipoint Grounding (MPG) Topology.

As discussed with respect to floating systems and single-point grounding, distributed stray (parasitic) capacitance is present from each circuit and along the return conductors to the signal reference. At high frequencies where the length of the return conductor exceeds 0.05 λ, single-point grounding essentially ceases to function in an optimal manner.

This reason subsequently leads to the logical postulate that "if you can't fight them, join them." In other words, if circuits were grounded at multiple points, it would be best to implement those connections in a controlled manner to a massive, solid conducting plane. This grounding method forms a homogenous low-impedance path between circuits. This plane also serves as a high-quality equipotential reference. This multipoint grounding topology is illustrated in Figure 4.30.

Multipoint grounding benefits from the simplicity of system or circuit construction owing to the fact that physically long conductors are no longer required for connection to the reference, provided that the quality of the signal reference is carefully implemented and maintained. Furthermore, multipoint grounding precludes standing-wave and resonance effects at higher frequencies.

Multipoint grounding is frequently the favored technique for higher frequency circuits, (e.g., digital, video, and RF circuits), permitting easy use of coaxial cables. This is because the shield of the coaxial cable does not have to be floated relative to the equipment cabinet or enclosure and can be grounded without violating the system's chosen grounding methodology.

In using multipoint grounding, it is presumed that the signal reference serving as the signal-current-return path exhibits extremely low impedance between any two points. As such, it approximates an equipotential structure at any frequency of concern. If this

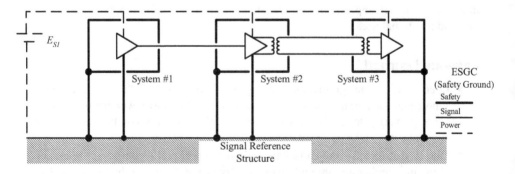

Figure 4.30. Multipoint grounding topology.

were not the case, common impedance coupling could occur between the circuits owing to the fact that the return currents from all circuits are flowing through the same path, introducing common impedance and offering no significant advantage for this topology over the "daisy chain" configuration depicted of Figure 4.22, as illustrated in Figure 4.32.

The belief that the signal reference exhibits a close-to-perfect equipotential behavior often leads to inferior designs when due consideration is not paid to practical limitations. Current from other sources may be sharing the same return path through the signal reference, whether intentionally or unintentionally. Observe, for instance the situation from Figure 4.13, illustrated in Figure 4.31 with modification. In this circuit, a PWM motor is located on the same printed circuit board used for the 3.3 V-powered logic circuitry, and its return current (e.g., 12 A) shares the same path through the ground reference back to the motor driver, developing potentially large high-frequency noise across the surface impedance. This interference voltage will couple into the digital circuits and may disrupt or degrade their performance. This situation constitutes a very poor system design!

Since a large conductive mass can easily be employed for the signal reference, it is often assumed that extremely low impedance between any ground-connection points will guarantee little voltage potential difference between the systems, thus ensuring proper performance. However, reality often dictates otherwise, mainly due to misinterpretation of the multipoint grounding concept and objectives. Returning to Figure 4.32, the source of this misapprehension is clearly revealed. When considering the ground impedance, merely taking into account the impedance of the ground structure itself (represented as Z_{G1}, Z_{G2}, and Z_{G3} in Figure 4.32), the impedances from the systems to the connection points on the ground reference structure (Z_1, Z_2, and Z_3 in Figure 4.32) are neglected. Actually, low impedance should exist between the reference points in the systems (marked A, B, and C in Figure 4.32).

Failing to consider the length of grounding leads from systems to grounding points on the signal reference will lead to increased (inductive) impedance between those points, contradicting the objectives of multipoint grounding. Merely connecting subsystems to different points on the signal reference, even if it exhibits absolutely zero impedance, does not constitute effective multipoint grounding, unless the spirit of this topology can be observed, that of maintaining low impedance between all reference points throughout the system at the frequencies of concern.

Care must be taken when implementing multipoint grounding, since numerous ground loops* are inherently created in this topology between each ground connection physically distant from other ground connections. These ground loops are prone to magnetic field coupling of RF energy, or may create radiated EMI, compromising performance of the system.

There is both truth and fiction in this view of multipoint grounding topology. The truth is that all evils ascribed to this design are quite possible and could happen when objectionable design measures are implemented, and also when all return connections are left undefined by the designer. The fiction lies in the fact that multipoint grounding

*Ground loops are discussed in Section 4.5.

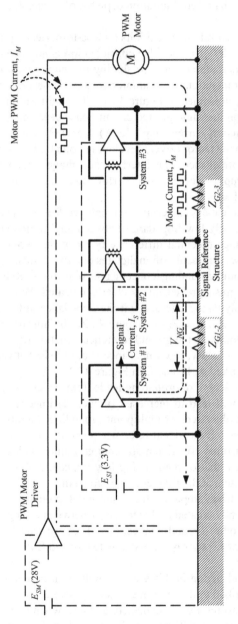

Figure 4.31. Poor design can produce interference coupling in multipoint ground configurations.

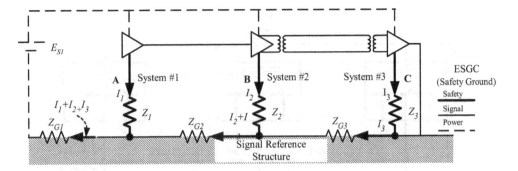

Figure 4.32. Common-impedance coupling is inherently generated in multipoint grounding topology.

is not optimal and must be avoided at any cost in favor of one of the previously described grounding topologies.

Where does the difference lie? It is simply a matter of how multipoint grounding is implemented,. Making low-impedance connections to a properly designed signal reference is what produces all the benefits associated with multipoint grounding. To prevent adverse effects resulting from the naturally resultant ground loops, the layout of the system and particularly its respective interconnects becomes extremely important. Keep in mind that high-frequency return currents will precisely follow the path of the signal conductors that exhibit the least impedance at high frequencies.* Consequently, multipoint grounding finds application in circuits operating at or above frequencies at which length of the return conductors would otherwise exceed 0.05 λ, such as RF circuits, high-speed digital circuits and systems, printed circuit boards (PCBs), and data networks, as well as for signal reference and cable shields.

In certain high-frequency applications, such as PCBs, multipoint grounding is implemented between the return planes in a PCB and the return plane in a backplane or motherboard to the equipment chassis. The implementation of such a configuration is schematically depicted in Figure 4.33.†

Lessons Learned

- Multipoint grounding is a preferred grounding topology for signal circuits and systems operating at higher frequencies (where length of the return conductors exceeds 0.05 λ), overcoming the shortcomings of floating systems and single-point grounding topologies.

- Multipoint grounding is the only way to avoid standing-wave and resonance effects in ground conductors at higher frequencies.

*Discussion of high-frequency return current propagation and the "path of least impedance" principle is found in Chapter 2.
†Grounding design and implementation in PCBs are discussed in Chapter 9.

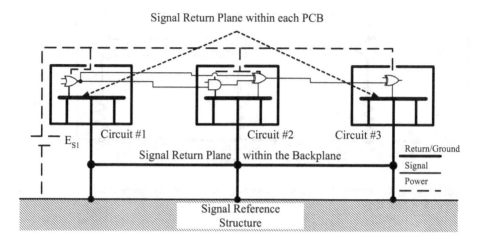

Figure 4.33. Multipoint grounding topology in a backplane.

- Effectiveness of multipoint grounding requires that the spirit of this topology be observed, namely, maintaining a low-impedance connection between all reference points throughout the system.
- Care must be taken to ensure that power frequency and other higher amplitude lower frequency return currents flowing through the common-signal reference structure do not conductively couple into sensitive signal circuits sharing the same plane, for preclusion of intolerable interference to susceptible circuits.

4.2.2.4 Composite (Hybrid) Grounding. Typically, single-point grounding is used in lower frequency analog systems in which low-level signal amplitudes are involved. In such systems, voltage drops in the signal reference on the order of millivolts and even microvolts can have significant effects on the system performance due to common-impedance coupling interactions.

Single-point grounding may also be required to be utilized for high-level "noisy" subsystems such as motor drives and solenoids, where the single-point ground is intended to preclude return currents of those systems from flowing through the signal-reference structure, producing high-voltage drops across the reference common to other sensitive circuits.

Digital circuits, on the other hand, exhibit an intrinsic immunity to externally introduced interference at reasonable levels. On the other hand, they may adversely respond to noise generated within the circuit.* In that respect, digital devices, particularly low-level, high-speed devices, are self-defeating, due both to internal noise and common-impedance coupling within the digital devices themselves. Multipoint grounding is best suited for this situation. They make use of large and "massive" ground planes, thereby effectively reducing the impedance of the return path.

*Chapter 9 discusses grounding considerations in printed circuits boards.

Mathematical modeling and predictions of EMI problems arising from grounding are exceptionally difficult to undertake for any but very simple and generic situations, or for single-point grounding configurations. Stray reactance, capacitance, and inductance radically affect high-frequency performance of the grounding system. Adding uncertainty associated with their value, which cannot, in practice, be appropriately estimated, interactions within the grounding system become exceedingly complex.

With the absence of mathematical tools to analyze this situation, a general selection criterion for determining whether single-point or multipoint grounding should be considered is, therefore, suggested. In most cases, the choice of the appropriate grounding topology was shown to abide by the following ground rules [3, 6, 7]:

$$\ell < 0.05 \ \lambda \ \text{or} \ 15 \, / \, f, \quad \text{Single-point grounding}$$
$$\ell \geq 0.05 \ \lambda \ \text{or} \ 15 \, / \, f, \quad \text{Multipoint grounding} \tag{4.30}$$

where:
ℓ = distance from the unit ground to the ground reference point (m)
f = highest operational frequency in the system (MHz)
λ = wavelength associated with the frequency f (m)

Practical systems may consist of both low-frequency (e.g., audio equipment) and high-frequency applications (e.g., RF and high-speed digital circuits). The best approach for grounding of both such systems is the application of a composite (hybrid) topology, offering the benefit of both single-point and multipoint grounding. Higher frequency components will benefit from multipoint grounding. From the standpoint of low-frequency components, they possess only a single common reference, whereas from the perspective of the higher frequency components, a system-wide multipoint grounding is implemented.

As observed in Figure 4.34, Systems #1 and #2 utilize single-point grounding, whereas Systems #3 and #4 are multipoint grounded. However, the configuration depicted in Figure 4.34, which at first glance seems to be an ideal solution for such situations, may be deceptive. Implementation of composite grounding topologies require due consideration to be paid to such items as:

- Power supply distribution
- Relative routing of signal and return path
- Type of signaling between the circuits (analog or digital)

Attention must be paid to ascertain that grounding performance is not compromised when such arrangements are applied.*

Figure 4.35(a) presents a circuit similar to that of Figure 4.34; however, the same power source is shared by the two system groups. Moreover, due to lack of foresight and poor design, the power source was placed adjacent to the digital circuits, whereby

---

*On PCBs, this dilemma is realized in mixed analog and digital circuits, comprising ADCs and DACs. This is discussed in Chapter 9.

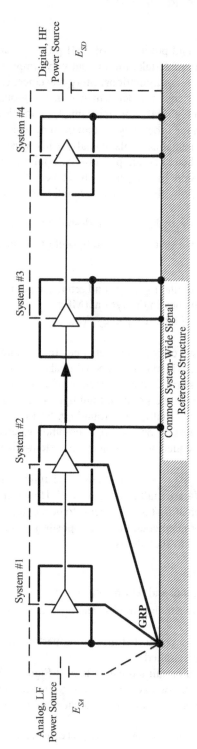

Figure 4.34. Composite (hybrid) grounding topology.

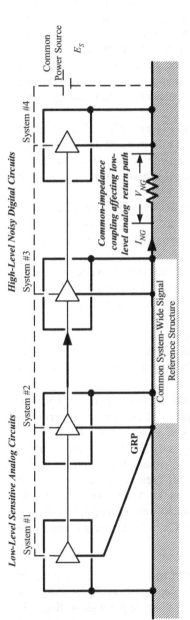

(a) Objectionable Implementation: Common-Impedance Coupling

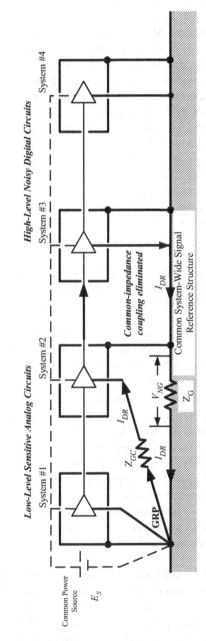

(b) Better Implementation: Common-Impedance Coupling Eliminated

Figure 4.35. Compromise of grounding performance due to system arrangement with composite grounding topology; common power supply and return path.

the sensitive, low-level analog circuits experienced the ground noise generated by the digital signal-return current in the signal-reference structure.

If both power and return conductors exhibit high impedance, common-impedance coupling will be developed across the signal reference and along the high-level digital return path. At first glance, this would not seem to constitute a problem since a voltage drop V_{NG} occurs in the digital current-return path only (high-speed signals return in the path of least inductance); however, a second glance reveals that the same path is shared by the low-level analog circuits. This situation, in fact, is equivalent to that depicted in Figure 4.21 and Figure 4.22 for the "daisy chain" SPG configuration, along with its disadvantages.

A solution to the undesired situation is provided in Figure 4.35(b). The circuit, essentially, did not change; however, the return conductor of the common power supply is now connected at a point common to the low-power analog return. In this case, little or no voltage drop occurs across the analog return current path but, rather, along the digital return path only.

In an attempt to eliminate the interference coupling mechanism due to the common-current-return path, an alternative approach is proposed, whereby dedicated power supplies for each of the circuits, analog and digital, are provided. This is illustrated in Figure 4.36.

Normally this configuration should function properly since the two circuits utilize a common low-impedance signal reference. However, other concerns may result, not necessarily related to EMC but, rather, to functionality concerns, including:

1. *Voltage difference between separate power supplies.* For the circuits to function properly, the voltage difference should be controlled. For instance, interfacing between 5 V and 3.3 V circuits may require buffering for level adjustment between circuits.

2. *Start-up order of the power sources:* Often, different power sources are generated from a multioutput switch-mode power supply or from separate independent power supplies. The order in which the power sources are turned on may be critical, since damage or improper functionality may result if different power sources do not start up in a predetermined order, either because of circuit start-up issues, or simply due to voltage differences present between two power sources. For instance, when interfacing 5 V and 3.3 V circuits, it is often required that the difference between the power sources not exceed 1.9 V. This requires that the order of and delay between the 5 V and 3.3 V power sources upon start-up be well controlled.

4.2.2.5 Distributed Single-Point Grounding. Distributed single-point grounding architecture constitutes a special case of composite/hybrid grounding, and could be considered as a combination of multipoint and single-point grounding. Distributed single-point grounding consists of multiple, isolated system grounding points common to an isolated set of equipment (i.e., single-point ground) referenced to a large common-conductive structure, such as "equipotential islands" on a nonconductive vehicle structure (Figure 4.37).

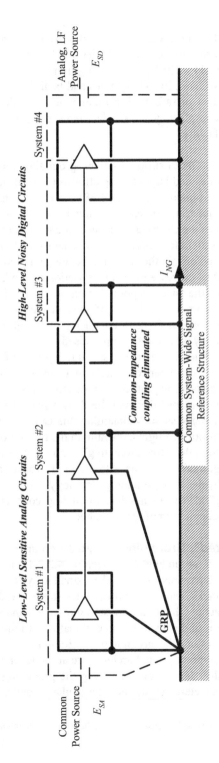

Figure 4.36. Composite grounding with power source separation.

197

Distributed single-point grounding finds it's most common application in primary power distribution networks of large vehicles and facilities in which different primary power sources (e.g., 115/220 V_{AC}, 24/28 V_{DC}) are used to feed separate clusters of equipment and subsystems. In such systems and facilities, performance of certain systems (such as mission- or safety-critical equipment) should not be dependent on the presence or absence of other systems, subsystems, equipment, or assemblies comprising the vehicles or facility.

4.2.2.6 "Tree" Grounding. Another grounding topology, more commonly found in large facilities and large-scale platforms (such as nonmetallic ships) is commonly called "tree grounding" (Figure 4.38),* owing to its tree-like representation, with a main trunk, limbs, and branches, eventually spreading out to the foliage, which, in practice, constitutes the electrical and electronic load equipment [1, 5].

One of the most salient factors requiring due consideration in the application of the "tree" grounding topology is the need for heavy-gauge conductors for the "trunk," with a possible gradual decrease in the cross section of conductors branching out from it. This is founded on the misconception that resistance, not impedance, of the conductors is of primary importance. However, if one considers only resistance while failing to consider the inductive reactance of conductors, the high-frequency performance of this grounding topology is severely compromised.

Generally, there should be a strongly stated requirement for total electrical galvanic isolation of the "tree" from other colocated conductors in the installation or facility (including piping, conduits, or any other metallic structures). Isolation precludes the development of "ground loops" that may be detrimental to the low-frequency performance of the grounding system. Maintaining this isolation, avoiding even accidental metallic contact, constitutes a tight constraint in the realization of such topologies.

It follows, therefore, that "tree grounding" is no more than a more complex variant of the single-point grounding scheme. Almost all drawbacks of single-point grounding are also possessed by the "tree grounding" topology.

"Tree grounding" also lacks a safety grounding connection that may violate the overall grounding scheme if improperly treated.

4.2.2.7 "Nested" Grounding. "Nested" grounding constitutes a blend of composite or hybrid grounding (see Section 4.2.2.4) and "tree grounding" (see Section 4.2.2.6) architectures as applied to smaller systems such as complex electronic assemblies, racks, cabinets, and consoles (these are discussed in particular in Section 4.8). In practice, this is the most common scheme in real-life electronic assemblies and systems and constitutes a combination of multiple and diverse grounding schemes. Very often, the grounding schemes of electronic assemblies are designed to form single-point ground architecture at the assembly level. When zooming into the subassemblies and circuits, however, the grounding scheme applied in those must match the characteristics of their particular circuits. This is similar to "tree

*Do not confuse "tree grounding" discussed here and the term "grounding tree," to be discussed in Section 4.3.

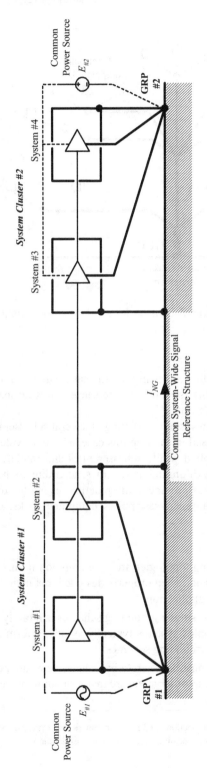

Figure 4.37. Distributed single-point grounding separation.

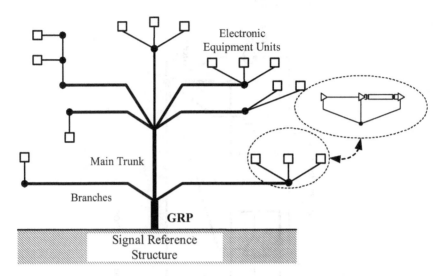

Figure 4.38. "Tree grounding" scheme for electronic equipment in large facilities and platforms.

grounding" in the sense that, normally, it still has a single common grounding point at the system level but differs from it in the sense that it normally does not utilize a large and massive "main trunk".*

Figure 4.39 illustrates a system consisting of several subassemblies, where different grounding schemes are used in each. In this case, a system-wide, single-point ("star") grounding scheme is applied, while subassemblies and modules "nested" within the system comprise a variety of grounding schemes, according to their respective applications. Sections 4.7 and 4.8 provide some practical examples of "nested" grounding schemes, as applied to equipment assemblies, equipment racks, and cabinets.

Lessons Learned

- Composite grounding topologies are very common in mixed analog–digital circuits and systems, and, when properly designed, maintain the necessary grounding design for each application.
- In complex systems, simple grounding schemes can rarely be used; common architectures for complex systems typically assume a form of "distributed SPG," "tree," or "nested" grounding schemes.
- Due consideration must be given to how the systems are powered on, grounded, and interconnected, in order to ensure that grounding objectives are not compromised or violated.

*The example to be presented in Section 4.3.2 results in practice in a "nested" grounding scheme when the grounding schemes in each of the assemblies are taken into consideration.

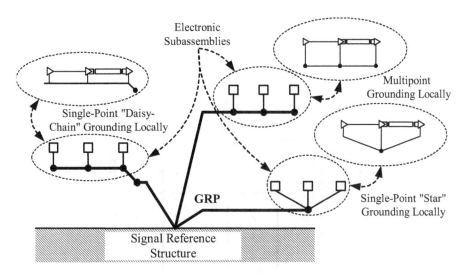

Figure 4.39. "Nested grounding" scheme for electronic equipment in complex assemblies.

- Maintaining a low-impedance signal reference and proper routing of signal conductors between circuits can overcome any problems that would otherwise result from this configuration.
- Using separate power sources in an attempt to overcome common-impedance coupling between circuits can further reduce the coupling; however, attention should be paid to the voltage build-up sequence and voltage difference between the power sources for preclusion of potential functionality concerns.

4.2.2.8 Frequency-Selective Grounding. In addition to the above fundamental grounding topologies, namely, single-point and multipoint grounding, situations occur in which a hybrid grounding topology may be utilized, constituting a combination of single-point and multipoint grounding simultaneously, depending on the functionality of the circuit and frequencies of concern.

Hybrid grounding is used in those special circumstances where:

1. Single-point grounding may be required at lower frequencies and multipoint grounding at higher frequencies
2. Multipoint grounding may be required at lower frequencies and single-point grounding at higher frequencies

Observe, for instance, Figure 4.40, which is a variation of Figure 4.31 depicting two systems that have a low-level, audio-frequency, single-ended I/O interface. In this circuit, both the source (System #1) and the load (System #2) appear to be internally

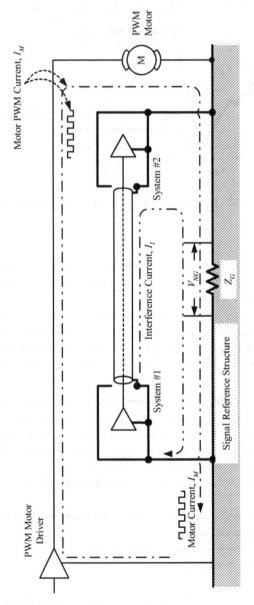

Figure 4.40. Common-impedance coupling in a low-frequency system using multipoint grounding.

202

grounded to their respective enclosures, which, in turn, are both bonded to the vehicle structure. The shield of the interconnecting cable (also intended to act as the return path for the single-ended interface) is grounding to the chassis of each of the enclosures.

Although this system must rely on multipoint grounding for its functionality, conflicting concerns due to common-impedance coupling may introduce unacceptable interference.

Due to the lower impedance of the signal-reference structure, most of the low-frequency return current will probably follow this path rather than through the intended return path, the shield, thus contributing to the flux cancellation and the higher noise immunity of the system. In fact, a system if better designed, for instance, by using a differential rather than a single-ended interface, would probably utilize a single-point signal grounding topology. The situation may be worse if a noise-source (e.g., a PWM motor) current shares the same return-current path through the signal-reference structure.

One possible solution is to float one of the enclosures* (Figure 4.41). In this case, a single-point grounding configuration has actually been reestablished. This will do for low-frequency environments (say, up to 300 kHz or so), but in a high-frequency radiated environment, RF currents will be induced on the cable shield. Thus, multipoint grounding may be desired for RF interference control purposes.

Figure 4.42 offers an excellent solution: Capacitively grounding the circuit will offer a controlled high-frequency multipoint ground connection while maintaining low-frequency single-point grounding.

Typical values of this high-frequency capacitance range from 1 nF to 10 nF depending on the frequency of concern. This capacitance may be implemented using discrete capacitors (in PCB-related applications),[†] or, for enclosure grounding, by implementing a capacitance by placing a dielectric compound between the enclosure and the signal reference (Figure 4.43).

Another approach to a solution is shown in Figure 4.44. In this case, the enclosure of both circuits is grounded. Low-frequency single-point grounding is realized by capacitively grounding the cable rather than the enclosure.[‡]

In summary, capacitive selective grounding eliminates low-frequency EMI ground loops while maintaining high-frequency grounding performance.

An opposite situation may occur when several systems are interconnected via a single-ended slow digital interface (e.g., RS-232); however, each of the systems must be grounded at one location (hence implying a multipoint grounding topology) at low frequencies for meeting electrical safety codes. However, the ESGC (the "green wire") often carries high-frequency RF current, which may penetrate the equipment through the single-ended RS-232 signal link (Figure 4.45).

---

*Note that when safety codes mandate grounding of the enclosure for electrical safety purposes, safety requirements must take precedence and should be followed. This comment applies to any of the examples discussed in this section associated with "floating" circuits and equipment at lower frequencies.
†Selective grounding on PCBs is discussed in Chapter 9.
‡Grounding of cable shields is discussed in Chapter 7.

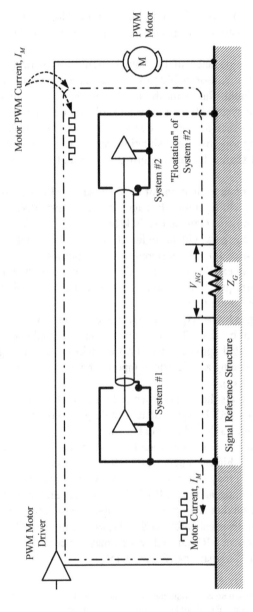

Figure 4.41. Floatation of system #2 for elimination of common-impedance coupling.

204

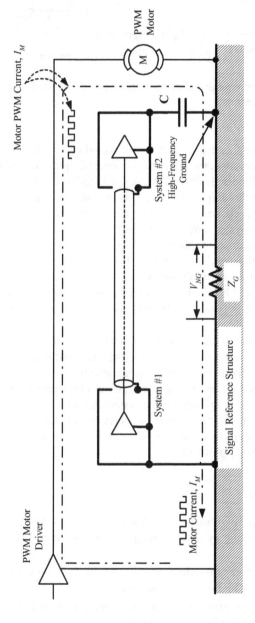

Figure 4.42. Capacitive selective grounding exhibits low-frequency single-point grounding and high-frequency multipoint grounding.

205

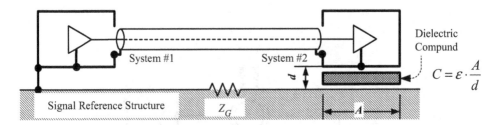

Figure 4.43. Dielectric spacer used to implement distributed capacitive selective grounding for an enclosure.

In this case, "floating" is impractical, as this would compromise the electrical safety of the system. Hence, adequate grounding at lower frequencies must be maintained. High impedance to RF noise is desired, however, in order to "block" interference current from propagating through the system. This high impedance is achieved by inserting an "AC block" RF choke (inductor) in series with the ESGC (Figure 4.46). At higher frequencies, the system now exhibits single-point grounding, while at lower frequencies, it exhibits multipoint grounding. High-frequency noise propagation is precluded while electrical safety is maintained. Typical values of inductance appropriate for this application range up to 100 nH, which at 50 Hz power frequency has an impedance of 30 mΩ, which is less than the maximum allowable resistance to ground for safety purposes, which is typically 100 mΩ; hence, electrical safety is not compromised.

In summary, inductive selective grounding isolates the signal return from the electrical safety ground conductor, thus eliminating high-frequency EMI ground loops while maintaining low-frequency safety grounding at the power frequency.

Incorporating capacitors or inductors in a ground topology allows us to steer RF currents in a manner that is optimal for the system design. One can take control of the system's performance at low and high frequencies by defining the path that RF return currents will take. Failure to locate the RF current-return path may result in either emissions or susceptibility problems.

Lessons Learned

- Frequency-selective grounding combines the characteristics of single-point and multipoint grounding by incorporating reactive elements (capacitors and inductors).
- Capacitive-selective grounding eliminates low-frequency EMI ground loops while maintaining high-frequency grounding.
- Inductive-selective grounding eliminates high-frequency EMI ground loops while maintaining low-frequency grounding.
- Hybrid grounding finds applications in circuit and cable grounding.

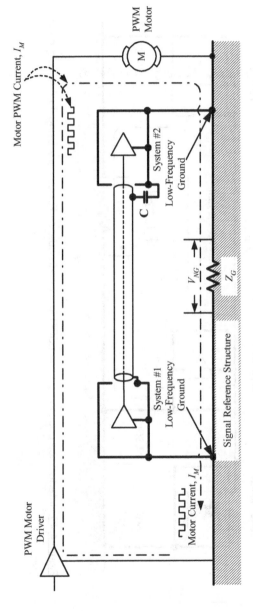

Figure 4.44. Capacitive selective grounding implemented in the grounding of a cable shield.

207

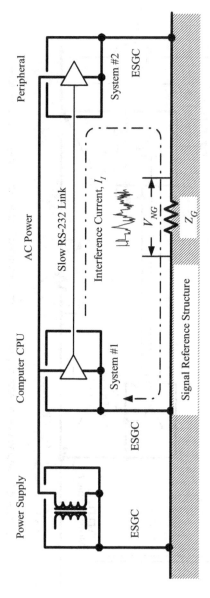

Figure 4.45 Interference coupling via the ESGC in a low-frequency RS-232 link using multipoint grounding.

208

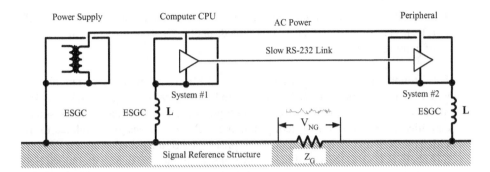

Figure 4.46. Inductive selective grounding provides high-frequency single-point grounding and low-frequency multipoint grounding.

4.3 GROUNDING TREES

4.3.1 Objectives and Basic Design Considerations

The discussion and examples presented in Section 4.2 demonstrate that a proper grounding topology is required for ensuring a design that does not compromise functionality, safety, or EMI concerns.

A complex electronic system may consist of a large number of circuit types, each exhibiting different operational characteristics. Separate power and signal-return conductors are, therefore, assigned to each subsystem identifiable as a distinct load according to its noise and signal characteristics. Examples of such dissimilar load categories include:

- DC and primary AC power returns
- Low-level analog signal returns
- High-level signal returns (e.g., relay drives)
- High-speed logic returns
- High-frequency RF returns

There are special cases in which additional dedicated return conductors may be required. Examples include:

- *Safety considerations of firing circuits of electrically initiated explosive devices (EIEDs),* such as squibs and pyrotechnic devices This dictates that special isolation measures be engaged for preclusion of chassis currents or pyro ground-fault currents during firing events, which could potentially cause magnetic field noise coupling into nearby sensitive circuits and, occasionally, even more severe outcomes [8].

- *Preserving red/black isolation ("TEMPEST")* between circuits processing "red" (clear, classified) and "black" (encrypted, unclassified) information, for control of compromising emanations [9].

The grounding complexity* of composite electronic systems is commonly acknowledged as a "grounding tree," demonstrated symbolically in Figure 4.47. This figure portrays examples of a number of circuit types, the characteristics of which should be carefully addressed in the system grounding architectural design, as related to the specific approach that should be applied for prevention of undesired coupling.

It is likely that some (or all) of these subsystems will have to share a common reference point for ensuring system functionality. However, in order to avoid circulating currents between circuits, it is often required that no more than one such point exist in the system.

The fixation that single-point grounding must religiously be applied often leads to inappropriate grounding schemes, particularly when high-frequency circuits are involved. However, in composite systems that consist of both low-frequency and high-frequency circuits, some variant of system-level, single-point grounding, at least as far as the low-frequency circuits are concerned, should be considered.[†]

Consequently, the "grounding tree" constitutes a simple (yet effective) method of ground architecture design by categories of circuits and signals. The "grounding tree" should depict different circuits and assemblies that constitute the systems and their respective grounding interconnections, particularly their common reference connection.

Inasmuch as the principles portrayed here are commonly addressed at the facility, platform, or system level, relatively complex assemblies could also be considered as a system. The grounding architecture considerations discussed here could be applied to them just as well.

4.3.2   Ground Tree Design Methodology

This section presents an example of a design procedure for a complex electronic assembly consisting of diverse circuits (e.g., analog, digital high-power, and sensitive RF circuits). The primary objective of this example is to present, in a step-by-step manner, particular considerations, conflicts, and compromises that exist, and the manner in which they can be addressed.

In practice, the design of the "grounding tree" is carried out in two tiers, namely:

1. Intercircuit "grounding tree," in which the grounding of an integrated system as a whole is carried out, considering all circuits and subassemblies as "black boxes."
2. Intracircuit "grounding tree," in which attention is paid to the subassemblies' or circuit's internal grounding topology. The following design example

---

*Further enhancing complexity of the grounding architecture is the need, particularly in facilities, to address not only the signal-return conductors, but also lightning protection system grounding, power earth connections, and so on. These aspects are discussed in Chapter 6.

[†]See Section 4.2.2.4 for "composite grounding" schemes.

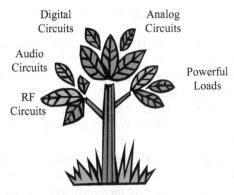

System Signal Reference Structure

Figure 4.47. Symbolic representation of the "grounding tree" concept.

clearly demonstrates the two tiers of a grounding architecture design of a complex system.

4.3.2.1 Step 1: Identify System Architecture. Examine the system portrayed in Figure 4.48. The first constraint in the grounding system is the physical layout. We begin by identifying the subassemblies (e.g., printed circuit boards).

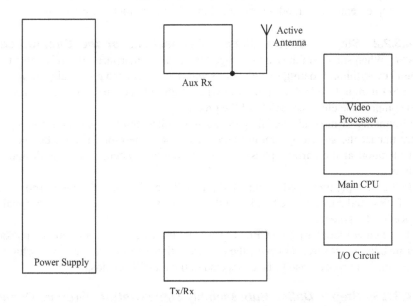

Figure 4.48. Step 1: Identify system architecture.

In this example, the system consists of six subassemblies, namely:

1. A switch-mode power supply (SMPS) accomplishes the power conditioning tasks and converts input AC power to multiple DC power outputs. The SMPS is implemented as a module packaged in a metallic enclosure.
2. A main central processing unit (CPU) circuit serves as the system's controller. This item consists primarily of digital circuits, but could also contain analog circuitry, principally audio. The CPU circuit is implemented as a separate printed circuit board.
3. The input/output (I/O) circuit serves as the external interface between the system and outside world. This circuit also serves as an interface to the transceiver module (Tx/Rx). The I/O circuit is implemented as a separate printed circuit board.
4. A video processor, processes video information and consists of high-speed digital circuits and analog video circuits (both low- and high-frequency analog). The video processor is implemented as a separate printed circuit board.
5. A radio transceiver (Tx/Rx) module provides high-power RF transmission and reception. Transmissions are pulsed, resulting in high-current pulses during transmission. The Tx/Rx is implemented as a module packaged in a separate metallic enclosure.
6. An auxiliary receiver (Aux Rx) module provides the capability of receiving a channel other than that to which the Tx/Rx was tuned. This is an extremely sensitive receiver utilizing an active antenna (i.e., an antenna containing an embedded low-noise amplifier, or LNA). DC power for operation of the LNA is superimposed on the RF signal fed into the coaxial transmission line. The Aux Rx is implemented as a module packaged in a dedicated metallic enclosure.

4.3.2.2 Step 2: Define Chassis Connections at the Circuit/Module Level.
When subassemblies are packaged in metallic enclosures, it is advisable to DC ground everything. Floating metallic parts within a system can potentially compromise the ESD immunity of the system, and when high voltages are concerned these may also compromise electrical safety of the product.

Grounding of internal metallic parts such as subassembly enclosures could potentially impact the system grounding architecture and, therefore, should be taken into consideration at the earliest phases of the grounding system design as design constraints.

In the example presented (Figure 4.49), metallic enclosures of modules (power supply, Tx/Rx, and Aux Rx) are bonded to the chassis, which also serves as the signal reference for the system.

Three other modules (main CPU, I/O circuit, and video processor) are not packaged in a metallic enclosure; however, their construction includes a heat sink, which, for heat-transfer purposes, had to be connected to the metallic enclosure.

4.3.2.3 Step 3: Define Subassembly Signal-Return (Ground) Requirements.
Based on the circuit/subassembly design, return conductors are allocated to

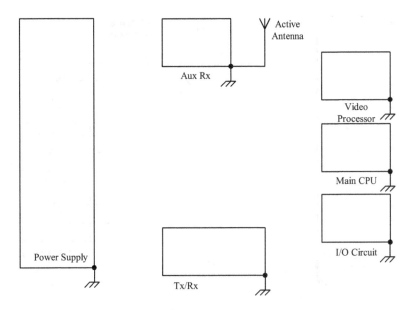

Figure 4.49. Step 2: Define chassis connections at the circuit/module level.

each circuit (Figure 4.51). The notation appearing in Figure 4.51 should be interpreted as in Figure 4.50.

Using this notation, for instance, the return conductor 5VD is interpreted as the return conductor for the digital circuit, whereas 15VA/RF can likewise be interpreted as the return of the analog RF circuit powered from a 15 V power supply. All return/reference conductors required for the operation of the different circuits and subassemblies are added into the diagram.

Note: As detailed in Chapter 9, with respect to grounding design for printed circuit boards, often a common ground plane serves as a common return path for both analog and digital circuits. In that case, a common return path would be drawn rather than separate analog and digital returns.

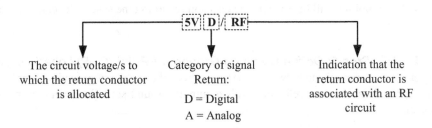

Figure 4.50. Notation for identification of return categories.

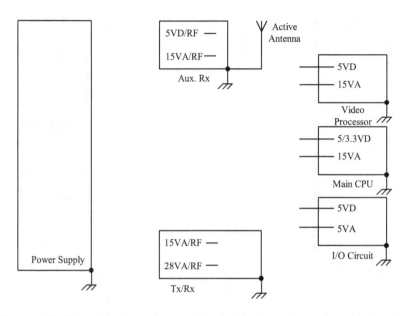

Figure 4.51. Step 3: Define subassembly signal returns (ground) requirements.

4.3.2.4 Step 4: Identify Chassis Isolation/Connection Requirements in Subassemblies.

In Figure 4.51, all return conductors are depicted without indicating their connection point to the chassis or signal-reference structure. However, it is well known that as well as high-speed digital circuits are commonly multipoint grounded at the circuit level to the chassis (Section 4.2.2.3).* The chassis connection, implemented locally at the subassembly level using the multipoint grounding topology (MPG), is shown in Figure 4.52. The analog return conductors are not connected to the chassis at the PCB level. This is also applied to the video return, which contains both low- and high-frequency signals. For these types of circuits, frequency-selective grounding by means of small capacitors (1 to 10 nF) could be implemented, providing high-frequency multipoint grounding and low-frequency single-point grounding (Section 4.2.2.8). For this example, this was not required.

As discussed in the following paragraphs, due consideration of the subassembly grounding topology will be given in the decision regarding the system-level grounding architecture.

4.3.2.5 Step 5: Define Common Grounding-Point (CGP) Location.

For signals that communicate between separate parts of the system, a grounding topology must provide a common reference with minimum ground shift (low common-mode noise).

---

*Note that in the figures and illustrations in this section, only a single chassis connection is shown for the purpose of simplicity.

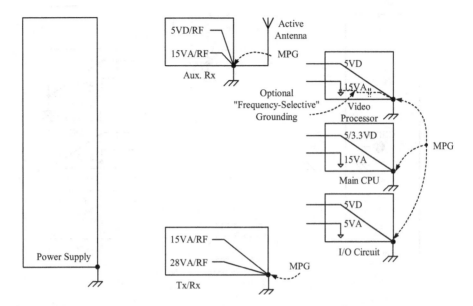

Figure 4.52. Step 4: Identify chassis isolation/connection requirements in subassemblies.

Once the manner of grounding connection is defined for each of the return conductors, a system-level common grounding point is defined. In Figure 4.53, the notation of different options for allocating the common grounding point (CGP), which serves as the system-level single point ground, is shown.

Note: There may be cases in which multipoint grounding is a better choice throughout the system, particularly if the system comprises many high-frequency/high-speed circuits. However, when mixed systems are concerned, a mixed ground topology is often applied, mandating that a CGP be defined and used. There may be cases in which other system-level topologies may be used, such as the "daisy chain" SPG architecture.

The CGP provides the common connection between all return conductors of the circuit or system.

The location of the CGP is determined based on several decisive factors, for instance:

1. Length of conductor
2. System layout
3. Sensitivity of system components
4. Functional safety grounding requirements

Three principal positions are normally considered for the CGP, namely:

1. Near the power supply
2. Within the power supply
3. Near the most sensitive subassembly

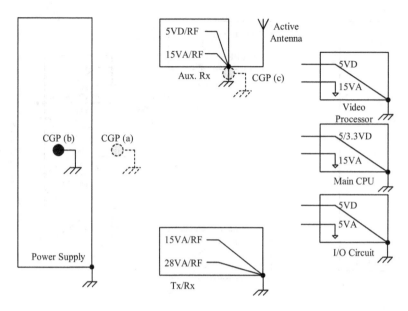

<u>Figure 4.53.</u> Step 5: define system-level common grounding point (CGP). Option (a): Near the power supply. Option (b): Within the power supply. Option (c): Near the most sensitive subassembly.

The following is a discussion of the above options, depicted in Figure 4.53.

(a) Location of CGP Near the Power Supply. The most common location for the CGP is as close as possible to or even within (see below) the power supply. Normally, all power leads branch out from the power supply and the return conductors would normally be connected to the respective return terminals. Therefore, locating the CGP as close as possible to the power supply actually guarantees short conductors, which are of great importance when considering the application of single-point grounding.

(b) Within the Power Supply. If confidence is gained regarding the location of the CGP in the vicinity of the power supply and, particularly, if the power supply is a multioutput assembly packaged in a metallic enclosure, the location of the CGP within the power supply is an even better choice, as the enclosure of the power supply can be easily bonded to the metallic chassis, if present. This, again, will ensure minimum grounding impedance for low- and high-frequency circuits.

Common to alternatives (a) and (b), and of particular importance when the equipment is not packaged in a metallic chassis, is the fact that grounding of the system can be attained only via the electrical safety ground conductor (ESCG, or "green wire"). In this case, placing the CGP within the power supply minimizes the length of the ground conductors to the system "safety ground" connection.

(c) Near the Most Sensitive Subassembly. Recall that when applying the "daisy chain" single-point grounding topology (Section 4.2.2.2.1), the actual system connection to the signal reference, serving as the CGP, is normally located as close as possible to the most sensitive subassembly owing to the fact that the lowest EMI voltages will develop on this short connection. It follows that another alternative for locating the CGP is near the most sensitive subassembly or circuit in the system. In this particular example, the receiver appears to be that most sensitive component and, hence, the CGP is placed at the point nearest to that assembly, and actually shares this chassis connection.

In Figure 4.53, the CGP is placed within the power supply, option (b), and is marked as such. From here, only reference to option (b) will be discussed.

4.3.2.6 Step 6: Determine Return Conductor Connections from the Circuits to the CGP. Once the location of the CGP is determined, all that remains is to connect the different return conductors to the CGP. As observed in Figure 4.54, the return conductors from most subassemblies are indeed connected to that point.

If the subassemblies consist of printed circuit boards and modules inserted into a motherboard or backplane, those interconnects would normally be realized in the form of wide planes.* If such a connection is not available, this architecture may suffer from unacceptably high ground-path impedance at higher frequencies, which may cause a functionality concern.

When a motherboard is not present, or if different system components actually comprise equipment assemblies themselves, such connection is often accomplished using a grounding conductor. In this case, care must be taken to ensure that the impedance of the grounding conductors be as low as reasonably achievable, using the techniques discussed in Section 4.1.2.1.

In Figure 4.54, the CGP serves as the reference point for the return terminals of the power modules, demonstrating that it truly serves as the system common reference point.

Note: Figure 4.54 depicts only the return conductors of the power system; the "hot" power leads are omitted from the drawing for clarity.

Noticeably, several return conductors are not connected to the CGP (as pointed out by the question mark "?"). Why is that? The answer lies in Step 7.

4.3.2.7 Step 7: Identify Potential Ground Loops. The common denominator between these two cases is that both are RF circuits, internally grounded within their metallic enclosure, which, in turn, is grounded to the assembly's chassis. That, by its own virtue, does not justify the preclusion of their connection to the CGP. Obviously, if they were to be connected to the CGP, that would, in practice, be accomplished using a relatively (electrically) long conductor, considering the very high frequency content used by these modules. Furthermore, by making that connection, these circuits would be grounded in two different locations in the system, forming ground loops.† As seen in Figure 4.55, the resulting ground loop could potentially bring about circulating interference current between the circuits.

---

*The exact manner of accomplishing such a connection in practice is discussed in Chapter 9.
†Ground loops are further discussed in Section 4.5.

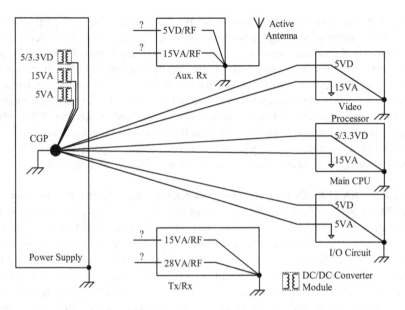

Figure 4.54. Step 6: Determine return conductors connections from the circuits to the CGP.

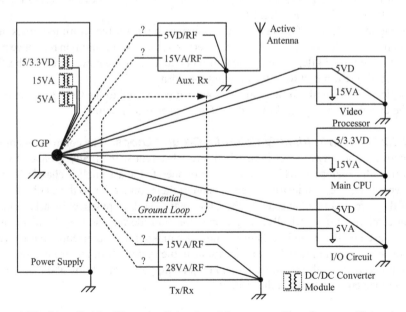

Figure 4.55. Step 7: Identify potential ground loops resulting from multiple chassis connections.

Normally, RF circuits such as the Tx/Rx and the Aux Rx, are multipoint grounded. However, in the particular case of this example, the Tx output (peak pulsed) power was +50 dBm (100 W), while the Rx as well as the Aux Receiver had a sensitivity better than –120 dBm. Isolation as high as 170 dB is desired in order to preclude unacceptable coupling between the circuits. It was feared that in this case, even though the high-frequency return current would mostly follow the power leads (path of least inductance), a small proportion of the return current may still circulate in the system and couple via the CGP to the chassis and the sensitive receivers, causing a functionality concern.

Therefore, the Tx/Rx and the Aux Rx cannot be grounded to the CGP and must be left to float at the input of those assemblies. That results in the need to allocate dedicated DC/DC power modules for this purpose (Figure 4.56). Observe that the outputs, which have dedicated terminals, are also attributed with a special designation ("/RF").

It is important that the DC/DC power modules exhibit DC isolation from input to output and also be floated from each other. That is easily accomplished when the DC/DC modules are implemented as "switch-mode power modules." Inasmuch as they do not provide absolute isolation at RF, due to parasitics, good ground loop control can be achieved by adequate filtering to support lower frequency isolation.*

Actually, another outcome of the grounding architecture design process discussed is the high-level characterization of the system power supply.

4.3.2.8 Step 8: Consider "Special Cases" Potentially Leading to Violation of the Grounding Architecture.
In some cases, system-specific isolation requirements may require particular attention. In our ongoing example, the I/O circuit is required to control the Tx/Rx module, whereas the main CPU controls the Aux Rx.

If single-ended signaling is used for this purpose (the dashed lines in Figure 4.57), resulting ground loops are unmistakably apparent. Furthermore, any system isolation requirement precludes any galvanic connection between the external circuits (i.e., circuits external to the system that interface with the I/O module) and internal circuits. This requirement would be violated by the manner in which the I/O circuit is connected to the Tx/Rx module. A similar concern lies in the connection between the Main CPU and the Aux Rx, due to the potential ground loop, particularly in light of the very high sensitivity of the Aux Rx.

4.3.2.9 Step 9: Incorporate "Isolation Measures" for Preclusion of Undesired Ground Loops.
In an attempt to overcome the situation identified in Section 4.3.2.8, measures for "opening" the ground loops are incorporated. Such measures, discussed in Section 4.5, include, for instance, isolation transformers, optical isolators/optocouplers, and balanced drivers.

In Figure 4.59 observe the Tx/Rx module and its connection to the I/O circuit while recalling that only return conductors are presented. Signaling between the I/O circuit and the Tx/Rx module is carried out through optocouplers, which facilitate the trans-

*Methods for "breaking" ground loops and their limitations are discussed in Section 4.5.5.

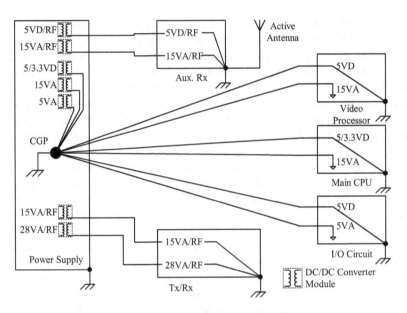

Figure 4.56. Dedicated DC/DC power modules are required for the RF circuits, for preclusion of ground loops.

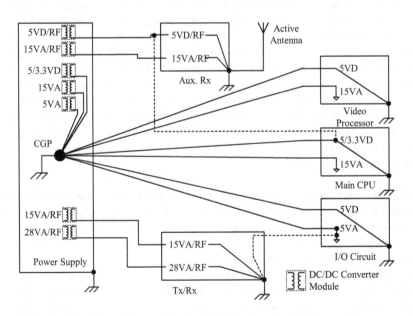

Figure 4.57. Step 8: "Special cases" potentially leading to violations of the grounding architecture.

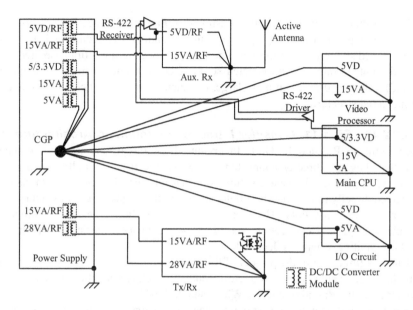

Figure 4.58. Step 9: Incorporate "isolation measures" to preclude undesired ground loops.

mission of an electrical signal in the form of an optical (light) signal, thereby eliminating the need for any common galvanic reference between the circuits.*

In the case of the link between the main CPU and the Aux Rx (Figure 4.59), a differential, balanced, and isolated high-speed RS-422 digital interface is used,† exhibiting high common-mode rejection and, thus, high immunity to ground loops.‡

Other approaches and techniques may be applied in other cases, as further elaborated in Section 4.5.5.

4.3.2.10 Step 10: Sketch the "Grounding Tree."

Once all the previous steps have been completed, sketch the system "grounding tree." This is normally performed by identifying the ground/return conductor connections beyond the CGP, located (in this case) within the power supply, which, by design, consists of isolated outputs.

Applying this to our example, Figure 4.58 can be transformed into the schematic of Figure 4.59. It is now easy to view all ground connections and identify potential ground loops or other abnormal ground connections. Even if loops were not identified in the previous steps, they will certainly be noticed in this schematic diagram, allowing timely implementation of corrective actions.

*Optocouplers are discussed in Section 4.5.5.1.3.
†In this particular case, optocouplers could not be used due to the high speed of the digital signal, which exceeded the bandwidth of the optocouplers.
‡Differential and balanced signaling is discussed in Section 4.5.5.1.5.

The resultant grounding scheme illustrated in Figure 4.59 clearly does not form a classis single-point grounding topology. This grounding architecture is better known as distributed single-point grounding, comprising a system of multiple, isolated system ground points common to an isolated set of circuits or equipment (i.e., single-point ground), referenced to a common, large, conductive structure such as the equipment equipotential structure.

4.3.2.11 Step 11: Consider Intracircuit Grounding Architecture. Now examine the Tx/Rx module. Up to this point, all circuits in this module are depicted such that they are directly connected to the chassis, using a multipoint grounding topology. In practice, however, it turns out that the digital circuits return cannot be connected directly to the chassis simultaneously with the analog circuits return, due to internal circuit constraints. Therefore, we have two alternatives for accomplishing module-level ground connection; see Figure 4.60, where a partial schematic of the Tx/Rx grounding scheme is depicted.

Due to a limitation in the pin-out of the Tx/Rx module, it is not possible to provide both digital and analog power and return conductors. Only one return conductor can be connected, either digital or analog. Which should it be?

Figure 4.60(a) depicts the first option, actually the one used up to now. In this case, the analog return conductor (AGND) is connected to the chassis of Tx/Rx directly (using the lowest impedance connection possible), whereas the digital return conductor (DGND) is also referenced to the chassis of the Tx/Rx via the AGND. The AGND–chas-

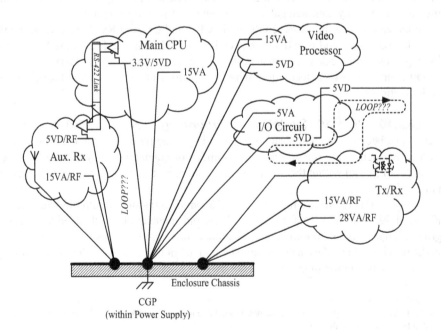

Figure 4.59. Step 10: Sketch the "grounding tree."

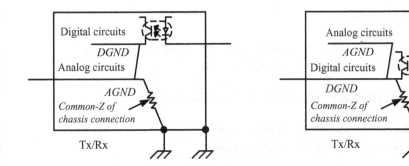

(a) Digital Return Referenced to Analog Return (acceptable)

(b) Analog Return Referenced to Digital Return (objectionable)

Figure 4.60. Step 11: The Tx/Rx grounding dilemma.

sis connection now forms a common impedance link (common-Z) between both the AGND plane and the chassis. The risk in this configuration is, of course, that any digital noise developed across the common impedance of the chassis connection will also appear across the analog circuit as "ground noise" and may disrupt the analog circuit performance. However, if the impedance (particularly, inductance) of this common conductor is kept to an absolute minimum, the outcome of such coupling will be minimized.

If, on the other hand, the connection portrayed in Figure 4.60(b) is used, the analog circuits will "ride" over the noisy digital ground plane to chassis connection. Even if the chassis connection is low impedance, a high probability exists for interference coupling due to the very noisy common impedance of the digital return.

In conclusion, it follows that in such cases in which a dilemma exists as to the interconnection between the analog and digital return conductors, the return conductor of the sensitive circuits (e.g., AGND) should always be the one directly connected to the signal reference.

Note: Observe that this constitutes a special case of "daisy chain" single-point grounding.

The interconnection between AGND and DGND within the Tx/Rx module (intracircuit architecture) should be included, accordingly, in the intercircuit grounding architecture.

4.3.2.12 Step 12: Define the Power Supply's Outputs Specification.
The definition of the power supply's outputs specification is directly derived from the grounding constraints of the system. In our example, it follows that the power supply will have to provide:

- A 3.3 V/5 V converter with a common digital return for all digital circuits
- A 5 V converter for the I/O circuit
- A ±15 V converter with a common analog return for analog circuits

However, Tx/Rx requires:

- A ±15 V converter with a common analog return for analog RF circuits
- A 28 V converter for analog RF (pulsed) circuits

Finally, Aux Rx requires two more dedicated converter outputs:

- A 5 V converter for the Aux Rx digital circuits
- A 5 V converter for the Aux Rx analog circuits

In summary, an optimal power distribution system architecture would be derived from use of a "grounding tree" and result in a seven-output switch-mode power supply.

Lessons Learned

- In any system design, care must be taken to ensure that undesired high-level signals do not conductively couple into sensitive signal circuits sharing the same reference plane, for preclusion of intolerable interference to susceptible circuits.
- Constructing the grounding architecture of a complex system must be performed in a systematic manner, considering system layout, circuit requirements, and special concerns in the system (e.g., isolation).
- The "grounding tree" should be expanded to include intracircuit interactions in addition to the intercircuit interactions depicted above. Such an extension may reveal yet other interactions not identified at this level of discussion.

4.4   ROLE OF SWITCH-MODE POWER SUPPLIES IN GROUNDING SYSTEM DESIGN

The grounding tree design example in Section 4.3.2 clearly demonstrated that isolated power supplies play an important role in the design of overall system grounding architecture. In particular, it was shown that the power supply architecture may be directly derived from the system grounding scheme. This section will now discuss in brief the unique features of switch-mode power supplies (SMPS), particularly those associated with grounding.*

Switch-mode power supplies have been used for many years in industrial and aerospace applications in which high efficiency, light weight, and small size are of prime concern. Today, SMPS are used extensively in almost all DC- or AC-powered electronic devices. A SMPS offers four main advantages over a conventional linear power supply:

---

*It is not the intention of this section to dwell on the actual topologies of switch-mode power supplies. Many excellent books have been written on this topic. The reader is encouraged to refer to the literature for a more detailed discussion of power supply design.

1. High efficiency (60% to 95%) compared to that of linear power supplies (40% to 50%), hence, less heat dissipation
2. Tighter and continuous regulation while producing a wide range of output voltages (often higher than the input voltage)
3. Smaller size and lower weight thanks to its higher frequency of operation
4. Possible input-to-output and output-to-output (in multiple-output SMPSs) isolation thanks to the incorporation of switching transformers in many power supply topologies

The main drawback of switch-mode power supplies is that of increased EMI, both common-mode and differential-mode emissions, due to the switching function fundamental to their performance.

4.4.1 Principle of Switch-Mode Power Supply Operation

Figure 4.61 shows a basic diagram of a "flyback" topology, switch-mode power supply consisting of a DC voltage source, a switching device, a step up/down transformer,* and a switching control stage.

The switching transistor and pulse transformer form the heart of all isolated switch-mode power supplies. Unregulated power is supplied to the switching transistor through the primary winding of the transformer. The switching transistor acts as a switch. When the switch is closed (the transistor is turned on, driven into saturation, and conducts) it provides a path for current to flow through the primary winding of the pulse transformer to the power return lead, producing an expanding magnetic field that couples and stores energy in the magnetizing inductance of the secondary winding of the transformer through the core. When the switching transistor turns off (the transistor is in cutoff and does not conduct), the stored energy is delivered through the rectifier into the load. Consequently, current does not flow simultaneously in the primary and secondary windings of the transformer in this topology.

Varying the momentary switching frequency (i.e., pulse-rate modulated[†] or PRM regulator) or duration (i.e., pulse-width modulated[‡] or PWM regulator), the switch remains closed per unit of time, resulting in duty cycle variation of the unregulated input voltage and a corresponding regulation of the level of the output voltage(s) (Figure 4.62).

---

*Not all switch-mode power supplies include transformers. Common topologies such as "buck" or "forward" regulators do not include a transformer and are not isolated.

[†]PRM regulators vary the rate (frequency) at which the switching transistor is turned "off" and "on," but pulse width remains unchanged. As the pulse rate increases, the "on time" decreases, again changing the duty cycle. As the rate of the "on" pulses is increased, more energy is delivered to the transformer per unit of time, increasing the output DC voltage and vice versa.

[‡]PWM regulators vary the duration in which the switching transistor is "on" (in conduction). The switching frequency is constant but the duty cycle varies. As the duration of the "on" pulse is increased, the switching transistor conducts for a longer duration and more energy is delivered to the transformer per unit of time, increasing the output DC voltage and vice versa.

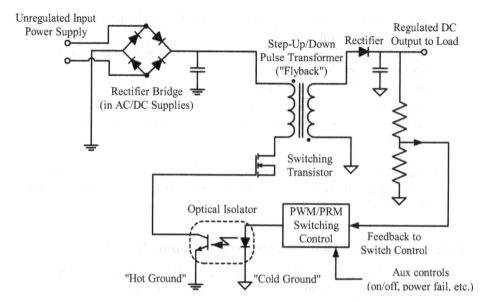

<u>Figure 4.61.</u> Basic configuration of a single output "flyback" switch-mode power supply.

Figure 4.63 illustrates one of the greatest advantages of an isolated switch-mode power supply, namely, the capability to produce multiple, isolated, and independent DC outputs by using multiple secondary windings of the switching transformer. As the transistor is switched on and off, the magnetic field alternately expands and collapses in all of the transformer secondary windings. By designing the transformer with multiple, different secondary turns' ratios, various amplitude pulses are produced at the output of each secondary winding. The output of each winding is applied to a high-speed switching diode rectifier and filter, which produces the regulated DC output voltages of the SMPS.

An isolation device (e.g., optical isolator or optocoupler) is required to maintain isolation between the "cold ground" (secondary side of the pulse transformer), and the "hot ground" (primary side), while coupling the DC feedback voltage. Signals or power delivered from one ground reference to another are often referred to as "crossing the isolation boundary." Grounding and isolation in SMPSs are now discussed.

4.4.2   The Need for Isolation

The need for isolation in a power supply is primarily an issue of safety, requiring that no direct conductive path exist between the source of input power to the power supply and its output terminals or load. The input voltage level usually defines the actual specification and can range from 3500 V for a universal line-voltage handling capability to 500 V for some telecommunications applications in which the maximum input

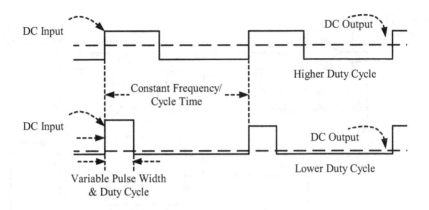

(a) Pulse-Width Modulation (PWM)

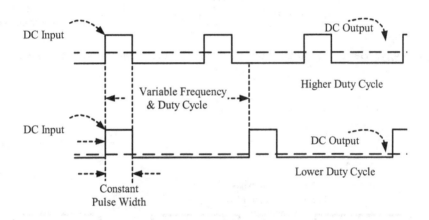

(b) Pulse-Rate Modulation (PRM)

Figure 4.62. PWM and PRM regulation.

line voltage might be only 80 V. In addition to safety, other driving forces for isolation are grounding issues associated with the control of EMI.

Isolation means that the net DC and AC extraneous or noise current is substantially reduced in the isolated interface. If signal, control, or power returns are not isolated at an assembly's interface, ground currents in the reference structure (chassis) may exist. Figure 4.64 illustrates both isolation of grounds between two subsystems and also lack of isolation (permitting a ground loop to exist). Return current can flow both in the return wire as well as through the chassis ground connections. When isolation is to be maintained, some form of AC coupling, power transformers in particular (in power supply systems), must be used to transmit AC power; no DC path exists between the

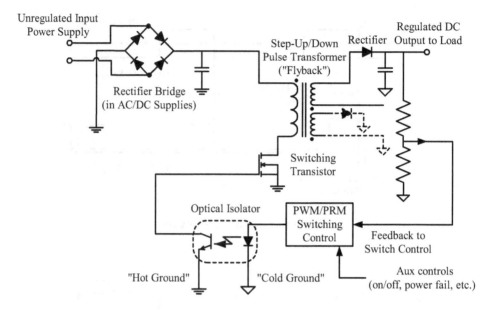

Figure 4.63. Basic configuration of a multiple (×2) output switch-mode power supply.

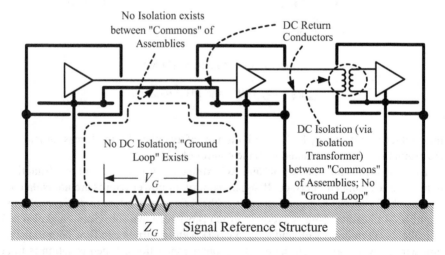

Figure 4.64. Illustration of isolation of grounds and lack of isolation (permitting a ground loop to exist).

interconnected assemblies. Note that isolation really means limiting of the flow of current. For instance, a typically recommended isolation impedance of 1 MΩ to the chassis in a 28 V DC power system implies that DC current lower than 28 μA to the chassis is permitted.

A particularly challenging situation may occur when several assemblies, which are powered from a single primary power source and share a common reference (and primary power-return path) interface with other assemblies, when no isolation exists in the input power supply of the assemblies.* This situation is illustrated in Figure 4.65.

In Figure 4.65(a), the electronic assemblies are powered by an isolated DC/DC switch-mode power supply (SMPS). Although single-ended signaling is used in the circuits, the only available signal-return path is through the controlled secondary power return[†] circuits; no path exists for the signal-return current through the primary power-return[‡] circuits thanks to the isolation of the DC/DC SMPS (as long as the layout of the assemblies is such that there is no physical common path between the assemblies that overlaps the primary power return).

In Figure 4.68(b), the electronic assemblies are powered by a nonisolated power source, whether via a nonisolated DC/DC switch-mode power supply (SMPS) or through direct power feed to the assembly. Note that the primary and secondary power-return circuits are now directly connected at multiple locations (producing potential "ground loops"). The single-ended signal current may now return through the uncontrolled primary power-return circuits, resulting in potential degradation of performance, increased EMI, and possible violation of safety objectives.

Note that when a grounding scheme that is totally isolated across all interfaces is used (and for that purpose an isolated DC/DC power supply was utilized), isolation throughout all circuits of the system (including power, control, signal, and data) must be maintained. Note that for any interface to be considered isolated, only one end of the interface (sending or receiving) needs to be isolated. Interface isolation techniques are discussed in Section 4.5.5.

4.4.3 Isolation and Grounding in Switch-Mode Power Supplies

Switch-mode power supplies are often categorized into two basic types with respect to their grounding schemes: isolated and nonisolated. A switch-mode power supply that generates a low-voltage isolated output from a primary (mains) source is often referred to as an "offline SMPS." In isolated power supplies, input and output circuits of the power supply are electrically separated (i.e., isolated) from each other, precluding electrically continuous current or current-return paths from existing be-

---

*An elaborate and extensive discussion of grounding schemes in power distribution schemes of integrated systems is provided in Section 4.7.3.

[†]Secondary electrical power is electrical power in a system that has been isolated from the primary electrical power before it is distributed to subsystems.

[‡]Primary electrical power is electrical power taken from power generation units without conditioning or isolation.

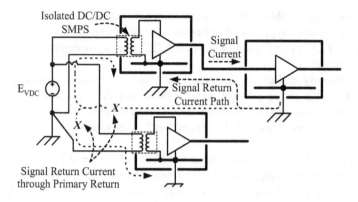

(a) Isolated Secondary Power Circuit: Signal-Return Current Flows

Only through Controlled Secondary Power-Return Circuit

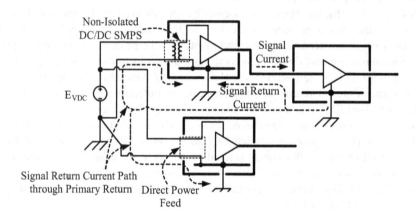

(b) Nonisolated Secondary Power Circuit: Signal-Return Current

May Flow through Uncontrolled Primary Power-Return Circuit

Figure 4.65. Effect of lack of isolation in the power supply interface on system functionality.

tween them. Requirements for electrical isolation often originate from various electrical safety codes. Isolation provides protection to equipment and personnel from hazardous voltages* but also improves common-mode rejection and interference-noise immunity. Isolation may, thus, eliminate potentially harmful ground loops.[†] In addition, isolated power supplies can be used to produce negative output voltages, for example, −5 V from +5 V.

---

*Hazardous voltages are voltage greater than 30 V_{RMS}, 42.4 V_{Peak}, or 60 VDC.
[†]Ground loops are discussed in Section 4.5.

Within all SMPSs, there are four major current loops. Two of the loops, the "switching transistor current loop" and "output rectified current loop," are rich with harmonics and noise due to the highly nonlinear switching operation (Figure 4.66). The current waveforms are typical trapezoidal pulses with high peak currents and rapid waveform di/dt transitions, thus comprising higher frequency spectral components. The two remaining current loops are the "input current loop" and "output current loop," which carry current of lower frequency content, and should remain interference-free. Uncontrolled flow of current through high interference to sensitive paths may adversely affect equipment performance and produce excessive EMI. Figure 4.66 illustrates the above-described current loops in nonisolated [Figure 4.66(a)] and transformer-isolated [Figure 4.66(b)] switch-mode power supply configurations.

The main drawback of nonisolated power supplies [Figure 4.66(a)] from the standpoint of EMI is self-evident. Any voltage potential difference present between the ground-reference points of the source and loads results in differential-mode interference current propagating between source and load or vice versa unless the power supply and its loads are floated (which may conflict with safety regulations). This could result in performance degradation and failure to comply with EMC requirements.

A transformer-isolated power supply overcomes this difficulty. Interference appearing across the ground reference now appears as common-mode interference across the windings of the transformer, minimizing the adverse effects of this interference (see Section 4.5.5.1.1). Furthermore, isolation in the power supply permits the application of an equipment internal grounding architecture independent of the host system, platform, or facility grounding architecture.

Notice in Figure 4.66(b) that a feedback signal must be isolated and brought over the isolation boundary from the secondary ("cold") ground to the primary ("hot") ground in order to maintain isolation. Figure 4.67 illustrates the effect of violation of this isolation on system grounding architecture. A conductive path exists, allowing current, I_{GL}, resulting from an interference potential difference between the "cold" and "hot" reference points, V_G, to flow through the feedback circuit between the primary and secondary circuits, intended to be isolated by the isolation transformer, making the transformer useless. The ground V_G will produce interference in the feedback circuit and, consequently, across the load.

The controller, the core of the converter, can be referenced to either the primary- or the secondary-side ground. Several methods can be used for deriving a feedback signal from the secondary side and bringing it across the isolation boundary. The best design approach for "crossing the isolation boundary" varies with application. Many factors such as performance, complexity, and cost need to be considered. Evaluation of the isolation circuit against the system objectives is necessary throughout each stage of the design.

Yet, nonisolated power supplies are often used in some applications such as in digital circuits where multiple nonisolated DC voltages (such as 5 V, 3.3 V, 2.5 V, and 1.8 V) are required. Circuits powered from such diverse voltage sources are commonly designed to share a common ground reference and thus isolation is neither required nor desired. Switch-mode power supplies are used in such applications simply for achieving higher efficiency of power conversion, particularly when high currents are drawn.

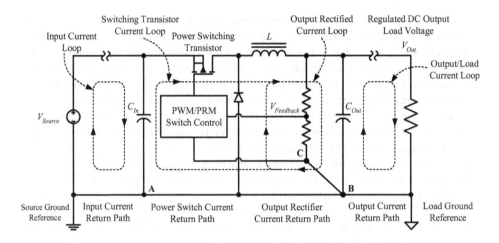

(a) Nonisolated ("Buck") Switch-Mode Power Supply

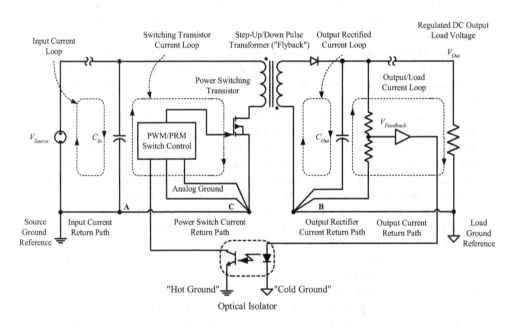

(b) Transformer-Isolated ("Flyback") Switch-Mode Power Supply

Figure 4.66. Current loops and return paths in major switch-mode power supply topologies.

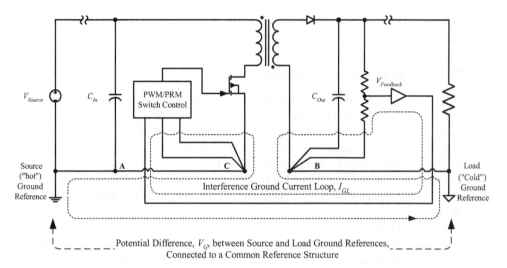

Figure 4.67. Effect of nonisolated feedback circuit on system grounding architecture.

Lessons Learned

- Switch-mode power supplies (SMPSs) offer many benefits including higher efficiency, tighter regulation, wide range of output voltages, smaller size, less weight, and input-to-output and output-to-output isolation.
- Input-to-output isolation is often required for ensuring safety and can also eliminate undesired ground loops. SMPSs thus constitute essential elements in grounding systems and architectures.
- Isolation must be carefully maintained and "crossing the isolation boundary" should be carefully controlled.
- EMI is intrinsically produced in SMPSs and must be carefully controlled.

4.5 GROUND LOOPS

One of the most problematic types of EMI to understand, diagnose, and resolve is the ground loop. Ground loops are a mystery to many. Most electronic engineers often fail to comprehend the concept of ground loops and typically ignore this aspect of system design.

Engineers generally concentrate on electrical power distribution or on utilizing equipment that hooks up to the power distribution network. Not much attention is paid normally to power distribution systems and equipment as a single entity in which ground loops arise.

All types of equipment are susceptible to ground loops: multimedia, medical, industrial, and data processing. Ground loops can cause data errors, component failure, lock-ups, and even safety hazards.

It is no wonder that addressing ground loops is one of the highest priorities in the quest for high-quality system performance.

4.5.1   Definition of a "Ground Loop"

Isolation of ground/return paths constitutes a central concept in a grounding systems' design. Isolation implies that the net DC and AC extraneous or noise current is substantially reduced. Frequently, however, it is necessary to transport signals between two or more physically separated subsystems associated with different grounding systems. In such cases, where more than one ground path exists between assemblies, for instance, a signal return is connected between circuits, each with a separate grounding connection to the signal reference, their signal interfaces will no longer be isolated from each other.

Physically, the grounding system may be realized as a wire, a trace on a printed circuit board, a metal chassis structure, or virtually anything that conducts electricity. Ideally, it should be a perfect conductor but in any practical system, it is not. As the complexity and size of the system are increased, the nonzero impedance of the ground system causes problems.

Figure 4.68 illustrates both isolation of grounds between two subsystems and the lack of isolation forming a ground loop. Signal-return current flows both in the return wire as well as through the chassis ground connections, resulting in an unbalanced system. An example of a DC-isolated interface is a transformer used to transmit AC signals or power, avoiding the need for a conductive DC path between the assemblies.

Two or more points in an electrical system that are nominally at ground potential and are interconnected by a conducting path such that either or all designated ground points are not maintained at an equal ground potential are commonly prone to the creation of ground loops [5, 10].

Whenever a ground loop is present, a potential for damage from intersystem ground noise may occur; the voltage-potential difference in the ground network causes inter-

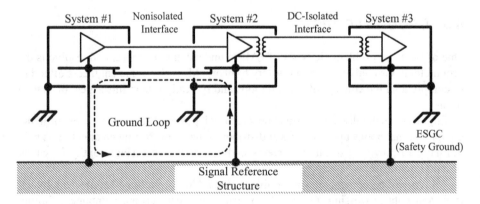

Figure 4.68. DC-isolated ground and nonisolated ground connections.

ference loop currents to flow into interconnects. These loop currents are transformed by the impedance of the ground interconnects into voltage fluctuations and, consequently, system ground can no longer serve as a stable system reference. The resultant ground noise is superimposed on the intended signals at the input of sensitive circuits and becomes part of the signal processed by the component. Figure 4.69 depicts a situation that could result in ground loops.

The AC-powered computer and peripherals are interconnected through their respective electrical safety ground wires within the building's power distribution network. A voltage drop across any impedance of the grounding system within the facility may result in a possible ground potential difference between equipment. Computers may also be connected to peripherals via unbalanced RS-232 data communications cables. Multiple ground paths frequently exist. Ground loops caused by RS-232 links between devices may cause computer lockups due to power line transients and electrical noise coupling onto the interface through the ground reference line. Similar results commonly occur within multimedia systems (audio and video).

Unbalanced interfaces are a common source of ground loops, often produced in shielded single-ended cable interconnects. Taking into account that some equipment assemblies are linked using shielded single-ended conductors, it is quite likely that some interference problems may arise. Return currents will almost certainly circulate from one assembly through the earth conductor into another grounded assembly and back to the first assembly via a shielded cable. This wire loop may also pick up interference from nearby magnetic fields and radio transmitters. Unfortunately, such interconnections are often overlooked when designers fail to realize that the shield actually serves as the signal's intended return path.

Figure 4.70 illustrates a typical ground loop situation. Two interconnected elements, Assemblies #1 and #2, are plugged into grounded AC outlets at different loca-

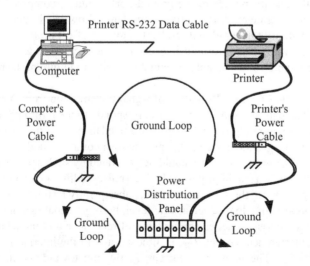

Figure 4.69. Ground loop situation in a computer system.

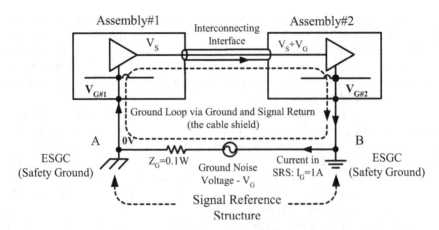

Figure 4.70. Ground loop between two circuits.

tions, denoted as A (serving as the system's 0 V reference) and B, respectively. The signal ground is also internally connected to the enclosure (safety ground) in each. The electrical safety ground path and the duplicate signal-return path formed by the interface shield form a loop that can pick up and conduct EMI throughout the system.

V_G represents the common-mode ground-noise voltage drop within the SRS due to ground current flowing through the ground conductors and reference structure. Two separate reference points, A and B, are provided for Assembly#1 and Assembly#2, respectively. Due to finite impedance of the signal reference, a reference potential difference, $V_G = |V_{G\#1} - V_{G\#2}|$, exists between the two subsystems' reference points.

Unwanted noise from one subsystem may be injected into the other through a common return path. The magnitude of the ground-noise voltage, compared to the signal level in each subsystem circuit, is of prime importance. Low-level interference, below the circuit's sensitivity threshold, will probably have no effect on the circuit's performance; however, when that threshold is exceeded, the received signals will become progressively corrupted and unusable, eventually leading to failure of the receiving circuit.

Figure 4.71 depicts a general situation of signal transmission from Assembly #1 to Assembly #2. The path A–B in Figure 4.71 corresponds to the return path through the signal reference structure ("ground"). Common-impedance coupling across the impedance Z_G converts the ground currents, I_{GI}, producing a common-mode interfering voltage, $V_{NG} = V_{CM}$. The ground currents could be generated by such sources as highly inductive circuits, such as a high-power PWM (pulse-width modulation) motor for hydraulic pumps, sharing the same return path, subsequently resulting in possible hum and noise within sensitive loads such as low-level, high-gain analog amplifiers.

Z_G normally comprises a series combination of resistance and inductance. The external ground interference current, I_{GE}, is almost equal to the internal ground current I_{GI} (i.e., $I_{GE} \cong I_{GI}$). The interfering current, I_I, flowing toward the interconnecting wiring, is, therefore, much smaller than the current flowing through the reference

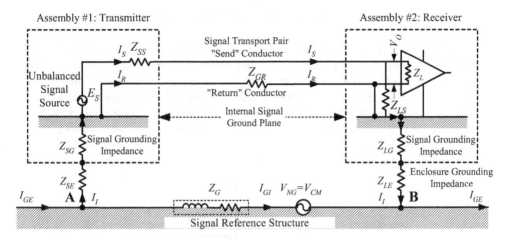

Figure 4.71. Conductive ground-loop interference-coupling mechanism in an unabalanced single-ended system.

structure. In spite of its low level, the current I_I is not negligible and may produce interference across sensitive loads. Once the current I_I arrives in Assembly #1, it splits into I_S flowing through the signal line, and I_R flowing through the signal-return wire (i.e., $I_I = I_S + I_R$). Note that the (common-mode) currents I_I, I_S, and I_R shown in Figure 4.71 are those produced by the external disturbance source and are independent of the signal source, E_S. The noise voltage across the receiver terminals in Assembly #2 is proportional to the voltage drop $V_{NG} = Z_G \times I_{GI}$ that generates the noise current I_I.

In single-ended unbalanced interfaces, the lower line depicted in Figure 4.71 serves as the intentional return conductor for the signal current. Therefore, a noise I_R, produced by the voltage drop V_{NG}, flows predominantly on the return conductor between Assembly #1 and Assembly #2 since its impedance is typically significantly lower than that of the signal line impedance to the reference structure. The intrinsic impedance associated with the return conductor, Z_{GR}, however, could be sufficiently large at higher frequencies so as to degrade the receiver performance due to the voltage drop, $V_{GR} = Z_{GR} \times I_R$. In fact, at all but very low frequencies, Z_{GR} can generally be assumed to be primarily inductive with partial inductance, L_{GR}; hence, Z_{GR} can be represented as $Z_{GR} \cong j\varphi L_{GR}$. In fact, for this single-ended connection, the noise voltage appearing across the input of the receiver can be expressed as

$$V_O = V_{GR} \cdot \left(\frac{Z_{LS}}{Z_{SR} + Z_{LS} + Z_G} \right) \cong V_{GR} \cdot \left(\frac{Z_{LS}}{Z_{SR} + Z_{LS}} \right), \text{ volts} \qquad (4.31)$$

Since, typically, $Z_{LS} \gg Z_{SS}$ and $V_O \cong V_{GR}$, almost all the noise is superimposed on the signal voltage produced by E_S and appears across the load, Z_{LS}. If, for instance, a current, I_{GI}, of 1 A flows through the impedance of the ground conductor, Z_G, of 0.1 Ω, results in a voltage difference between the two reference points A and B, V_{NG}, of 100 mV. This voltage drop will thus appear across the load as interference superimposed on the signal voltage, E_S.

Assume that the input to the receiver is in the form of a "Darlington" stage [11], as illustrated in Figure 4.72(a). Considering the 600 mV base-emitter voltage of both transistors and a 500 mV noise margin of the TTL gate, overall system noise immunity is 2×600 mV $-$ 500 mV $=$ 700 mV, which provides sufficient noise immunity for achieving optimal performance. If, however, a Darlington stage is not used and a single transistor only is used at the receiver's interface, the noise immunity reduces to 600

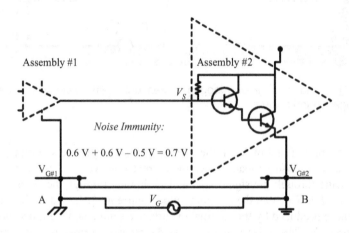

(a) Darlington Stage Input Circuit

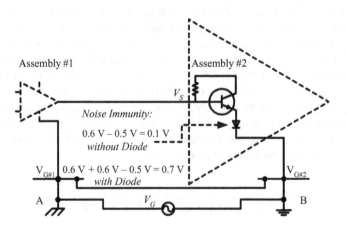

(b) Single-Transistor-Stage Input Circuit

Figure 4.72. Effect of the input stage of subsystem #2 on the common-impedance interference coupling as a result of the ground loop.

mV – 500 mV = 100 mV. This margin may be too small to guarantee sufficiently reliable performance.

With a single-transistor interface [Figure 4.72(b)], the noise immunity drops to 100 mV if the diode were not present in series with the emitter, which is a very small immunity margin to guarantee sufficiently reliable performance. Inserting the diode increases the noise immunity margin back to 700 mV.

4.5.2 Ground Loop Consequences ("Who's Afraid of the Big Bad Loop?")

Do ground loops invariably constitute a problem? They are often demonized and relentless efforts are exerted to have them eliminated at any cost. Why?

The following demonstrates why (and when) ground loops constitute a problem, and when they do not.

4.5.2.1 Why Are Ground Loops a Problem? When ground loops exist, the current that flows in the system reference structure becomes unpredictable.

The return paths formed by multiple connections are the equivalent of a loop, which not only conductively couples noise between different assemblies, but also acts as a loop antenna that may very efficiently pick up interference noise. Loop currents are produced by voltage differences, induction from other cables or devices, wiring connection errors, ground faults, and normal equipment leakage current. These currents can be at DC, 50/60 Hz power frequency, or, in reality, at any frequency.

Line-to-structure current leakage via capacitors in power line filters, for instance, could introduce RF current in the return path. This leakage current is typically on the order of milliamperes per unit (typically less than 1 mA in computer equipments). However, with many such sources, each contributing small leakage currents, several amperes can easily be observed and measured.

The line-to-ground capacitance of large machines such as motors can exceed safety levels internal to power line filters. Currents from such sources are usually of the order of 1 A.

Even a small induced voltage can cause a large amount of current in a ground-conductor loop, as the resistance and inductance are generally low. These loop currents can be as high as tens of amperes. Cables carrying such high currents can create high-level interference current induction.

As a result of the above, specific equipment problems can be of three distinct types (Figure 4.73):

1. **Low-level currents** in the ground system generate interference voltages that cause signal and data errors. These can be low frequency (such as power frequency hum on analog systems) or high frequency (electrical noise).

    Ground loops are the most common cause of AC-power frequency interference to audio and video signals when multiple multimedia system elements are interconnected. A typical indication of a ground-loop problem is the annoying audible "hum" and "buzz" noise (50 Hz/60 Hz power frequency and their har-

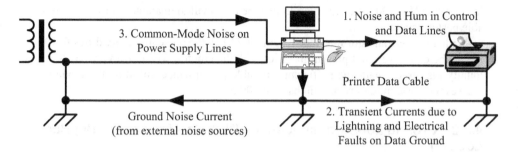

Figure 4.73. Common interference problems associated with ground loops.

monics) superimposed on the signals. In video or television systems, this hum may be perceived as stationary or moving bars across the screen.

Audio-frequency ground loop problems are typically in the low-millivolt range. It does not take much interference in a grounding system to introduce problems.

2. **High-energy transients** can cause serious problems as transient currents flow to earth through return conductors rather than through the electrical safety ground. Transients can be generated internally (switching or inrush currents) or externally (lightning transients).

Transients may cause equipment damage such as sparking in connectors, heating, and burned wiring, as well as damages to driver and receiver communication ports as well as to the digital signal-processing circuits themselves.

3. Ground loops constitute one source of **common-mode noise** in power and signal circuits. This noise is subsequently injected into circuits and can cause equipment disruption.

Industrial and process-control applications, for instance, which are concerned with sensitive and accurate measurements of voltage, current, temperature, pressure, strain, and flow, often involve environments with hazardous voltages, transient signals, common-mode voltages, and fluctuating ground potentials. Such phenomena are capable of damaging measurement systems and significantly degrading measurement accuracy.

4.5.2.2 When Are Ground Loops Not a Problem? Ground loops do not cause problems when both the following provisions hold:

1. **None of the wires in the loop carries any current.** This, however, is impractical in real systems in which nonzero impedances of conductors are virtually nonexistent.

2. **The loop is not exposed to external time-variant electromagnetic fields,** which implies a zero-ambient electromagnetic field environment. However, there is practically no situation in which such a condition can be satisfied.

4.5.3 Ground-Loop Interference-Coupling Mechanisms

Most ground-loop interactions, particularly where higher frequencies are concerned, result in common-mode currents flowing within the system. Elimination of ground-loop problems (the cause) has often become, therefore, synonymous with common-mode interference (the outcome) suppression.

As will be shown in Section 0, ground-loop interference is typically associated with common-impedance coupling. Interference currents flow through common impedance introduced by the signal reference structure shared by the victim and culprit circuits. Interference currents may also be produced through radiated electromagnetic fields due to cable interaction, whereby interference currents are induced into loops formed by the circuit wiring and the common reference structure.

In this section, the manner in which common-mode currents are produced in electrical and electronic systems due to electromagnetic fields is discussed. The effect of such ground loops in differential circuits due to load imbalance is shown. The discussion is concluded with an introduction of the concept of circuit transfer impedance, serving as a figure of merit for common-mode to differential-mode interference conversion.

4.5.3.1 Coupled Ground-Loop Interactions. Electromagnetic fields can efficiently couple undesired signals into interconnected circuits grounded to a common signal reference, particularly if little or no attention has been paid to minimizing field-to-cable interaction. Even in situations in which a high-quality grounding system is present, large loop areas bounded by signal conductors and the signal reference will exacerbate interference coupling [1, 12].

4.5.3.1.1 MAGNETIC FIELD INTERACTIONS. Magnetic (inductive) coupling involving near-field magnetic flux often constitutes the most severe offender in certain situations (Figure 4.74). The magnetic field produced by an adjacent current-carrying circuit induces electromotive force (emf) sources into the signal transport conductors, producing common-mode current flow between both conductors and the signal reference.

The level of induced current is directly related to the coupling efficiency, which, in turn, depends on the circuit's layout and the interference field characteristics (field intensity and frequency).

Figure 4.75 illustrates that H-field coupling can introduce common-mode interference into a circuit by coupling into a transmission line grounded both at the source and load ends. Coupling is enhanced by an increase in the loop area between each of the circuits to the signal reference plane.

The key to minimizing the effects of inductive coupling into ground loops is to minimize the enclosed loop area around which current flows (Faraday's Law of Induction"):

$$V_{NG} = \oint_{C_{Loop}} \vec{E} \cdot d\vec{l} = -\frac{\partial}{\partial t} \int_{S_{Loop}} \vec{B} \cdot d\vec{s} \approx -\frac{\partial \vec{B}}{\partial t} \cdot S_{Loop} \qquad (4.32)$$

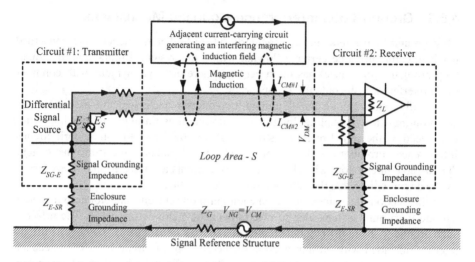

Figure 4.74. Magnetic (inductive) interference coupling producing ground-loop interference.

In practice, minimizing inductive coupling can be accomplished by reducing spacing between the conductors in the victim circuit, increasing the spacing between the source and victim circuits, and keeping both conductors adjacent to a signal reference as much is possible. As a result, the area of the loops formed between both conductors in the victim circuits (the "small loop"), as well as between the victim circuit conductors and the signal reference (the "large loop") are reduced.

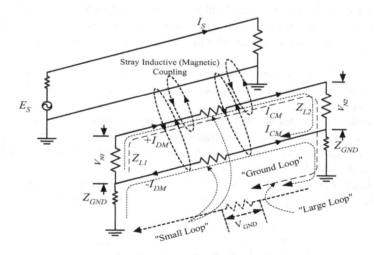

Figure 4.75. H-field common-mode interaction in a victim signal-transmission circuit.

4.5.3.1.2 ELECTRIC FIELD INTERACTIONS. Interactions of E-fields with conductor loops are similar in nature to the H-field interaction. Both may introduce common-mode currents and voltages into victim circuits. E-field interaction is shown in Figure 4.76. Often disregarded by system designers is the fact that E-field interactions are associated with stray reactance paths that normally exist within a product design. E-field interaction is barely affected by minimizing or maximizing the quality of "ground connections," "ground conductors," and the like. On the other hand, source-to-victim separation, mutually exposed circuit areas, relative orientation between circuits, and so on are of utmost significance. The fundamental mechanisms of E-field interaction between circuits are shown in Figure 4.77.

It is importance to realize that a common reference between the circuits may be the chassis of the enclosure, the facility ground, a signal reference structure, a return plane in a printed circuit board, or any combination thereof. The level of the coupled current is directly related to coupling efficiency, which, in turn, depends on the mutual circuits' layout and on the interference field characteristics (field intensity and frequency).

The key to minimizing the effects of capacitive coupling into ground loops is to minimize displacement current $\partial \vec{D}/\partial t$ between circuits, which, in turn, depends on the stray capacitance between the circuits (Ampere's Law):

$$\oint_C \vec{H} \cdot d\vec{l} = \int_S \left(\vec{J} + \frac{\partial \vec{D}}{\partial t} \right) \cdot d\vec{s} \qquad (4.33)$$

Electric coupling is enhanced with increased conductor length and proximity between source and victim circuits. Hence, minimizing interaction can be accomplished by in-

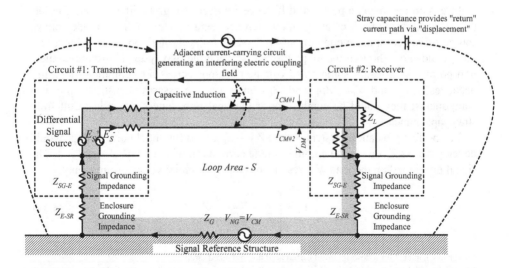

Figure 4.76. Electric (capacitive) interference coupling producing ground-loop interference.

creasing separation and minimizing the parallel common length between adjacent conductors. Reducing loop area in the circuits will yield little improvement since, unlike magnetic coupling, interaction is not dominated by loop area. Electric (capacitive) interactions are commonly dominant in high-frequency circuits as they are limited by the value of the stray capacitance between the circuits.

It is also evident from Figure 4.77 that even in cases in which grounding impedance Z_{GND} in the victim circuit is nonconductive, such as when it is "floated" at one or both ends, there will be little reduction in coupling at higher frequencies. Stray capacitance will "complete the loop" for induced common-mode currents on the victim circuits' conductors.

4.5.3.2 Ground-Loop Interference Due to Load Imbalance.

Differential and balanced circuits* are erroneously considered a "magic solution" to ground-loop coupled-interference problems. Perfectly balanced circuits would be indifferent to externally introduced common-mode interference, since the effect of common-mode currents present on both the "send" and "return" conductors would cancel across the receiver's balanced load. No differential-mode interference signal subsequently develops across its load.

If the two lines are not identical ("unbalanced lines"), however, "common-mode to differential-mode conversion" occurs, resulting in differential-mode noise across the differential receiver input.

Figure 4.78 depicts undesired common-impedance (conductive) coupling between two interconnected assemblies [12]. Assembly #1 is a differential transmitter represented as the signal source E_S[†] with source impedance Z_S, whereas Assembly #2 is a differential receiver, represented by the differential load impedance Z_L. Both circuits are grounded to the signal reference structure (SRS).

In this arrangement, a potential difference between two grounding reference points ($V_{NG} = V_{CM}$) is assumed, creating an interference voltage. This common-mode voltage source drives common-mode currents $I_{CM\#1}$ and $I_{CM\#2}$ through both "send" and "return" conductors through the loops formed between the two systems' ground-connection points. As a result, some noise voltage develops across the "common-mode impedances" Z_{C1} and Z_{C2}. Z_{C1} and Z_{C2} may not necessarily constitute physical components; they could be conductors or physical impedances or could result from stray capacitance.

In a perfectly balanced system, where $Z_{C1} = Z_{C2}$, no differential noise voltage would develop across Z_L due to the common-mode currents flowing though Z_{C1} and Z_{C2}. The signal developed on the load will result from the differential voltage source only:

$$V_L \approx \left(\frac{E_S}{Z_L + 2Z_S} \right) \cdot Z_L \qquad (4.34)$$

---

*Refer to Chapter 2 for a description of differential and balanced circuits.
[†]In a differential and balanced circuit, in practice, the source voltage E_S is a differential source, appearing between the two lines. For illustrating the balance of the source, E_S is shown as a split source between the signal and return lines.

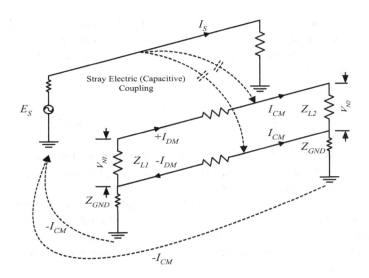

Figure 4.77. E-field common-mode interaction in a victim signal transmission circuit.

In realistic circuits, this is seldom the case. It would be reasonable to assume that Z_{C1} $\neq Z_{C2}$. Any difference, $\pm\Delta Z$ ($\Delta Z = |Z_{C1} - Z_{C2}|$), represents a degree of the "load imbalance" that exists between them to a "differential load" Z_L such that

$$Z_{C1} = \frac{Z + \Delta Z}{2} \qquad \text{and} \qquad Z_{C2} = \frac{Z - \Delta Z}{2} \qquad (4.35)$$

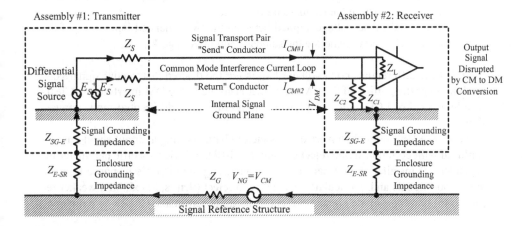

Figure 4.78. Conductive ground-loop interference coupling in a differential system introduces common-mode ground-to-differential-mode noise due to circuit imbalance.

Consequently, some differential voltage, $V_{DM} = V_L$, develops across the differential load Z_L, introducing interference into the load circuit. Ignoring transmission-line impedance (assuming electrically short conductors), the resultant total differential voltage across the load is

$$V_L \approx I_{CM\#1} \cdot \left(\frac{Z + \Delta Z}{2} \right) - I_{CM\#2} \cdot \left(\frac{Z - \Delta Z}{2} \right) + \left(\frac{E_S}{Z_L + 2Z_S} \right) \cdot Z_L \qquad (4.36)$$

It is assumed that $Z_L \ll Z_{C1} + Z_{C2}$. By definition, in a perfectly balanced circuit ($\Delta Z = 0$), common-mode currents are equal ($I_{CM\#1} = I_{CM\#2} = I_{CM}$) resulting in the desired response, expressed in Equation (4.36).

It follows that in "real-world" (nonideal) balanced systems, common-mode noise voltage present across the signal reference structure will result in undesired differential-mode interference introduced across the receiver's load. This differential-mode voltage appears as a "legitimate" input across the load and may cause an undesired response in the receiver.

The degree of unwanted differential-mode interference signal V'_{DM} produced at the terminals of the network, due to the presence of a common-mode (or longitudinal) signal on the connecting leads is defined as the longitudinal conversion loss factor or LCL.* LCL represents the extent of common-mode to differential-mode conversion in the system [14].

LCL is expressed, in dB, as

$$LCL_{dB} = 20 \cdot \log \left(\frac{V_{CM}}{V_{DM}'} \right) \Bigg|_{V_O = \text{Constant}} \qquad (4.37)$$

where:

V_{CM} = Common-mode voltage applied to the two balanced input ports of the system, required to generate the standard output voltage, V_O

V'_{DM} = Differential-mode voltage applied between the two balanced input ports of the system, required to generate the standard output voltage V_O

Note that the differential-mode voltage V'_{DM} is *not* the desired differential signal voltage but rather the portion of the CM noise that is converted to DM interference. Clearly, a high level of LCL indicates a lower common-mode to differential-mode conversion.

Note: The longitudinal conversion loss factor characterizing a circuit should not be confused with common-mode rejection ratio (CMRR), a device-specific parameter independent of installation and parasitics characterizing the circuit's response to common-mode signals and, particularly, its ability to reject (that is, to exhibit a weaker response to) common-mode signals present at its input terminals. CMRR is defined as

---

*The longitudinal conversion loss or LCL is a parameter originally used for telecommunication equipment. See Chapter 2 for a discussion on common-mode to differential-mode conversion and vice versa.

ratio of the common-mode voltage V_{CM} required to produce a standard output voltage V_O across the differential load to the nominal differential voltage V_{DM} that would produce the same differential output voltage, V_O:

$$CMRR_{dB} = 20 \, \log \left(\frac{V_{CM}}{V_{DM}} \right) \Bigg|_{V_O = constant} \tag{4.38}$$

A CMRR figure is typically provided by manufacturers in the devices' data sheets. Keep in mind, however, that the CMRR is frequency-dependent: It can reach very high values (as high as 80 dB) at lower frequencies, falling rapidly as the frequency increases beyond the operational frequency of the component. At higher frequencies, achieving a large CMRR value may be difficult to accomplish.

Combining Equations (4.36) and (4.38) while assuming $\Delta Z \ll Z$, we obtain

$$LCL_{dB} = 20 \, \log \left(\frac{V_{CM}}{V'_{DM}} \right) \approx 20 \, \log \left(\frac{Z + 2Z_S}{2\Delta Z} \right) \tag{4.39}$$

The extent of circuit imbalance will determine the extent of mode conversion and, hence, ground-induced interference across the differential load.

Unfortunately, a system designer has little, if any, control on the amount of common-mode current flowing through the system. It is often determined by factors external to the system itself, such as fault-current surges or lightning impulses. Therefore, for precluding the common-mode to differential-mode conversion, the best design practice is that of minimizing the common impedance across the broadest frequency band achievable. The second factor under the designer's control is circuit balance. Common-mode to differential-mode conversion and the effect of circuit balance on system performance are further discussed in Section 4.5.4.

Would "floating" of the circuits offer an adequate solution and eliminate ground loops? In all but very low-frequency applications, "floating" of either circuit, whether between the internal signal ground plane to the enclosure chassis (Z_{SG-E}) or from the enclosure chassis to the signal reference structure (Z_{E-SR}), will be of little benefit in eliminating the problem. Higher frequency displacement current can still flow through these loops owing to the large parasitic capacitance present between the surfaces, effectively completing the loop circuits. Floating will, therefore, significantly suppress only low-frequency ground-loop currents while having little, if any, effect on the high frequency currents, a fact regretfully commonly overlooked by system designers.

Common-mode currents flow through potentially large loops, between either of the signal transmission line pairs or the signal reference (note the arrows in Figure 4.78). Each of these common-mode currents acts as two differential-mode currents from the standpoint of producing radiated fields, the level of which are proportional the loops' dimensions.

4.5.3.3 Application of the Transfer Impedance Concept to Ground-Loop Interference Coupling. In previous discussions, it was demonstrated that

common-impedance is the immediate cause for ground-loop interference coupling, but no insight was provided as to the nature of this impedance. It is evident that in single-ended unbalanced systems, the interference voltage present across the impedance of the reference structure between the interconnected assemblies appears in its entirety as noise superimposed on the intended received signal across the load. From the examples presented above, it may be deduced that as long as low impedance of the ground-reference structure is maintained, no interference coupling problems will occur. Unfortunately, the situation is not that simple, particularly if differential and real-world balanced interfaces are used.

In this section, a quantitative approach for computation of ground-loop interference coupling (GLIC) mechanisms is presented, making use of the concepts of transfer impedance and partial inductance. This technique sheds light on the principal factors associated with the GLIC and practical approaches for effectively addressing such challenges.

Figure 4.79 depicts a general situation of the balanced signal-transmission interface from Assembly #1 to Assembly #2. The path A–B in Figure 4.79 corresponds to the return path through the signal-reference structure ("ground"). As previously shown, it is anticipated that common-impedance coupling across the impedance Z_G converts the ground currents I_{GI} to a common-mode interfering voltage, $V_{NG} = V_{CM}$. Here again, ground currents flowing through Z_G may result in interference to sensitive loads such as low-level, sensitive, high-gain analog amplifiers. As will be shown herein, common noise produced across the reference structure will still produce some interference across the differential load.

In the case of balanced, differential-mode signaling, the Thevenin equivalent circuit of the source driver typically has low values of $Z_{SS} = Z_{SR}$ and the source voltage is represented as a split and complementary pair of sources $E_{SS} = -E_{SR}$ in series with Z_S and

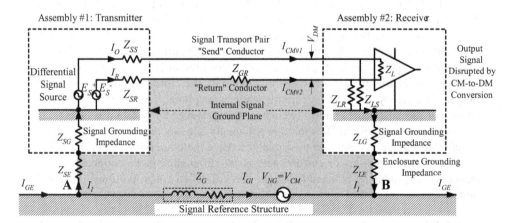

Figure 4.79. Conductive ground-loop interference current introduces common-mode to differential-mode noise due to circuit imbalance in a differential system.

Z_R, respectively, in order to excite a differential-mode signal (Figure 4.79).* At the end of the transmission line, the differential load consists of a differential load, Z_L, in addition to two impedances to the local reference, $Z_{LS} = Z_{LR}$, which may be present and are ideally maintained equal for the purpose of circuit balancing. Also, in high-speed transmission circuits, it is desirable to ensure that $Z_{LS} = Z_{LR} = Z_0$, where Z_0 is the differential-mode characteristic impedance of the transmission line between the two circuits. The interfering current, I_I, produced by the voltage drop, V_{NG}, subsequently appears as common-mode current for the system comprising the two conductors between the driver and the receiver.

The concept of transfer impedance is now applied to quantitively characterize the common-impedance coupling between a loop carrying high current that produces unwanted current in a second loop. The two loops share a segment of conductor with a low effective impedance [13].

Consider a simplified case of the circuit in Figure 4.79 with the two circuits directly interconnected to the signal reference structure ("ground"). This case is depicted in Figure 4.80(a) [13]. The interference voltage, V_{NG}, could be produced by an external noise current flowing through the reference structure shared with some noisy circuit (e.g., a PWM-operated motor), as described above, or by an external electromagnetic field coupling into the loop formed by the return conductor and the reference structure. That loop area is depicted as the grey area in Figure 4.79.

Since $I_O \ll I_R$, then $I_R \cong I_I$, where I_I is the current produced by the EMI voltage source, V_{NG}, acting as a forcing current around the path formed by the reference structure (A–B) and the return conductor in the transmission line, as shown in Figure 4.79.

The combined impedances ($Z_{GS} = Z_{SG} + Z_{SE}$) and ($Z_{GL} = Z_{LG} + Z_{LEE}$) are associated with the paths between the source Assembly #1 (subscript "$_S$") and load Assembly #2 (subscript "$_L$") to the reference structure, respectively. The induced noise voltage at the interconnect output is the voltage V_O occurring at the input of the receiver. To quantify the interference, the ground-loop interference coupling (GLIC) parameter is introduced, defined as[†]

$$GLIC = 20 \cdot \log \cdot \left(\frac{V_O}{V_{NG}} \right), \ dB \qquad (4.40)$$

When the conductors between source and load assemblies are electrically short, the equivalent circuit of Figure 4.80(b) can be used to compute the GLIC across the frequency range of interest. In this circuit, capacitive effects are neglected because the impedances of the circuit, Z_{GS} and Z_{GL}, are normally of low value for digital signaling, so inductive effects prevail. In the circuit of Figure 4.80(b), the inductances, L_{SW} and L_{RW}, of the signal and return wires, respectively, represent the self-partial inductances,[‡]

---

*Note that for simplicity of the image, this split source is not illustrated in Figure 4.80. As it can be assumed to be an ideal source (with the source impedances considered in the circuit), it has no effect on the discussion herein.

[†]Note that the GLIC, as defined, is a special case of the LCL (longitudinal conversion loss) defined above.

[‡]Partial inductance is discussed in detail in Chapter 2.

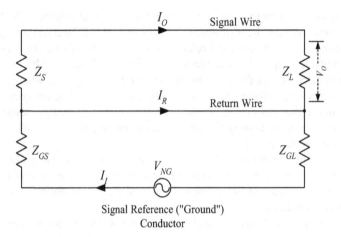

(a) Simplified Model Structure

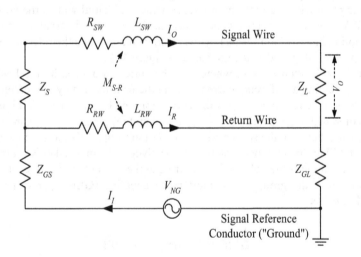

(b) Equivalent Circuit for Analysis

<u>Figure 4.80.</u> Model of single-ended interface for analysis. (Courtesy of Spartaco Caniggia and Francescaromana Maradei.)

whereas M_{S-R} symbolizes the mutual partial inductance between the signal and return conductors. R_{SW} and R_{RW} stand for the resistances of the signal and return conductors, respectively. The impedances Z_{SG} and Z_{LG} are associated with the connections to the reference structure (ground) and, therefore, are of very low value. The parameters Z_S and Z_L correspond to the circuits' source and load impedances, respectively.

To facilitate the calculation of the GLIC, the concept of partial inductance is now introduced. A closed current loop (of contour C) is correspondingly divided into a number of segments C_i (where the index i represents the number of segments), each of which is assumed to have an unequivocal value of self-partial inductance regardless of the character of the overall loop.* In addition, mutual partial inductance between any two segments of the loop is introduced (Figure 4.81).

Recalling that the magnetic flux density, $\vec{B}$, can be expressed as a function of a vector potential, $\vec{A}$, such that $\vec{B} = \nabla \times \vec{A}$, and applying Stokes' Theorem, the inductance of the entire loop of contour C can be expressed as

$$L_C = \frac{\phi_A}{I} = \frac{1}{I} \int_A \vec{B} \cdot d\vec{a} = \frac{1}{I} \oint_C \vec{A} \cdot d\vec{l} = \sum_{i=1}^{4} L_i \qquad (4.41)$$

Note that $I_1 = I_2 = I_3 = I_4 = I$. As the entire contour C forming the loop comprises the sum of all segments C_i of the loop, the effective inductance L_i of each ith segment of the loop is thus defined as a sum of all partial inductances associated with this segment. Accordingly, the total inductance of the loop is the sum of all self- and mutual partial inductances associated with each segment (Figure 4.81):

$$L_i = \sum_{j=1}^{4} \pm L_{p_{ij}} \qquad (4.42)$$

The partial inductance associated with the ith segment of the contour C_i is thus

$$L_{p_{ij}} = \frac{1}{I_j} \int_{C_i} \vec{A}_{ij} \cdot d\vec{C}_i = \frac{1}{I_j} \int_A \vec{B}_{ij} \cdot d\vec{a}_i \qquad (4.43)$$

The self-partial inductance L_{pii} and partial mutual inductance L_{pij} for circular wire conductor(s) of length ℓ and radius r_w, separated by distance d, are expressed, respectively, as [13]

$$Lp_{ii} \approx \frac{\mu_0}{2\pi} \ell \cdot \left[\ln\left(\frac{2\ell}{r_w}\right) - 1 \right], \text{ henry} \qquad (4.44)$$

$$Lp_{ij} \approx \frac{\mu l}{2\pi} \ell \cdot \left[\ln\left(\frac{2\ell}{d}\right) - 1 \right], \text{ henry} \qquad (4.45)$$

These expressions are valid as long as $d \ll \ell$ and $r_{w<s>} \ll \ell$.

This concept can now be applied to the circuit in Figure 4.80(b), which is expanded to Figure 4.82, in which each of the circuit conductors, the signal (S), return (R), and reference (G) are replaced by the corresponding partial inductance, L_{PSW}, L_{PRW}, and

---

*The concept of partial inductance is discussed in detail in Chapter 2, and is only partially repeated here for clarity.

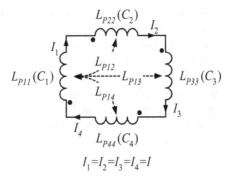

$$I_1 = I_2 = I_3 = I_4 = I$$

Figure 4.81. Conceptual illustration of partial inductance associated with the ith segment of a contour, C_i.

L_{PG}, respectively. The effect of the current flowing on each of the conductors introduces an emf (electromotive force) or voltage source across the neighboring conductors, due to the mutual partial inductance. In general, the current in the ith conductor, I_i, produces a voltage drop across the jth conductor, due to the partial mutual inductance between them, equal to

$$V_{ij} = -j\omega L_{Pi-j} \cdot I_i, \text{ volts} \tag{4.46}$$

For instance, the current in the return conductor, I_R, introduces an emf source, $I_{S\text{-}R}$, across the signal conductor, equal to $-j\omega L_{PS\text{-}R}(I_R$ (note the negative sign from Fara-

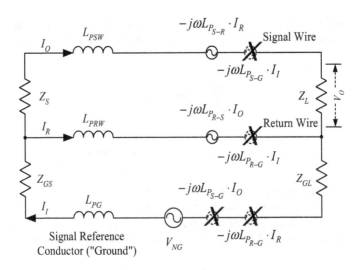

Figure 4.82. Partial inductance equivalent circuit of a single-ended interface for GLIC analysis. (Courtesy of Spartaco Caniggia and Francescaromana Maradei.)

day's Law of Induction). In a similar manner, partial mutual inductances with the reference (ground) conductor are depicted.

The impedances Z_{SG} and Z_{LG}, associated with the connections to the reference structure are assumed to be very small and are, therefore, neglected. Capacitive effects are likewise neglected as well as the partial mutual inductances between the ground reference conductor and any of the other wires, signal and return (observe the X notation), both due to the relatively large separation between the conductors and the reference conductor. It can also be assumed that $I_I \gg I_R \gg I_O$ based on reasoning presented above; therefore the source ground interference voltage, V_{NG}, can be approximated as

$$|V_{NG}| \cong j\omega L_{PG} \cdot I_I, \text{ volts} \qquad (4.47)$$

Through the application of Kirchhoff's loop equations to the circuit of Figure 4.82, a simplified equivalent circuit shown in Figure 4.83 is derived. Each of the dependent current sources (i.e., current sources that are coupled and result from current in the oth-

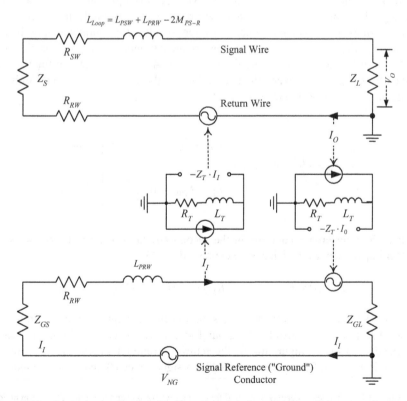

Figure 4.83. Equivalent circuit with the transfer impedance, Z_T, concept applied. (Courtesy of Spartaco Caniggia and Francescaromana Maradei.)

er circuit) illustrated in Figure 4.83 represents the corresponding currents I_I and I_O flowing through the transfer impedance, Z_T. The "transfer" feature of this impedance is revealed from observation of its definition for our application:

$$Z_T \triangleq \left.\frac{V_O}{I_I}\right|_{I_O=0} \text{, ohms} \tag{4.48}$$

The transfer impedance Z_T comprises a series combination of transfer resistance R_T and transfer inductance, L_T. R_T represents the frequency-dependent* resistance of the return conductor, whereas L_T symbolizes the equivalent inductance, which corresponds to the difference between the self-partial inductance L_{PSW} and the mutual-partial inductance M_{S-R} in the signal and return loops. The dependent voltage sources $Z_T \cdot I_I$ and $Z_T \cdot I_O$ determine the current induced in each of the respective loops.

The transfer impedance Z_T signifies, therefore, the extent of coupling between the circuits through the common impedance in the reference structure. From an EMC perspective, the transfer impedance Z_T should thus be minimized.

Typically $I_I \gg I_O$; thus, the contribution of the dependent voltage source $Z_T \cdot I_O$ in Figure 4.83 is negligible with respect to the voltage developing across the load impedance Z_L. The shape of the signal and return conductors strongly influence Z_T, and closed-form expressions are available for a few specific configurations only. It can be measured indirectly, however, when low-frequency approximation (line electrically short) is valid as shown in the circuit in Figure 4.84. The schematic in Figure 4.84 is derived from Figure 4.83 when the bottom loop of the circuit is open, and can be used to calculate the transfer inductance $L_T = L_{RW}$:

$$Z_T = \frac{V_{RW}}{I_S}\text{, ohms} \Rightarrow \begin{cases} R_T = R_{RW} = \text{Re}\left\{\dfrac{V_{RW}}{I_S}\right\}\text{, ohms} \\[2mm] L_T = L_{RW} = \text{Im}\left\{\dfrac{V_{RW}}{\omega\,I_S}\right\}\text{, henry} \end{cases} \tag{4.49}$$

Clearly, the interference coupling in the second circuit is directly determined by the transfer impedance, Z_T, which can be expressed as [13]

$$Z_T(\omega) = R_T + j\omega L_T = R_{RW} + j\omega\left(L_{RW} - M_{S-R}\right)\text{, ohms} \tag{4.50}$$

Figure 4.85 illustrates the effect of transmission-line configuration on its transfer inductance L_T (and, subsequently, the ground-loop interference coupling), with particular attention to the configuration of the return ("ground") structures. The structures depicted in Figure 4.85 are characterized by a signal trace (t) and a corresponding re-

---

*The value of R_T is frequency dependent beyond frequencies where skin and proximity effects can no longer be neglected.

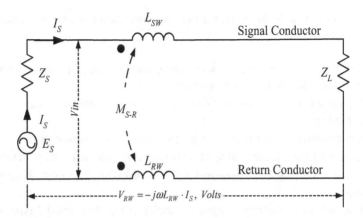

Figure 4.84. Equivalent circuit used to calculate transfer inductance L_T. (Courtesy of Spartaco Caniggia and Francescaromana Maradei.)

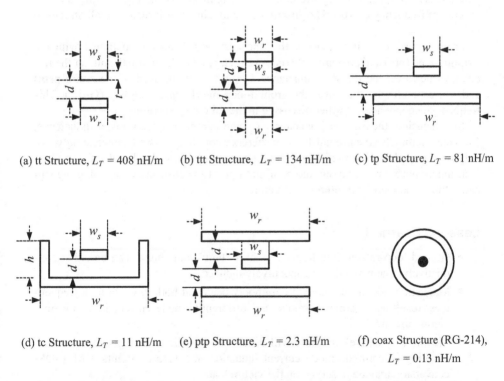

(a) tt Structure, $L_T = 408$ nH/m   (b) ttt Structure, $L_T = 134$ nH/m   (c) tp Structure, $L_T = 81$ nH/m

(d) tc Structure, $L_T = 11$ nH/m   (e) ptp Structure, $L_T = 2.3$ nH/m   (f) coax Structure (RG-214),
$L_T = 0.13$ nH/m

Figure 4.85. Transfer inductance, L_T, of various signal and return conductor structures ($t = 0.1$ mm, $w_s = 0.25$ mm, $w_r = 10 w_s$, $d = 0.5$ mm, $h = 6 \times w_s$). (Courtesy of Spartaco Caniggia and Francescaromana Maradei.)

turn conductor(s), which, likewise, may be in a form of a trace (*t*) or a plane (*p*), as follows* [13]:

a) Broadside-coupled traces: traces having another trace of equal size as return conductor, denoted trace–trace structure (tt)

b) A trace embedded between two return traces of equal size: trace–trace–trace structure (ttt)

c) A trace above a finite return plane: microstrip-type structure (tp)

d) A trace within a finite U-shaped return structure: conduit-type structure (tc)

e) A trace embedded between two finite return planes: stripline-type structure (ptp)

f) A round wire enclosed within a tubular conductor: coaxial-type structure (coax), used as reference

Using the values of the transfer inductance L_T (which dominates the transfer impedance, Z_T) from Figure 4.85, the ground loop interference coupling (GLIC) was computed versus frequency using the equivalent circuit shown in Figure 4.83 up to a maximum frequency of 10 MHz (in order to maintain an electrically small structure) [13].

From Figure 4.86, it is evident that the best performance is obtained with the stripline-type (ptp) and coaxial cable (coax) structures. Furthermore, the higher the frequency, the lower the transfer inductance L_T (or L_{RW}, inductance of the return path) and, correspondingly, the lower the ground-loop interference coupling (GLIC). This frequency dependence is further discussed in the following section.

In conclusion, the transfer impedance, Z_T, and particularly the transfer inductance, L_T, which dominates the ground-loop interference coupling, depends, interestingly, not only on the quality of the reference structure, as is often assumed, but rather and, in fact, much more, on the layout and configuration of the structure comprising the signal, return conductor, and reference structure.

Lessons Learned

- Ground-loop-related interference such as "hum" and "buzz" can be either conductively, magnetically, or capacitively induced.

- Magnetic (inductive) coupling commonly occurs at both low and high frequencies, resulting in common-mode and differential-mode current induction into victim circuits.

- Electric (capacitive) coupling will usually be significant at higher frequencies, resulting in common-mode current induction into victim circuits. CM-to-DM conversion may be observed on the victim loads.

---

*The transfer inductance, L_T, of the various configurations were computed numerically, using the method of moments (MoM) and the nodal method [13].

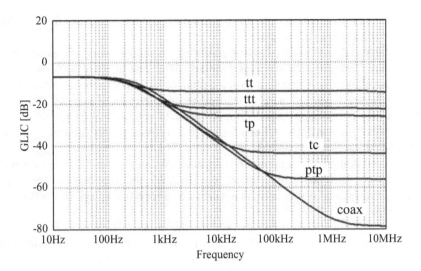

<u>Figure 4.86.</u> Computed GLIC (dB) for the various signal and return conductor structures. (Courtesy of Spartaco Caniggia and Francescaromana Maradei.)

- Ground-loop interference coupling (GLIC) represents the translation of ground-induced interference voltage to circuit interference voltage.
- GLIC is dominated by inductance associated with the signal return path; this inductance is primarily geometry dependent.
- GLIC is frequency dependent and typically falls with increase of frequency.

4.5.4 Ground-Loop Interactions: Frequency Considerations in CM-to-DM Interference Conversion

Section 4.5.3.2 demonstrated the importance of circuit balance for enhanced circuit performance. For instance, perfectly balanced circuits exhibit high immunity to externally introduced common-mode interference, since the effect of common-mode currents induced on both conductors cancels out at the receiver's input, leaving a net zero differential-mode interference signal at the circuit input.

Consider again the wire pair interconnecting two circuits. It was shown that a poorly balanced, loosely coupled pair has more of a propensity to cause interference than a well-balanced, tightly coupled one. This should be of no surprise, since the capture area of a poorly balanced, loosely coupled wire pair will allow greater amplitudes of energy to be converted to and dissipated as common-mode energy into the environment or accepted into the poorly balanced pair as differential-mode energy from external stimuli.

There is no magic associated with common-mode to differential-mode conversion in coupled (e.g., differential) transmission lines. Differential transmission lines are simply two individual electromagnetically coupled lines. Typically, designers try to make both lines identical. One can excite these lines in the odd mode (equal and opposite voltages on each line, or differential signaling) or in the even mode (equal voltages on each line, or common-mode signaling).

A not perfectly balanced excitation results in a combination of odd and even electromagnetic-field propagation modes. Similarly, if the two lines are not identical, often called "unbalanced lines," both modes are produced in order to satisfy boundary conditions.* Hence, when common-mode interference is introduced into an unbalanced or asymmetrical pair of lines, some differential-mode voltage will be produced by the asymmetry of the transmission line or loads. This is usually what is meant by the term "CM-to-DM interference conversion."

All discussion presented up to now concentrated on the effect of circuit configuration on CM-to-DM interference conversion. However, there has been no detailed discussion of the effect of frequency dependence of circuit balancing. This is now illustrated (Figure 4.87).

Assume two circuits, Circuit #1 and #2. The circuits are interconnected by a two-conductor transmission line. Circuit #1 contains a differential line transmitter (source) represented as an ideal signal voltage source V_S and source impedance Z_S (illustrated as a split impedance at the source, as would be the case for a truly balanced source), whereas Circuit #2 consists of a differential line receiver (load), represented as Z_L. Both circuits are mounted in metallic enclosures, grounded to the signal reference structure (SRS). Both circuits are initially floated from their respective enclosure structures.

The circuit area A with separation distance d from the SRS forms capacitance C_G ($C \approx \varepsilon \cdot A/d$, where $\varepsilon = \varepsilon \cdot \varepsilon_r$, the permittivity of the medium between surface A and the SRS). For simplicity of discussion, it is assumed that C_G is identical for both Circuit #1 and #2.

The conductors (transmission line) are placed at height h above the SRS, and have length S. At low frequencies, these conductors can be treated as a low-pass filter with an equivalent series inductance L_C and interconductor capacitance C_D. For the sake of clarity, common-mode capacitance existing between the conductors and input ports to the SRS are omitted.

It is noted that the "CM impedances" Z_1 and Z_2 may not be physical components; they could be conductors, physical impedances, or stray capacitance. It is also reasonable to assume that Z_1 and Z_2 are rarely identical; thus, some small difference, $\pm\Delta Z$ representing the degree of the "load imbalance" yielding

$$\oint_C \vec{H} \cdot d\vec{l} = \int_S \left(\vec{J} + \frac{\partial \vec{D}}{\partial t} \right) \cdot d\vec{s} \qquad (4.51)$$

$$Z_1 = \frac{Z + \Delta Z}{2} \qquad \text{and} \qquad Z_2 = \frac{Z - \Delta Z}{2} \qquad (4.52)$$

An arbitrary CM voltage source $V_{SRS} = V_{CM}$ is now assumed present in the SRS due to one of the mechanisms described in Section 0, namely, electric, magnetic, or conduc-

---

*Boundary conditions are defined and discussed in Chapter 2.

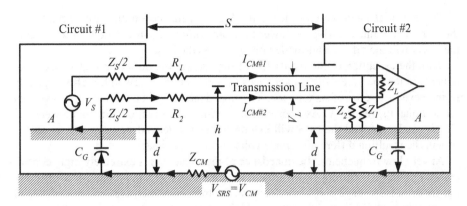

<u>Figure 4.87.</u> Typical ground-loop path generating common-mode to differential-mode conversion.

tive coupling. This interference voltage source will cause current to flow in the circuit loop, limited by the loop impedance, developing some noise voltages on the unequal impedances Z_1 and Z_2. Due to this load imbalance, some differential voltage V_L (or V_{DM}) will develop across the differential load Z_L, introducing potential interference into the load circuit. Ignoring the reactive impedance of the transmission line, the total differential voltage developed across the load is equal to

$$V_L = I_{CM\#1} \cdot \left(\frac{Z + \Delta Z}{2} \right) - I_{CM\#2} \cdot \left(\frac{Z - \Delta Z}{2} \right) + I_S \cdot Z_L \qquad (4.53)$$

where I_S stands for the differential current circulating in the circuit due to the differential voltage source, V_S.

By definition, common-mode currents $I_{CM\#1}$ and $I_{CM\#2}$ are equal; hence, it follows that when $\Delta Z = 0$, representing a perfectly balanced load, Equation (4.53) simply reduces to

$$V_L = I_S \cdot Z_L \qquad (4.54)$$

If the circuit is perfectly balanced, implying $Z_1 = Z_2$, no voltage will develop across Z_L due to the common-mode currents flowing though Z_1 and Z_2, and the signal developed on the load will result from the differential voltage source. However, this is seldom the case, thus $Z_1 \neq Z_2$. An interference voltage difference will be present proportional to the difference in impedance ΔZ.

For the purpose of the following discussion, an equivalent circuit for this problem corresponding to Figure 4.87 is depicted in Figure 4.88.

Note that differential input V_S is not included. It is assumed to be an ideal voltage source; the source internal impedance Z_S is depicted as being equally split between both lines, representing a balanced circuit. From here, the frequency-dependent characteristics of the system can now be investigated.

The essential factor to consider is the loop impedance limiting the current flow through the circuit. Clearly, the lower the common-mode current flow in the circuit, the lower will the differential-mode voltage across the load.

With the exception of the (split) source impedance, the dominant impedance components in the loop are stray capacitances C_G between the PCBs to the SRS, typically in the several to tens of nanofarads. The impedance of this capacitance is inversely proportional to frequency. As a result, common-mode current circulating between the circuit conductors and the SRS will exhibit frequency dependency and, subsequently, so will the resulting differential-mode voltage across the load.

At very low frequencies, the impedance of the capacitors is extremely high, considerably limiting RF current flow through the circuit, thus maintaining a low-differential-mode noise voltage across the load. As the frequency increases, the impedance of the capacitances C_G decreases, allowing higher current flow through the circuit, resulting in higher differential-mode noise voltage across the load. However, as frequency further increases, the impedance of the capacitors becomes negligible as the low-pass characteristics of the conductors become gradually influential in dominating the loop impedance. Consequently, significant high-frequency current limiting occurs after peaking, followed by a droop in the load voltage curve at high frequencies.

Three distinct cases are now investigated, originating from the circuit depicted in Figure 4.88.

Case A: Floating Circuit
Figure 4.88 depicted a circuit that is totally floated with respect to the SRS, creating a true differential and balanced system. The circuit is balanced at lower frequencies. Due to the high impedance of the capacitance to the SRS, C_G, the impedance of the loop formed by the SRS and the conductors is dominated by this capacitance. Only a negli-

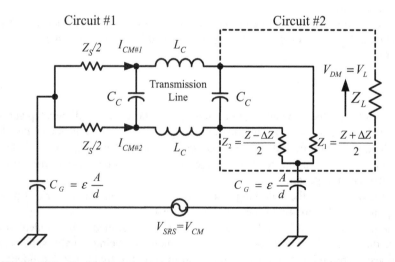

Figure 4.88. Equivalent circuit for analysis: "Floated" circuit (initial situation).

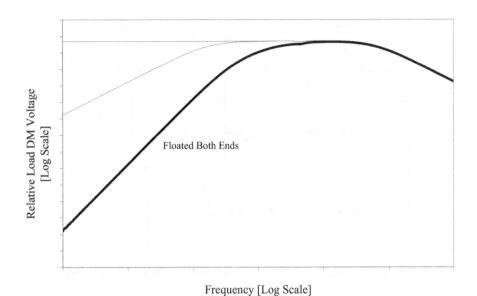

Frequency [Log Scale]

Figure 4.89. Relative DM interference from CM noise coupling in a floating circuit (thick line).

gible fraction of the CM current may, therefore, flow in this loop. As frequency increases, however, the capacitive impedance drops progressively, allowing more current to flow in the loop. This is displacement current flowing through the stray capacitance, resulting in an increase in the CM voltages developed across the common-mode load impedances Z_1 and Z_2, $V_{CM\#1}$ and $V_{CM\#2}$, respectively. As these voltages are not equal due to the imbalance in the circuit, they result in a differential-mode voltage V_{DM} = V_L across load Z_L corresponding to the system CMRR. The thick curve in Figure 4.89 depicts the frequency dependence of the voltage across the load.

Case B: Circuit Connected to SRS ("Grounded") at One End
Figure 4.90 depicts a situation in which the circuit is connected to the SRS at one end, with the other end remaining floated. For this analysis, it is immaterial whether the source or the load side of the circuit is grounded. In this situation, the circuit may still be considered balanced, since the signal current can return only through the return conductor.

In this case, the low-frequency loop impedance, dominated by the impedance of the capacitance to the SRS, C_G, has been significantly reduced (when the capacitances dominate the circuit loop impedance, the loop impedance will be approximately halved). From the two capacitances existing between the circuits to the SRS, only one remains, due to the grounding at one end, shorting out the capacitance between the PCB and the SRS. As the loop impedance is reduced, the common-mode current flowing through the loop formed by the conductors and the SRS is accord-

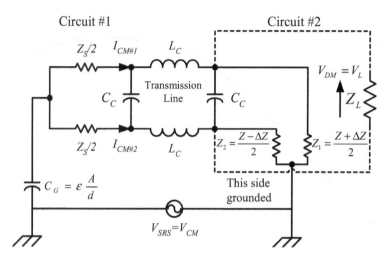

Figure 4.90. Equivalent circuit for analysis: grounded at one end (at load end).

ingly increased, resulting in an increase of the low-frequency differential-mode voltage across the load. No significant change is observed in the high-frequency curve since it is dominated by factors other than the capacitance to the SRS, primarily, the transmission-line low-pass characteristics. The thick curve in Figure 4.91 depicts the frequency dependence of the voltage across the load in this situation. It is noteworthy that when the loop impedance is dominated by the capacitances at both ends, the transition from a floated circuit (Case A) to the single-point grounded system (Case B) will result in an increase of coupled interference at the lower frequencies. Note that when the capacitive reactance between the circuits' ground and the reference structure dominates the impedance of the circuit's CM loops, elimination of the one capacitance due to grounding at one end will result in about 6 dB increase of coupled noise level at lower frequencies, due to the approximately 50% (or 6 dB) reduction in loop impedance.

Case C: Circuit Connected to SRS ("Grounded") at Both Ends
Figure 4.92 illustrates the circuit when it is connected to the SRS at both ends. With both sides grounded, an unbalanced system is produced. The capacitance to the SRS, CG, is bypassed at both ends of the circuit to the SRS. As a result, the loop impedance is dominated at lower frequencies only by the termination impedances and by transmission-line reactance (which is influential at higher frequencies only). Consequently, the circuit introduces little loop-current limiting at lower frequencies, resulting in high differential-mode voltage across the load even at lower frequencies. At higher frequencies, however, no difference is observed as the high-frequency performance is dominated by other factors, no different from than in both previous cases (Figure 4.93).

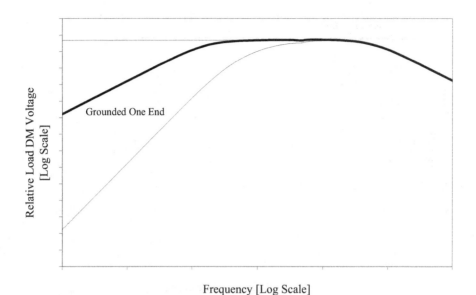

Figure 4.91. Relative DM interference from CM noise coupling in a circuit grounded at one end (thick line).

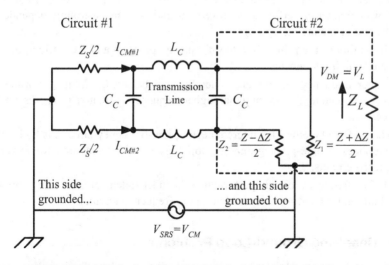

Figure 4.92. Equivalent circuit for analysis: Grounded both ends.

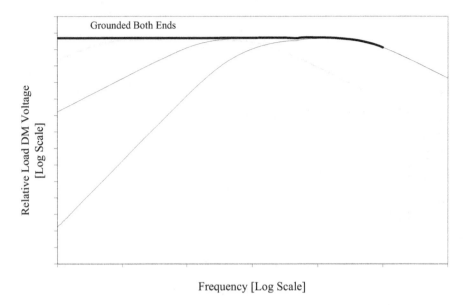

Figure 4.93. Relative DM interference from CM noise coupling in a circuit grounded at both ends (thick line).

Lessons Learned

- Imbalance in differential circuit will yield CM-to-DM conversion, introducing DM interference across the load.
- The effects of circuit imbalance are primarily a concern at lower frequencies. Lower-frequency performance of circuits and systems is strongly dependent on the circuit grounding topology.
- Ground loops are primarily a low-frequency problem and are difficult to eliminate at high frequencies.
- Higher frequency circuits are generally well behaved. Circuit performance is very predictable, considering that higher frequency current-return paths are foreseeable.
- The preferred method for interconnecting low-frequency circuits is by balanced interfaces. Single-point grounding topology should be favored for lower frequency circuits.
- Higher frequency circuit performance is independent of grounding topology. Multipoint grounding is favored in higher frequency circuits.

4.5.5 Resolving Ground-Loop Problems

The previous discussion clearly demonstrated that "ground loops" arise when multiple connections to the same signal reference occur, and are a direct result of:

- Potential differences across the signal reference structure
- External interference sources introducing conductively, inductively or capacitively undesired currents into the circuit formed between circuit conductors and the signal reference

Avoiding (or controlling) the ground-related interference resulting from ground loops or "living with it" are the essence of optimal grounding design for the purpose of EMI control.

There are many ways to address ground loop problems. So far, two primary methods have been proposed for addressing the challenge of common impedance or ground loop interference coupling via the return conductors (ground system), namely:

1. Improving the Quality of the Grounding Return System (Section 0)
2. Avoiding Common Return Connections by reverting to a single-point grounding architecture (Section 0)

In practice, many electronic systems have specific sensitivities and configurations that prohibit or make floated or single-point grounding topologies difficult to implement. Fortunately, a third approach is available to reduce or eliminate ground loops and their effects on systems: the elimination of interference current circulating between the circuit conductors and the signal reference structure. This approach is commonly known as "breaking ground loops."

4.5.5.1 Breaking Ground Loops. The key to resolving ground-loop problems lies in the opening or breaking of the loops. In an effort to eliminate ground loops, driving a load with a differential driver* is recommended [1], as illustrated in Figure 4.94. In logic state "1," transistors T_1 and T_4 are "on" while T_2 and T_3 are "off." The capacitance C is assumed to exhibit a short circuit at high frequencies. The transistors are assumed to represent very low forward saturation impedance, typically less than 5 Ω. From observation, the impedance to "GND" from each of the transistors T_1 and T_4 is *not* identical. This "totem pole" configuration does not exhibit a balanced source! The difference between the impedance to "GND" from each of the transistors T_1 and T_4 is denoted ΔZ_S.

Figure 4.95 depicts a noise-equivalent circuit. As the GND return connection points A and B are remotely spaced, a "ground noise" source V_N is assumed to be present across some noise impedance Z_N. This difference will introduce a differential noise across the differential load, Z_L, comprised of Z_{L1} and Z_{L2}.

The noise voltage across the differential load V_L, due to the ground-noise source V_N, is expressed as

---

*Note, however, that so-called differential networks are not truly differential, and the term "pseudo-differential" is more appropriate. A differential driver comprises two complementary single-ended drivers, referenced to a common return (or "ground"), compromising the network's differential nature. The receiver, however, is truly differential, as it senses the difference between the two nets, independent of the common "ground," thus providing high common-mode rejection and noise immunity.

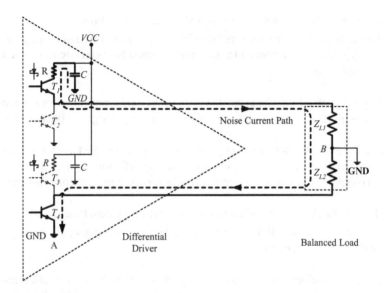

Figure 4.94. A Presumed differential "totem pole" source driving a balanced load.

$$V_L = \left(\frac{Z_{L1}}{Z_{L1} + \Delta Z_S + Z_N} - \frac{Z_{L2}}{Z_{L2} + Z_N} \right) \cdot V_N \cong \left(\frac{\Delta Z_S}{Z_{L1} + \Delta Z_S} \right) \cdot V_N \; ;$$

$$Z_{L1} = Z_{L2}, \; Z_N \ll Z_{L1}, Z_{L2} \qquad (4.55)$$

Note that a balanced load is assumed, and that the impedance between the driver and load return connections (A and B) is small compared to the high load impedance.

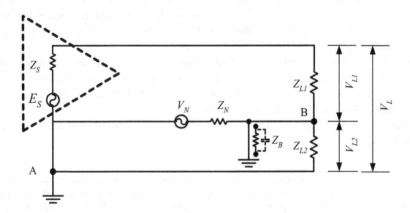

Figure 4.95. A noise equivalent circuit of the differential "totem pole" source driving a balanced load.

From Equation (4.37), combined with Equation (4.55), we can now determine the LCL expected from this "balanced" circuit [14]:

$$LCL_{dB} = 20 \ \log \left| \frac{V_{CM}}{V_{DM}} \right|_{V_0 = constant} = 20 \ \log \left| \frac{V_N}{V_L} \right|_{V_0 = constant} \cong \left(\frac{Z_{L1} + \Delta Z_S}{\Delta Z_S} \right) \quad (4.56)$$

Obviously, matched transmission lines, where Z_{L1} is represented by characteristic impedance and Z_0 results in high LCL, that is, high common-mode to differential-mode conversion or little common-mode rejection implies significant ground-loop interference! We may conclude, therefore, that a differential driver does not necessarily guarantee high common-mode (and ground-loop interference) suppression.

A hint to a possible solution results from an interesting observation emerging from Equation (4.56): The differential interference voltage V_L, developed across load Z_{L1} (and Z_{L2}) by the common-mode source V_N, is a function of the circuit balance as well as the circuit load impedance. The better the source balance and/or the higher the load impedance, the lower will be the interference voltage across the load.

One possible approach for reducing source impedance imbalance, ΔZ_S, is to replace the current limiting resistor R in the collectors of T_2 and T_4 by Schottky diodes. This results in a close to perfect matching of the differential driver output.

High impedance inserted in series between point B and the load ground, instead of a short circuit (Figure 4.95) will have no effect on the differential signal path between the driver and the load. It will, however, have a substantial effect on the suppression of ground-loop current flowing between points A and B. Equation (4.56) is now converted to

$$\frac{V_L}{V_N} = \left(\frac{Z_{L1}}{Z_{L1} + \Delta Z_S + Z_N + Z_B} - \frac{Z_{L2}}{Z_{L2} + Z_N + Z_B} \right) \cong \left(\frac{\Delta Z_S}{Z_{L1} + Z_B} \right) \to 0; \ Z_B \gg Z_N, \ \Delta Z_S$$

$$(4.57)$$

Hence, if Z_B is large compared to the ground noise impedance Z_N and the source imbalance resistance ΔZ_S, the voltages across Z_{L1} and Z_{L2} will approach an equal value. The two fractions in the left-hand part of Equation (4.57) become equal and the entire expression vanishes. A practical value of such resistors would be in the tens of kΩ. Often, this resistor is bypassed with a 10 nF capacitor in order to decouple high-frequency interference that may develop at point B due to unrelated sources.

Four elementary approaches can, therefore, be employed for breaking ground loops, easily remembered as B^2-I-D:

- **Blocking**
- **Balancing**
- **Isolation**
- **Diversion/Decoupling**

All of the above, with the exception of the last, have to do with the utilization of electrical isolation techniques. The different techniques for implementing each of the above approaches are now discussed.

4.5.5.1.1 ISOLATION TRANSFORMERS. An isolation transformer (see Figure 4.96) is an effective device for blocking common-mode interference coupling and eliminating undesired effects of ground loops. The ground-loop interference currents and voltages are still present, but will be observed only at the input terminals of the transformer.

The transformer accomplishes this function by magnetically coupling the desired signal from its primary to its secondary. This does not require any galvanic connection or common reference between source and load.

In an ideal transformer, which has no parasitics, leakage, or losses and exhibits an infinitely large inductance in the primary and secondary windings, the voltage transfer ratio is directly related to the windings ratio, n, resulting in a secondary-to-primary impedance ratio of n^2.

A practical transformer, however, exhibits a leakage inductance factor, $(1 - k)$; thus, the inductance of each winding $(1 - k)L_1$ and $(1 - k)L_2$ results in an effective coupling coefficient, k. Using M for the mutual inductance between the two windings,

$$k = \frac{M}{\sqrt{L_1 L_2}} \leq 1 \tag{4.58}$$

When an ideal transformer ($k = 1$), exhibiting no parasitics, is inserted in the transmission line (Figure 4.97), ground (common-mode) noise appears only across the windings of the transformer and not across the load. The desired signal current, on the other hand, couples perfectly from source to load through the transformer to loads Z_{L1} and Z_{L2}.

Any noise coupling that occurs is a result of parasitic capacitance, C_P, between the primary and secondary windings of the transformer (Figure 4.96). Noise developed across the load Z_L ($Z_A + Z_B$) is

$$V_L = \frac{Z_L}{Z_L + \dfrac{1}{j\omega C_P}} = \frac{j\omega C_P Z_L}{1 + j\omega C_P Z_L} \tag{4.59}$$

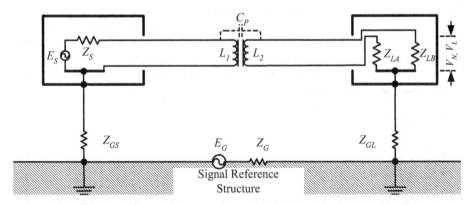

Figure 4.96. Isolation transformers block the common-mode current path, balance the transmission line, and galvanically isolate the source and load circuits.

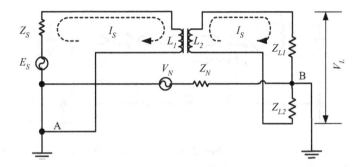

Figure 4.97. Effect of the isolation transformer in the circuit of Figure 4.95.

The degree of common-mode isolation achieved by transformers is limited by the interwinding capacitive parasitics. Capacitance is frequency dependent, which limits the high-frequency performance of the transformer. Improvement of high-frequency performance can be achieved by introducing a shield between the windings (Figure 4.98). In this case, the shield should be connected to the signal reference (the chassis). If shielded transformers are used, both the interwinding shield as well as the shielded enclosure of the transformer must be grounded for electrical safety. This shield is commonly called a "Faraday shield."

Grounding of the shield is not arbitrary and should be connected to the output of the transformer at point (2). If the shield were grounded at point (1), the potential across C_{P1} would be zero and the ground noise potential E_G would have been coupled directly to capacitance C_{P2}, facilitating noise coupling to the transformer's output.

In certain cases (Figure 4.99), particularly when isolation transformers are used at both ends of the line, such as in certain data bus applications, further improvement of the transformer's RF isolation can be achieved. This is accomplished by introducing an RF

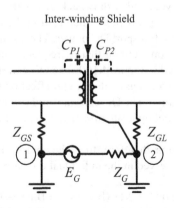

Figure 4.98. An interwinding electrostatic shield in an isolation transformer reduces common coupling.

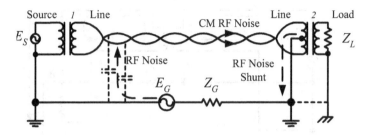

Figure 4.99. An RF short from the transformer's primary winding center tap to the local ground further improves RF isolation.

short from the line-side center tap of the transformer T_2 primary winding (if transformers are placed at both ends, such shorts can be implemented at both ends). If load Z_L is isolated from source grounding, the center tap should be grounded to the source ground and also interconnected to the load grounding point (indicated by a dashed line in Figure 4.99), or the center tap should be connected to the enclosure chassis.

Isolation transformers exhibit excellent performance at lower frequency AC applications and, in particular, find use in audio and AC-power circuits, where common-mode rejections of 100 to 140 dB can be achieved at $f = 1$ kHz. Isolation transformers can also be found in data bus and local area network (LAN) applications such as the MIL-STD-1553 multiplex bus and 10/100/1,000BaseT Ethernet.

The primary disadvantages of isolation transformers are their physical size, requiring costly real estate on printed circuit boards, and additional wiring to transport both the signal and return between source and load, as well as relatively high cost. In addition, when isolation of multiple circuits is required, a dedicated transformer is required for each circuit.

From the EMC point of view, isolation transformers may be intrinsically susceptible to interference from stray external magnetic fields because inductive coupling involves the use of magnetic fields for signal transmission.

Frequency limitation is yet another drawback of isolation transformers: their high-frequency limitation is due to parasitics, which restrict their usage to several hundred MHz. They cannot be used to transport DC power. The last drawback, however, can be overcome by using a "phantom" circuit that is frequently used in balanced telephone and data networks, illustrated in Figure 4.100. The DC current flowing through the circuit (and the transformers' windings) will have no effect on the signal itself, nor will it saturate the transformers' cores. The DC current is fed symmetrically into the line-side windings, producing equal but opposite magnetic flux, resulting in zero total magnetic flux in the cores.

The term "phantom" stems, therefore, from the obvious fact that, due to symmetry, the DC power cannot be observed in the individual transformers.

4.5.5.1.2 COMMON-MODE CHOKES (BALUNS OR BIFILAR CHOKES). One of the most effective techniques for controlling common-mode current flow is the common-mode choke, also known as a "balun" (BALanced–UNbalanced transformer) or "bifilar

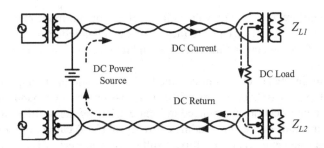

Figure 4.100. "Phantom" circuits can be used to transport DC power through an isolation transformer system.

choke." One advantage of the common-mode choke is that the device exhibits high series impedance to common-mode interference current propagating in the circuit, without affecting propagation of the functional differential-mode signal. Figure 4.101 illustrates the application of the common-mode choke.

The common-mode currents applied to the circuit, I_{CM1} and I_{CM2}, will develop a voltage across the inductors of the common-mode choke (ignoring the wiring resistance and all parasitics):

$$V_{L_1} = j\omega L_1 I_{CM1} + j\omega M I_{CM2}$$
$$V_{L_2} = j\omega L_1 I_{CM2} + j\omega M I_{CM1}$$

(4.60)

With tightly coupled inductors, that is, $k = 1$ and $M = L_1 = L_2 = L$, and assuming a perfectly balanced system, implying $I_{CM1} = I_{CM2} = I_{CM}$,

$$Z_{L_1} = Z_{L_2} = 2j\omega L$$

(4.61)

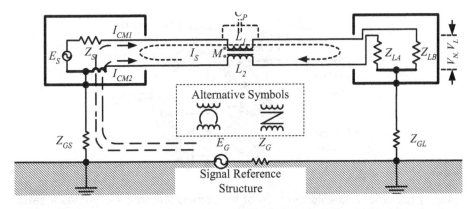

Figure 4.101. Common-mode chokes block common-mode interference current but maintain the metallic path for uninterrupted differential-mode current.

It follows therefore, that the equivalent circuit of a common-mode choke with respect to common-mode currents is equal to the choke's winding inductance for each branch. For $k = 0$ or $M = 0$ (uncoupled inductors), each branch will represent a separate inductor of value L.

The key to this behavior of common-mode chokes lies in their unique construction, depicted in Figure 4.102: Two wires of the transmission line are wound on the choke's core in the same direction, symbolized in Figure 4.101 by the position of the dot present at the same side of the two windings. The signal (DM) current introduces into the core equal but opposing magnetic flux, resulting in near perfect flux cancellation. A common-mode choke has no effect on differential-mode functional signal current. Common-mode currents, on the other hand, flow in the same direction in the core's windings by virtue of their common source E_G, introducing equal and additive magnetic flux in the choke's core, which, with the high permeability of the core, will represent a very high series inductance. This effectively inhibits the CM current, suppressing its effect on the circuit's differential load, Z_L.

A simple mathematical derivation [15] can be very helpful in understanding how common-mode chokes work. Examine Figure 4.103, in which the equivalent circuits for differential-mode and common-mode currents are identified. Applying Kirchhoff's voltage rule for the differential-mode current case (a), we obtain, neglecting Z_G,

$$E_S = j\omega(L_1 + L_2) \cdot I_S - 2j\omega M \cdot I_S + (Z_S + Z_L + R_{C1} + R_{C2}) \cdot I_S$$

$$\Rightarrow I_S = \frac{E_S}{Z_S + Z_L + R_{C1} + R_{C2}} \approx \frac{E_S}{Z_S + Z_L}; \ L_1 = L_2 = M; \ (Z_S + Z_L) \gg (R_{C1} + R_{C2})$$

$$(4.62)$$

Based on the assumption of $k = 1$ and $M = L_1 = L_2 = L$, and neglecting the wiring resistances, R_{C1} and R_{C2}, it is clear that a common-mode choke has no effect whatsoever on the differential signal.

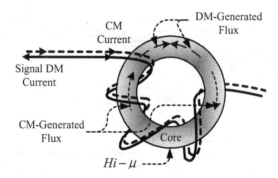

Figure 4.102. The unique construction of the common-mode choke makes it seem "transparent" to differential-mode signals and highly inductive to common-mode currents.

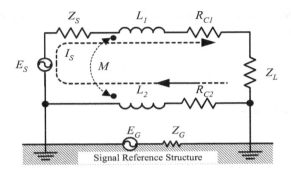

(a) Differential-Mode Current Case

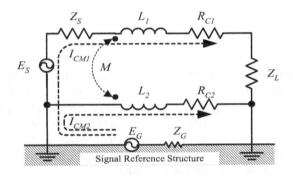

(b) Common-Mode Current Case

Figure 4.103. A simple CM and DM circuit analysis of common-mode choke embedded circuits.

As for the common-mode current in case (b), we find

Upper Loop: $\qquad E_G = j\omega L_1 \cdot I_{CM1} + j\omega M \cdot I_{CM2} + \left(Z_S + Z_L\right)\cdot I_{CM1}$

$$\Rightarrow I_{CM1} = \frac{E_G \cdot R_{C2}}{j\omega L \cdot \left(Z_S + Z_L + R_{C2}\right) + \left(\left(Z_S + Z_L\right)\cdot R_{C2}\right)}$$

Lower Loop: $\qquad E_G = j\omega L_2 \cdot I_{CM2} + j\omega M \cdot I_{CM1} + R_{C2}\cdot I_{CM2}$

$$\Rightarrow I_{CM2} = \frac{E_G - j\omega M \cdot I_{CM1}}{j\omega L_2 + R_{C2}}$$

$$(4.63)$$

In this case, only R_{C1} is neglected, as it is in series with $(Z_S + Z_L) \gg R_{C1}$, whereas R_{C2} is in series with inductor L_2 only.

Since load noise voltage V_N is a result of I_{CM1} flowing through it and since $(Z_S + Z_L) \gg R_{C2}$, we derive

$$V_N = I_{CM1} \cdot Z_L \approx \frac{E_G \cdot R_{C2}/L}{j\omega + R_{C2}/L} \qquad (4.64)$$

Rearranging Equation (4.64):

$$\frac{V_N}{E_G} \approx \frac{R_{C2}/L}{j\omega + R_{C2}/L} \qquad (4.65)$$

This indicates that V_N/E_G is minimized when $L \gg R_{C2}/\omega$. From here, the condition for the effectiveness of the common-mode choke is derived:

$$\omega \geq 5 \cdot \frac{R_{C2}}{L} \qquad (4.66)$$

Figure 4.104 illustrates graphically the situation described mathematically in Equation (4.65) and the conditions of Equation (4.66). For achieving high effectiveness of the common-mode choke, high-permeability (μ) cores are required. Typically, values of relative permeability are $\mu_r \approx 2000$ for low-frequency applications and 100 to 200 for high-frequency applications, resulting in high inductance. In terms of suppression, common-mode rejection exceeding 80 to 100 dB can be achieved beyond the cutoff frequency of the functional portion of the choke.

The primary advantage of the common-mode choke is its capability to reject high-frequency common-mode signals propagating along a transmission line with little effect on the functional differential signals. With care, multiple common-mode chokes

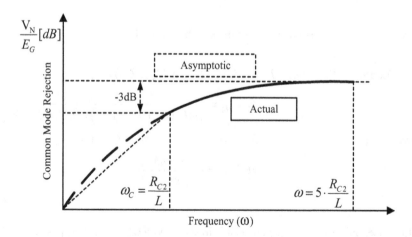

Figure 4.104. Conditions for common-mode chokes effectiveness.

can also be wound on the same core structure, increasing the density or number of signal lines that the choke can handle. Also, common-mode chokes can be used on circuits employing more than two wire transmission lines, such as three-phase AC-power circuits. In this situation, the phase lines and neutral must all be wound on the same core to ensure optimal performance and preclude saturation of the core.

Another advantage in using common-mode chokes is their capability, unlike isolation transformers, to pass DC power without running the risk of saturation of a high-μ core, thanks to DM current-induced flux cancellation. Practical common-mode chokes will exhibit some leakage inductance due to imperfections in the symmetry of the windings, or $k < 1$, thus $L'_1 = (1 - k)L_1$ and $L'_2 = (1 - k)L_2$. With $L_1 = L_2$, the total leakage inductance affecting the differential signal path is

$$L_T = L'_1 + L'_2 = 2L_1 \cdot (1 - k) \tag{4.67}$$

Using Equation (4.62) while neglecting R_{C1}, R_{C2} but inserting L_T in series with the source and load impedance yields

$$E_S = j\omega L_T \cdot I_S + (Z_S + Z_L + R_{C1} + R_{C2}) \cdot I_S \Rightarrow I_S \approx \frac{E_S}{(Z_S + Z_L) + j\omega L_T} \tag{4.68}$$

Even with low winding losses (R_{C1} and R_{C2}), some suppression of differential functional signal occurs. For $k \gg 0.9$, for instance, suppression can be significant. Typical common-mode chokes exhibit $k = 0.98$, whereas for high quality chokes, k may be as high as 0.999.

Similar to the isolation transformer, parasitic capacitance limits the high-frequency performance of common-mode chokes, particularly due to parasitic capacitance which, at higher frequencies, shorts out the windings of the choke. Common-mode chokes may also be somewhat susceptible to interference from stray external magnetic fields, although to a lesser extent compared to isolation transformers. Also, like isolation transformers, physical size and cost are the main disadvantages of using common-mode chokes.

4.5.5.1.3 OPTOCOUPLERS AND OPTICAL ISOLATORS. Another technique to prevent ground loops and minimize common-mode current flow is optical isolation. Optocouplers, one of the oldest and most commonly used isolation devices, are digital isolators based on optical coupling principles. Light-emitting diodes (LEDs) produce light signals when a voltage is applied across them. Optical isolation uses a LED along with a photo-detector device to transmit signals across an isolation barrier using light as the method of data transmission. A photo detector receives the light transmitted by the LED and converts it back to the original signal. As observed in Figure 4.105, the optical isolator interrupts the direct metallic transmission path between two circuits completely. Ground noise voltage appears between the input LED and the output phototransistor terminals rather than across two input terminals.

Note: A fiber-optic link could be considered an extension of the optical isolator. The source will drive a light-emitting diode converter, which transforms electrical to optical signals, whereas, at the end of the fiber-optic link, the phototransistor converts the light signals back to electrical signals.

Optic isolators perform best when a large voltage potential exists between circuits. In particular, optical isolators are best suited for digital designs owing to their nonlinearity. Nonlinearity may introduce distortions in analog circuits, but in certain cases could be used in small-signal analog circuits. For low-level analog (< 10 mV) circuits and precise tracking applications, optic isolators and fiber optics can be used at the cost of increased complexity of the circuitry.

Optical isolation is one of the most commonly used methods for isolation. One benefit of using optical isolation is its immunity to electrical and magnetic noise. With respect to control of EMI, optical isolators constitute a close to perfect solution for preclusion of ground loops, offering isolation exceeding 120 dB at DC but dropping to 20 to 30 dB at 30 MHz, approximately. This higher frequency limit stems from the optic isolator's relatively large intrinsic parasitic capacitance, typically on the order of ~2 pF between the LED and the phototransistor. This capacitance limits high-frequency isolation performance.

To enhance common-mode rejection capability of optocouplers, an internal shield is sometimes inserted between the input LED and the phototransistor. The internal shield is a transparent conductive shield that allows optical coupling to the phototransistor but diverts electrically coupled current to the ground pin. The shield, shown in Figure 4.106, reduces common-mode current coupling by at least an order of magnitude, improving the common-mode rejection of the optical isolator. A typical unshielded optocoupler might have a specified common-mode rejection of 10 or 50 V peak and a slew rate of 1000 V/msec, whereas by using a shielded optocoupler, common-mode rejection may surge to ratings as high as 1500 V peak and a slew rate of 30,000 V/msec [16].

A main advantage of optical isolators is found in their use in digital circuits, even when relatively high potential differences exist, making them valuable when surge immunity requirements are applied. They are also physically small and relatively inex-

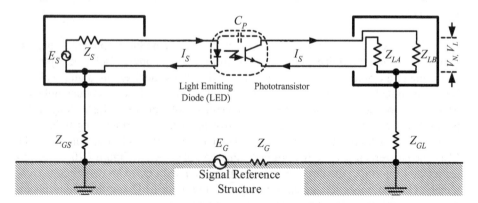

Figure 4.105. Optocouplers and optical isolators interrupt the direct common-mode interference current while optically transmitting the differential-mode signal.

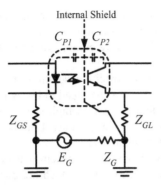

Figure 4.106. An Electrostatic shield between the light-emitting diode and phototransistor.

pensive. A primary disadvantage, however, lies in the limited use of optical isolators in analog circuits and, of course, their incapability to transfer DC or AC power. Additionally, the intrinsic low conversion efficiencies for electrical light conversion, low LED switching speed, and slow photodetector response lead to optocoupler limitations in terms of lifetime, transmission speed, and power consumption and dissipation.

4.5.5.1.4 CAPACITIVE COUPLERS/ISOLATORS. Capacitive coupling facilitates the transfer of electrical energy within a network by means of capacitance between circuit nodes. Capacitive coupling is typically achieved by placing a capacitor in series with the signal path to be coupled, such that only the AC signal from the one circuit can propagate to the other while DC and lower frequency signals are blocked (Figure 4.107). This technique thus helps to isolate the DC bias settings of the two coupled circuits. Capacitive coupling is also known as AC coupling and the capacitor used for the

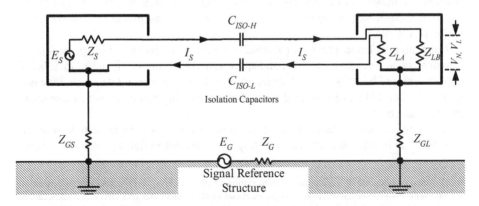

Figure 4.107. Capacitive isolators interrupt DC and lower frequency common-mode interference current while capacitively coupling higher frequency signals.

purpose is known as a coupling or DC blocking capacitor. Used in conjunction with isolated power supplies, capacitive isolators prevent noise currents on a data bus or other circuits from entering the local reference ground and interfering with or damaging sensitive circuitry.

Compared to optical isolation, capacitive isolation can support faster data transmission rates because there are no LEDs that need to be switched. Consequently, capacitive isolation is commonly found today in high-speed data acquisition systems and computer peripheral interfaces, such as Fast Ethernet [17] (e.g., 10BaseT and 100BaseT) and IEEE 1394 [18] data buses.

Performing isolation through the use of capacitors is not so obvious until the spectral content of signal and noise currents involved are considered. Signal frequencies between the data communication link and PHY* chips are typically in the tens to hundreds of MHz range. The signals are high-slew-rate analog or digital in nature with typical rise times of 1 ns or so. The ground noise spectrum, on the other hand, normally comprises much lower frequencies. Common frequency components result from the use of switching power supplies; still more common are the fundamental and harmonics of the power line frequency (e.g., 50 or 60 Hz). In reality, measured spectral content of ground noise currents seldom exceeds 1 MHz.

At higher frequencies, the impedance observed across isolation capacitors C_{ISO} (typically in the range of 10pF to 1nF) is much lower than the isolation resistance R_{ISO}, usually exceeding many MΩs. At 10 MHz, for instance, a barrier capacitance of 1nF has an equivalent impedance of

$$X_C = \frac{1}{2\pi f C_{ISO}} \approx \frac{1}{2\pi \cdot (10\times10^6)\cdot(1\times10^{-9})} \approx 16\,\Omega \qquad (4.69)$$

The higher frequency PHY to twisted pair link interface signals can thus "propagate through" a properly selected isolation capacitance, whereas lower frequency ground structure noise will be blocked or significantly attenuated. In fact, the most harmful frequencies, those near the power-line fundamental frequency, will be very greatly attenuated; a quick look at the ratio 50 Hz/100 MHz, for instance, yields a relative attenuation of 126 dB.

Capacitive isolation provides galvanic isolation up to 50 volts DC (or peak), approximately, and is a cost-effective and space-efficient solution. Thanks to their broad applications and benefits, several device manufacturers have integrated capacitive isolators into their PHY or interface devices, further reducing part count and, subsequently, space and cost.

Capacitive isolation also exhibits high immunity to stray magnetic field interference but may be adversely affected by stray external electric fields as capacitive isola-

*PHY (often pronounced "fi") is a common abbreviation for physical layer. PHY is also a generic electronics term referring to a special electronic integrated circuit or functional block of a circuit that takes care of encoding and decoding between a pure digital domain and a modulation in the analog domain. PHYs are often used to interface a field-programmable gate array (FPGA) or complex programmable logic device (CPLD) to a specific type of interface.

tion involves electric field coupling for data transmission. Capacitive isolation devices also exhibit lower transient immunity performance since some electrical fast transient (EFT) common-mode pulses propagate across the coupling capacitor and are not blocked or filtered out.

Capacitive isolation is suitable for many applications in the information technology (IT) world. However, certain industrial and commercial applications may require isolation voltage levels in excess of those provided by the capacitive scheme. Transformer isolation may be preferable in such cases. In other cases, combined capacitive and transformer isolation schemes are employed.

Capacitive isolation suffers the disadvantage of degrading lower frequency performance of systems containing capacitively coupled subsystems. Each coupling capacitor combined with the input impedance of the following stage forms a high-pass filter and each successive filter results in a cumulative filter exhibiting a 3dB corner frequency that may be higher than in each individual filter. So for adequate response at the lowest frequency of interest, the capacitors used must have high capacitance ratings.

4.5.5.1.5 HIGH-SPEED DIGITAL ISOLATORS. In many information technology (IT) and industrial applications, such as process control systems or data acquisition and control systems, high-speed digital signals must be transmitted from various sensors to a central controller for processing and analysis. When galvanic isolation is required on such interfaces, transformer-based isolators, optocouplers, and capacitively coupled isolators, presented in previous sections, may be inappropriate for such high-speed applications due to limited bandwidth and subsequent response time. Use of these isolators is further restricted owing to limitations on their incorporation in integrated circuits and the fact that they often require expensive hybrid packaging.

For addressing these objectives, monolithic high-speed digital isolators have recently been introduced as drop-in replacements for most optocouplers and magnetic isolators. High-speed digital isolators are divided into two main technologies: transformer-coupled and capacitively coupled isolators.

In addition to the fact that thousands of volts of isolation (2.5 kV to 4 kV, typically) can be achieved on-chip, these novel devices also make it possible to efficiently, accurately, and reliably transmit very high bandwidth signals, providing for signaling rates exceeding 100 Mbps.

Both isolation technologies offer the advantage of practically impervious immunity to external stray magnetic fields that frequently occur in the industrial environment and can distort signal integrity. On the other hand and in spite of their small size, these devices offer enhanced immunity to data corruption, providing a minimum protection level of 25 kV/μs due to common-mode fast voltage transients, when judged against that achievable in present-day high-performance optocouplers, typically 10 kV/μs.

The transformer-coupled* technology is based on small, chip-scale transformers [20]. The coupler has three main parts: a transmitter, transformers, and a receiver. The transmitter encodes and transmits the signal across the isolation barrier using the trans-

---

*See, for instance the Texas Instruments ISO721, ISO722 3.3-V/5-V High-Speed Digital Isolators data sheet, SLLS629D-January 2006, Revised February 2007.

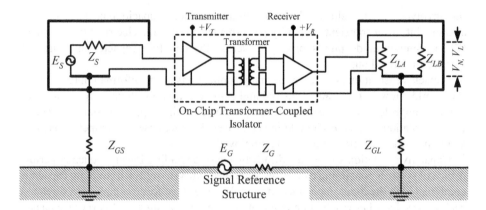

<u>Figure 4.108.</u> Simplified functional schematic of on-chip transformer-coupled, high-speed digital isolators.

former. The received signal is decoded on the other side by the receiver circuitry, as shown in Figure 4.108.

In addition to the enhanced performance discussed previously, this technology also offers advantages in terms of integration in the form of multichannel isolated devices. The bidirectional nature of inductive coupling facilitates bidirectional and multilevel signal transfer. One data channel can transmit signals in the one direction while another channel transmits signals in the opposite direction. One side of the channel could be at one level, say 2.7 V, for instance, while the other could be at 5.5 V.

The capacitively coupled* technology is based on the use of on-chip logic input and output buffers separated by a high-voltage capacitive isolation barrier employing a semiconductor-grade silicon oxide (SiO_2) dielectric substrate. A binary input signal is conditioned and translated to a balanced signal that is then differentiated by the capacitive barrier, as shown in Figure 4.109. Across the isolation barrier, a differential comparator receives the logic-transition information, accordingly setting the output circuit. These devices too can operate from two supply voltages of 3.3 V and 5 V or any combination thereof.

Both types (transformer- and capacitively coupled) of isolators may also be integrated into larger isolated data bus line transceivers (e.g., RS-422/RS-485[†]) and buffers or isolation amplifiers (see Section 4.5.5.1.6). A limited number of such devices are commercially available, exhibiting standard interface characteristics.

4.5.5.1.6   ANALOG DIFFERENTIAL, INSTRUMENTATION, AND ISOLATION AMPLIFIERS. Amplifiers are generally used in the analog front end of sensitive data-acquisition systems instrumentation such as that present in process-control systems and, particularly,

*See, for instance, the Analog Devices *i*Coupler® Digital Isolators, ADuM1100 data sheet C02462-0-10/03(E), Revision E, October 2003.
[†]See, for instance, the NVE Corporation IL422 3.3-V Isolated RS422/RS485 Interfaces, data sheet ISB-DS-001-IL422-M, September 2007.

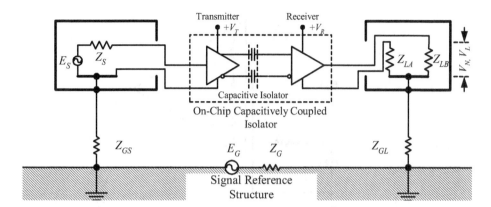

<u>Figure 4.109.</u> Simplified functional schematic of on-chip capacitively coupled high-speed digital isolators.

in medical measurement and monitoring equipment. In medical applications, amplifiers exhibiting a high degree of isolation usually serve to provide sufficient protection against electric shock to the staff and patients, for instance, when using defibrillators. They are also used for measurement of neurophysiological signals, such as ECGs and EEGs, as well as in microwave therapy, patient-monitoring devices, and so on [21].

Amplifiers used as isolators are unique in \ that they are gain devices, compared to all other techniques described above, and are thus of particular value for such measurement applications. Many measurements of neurophysiological signals, for instance, involve voltages at very low levels, typically ranging between 1 μV and 100 mV, superimposed with noise and interference from different sources. Amplifiers are commonly used to couple these low-level signals from high-impedance sources while making them compatible with devices such as recorders, displays, and analog-to-digital converters (ADCs) for digital data processing.

Amplifiers adequate for measuring neurophysiological signals must, therefore, satisfy several very specific requirements. They have to provide amplification specific to the signal, reject superimposed noise and interference signals, and guarantee protection against hazards or damage from voltage and current surges to patients and electronic equipment (i.e., isolation).* Three types of amplifiers, with increasing performance with respect to the above objectives, are described herein: differential amplifiers, instrumentation amplifiers, and isolation amplifiers.

Differential Amplifiers
In differential amplifiers, the differential measuring technique is used to minimize interference and noise coupling, which are usually superimposed on these low-level signals. A typical amplifier configuration is shown in Figure 4.110. Three leads, two of them collecting the measured signal V_{mea} and the third providing the reference potential V_{ref}, connect the sensor or object to the input of the amplifier. The output of the dif-

---

*Amplifiers featuring these specifications for medical applications are known as biopotential amplifiers.

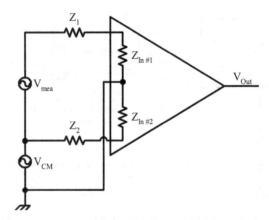

Figure 4.110. Typical configuration of a differential amplifier.

ferential amplifier is, therefore, a result of the difference between the two input signals times a gain factor.

Strong rejection of common-mode signals (i.e., high common-mode ejection ratio or CMRR) is one of the most important features of good differential amplifiers. Rejection of the common-mode signal is a function of both the amplifier CMRR and the source impedances Z_1 and Z_2. Desired values are on order of at least 100 dB.

Instrumentation Amplifiers

Instrumentation amplifiers exhibit superior performance (i.e., enhanced signal-voltage amplification and rejection of the common-mode signal) when weighed against standard differential amplifiers. Crucial to the common-mode rejection (or isolation) performance of the preamplifier is the input impedance, which should be as high as possible. The input stage of an instrumentation amplifier usually consists of two voltage followers, which have the highest input impedance of any common amplifier configuration, compared to standard single operational amplifier (op-amp) design without the requirement of close resistor matching (Figure 4.111).

The differential output of the first stage represents a signal with substantial relative reduction of the input common-mode signal and is used to drive a standard differential amplifier that further suppresses the common-mode signal. Complete instrumentation amplifier-integrated circuits based on this instrumentation amplifier configuration are commercially available.

Instrumentation amplifiers have some limitations, including offset voltage, gain error, limited bandwidth, and settling time. The offset voltage and gain error can be calibrated as part of the measurement, but the bandwidth and settling time are parameters that limit the frequencies of amplified signals.

It is noteworthy that standard differential interface amplifiers (including both differential and instrumentation amplifiers) are generally designed to allow low-to-zero chassis ground currents flow and may violate isolation needs when deenergized. Non-

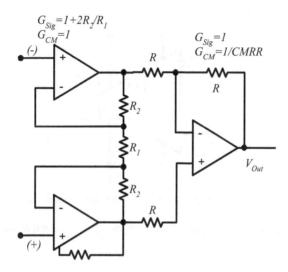

Figure 4.111. Typical configuration of an instrumentation amplifier.

intuitive paths in the devices may exist in this situation and may allow flow of unintentional ground currents. Also, both differential and instrumentation amplifiers (to a different extent) exhibit lower common-mode noise rejection compared to isolation devices such as transformers or optical isolators. Figure 4.94 and Figure 4.95 in Section 4.5.5.1 illustrate the shortcomings of real-world differential drivers as truly balanced sources.

Isolation Amplifiers

Isolation amplifiers are superior to standard differential and even instrumentation amplifiers. They are designed to provide isolation between two circuits powered from an isolated power and return system. This device includes a built-in amplifier stage that is decoupled from the input circuit by a high-impedance, low-capacitance stage (Figure 4.112).

In medical applications, for instance, amplifiers must provide sufficient protection against electric shock to the user and patient in the case of patient-coupled medical instrumentation. To that end, isolation amplifiers may be used to break detrimental ground loops and to provide isolation of the patient and electronic equipment.

An isolation barrier (labeled "isolation" in Figure 4.112) provides a complete galvanic separation of the input—patient and preamplifier—from all equipment on the amplifier output. The isolation barrier is realized by using one of the different techniques discussed in the previous sections, namely, transformer, capacitor, or optical isolation).

Isolation amplifiers are particularly useful in preventing noisy signals present in the source circuit signal reference from entering the output amplifier through common-mode paths inside the amplifier. Ideally, no flow of electric current should occur

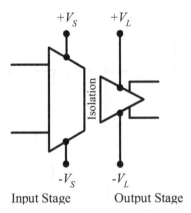

Figure 4.112. Typical configuration of an isolation amplifier.

across the barrier. A figure of merit for the extent of isolation in the analog isolation amplifier is the isolation-mode voltage, which is the voltage appearing across the isolation barrier, that is, between the input common and output common. The isolation-mode rejection ratio (IMRR) is the ratio between the isolation voltage and the amplitude of the isolation signal appearing at the output of the isolation amplifier. IMRR levels of 60 dB to 100 dB are commonly provided by isolation amplifiers; however, values as high as 120 dB may be achieved in particular models. Since the IMRR in practical devices is finite, some leakage will always be present across the isolation barrier. Figure 4.113 illustrates the use of an isolation amplifier in a circuit.

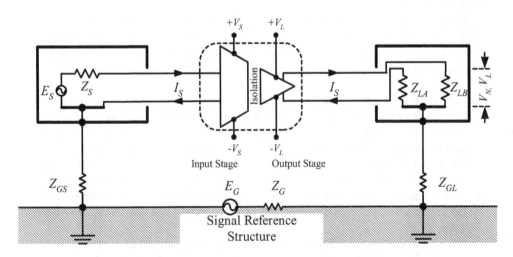

Figure 4.113. Isolation amplifiers include built-in amplifier stages separated by a high-impedance isolation barrier.

The main disadvantage of the analog isolation amplifier is cost, which includes the need for double the number of isolated power supplies, in addition to the high cost of the device itself.

4.5.5.1.5 CIRCUIT BYPASSING. Better classified as a filtering technique,* common-mode (i.e., line-to-chassis) capacitors may also serve in common-mode signal suppression as they decouple and divert common-mode current present on the lines away from sensitive circuits that that need to protected.

Unlike previous techniques (isolation transformer, common-mode choke, optical isolator, or buffer amplifier), bypass capacitors do not block the common-mode current circulation path. However, with capacitors, as illustrated in Figure 4.114, common-mode current is diverted away from the source and load impedance, as long as the capacitors exhibit sufficiently low impedance compared to the load's common-mode impedance. Recalling the discussion on "real-world capacitors" in Chapter 2, parasitics internal to the capacitors, particularly series-equivalent inductance, cannot be overlooked, as parasitics will significantly degrade the high-frequency performance of the nonideal bypass capacitor.

For ensuring adequate decoupling by the bypass capacitors, they must be properly selected, considering type, dielectric material, and package. Installation also plays a significant role in their performance, particularly the need to eliminate any lead inductance due to excessive wiring, trace length on the PCB, vias, and so on.

On the other hand, overbypassing can cause other undesired situations. Large capacitors on signal lines may distort the signal integrity of the circuit due to high-frequency insertion losses. Too much capacitance on power lines is of particular concern as it may compromise safety due to excessive leakage current, which must be kept to a minimal level to prevent a hazard of electric shock.

Bypass capacitors enjoy, therefore, the advantage of being a simple and inexpensive solution, in addition to being efficient devices from the standpoint of EMI mitigation and control.

One technique to overcome disadvantages of using bypass capacitors while providing adequate ground-loop decoupling is by combining several techniques, such as a common-mode choke with bypass capacitors. In this case, excessive use of capacitance can be compensated by large inductance of the choke, precluding a potential safety hazard or undesired signal degradation. An example of this application is in the design of a data line interface isolator for 10/100BaseT "Ethernet" twisted wire pair links, presented in the next Section.

4.5.5.1.8 SUMMARY OF INTERFACE ISOLATION TECHNIQUES. The isolation techniques presented in the above sections constitute the most common and effective techniques. Other techniques are occasionally used and all are intended to maintain the integrity of a grounding scheme by ensuring that little or no current flows through the signal reference structure path. Cost, performance, and other constraints will determine the technique most suitable for the application. Table 4.1 presents a summary of typi-

---

*Filtering and grounding of filters are presented in Chapter 8.

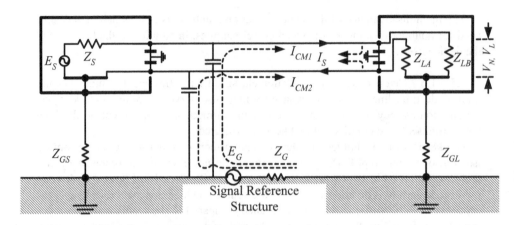

<u>Figure 4.114.</u> Common-mode bypass capacitors divert common-mode currents away from the load.

cal totally ("hard") isolated circuits, including those discussed above, whereas Table 4.2 shows circuits that provide a lesser degree of isolation [21]. Of course, "hard" isolation techniques are preferred but the "soft" techniques shown in Table 4.2 are better than having no isolation at all.

4.5.5.2 Example: Data Line Interface Isolation Design (10/100BaseT).
Data line isolation effectively breaks ground loops and prevents noise currents from affecting circuit performance. Some data transmission paths incorporate built-in isolation circuitry; others can easily be isolated by addition of proper isolation devices [17, 18].

The Ethernet 10/100BaseT Interface illustrates one manner of using data line interface isolation, which consists of several of the techniques discussed above.

4.5.5.2.1 DEFINITION OF THE 10/100BASET INTERFACE. Ethernet is the most widely used standard for connecting computers into local or wide-area networks (LAN/WAN). Various forms of Ethernet are known as 10BaseT, 100BaseT, and 1GBaseT.

This designation is an IEEE (Institute of Electrical and Electronic Engineers) identifier. The "10/100/1000" in the media type designation refers to the transmission speeds of 10, 100, or 1000 Mbps. The "Base" refers to baseband signaling. This means that only Ethernet signals are carried within the transmission medium. The "T" represents twisted pair and the "F" represents fiber optic cable.

4.5.5.2.2 TWISTED-PAIR INTERFACE. The Ethernet interface is required to exhibit close to perfect balance within the interface cable, both for preclusion of interference due to common-mode interference in the wiring and for elimination of radiated EMI from unshielded twisted wire pairs.

Table 4.1. Summary of techniques for total ("hard") interface isolation

Isolation scheme	Schematic	Advantages	Drawbacks	Common applications
Isolation transformer	GND_S GND_L	• Galvanic isolation • Differential signal transmission	• Large and bulky, heavy and costly • Limited frequency range* • Inappropriate for DC circuits	• Differential digital data buses (e.g., Ethernet) • Audio circuits • AC power circuits and isolated switch-mode power supplies
Optical isolators	GND_S GND_L	• Galvanic isolation • Small, SMT devices	• Active, power-consuming devices • Relatively limited frequency range	• Digital data and control circuits • Small-signal analog circuits
Capacitive couplers/ isolators	~1-100nF ~1-100nF GND_S GND_L	• Cost-effective and space efficient • Power-off isolated • Higher speed	• Degraded lower frequency performance • Extra part count	• High-frequency analog/digital links • "DC blocking"
High-speed on-chip digital isolators	Isolators GND_S GND_L	• Higher speed • High galvanic isolation • Cost-effective and space efficient • Power-off isolated	• Novel, nonstandard technique	• High-speed digital channels
Isolated analog operational amplifier (op-amp)	~1M ~1M GND_S GND_L	• Easy to implement • Power-off isolated	• Low bandwidth • Extra part count	• Analog measurement circuits
Analog isolation amplifier	V_S V_L Isolation GND_S GND_L	• True galvanic isolation • SMT devices	• Low bandwidth • Relatively costly	• Precision measurement circuits • Medical devices and instrumentation
Galvanically isolated line driver/receiver	Driver Receiver Isolators GND_{SI} GND_{SOt} GND_L	• Established interface standard • True galvanically isolated interface • Lower power consumption	• Current availability to few interfaces only	• RS-422 and RS-485 data interfaces

(continued)

Table 4.1. *Continued*

Isolation scheme	Schematic	Advantages	Drawbacks	Common applications
Remotely referenced (e.g., temperature sensors/ transducers)	GND_S GND_L	• Simple	• Electrical insulators often do not provide thermal conductivity • Low-frequency solution	• Commonly used in sensing circuits • In practice, single-point grounding
Relays (coils to contacts)	GND_S GND_L	• Excellent DC and AC isolation • High current capability	• Lower reliability and useful life • Low-frequency response • Large packages • Binary operation • High stray capacitance in large power-switching relays	• Power circuits • High current control circuits

*Note that in each of the above isolation devices and circuits, the input and output are isolated from each other by high DC impedance, usually on the order of 1 MΩ or more. AC isolation generally deteriorates at higher frequencies (above 1 MHz approximately), due mainly to stray capacitances that do not show on the schematics, particularly in operational amplifiers in which their common-mode rejection ratio (CMRR) decreases with increasing frequency.

To accomplish these objectives, the Ethernet Physical Layer interface consists of magnetics and other discrete components. Within the magnetics portion of the interface, we have an isolation transformer and common-mode choke, as well as bypass capacitors and termination networks for both receiver and transmitter.

4.5.5.2.3   RECEIVE INTERFACE CIRCUIT. The receive interface circuit (Figure 4.115) consists of magnetics, including an isolation transformer and a common-mode choke as well as series 270 pF capacitors for enhanced galvanic isolation. Some vendors place this filter on the line (primary) side of the main winding; others place it on the device (secondary) side. Either approach is acceptable. The center-tap of the isolation transformer at the line side is tied to chassis via a 10 nF capacitor, which is mandated within the interface specification.

However, when using this single device package with the common-mode choke adjacent to the PHY (physical layer device driver), the 10 nF bypass capacitor should not be connected from the device-side center tap to the chassis. RF noise from the chassis can efficiently couple through the capacitor into the center tap, bypassing the common-mode choke and resulting in potential EMI problems.

4.5.5.2.4   TRANSMIT INTERFACE CIRCUIT. The transmit interface circuit of Figure 4.116 also consists of magnetics that include an isolation transformer and a common-mode choke. Some vendors place the choke on the line (primary) side of the main wind-

Table 4.2. Summary of techniques for partial ("soft") interface isolation

Isolation scheme	Schematic	Advantages	Drawbacks	Common applications
Common-mode choke (balun, bifilar choke)	GND_S GND_L	• DC continuity • Rejects common-mode noise	• DC and lower frequency ground loops not precluded • Saturation of cores possible at high DC currents • Stray capacitance limits higher frequency isolation	• Power-line filters • Differential digital data buses (e.g., Ethernet)
Analog op-amp, differential amplifier	GND_S GND_L	• Easy to implement	• Not isolated when op-amp is not powered • Common-mode rejection may be insufficient	• Analog circuits
Balanced differential and instrumentation amplifier	GND_S GND_L	• Noise couples equally to both leads • Rejection through CMR	• AC isolation limited • No DC power-off isolation • Common-mode rejection degrades with frequency increase and with imbalance	• Generally not recommended
Line driver/receiver	Driver Receiver GND_S GND_L	• Established interface standard	• Possible lack of common-mode noise immunity • Possible lack of DC power-off isolation	• RS-422 interfaces
Circuit bypassing	GND_S GND_L	• Shunts higher frequency ground loop noise	• Provides no galvanic isolation • May affect performance of high-frequency circuits	• Common for analog and digital interfaces

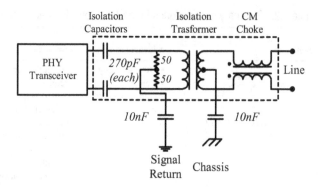

Figure 4.115. Typical 10/100BaseT receive interface circuit. (Note the different ground nodes.)

ing; others place it on the device (secondary) side. A few vendors include two transmit chokes, one on each side of the isolation transformer. Any of these approaches is acceptable.

The line-side center tap can be bypassed to the chassis, but this should be carefully evaluated in the system application. Bypassing both center taps of the transmit winding to the same reference may produce undesirable results by creating a low-impedance AC coupling between the chassis and the circuit's signal return.

Lessons Learned

- Avoiding (or controlling) ground loops or "living with them" is the essence of grounding design for EMC.
- Resolving ground-loop problems is accomplished by blocking, balancing, isolating, and decoupling (B^2-I-D).

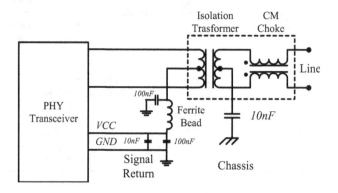

Figure 4.116. Typical 10/100BaseT transmit interface circuit. (Note the different ground nodes.)

- Differential drivers are desirable for eliminating ground loops; however, differential drivers comprise two complementary single-ended drivers referenced to a common "ground," compromising the network's differential nature.
- Perfectly balanced circuits rarely exist; appropriate rejection devices and circuits must be used.
- "Hard" isolation techniques for preclusion of ground loops and their effects are preferred but "soft" techniques are better than having no isolation at all.
- Isolation transformers provide high low-frequency, common-mode rejection but cannot transport DC signals (except in the "phantom" configuration), whereas common-mode chokes provide excellent high-frequency isolation and can also transport DC signals.
- Optic isolators and optocouplers are best suited for digital applications, whereas buffer amplifiers are mostly suited for low-level, low-frequency analog circuits.
- Capacitive couplers/isolators provide DC and lower frequency isolation; however, they may affect lower frequency components of the transmitted signals.
- The advantages of on-chip, high-speed digital capacitive and inductive isolators over traditional isolation techniques in terms of power consumption, signal bandwidth, robustness, and ease of integration will make them ideal choices for future demanding isolation applications.
- Isolation amplifiers are of particular benefit for low-level analog signals thanks to their intrinsic gain. Isolation amplifiers are superior to differential amplifiers and instrumentation amplifiers thanks to their intrinsic galvanic isolation.
- Bypassing signal and power lines provides a complementary solution but may cause signal degradation or compromise safety.
- Ground-loop problems can be reduced or eliminated with a coordinated grounding policy; no single technique can address all problems.

4.6 ZONED GROUNDING

Up to this point, fundamental concepts of grounding were discussed. Applications of grounding, ranging from assembly, system, and facility, are discussed in the following sections. However, before discussing grounding applications in large-scale, complex systems and facilities, the useful concept of "zoning" and its application to grounding is introduced.

4.6.1 The Zoning Concept as Applied to Grounding

Complex systems, ranging from critical-equipment assemblies to essential installations such as power plants, information technology (IT) centers, and military command, control, and communication (C^3) facilities, require interference-free operation, even in the presence of severe electromagnetic phenomena such as high-intensity radiated electromagnetic fields, nuclear electromagnetic pulses (NEMP), and lightning. The disruption or destruction of such systems and facilities can have far-reaching consequences. In order to safeguard the continuous availability of critical complex systems and facilities, protective measures must be implemented. For this purpose, the electromagnetic (EM) protection zones concept, initially developed for lightning protection,* is applied. The zoning approach enables designers and operators to plan, implement, and control cost-effective protective measures within a modern infrastructure.

According to the EM protection zone concept, a structure is separated into different risk zones. Within each zone, a comparable electromagnetic environment can be described. A significant difference in this environment occurs at the boundary of a protection zone. A zone boundary may be established by a structure exhibiting a certain degree of electromagnetic shielding or by installation of terminal protective devices (e.g., filters and transient-suppression devices). Zoning is of particular advantage in large systems and facilities in which multiple shielded boundaries may be used (note that metallic enclosures of equipment assemblies installed within a facility can also be treated as boundaries between zones). Appropriate "zoned grounding" architecture, consistent with the system and facility zoning, constitutes an inseparable element in the implementation of the electromagnetic protection zones concept.

Observe the grounding architecture depicted schematically in Figure 4.117. Metallic structures within each zone, such as equipment enclosures, cable shields, and conduits, which are not intentionally maintained at a potential different from the boundary shield, including the outer surface of the boundary with the higher-order zone,[†] are all grounded to the inner surface of the zone boundary through a single grounding conductor.

Within each zone, a grounding architecture should be maintained such that ground loops are precluded. In most cases, zone-wide grounding schemes are recommended (as illustrated in Figure 4.117). If application of multipoint grounding topology is required, for instance, in the case of high-frequency RF or digital circuits, composite (hybrid) grounding may be applied. The equipotential reference structure will then be enclosed within a single zone and grounded through a single conductor to the inner boundary of this zone.

4.6.2   Zoning Compromises and Violations

Grounding conductors should never traverse the shielded boundaries between zones. Going across shielded boundaries severely compromises integrity of the shields between zones. If such connections are necessary, these leads should be treated in a manner similar to any other penetrating conductors.

Figure 4.118 depicts two examples of EM protection zoning compromises. In Figure 4.118(a), components within Shield #2 are grounded to Shield #1 through an aperture in Shield #2. Consequently, Shield #2 is ineffective, because the penetrating grounding conductor "imports" Zone 1 into the area enclosed by Shield #2 (which

---

*The concept of the zoning for lightning protection is described in IEC 62305-4/FDIS, Protection against Lightning, Part 4: Electrical and Electronic Systems within Structures.
[†]EM protective zones are typically numbered from the external zone ("Zone 0") to inner zones ("Zone 1," "Zone 2," etc.). The higher the zone number, the higher is the order of the zone.

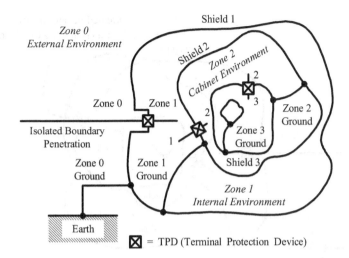

Figure 4.117. Zoned grounding in a complex shielded facility.

should have been Zone 2). In Figure 4.118(b), an even more severe violation is shown. Both Shield #1 and Shield #2 are violated due to the penetration of a grounding conductor from the external electromagnetic environment (Zone 0). Shield #1 and Shield #2 now form, in practice, a single shielded boundary. This shield now only includes the area between the shields and excludes the area within Shield #2. These two examples demonstrate a most important rule for grounding in multizoned facilities: Grounding conductors should never penetrate shielded boundaries.

When grounding conductors are connected to the boundary shield, the interference current flowing on the conductor is transferred to the shield surface. This current will typically flow on the surface to which the grounding conductor is connected, as long as the thickness of the shield is larger than the material's skin depth. In this manner, currents flowing on external conductors may be diverted to the external surface of the shield. Figure 4.119 shows grounding conductor connections implemented in an acceptable (a) or compromising (b) manner and in serious violation of the zoning scheme (c).

4.6.3 Impact of Zoning on Subsystem Grounding Architecture

Consider a sensitive system consisting of several assemblies that are interconnected by cables as illustrated in Figure 4.120. Each of the system's assemblies is packaged within a metallic (shielded) enclosure (Enclosure #1), creating a local zone for each (designated "Zone 2"). In addition, all assemblies but Assembly #4 are installed within a larger shielded enclosure, such as a shielded rack, a shielded room, a tactical shelter, or an aircraft (forming "Zone 1"). This enclosure provides a boundary between the external electromagnetic environment ("Zone 0") and the internal volume of the enclosure. Assemblies #1 and #2 as well as the power supply are interconnected by cables. In addition, an electrical connection exists between Assembly #2 (enclosed within the larg-

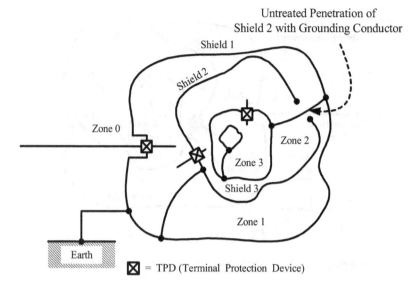

(a) Compromise of Shield #2 by Untreated Ground Conductor Penetration

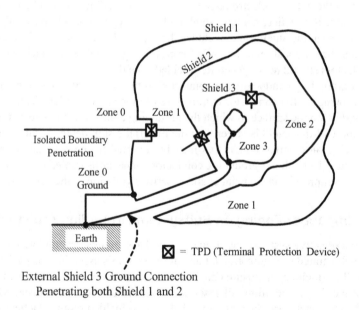

(b) Compromise of Both Outer Shields by External Ground Connection

Figure 4.118. Examples of EM protection zone compromises due to incorrect grounding connections.

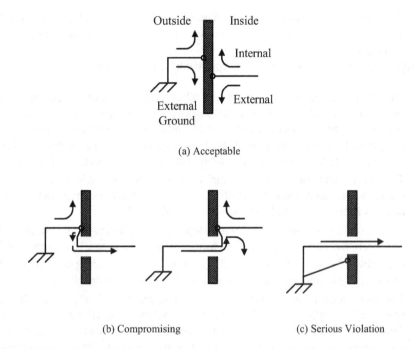

(a) Acceptable

(b) Compromising   (c) Serious Violation

Figure 4.119. Grounding connections maintaining and violating the zone boundary integrity.

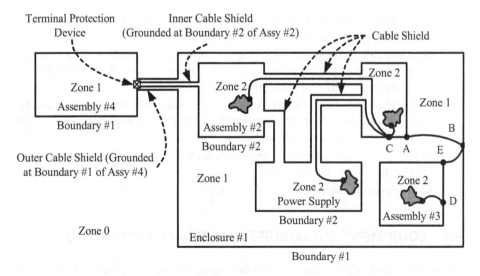

Figure 2.120. Application of zoned grounding to subsystems.

er shielded enclosure, "Zone #1") and Assembly #4, which is placed outside the boundary of the shielded enclosure. Assembly #3 is not connected to any other assembly within the system.

Proper application of the zoning concept mandates the following:

1. The enclosures of Assemblies #1 and #2 as well as that of the power supply should be isolated from the external enclosure; the only grounding connection between the enclosures of Assemblies #1 and #2 to the external enclosure is accomplished through conductor A–B. This connection may be implemented in a single-point or multipoint scheme, as applicable to the respective subsystems.
2. The cables interconnecting the various assemblies are shielded. The cable shields are bonded circumferentially at all ends to the respective assemblies in order to form a single and continuous zone ("Zone 2").
3. The grounding scheme within each assembly is implemented according to the characteristics of each, typically in a form of a "grounding tree" (see Section 4.3).
4. Assembly #3 is internally grounded according to its "grounding tree" at point D. Externally, Assembly #3 is grounded using bonding conductor E–B.
5. The connection between Assembly #2 and Assembly #4 is accomplished by means of a double-shielded cable, with insulation placed between the two shields. The inner shield is grounded to the enclosure of Assembly #2, whereas the external shield is terminated to the external enclosure. Terminal protection devices (including filters and/or surge protection devices) should also be installed on the interfacing ports of this cable.

Lessons Learned

- Grounding architectures in complex systems should be carried out based on the electromagnetic protection zones concept.
- Structures are separated into different risk zones, each characterized by a comparable electromagnetic environment separated by a zone boundary established by shielding and/or terminal protective devices.
- Appropriate "zoned grounding" architecture, consistent with the system and facility zoning, constitutes an inseparable element in the implementation of the electromagnetic protection zones concept.
- Within each zone, a consistent grounding architecture should be maintained according to system characteristics.
- Grounding conductors must never penetrate shielded zone boundaries.

4.7   EQUIPMENT ENCLOSURE AND SIGNAL GROUNDING

Up to this point, signal return conductors were discussed separately from safety grounding. Whereas a signal ground (for "noisy" and "sensitive" signals) is intended to carry signal-return currents, the safety ground conductor is not intended to carry any type of current, except in cases of a fault hazard or for diverting electrostatic discharge currents away from susceptible components to preclude damage. The key to understanding the need for these different and distinct grounding systems has been shown to be important because of the necessity to prevent common-impedance coupling between circuits [1].

Somewhere in the overall system, a common point or galvanic connection between the signal circuits and the electrical system grounding network is typically identified. Such connection is essential for situations in which single-ended interconnects between circuits exist. The connection of separate grounding systems, inasmuch as it addresses some functionality objectives, may also be the source of problems (Figure 4.121).

4.7.1 External Signal and Safety Grounding Interconnects between Enclosures

Figure 4.121 illustrates a common situation in which, in the effort to avoid common-impedance coupling, no connection is provided between the signal-ground conductors of the circuits internal to Circuits #1 and #2 and their respective chassis; however, the enclosures of Circuits #1 and #2 are interconnected by "chassis ground" conductors for safety or ESD control purposes.

The primary drawback of this configuration lies with signal return (ground) wires routed outside the equipment assemblies' shielded enclosures. A severe violation of

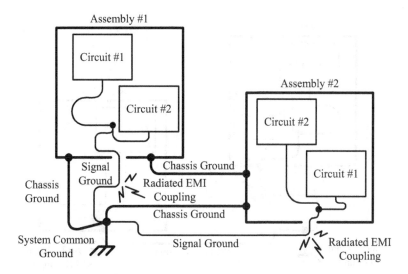

Figure 4.121. Signal ground conductors not grounded internally to the assemblies could result in radiated EMI.

the shielding integrity that would result, producing potential EMI interactions (Figure 4.121).

Connecting the circuits' return internally to the enclosures of the different assemblies (Figure 4.122) eliminates potential radiated emissions from the ground conductors. The enclosures can now be grounded to the system reference point using one conductor, serving both as the intended current-carrying signal and a noncurrent-carrying safety. The chassis ground connections to the system common ground were deleted for clarification purposes. If, however, a chassis ground connection between different assemblies in addition to the system common ground (marked as a dashed line) were to remain, a detrimental ground loop would result, potentially introducing interference into the system circuits. The secret to success lies in maintaining low-impedance interconnections between different assemblies and the system common ground.

4.7.2 Equipment DC Power, Signal, and Safety Grounding

In most of the figures accompanying the discussion on the various grounding topologies, the signal reference terminal of the circuit was shown to be isolated from the equipment's enclosure chassis (e.g., Figure 4.30).

System designers often do not recognize or, rather, tend to ignore the fact that according to common practice, electronic and electronic equipment and assemblies contain a deliberate galvanic connection between the DC power supply system's common terminal and the equipment's enclosure at the input terminal, also serving as the electrical safety ground conductor (ESGC, or simply, safety ground).

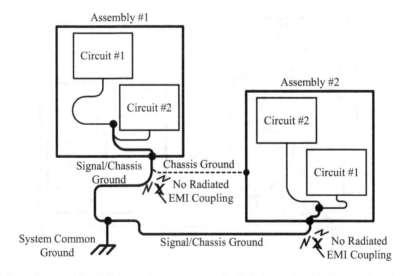

Figure 4.122. Interconnection of grounding systems of several assemblies may result in undesired outcomes.

In reality, an electrical connection is established between the DC power supply return/reference (or common) terminals and the equipment chassis. The advantage of such a connection was previously presented. The connection depicted in Figure 4.123 serves as the rule, rather than the exception, in modern electronic equipment.

In Figure 4.123(a), all outputs of the power supply share a common return (hence the name "DC Common"), whereas in Figure 4.123(b) each DC output has its dedicated return conductor, and all return conductors are galvanically connected at common point to the equipment chassis, serving as the signal reference structure of the equipment.

The DC return conductors also serve as signal return conductors. Any interference coupling onto those conductors will appear also in the signal circuit itself, potentially introducing interference. The advantage that (b) has over (a) lies in the fact that when a

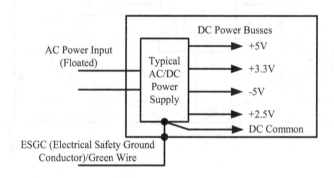

(a) Multiple-output Power Supply, Sharing a

Single Common Terminal

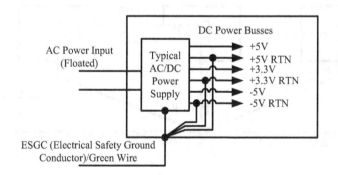

(b) Multiple-output Power Supply, with

dedicated Return Terminals

Figure 4.123. Signal DC reference terminal(s) is typically grounded to the equipment chassis in most modern electronic equipment.

single common return is used, it serves as a "common impedance" between the different load circuits utilizing the different power outputs, potentially producing "common-impedance coupling," which may itself introduce interference to sensitive loads. In (b), each power output terminal is allocated a dedicated return conductor. If sensitive circuits are provided a dedicated power output (and a dedicated power return conductor), common-impedance coupling is not possible.

In all grounding topologies except multipoint, particularly when "floating" or single-point is desirable, or in any case in which a special grounding topology is to be employed in systems where such equipment is used, it is imperative that the internal grounding configuration be taken into consideration. This is demonstrated for single-point ("star") grounding in Figure 4.124.

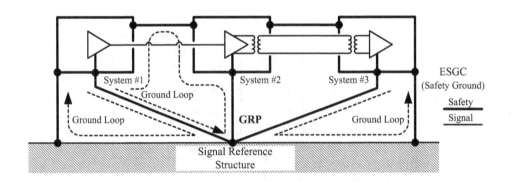

(a) Enclosures "Daisy-Chained" Produces Multiple "Ground Loops"

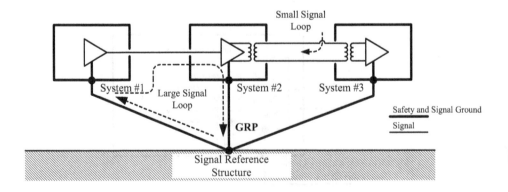

(b) Enclosures Single-Point Grounded Precludes "Ground Loops" and Maintains Single-Point

Grounding Topology, but Creates Large Signal loops

Figure 4.124. Internal grounding of the equipment requires due consideration in the system grounding implementation to preclude violation of the system grounding topology.

Figure 4.124(a) clearly demonstrates that when the equipment is internally grounded to the enclosure, which in turn is grounded to the signal reference for safety purposes, multiple ground loops* are consequentially produced, some of which are illustrated in the figure.

In Figure 4.124(b), multiple connections of the ESGC are avoided. The same grounding conductor serves as both DC, signal, and safety grounding. Note that between System #2 and System #3, differential and balanced low-frequency signaling prevents large signal loops.

Note: In high frequency signaling, coaxial interconnects will serve a similar role. This is discussed in Chapter 7.

It follows therefore, that in large and complex systems,[†] constituting several equipment assemblies, knowledge of the grounding topology design of the equipment must be obtained before determining the system-level grounding topology. Particular attention must be paid to the continued preservation of the equipment electrical safety grounding connection, regardless of the system-wide grounding topology chosen to prevent electrical shock and fire hazards.

4.7.3 Power Distribution Grounding Schemes in Integrated Clustered Systems

The situation described in Section 4.7.2 regarding the DC power grounding scheme within an equipment assembly is now further extended to integrated systems containing several assemblies, also known as "clustered systems" (see Section 4.9.3). Grounding in integrated systems such as equipment racks, airborne pods, and satellites may be of special concern, above all with respect to the power distribution system (PDS) grounding architecture, in particular with regard to secondary and tertiary power distribution.

Well-designed clustered system (e.g. equipment racks) are typically operated from a single AC (e.g. 115 V/230 V/380 V) or DC (e.g., 12 V/24 V/–48 V) primary power source (known as "mains"). This power source typically does not directly feed low voltage (and sensitive) electrical and electronic circuits, but, rather, is converted by AC/DC or DC/DC power supplies to lower-level secondary and tertiary power outputs.

Secondary and tertiary power distribution schemes are particular concern, as these are used for direct feed of the system's electronic assemblies and circuits. The potential for inadvertent violation of the intended grounding scheme at the system level is, therefore, particularly high, especially when distinction is made by the system designer between the power-return and the signal-return functions within the assembly. Commercial off-the-shelf (COTS) or nondevelopment items (NDI) assemblies introduce further complication when the user is not knowledgeable of their internal signal (and power) grounding scheme. Incorporating such equipment in a composite system under such circumstances may, in certain cases, compromise overall system performance.

---

*Ground loops and their significance in are discussed in Section 4.5.
†The strategy for grounding architecture expanded to large-scale systems is discussed in Section 4.9.

This section addresses this question and introduces commonly used power distribution schemes in such clustered systems, with particular emphasis on the impact of the particular scheme on grounding and, hence, system performance.

Several approaches discussed below are commonly found in clustered systems power distribution arrangements, each enjoying advantages and suffering disadvantages with respect to the system's overall grounding performance.

a) Centralized Power Scheme with Secondary Power Supplies

In this scheme, primary power (e.g., 115/230 VAC or 24/28/–48 VDC) is supplied to a central power supply only (denoted "primary power supply"), which generates all necessary secondary power outputs (e.g., 24 VDC/AC, ±15 VDC, 12 VDC) for all of assemblies incorporated in the system [Figure 4.125(a)]. The return lead of the primary input power is commonly "floated" at the power supply input (see Section 4.4.2), whereas the electrical safety ground conductor (ESGC) or safety ground lead is connected to the system chassis (e.g., the rack's frame) at its designated grounding point, normally adjacent to the power entry to the system.

Secondary DC (and, occasionally, low-voltage AC) outputs may be floated at the output of the power supply and power distribution system and, thus, are mutually isolated (if their return leads are not connected to a common reference point). The return leads of the secondary power outputs are typically grounded to a common grounding point, however, at the host system and distributed thereafter to the different assemblies; thus, full isolation is not maintained between the respective outputs when incorporated in the rack.

Each secondary output power lead is typically accompanied by a dedicated power-return conductor. This return conductor may be grounded to the respective assembly's enclosure at the local common point. Good design practices dictate, however, that each of the rack assemblies powered from the power distribution unit incorporate a local, isolated DC/DC power supply in order to generate the necessary tertiary voltages for their associated circuits, for example, 5 V, 3.3 V, and 1.2 V. With such an isolated power supply, the power input from the main rack-level power supply should be maintained floating at the input to the assemblies' respective power supply inputs. Potentially disturbing galvanic ground loops between circuits of different assemblies are thus precluded.

Single-ended links between assemblies contained in the system could potentially lead to ground loops at the low-voltage outputs. Thanks to the isolation within the assemblies' power supplies, however, this will often not constitute a source of concern to performance or functionality.

If nonisolated DC/DC power supplies are used in the system's assemblies, it is best to DC-float the secondary DC outputs at the primary power supply (capacitive, frequency-selective grounding may be appropriate for AC grounding at the output of the primary power supply). This scheme limits potentially harmful ground loops to the secondary power circuits of the system at most [Figure 4.125(b)].

The main advantage of this scheme is the elimination of multiple high-voltage power supplies at each of the assemblies. This is particularly advantageous when compli-

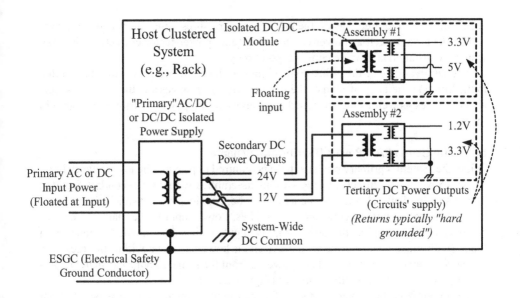

(a) Isolated Secondary DC/DC Power Supplies with Extensions

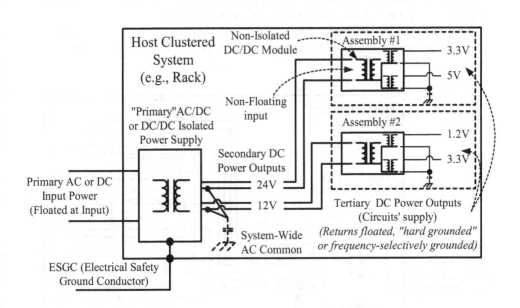

(b) Nonisolated Secondary DC/DC Power Supplies

Figure 4.125. Grounding in a centralized power distribution with secondary power supplies scheme.

ance with EMC requirements on the power mains requires large filters and surge pro-tective devices, as well as power conditioning (e.g., power-factor correction circuits), which are installed at a single common point only.

The disadvantage of this scheme is, of course, the increased power dissipation due to the limited efficiency of each of the power conversion stages and the need for sever-al such stages as well as the increased complexity of the power distribution system in the rack.

b)   Fully Centralized Power Distribution Scheme

Often, the system's primary power supply is required to provide the low voltages (e.g., ±15 V, 5 V, 3.3 V) directly to the various assemblies without a secondary isolated power supply. This scheme, simple as it may be, provides for possible ground loops between signal circuits of collocated assemblies. Power input to the main power sup-ply may be "floating" at the power supply input, which does preclude intersystem loops (with other racks of interfacing systems); however, loops within the rack, be-tween collocated assemblies that are powered from the same outputs and share a com-mon reference point, will still be present (Figure 4.126).

For minimizing the adverse effects of system-wide ground loops, it is best to float the secondary DC outputs at the primary power supply, as those may be grounded at the various assemblies. However, if those are floated (it is easy to validate that through bonding impedance measurements), the secondary outputs of the primary power sup-ply should be grounded at a central point near the power supply.

Lower frequency secondary power outputs at the various assemblies, if "floated," may be capacitively grounded for elimination of high-frequency noise pickup while

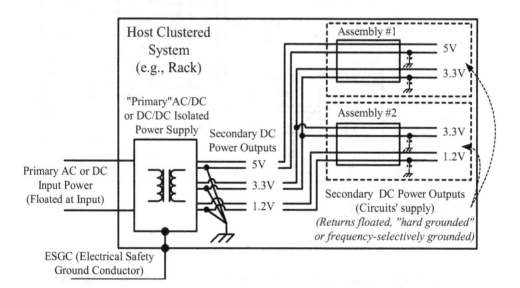

Figure 4.126. Fully centralized power distribution scheme.

providing single-point grounding at the lower frequencies (5 V output in Figure 4.126). Outputs powering high-frequency circuits (e.g., RF or high-speed digital circuits) on the other hand, may be "hard" grounded, effectively providing multipoint grounding (1.2 V and 3.3 V outputs in Figure 4.126).

c) Decentralized (Distributed) Power Distribution Scheme

Finally, a scheme is also often used whereby no common secondary power supply is present. A common primary AC or DC power input may still be present, incorporating the EMI filters and surge protection devices. The primary power distribution unit is used merely to deliver mains power to the system's assemblies. For generating the necessary secondary power for the various assemblies, each should include an isolated AC/DC or DC/DC power supply, as appropriate.

The main advantage of this architecture is the fact that through proper design, potential ground loops between the electronic circuits (powered from the secondary power generated by the power supplies in each assemblies) are contained and limited to the secondary circuits alone. Thanks to the isolation provided by each such power supply, ground loops, involving the primary power circuits cannot occur.

Note that in this architecture, the grounding scheme of each of the secondary circuits in each of the assemblies (when isolation is maintained in the power supplies of each assembly) has little impact on the system-wide grounding system performance, also thanks to the isolation provided.

If, however, no isolation exists at the respective power supplies in each of the assemblies, this architecture could become the worst of all three, as a direct conductive path exists between the primary and the secondary circuits. Return paths in both the secondary and tertiary power distribution systems may be mixed with the primary power return, consequently resulting in multiple ground loops. Extremely careful system layout and EMI control measures in both secondary and tertiary power and signal circuits would be necessary to preclude the adverse outcome of this situation.

Of course, practical systems may consist of a combination of the above schemes or variations of them. The principles outlined for the above three fundamental schemes may be applied to variants of these configurations, while keeping in mind (a) the vulnerability of the secondary circuit to EMI and (b) the potential risk of potential ground loops throughout the system.

4.7.4 Grounding of Equipment Enclosure Shield

A question often asked is whether an equipment-shielding enclosure should be grounded. The reply to that question lies in the physics of shielding theory, namely reflection and absorption.* None of these properties is associated with the electrical potential of the shielded enclosure. Self-contained shielded equipment will function even if the equipment shield is ungrounded. Consider, for instance, shielded enclosures of hand-held portable devices, satellites, and airborne vehicles as typical examples. In all these cases, the shielded enclosure provides effective suppression of incident electromagnet-

---

*Excellent coverage of shielding theory can be found, for instance, in [3, 11, 12, 15].

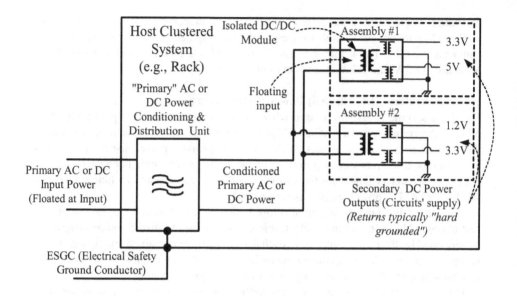

(a) Isolated DC/DC Power Supplies in Assemblies

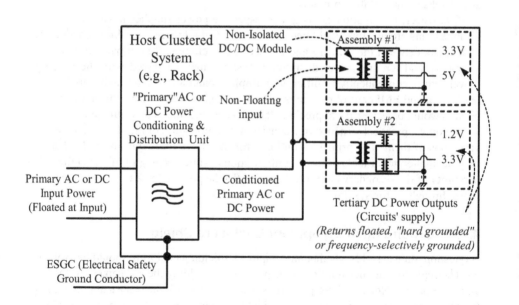

(b) Nonisolated Power Supplies in Assemblies

Figure 4.127 Noncentralized (distributed) power distribution scheme.

ic fields even when not grounded. It follows that grounding of equipment-shielding enclosures is dictated by objectives other than shielding performance:

1. **Electrical Safety.** When equipment is powered from a 60 V_{RMS} or 42.4 V peak AC or DC source, electrical safety regulations dictate that exposed metallic parts of the equipment be earthed through an electrical safety ground conductor to avoid electrical shock in the event of electrical power fault or lightning.

2. **Single-Ended Interface Performance:** When single-ended interfaces such as TTL or RS-232 are used, a common reference (and current-return) path must be provided between source and load circuits. This can be accomplished either by providing a return conductor (virtually transforming the circuit to a differential, even if not balanced, circuit) or by the two systems sharing a common signal reference structure. When the circuits in each of the assemblies are grounded internally to chassis, a common reference and current-return path is provided through the grounding of equipment assemblies to a common reference structure.

3. **Electrostatic Discharge Control (ESDC)*.** Most electronic circuitry, whether single-ended, balanced and floating, or isolated, are related to the chassis of the equipment either through the power supply return ground or, even when floated, through stray capacitance between the circuit boards and interconnecting cables to the metallic structure of the equipment enclosure. A totally isolated equipment enclosure may exhibit an electrostatic charge buildup with no path for controlled discharge, eventually resulting in arcing to the electronic circuits at some weak point, resulting in possible upset and even damage to circuits. Grounding the equipment chassis directly ("hard grounding") or even through a large (e.g., 100 kΩ to 1 MΩ) "bleeding" resistor ("soft grounding") can alleviate such a problem.

4. **Cable Shielding Integrity†:** When equipment constitutes part of a system and interconnects with some other assemblies through shielded cables, continuity of the cable shield is mandatory to ensure its effectiveness. When terminated to a nongrounded enclosure, the cable shield will be effective until such frequency at which it becomes electrically long (when its physical length exceeds λ/10, where λ is the wavelength associated with the frequency of interference). Above that frequency, standing current and voltage waves on the shield may result in increased coupling to the shielded wires. Grounding the enclosure provides effective shield termination to the common reference structure.

5. **Performance of Terminal Protection Device:** Most power and signal circuits use terminal protection devices and filters that contain capacitors or surge protection devices (SPDs) between each of the lines interfacing with other assemblies to equipment chassis (for common-mode suppression/protection). If not grounded, performance of the protection devices in common-mode suppression will be significantly compromised.

---

*Grounding for control of electrostatic discharge effects are described in Chapter 10.
†Grounding of cable shields is discussed in Chapter 7.

In conclusion, with the exception of electrical safety considerations, reasons for grounding of shielded enclosures in all other cases are related to performance of interface circuits and their associated shields or terminal protection devices. Under no circumstances is grounding required for the actual performance of the shielding enclosure itself. A perfect, continuous, and self-contained shielded enclosure, having no external interfaces, acts like an ideal "Faraday cage" and, even if charged with high electrostatic charge, no discharge could occur to the interior of the enclosure nor would any external incident electromagnetic fields penetrate the shield (Gauss's Law), hence, no interference or damage could occur to internal circuitry. Figure 4.128 illustrates a few of the cases discussed above.

4.8 RACK AND CABINET SUBSYSTEM GROUNDING ARCHITECTURE

4.8.1 Grounding Ground Rules in Racks and Cabinets

A special yet very common case of grounding system implementation is found in subsystems comprising equipment racks and cabinets. Frequently, equipment racks are specified with explicit requirements for AC power and signal ground (Figure 4.129). Obviously, what the system designer had in mind was that the equipment installed in

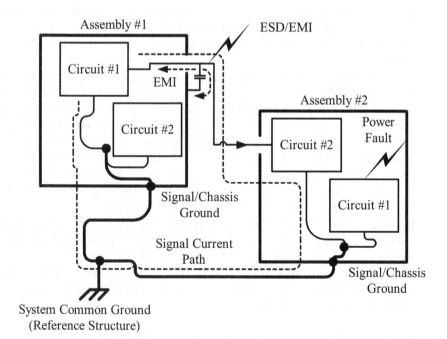

Figure 4.128. Reasons for grounding equipment enclosures.

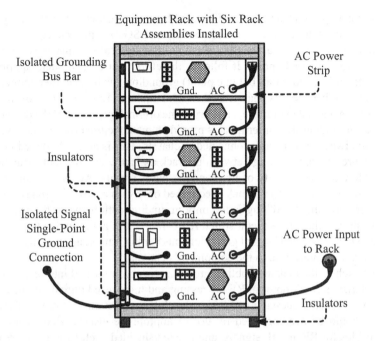

Figure 4.129. Commonly specified power and grounding connection implementation in typical equipment racks, using an isolated ground bus bar arrangement.

the rack would benefit from isolation of a grounded bus bar from the "noisy" rack, believing that in this manner a "quiet," interference-free grounding system would be maintained. This designer's conviction could not be further from the truth.

Indeed, in a well-designed rack, the primary AC or DC input power feeds only the main AC/DC or DC/DC power supply and is "floating" at the power supply input. The electric safety ground conductor (ESGC), or safety ground wire, is thus connected to the rack's frame at its designated grounding point, normally adjacent to the power entry terminal. In spite of this, however, other constraints and practices dictate that the intentions of the designer may not be fully accomplished.

First and foremost, implementation of an isolated AC power return bus bar in an AC-powered rack could be in severe conflict with electrical safety codes [e.g., NEC (National Electric Code, USA)] and, thus, may constitute a safety hazard (e.g., shock or fire).

Second, typical electronic assemblies installed in equipment racks are normally delivered as self-contained apparatus, equipped with all provisions for meeting product safety objectives at the assembly level, in particular, grounding. Such grounding provisions typically include a grounding lug (marked as "Gnd" in Figure 4.129) on the assembly's enclosure or chassis in order to achieve a low-impedance multipoint grounding connection (as opposed to the "daisy chain" single-point grounding), between interconnected rack assemblies to the grounding bus bar.

The grounding lug is normally not isolated from the chassis of the assembly and is common to the electrical safety ground conductor (ESGC) within the equipment's AC power cord. The rack's grounding bus bar, in turn, is typically connected to the rack's frame, at least at a single point. It follows, therefore, that even if such equipment is grounded to an isolated (at all but one point) grounding bus bar, another connection is actually present between its chassis to the rack structure. Thus, isolation of the signal ground is not achieved in practice, contrary to the original intentions of the designer [7].

When a grounding stud or terminal is provided, manufactured by unrelated vendors, it is almost always galvanically connected to chassis. This is normally done in order to obtain a more reliable bonding between the rack assembly and the rack structure, not relying solely on the integrity of its mounting hardware, the AC power cord's safety ground wire, or both. When the rack is equipped only with a two-wire (phase and neutral) power cord, the "Gnd" lug is the only grounding path between the rack assembly's chassis and the rack's metal structure or frame. It is not intended to be used for connection to an "isolated" signal grounding bus bar but, rather, as an easily accessible direct grounding connection to the equipment's chassis.

Within each of the rack assemblies, grounding is accomplished internally, befitting its characteristics (multipoint in RF processing and high-speed digital circuits, single-point in assemblies processing audio signals). Interconnections between different assemblies should be accomplished by use of appropriate electrical conductors (e.g., coaxial cables for RF and IF signals, and twisted/shielded cables for audio frequency signals).* Multiple connections, however, may produce large loops, which may be of particular concern at lower frequency interfaces (e.g., at audio frequencies) if adequate precautions and measures (such as isolated interfaces) are not put into practice.

Figure 4.130 [7] depicts a recommended rack power and grounding connection implementation in a nonisolated ground bus bar arrangement, as previously discussed. In addition to the ground bus bar arrangement, the enclosure for rack- or cabinet-mounted equipment must be directly bonded to the rack. The rack, as a whole, regardless of the internal grounding scheme, should also be grounded to the nearest point on the fault protection subsystem within the facility via the rack's grounding provisions, resulting in minimum potential differences between collocated racks in the installation or facility. Figure 4.131 provides a typical equipment rack grounding detail [2].

4.8.2 Ground Loops and their Mitigation in Racks and Cabinets

From the discussion in Section 4.8.1, it is evident that ground loops are inherent to equipment installed in racks and cabinets, due to the multiple connections between the rack assemblies and the chassis, through the grounding bus bar and the electrical safety ground conductor present in the power outlet. Both are typically joined at the chassis of the rack.

Racks and consoles lend themselves to grounding thanks to the fact that impedance between various rack assemblies is controlled through the grounding bus bar or the

---

*Cable shielding and shield grounding are discussed in Chapter 7.

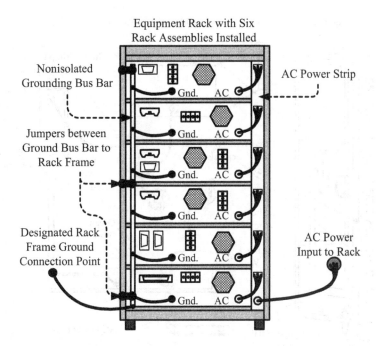

Figure 4.130. Recommended rack power and grounding connection implementation using a nonisolated ground bus bar arrangement.

chassis (depending on the particular grounding scheme used). However, with poorly designed rack assemblies, severe grounding problems may arise.

Consider the objectionable situation depicted in Figure 4.132. A common (signal) reference point (CRP) is created within each rack assembly, interconnecting all grounds of the circuits inside the rack assembly. This reference point, however, is floated internally to the assembly, counting on the rack's "signal ground bus bar" as the reference and current-return path. The chassis (or electrical safety ground conductor) of each assembly is connected to the rack's safety grounding bus bar. Consequently, a large signal current return loop is created between the rack assemblies' common reference point, potentially creating severe EMI and functionality problems, particularly when single-ended interconnects are used in the rack. Note that any structural noise on the rack's chassis will directly couple into the internal circuits in this configuration, even if those circuits do not interface with other assemblies.

When single-ended unbalanced interconnects (with the exception of coaxial cables in high-frequency or RF application) are used, the severity of this situation may further increase. The rack grounding bus bar serves as the signal return path and common-impedance coupling may occur through this structure if noise is present.

The only prospect for such a scheme to function properly is if separate grounding bus bars are provided for the signal ("signal grounding bus bar") and for electrical safety ("safety grounding bus bar"). Furthermore, the "signal grounding bus bar" must

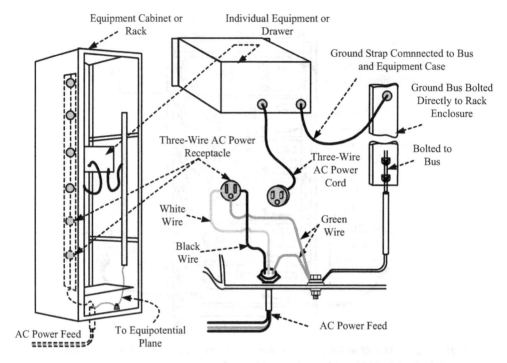

Figure 4.131. Typical equipment cabinet grounding details.

be bonded at a single point only to the rack (a "daisy chain" single-point grounding topology) for preclusion of any undesirable coupling through the rack structure.

An enhanced situation is illustrated in Figure 4.133. Now, the common (signal) reference point (CRP) is internally bonded to the chassis of each rack assembly. Internal circuitry is no longer affected by rack structural noise and does not depend on the signal grounding bus bar of the rack. No noise will now couple into any of the internal circuits of the rack assemblies. If single-ended interconnects are used, a separate signal grounding bus bar is still required, however, as common-impedance between the signal return current and any structural noise current can still occur. However, if balanced interconnects are utilized, separation of grounding busses may not be required and one common bus bar may be sufficient.

Lessons Learned

- A galvanic electrical connection is commonly present in electronic equipment between the DC power supply (and signal) return (or common) terminals and the equipment chassis ("safety ground").

- In systems constituting several assemblies, knowledge of the equipment's internal grounding topology must be obtained before determining the system-level grounding architecture.

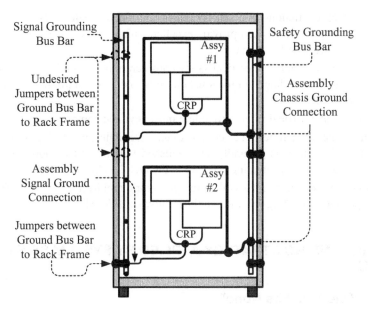

Figure 4.132. Objectionable design: signal ground loop involves assemblies' internal grounding. Note signal conductors between assemblies not shown.

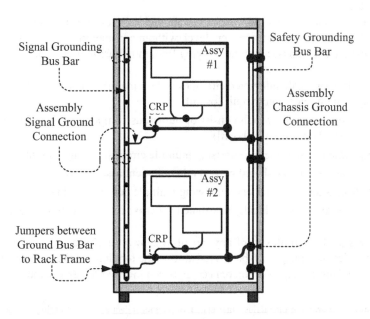

Figure 4.133. Acceptable design: Rack grounding and assemblies' internal grounding separated and small signal and return path created. Note: signal conductors between assemblies not shown.

314 FUNDAMENTALS OF GROUNDING DESIGN

- A perfect, continuous, and self-contained equipment shielding enclosure having no external interfaces can provide excellent shielding performance whether the shield is grounded or is not.
- With the exception of electrical safety considerations, grounding of shielded enclosures is only related to functional interface and EMI control measures through interconnects.
- Equipment electrical safety grounding must never be compromised!
- Due to the internal design of complex rack assemblies, grounding of the circuits enclosed in them is generally accomplished within the assembly, resulting in mixed/composite grounding schemes at the rack level.
- Signal interconnects may violate the grounding architecture of the rack if their respective grounding scheme is not carefully considered.

4.9 GROUNDING STRATEGY APPLIED BY SYSTEM SIZE AND LAYOUT

4.9.1 "One Size Fits None"

Section 4.8 discussed the grounding architecture of subsystems arranged in equipment racks or cabinets. This section is intended to generalize the discussion to the approach to be employed for implementation of the grounding architecture, whether large or small scale [7].

There is no one "correct" approach to be applied in the design and implementation of a grounding architecture at the system level, without consideration of the term system.* A system may encompass anything, ranging from miniature to huge. For instance:

- A very large scale integrated circuit (VLSI) containing manifold gates on a die (a "system on integrated circuit" or SoIC)
- Electronic circuits, which consist of numerous components on a printed circuit board (PCB) (system on board)
- Equipment assemblies, comprising multiple circuits, wiring, and interconnects
- Consoles and racks shared by several equipment assemblies
- Platforms and facilities encompassing multiple consoles and racks
- Large-scale, geographically distributed networks interconnecting several facilities

Regardless of the complexity of any tier of systems, a useful clue to the appropriate grounding design strategy can be arrived at from the topology of the power distribution system and from the signal interconnects between the next lower assemblies. This ap-

---

*A system is (1) a group of interacting, interrelated, or interdependent elements forming a complex whole; (2) a functionally related group of elements, such as a group of interacting mechanical or electrical components, a network of structures and channels as for communication or distribution, or network of related computer software, hardware, and data transmission devices.

proach can be easily warranted by considering the conclusions from the discussion in Section 4.7 regarding equipment enclosure and signal grounding. As such, system-level grounding schemes can be classified into one of the following categories [7]:

1. Isolated systems
2. Clustered systems
3. Distributed systems
4. Nested-distributed systems
5. Central system with extensions

The grounding strategy for each of these system tiers is now discussed.

4.9.2 Isolated System

An isolated system is one in which all functions are accomplished within one assembly enclosure. In other words, it represents a "self-sufficient" or "self-contained" unit with no external signal interfaces. Isolated systems are typically powered from a single power source such as a battery or an AC or DC power distribution network.

The grounding strategy in isolated systems is driven by safety considerations only. In systems powered from AC or high-voltage DC,* whether directly or through a switch-mode power supply,† an electrical safety ground connection (ESGC) to the signal reference structure (vehicle/platform body or earth) must exist, unless the equipment is exempt from an earth connection, through double-insulated protection schemes. When not mandated for electrical safety or for lightning arc prevention, such grounding connections may not even be present (Figure 4.134 [7]). The equipment's internal grounding system (single-point or multipoint) architecture should be determined by the designer based on the particular equipment characteristics.

Common examples of isolated systems include notebook or desktop personal computers, certain home appliances (e.g., refrigerator, electrical stove, washing machine), and hand-held devices (e.g., pocket calculator, CD/MP3 player).

4.9.3 Clustered System

A clustered system comprises multiple elements (e.g., 19 inch (48.3 cm) equipment racks or consoles) located at a central area. Clustered systems are typically characterized by a relatively compact layout of their assemblies. Clustered systems are typically operated from a common AC or DC power source, and connected to a common signal reference for the purpose of ensuring electrical safety and lightning protection (Figure 4.135) [7].

Examples of clustered systems are home entertainment/multimedia systems, minicomputer systems, and their peripherals.

---

*Voltages below 60 V_{RMS} or 42.4 V peak, AC or DC, are generally not regarded as unsafe.
†Grounding issues related to switch-mode power supplies are discussed in Section 4.4.

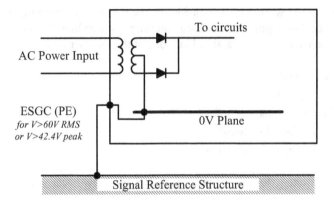

(a) AC Powered System

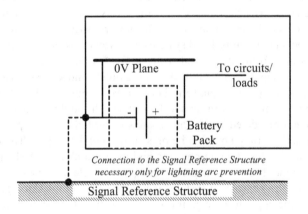

(b) Battery Powered System

Figure 4.134. Grounding strategy in an isolated system.

Interfacing through multiple interconnects within the system assemblies is likely to occur. These interconnects typically include power, signal, and control lines and their associated return conductors. However, no interconnects to other systems, whether directly or through a local/wide area network (LAN/WAN), are present. Interconnects may be single-ended (e.g., TTL lines), balanced (e.g., twisted pairs), or coaxial, according to the interface characteristics. Depending on type of interfacing circuit, the signal return may consist of return wires or the shield of a coaxial cable. When balanced interfaces are not implemented, these return conductors can introduce potential difficulties where low-frequency circuits are concerned due to the potential detrimen-

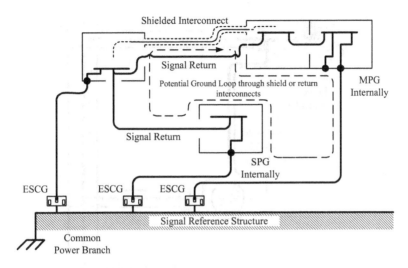

Figure 4.135. Grounding strategy in a clustered system.

tal ground loops that may come about when multiple electrical safety ground connections in addition to signal return or coaxial shield interconnects exist simultaneously.* For preclusion of the harmful effects of such ground loops, coaxial cables may need to avoided, or their shield may have to be isolated from the (grounded) equipment enclosures at all but one end to prevent such ground loops, which might consequently result in severe EMI interactions in a harsh electromagnetic environment.[†] Owing to possible variations in installations as well as site- or system-specific constrains, it is good practice to include an optional ground connection terminal on each assembly. This will provide some flexibility in the selection of the actual ground connection architecture when integrating systems on-site.

Figure 4.136 depicts a typical installation. In production facilities, process control computers often interconnect to their respective machinery through a single-ended RS-232 interface, thereby forming a clustered system. Even when the systems are within a small distance from each other, a significant potential difference may occur between power outlets along with their corresponding ESGC connections due to power frequency and transient currents flowing through the nonzero impedance of the conductors between the outlets, even when connected to the same power and ground bus. Unfortunately, in single-ended communication links, a voltage potential difference also appears across the control link, resulting in interference and possible malfunction of the network. Single-ended interfaces should, therefore, be avoided or, alternatively, adjacent power outlets (connected to the same bus) should be utilized.

The grounding strategy in clustered systems should emerge from the particular installation and layout constraints, the frequency of the intrasystem interface circuits,

---

*Ground loops and their preclusion are discussed in Section 4.5.
[†]The importance of proper cable shield termination is discussed in Chapter 7.

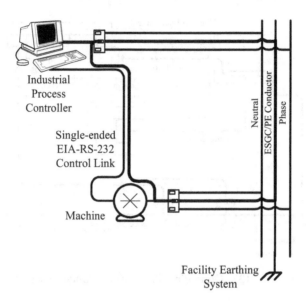

Figure 4.136. Ground loops in a clustered system due to large voltage potential difference between outlets in an industrial process control system.

and the characteristics of the ambient electromagnetic environment. As discussed in Section 4.2, single-point grounding topologies are best suited for low-frequency circuits. Coaxial cables should be avoided to eliminate multiple grounding of signal returns. On the other hand, when higher frequencies are of concern, multipoint grounding (using ground meshes or planes) is more appropriate. In intermediate or mixed low-/high-frequency situations, where coaxial cables may need to be used, either multipoint, hybrid, or frequency-selective grounding topologies may be more appropriate and should be considered.

4.9.2 Distributed System

A distributed system comprises of multiple subsystems that are physically separated so different components are powered from separate power outlets, branch circuits, or power substations. Each of the subsystems constituting the distributed system should be considered an isolated or clustered system as appropriate. Excellent examples of distributed systems are local and wide-area networks (LANs and WANs) (Figure 4.137) [7].

Owing to their large area, multiple sites, different power substations and feeds, and separate and independent electrical safety and lightning protection, grounding systems are required on each site. Interconnection between system components is commonly accomplished by means of electrically long signal and control wiring networks. Multipoint grounding is normally the most appropriate architecture for such systems when a common signal reference is available, albeit large, for example, on board large plat-

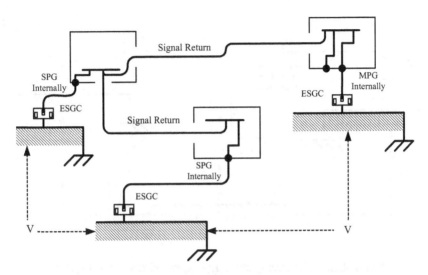

Figure 4.137. Grounding strategy in a distributed system.

forms, such as ships or aircraft. However, in most practical cases a common low-impedance signal reference connection is not available, and the system should be designed to perform its intended function under far less than optimal conditions. As a result, all signal and power ports of distributed subsystems should be treated as if they are exposed to and interfaced with a noisy conducted and radiated electromagnetic environment, including lightning. Ensuring interference-free performance of distributed systems mandates the application of terminal-protective measures such as filters and transient protection devices on signal and power lines.* Single-ended interfacing must be avoided and balanced, transformer-coupled, optically isolated, or fiber-optic interconnects are usually recommended and should be considered.

4.9.5   Nested-Distributed System

A nested-distributed system is similar to a distributed system. At each general area, several components of the system are operating. On each site, it would be expected that the systems utilize power from the same primary power source at each location. Examples of nested-distributed systems are wide-area networks (WANs) (Figure 4.138) [7].

Treatment of nested-distributed systems is similar to that of distributed systems; however, due to increased complexity and the large interference potential in such systems, both intersystem and intrasystem, due consideration to enhanced protective measures (e.g., shielding and terminal protection) is necessary. At each location, the grounding scheme follows the characteristics of the local system layout, as was discussed with respect to distributed systems.

*Grounding considerations in terminal protection devices are discussed in Chapter 8.

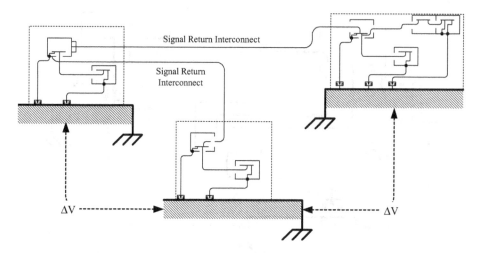

Figure 4.138. Grounding strategy in a nested-distributed system.

4.9.6 Central System with Extensions

A central system with extensions is actually an extension of the isolated or clustered system, where integral elements extend out from a central element with long physical and electrical interconnects. With this system architecture, extended system components normally branch out of the central element, hence, their respective connections to the primary power source is realized via the central element. A common example of a central system with extensions is an industrial process controller along with its sensors and associated instrumentation located remotely from the controller (Figure 4.139) [7].

The best approach for implementing grounding in a central system with extensions is to treat it like a clustered or isolated system, as discussed above. In accordance with the frequency band and dimensions of the system single, multipoint, hybrid or frequency-selective grounding should be used. Following the general system architecture, its ground node should be located at the central element, whereas the extended elements should be floated or balanced. This mandates that balanced signal transmission lines, such as those implemented by using twisted pairs, should be used between the central and extended subsystems.

Lessons Learned

- Grounding architecture of complex and large systems or installations is dependent on the system topology and layout.
- In all but isolated and compact clustered systems, interconnections between remotely installed systems will necessitate careful consideration of the interface circuits, particularly when a severe electromagnetic environment is of major concern.

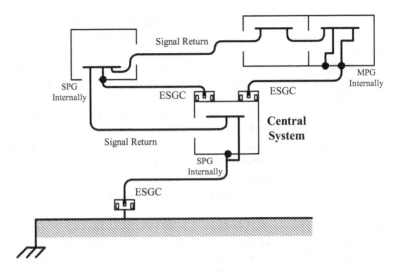

Figure 4.139. Grounding strategy in a central system with extensions.

- Multipoint grounding in wide-layout systems (e.g., distributed and nested-distributed) is normally favored, but may introduce low-frequency interference coupling through ground loops interface control is inadequate (e.g., coaxial cable shields grounded at both ends).
- When designing such large-scale installations, whether terrestrial or on large platforms (e.g., ships and aircraft), all grounding connections of the system assemblies and subsystems must be considered and necessary interference mitigation measures should be incorporated in the system design.

BIBLIOGRAPHY

[1] MIL-HDBK-1857, *Department of Defense Handbook, Grounding, Bonding and Shielding Design Practices,* Washington, DC: U.S. Government Printing Office, March 1998.

[2] MIL-HDBK-419A, *Military Handbook, Grounding, Bonding, and Shielding for Electronic Equipments and Facilities,* Washington, DC: U.S. Government Printing Office, 1987.

[3] Hartal, O., *Electromagnetic Compatibility by Design,* West Conshohocken, PA: R&B Enterprises, 1994.

[4] Jordan, E. C., and Balmain, K. G., *Electromagnetic Waves and Radiating Systems,* 2nd Ed., Englewood Cliffs, NJ: Prentice-Hall, 1968.

[5] MIL-STD-1310G (NAVY), *DoD Standard, Practice for Shipboard Bonding, Grounding, and other Techniques for Electromagnetic Compatibility and Safety,* Washington, DC: U.S. Government Printing Office, 28 June 1996.

[6] Montrose, M. I., *EMC and the Printed Circuit Board, Design, Theory and Layout Made Simple,* New York: IEEE Press/Wiley, 1998.

[7]   Mardiguian, M., *A Handbook Series on Electromagnetic Interference and Compatibility, Volume 2, Grounding and Bonding,* Gainesville, VA: Interference Control Technologies, Inc., 1988.

[8]   NASA-HDBK-4001, *Electrical Grounding Architecture for Unmanned Spacecraft, NASA Technical Handbook,* Washington, DC: U.S. Government Printing Office, February 18, 1998.

[9]   MIL-HDBK-232A, *Military Handbook, Red/Black Engineering, Installation Guidelines,* DoD, Washington, DC, March 1987.

[10] IEEE-1100, *IEEE Recommended Practice for Powering and Grounding Electronic Equipment*(IEEE Emerald Book), New York: IEEE Press, 1999.

[11] Paul, C. R., *Introduction to Electromagnetic Compatibility,* New York: Wiley, 1992.

[12] Perez, R. (Ed.), *Handbook of Electromagnetic Compatibility,* New York: Academic Press, 1995.

[13] Caniggia S., and Maradei, F., "Investigation on the Ground Loop Coupling by Simulation Tools Based on the Partial Inductance Concept," in *Proceedings of the 2008 Asia-Pacific Symposium on Electromagnetic Compatibility and 19th International Zurich Symposium on Electromagnetic Compatibility,* May 2008, Singapore.

[14] ITU-T Telecommunication Standardization Sector of ITU Series O: *Specifications of Measuring Equipment—General Measuring Arrangements to Assess the Degree of Unbalance About Earth,* ITU-T Recommendation 0.9, March 1999.

[15] Ott, H. W., *Noise Reduction Techniques in Electronic Systems* (2nd Ed.), New York: Wiley Interscience, 1988.

[16] Agilent® Technologies Application Note 1043, *Common-Mode Noise: Sources and Solutions* (published on line at URL: http://cp.literature.agilent.com/litweb/pdf/5965-5979E.pdf), 2006.

[17] Intel® Application Note AP-438, *Interfacing Intel(r) 8255x Fast Ethernet Controllers without Magnetics,* Revision 1.0, November 2005.

[18] Intel® Application Note, *LXT971A/972A 3.3V PHY Transceivers Design and Layout Guide,* Document #249016, Revision #:003, November 2001.

[19] Parker, R., Phillips® Semiconductors Application Note, *AN2452, IEEE 1394 Bus Node Galvanic Isolation and Power Supply Design,* February 13, 2001.

[20] Kliger, R., "Integrated Transformer-Coupled Isolation," *IEEE Instrumentation and Measurement Magazine,* Volume 6, Issue 1, pp. 16–19, March 2003.

[21] Dehmelt F., "Galvanic Isolation Using Capacitive Coupling—CMOS-technology Allows 100Mbps and high integration," in *Proceedings of the Euro DesignCon 2005.*

[22] NASA-HDBK-4001, *NASA Technical Handbook, Electrical Grounding Architecture for Unmanned Spacecraft,* Washington, DC: U.S. Government Printing Office, February, 1998.

5

BONDING PRINCIPLES

A bond is an electrical union between two metallic structures, intended to provide a low-impedance path for the flow of electric current between them. Bonding is, therefore, the process of establishing a required degree of electrical continuity between conductive surfaces of members to be joined [1]. Structures (or members) involved may be housings, subassemblies, or components such as the chassis of a platform, an electronic assembly, or the frame of an electrical machine.

Note: Often, bonding is confused with grounding, which is associated with establishment of an electrical conductive path between a circuit and some reference structure/point. Good grounding performance depends on high-quality bonds. The two terms are complementary, not synonymous.

In electrical schematics, a bond frequently appears as a black dot, where the line (wire, trace, or circuit) interconnects with a mating structure (return plane, reference structure, etc.). However, there is more to this black dot than initially meets the eye...

5.1 OBJECTIVES OF BONDING

In any realistic electronic system, whether it is only one piece of equipment or an entire facility, numerous interconnections must be established in order to provide electric power and to preclude the development of electric potential differences between metallic assemblies. Low-impedance bonding minimizes electric shock hazards and electromagnetic interference, provides lightning protection, and establishes references and return paths for electronic signals. Ideally, each of these interconnections should be made so that both mechanical and electrical properties of the path are determined by the connected members and not by the interconnection junction. Further, the joint must maintain its properties over an extended period of time in order to prevent progressive degradation of the performance initially established by the interconnection.

Bonding is concerned with the techniques and procedures necessary to achieve a mechanically resilient, low-impedance interconnection and to prevent deterioration through corrosion or mechanical looseness of this connection.

In terms of the results to be achieved, bonding is necessary for:

1. **Providing electrical power and signal current return paths** for systems using a metallic structure for power and/or signal return currents, while minimizing the total voltage drop along the path

2. **Attaining adequate antenna patterns and required antenna gain** when ground planes or counterpoises are required for generating the necessary radiation patterns

3. **Protection from hazards of shock,** ensuring that sufficient fault current is allowed to flow or trip circuit protection devices in a timely manner

4. **Preclusion of ignition of flammable vapors** due to arcing, sparking (characterized by hot particles and voltage breakdown) and thermal hot spots resulting from the flow of electrical fault currents

5. **Protection of people and equipment from hazard of lightning discharges,** by providing high discharge currents an alternative, low-impedance path, and diverting them from sensitive equipment, precluding damage to apparatus and shock hazards to people

6. **Preclusion of coupling of interference signals into the equipment,** providing electrical continuity across external mechanical and electrical interfaces, both within the equipment and between system elements, minimizing potential differences across the metallic interfaces, and, in particular, maintaining shielding integrity of the equipment enclosure by ensuring the continuum of metallic coverage over the enclosure, so as to control the coupling of electromagnetic fields, whether from within the equipment housing or external to it

7. **Prevention of static charge build-up,** controlling and dissipating the buildup of stray voltages due to an electrostatic charging processes, for protecting people from shock hazards, and to prevent performance degradation or damage to electronic equipment, as well as preclusion of fuel ignition and ordnance hazards

Essentially, the role of bonding is to minimize voltage drops throughout the system, providing low-impedance paths for currents to flow.

With proper design and implementation, electrical bonds minimize differences in potential between points within the fault protection limits required in addition to circuit signal return, referencing, shielding, and lightning protection. Conversely, poor bonding leads to a variety of hazardous and interference-producing situations. For instance, loose connections in AC power lines can produce unacceptable voltage drops at the load. The heat generated by the load current through the increased resistance of a poor joint can be sufficient cause insulation of the wires to melt, subsequently producing a potential fire hazard. Loose or high-impedance joints in signal lines are particularly annoying because of intermittent signal behavior such as a decrease in signal amplitude, increase in noise and distortion levels, or both. Poor joints in lightning protection networks can also be particularly dangerous. The high lightning discharge current may generate several thousand volts across a poor joint. Arcs produced thereby present both a fire and explosion hazard and may possibly be a source of electromagnetic interference to equipment. The additional voltage developed across the joint also increases the likelihood of flashover arcing to other circuits and equipment in the vicinity of the discharge path.

Good electrical bonding practices have long been recognized as a key element of successful system design. An indicator of the importance of electrical bonding is that the first item often considered when EMC problems occur is whether bonding is adequate. Frequently, degradation in system performance due to high interference levels is, in fact, traceable to poorly bonded or deteriorating joints in circuit returns and signal referencing networks due to aging, corrosion, and mechanical or other environmental stresses. As a reference network is required to provide low-impedance interconnects between potentially incompatible signals, such deterioration will further exacerbate system performance. The increased impedance between elements of the reference network as a result of such poor connections will allow interference voltages to develop due to the currents flowing through them, preventing circuit and equipment signal references from being at the same potential. When such circuits and equipment are interconnected, the voltage differential represents an unwanted signal within the system.

Bonding is also important to the performance of other interference control measures. For example, adequate bonding of connector shells to equipment enclosures is essential to the integrity of cable shield. This is in addition to the retention of the low-loss transmission properties of the shielded cable. Careful bonding of seams and joints in electromagnetic shields is essential to the achievement of a high degree of shielding effectiveness.

Interference control components and terminal protection devices (comprising EMI filters and transient suppressors) must also be well bonded for optimum performance, particularly when common-mode interference control is concerned.* Consider a typical power line filter shown in Figure 5.1.

If the return side of the filter, usually the chassis of the enclosure, is inadequately bonded to the signal reference structure (i.e., the equipment case or chassis), the bond impedance, Z_B, may be sufficiently high to impair the filter's performance. The filter

---

*Refer to Chapter 8 for a discussion of grounding in terminal protection devices and circuits. Common-mode EMI control elements such as capacitors and transient suppression devices rely on their proper bonding to the signal reference or chassis for reducing the common-mode path impedance required for their effective performance.

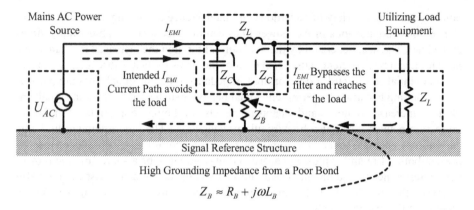

Figure 5.1. Due to poor filter bonding, I_{EMI} bypasses the EMI filter and reaches the load.

shown is a low-pass filter intended to remove high-frequency interference components from the power lines of equipment. This filter is intended to achieve its goal in part by the fact that the reactance, X_C, of the shunt capacitors is low at the frequency of the interference signal, compared to the series reactance, X_L. Interfering signals present on the AC line are, therefore, shunted to ground, bypassing the load. However, a poor bond can result in the following:

$$\left(2 \cdot \frac{1}{j\omega C} + Z_L \right) < \left(j\omega L_B + R_B \right) \tag{5.1}$$

When the impedance to the reference structure Z_B ($Z_B = Z_B + j\omega L_B$) is higher than the series impedance of the circuit comprised primarily of the series combination of the two capacitors $2(X_C$ (where $X_C \approx 1/j\omega C)$* and the load impedance, Z_L, due to a poor bond, interference currents will bypass the filter's inductor as well as the reference connection and will follow an undesired path to the load. The effective performance of the filter is, hence, compromised.

If a bond in a current path is not securely made or works loose through vibration, it can behave as a set of intermittent contacts. Even if the current through the joint is at DC or at AC power frequency, sparking may occur across the bond, generating broadband interference signals with frequency components ranging up to several hundred megahertz.

Poor bonds in the presence of high-level RF fields, such as those in the immediate vicinity of high-power transmitters, can produce a particularly troublesome type of in-

---

*This expression is valid as long as the equivalent series inductance (ESL) and resistance (ESR) of the capacitor can be ignored. If, due to poor selection of the capacitors and their respective installation, this condition does not prevail, the current will split between the series inductor, X_L and the two capacitors in series, $2 \cdot X_C$, again resulting in unacceptable performance of the filter.

terference. Poorly bonded joints and nonlinear junctions are known to produce inter-modulation products when irradiated by two or more high-level signals [2, 3]. Metal oxides behave as nonlinear junctions to provide mixing action between incident signals in a manner similar to semiconductor P–N junctions. Interference thus produced can couple into nearby susceptible equipment, particularly receivers collocated in dense RF environments such as shipboard applications (where this phenomenon is well known as the "rusty bolt effect" [2, 4]).

Note: Electrical bonding involves obtaining good electrical contact between metallic surfaces while corrosion control measures are taken into consideration in the attempt to avoid electrical continuity between dissimilar materials (Section 0). It is essential, therefore, that potentially conflicting requirements be carefully considered during system design.

Experience reveals that the vast majority of 90% of EMC life cycle problems can be attributed to deterioration of bonding. This is an encroaching effect as it slowly builds up with time. It can be almost unnoticeable while users "learn to live with it" unnecessarily until severe outcomes result. Consequently, maintaining lifetime EMC hardness of equipment almost entirely lies in the preservation of high-quality bonds.

The primary electrical and mechanical properties of concern associated with bonding characteristics include:

1. Impedance, both resistive and reactive
2. Mechanical stress endurance
3. Electrochemical corrosion resistance due to dissimilarity of mating materials
4. Current carrying capability for high-current stresses, such as fault currents, lightning, and NEMP (nuclear electromagnetic pulse) induced surges
5. Simplicity of inspection and maintenance

5.2 BOND IMPEDANCE REQUIREMENTS

The chief requirement for effective bonding is that a low-impedance connection be created between the two joined objects. Hence, the bond impedance serves as a well-established figure of merit for the bonding quality.

Often, bonding impedance is defined in terms of resistance; however, it should be recognized that a low DC bond resistance is not a reliable indicator of the performance of the bond at higher frequencies as the low- and high-frequency characteristics of most bonding techniques are poles apart. Impedance, rather than resistance, serves as the governing factor: Intrinsic conductor inductance and stray capacitance, along with associated standing wave effects and path resonances, will determine the impedance of the bond. Accordingly, at higher frequencies, which are of particular concern for most sensitive electronic equipment, reactance must be considered along with the DC resistance.

Note: In many instances, however, resistance rather than impedance is quoted for practical reasons. Unlike reactance, measurement of resistance is straightforward.

The bond impedance must remain low with use and time. The specific bonding values needed at a particular junction to obtain required performance are application-specific, and are a function of the current, actual or anticipated, through the path.

For example, where the bond serves only to prevent static charge buildup, a very high resistance (50 kΩ or higher) is acceptable. When lightning discharge or heavy fault currents are of concern, on the other hand, the total bond impedance must be very low to minimize heating and arcing effects and preclude dielectric breakdown due to high voltages developing across highly inductive bond connections.

Interference control requires that path impedances, typically less than 50 mΩ, be ensured at all times. Rarely will noise control require bond resistances as low as those necessary for fault and lightning currents. Bond resistance based strictly on noise minimization requires information on what magnitude of voltage constitutes an interference threat, and the magnitude of the current through the junction.

Historically, MIL-B-5087* [5] first established electrical bonding requirements for aircraft in 1949. Several electrical bonding classes were then defined and eventually designated in subsequent revisions as follows:

1. **Class A** for antenna installation, no bonding resistance specified
2. **Class C** for current return path, fault current versus resistance table provided
3. **Class H** for shock hazard, 100 mΩ
4. **Class L** for lightning protection, control internal vehicle voltages to 500 volts
5. **Class R** for RF potentials, 2.5 mΩ from electronic units to structure
6. **Class S** for static charge, 1.0 Ω

Over the years, the 2.5 mΩ "Class R" became universally accepted as a design requirement for electrical connections established across a metallic interface, particularly aluminum[†] [1, 6]. This value originally emerged from the need to protect aircrew when their aircraft is struck by a severe lightning discharge (i.e., peak return strike current of 200 kA). With 200 kA flowing through metallic joints on the aircraft structure exhibiting an impedance of 2.5 mΩ across the interfaces, the resultant momentary potential difference (200 kA · 2.5 mΩ = 500 V) would not put the pilots at risk if they came into contact with both surfaces.

No scientific basis has been found for this requirement from the standpoint of EMI control, however, other than the fact that it is a value that can be achieved with good metal-to-metal contact. There is no technical evidence that it must be strictly met to avoid interference. Higher resistance values do tend to indicate, though, that a quality or workmanship problem may be present and that the bonding may be degrading. It provides, thus, an excellent good figure of merit to ensure that adequate electrical bonding measures are implemented in the design.

Previously, the rationale behind "Class R" bonding was to ensure that impedances of the signal current return circuits were kept very low due to the extensive use of sin-

---

*Superseded by MIL-STD 464 [6].

[†]Higher bonding resistance values may be more appropriate for other metals such as stainless steel or titanium, as well as composite materials, which will inherently exhibit much higher levels.

gle-ended circuits back in 1949. Modern electronics now use primarily balanced circuits and the need for such strict bonding requirements is less obvious. Yet, the 2.5 mΩ still serves as a desired design goal for some application bonds such as cable shield terminations to connectors and bonding a connector to the equipment case.* The other bonding values of MIL-B-5087 for shock protection, current return paths, and static charge are still valid for common use today.

A similarly low resistance between widely separated points on a signal reference guarantees that well-established electrical connections are achieved and that reasonably adequate quantities of conductors are provided throughout the network. In this fashion, resistive voltage drops are minimized, resulting in improved interference control. In addition, the need for reaching a low resistance tends to force the use of reasonably sized conductors, which also helps minimize path inductance.

A much lower resistance value could provide greater protection against very high currents but would be more difficult to accomplish at many common types of bonds such as at connector shells and across shock mounts. However, there is little benefit in achieving a junction resistance that is substantially lower than the intrinsic resistance of the conductors being joined themselves.

The actual need for certain bonding in a particular application is not easily ascertained. It is dependent on various items such as the shielding topology, type of circuit interfaces utilized, and the use of the enclosure as a ground reference for circuits and filters. For example, a subsystem that is wholly contained, all enclosures and cable interfaces in a continuous unbroken shield, typically does not require bonding for RF potential control. External interference surface currents will remain outside the shield while internal return currents, if allowed to flow on the enclosure structure, will remain on its inner surface, both owing to the skin effect.

Other values of bonding impedance, initially specified in MIL-B-5087 for purposes of shock protection, current return paths, and static charge, are still valid and are commonly applied.

Lessons Learned

- Bonding must be designed into a system for meeting electrical safety, functional performance, and EMI control objectives.
- Bonding must be controlled, qualitatively and quantitively.
- Bonding must not be haphazard or erratic; repeatability of performance from system to system and over time is critical.

5.3 TYPES OF BONDS

The term bond relates both to the mechanical interface between mating surfaces as well as to the conductor bonding jumpers or straps that may be used to interconnect

---

*Cable shield grounding and bonding requirements are discussed in Chapter 7.

two separate structures. When the mating surfaces come into immediate contact, the bond is called a direct bond; otherwise it is an indirect bond.

5.3.1 Direct Bonds

Direct bonding constitutes the establishment of a desired electrical path between interconnected members without use of an auxiliary intermediate conductor, using direct metal-to-metal contact between the mating surfaces.

Direct bonds may be either permanent or semipermanent. Permanent bonds may be defined as those intended to remain in place for the expected life of the installation and not required to be disassembled for inspection, maintenance, or system modifications. Such contact can constitute a permanent joint formed of machined metal surfaces. Joints that are inaccessible by virtue of their location should be permanently bonded with appropriate measures employed to protect the bond against deterioration due to galvanic corrosion (discussed in Section 5.5).

In many applications, permanent bonds are not desired. For example, equipment may have to be removed from enclosures, or moved to other locations, for implementing system modifications, for network noise or resistance measurements, and for other related reasons, all of which require that electrical joints be disconnected without destroying or significantly altering the bonded members.

Junctions that should not be permanently bonded are defined as semipermanent bonds. Semipermanent bonds include riveted or bolted joints, screws, clamps, or pinned fittings driven tightly, and other types of fasteners, not being subject to wear or vibration. Besides offering greater flexibility and lower cost, semipermanent bonds may be easier to use, require less operator training and require fewer specialized tools.

Regardless of the type of bonding technique utilized, high-quality, metal-to-metal contact must be maintained throughout the useful life of the joint. Precautions must be taken to protect the joint against moisture that could instigate galvanic corrosion.

Examples of direct bonds are splices between bus bar sections, the connections between lightning down-conductors and the earth electrode subsystem, the mating of the equipment front panel to equipment racks, and the mounting of connector shells to equipment panels.

Properly constructed direct bonds exhibit low DC resistance and provide RF impedance as low as the configuration of the bond permits. Direct bonding is always preferred; however, it can be used only when two members can be connected together and can remain so without relative movement. The establishment of electrical continuity across joints, seams, hinges, or fixed objects that must be spatially separated requires indirect bonding with straps, jumpers, or other auxiliary conductors.

Current flow through two configurations of a direct bond is illustrated in Figure 5.2.

As shown in Chapter 4, the DC resistance of the path through the conductors on either side of the bond R_C is proportional to the resistivity of the material ρ and to the length in the direction of the current flow ℓ, and inversely proportional to the cross-sectional area (assumed to be equal for both conductors) perpendicular to the current flow:

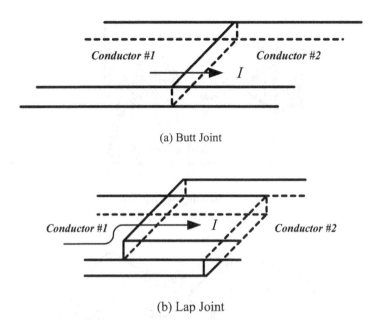

(a) Butt Joint

(b) Lap Joint

Figure 5.2. Current flow through direct bonds.

$$R_C = \rho \frac{\ell}{A} \qquad (5.2)$$

Any bond resistance at the junction will increase the total path resistance. Therefore, the objective in bonding is to reduce the bond resistance to a value negligible in comparison to the conductor resistance so that the total path resistance is primarily determined by the intrinsic resistance of the conductors.

Some examples of implementation of direct bonds are depicted in Figure 5.3 [7].

Optimally bonded joints are formed by metal flow processes such as welding, brazing, or soldering. With such processes, the resistance of the joint is determined by the resistivity of the weld or filler metal. The bond members are raised to temperatures sufficient to form a continuous metal bridge across the junction.

When metal flow processes are not desirable, for reasons of economy, future accessibility, or functional requirements, high-quality semipermanent bonding techniques can also be established by bringing the mating surfaces together under high pressure using auxiliary fasteners such as bolts, screws, rivets, or clamps. The resistance of these auxiliary bonding methods is determined by the kinds of metals involved, the surface conditions within the bond area, contact pressure, and cross-sectional area of the mating surfaces.

Take note that screw threads themselves do not provide adequate bonding, and should not be considered as such. The contact impedance achieved is too high. All of these techniques can ensure that adequate pressure is impressed on the mating sur-

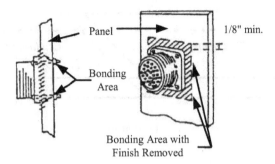

(a) Bonding of Connector

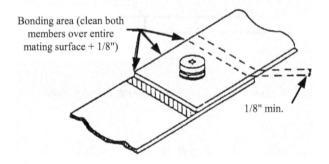

(b) Bracket Installation (Rivet or Weld)

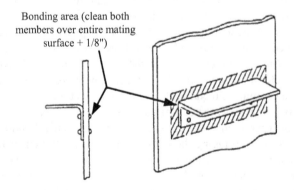

(c) Bolted Members

Figure 5.3. Examples of implementation of direct bonds.

faces, bringing them together in direct surface contact, however, screws, rivets and the like, by themselves, serve only as localized contact points, prone to corrosion.

In order of preference, the acceptable direct bonding techniques are discussed below.

1. Welding

In terms of electrical performance, welding is the ideal method of bonding. The intense heat (in excess of 4000°F or 2200°C, approximately) is sufficient to boil away contaminating films and foreign substances. A continuous metallic bridge is thus formed across the joint. The conductivity of this bridge typically approximates that of the bond members. The net resistance of the bond is essentially zero because the bridge is very short relative to the length of the bond members. The mechanical strength of the bond is high, approaching or exceeding the strength of the bond members themselves. Since no moisture or contaminants can penetrate the weld, corrosion is minimized. Therefore, lifetime use of the bond should be as great as that of the bond members themselves.

Welds should be utilized whenever practical for permanently joined bonds. Although welding may be more expensive, the reliability of the joint makes it very attractive for those bonds that will be inaccessible once construction is completed. Most metals encountered in normal construction can be welded using one of several standard welding techniques.

2. Brazing

Brazing, which includes silver soldering, is another metal flow process for permanent bonding and constitutes a process for joining similar or dissimilar metals using a filler metal that typically includes a base of copper combined with silver, nickel, zinc, or phosphorus. Brazing covers a temperature range of 900°F to 2200°F (470°C to 1190°C).

Metal with an appropriate flux is applied to the heated members, which wets the bond surfaces to provide intimate contact between the brazing solder and the bond surfaces. As with higher temperature welds, the resistance of the brazed joint is essentially zero.

Brazing differs from welding in that brazing does not melt the base metals; therefore, brazing temperatures are lower than the melting points of the base metals. For the same reason, brazing is a superior choice for joining dissimilar metals; however, since brazing frequently involves the use of metal different from the primary bond members, additional precautions must be taken to protect the bond from deterioration through corrosion. Brazed joints are strong. A properly-made joint (like a welded joint) will in many cases be equally strong or even stronger than the base metals being joined.

3. Soft Soldering

Soft soldering is an attractive metal flow bonding process because of the ease with which it can be applied. Relatively low temperatures are involved and it can be readily employed with several of the high-conductivity metals such as copper, tin, and cadmium. With appropriate fluxes, aluminum and other metals can be bonded together.

Properly applied to compatible materials, this bonding process is nearly as low in resistance as one formed by welding or brazing. Because of its low melting point, soft soldering should not be used as the primary bonding technique where high currents may be expected. For this reason, soldered connections are not permitted in grounding circuits for fault protection and for interconnections between elements of lightning protection networks.

In addition to its temperature limitation, soft solder exhibits low mechanical strength and tends to crystallize if the bond members physically move while the solder is cooling. Soldered bonds also tend to crack microscopically under mechanical stress, creating a disconnect that is hard to identify by visual inspection. Therefore, soft solder should not be used if the joint must withstand mechanical stresses.

Soft solder can be used effectively in a number of ways. For example, it can be used to tin surfaces prior to assembly to assist in corrosion control. Soft solder can be used effectively for the bonding of seams in shields and for the joining of circuit components together and to the signal reference subsystem associated with the circuit. Soft solder is often combined with secondary mechanical fasteners. By sufficiently heating the joint to melt the solder, a low-resistance filler metal augments the path established by the second fastener. In addition, solder provides a barrier to keep moisture and contaminants from reaching the mating surfaces.

4. Bolts

The most common semipermanent bond is the bolted connection, or one held in place with machine screws, lag bolts, or other threaded fasteners. This type of bond provides the flexibility and accessibility that is frequently required. The bolt (or screw) should serve only as a fastener to provide the necessary force to maintain the pressure required between the contact surfaces for satisfactory bonding. Hence, although metals are generally required to provide tensile strength, the fastener itself need not be conductive. Although bolt or screw threads may provide an auxiliary current path through the bond, the primary current path should be established across the metallic interface. A primary consideration in bolted bonds constitutes reliability, particularly under mechanical stresses such as vibration of the joint.

The size, number and spacing of the fasteners should be sufficient to establish the required bonding pressure over the entire joint area under the required dynamic stresses.

5. Rivets

Riveted bonds are less desirable than bolted connections or joints bridged by metal flow processes. Rivets lack the flexibility of bolts without offering the degree of protection against corrosion achieved by welding, brazing, or soldering. The chief advantage of rivets is that they can be rapidly and uniformly installed with automatic tools.

6. Conductive Adhesive

Conductive adhesive is a silver-filled, two-component thermosetting epoxy resin that when cured produces an electrically conductive material. It can be used between mating surfaces to provide low-resistance bonding. It offers the advantage of providing a direct bond without the application of heat required by metal flow processes. In many locations, heat necessary for metal flow bonding may pose a fire or explosion threat. When used in conjunction with bolts, conductive adhesives provide an effective metal-like bridge with high corrosion resistance along with high mechanical strength. In its cured state, the resistance of the adhesive may increase through time. It also tends to adhere tightly to the mating surfaces. On the other hand, conductive epoxy has been known to crack under mechanical stress; hence, mechanical means to secure rigidity of the bond must be used in conjunction with the glue/paste. An epoxy–bolt bond is also less convenient to disassemble. In some applications, the advantages of conductive adhesive may however outweigh this inconvenience.

5.3.2 Indirect Bonds

The preferred method of bonding is direct bonding, joining objects together with no intervening conductor. Unfortunately, there are situations in which operational requirements or equipment locations often preclude the implementation of direct bonding; for instance, parts that exhibit a relative motion, such as equipment installed on shock mounts. Under such circumstances, auxiliary bonding conductors such as bonding straps or jumpers must be used for bonding such equipment to a structural ground reference. Bonding straps are also used for bypassing structural elements, such as the hinges on distribution box covers or on equipment covers, to eliminate the wideband noise when illuminated by intense radiated fields or when carrying high-level currents. Bond straps or cables are also used to prevent static charge buildup and to connect metal objects to lightning down-conductors to prevent flashover. Typical applications of implementation of indirect bonds are shown in Figure 5.4 [8]. A common application of bridging across a shock absorber is shown in Figure 5.5 [7].

A good indirect bond is one that exhibits a low impedance across the entire frequency spectrum of interest and retains it effectiveness for an extended period of time. Indirect bonds are normally implemented by means of jumpers or bonding straps. Bonding

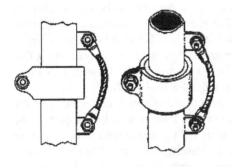

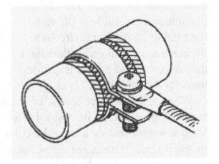

(a) Bonding Tubing across Clamps (b) Clamp Connection, Jumper to Tube

Figure 5.4. Typical applications of indirect bonds.

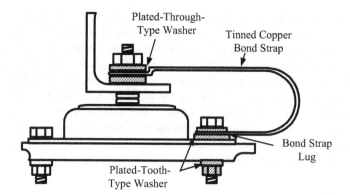

Figure 5.5. Example of implementation of an indirect bond—bonding across a shock absorber.

jumpers are short round or braided conductors. Jumpers normally find use in low-frequency applications, below a few MHz, or where the buildup of static electricity is to be avoided. Bonding straps, on the other hand, are either solid, flat metal straps or woven braid. Solid metal straps are generally preferred for most applications.

Typical bond straps may be manufactured as solid or braided flat straps or solid wire. The material used is generally copper or aluminum. In the case of flat straps, those made of phosphor bronze are commonly used. Braided or stranded straps are normally not recommended, particularly due to their vulnerability to corrosion, which occurs in a nonuniform manner along the jumper strands. Such corrosion can generate RF interference due to the nonlinear nature of oxidized/corroded junctions (the "rusty bolt effect"), whereas the strap itself will act as an efficient high-frequency antenna transmitting that interference. Stranded jumpers also exhibit a significantly higher self-inductance than the flat and solid bonding straps and may not be an optimal choice.

Indirect bonds are only a substitute for direct bonds and should be used only when direct bonds are impractical. When jumpers are used, they should be kept as short as possible and exhibit low DC resistance and low inductance-to-capacitance (L/C) ratio (increasing its resonance frequency, discussed in Section 5.3.3), and not be lower in the electrochemical series than the bonded members.* A good rule to maintain is that the jumper should have a length-to-width ratio of less than 5:1 or, better still, 3:1 (the basis for the "good rule" is also further elaborated in Section 5.3.3).

When using indirect bonds, care must be paid to the quality and durability of the bond. Jumpers should be bonded directly, rather than through an adjacent part, to the basic structure and should not be connected with self-tapping screws or by any other means if screw threads are the primary means of bonding. Also, bonding straps should be continuous and not be broken along the bonding path. Figures 5.6 and 5.7 illustrate the objectionable and acceptable manners of implementing indirect bonds.

---

*Refer to Section 5.5 for a discussion on dissimilar metals in electrical bonds and joints.

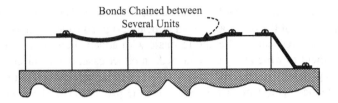

(a) Objectionable: Jumpers not Bonded Directly

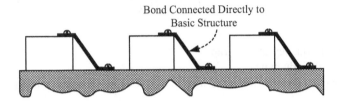

(b) Acceptable: Jumpers Bonded Directly to Basic Structure

Figure 5.6. When using indirect bonds, jumpers should be bonded directly to the basic structure.

5.3.3 Bonding Impedance and Effectiveness

Bonding jumpers possess the usual electrical parameters of resistance, inductance, and capacitance. The use of bonding jumpers in indirect bonding is equivalent to the problem of maintaining low-impedance paths, briefly discussed in Chapter 4. At lower frequencies, bonding jumpers do not present any particular concern, with the exception of

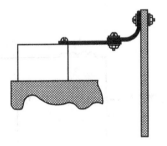

(a) Objectionable: Noncontinuous Bonding   (b) Acceptable: Continuous Bonding Strap
Strap, Broken Along the Bonding Path

Figure 5.7. When using indirect bonds, jumpers should be bonded directly to the basic structure.

resistance. Any reasonable length jumper can be used. At higher frequencies, however, the RF impedance of the bond becomes a critical design consideration. Bonding straps may become self-resonant and could resonate with added stray or parasitic reactance of the equipment to which they are attached. In either case, the occurrence of parallel resonance is generally undesirable as it results in a remarkable increase of the bonding path's impedance. The general expression for the primary (i.e., the lowest) resonant frequency in these cases, ignoring the AC resistance, R_S, is

$$f_r = \frac{1}{2\pi\sqrt{L_s C_c}} \tag{5.3}$$

where:
f_r = resonant frequency of the strap
L_S = strap's self-inductance
C_S = stray capacitance between the strap and the two items being bonded

This expression represents a sufficient approximation of the equivalent circuit of a bond strap when the length of the strap is short compared to the wavelength, illustrated in Figure 5.8(a).

Of these parameters, resistance is an inherent property of jumper resistivity, depending on the material selected. Capacitance is dependent upon the physical configuration and separation from the bonded units, and inductance is dependent upon the physical dimensions of the bonding jumper. The inductance and capacitance involved in creating this simplified parallel-tuned equivalent circuit are not confined to the bonding conductor itself. They also involve the structures such as the equipment enclosure being connected across the bonding path. The latter can significantly affect bond strap performance. Figure 5.9 [7] illustrates the application of the simplified equivalent circuit to a bonding strap across a shock absorber.

The equivalent circuits of Figure 5.8(a) and Figure 5.9 do not take into consideration the effects of the equipment enclosure or other objects being bonded. In many cas-

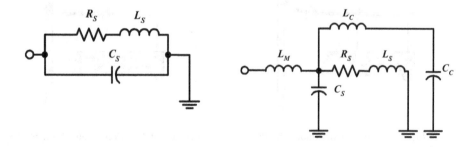

(a) Simplified Low-Frequency Equivalent Circuit   (b) High-Frequency Equivalent Circuit

Figure 5.8. Equivalent circuit of a bonding strap.

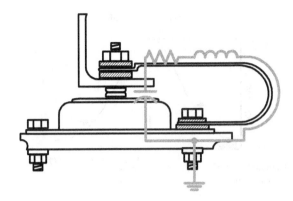

Figure 5.9. Simplified equivalent circuit of a bonding strap, related to the physical structure.

es, the high-frequency effects of this form of coupling will be predominant in comparison with other paths. Figure 5.8(b) depicts the true equivalent circuit when taking into consideration the system being bonded. In Figure 5.8(b), the intrinsic inductance of an equipment case is represented by L_C. Capacitance, predominantly between the equipment bottom surface mounted on the shock absorber and the reference plane, is indicated by C_C.* The inductance contributed by the insertion loss measuring circuit is represented by L_M.

Except for extremely short straps, the magnitude of the inductive reactance of the strap will be significantly larger than the AC resistance, R_S,. At frequencies above approximately 100 kHz, resistive effects can again be ignored as reactance dominates the resonance effects. In practical situations, $L_S \gg L_C$ and $C_C \gg C_S$, thus, the magnitude of impedance, Z_S, of the equivalent circuit at higher frequencies is expressed as

$$|Z_S| \approx \frac{1}{\left(\dfrac{1}{\omega L_S} - \omega C_S\right)} = \frac{\omega L_S}{1 - \omega^2 L_S C_S} \tag{5.4}$$

The impedance characteristic of the bonding strap is depicted in Figure 5.10. The resonance characteristics depicted in Figure 5.10 occur at surprisingly low frequencies, as low as 10 to 15 MHz in typical configurations. In the vicinity of these resonances, bonding path impedances of several hundred ohms are common. As a result, the bonding effectiveness provided by the strap is significantly compromised. In fact, in these higher impedance regions, the bonded system may act as an effective antenna thath may exacerbate the pickup of the same signals that bond straps are intended to reduce.

---

*Refer to Section 5.3.3.3 for a detailed discussion of the effect of the capacitance between equipment and reference structure.

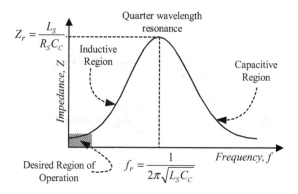

Figure 5.10. Bonding strap impedance characteristics.

5.3.3.1 Bond Resistance.

The DC resistance of a bonding strap per unit length can be obtained from

$$R = \frac{\rho}{A}; \qquad \text{ohm}/\text{m} \tag{5.5}$$

(refer to Chapter 4 for details), where ρ is the specific resistivity of the material and A is the cross-sectional area of the bond. If ρ is in units of $\Omega \cdot \text{cm}$ (for copper, $\rho = 1.724$ $10^{-6}\ \Omega \cdot \text{cm}$) and A is in cm^2, then R_{DC} is in Ω/cm of bond length.

At higher frequencies, at which the skin effect becomes significant, the AC resistance of the bond differs significantly from its DC value. Skin depth was defined in Chapter 4 and is given by

$$\delta = \frac{1}{\sqrt{\pi f \sigma \mu}} \tag{5.6}$$

where f is given in Hz and σ and μ are the specific conductivity and permeability of the strap, respectively.

The assumption that all current flowing through a conductor is contained within the first skin depth defined by Equation (5.6) results, in the case of circular conductors, in the following equation for AC resistance:

$$R_{AC} \approx \rho \frac{\ell}{2r} \cdot \sqrt{\frac{f \sigma \mu}{\pi}} = R_{DC} \cdot \frac{r}{2\delta}; \qquad \text{ohms} \tag{5.7}$$

where r is the conductor radius [in the same units of length as ℓ and δ (e.g., m and cm)].

Figure 5.11 [15] illustrates the ratio of AC-to-DC resistance for several sizes of a single strand of copper wire over the frequency range of 1 to 1000 MHz.

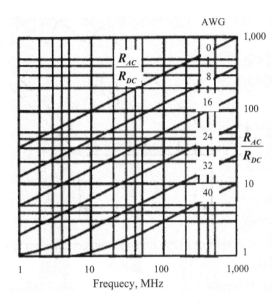

<u>Figure 5.11.</u> Ratio of AC-to-DC resistance for several sizes of copper wire.

5.3.3.2 Bond Reactance. The geometrical configuration of a bonding conductor and the physical relationship between objects being bonded introduce reactive components into the impedance of the bond. The strap itself exhibits an internal inductance that is related to its physical dimensions. For a straight flat strap of nonmagnetic metal, inductance L is given by

$$L = 0.002 \cdot \ell \cdot \left[2.303 \cdot Log\left(\frac{2 \cdot \ell}{w+t} \right) + 0.5 + 0.2235 \cdot \left(\frac{w+t}{\ell} \right) \right]; \qquad \mu H \quad (5.8)$$

If the bond is made with a straight piece of wire of circular cross section, L is given by*

$$L_{HF} = 0.002 \cdot \ell \cdot \left(2.303 \log \left(\frac{4\ell}{d} \right) - 1 \right); \qquad \mu H \qquad (5.9)$$

In the last two equations,
ℓ = length of the strap (cm)
w = width of the strap (cm)
t = thickness of the strap (cm)
d = wire diameter (cm)

---

*For practical purposes and ease of data retention, the internal partial inductance of any practical circular conductor can be approximated as 1 μH/m, almost irrespective of the conductor's cross section area.

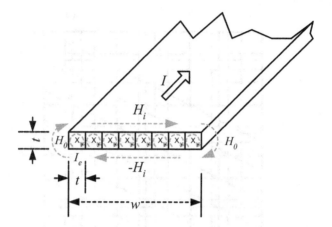

<u>Figure 5.12.</u> Characteristics of a flat bonding strap contributing to reduction of its self-internal inductance.

An interesting feature of flat bonding straps with respect to their advantage over a circular wire is illustrated in Figure 5.12 [1]. When current flows through the cross section of a bonding strap, the current tends to be distributed along the perimeter of the strap. When the thickness of the strap is comparable to the skin depth,* the current actually distributes itself as depicted in Figure 5.12 [10].

The current shown in Figure 5.12 can, therefore, be considered as infinitesimal "current elements," each carrying an equal fraction of the total current flowing along the strap and each generating a magnetic field that surrounds it, H_i, in accordance with Ampere's Law (or the "Right-Hand Rule"). At the boundaries between each current element, magnetic flux opposes and, hence, cancels it out. The resulting net magnetic field exists only above and below the strap, H_i and H_j. Since the strap is assumed to be very thin, those portions of the magnetic flux tend to cancel out, leaving only the flux at both edges of the strap, H_0, as the only contributors to the total flux surrounding the strap, resulting in very little total flux.

Inductance, as defined in Chapter 2, is

$$L \triangleq \frac{\Phi}{I}; \qquad \text{henry} \tag{5.10}$$

where Φ is the total magnetic flux surrounding the conductor and I is the current flowing through that conductor. This leads us to

$$\Phi = \int_A Bda = \mu \cdot \int_A Hda \approx \mu \cdot H \cdot A; \qquad \text{weber} \tag{5.11}$$

*A discussion of skin effect and skin depth is provided in Chapter 2.

where A is the cross section of the flux area, B is the magnetic flux density, μ, the permeability of the space surrounding the strap, and H is the magnetic force.

Since H is produced by the end current elements only, we derive (for $w \ll r$) from Equation (5.11) for the flux at an arbitrary distance r from the strap,

$$H_{i(Flat)} = 2 \cdot \frac{I_e}{2\pi r} = I \cdot \frac{t/w}{\pi r}; \qquad \text{henry} \qquad (5.12)$$

For an equivalent bonding wire with a round cross section, the internal inductance due to symmetry is (see Figure 5.13):

$$H_{i(Round)} = \frac{I}{\pi r}; \qquad \text{henry} \qquad (5.13)$$

Comparing Equations (5.12) and (5.13), the advantage of using a flat strap over a circular bonding wire having the same cross section is demonstrated [Equation (5.14) and Figure 5.14]:

$$\frac{H_{i(Flat)}}{H_{i(Round)}} = \frac{I \cdot \dfrac{t/w}{\pi r}}{I \cdot \dfrac{1}{2\pi r}} = 2 \cdot \frac{t}{w} \qquad (5.14)$$

Equation (5.14) demonstrates, therefore, that a flat strap's inductance is approximately $2 \cdot t/w$ times less than the inductance of a circular wire carrying the same current, I. This effectively implies that the wider and thinner the strap, the lower will be its inductance. As the length of the strap is increased, its impedance increases nonlinearly for a given width; however, as the width, w, increases, there is a nonlinear decrease in strap impedance, depicted in Figure 5.15.

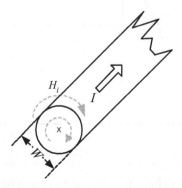

Figure 5.13. Model for internal inductance of a round bonding wire.

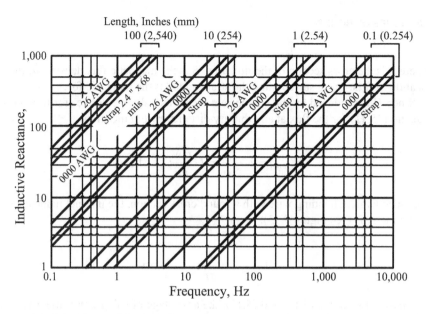

Figure 5.14. Advantage of flat bonding strap over a circular bonding wire with respect to their internal partial inductance.

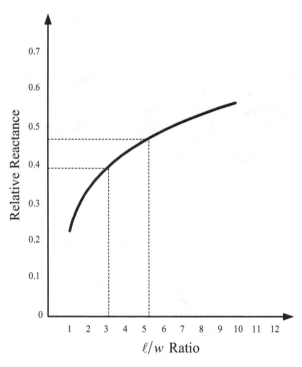

Figure 5.15. Relative internal inductive reactance versus length-to-width ratio of flat straps.

Figure 5.15 shows that the relative reactance of a strap decreases significantly as the ℓ/w ratio decreases. The curve illustrates that a strap with an ℓ/w ratio of 5:1 has an inductive reactance that is 45% that of a thin wire (i.e., very high ℓ/w ratio); a 3:1 ratio decreases this ratio to 38%.

Table 5.1 details the calculated inductance, corresponding to Equation (5.8), of a nonmagnetic rectangular strap 6 inches (15.2 cm) long. Table 5.2 compares the inductance of 6, 12, and 36 inch lengths (15.2, 30.5, and 91.4 cm, respectively) of 0.05 inch (1.27 mm) thick straps and Table 5.3 tabulates the inductance of 6, 12, and 36 inch lengths of selected standard size cables from No.14 AWG to 4/0 AWG, corresponding to Equation (5.9) [1].

Even at relatively lower frequencies, the reactance of the inductive component of the bond impedance becomes much larger than the resistance. Thus, in the application of bonding straps, the inductive properties as well as the resistance of the strap must be considered. Bear in mind, however, that the bonding conductor's total inductance comprises internal inductance (which is a function of the conductor's cross section and length only) and external inductance (which is a function of the total loop area formed by the bonding conductor and the bonded objects). In applications such as U-shaped bonding straps bypassing shock absorbers (Figure 5.9), the total external inductance associated with this arrangement is significantly higher than the internal inductance associated with the bonding strap. Consequently, flat straps in practice offer little if any appreciable benefit over circular conductors for higher frequency applications in which inductance dominates impedance.

In addition to inductance, a certain amount of stray capacitance is inherently present between the bonding jumper and the objects being bonded as well as between the bonded objects themselves. This capacitance, along with the inductance, explains the resonant behavior of the bonding strap.

Minimizing impedance between items being bonded is of utmost importance. Use of a single bonding strap may make this objective almost unattainable. Multiple, parallel straps may lower the total bonding impedance because of the reduced total AC and DC resistance, and, of course, above all, total inductance. As to inductance, care must be taken to control the mutual inductance between adjacent bonding straps. Spacing

Table 5.1. Calculated inductance of a rectangular strap, 6, 12, and 36 inch long

Width, w, inch (mm)	Thickness, t, inch (mm)	Inductance, L, µH
0.5 (12.7)	0.01 (0.25)	0.112
0.5	0.05 (1.27)	0.110
0.5	0.10 (2.54)	0.1017
1.0 (25.4)	0.01	0.092
1.0	0.05	0.091
1.0	0.10	0.089
2.0 (50.8)	0.01	0.072
2.0	0.05	0.071
2.0	0.10	0.071

Table 5.2. Calculated inductance of a rectangular strap, 6 inch (15.2 cm) lengths of 0.05 inch (1.27 mm) thick straps

Width, *w*, inch	Length		
	6 Inch (15.2 cm)	12 Inch (30.4 cm)	36 Inch (91 cm)
0.5 (12.7 mm)	0.110	0.261	0.984
1.0 (25.4 mm)	0.091	0.222	0.866
2.0 (50.8 mm)	0.071	0.183	0.745

must thus be maintained between straps at least as long as the length of the longest strap being employed, if coupling is to be avoided.

5.3.3.3 *Bond Effectiveness.* The representation of the equivalent circuit of a bond as a parallel-tuned circuit, considering the effects of the equipment enclosure or other item being bonded, has been confirmed by measurements. Figure 5.16 details the measured effectiveness of a bonding strap 9.5 inch (24.1 cm) long in the reduction of the RF voltage induced by a radiated field on an equipment cabinet above a ground plane [1]. The proximity of the bond strap to the ground plane is also a parameter within the plots.

The bond effectiveness [1, 7] indicates the amount of voltage reduction achieved by addition of the bonding strap. Positive values of bonding effectiveness indicate a lowering of the induced voltage, whereas a negative value of bonding effectiveness in the plots indicates that the bond strap increases the amount of voltage developed on an equipment case at the indicated frequency. At frequencies near network resonances, induced voltages are higher with the bonding straps than without the straps.

Under certain circumstances, the dip in bonding effectiveness caused by resonant effects in fact goes below 0 dB (dips as low as 10 to 20 dB have been noted), indicating that the induced voltage on the unit being bonded can be increased by attachment of a bonding strap.

Table 5.3. Calculated inductance (μH) of a standard size cable, 6 inches (15.2 cm) long

AWG No.	Length		
	6 Inch (15.2 cm)	12 Inch (30.4 cm)	36 Inch (91 cm)
4/0	0.098	0.238	0.914
1/0	0.108	0.259	0.997
2	0.115	0.273	1.020
4	0.122	0.287	1.063
6	0.129	0.301	1.105
10	0.144	0.329	1.189
14	0.158	0.358	1.274

Figure 5.16 shows a wide frequency range for each bond when the bond creates the negative bonding effectiveness condition cited earlier. In particular, it demonstrates that:

1. At low frequencies at which the reactance of the strap is low, bonding straps will provide effective bonding.
2. At frequencies at which parallel resonances exist, bonding straps may severely enhance the pickup of unwanted signals.
3. Above the parallel resonant frequency, bonding straps have little effect on the pickup of radiated signals, either positively or negatively.
4. The frequency range over which effectiveness of the bond is compromised can be extended by employing shorter bonds located away from the reference structure.

In conclusion, bonding straps should be designed and used with care. In selecting a bond strap, the resonant frequency of the strap must be well above the highest interfering frequency that is expected to be encountered. This resonant frequency corresponds closely to the frequency at which the worst bonding effectiveness value is obtained. To reduce the RF impedance of the bond and thus increase its bonding effectiveness, a high C/L ratio should be achieved by minimizing the case-to-ground spacing or the

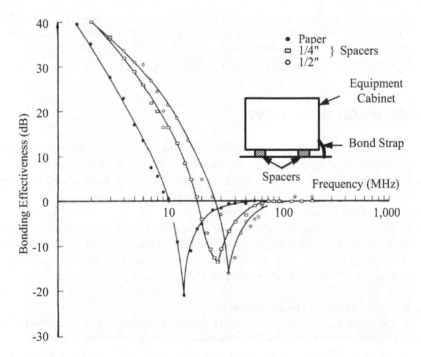

Figure 5.16. Measured bonding effectiveness of a 9.5 inch (24.1 cm) bonding strap.

length-to-width (ℓ/w) ratio of the bonding strap. Because of this reduction in reactance, bonding straps that are expected to provide a path for RF currents are frequently recommended to have a length-to-width ratio of 5 to 1 or less, with a ratio of 3 to 1 preferred. A good bond should have a total inductance of less than 25 nH.

Lessons Learned

- A direct bond is always preferable to an indirect bond. The best bond is obtained by metal-to-metal contact between overlapping surfaces, without a dielectric insulator between the two mating pieces of metal.

- Bonding jumpers are only a substitute for direct bonds. Jumpers should be kept as short as possible and have a low resistance and low ℓ/w ratio. The bonded items and jumpers should be as close together as possible within the electrochemical series.

- The length-to-width ratio of bonding straps, if used, should not exceed 5:1 or, better still, 3:1.

- Bonding must achieve and maintain intimate contact between metal surfaces. Fasteners must exert sufficient pressure to hold the surfaces in contact in the presence of the deforming stresses, shocks, and vibrations associated with the equipment and its environment. Fasteners, screws, or rivets by themselves should not be counted on, however, to provide an acceptable bond.

- It is of utmost importance, in the broadest types of bonding applications, that the direct or indirect bond be sufficient to carry all currents that may flow through it. This includes signal return path, personnel safety, and lightning protection along with mechanical and operational needs.

- Bonds must be maintained for the life of the joint.

5.4 SURFACE TREATMENT

Both direct and indirect bonding connections require a high-quality, metal-to-metal contact between conductive surfaces. In practice, such a metal-to-metal contact is not easy to achieve and requires proper planning before designing any mechanical assembly.

Factors that could potentially compromise the quality of a bond include: mating surface conditions, particularly the roughness/smoothness within the bond area; the cross-sectional area occupied by the bond; the presence of contaminates and nonconductive compounds on the mating surfaces; contact pressure on the mating surfaces; and electrochemical compatibility between metals.

Roughness of Mating Surface Conditions
No metallic surface is perfectly smooth. In fact, surfaces consist of many peaks and valleys. Even the smoothest commercial surfaces exhibit an RMS roughness of 0.5 to 1 millionth of an inch [1]; the roughness of most electrical bonding surfaces will even be

several orders of magnitude greater. When two such surfaces are brought into contact, they touch only at the tips of the peaks, called asperities. As a result, the actual effective contact area for current flow is much smaller than the apparent area of metallic contact.

An exaggerated side view of the actual contact surfaces at a bond interface is depicted in Figure 5.17. Theoretically, two infinitely hard surfaces would touch at only three asperities. In practice however, pressure, elastic deformation, and plasticity allow other asperities to come into contact.

Current passes between mating surfaces only at those points where asperities have been crushed and deformed to establish true metal contact. The total effective area of electrical contact is equal to the sum of the individual areas of contacting asperities. This area of contact will, therefore, be on the order of magnitude of as little as one millionth of the apparent contact area.

Surface Contaminants

It is frequently necessary to remove protective coatings from metals to provide a satisfactory bond. The area cleaned for bonding should be slightly larger than the area to be bonded. Oxidized surface films, protective coatings, paint, grease, or simply dirt will be present on practically any mating surface.

The more electrochemically active metals such as iron and aluminum readily oxidize, allowing surface contamination to develop, whereas noble metals such as gold, silver, and nickel are less affected by oxide films. Of all metals, gold is the least affected by oxide films. Although silver does not oxidize severely, silver sulfide forms readily in the presence of sulfur compounds.

If the surface films are much softer than the contact material, they can be squeezed out from between the asperities to establish a quasimetallic contact. Harder films, such as those provided by anodization, may support all or part of the applied pressure, reducing or eliminating useful conductive contact area. Foreign particulate

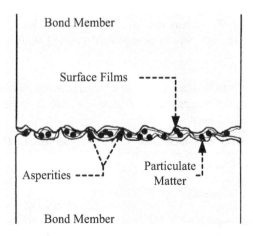

Figure 5.17. Nature of practical contact between mating bond members.

matter on the bond surfaces, such as dirt and other solid matter, can further impair bonding.

For ensuring a high quality bond between mating surfaces, nonconductive films, paint, grease, and dirt must be thoroughly removed prior to the joining of the bond members, or they would compromise contact resistance. The comprehensive application of metal flow bonding does not guarantee that an acceptable metal-to-metal contact is established if contamination is present, unless both surfaces have been cleaned beforehand.

Surface Hardness

The hardness of the bond surfaces affects contact resistance. Under a given load, the asperities of softer metals will undergo greater plastic deformation and establish superior metallic contact. Likewise, at a junction between a soft and a hard material, the softer material will tend to conform to the surface contours provided by the harder material and will provide a lower contact resistance than would be afforded by two hard materials themselves. Table 5.4 illustrates the variation of contact resistance of 1 inch^2 (6.45 cm^2) bonds [1].

Contact Pressure

The influence of mechanical pressure on bond resistance is detailed in Figure 5.18 [1]. This figure shows the resistance variation of a 1 in^2 (6.45 cm^2) bond held in place with a 1/4-20 steel bolt as a function of torque. The resistance variation for brass is lowest due to its softness and the absence of insulating oxide films. Even though aluminum is relatively soft, the insulating properties of aluminum oxide cause the bond resistance to be highly dependent upon fastener torque, which can be up to approximately 40 inch-pound (in-lb). This corresponds to a contact pressure of about 1200 pounds per square inch (psi). Mild steel, being harder and also susceptible to oxide formations, exhibits a resistance that is dependent upon load compression, below 80 in-lb, or about 1500 psi. Above these pressure levels, no significant improvement in contact resistance can be expected.

Bond Area

Large bond mating surface areas maximize the cross-section of the path for current flow. In addition to the obvious advantage of decreased bond resistance, the current

Table 5.4. DC resistance of direct bonds between selected mating metals (apparent bond area: 1 in^2 (6.45 cm^2); fastener torque:100 in-lb)

Bond metallic composition	Resistance, $\mu\Omega$
Brass–Brass	6
Aluminum–Aluminum	25
Brass–Aluminum	50
Brass–Steel	150
Aluminum–Steel	300
Steel–Steel	1,500

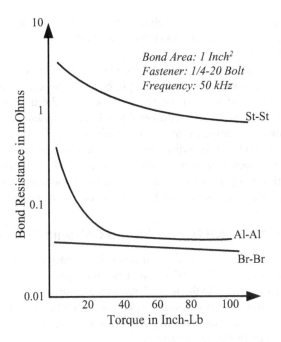

Figure 5.18. Resistance of a test bond as a function of fastener torque.

crowding that may occur during power fault situations or under a severe lightning discharge is decreased. Such current crowding produces a higher effective bond resistance than is present during low current flow, resulting in a raised voltage drop across the junction to even higher values and adding to the heat generated at the junction by the heavy current flow. Large bond areas not only lessen the factors that contribute to heat generation, they also distribute the heat over a larger metallic area, which facilitates its removal.

Another advantage of a larger size bond is that it will provide greater mechanical strength and will be more resistant to long-term erosion by corrosion.

5.5 DISSIMILAR METALS CONSIDERATION: CORROSION CONTROL

In almost every application involving metals in the design of an electronic system, the designer is faced with the problem that dissimilar metals come into contact through the process of bonding. For instance, copper-based printed circuit boards are often bonded via steel screws, washers, and nuts to an aluminum chassis. Direct contact of dissimilar metals in an atmosphere with some degree of humidity will result in corrosion, which degrades the integrity of the electrical joints, thus impairing the effectiveness of and physically weakening the bond. In addition, corrosion may even generate electromagnetic interference (EMI) due to the nonlinear nature of the corroded junction, resulting in interference effects such as intermodulation between incident RF signals present in the environment, known on ships as the "rusty bolt effect."

It is of prime importance that consideration be given to some appropriate measure of protection from corrosion. In order to implement this protection, the designer must be aware of the application requirements of the design, the intended environmental conditions that the product will encounter throughout its service, the materials available, and the protective measures that can be employed.

For each design problem, it is difficult to achieve a solution that meets all requirements. It, thus, becomes necessary to consider design trade-offs, balancing corrosion resistance qualities of a particular metal against cost, workability, mechanical properties, manufacturability, availability, and, of course, electrical properties.

5.5.1 Electrochemical Basis of Bond Galvanic Corrosion

Corrosion is the process involving the transformation of metals to their elemental compounds through an electrochemical reaction whereby cathodic metals transfer electrons to more anodic metals in the joint. Thus, electric current flowing through a joint comprising two or more electrochemically dissimilar metals can cause corrosion. Most environments are corrosive to some degree, yet those containing salt sprays and industrial contaminants are particularly harmful.

Bonds exposed to these and other environments will normally experience deterioration of the metal and must be protected to prevent degradation of the low-resistance connection. Four factors are necessary for causing corrosion to occur:

1. Electrolyte (continuous liquid path, typically water, in the form of condensate, salt spray and similar liquids)
2. Cause (electrolytic action between dissimilar metals)
3. Electron conductor (in a structure, usually a metal-to-metal contact, e.g., rivets, bolts, or spot welds)
4. Effect (corrosion caused by a galvanic action)

Figure 5.19 illustrates the corrosion process for most common types of metal. It demonstrates that for this process to take place, first an anode with a positive potential and a cathode with a negative potential must be present to form an electrochemical cell and, second, there must be a complete path for the flow of current. These conditions occur readily in many environments, allowing current to flow from anode to cathode, resulting in corrosion of the bond.

Even on the surface of a single piece of metal, anodic and cathodic regions are present owing to impurities, grain boundaries and orientations, or localized stresses. These anodic and cathodic regions come into electrical contact through the body of metal. The presence of an electrolyte or conducting fluid completes the circuit and allows the current to flow from the anode to the cathode of the cell.

Anything capable of preventing the above conditions will prevent corrosion. For example, in pure water, hydrogen will accumulate on the cathode to provide an insulat-

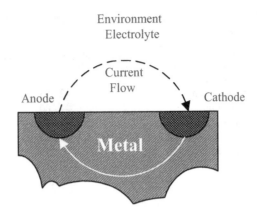

Figure 5.19. Basic diagram of the corrosion process.

ing blanket to prohibit current flow. Most water, however, contains dissolved oxygen that combines with hydrogen to form additional molecules of water. The removal of this hydrogen permits corrosion to proceed. This principle of insulation is employed in the use of paint as a corrosion-preventive measure. Paint prevents moisture from reaching the metal and, thus, prevents the necessary electrolytic path from being established.

5.5.2   Electrochemical Series

The oxidation of metal involves the transfer of electrons from a metal to an oxidizing agent. In the process of oxidation, an electromotive force (EMF) is established between the metal and the solution containing the oxidizing agent. A metal in contact with an oxidizing agent containing its own metal ions establishes a fixed potential difference with respect to every other metal.

The set of potentials determined under a standardized set of conditions, including temperature and ion concentration in the solution, is known as the electrochemical series. The electrochemical series, with hydrogen as the referenced potential of 0 volts for the more common metals is listed in Table 5.5 [1, 13]. The importance of the electrochemical series is that it shows the relative tendencies of metals to acquire an electric potential (and corrode) when immersed in an aqueous solution. Metals high in the electrochemical series, for example, magnesium, aluminum, beryllium, and zinc, exhibit a greater tendency to form ions in solutions and corrode more readily than metals that are located at the lower, nobler end of the electrochemical series (e.g., platinum and gold). The electrochemical series also indicates the magnitude of the potential established when two metals are coupled to form a cell. The farther apart the metals are in the series, the higher the potential difference between them. Metals higher in the series will act as the anode and the one lower in the series will act as the cathode. When two metals are in contact, loss of metal at the anode will occur through oxidation to supply the electrons to support current flow. This type of corrosion is defined as gal-

Table 5.5. Electromotive force (EMF) series of commonly used metals

Metal	Electrode potential (V) at 25°C
More active (anodic)	
Magnesium	−2.375
Beryllium	−1.700
Aluminum	−1.670
Zinc	−0.7628
Chromium	−0.740
Iron	−0.441
Cadmium	−0.402
Cobalt	−0.277
Nickel	−0.230
Tin	−0.1406
Lead	−0.1263
More noble (cathodic)	
Hydrogen	0.000
Copper	+0.345
Silver	+0.7996
Mercury	+0.854
Palladium	+0.987
Platinum	+1.200
Gold	+1.420

Note: The EMF values are obtained when a specific metal is immersed in a solution containing one equivalent weight of its ions per liter.

vanic corrosion. The greater the potential difference in the cell, that is, the greater the dissimilarity of the metals, the greater the rate of corrosion of the anode.

5.5.3 Galvanic Series

The EMF (electromotive force) series is based on metals in their pure state, free of oxides and other films, in contact with a standardized solution. Of greater interest in practice, however, is the relative ranking of metals in a typical environment with the effects of surface films included. This ranking is referred to as the galvanic series. The most commonly referenced galvanic series is provided in Table 5.6 [1]. A "galvanic series" applies to a particular electrolytic solution. For each specific solution that is expected to be encountered in actual use, a different order or series will ensue. This series therefore, based on tests performed in sea water, should be used only as an indicator where other environments are of concern.

Galvanic series relationships are useful as a guide for selecting metals to be joined. They will help in the selection of metals having minimal tendency to interact galvanically or can indicate the need or degree of protection to be applied to lessen expected potential interactions.

Table 5.6. Galvanic series of common metals and alloys in seawater

Anodic (active) end	
	Manganese bronze
Magnesium	Brasses
Magnesium alloys	Aluminum bronze
Zinc	Copper
Galvanized steel or iron	Silicon bronze
1100 Aluminum	Monel
Cadmium	Silver solder
2024 Aluminum	Nickel
Mild steel or wrought iron	Inconel
Cast iron	Chromium steel
Chromium steel (active)	18-8 Stainless steel
Ni-Resist (high-Ni cast iron)	18-8 Mo Stainless steel
18-8 Stainless steel (active)	Hastelloy C
18-8 Mo stainless steel (active)	Chlorimet 3
Lead–tin solders	Silver
Lead	Titanium
Tin	Graphite
Nickel (active)	Gold
Inconel (active)	Platinum
Hastelloy B	
	Cathodic (most noble end)

Galvanic corrosion in the atmosphere is dependent largely on the type and amount of moisture present. For example, corrosion will be more severe near the ocean and in polluted industrial environments than in dry rural settings. Condensation in such environments is more conductive even under equal humidity and temperature conditions, due to increased concentration of sulfur and chlorine compounds. The higher conductivity means that the rate of corrosion is increased.

5.5.4 Galvanic Couples

The most effective approach for avoiding adverse effects of galvanic corrosion is to use noble metals, that is, those low in the electrochemical series. However, due to practical considerations, this is not always possible. In aerospace and aviation systems, for instance, where weight constitutes a significant factor, lighter (but more active) metals such as aluminum and titanium are commonly used.

Since the rate of (galvanic) corrosion depends on the separation between the mating metals in the EMF series, the closer one metal is to another in the series, the smaller the potential difference between them will be, the more compatible they will be, and the adverse galvanic effects of corrosion will be minimized. Conversely, the farther one metal is from another, the greater will be these effects. In a galvanic couple, the higher metal in the series represents the anode, which will corrode preferentially in the given environment.

Normally, corrosion is minimized if the combined potential difference between the two metals does not exceed 0.3 V in harsh environments (e.g., exposure to salt spray or weathering) and 0.5 V in benign environments (e.g., interior, salt-free condensation only).

Aluminum and silver, present in many conductive gaskets, form a galvanic couple of high potential difference (about 2.5 V), which tends to produce rapid corrosion of the aluminum. Figure 5.20 illustrates corrosion produced from an electrochemically compatible conductive gasket (left) and a pure-silver-filled gasket (right) mated with an aluminum disk, after 168 hours of corrosive salt fog.

Various handbooks and design guides [15, 8] provide details on bonding practices for corrosion prevention. Based on these recommendations, Tables 5.7 and 5.8 present a compatible grouping of the most commonly used metals in electronic systems. If the mated metals belong to different groups, the metal coming first in the list in Table 5.7 will constitute the anode and will be relatively heavily corroded, whereas the metal coming from the lower group will form the cathode and will be relatively free from corrosion.

The greatest degree of corrosion occurs when dissimilar metals are openly exposed to saltwater, salt spray, rain, gasoline, jet fuel, or other liquids that may act as an electrolyte. The least amount of corrosion occurs when metals are kept dry and completely free from exposure to moisture. The following three exposure conditions are defined (Table 5.8):

1. **Exposed:** Metal has an open, unprotected exposure to weather.
2. **Sheltered:** Milder exposure than above. The metal surfaces receive limited protection from direct action of weather.
3. **Housed:** Metal surfaces of equipment are housed in weatherproof buildings.

For any given pair of dissimilar metals in contact under these three exposure conditions, the extent of corrosion can be minimized. Table 5.8 may be used for this purpose. The table indicates the protection to be applied to a bond for the expected environmental conditions in which the particular type of bond is to be used. Table 5.8 must be used in conjunction with Table 5.7 to identify the group associated with each bond metal.

Figure 5.20. Corrosion on a corrosion-resistant (CHO-SEAL 1298) EMI shielding gasket (left) and a pure-silver-filled gasket (right) mated with an aluminum disk, after 168 hours of salt-fog exposure. (Courtesy of Chomerics, a Parker Company.)

Table 5.7. Electrochemical series grouped and ordered by decreasing sensitivity to corrosion

Anodic end (most active and easily corroded)	
Group I	Magnesium
Group II	Aluminum, Aluminum Alloys, Zinc, Cadmium
Group III	Carbon Steel, Iron, Lead, Tin, Tin–Lead Solder
Group IV	Nickel, Chromium, Stainless Steel
Group V	Copper, Silver, Gold, Platinum, Titanium
Cathodic end (most noble and least corroded)	

Another representation of galvanic coupling is provided in Table 5.9, where groups of metals connected by lines are considered to be galvanic couples. Table 5.9 lists the order of their relative activity within a seawater environment, which can serve as a representative corrosive atmosphere. The list begins with the more active (anodic) metals and proceeds down to the least active (cathodic) metals in the galvanic series.

In Table 5.9, allowable combinations of mating metal members are shown [16]. Incompatible combinations (designated "I") should be avoided. "Allowable combina-

Table 5.8. Groups of metals recommended for providing protective bonds between two dissimilar metals used as anodes and cathodes

Condition of Exposure (see below)	Anode				Cathode
	I	II	III	IV	
Exposed	A	A			II
Sheltered	A	A			
Housed	A	A			
Exposed	C	A	B		III
Sheltered	A	B	B		
Housed	A	B	B		
Exposed	C	A	B	B	IV
Sheltered	A	A	B	B	
Housed	A	B	B	B	
Exposed	C	C	C	A	V
Sheltered	A	A	A	B	
Housed	A	A	B	B	

A: The mating metals must have a protective finish after metal-to-metal contact has been established so that no liquid film can bridge the two elements of the couple.

B: The metals may be joined with bare metal exposed at junction surfaces. The remainder must be given an appropriate protective finish.

C: This combination cannot be used except if short life expectancy can be tolerated, or when the equipment is normally stored and exposed for only short intervals. Protective coatings are mandatory.

Table 5.9. Guide for selection of permissible galvanic couples and protective systems for service in seawater, marine atmosphere, and industrial atmosphere

First metal/alloy → ⟶ Active (anodic) ——— Noble (cathodic) ⟶

In each cell the two numbers give the surface-treating/finishing system group (from Table 5.11) for Metal 1 and Metal 2. The code letters give the recommended system by environment (Sea Water; Marine Atmosphere; Industrial Atmosphere): C = coating/treatment recommended, – = not applicable, G = optimum (same-metal) couple.

Second metal/alloy ↓	A	B	C	D	E	F	G	H	I	J	K	L	M	N	O	P	Q	R	S	T
A Magnesium, Mg Coating	1 / G	1 2	1 3	1 3	1 4	1 5	1 6	1 6	1 7	1 8	1 9	1 9	1 9	1 9	1 9	1 9	1 10	1 11	1 12	1 13
B Zinc (Zn), Zn Coating		2 / G	2 3	2 4	2 4	2 5	2 6	2 6	2 7	2 8	2 9	2 9	2 9	2 9	2 9	2 9	2 10	2 11	2 12	2 13
C Cadmium, Beryllium			3 / G	3 4	3 4	3 5	3 6	3 6	3 7	3 8	3 9	3 9	3 9	3 9	3 9	3 9	3 10	3 11	3 12	3 13
D Aluminum (Al), Al–Mg, Al–Zn				4 / G	4 4	4 5	4 6	4 6	4 7	4 8	4 9	4 9	4 9	4 9	4 9	4 9	4 10	4 11	4 12	4 13
E Al–Copper (Al–Cu)				4 / G	4 / G	4 5	4 6	4 6	4 7	4 8	4 9	4 9	4 9	4 9	4 9	4 9	4 10	4 11	4 12	4 13
F Steels, Low Alloy, Carbon					5 / G	5 / G	5 6	5 6	5 7	5 8	5 9	5 9	5 9	5 9	5 9	5 9	5 10	5 11	5 12	5 13
G Lead (Pb)						6 / G	6 / G	6 6	6 7	6 8	6 9	6 9	6 9	6 9	6 9	6 9	6 10	6 11	6 12	6 13
H Tin (Sn), Sb–Pb, Indium							6 / G	6 / G	6 7	6 8	6 9	6 9	6 9	6 9	6 9	6 9	6 10	6 11	6 12	6 13
I Stainless Steel, Ferric, Martensitic								7 / G	7 / G	7 8	7 9	7 9	7 9	7 9	7 9	7 9	7 10	7 11	7 12	7 13
J Chromium, Tungsten									8 / G	8 / G	8 9	8 9	8 9	8 9	8 9	8 9	8 10	8 11	8 12	8 13
K Stainless Steel, Austenitic, Superstrength										9 / G	9 / G	9 9	9 9	9 9	9 9	9 9	9 10	9 11	9 12	9 13
L Brass–Lead, Bronze											9 / G	9 / G	9 9	9 9	9 9	9 9	9 10	9 11	9 12	9 13

Information on this chart is presented for equal exposed areas of each metal comprising the galvanic couple.

Numerical notations refer to surface treating and finishing systems listed in Table 5.11 for each metal group for ameliorating corrosion of joint metals. The systems are arranged in Table 5.11 in decreasing order of effectiveness. An optimum system is presented in each case for use with joint similar or dissimilar metals, intended for service in severe environ-

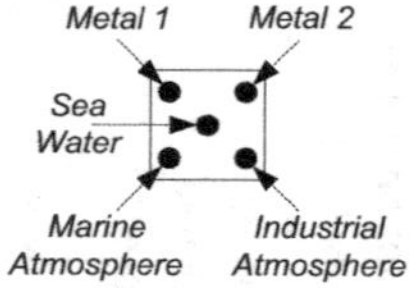

Legend (cell diagram):

```
Metal 1  •        •  Metal 2
          Sea Water
         •
Marine   •        •  Industrial
Atmosphere          Atmosphere
```

The following is a galvanic‑compatibility matrix (rotated 90° on the page). The material groups are listed in the left column; the remaining cells give the group numbers (top) and the compatibility notations (bottom).

Material	M	N	O	P	Q	R	S	T
M Brass–Low Cu, Bronze–Low Cu	9 G	9,9 C,C	9,9 C,C	9,9 C,C	9,10 I,C	9,11 C,C	9,12 I,C	9,13 I,C
N Brass–High Cu, Bronze–High Cu		9 G	9,9 I,C	9,9 C,C	9,10 I,C	9,11 C,C	9,12 I,C	9,13 I,C
O Copper–High Nickel, Monel			9 G	9,9 C,C	9,10 C,C	9,11 C,C	9,12 I,C	9,13 I,C
P Nickel, Cobalt				9 G	9,10 C,C	9,11 C,C	9,12 I,C	9,13 I,C
Q Titanium					10 G	10,11 C,C	10,12 C,C	10,13 C,C
R Silver						11 G	11,12 C,C	11,13 C,C
S Gold, Platinum, Rhodium							12 G	11,13 C,C
T Graphite								13 G

ments. Alternative systems are provided for use in service situations that preclude the maximum protective system or for milder environment service situations.

Letter notations "C" or "I" signify compatibility or incompatibility, respectively, of joint metals in the specific environment. Occasionally, "C" or "I" is not clearly resolvable, and in such borderline cases "I" is indicated. Further, "C" indicates negligible galvanic interaction between bare, dissimilar metals, when joined and subjected to the specific environment, whereas "I" implies significant galvanic corrosion of bare, dissimilar metals, when joined and exposed to the specified environment.

"G" signifies compatibility of same-metal couple, bare, in seawater, marine atmosphere, or industrial atmosphere.

tions" should not be construed as being devoid of any galvanic action: Permissible couples only represent a low and tolerable galvanic action. Several factors further influence and control galvanic action, namely, the effectiveness of the electric circuit, the ratio of the anode-to-cathode areas, and the polarization of the electrodes.

5.5.5 Corrosion Protection

Special attention should be given to the interdependent relationship between electrical bonding and corrosion control. The need for mating bare metal to bare metal for achieving a satisfactory bond brings about a frequent conflict between bonding and surface finishing specifications. Generally speaking, oxides that form on metal surfaces are nonconductive. It is desirable that the oxides be softer than the base metal and as thin as possible. Unfortunately in most cases, such as aluminum, oxides and other corrosion products are often much harder than the base metal and, thus, impenetrable, making bonding of bare surfaces a "mission impossible."

Obtaining a good electrical bond can lead to potential corrosion problems if the bonding is not properly implemented. Conversely, design techniques for effective corrosion protection, such as the use of surface finishes that are not electrically conductive, can be self-defeating as they result in lack of bonding.

The following section discusses techniques that may be employed for corrosion control and effective bonding.

5.5.5.1 Protective Conductive Coatings.
The most effective means of preventing corrosion is the application of protective conductive films or coatings. Protective coatings are categorized as follows:

1. **Chemical conversion treatment or anodic films.** Metals and alloys are coated with suitable solutions of chemicals under controlled conditions to form a protective surface coating. This coating is physically integrated with the underlying metal and serves as a barrier against corrosive attack. Coatings commonly used are oxides, phosphates, chromates, or complex compounds of the substrate metal. These coatings are commonly applied on iron, steel, aluminum, cadmium, and other metals.

2. **Metallic coatings.** Selected for suitability and application involved, with attention paid to problems of aging, cracking, diffusion, and corrosion. Recommendations for the prevention of corrosion that should be considered are provided in Table 5.10 [14].

    Metallic coatings are also applied to some metals by the process of hot-dipping, largely confined to the coatings of ferrous alloys with metals and alloys of low melting points. Typical hot-dipping coating materials are zinc (Zn) and tin (Sn) as well as lead (Pb) alloys. Tinned steel and zinc-coated or galvanized iron and steel are the most common hot-dipped products.

    The noble metals [gold (Au), palladium (Pd), platinum (Pt), and rhodium (Rh)] and the corrosion-resistant metals [chromium (Cr), nickel (Ni), tin (Sn), tin-lead solder, and titanium (Ti)] require no finish other than cleaning.

Table 5.10. Selection of metallic coatings for minimum corrosion

Purpose/criterion	Recommended	Not recommended
Contact with aluminum or magnesium	Cadmium or tin	Chromium, copper, silver, gold
Prepaint coating	Cadmium or tin	Chromium, copper, nickel, silver, gold
Tarnish prevention	Rhodium over silver Gold over silver, copper, or nickel Nickel between copper and silver	
Marine exposure	Heavy gold, 0.00030 inch (0.3 mil) minimum	
Solderability	Tin, gold, or tin–lead	Nickel, chromium, rhodium
Storage	Gold, rhodium, or reflowed heavy tin	Cadmium, silver, copper
Wear	Chromium, nickel, rhodium, or hard gold	Cadmium, tin
Easy etching (for printed circuit boards)	Cadmium, nickel (in ferric chloride only), indium, tin	Rhodium, silver, tin–lead, gold

Applications of aluminum (Al), copper (Cu), and magnesium (Mg) require special treatment unless they are used in hermetically sealed enclosures. Aluminum should be anodized, and where this is impossible, chemical conversion treatment (i.e., alodine and irridite) should be used. Copper and copper alloys may be black-oxide treated, plated, or painted. If bare copper is required, a tarnish-preventive, thin silicone-cured resin film is commonly used. Magnesium exhibits very poor corrosion resistance and, therefore, should be anodized or otherwise treated with several coats of alkali-resistant primer or other moisture-proofing coatings. If used with any other metal, extreme precautions should be employed to prevent destructive corrosion.

The list of treatments provided in Table 5.11 [16] represent decreasing order of protectiveness for the metals to which they apply. Where a choice of treatments can be carried out and long-term economics permit, the selection of treatments should be made accordingly. Specific enhancing effects can be accomplished by selecting the higher level treatments for each metal. Alternatively, higher degree protection is frequently achieved if an optimum treatment is selected for one metal, and a second or third protection option taken for the second metals. In atmospheric corrosion considerations when cost must be taken into consideration, a recommended strategy would be to select a higher level treatment for the more active metal and an alternate treatment for the less active metal. This selection takes into account the fact that the more active metal is likely to undergo more corrosion initially, even under mild conditions when

Table 5.11. Recommended protective treatment in order of protective effectiveness

Order of effectiveness	Treated metal/alloy	Recommended treatment
1	Magnesium (Mg)	a) Anodic coating plus alkali-resistant paint system or resin seal. b) Chromate conversion coating plus alkali-resistant paint system or resin system (alternate for use in nonpersistent wet or marine atmosphere, or anodic coating without organic system). c) Metallic coating, electroless nickel plus cadmium overplating. For electrical and thermal conducting purposes, in the absence of wet, saline, or acidic atmospheric conditions. d) Chromate treatment, suitable for assured condensation and acid-free conditions. *Note:* Do not use bare magnesium.
2	Zinc (Zn) and zinc coatings	a) Anodic coating plus paint or resin coating system primarily used for coatings. b) Chromate conversion coating plus paint or resin system; or anodic coating without organic systems. For use in nonpersistent wet or marine atmosphere, for electrical and thermal conducting purposes in mild atmospheres in the absence of wet, saline, or acidic conditions. c) Chromate conversion coating, without paint or resin coating system.
3	Cadmium (Cd) or beryllium (Be)	a) Chromate conversion coating plus paint or resin coating system. b) Chromate conversion coating, without organic system. For use in nonpersistent wet or marine atmosphere; for electrical, thermal conducting purposes in mild atmospheres in the absence of wet, saline, or acidic conditions. Recommended for beryllium in high-temperature applications to forestall catastrophic oxidation in oxygen-containing atmosphere.
4	Aluminum (Al) and aluminum alloys	a) Anodic coating plus paint or resin coating system. b) Chromate conversion coating plus paint or resin system; or anodic coating, sealed with resin seal (when porous castings are used, impregnated with resin prior to surface treatment and finishing). c) Chromate conversion coating without paint or resin coating. For electrical and thermal conducting purposes in mild atmospheres in the absence of saline, alkaline, or acidic conditions.

Table 5.11. Recommended protective treatment in order of protective effectiveness, *continued*

Order of effectiveness	Treated metal/alloy	Recommended treatment
4	Aluminum (Al) and aluminum alloys (*cont.*)	d) Bare aluminum may be used when surface treatment would interfere with the application, under conditions free of salinity or extended wetness, or when highly corrosion resistant alloys are used. Faying edges should be sealed to prevent crevice corrosion.
5	Carbon (C) and low-alloy steel	a) Metallic coating (e.g, sacrificial Zn, Cd plus chromate treatment, or nonsacrificial Cu, Ni) plus paint coating system. For steels of strengths greater than 220 ksi, metallic coatings should be applied by non-electrolytic methods; zinc or cadmium are prohibited. For steels of strengths up to 220 ksi metallic coatings, coatings may be applied electrolytically, but the steel should be stress relieved before plating and hydrogen embrittlement is required after plating.
		b) Metallic coating, for example, sacrificial Zn, Cd with supplemental surface treatment, or nonsacrificial Cu or Ni, without paint coating system for direct metallic contact or for achieving the least potential difference between the joint metals. For metals of strengths greater than 220 ksi, metallic coatings, if required, should be applied by nonelectrolytic methods; zinc or cadmium are prohibited.
		c) Zinc-phosphate conversion coating plus paint coating system. Caution is required if phosphate coating is used on steels of strengths between 150 to 220 ksi, hydrogen relief required; stress relief required prior to phosphating and hydrogen embrittlement required after phosphating.
		d) Pretreatment primer plus paint coating system.
		e) Heavy phosphate conversion coating plus supplemental treatment. Not for steel of strengths greater than 220 ksi. *Note:* Bare steel is not recommended.
6	Treatment for lead (Pb), tin (Sn), solders, and indium (In)	Coating is applied by hot-dipping, fusing or electroplating. a) Coat with paint or resin coating system. Electroplating should be "flowed" prior to applying the coating system.
		b) Electroplate with other metal to reduce the electropotential difference of metals being joined, where direct contact of metals is required for electrical purposes.

(*continued*)

Table 5.11. Recommended protective treatment in order of protective effectiveness, *continued*

Order of effectiveness	Treated metal/alloy	Recommended treatment
7	Steels: carbon, low-alloys, martensitic, and ferritic stainless steel	Steels with chromium contents in the region of 12% will undergo considerable surface staining and limited rusting in corrosive environments, but, on the whole, are appreciably less corroded than carbon steels. a) Paint or apply resin coating; zinc phosphate carbon steel prior to application of paint or resin coating. b) May be electroplated or used bare for use in nonpersistent wet or marine atmosphere, and for electrical or thermal conducting purposes.
8	Chromium (Cr) (plate), tungsten	a) Paint or apply resin coating to reduce corrosion at voids in chromium plating, or staining of tungsten surfaces. b) Normally may be used bare for electrical wear resistance, or thermal conducting purposes. Seal faying edges to mitigate crevice attack of metal to which it is joined.
9	Steel: stainless-austenitic, super strength, brass-leaded, bronze, brass–low-copper, and copper–high-nickel	a) Apply metallic coating as may be required to minimize electrical potential difference between the metals to be joined. Apply paint or resin coating, primarily to diminish ion contamination from metals of this group onto more anodic metals to which they might be joined, thereby diminishing potential damage to the more anodic metal. b) Apply metallic coating (as in "a" above) without paint or resin coating for electrical or thermal conducting purposes. It may be expedient to overcoat completed assembly with paint or resin. c) Apply paint or resin coating system and seal faying edges. d) Use bare and seal faying edges for electrical and thermal conducting purposes if more anodic metals are not directly joined or in close proximity to receive a rundown of surface condensate. e) Select galvanically compatible metals required to be coupled for high-temperature applications, where metallic coatings may not be useful and paint or resin coatings are impractical.
10	Titanium (Ti)	a) Anodize for antigalling and wear resistance. b) Apply metallic coating (Cd, Zn prohibited, Ag over Ni acceptable) plus paint or resin coating. c) Apply metallic coating (Cd, Zn prohibited; Ag over Ni acceptable), seal faying edges. For electrical or thermal conducting purposes.

Table 5.11. Recommended protective treatment in order of protective effectiveness, *continued*

Order of effectiveness	Treated metal/alloy	Recommended treatment
10	Titanium (Ti) (*cont.*)	d) May be used bare with faying edges sealed in contact with metals other than magnesium, zinc, or chromium; for electrical or thermal conducting purposes.
11	Silver (Ag)	a) Silver or silver-plated parts to be used as electrical, open/close contact points, plugs, and receptacles should be plated over with rhodium, palladium, or gold. b) May be used in stationary components or electrical assemblies, for example, connectors and printed circuit boards, but should be enveloped by sulfur-free conformal coatings. c) Apply chromate conversion coating plus corrosion inhibiting fluid film to parts of electrical plugs, receptacles, and so on.
12	Gold (Au), rhodium, platinum (Pt), and alloys	Use bare, with compound sealant at edges of dissimilar metal joint, or by enveloping dissimilar metal joint in conformal coating, where feasible.
13	Graphite	a) Plate graphite to minimize electrical potential difference between graphite and metal to be joined to it. Seal faying edges to preclude corrosion at contacting surface of the metal member if service is electrical, or apply conformal coating. b) May be used bare in electrical or thermal conducting service, conditions permitting. Seal faying edges.

galvanic effects would be minimal. Therefore, cathodic control of corrosion, frequently useful in electrolytic solutions, is virtually inoperative under regular atmospheric exposure conditions. Table 5.10 [14] similarly provides guidance on the selection of metallic coatings for minimum corrosion between metals, with reference to several application criteria.

For the purpose of achieving EMI control, it is required to remove all nonconductive finishes on all mated surfaces on which bonding effectiveness would otherwise be severely compromised. There are certain corrosion-resistant chemical conversion finishes that offer low electrical resistance, such as iridite or alodine for aluminum or its alloys, as well as metallic plating, such as cadmium, tin, or silver, which generally need not be removed, thanks to their conductive nature. Most other coatings however, such as paints and anodize, are nonconductive and must be removed prior to forming a low-impedance bond. Anodized aluminum, for instance, would appear at first glance to be a good conductive surface for the purpose of bonding, yet, in practice, anodizing forms highly insulating hard surfaces, severely degrading the bonding effectiveness.

Figure 5.21 [1] illustrates the extent of degradation of shielding effectiveness of metal-to-metal joints on a shielded enclosure, caused by anodizing aluminum, compared to the performance with various conductive finishes on the aluminum surfaces. The superiority of the bare metal and even the conductive conversion treatments over the nonconductive anodized surfaces is self-evident.

Bonds should be kept tight and well coated after forming the bond. Paint, protective coatings, or metallic plating used for the purpose of excluding moisture or for providing a third, intermediate metal compatible with both bond members should be applied with caution. When such surface treatments are used, both members should be treated as detailed in Figure 5.22 [1]. Coating the anode alone [Figure 5.22(a)] must be avoided. If only the anode is coated, corrosion will be severe at imperfections and breaks in the coating, owing to the relatively small effective (exposed) area of the anode. All other cases will yield acceptable [Figure 5.22(b)] or optimal [Figure 5.22(c)] protection against corrosion. For maintaining the bond's endurance to corrosion during the equipment's life cycle, protective coating, metallic plating, or paints must be maintained in good condition. In particular, when maintenance actions involving disconnection of the bond were carried out, any protective coating, plating, or paints must be reapplied, requiring that appropriate maintenance procedures, instructions, and verification measures be put into practice.

Another point of concern in the application of protective coating is associated with ridges of paint around the periphery of the bonding area, which can prevent good metal-to-metal contact. Immediately prior to bonding, the surfaces to be bonded must be cleaned of any nonconductive substances and objects. All chips, paint, grease, or other

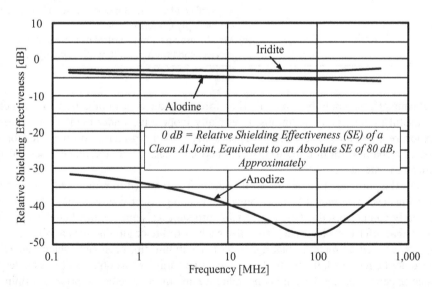

Figure 5.21. Degradation of shielding effectiveness of a shielded enclosure caused by surface finishes on aluminum, used for treatment of the bonded metal members of the enclosure.

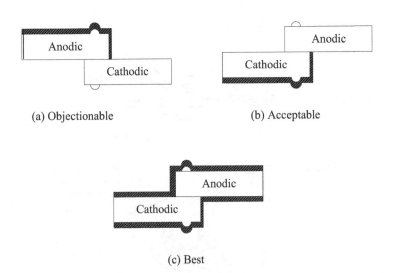

(a) Objectionable   (b) Acceptable

(c) Best

Figure 5.22. Techniques for protecting bonds between dissimilar metals.

foreign matter must be removed with a proper cleaning solution. Washers or fittings must fit inside the cleaned area to ensure the quality of the bond and the continuous smoothness of the bonded surfaces. After bonding, the exposed areas should be refinished as soon as possible with an appropriate finish, using the most appropriate technique from Figure 5.22.

Figure 5.23 illustrates the manner of implementation of an indirect bond at the contact area to the mating surface.

5.5.5.2 Interposing/Sacrificial Metals. Consideration should be given to the electrochemical series to assure that, when possible, corrosion occurs only to replaceable elements of a bond, such as bonding jumpers, washers, bolts, or clamps, rather than structural members or equipment enclosures. When two dissimilar metals are brought in contact, the one higher in the electromotive force series will be more affected by corrosion rather than the other. The smaller mass (generally the more easily replaceable) should, therefore, be made of the higher series metal; for instance, cadmium-plated washers are recommended for use with steel surfaces since cadmium is lower than steel in the galvanic series.

Metals widely separated in the galvanic series must be protected if they are to be joined. One manner of addressing that is by making use of an interposing metal between the two. Improved protection can be accomplished by applying to the anodic member a sacrificial metal coating having a potential similar to or near that of the anodic member and by sealing to ensure that the faying surfaces are water or humidity tight and by painting or coating all surfaces. For example, referring to the grouping in Table 5.7, if an aluminum (Group II) equipment case is to be bonded to a stainless steel (Group IV) frame, it is good practice to interpose a tin (Group III) or cadmium (Group

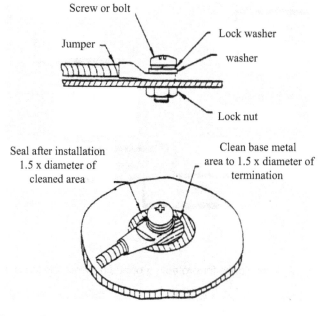

Figure 5.23. Recommended bonding strap bolting installation.

II) plated washer between the two metal surfaces. If the protective coating is then chipped, the washer instead of the aluminum case will be attacked by corrosion.

5.5.5.3 *Relative Area of Anodic Member.* When joints between two dissimilar metals are unavoidable, a small anodic area should be avoided and the anodic member of the pair should be the larger of the two. The same metal or more noble (cathodic) metals should be utilized for small fasteners and bolts. For a given current flow in a galvanic cell, the current density is higher for a small electrode than for a larger one. The greater the current density of the current leaving an anode, the greater will be the rate of corrosion (Figure 5.24). Therefore, the larger the relative anode area, the lower will be the galvanic current density on the anode and the lesser will be the corrosive attack. The galvanic corrosion effect may thus be considered as inverse to the anode–cathode area ratio.

As an example, if a copper strap or cable is bonded to a steel column, the rate of corrosion of the steel will be low because of the large anodic area. On the other hand, a steel strap or bolt fastener in contact with a copper plate will corrode rapidly because of the relatively small area of the anode of the cell.

Lessons Learned

- The mating surfaces must be treated in order to ensure minimum contact resistance.

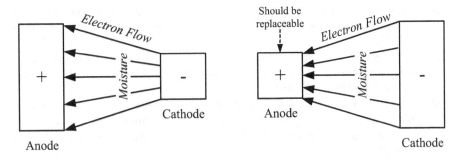

(a) Acceptable: Small Cathode   (b) Objectionable: Large Cathode

Figure 5.24. When joints between two dissimilar metals are unavoidable, the anodic member of the pair should be the larger of the two.

- Bonding must achieve and maintain intimate contact between metal surfaces. The surfaces must be smooth, clean, and free of nonconductive finishes.

- The compatibility of the mating metals must be considered for corrosion control. If this is not possible, special attention must be paid to the control of bond corrosion through the choice of the materials to be bonded, the selection of supplementary components (such as washers) to assure that corrosion will affect replaceable elements only, and the use of protective finishes and surface treatments.

- Protection of the bond from moisture and other corrosive elements must be provided.

- Finally, throughout the lifetime of the equipment, system, or facility, the bonds must be inspected, tested, and maintained to assure that they continue to perform as required.

BIBLIOGRAPHY

[1] MIL-HDBK-419A, *Military Handbook, Grounding, Bonding, and Shielding for Electronic Equipments and Facilities,* Washington, DC: U.S. Government Printing Office, 1987.

[2] Law, P. E., Jr., *Shipboard Electromagnetics,* Boston, MA: Artech House, 1987.

[3] MIL-STD-1310G (NAVY), *Department of Defense Standard, Practice for Shipboard Bonding, Grounding, and Other Techniques for Electromagnetic Compatibility and Safety,* Washington, DC: U.S. Government Printing Office, 28 June 1996.

[4] MIL-STD-1605(SHIPS), *Department of Defense Standard Practice, Procedures for Conducting a Shipboard Electromagnetic Interference (EMI) Survey (Surface Ships),* Washington, DC: U.S. Government Printing Office, 20 April 1973.

[5] MIL-B-5087B (ASG), *Bonding, Electrical and Lightning Protection for Aerospace Systems,* Washington, DC: U.S. Government Printing Office, 15 October 1964.

[6]   MIL-STD-464A, *DoD Interface Standard, Electromagnetic Environmental Effects, Requirements for Systems,* Washington, DC: U.S. Government Printing Office, 19 December 2002.

[7]   NAVSEA OD 30393, *Design Principles and Practices for Controlling Hazards of Electromagnetic Radiation to Ordnance (HERO Design Guide),* Second Revision, Naval Sea Systems Command, April 2001.

[8]   AFSC DH 1-4, *Air Force System Command Design Handbook, Electromagnetic Compatibility,* 4th Ed., Revision 1, Wright-Patterson AFB, Ohio, January 1991.

[9]   Perez, R. (Ed.), *Handbook of Electromagnetic Compatibility,* New York: Academic Press, 1995.

[10] Hartal, O., *Electromagnetic Compatibility by Design,* W. Conshohocken, PA: R&B Enterprises, 1994.

[11] MIL-HDBK-1857, *Department of Defense handbook, Grounding, Bonding and Shielding Design Practices,* Washington, DC: U.S. Government Printing Office, March 1998.

[12] Mardiguian, M., *A Handbook Series on Electromagnetic Interference and Compatibility, Volume 2, Grounding and Bonding,* Gainesville, VA: Interference Control Technologies, 1988.

[13] DARCOM-P 706-410, *DARCOM Pamphlet, Engineering Design Handbook, Electromagnetic Compatibility,* HQ, Alexandria, VA: USA Material Development and Readiness Command, March 1977.

[14] AMC 706-235, *AMC Pamphlet, Engineering Design Handbook, Hardening Weapon Systems Against RF Energy,* HQ, Alexandria, VA: USA Material Development and Readiness Command, February 1972.

[15] NAVAIR AD 1115, *Electromagnetic Compatibility Design Guide for Avionics and Related Ground Support Equipment,* 3rd Ed., Naval Air System Command, June, 1988.

[16] MIL-STD-889B, Notice 3 (USAF), *Military Standard, Dissimilar Metals,* Washington, DC: U.S. Government Printing Office, May 1993.

[17] IEEE-1100, *IEEE Recommended Practice for Powering and Grounding Electronic Equipment* (IEEE Emerald Book), New York: IEEE Press, 1999.

6

GROUNDING FOR POWER DISTRIBUTION AND LIGHTNING PROTECTION SYSTEMS

6.1 INTRODUCTION

The objectives of grounding have been discussed in several sections of this book. In the case of electronic circuits, the ground conductor or the ground plane normally functions as a return path for the signal and power supply. The ground also performs the important task of a common-signal reference for electronic circuits. A good ground design is crucial for electromagnetic compatibility (EMC). Functionally, the ground for electronic circuits does not have to be physically connected to the general mass of the earth (i.e., "earthed").

The grounding scenario is rather different for power and lightning protection where the ground invariably implies a physical connection to the earth mass. Electrical safety and lightning and system protections are often the only considerations of power engineers when dealing with power system grounding.

Electronic or EMC engineers know very well that the signal reference ground is somewhere connected to the safety ground and ultimately to the mass of the earth. They are also aware that the power or the electrical safety grounds often contain electrical noise that is detrimental to EMC performance. In general, however, they may not appreciate that improper equipment grounding and connection to a ground electrode subsystem is often the root cause of performance, safety, and equipment damage problems under ground-fault conditions, power system switching, and lightning transients.

Grounds for Grounding. By Elya B. Joffe and Kai-Sang Lock

Grounding designs for electronic circuits and equipment are mainly determined by considerations of performance and EMC compliance. The design and installation practices of grounding for power system [1–9] and lightning protection [10–28] are governed not only by sound engineering principles but are also dictated by relevant standards and codes of practice, which vary substantially from country to country.

In this chapter, the principles and practices of grounding for power systems and lightning protection are presented. To avoid confusing terminology, the term earthing is used in place of grounding when we refer to connection to the earth.

6.2   POWER SYSTEM EARTHING

Power system earthing is a standard practice throughout the world nowadays [29, 30]. However, early electrical power systems were unearthed, with electrical power being delivered with noninsulated lines supported by insulators and placed out of reach on tall wooden poles. It was only in the early 1920s that United Kingdom and France introduced the requirements for earthing of metalwork of appliances or frames for safety reasons.

6.2.1   Objectives of Power System Earthing

An earthed power system usually refers to a system in which the neutral point of transformer or generator windings is intentionally grounded, either solidly or through impedance.

Power system earthing should provide a sufficiently low impedance path, via the return conductors, back to the supply source to facilitate the operation of protective relays under fault conditions.

Grounding and bonding are also implemented to limit the potential rise to a safe value on all noncurrent-carrying metalwork, with which persons and animals may come into contact, under normal and abnormal circuit conditions. Bonding together all normally exposed conductive parts, such as gas, water, heating, and air conditioning metallic piping, and the connection of that bond to the earth terminal will prevent the occurrence of a dangerous potential difference between adjoining metalwork under abnormal conditions.

The main objectives for power system earthing are summarized as follows:

1. Ensure that living beings are not subject to dangerous electric shock hazards under normal or fault conditions.
2. Provide a path for the flow of fault current that allows detection of an unwanted connection between system conductors and ground so as to facilitate automatic operation of protective devices to remove the fault from the system.
3. Maintain system voltage within reasonable limits under fault conditions so that system insulation breakdown voltages are not exceeded.
4. Control the voltage to earth, or ground, within predictable limits so that graded insulation may be used in design and construction of power transformers.

The reasons for achieving these objectives and how these objectives can be realized are explained in the various sections of this chapter.

6.2.2 Faults in Power Supply Systems

It is inevitable that electrical systems experience occasional faults or short circuits due to insulation failure or a bridging of insulation in the power apparatus, the transmission and distribution lines, and the associated switchgear and equipment. A fault may be described as a circuit condition in which current flows through an abnormal or unintended path. There are generally two types of faults: a short-circuit fault or an earth fault. A short-circuit fault occurs where the current flows between live parts, which may include the neutral. In an earth fault, the current flows between a live part and earth, including any conductive part or conductor that is connected to the earth terminal or is substantially in contact with the general mass of the earth.

The current flow during a short circuit or fault at any point in an electrical system is limited by the impedance of circuits and equipment from the source or sources to the point of the fault. This point is explained more clearly by referring to Figure 6.1, where the path of fault current is indicated for the case of a phase-to-neutral short circuit.

Generally, the impedance between live conductors or between live conductors and earthed metal parts at the fault position may be considered negligible. In the case of short-circuit faults the current, I_F, is limited primarily by the equivalent source impedance, the transformer impedance, and the impedance of the transmission and distribution lines, which collectively form the fault loop impedance as seen from the

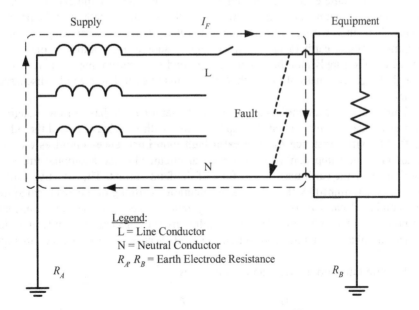

Figure 6.1. Short circuit between phase and neutral conductor.

point of the short circuit. The magnitude of short-circuit current in an electrical sup-ply system can be in the range of tens of thousands of amperes. This very high fault current can cause excessive thermal and mechanical stresses on the electrical system components and must be removed immediately by the automatic operation of protec-tive devices.

The fault current, I_F, is approximated by

$$I_F = \frac{U_0}{Z_F} \approx \frac{U_0}{Z_S + Z_T + Z_{Ph} + Z_N} \tag{6.1}$$

where:
U_0 = Phase-to-neutral voltage, volts
Z_F = Fault-loop impedance, Ω
Z_S = Effective source impedance as seen from the transformer, Ω
Z_T = Transformer impedance per phase, Ω
Z_{Ph} = Impedance of the phase conductor from transformer to the point of fault, Ω
Z_N = Impedance of the neutral conductor from transformer to the point of fault, Ω

The short-circuit current can be very high because of the low fault-loop impedance. It is not uncommon to have short circuit current in the region of 30 to 40 kA even in low-voltage (LV) power distribution systems. The energy in a short-circuit fault can be-come sufficiently high to vaporize circuit conductors or may lead to fire hazard.

A ground fault is an unintentional, electrically conducting connection between an ungrounded conductor of an electrical circuit and the normally noncurrent-carrying conductors, metallic enclosures, metallic raceways, metallic equipment, or earth. The ground-fault current path comprises an electrically conductive path from the point of a ground fault on a wiring system through normally noncurrent-carrying conductors, equipment, or the earth to the electrical supply source. Depending on the earthing schemes of the supply and the facility, the ground fault current may or may not flow through the general mass of the earth. More details of earth faults will be discussed in Section 6.3.

Figure 6.2 shows an earth fault in which the fault current has to flow though two earth electrode systems, one at the supply source with resistance R_A and the other at the facility with resistance R_B. The earth fault loop impedance comprises the short-circuit fault loop impedance of the electrical circuit plus the additional impedance contributed by the earth mass to the flow of the fault current. The effective earth re-sistance is determined by the schemes of intentional earthing of the electrical system. This combined earth resistance is normally much larger than the other impedance components in the fault loop. The earth fault current in this case is significantly low-er than the phase-to-neutral short-circuit current discussed with reference to Figure 6.1.

The earth fault current, I_F, is approximated by

$$I_F = \frac{U_0}{Z_F} \approx \frac{U_0}{Z_S + Z_T + Z_{Ph} + Z_{PE} + R_A + R_B} \tag{6.2}$$

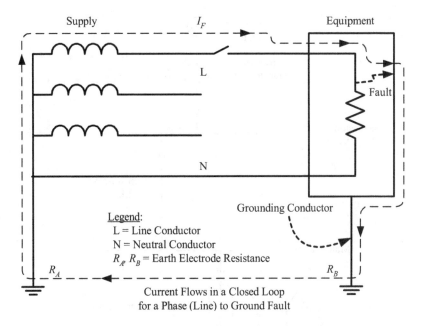

Figure 6.2. Phase-to-earth fault.

where:

U_0 = Phase-to-neutral voltage, volts

Z_F = Fault-loop impedance, Ω

Z_S = Effective source impedance as seen from the transformer, Ω

Z_T = Transformer impedance per phase, Ω

Z_{Ph} = Impedance of the phase conductor from transformer to the point of fault, Ω

Z_{PE} = Impedance of the protective earth (PE*) conductor, Ω

R_A = Earth resistance of the supply earthing system, Ω

R_B = Earth resistance of the facility earthing system, Ω

For effective fault protection, it is desirable to have low fault loop impedance so that the resulting high fault current can rapidly trip circuit breakers or blow fuses.

6.2.3 General Configuration of a Power Distribution System

The earthing schemes in a simplified transmission and distribution system are now illustrated with reference to Figure 6.3 to facilitate a better understanding of the functions of earthing in power systems.[†]

---

*The PE conductor, commonly used by power engineers, is also known as ESGC (electrical safety ground conductor), used in other chapters in this book. The two terms are equivalent and interchangeable.

[†]Note that in subsequent drawings, the cutout fuse and other protective devices are not shown in order not to clutter up the circuit diagrams.

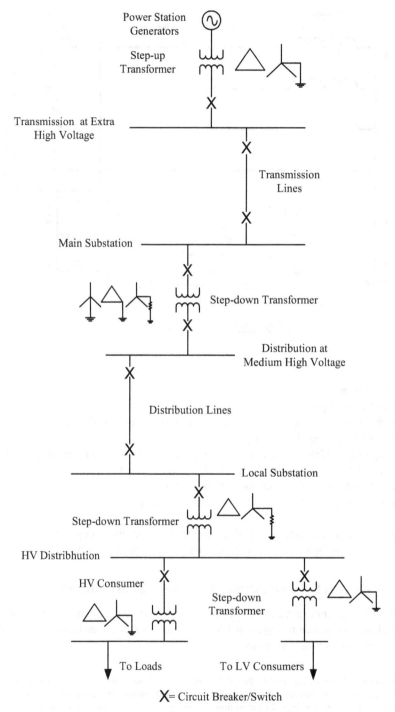

Figure 6.3. Example of transmission and distribution earthing schemes.

In the particular example given in Figure 6.3, the step-up transformer at the power station is solidly earthed at the neutral point of the extra-high-voltage* (EHV), wye-connected secondary windings. The main reason to use direct connection to earth is to ensure that the neutral point is at earth potential under normal and abnormal conditions. Different parts of the high-voltage windings will be subjected to different level of voltage stress; the further that part of the winding is from the neutral point, the higher will be the voltage stress. It is, therefore, unnecessary to design a uniform insulation for the whole transformer to withstand the highest voltage level. This graded insulation design, as illustrated in Figure 6.4, allows substantial savings in the cost of high-voltage transformers.

The drawback of solidly earthing the neutral of a high-voltage transformer is the excessively high earth fault current. In practice, the earth fault current is kept to a manageable level by selecting high-impedance transformers or earthing through a resistor.

At the main substation, the wye-connected primary winding of the transformer is again solidly earthed to take advantage of graded insulation. The neutral of the medium-high-voltage, wye-connected secondary windings is intentionally earthed through impedance (see Figure 6.3). At lower voltage levels, the consideration of graded insulation is less compelling compared with the desire to restrict fault current.

At a load center distribution substation, the neutral point of the wye-connected secondary windings is again earthed via low impedance in order to limit the fault current. At low-voltage† distribution to consumers, the neutral point of the transformer windings is solidly earthed. An example of power supply earthing is shown in Figure 6.5. Various earthing methods for low-voltage power are further discussed in greater detail in Section 6.3.

Grounding for electrical systems can be divided into system earthing and equipment grounding. Figure 6.5 shows an earthing system in which the supply earth and the consumer equipment earth are independent of each other. In some earthing schemes, these two earthing systems are kept separate from each other except at the point where they receive their source of electrical power.

6.2.4 Electric Shock Hazards

During an earth fault on a high-voltage system, the flow of current to earth will produce a dangerously high ground potential rise (GPR) at the earthing system with respect to a remote earth [2]. This GPR is equal to the maximum fault current times the earthing grid resistance. When earth-fault current flows from an earthing electrode to the surrounding soil, potential gradients exist in the soil and on the ground surface. The distribution of ground surface potential gradient due to earth-fault current flowing through a vertical earth electrode is shown in Figure 6.6. The potential gradients close to the location of the buried electrode, both in the soil and on the ground surface, are generally at their maximum values and are, therefore, the most dangerous. The potential gradient decreases rapidly with increased distance from the earth elec-

---

*Extra-high voltage refers to voltage in the range of hundreds of kV.
†Low-voltage distribution refers to voltage below 1000 V.

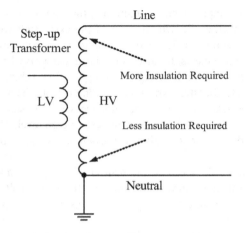

Figure 6.4. Graded insulation design for high-voltage transformer.

trode. The reasons for such potential gradient distribution will be discussed further in Section 6.7.5.

6.2.4.1 Step and Touch Voltage and Transferred Potential Arising from Ground Faults.

During a HV ground fault, a person walking on the ground surface adjacent to the earth electrode system may be subjected to a ground potential difference that can cause current to flow from one foot to the other foot through the hu-

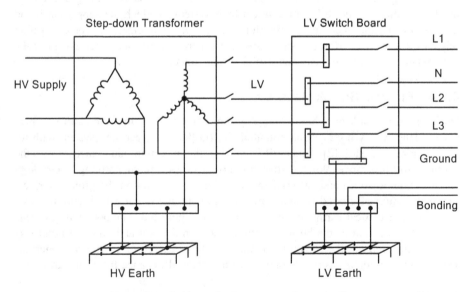

Figure 6.5. Scheme for power supply grounding.

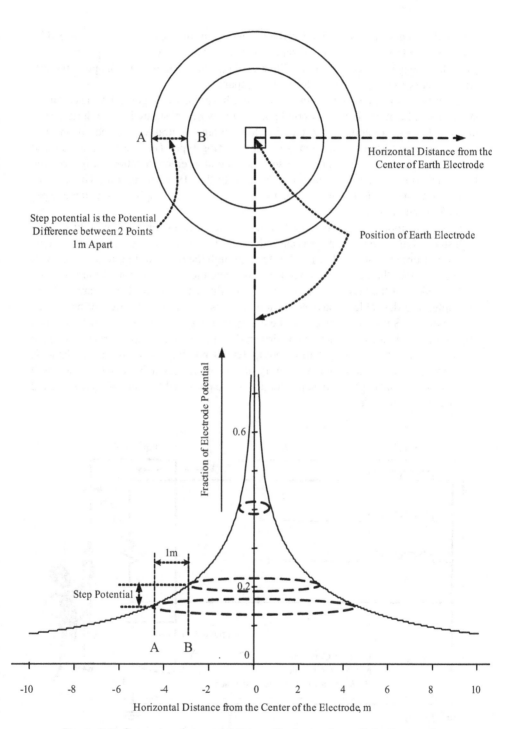

Figure 6.6. Ground surface potential gradients due to earth-fault current.

man body. Step-voltage is defined as the difference in surface potential experienced by a person with the feet bridging a distance of 1 m without contacting any grounded object. This concept of step potential is illustrated in Figure 6.6 as being the potential difference between points A and B located 1 m apart.

Earth faults on HV systems can also transfer hazardous voltages to LV installations at an outside location. This transferred potential may be transmitted by conduits, pipes, metallic fences, communication circuits, low-voltage neutral wires, and so on. The danger normally arises from contact of the touching type. The hazard of transferred potential generally occurs when a person standing at a remote location away from the HV installation touches a conductor connected to the HV earthing grid. This type of transferred potential is a serious problem because of the very high touch voltage experienced by the victim.

Hazardous touch voltage is more frequently encountered compared with hazardous high step voltage. Touch voltage is caused by a potential difference that can cause current flow from hand to hand or hand to foot through the body. In Figure 6.7, the touch voltage, V_T, that the equipment operator is subjected to during a ground fault is equal to the product of the fault current, I_F, and the facility earth electrode resistance, R_B. In principle, it is desirable to have R_B as low as possible in order to prevent hazardous touch voltage. Such an approach can be costly in areas of poor soil conductivity, as a large number of electrodes have to be driven down into the ground in order to achieve a low earth resistance. A system of equipotential bonding, to be discussed later, is much more effective, and often mandatory, in achieving acceptable touch voltage with minimal additional expense. In some countries, the permissible value of R_B is specified by local electrical code.

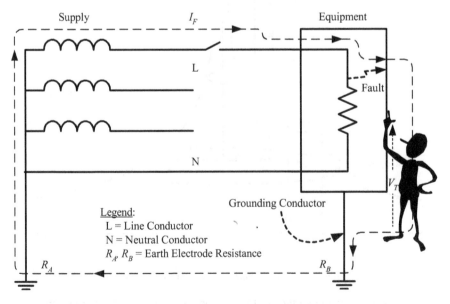

Figure 6.7. Touch voltage arising from a ground fault.

6.2.4.2 Leakage via Power Line Filter Capacitance.

Every LV installation has a permanent leakage current to earth, which is mainly due to the imperfect insulation and intrinsic capacitance between live conductors and earth. A large installation will have lower insulation resistance and greater capacitance. Consequently, leakage current is higher in larger installations.

The capacitive current to earth is often increased significantly by the filtering capacitors of electronic equipment. A simple circuit of a power line filter is shown in Figure 6.8. For the purpose of suppressing common-mode noise, the junction between C_2 and C_3 is bonded to the chassis ground. As a result, there is always a steady-state power frequency leakage current flowing to ground via C_2. For a filter design with C_2 = C_3, half the supply line voltage will be applied across C_2. For a supply voltage of U_0, the leakage current to chassis, I_L, is easily calculated as:

$$I_L = \frac{U_0}{2X_C} = \frac{U_0}{4\pi f C_2} \tag{6.3}$$

where X_C is the capacitive reactance and f is the frequency of the supply.

Safety aspects for installation with high leakage current will be discussed in Sections 6.3.6 and 6.3.7.

6.2.4.3 Shock Protection by Earthed Equipotential Bonding and Automatic Disconnection of Supply.

It is commonly thought that any earthed metal is free from electric shock hazard, even under fault conditions. Unfortunately, this is not always true. There are many examples of people having been killed or injured by hazardous shock potential when they were in indirect contact to energized exposed conductive parts that were properly earthed. Electric shock hazard may occur through indirect contact by touching a conductive part made live by a fault on equipment when the conductive part is connected via a protective conductor to the earth terminal. A safe electrical installation should be designed and constructed to have effective earthing together with a system of equipotential bonding and automatic disconnection of supply when a fault occurs.

Refer to Figure 6.9 for an illustration of the concept of earthed equipotential bonding for reducing the prospective touch voltage. This figure is basically an adaptation of Figure 6.7. In the figure, the person has one hand in contact with the exposed conductive part of the faulty equipment and his other hand is holding a metallic pipe in the facility.

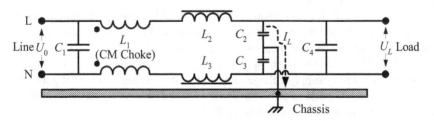

Figure 6.8. Circuit of a simple power line filter.

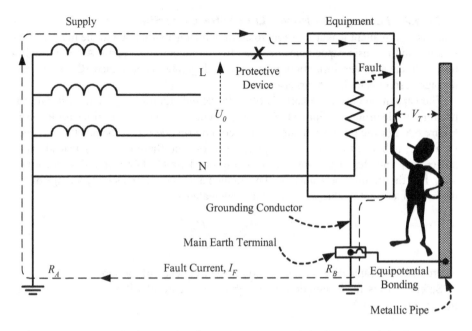

<u>Figure 6.9.</u> Earthed equipotential bonding and automatic disconnection of supply.

He is subjected to a hand-to-hand touch voltage, V_T. The fault current, I_F, has previously been determined as Equation (6.2), which is reproduced as Equation (6.4) for easy reference:

$$I_F = \frac{U_0}{Z_F} = \frac{U_0}{Z_S + Z_T + Z_{Ph} + Z_{PE} + R_A + R_B} \tag{6.4}$$

In a practical LV power supply system, R_A is normally below 1 Ω and R_B has a range of values varying from about 1 Ω to 50 Ω depending on the requirements of the local codes and regulations. The other components of the fault loop impedance in Equation (6.4) significantly lower than the combined value of R_A and R_B and may be neglected in estimating the fault current. R_A and R_B essentially act as potential dividers, with the supply voltage U_0 approximately divided between R_A and R_B, proportional to their respective resistance value.

Consider first the situation where the equipotential bonding is not implemented. For the case where R_B and R_A are both 1 Ω, the person in Figure 6.9 is subject to a touch voltage equal to half the supply phase voltage, that is, ½U_0. If R_B is close to the upper range of 50 Ω, almost the entire supply voltage U_0 will appear as the touch voltage. For normal LV utility supply voltage, the touch voltage is higher than the generally accepted safe touch voltage of 50 V,* even when facility earth resistance R_B is about the same value as the supply earth resistance R_A.

---

*A safe touch voltage limit of 50 V is implied in BS 7671:1992, Amendment 2.

This hazardous touch voltage is normally resolved satisfactorily by implementing a system of equipotential bonding. The bonding conductor shown in Figure 6.9 creates an equipotential between the metallic pipe and the main earth terminal. The latter is a bar or plate provided for the connection of protective conductors, including equipotential bonding conductors, and conductors for functional earthing, if any, to the means of earthing. With the equipotential bonding implemented as in Figure 6.9, the touch voltage V_T during an earth fault is reduced to that of the voltage drop created by the flow of fault current through the impedance of the grounding conductor:

$$V_T = I_F \times Z_{PE} \tag{6.5}$$

Since the impedance of the grounding conductor at power frequency is very much lower than R_B, the prospective touch voltage is well within the range of safe touch voltage for the type of earthing scheme shown in Figure 6.9.

Apart from using earthed equipotential bonding to prevent hazardous touch voltage, it is also required for safety consideration to incorporate protective devices in the electrical system so that fault conditions are detected and the faults are automatically disconnected immediately. For socket outlet circuits, for example, a common requirement is that disconnection of supply must occur within a time of 0.4 seconds [7, 8].

6.2.5   Methods of Power System Earthing

Most power systems are earthed either solidly or through impedance. Impedance earthing may be either resistance earthing or reactance earthing. Broadly, the types of system earthing normally used in industrial and commercial systems are:

1. Solid earthing
2. Resistance earthing
3. Reactance earthing
4. Unearthed

Each type of earthing has its advantages and disadvantages, and factors that influence the choice include a combination of the following:

1. The national earthing code
2. Voltage level of the power system
3. Transient overvoltage consideration
4. Fault current acceptable
5. Power quality requirements for continuity and voltage dip minimization
6. Existing method used on the system
7. Type of equipment on the system
8. Overall earthing and system equipment cost
9. Shock hazard and fire safety

6.2.5.1 Solid Earthing. Solid earthing, as shown in Figure 6.10, is the most common form of earthing for low-voltage supply systems. It refers to the direct connection of the neutral of a generator or transformer to the earthing system of the substation or facility. In the event of an earth fault, the fault current is limited by the earth fault loop impedance, which comprises the reactance of the grounded generator or transformer in series with the grounding conductor and the effective resistance of the earth electrode system. Solidly earthed systems provide the greatest control of overvoltage, both under steady-state and transient conditions. A disadvantage of the solidly earthed system is the high magnitude of earth-fault currents that can occur, and the destructive nature of the arcing earth faults. The high magnitude of earth-fault current requires that the supply circuit be immediately tripped off, inadvertently causing disruptive forced outages.

6.2.5.2 Impedance Earthing. A disadvantage of the solidly earthed power system is the high line-to-ground fault current that may cause a ground fault escalating to a phase-to-phase or three-phase fault. There is also the associated safety hazard of severe flash or arcing arising from the high ground fault current.

Impedance earthing, as shown in Figure 6.11, is normally implemented in high-voltage power systems to limit the fault current to an acceptable level. In impedance earthing, the neutral of the generator or transformer windings is connected to earth through either a resistor or a reactor. Resistance earthing is more commonly used.

In the low-resistance method, the ground-fault current is large enough to activate the ground-fault protective relay for immediate and selective fault clearance. Sometimes, high-resistance earthing is used to limit the ground fault current to a low value such that there is no immediate requirement of clearing a ground fault. The protective scheme is usually in the form of fault detection and alarm instead of immediate supply interruption.

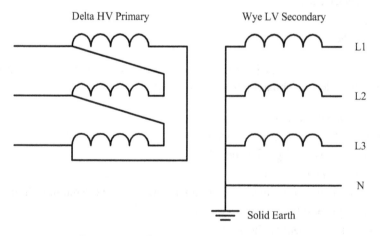

Figure 6.10. Solidly earthed power supply.

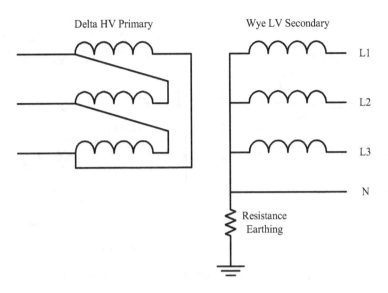

Figure 6.11. Resistance earthing of a power supply.

6.2.6   The Ungrounded System

Ungrounded systems are normally confined to installations supplied by private low-voltage transformers. Independent or dedicated power systems such as those on ships* and offshore structures often adopt ungrounded electrical systems. The major advantage of using an ungrounded electrical system is continuity of supply because the electrical supply need not be interrupted immediately on the occurrence of the first ground fault. This continuity of power supply is particularly important, for examples, for critical life-supporting circuits in hospitals and for some critical industrial processes. In a hospital facility, an ungrounded power supply reduces leakage current of medical devices to a low value, thereby enhancing the safety of patients. The second benefit of using an ungrounded electrical system is an economic factor—the absence of grounding conductors and ground fault protection devices.

An ungrounded power supply system is shown in Figure 6.12. The supply system can continue in operation with the first ground fault, fault A, because there is no complete fault current path. If the first fault is not removed, the supply system is now effectively an earthed supply system, connected to earth through the fault. There is a complete fault path, however, on the occurrence of the second fault, fault B. The flow path of the fault current is indicated in Figure 6.12. The resulting fault current can be very high as it is only limited by the supply circuit impedance. The protective devices must act immediately to remove the fault(s) to prevent any damage to the electrical system.

An ungrounded supply system should have an adequate ground detection scheme with a well-structured program for removing ground faults when they occur. The oc-

*Grounding on surface ships is discussed in Chapter 10.

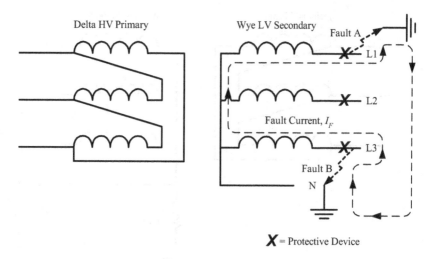

Figure 6.12. Faults on an unearthed supply system.

currence of the first ground fault changes the system to a grounded system and, consequently, the supply system must be interrupted immediately when the second fault occurs. Ground-fault detection schemes include insulation monitoring devices and leakage current detection systems with features to alert the system operator upon detection of the first insulation fault.

In practice, an ungrounded system has stray or leakage impedance to earth, as shown in Figure 6.13, formed by the parallel combination of the distributed resistive

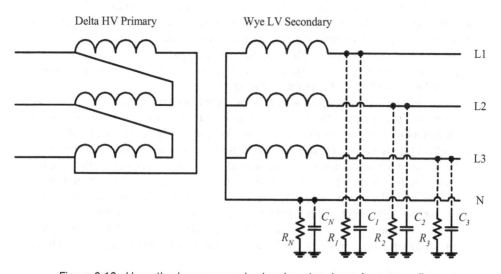

Figure 6.13. Unearthed power supply showing stray impedance coupling.

path and the distributed capacitive path. At 50 Hz or 60 Hz power frequency, this is extremely high impedance, and the capacitive coupling to ground is weak and changes with circuit conditions. Because of the capacitance coupling to ground, the ungrounded system is subjected to dangerous transient overvoltage. Overvoltage of magnitude five times normal or more may occur as a result of an intermittent contact ground fault (arcing ground) or where the fault circuit has a high inductive reactance connected from one phase to ground or phase to phase [30].

Lessons Learned

- Power distribution systems are earthed mainly for electric shock and fault protection.
- In an earthed power system, the neutral of a three-phase power supply system may be directly connected to an earth electrode system ("solid earthing") or through impedance ("impedance earthing").
- For continuity of supply, a power system may be unearthed if adequate precautionary measures are taken to monitor the insulation integrity.
- The earthing and bonding schemes affect the touch and step voltage in the event of an earth fault.

6.3 EARTHING FOR LOW-VOLTAGE DISTRIBUTION SYSTEMS

Different countries in the world have significant differences in earthing practices for their low-voltage supply systems. The earthing requirements to the consumers are well-specified in the national and international standards. IEC 60364-1 0 defines three categories of system grounding: the TN, TT, and IT systems. All the three earthing systems satisfy the primary function of safety, particularly protection against indirect-contact hazards. The selection criteria depend on regulatory and functional requirements, the type of supply network, the required continuity of service, and the operating conditions. Additional considerations include protection against fire of electrical origin and protection against electromagnetic disturbances.

Note: In IEC standards, notations for earthing schemes have the following meanings:

1. The First letter depicts the relationship of the power system to earth:

   T: Direct connection of one point to earth

   I: All live parts isolated from earth, or one point connected to earth through impedance

2. The second letter depicts the relationship of the exposed conductive parts of the installation to earth:

   T: Direct electrical connection of exposed conductive parts of the installation to earth

N: Direct electrical connection of the exposed conductive parts to the earthed point of the power system. (In AC systems, the earthed point of the power system is normally the neutral point or, if a neutral point is not available, a phase conductor.)

3. Subsequent letter(s), if any, depict(s) the arrangement of neutral and protective conductors:

S: Protective function provided by a conductor separate from the neutral or from the earthed line (or in AC systems, earthed phase) conductor

C: Neutral and protective functions combined in a single protective earth + neutral (PEN) conductor

Most industrial and commercial power systems are supplied from power transformers typically having a delta-connected primary and a wye-connected secondary. The neutral point of the secondary windings is normally solidly earthed. The wye secondary permits flexibility in connecting loads from phase to phase and from phase to neutral.

The different earthing methods specify how to earth the secondary winding of a HV/LV transformer and the methods used for earthing the exposed conductive parts of the LV installation supplied from it. These methods are now discussed in more detail.

6.3.1   TN System

In TN power supply systems, only one point of the electrical system is connected directly to earth—the neutral of the transformer secondary windings. The exposed conductive parts of the installation are connected to this earth point by protective conductors. Three types of TN systems are provided in IEC 60364-1 [4]; each varies according to the arrangement of neutral and protective conductors.

6.3.1.1   TN-S System. Figure 6.14 illustrates the configuration of a TN-S system in which the neutral and protective earth (PE) conductors are separate throughout, except at the supply transformer where they are bonded. All exposed conductive parts of an installation are connected to the PE conductor. The PE conductor can be a separate conductor or the metallic covering of the cable supplying electrical power to the installation.

When a ground fault occurs in an electrical installation supplied from a TN system, as shown in Figure 6.14, the fault current flows in a path made up entirely of conductors. Consequently, the fault return path does not include the general mass of the earth. The resulting fault current is limited by the fault loop impedance which comprises the series combination of the transformer impedance, the impedance, of the service cable, and the circuit protective conductor (the ground conductor). The earth fault loop impedance (Z_E) varies with the length of the service cable that supplies the electrical installation. Z_E is comparatively low, particularly when the fault occurs close to the transformer or the service panel. The TN system is characterized by high fault current, which may reach 30 to 40 kA when the fault occurs close to the main switchboard. The high fault current may pose a fire risk in premises with readily ignitable materials. The

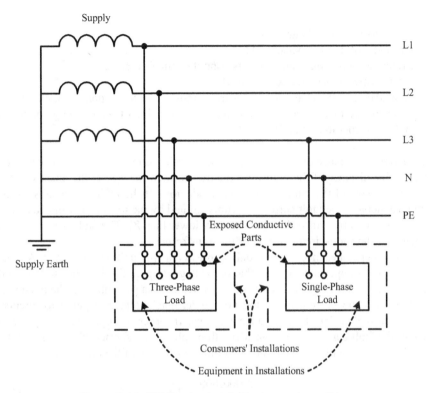

Figure 6.14. TN-S scheme of power supply earthing.

high fault current is undesirable from the standpoint of EMC as it may lead to a system voltage dip until the fault is cleared or isolated. Besides, the high impulse of the fault current can cause interference to other sensitive circuits via magnetic coupling.

It is of interest to note that the short-circuit fault current has a magnitude as high as the median lightning discharge current. The major difference is that the short-circuit current is a voltage-source phenomenon, whereas the lightning current is a current-source phenomenon with exceedingly high voltage. The fault energy to be dissipated in a power circuit may be higher than that associated with a lightning discharge because the fault current may flow for a duration several thousands time longer than the typical duration of lightning current. Electrical supply circuits and power equipment must be designed to withstand abnormal transient voltage, current, and energy during fault conditions.

A formal calculation of fault currents in the power system is rather involved and complicated [32]. For a good approximation, the ground fault current, I_F, may be estimated by

$$I_F = \frac{U_0}{Z_E} \approx \frac{U_0}{Z_S + Z_T + Z_{Ph} + Z_{PE}} \qquad (6.6)$$

where:

U_0 = Phase-to-neutral voltage, volts
Z_E = Ground fault loop impedance, Ω
Z_S = Effective source impedance seen beyond the transformer, Ω
Z_T = Transformer impedance per phase, Ω
Z_{Ph} = Impedance of the phase conductor from transformer to the point of fault, Ω
Z_{PE} = Impedance of protective earth conductor, inclusive of the earthing conductor, from transformer to point of fault, Ω

The relative importance of Z_{Ph} and Z_{PE} to the overall ground fault loop impedance depends on the distance of the fault from the transformer, the size of the phase conductors, and the size of the circuit-protective conductors. If the fault occurs close to the main switchboard, the fault current is mainly limited by the transformer impedance Z_T since the source impedance Z_S is substantially lower than Z_T. The fault current may be estimated as the short-circuit current of the transformer.

Transformer impedance is commonly expressed as per unit or percentage. A transformer with 5% impedance implies that the voltage drop on full load due to the winding resistance and leakage reactance is 5%, expressed as a percentage of the rated voltage. It is also the percentage of the normal terminal voltage required to circulate full-load current under short-circuit conditions. In order to appreciate the practical range of the fault current, let us consider a three-phase transformer with ratings of 400 V, 2 MVA, and 7% impedance. The rated full load current of the transformer is

$$I_{FL} = \frac{2,000,000}{\sqrt{3} \times 400} = 2890\,\text{A} \tag{6.7}$$

The ratio of fault current to rated full load current is given by the reciprocal of the percentage impedance. The fault current I_F for this transformer is therefore equal to

$$I_F = \frac{2890 \times 100}{7} = 41.3\ \text{kA} \tag{6.8}$$

This is obviously an overestimation since other components of the fault loop impedance are not considered. Nevertheless, this sample calculation reveals the order of magnitude of fault current that can occur in a TN supply system. Magnitude of fault currents is important information for power system protection engineers for sizing of switchgears and for other system-protection aspects.

If equipotential bonding is not implemented, the touch voltage, V_T, is due to the fault voltage over the length of the protective earth conductor ABCD, as shown in Figure 6.15:

$$V_T = I_F \times Z_{PE} \tag{6.9}$$

When main bonding is implemented, as shown in Figure 6.15, the touch voltage is reduced to the fault voltage drop over the length of the PE conductor from the equip-

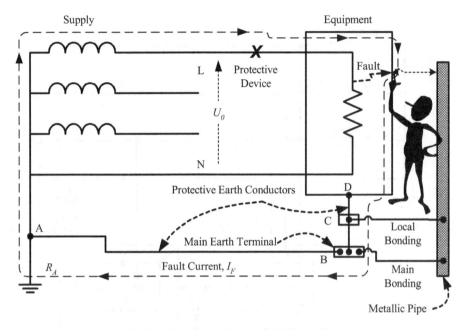

Figure 6.15. Earth fault in a TN-S system.

ment, marked as point D, to the main earth terminal, marked as point B. In areas with higher shock risk, such as a bathroom, local equipotential bonding is recommended. The touch voltage is thus further reduced to the fault voltage drop over a short length of the CPC, between points C and D in Figure 6.15.

TN-S system has the best EMC performance among the variations of TN systems. With the separation of neutral and circuit-protective conductor, the normal flow of load current is confined to the respective conductors. Consequently, this system of earthing does not encounter the "net-current" problem that creates magnetic field interference in the TN-C system.

6.3.1.2 TN-C System. In the TN-C system of earthing scheme shown Figure 6.16, the neutral and protective functions are combined in a single conductor, the PEN conductor, throughout the system. It is, therefore, a cost-effective scheme. All exposed conductive parts of an installation are connected to the PEN conductor. The neutral current not only flows via the PEN conductor, but also through the exposed conductive parts, the signal reference conductors, and extraneous conductive parts such as the structural metalwork.

In the TN-C system, an earth fault also constitutes a phase-to-neutral short circuit. The earth fault current flows through the PEN conductor instead of a dedicated protective conductor. Since the PEN conductor has to carry the normal neutral current, it has

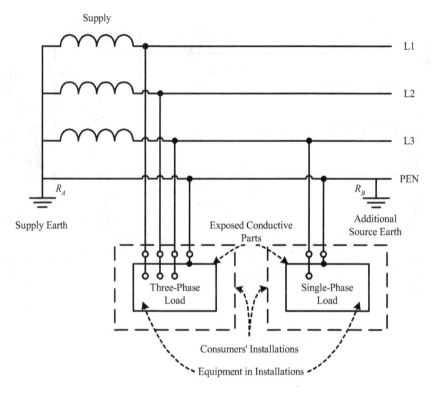

Figure 6.16. TN-C scheme of power supply earthing.

a larger cross-sectional area compared with a normal PE conductor. Consequently, the total earth fault loop impedance in a TN-C system is somewhat lower than that of a TN-S system. The resulting fault current is, therefore, higher. Due to the high fault current, automatic disconnection is mandatory in the event of faults. This disconnection should be provided by circuit breakers or fuses. Residual current devices are not suitable because part of the neutral current returns as stray current. The absence of protection using residual current devices may pose a fire risk for high-impedance faults when the fault current is insufficient to operate the overcurrent protection devices.

The TN-C system should not be adopted for serving buildings having significant IT or other sensitive equipment in order to avoid electromagnetic interference problems. In Figure 6.17 the "stray" paths taken by the return neutral current are shown. Separate protective earth (PE) and neutral (N) conductors should be used after the power entry point, such as in the TN-C-S scheme (see below), in order to minimize the possibility of EMI problems. Electrical conductors are normally laid closely together so that the incoming and outgoing currents sum to zero under normal operating conditions. As a result, the magnetic fields produced by the conductors tend to cancel out. In the case of the TN-C power supply system shown in Figure 6.17, the neutral return current follows many different paths. Along the cable route, the total

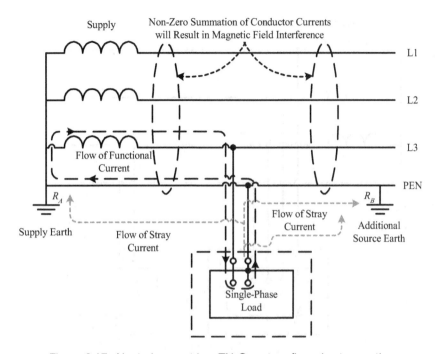

Figure 6.17. Neutral current in a TN-C system flows in stray paths.

of the conductor currents no longer sums to zero, and this "net" current is a source of electromagnetic interference.

The TN-C system is not permitted in a facility where there is a high risk of fire or explosion because of the flow of stray currents arising from connection of the extraneous conductive parts of the facility to the *PEN* conductor. During fault conditions, these stray currents in building structures and extraneous conductive parts are significantly increased, resulting in a risk of fire and electromagnetic interference.

6.3.1.3 TN-C-S System. Figure 6.18 shows the configuration of a TN-C-S system. Neutral and protective functions are combined in a single conductor (PEN conductor) in a part of the system. Normally, the electrical power supply is TN-C and the arrangement in the installation is TN-S. This type of distribution is also known as protective multiple earthing (PME), and the PEN conductor is referred to as the combined neutral and earth conductor. The supply PEN conductor is earthed at several points and an earth electrode may be necessary at or near a consumer's installation. All exposed conductive parts of an installation are connected to the PEN conductor via the main earthing terminal and the neutral terminal; these terminals are linked together. The magnitude of earth fault current is similar to that of the TN-C system.

In order to prevent any EMC problems, the TN-C system should be avoided for distribution within the building. TN-S system with separate neutral and PE conductors should be maintained within the building from the supply origin onward.

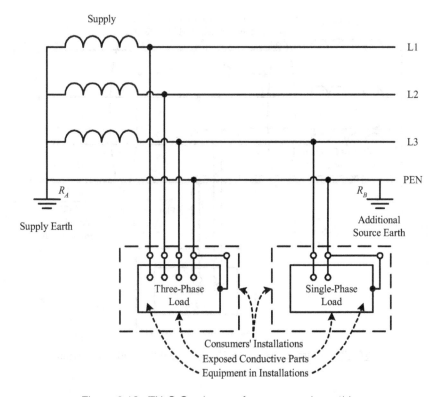

Figure 6.18. TN-C-S scheme of power supply earthing.

6.3.2   TT System

In this scheme, the neutral of the electrical power supply is directly earthed, as shown in Figure 6.19. All exposed conductive parts of an installation are connected to an earth electrode, which is electrically independent of the source earth. Because of its simplicity of design and installation, the TT system is the most common public LV distribution system in the world.

For the purpose of system protection, low earth fault loop impedance is desired so that the resulting fault current is sufficiently high to activate the circuit-protective devices. The earth fault loop impedance comprises the transformer impedance, the relevant cable impedance and the impedance of two earth electrodes, the transformer neutral earth electrode, and the installation earth electrode. The earth electrode resistances generally have much higher values than those of the other components of the earth fault loop impedance.

For the purpose of estimating the likely range of earth fault current in a TT system, references are made again to Figure 6.2 and Equation (6.2). The latter is reproduced as

$$I_F = \frac{U_0}{Z_F} = \frac{U_0}{Z_S + Z_T + Z_{Ph} + Z_{PE} + R_A + R_B} \tag{6.10}$$

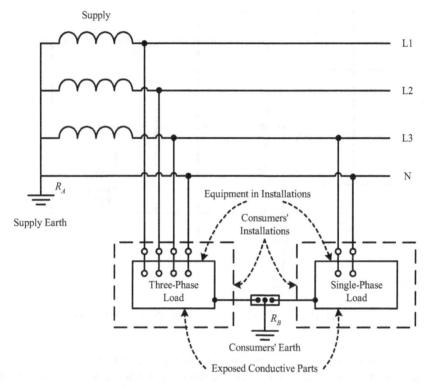

Figure 6.19. TT scheme of power supply earthing.

As discussed earlier, the earth fault loop impedance is predominantly constituted by the combined value of R_A and R_B. R_A, the earth resistance associated with the supply transformer, normally has a value not exceeding 1 Ω, whereas R_B, the facility earth resistance, can have a practical value ranging from under 1 Ω to 50 Ω. The earth fault current is normally in the range of 5 to 200 A.

The earth fault current of LV supply with phase voltage of 230 V is expected to be above 200 A for an effective TT earthing scheme that has a combined $(R_A + R_B)$ value of about 1 Ω. When $(R_A + R_B)$ is about 50 Ω, the earth fault current is less than 5 A.

The National Electrical Code (NEC) [6] specifies an upper value of 25 Ω for R_B. For a typical supply system of 120 V (Figure 6.20), the expected earth fault current is about 5 A or more.

Due to limited fault current, overcurrent protective devices are not effective for the protection of earth fault in a TT system. Earth faults relays with setting substantially lower than the overcurrent protection setting are generally installed to provide the appropriate sensitivity to detect and remove the earth faults. Sensitive residual current devices (RCDs) are commonly used for supplementary electrical protection. In some countries, the application of RCD is mandatory for circuits that supply portable equipment.

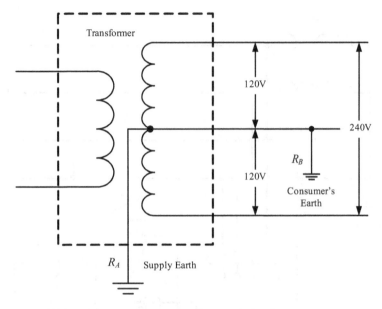

Figure 6.20. 240/120 V, single-phase, three-wire system.

The TT system places the responsibility of assuring an effective ground in the hands of the consumers themselves as they are the most appropriate ones to maintain the integrity of the earthing system within their own premises. In the case of the TN system, particularly the TN-S system, the supplier has to ensure the integrity and effectiveness of the protective conductor (ground conductor), which must be earthed at the service transformer end. This discussion of shared responsibility is based on the assumption that the service transformer is the responsibility of the electricity power supplier.

TT systems are characterized by two separate earth electrode systems. They are, therefore, more prone to transient overvoltage arising from lightning activities. Overvoltage protection is commonly provided [33].

6.3.3 IT System

In the IT system, the supply is either isolated or has one point earthed through a high impedance. As shown in Figure 6.21, all exposed conductive parts of the electrical installation are connected to an earth electrode independent of the source. The IT electrical supply can be derived from a normally earthed electrical system using an isolation transformer.

The IT system is adopted in applications that require very high availability of supply during operation. Similarly to the ungrounded system discussed in Section 6.2.6, the supply system can remain in operation after a first earth fault. The supply must be tripped immediately on occurrence of a second fault if the first fault has not been already cleared. Examples of IT system use include hospital operating theaters and

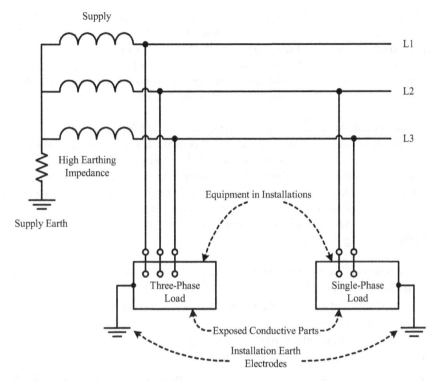

Figure 6.21. IT scheme of power supply earthing.

mines, where continuity of supply under a first earth fault condition is of critical importance.

Like the ungrounded system, the IT system requires the installation of insulation monitoring devices that will detect the occurrence of the first fault, which will then be immediately located and removed. The IT system is normally implemented only when trained maintenance and operational personnel are present for prompt action to locate and isolate the first insulation fault.

6.3.4 Temporary Overvoltage in Low-Voltage Installations Due to Faults between High-Voltage Systems and Earth

When an earth fault occurs in the HV system in a substation, severe temporary overvoltage may be transferred to the LV system, introducing a potential safety hazard to personnel and equipment in the LV system. Such faults cause a current to flow in the substation earth electrode to which the exposed conductive parts of the substation are connected. The fault current causes a general rise of the potential of the exposed conductive parts of the low-voltage system with respect to earth. This potential rise leads to an unsafe touch voltage.

The fault voltage, which may be as high as several thousand volts, can cause an insulation breakdown in the low-voltage equipment. The magnitude of the fault voltage depends on the fault-current magnitude and the resistance of the earth electrode of the exposed conductive parts of the substation. These temporary overvoltages (TOV) have implications for the design, including the ratings, of grounded common-mode protective devices such as filters, capacitors, avalanche diodes, and metal-oxide varistors (MOVs).

The magnitude of the fault current depends on the fault-loop impedance on the HV circuit, that is, on how the high-voltage neutral is earthed. The resistance of the ground electrode system in the substation is normally below 1 Ω [5].

The fault voltage at the substation is transferred to the consumer LV installation, causing unsafe contact voltage and the risk of insulation breakdown of equipment. The effects of overvoltage depend on the interconnection of the exposed conductive parts of LV loads with neutral earth connection of the supply transformer.

In the following discussion, the symbols used are:

I_m = Earth fault current in the high-voltage system that flows through the earth electrode of the exposed conductive parts of the transformer substation

R_A = Resistance of the earth electrode of the exposed conductive parts of the transformer substation

V_0 = Line-to-neutral voltage of the low-voltage system

V = Line-to-line voltage of the low-voltage system

V_f = Fault voltage in the LV system between exposed conductive parts and earth

V_1 = Stress voltage in the LV equipment of the transformer substation

V_2 = Stress voltage in the LV equipment of the consumer's system

In the TN system, the neutral and the ground protective conductors are effectively at the same potential. Figure 6.22 illustrates the case of the fault voltage causing the exposed conductive parts of the consumer's installation to rise to a dangerously high value, V_f, if the neutral of the transformer LV winding is bonded to the transformer frame. To avoid the risk of indirect contact with high fault voltage, it is necessary to ensure that the building is equipotentially bonded. The equipment at the consumer's installation is not subject to excessive overvoltage.

To avoid the HV fault voltage being transferred to the consumer's LV installation, the neutral conductor of the transformer should be separately grounded and no bonding should be made to the transformer frame, as shown in Figure 6.23.

In the TT system shown in Figure 6.24, the neutral of the LV winding of the substation transformer is bonded to the transformer frame. The exposed conductive parts at the consumer's installation are bonded to the consumer's ground electrode, which is electrically independent of the substation supply earthing electrode. Therefore, there is no risk of indirect contact with dangerously high transferred voltage from the substation arising from HV faults. The consumer's equipment, however, is subjected to stress voltage between the live conductors and the grounded equipment metallic enclosure, the exposed conductive parts.

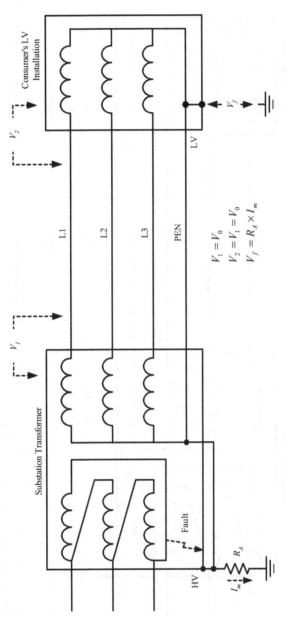

Figure 6.22. TN-S scheme of power supply earthing: LV neutral bonded to frame, HV ground fault, high transferred voltage to LV installation.

$V_1 = V_0$

$V_2 = V_1 = V_0$

$V_f = R_A \times I_m$

399

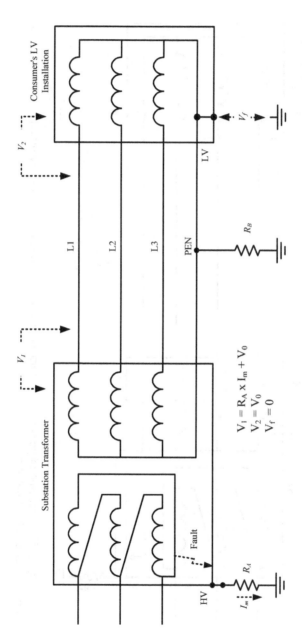

$V_1 = R_A \times I_m + V_0$
$V_2 = V_0$
$V_f = 0$

Figure 6.23. TN scheme of power supply earthing: LV neutral not bonded to frame, HV ground fault, and no transferred voltage to LV installation.

400

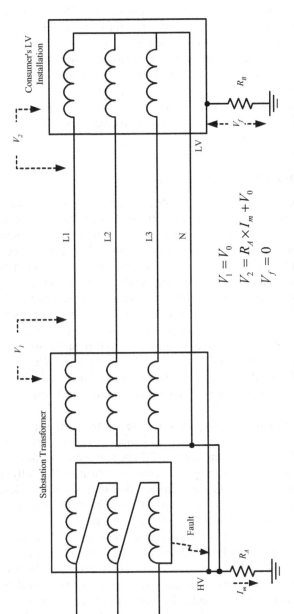

Figure 6.24. TT Scheme of power supply earthing: LV neutral winding bonded to transformer frame, consumer's equipment subject to high stress fault voltage, no excessive touch voltage.

$$V_1 = V_0$$
$$V_2 = R_A \times I_m + V_0$$
$$V_f = 0$$

401

In Figure 6.25, the neutral of the substation transformer LV winding has an independent ground. The consumer's LV installation is not affected by the transformer HV fault. There is no transfer of high voltage to the LV installation.

Regulations in some countries require that where the HV equipment earth and the LV neutral earth are connected together the combined resistance to earth must not exceed 1 Ω [29]. Otherwise, the HV equipment earth and the LV system earth have to be separated so that the electrodes are outside the respective areas of influence.

From the EMC point of view, transient overvoltage can affect the earth/signal reference potential. Besides, the high-frequency spectrum associated with the voltage surges can be capacitively coupled to sensitive electronic equipment and impact its proper operation. HV faults in power system are normally cleared in less than a second by high-speed protective relays.

6.3.5   Earthing Systems and EMC

The power supply earthing system topology has a profound influence on the EMC environment for electronic equipment. Electronic equipment is now commonly interconnected with other information technology and communication equipment, sensors, surveillance devices, and other accessories. Figure 6.26 illustrates the case of electronic equipment connected to a sensor via a cable link. An insulation breakdown between the live and ground conductors can result in a large current flowing through the PE conductor, causing a voltage drop V between points B and C. This common-impedance coupling between the equipment and the sensor can upset signal transmission. Additionally, the ground loop, ABCD, formed by the PE conductor, the bonding cables, and the cable shield, can create further problems. This interference mode only applies to single-ended signaling. In balanced signaling, this interference will not occur due to the common-mode rejection of balanced interfaces. Detailed discussions of balancing and ground loops are covered in Chapter 2 and Chapter 4, respectively.

Besides the common-mode coupling effect, severe transient magnetic field interference is also produced by the fault current. At junction C, the fault current I_{F1} splits into two paths, one flowing as I_{F2} through the PE conductor and the other flowing as I_{F3} through the grounding and bonding conductors as indicated. Normally, the phase (L) and protective earth (PE) conductors are bundled together rather closely to ensure good EMC performance. The ground loop ABCD with an effective loop current of I_{F3} can produce significant interference during fault conditions.

In the TN system, the fault current that flows through the PE conductor can be as high as 30 kA. The resulting EMI problems can be very severe. TT system generates comparatively much less EM disturbances because the magnitude of the earth fault current is much lower. It is possible to continue to mix both functional grounding conductors and electrical exposed conductive parts closely and to make the most of the meshing and equipotent bonding. In IT systems, usually much less than 1 A will flow on the first earth fault since the value of the neutral earthing resistor is generally more than 1000 Ω. In the event of a double fault, as discussed in Section 6.2.6, the situation is the same as in the TN system.

403

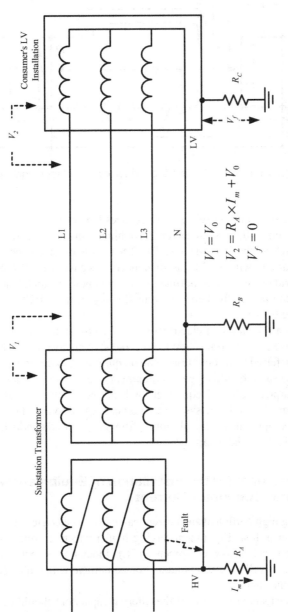

$V_1 = V_0$

$V_2 = R_A \times I_m + V_0$

$V_f = 0$

Figure 6.25. TT scheme of power supply earthing: LV neutral winding not bonded to transformer frame, no transfer of high voltage to consumer's installation.

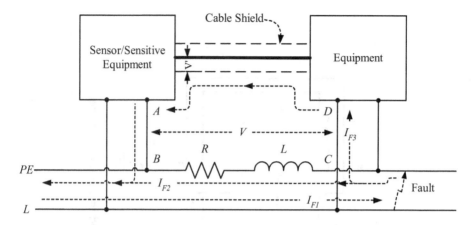

<u>Figure 6.26.</u> Current flowing in PE conductor may affect signal integrity.

In order to avoid grounding-related disturbances between communicating devices, it is prudent to avoid earthing systems that have high fault current flowing through the PE conductor, such as the TN systems. The TN-C system should never be used for supplying installations where EMC is of concern because the PEN conductors are meant to carry load current under normal operating conditions and, hence, there is always a steady-state power frequency current flowing in the ABCD ground loop referred to in Figure 6.26.

To prevent hazard of indirect electrical shock, electrical and electronic equipment are required to have ground connection to their metallic enclosures or chassis. Class II equipment, also referred to as double-insulated equipment, is designed with sufficient insulation features such that this type of equipment does not require a ground connection for safety purpose. Class II equipment will not suffer the ground loop problem when supplied from TT or TN-C systems. Another method to avoid the EMI problem is to insulate the equipment functional ground from the exposed conductive parts using an isolation transformer where appropriate.

6.3.6   Requirements for the Installation of Equipment with High Protective Earth Conductor Current

Equipment having high earth leakage current can cause an electric shock hazard if the ground connection is lost. Equipment having high protective conductor current includes information technology equipment (ITE); industrial, scientific, and medical (ISM) equipment; as well as telecommunication equipment with radio-frequency interference-suppression filters.

The maximum capacitance value of the filtering capacitors should be limited in order to control the magnitude of leakage current within safe levels. Product safety standards impose permissible limits on leakage current for electrical and electronic equipment. Due to the proliferation of ITE and their large number in workplaces and data centers, guidelines and standards have been established by various organizations [7,

34, 35] to address the earthing and bonding requirements. Apart from addressing safety from electrical hazards, a proper system of earthing and bonding will help to establish reliable signal reference and ensure satisfactory electromagnetic performance of the entire information technology installation.

For installation of equipment having high earth leakage currents in normal service, there should be a high-integrity protective conductor or equivalent arrangement. To enhanced physical integrity of a protective conductor, a good practice is to use a single protective conductor having a cross-sectional area larger than normal requirements. Alternatively, as shown in Figure 6.27, two protective conductors are used, the ends of the protective conductors being terminated independently of each other at all connection points throughout the circuit, for example, at the distribution board and socket outlets.

To ensure electrical safety, residual current devices of 30 mA trip setting are commonly used to protect electrical circuits with high-protective conductor current. Care must be taken to avoid nuisance tripping of these protective devices arising from excessive residual current of the connected equipment.

6.3.7 Application of Residual-Current Devices for Shock Protection

A residual-current device (RCD) is a protective device used to automatically disconnect the electrical supply when an imbalance is detected between live conductors. This type of device is also known as ground fault circuit interrupter (GFCI) in the United States and Canada.

A typical circuit of a single-phase RCD is shown in Figure 6.28. The device monitors the difference in currents between the phase (L) and neutral (N) conductors. In a healthy circuit, where there is no ground fault current or protective conductor current, the sum of the currents in the L and N conductors is zero. If a phase-to-ground fault develops, a portion of the phase conductor current will not return through the neutral

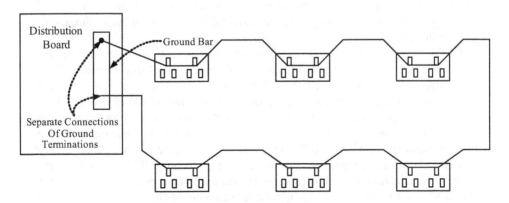

Figure 6.27. Ring final circuit supplying twin socket outlets.

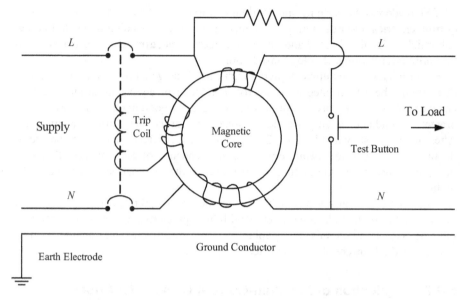

Figure 6.28. Single-phase residual current device (RCD).

conductor. When this unbalance or residual current reaches a preset limit, the induced current in the trip coil will operate and disconnect the circuit.

RCDs are intended to provide protection against electric shock that can result from a person touching a live exposed conductor (direct contact) or touching exposed conductive parts (indirect contact) that have become energized under fault conditions. If the exposed conductive part is connected to a circuit-protective conductor, the leakage current is basically the fault current and has a magnitude that far exceeds the operating residual current required to cause immediate tripping of supply.

If the energized exposed conductive part is not connected to a circuit-protective conductor, then the fault will not be cleared even when the circuit is protected by an RCD. A person touching this energized metalwork and in contact with the ground will suffer an electric shock due to passage of current from the live conductor through the body to the ground.

RCDs alone cannot be relied upon for protection against direct contact with live parts. They are intended for supplementary protection against direct contact. RCDs also provide protection against risk of fires and thermal effects likely to arise from ground fault currents. Industrial practice has an upper limit of 300 mA for RCDs intended for fire protection. An RCD on its own does not provide protection against overcurrent. Overcurrent protection is provided by a fuse or a miniature circuit breaker (MCB). However, combined RCDs and MCBs are available and are called RCBOs (residual current circuit breaker with overcurrent protection). RCBOs combine in a single device the residual current function and the overcurrent protection function typical of MCBs. RCBOs are tripped by both residual current and overcurrent. They are self-protecting up to a maximum short-circuit current value designed for the device.

Selection of a RCD for a particular application is based on three main characteristics:

1. Rating of the device in amperes
2. Rated residual operating current in amperes, $I\Delta n$
3. Tripping-time characteristics

The rating of the device is selected with considerations of the normal full load current together with the requirement of overcurrent protection. The tripping time is either instantaneous or time delayed to permit discrimination.

The use of RCDs with a rated operating residual current not exceeding 30 mA is recognized as additional protection in case of direct contact in the event of failure of other measures of protection or carelessness by users [36]. Where protection is provided by automatic disconnection of supply, RCDs with rated residual operating current not exceeding 30 mA should be used to protect circuits supplying portable equipment and socket outlets intended to supply portable or transportable equipment. RCDs with $I\Delta n$ of 10 mA provide a high degree of sensitivity that is sometimes used to protect laboratory benches in schools or in healthcare environment.

General purpose RCDs provide almost instantaneous tripping in the event of a ground fault. In accordance with RCD standards, the device operates in less than 200 ms at the rated residual operating current $I \times \Delta n$. At a current of $5 \times I\Delta n$, the device would operate in less than 40 ms. For the purpose of discrimination, S-type RCDs, which have intentional an time delay incorporated, may be used in series with general type RCDs.

For circuits that are protected by transient-protective devices (TPD), the RCD should be installed after the TPD in order to prevent nuisance tripping when transient/surge currents are diverted to ground. Likewise, RCDs for EMC laboratories should be installed after the big leaky filter, else the leakage current will activate the cutout fuse.

Lessons Learned

- There are three common earthing schemes for low-voltage power distribution systems: the TN, TT, and IT systems.
- The three earthing schemes depend on how the supply is connected to earth and how the exposed conductive parts of the installation or equipment are referenced to earth.
- The type of earthing schemes affects the magnitude of the fault current and the touch and step voltages.
- A high-voltage earth fault can create temporary overvoltages in low-voltage installations.
- The EMC environment is affected by the power supply earthing scheme.
- Residual current devices are recommended for protection of equipment with high leakage current.

6.4   LIGHTNING PROTECTION

6.4.1   An Overview of the Lightning Phenomenon

Lightning is the result of the natural buildup of regions of opposite electrical charges in storm clouds [9, 36–38] due to interaction of particles of raindrops, snowflakes, and ice crystals. Breakdown can occur between the charged regions within the cloud to produce intracloud lightning. When the potential difference between clouds is sufficiently high, cloud-to-cloud discharge occurs, giving rise to intercloud lightning. Intracloud and cloud-to-cloud discharges do not pose a direct threat to personnel or structures on the ground. However, voltages that could be induced in long cables by such discharges indicate that they present a definite threat to signal and control equipment, particularly those employing electronic or semiconductor devices. Cloud-to-cloud discharges pose severe EMI threats to aircraft and must be addressed by the designers of aircraft electronic (e.g., avionic) systems.

The normal distribution of charged particles gives rise to predominant negative charges at the base of the thundercloud, which, in turn, causes buildup of positive charges on the ground through electrostatic induction. When the concentration of a charge center is sufficiently high, the air in the vicinity is ionized by the strong electric field. Figure 6.29 illustrates the formation of a cloud-to-ground lightning discharge. A column of ionized air, called a pilot streamer, begins to extend toward earth. After the pilot streamer has moved perhaps about 30 to 45 meters, a more intense discharge called a stepped leader takes place. This discharge lowers additional negative charge into the region around the pilot streamer and allows the pilot streamer to advance for another 30 to 45 meters, after which the cycle repeats [9]. The stepped leader progresses toward the earth in a series of steps with a time interval between steps on the order of 50 μs.

Although the general direction of the streamer is toward the earth, the specific angle of departure from the tip of the previous streamer that the succeeding streamer takes is rather unpredictable. Therefore, each 40 to 45 m segment of the discharge will likely approach the earth at a different angle. This changing angle of approach gives the overall flash its characteristic zigzag appearance. Being a highly ionized column, the stepped leader is essentially at the same potential as the charged area from which it originates. Thus, as the stepped leader approaches the earth, the voltage gradient between the earth and the tip of the leader increases. The increasing voltage further encourages the air between the two to break down.

When the negatively charged leader is close to the ground, induced positive electrical charges on the ground propagate upwards as streamers and are likely to be initiated at protruding earthed objects. A conducting path through ionized air is completed when the downward leader meets an upward streamer, leading to the high-current return strike. A complete ground flash consists of a sequence of one or more high-amplitude, short-duration current impulses or strikes. As many as twenty restrikes may follow the first return strike, lasting for about 1–2 seconds.

Cloud-to-ground flashes are normally of greatest concern with respect to lightning protection because of the associated hazards. The high current flowing during the charge equalization process of a lightning flash can melt conductors, ignite fires

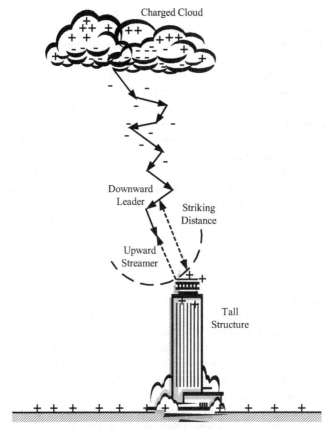

Figure 6.29. Formation of a lightning strike.

through the generation of sparks or the heating of metals, damage or destroy components or equipment through burning or voltage stressing and produce voltages well in excess of the lethal limit for people and animals. The objective of lightning protection is to preclude hazards to persons, structures, or buildings and their contents attributable to the effects of lightning. The lightning protection system accomplishes this goal by diverting the high currents associated with the lightning strike away from susceptible elements or by limiting the consequent voltage gradients to safe levels.

6.4.2 Lightning Attachment Point and Zones of Protection

Lightning strike attachment point is a term used to describe the point on the ground or on a structure where the lower end of the lightning discharge channel connects with the ground or structure. The attachment point is determined by the path of the successive streamers. The distance between the tip of the downward leader and the eventual strike attachment point at the moment of initiation of an upward intercepting leader is called the striking distance. The striking distance of a lightning strike is dependent on the magnitude of the lightning discharge current; the larger the peak current, the longer the striking distance.

Many national lightning protection codes adopt the "rolling sphere" method* for determining lightning strike attachment points [12, 19–21]. In this method, a sphere of specified radius is imagined to be rolled across the ground toward the building, up the side, over the top of the building, and down the other side to ground. Any point on the building touched by the sphere is a possible lightning strike attachment point. A shorter sphere radius is used when a higher degree of protection is required. This rolling sphere technique is illustrated in Figure 6.30 for determining the zone of protection of a structure. The area within the protected zone is unlikely to be struck by lightning as the attachment point is toward the top edge of the structure. The rolling sphere method has to be applied with electric-field enhancement effects in mind so that high priority is given to providing air termination electrodes at the more probable attachment points.

The advantage of the rolling sphere method is that it is relatively easy to apply, even to buildings of complicated shape. The method assumes that when the downward leader makes the last jump, the point of attachment can be any point on the ground or the structure equidistant from the end of the leader. The limitation of the method is that no account is taken of the influence of electric fields in initiating return streamers, and the method, therefore, does not distinguish between likely and unlikely attachment points. In particular, the enhancement of the electric field at the upper outer corners of a building makes these corners the most probable strike attachment points, whereas return streamers are unlikely to be initiated from a flat surface away from a corner or edge, even if on the roof and touched by the sphere.

The striking distance, d_S, is the distance between the leader tip and the eventual strike attachment point. The following relationship has been proposed [21]:

$$d_S = 10 \times i_{max}^{0.65} \qquad (6.11)$$

where:
d_S = Striking distance, m
i_{max} = Peak current of the return stroke, kA

For achieving an acceptable degree of protection at reasonable cost, a number of codes recommend a radius of 45 m for general protection of a structure [12, 19–22]. Using Equation (6.11), it can be determined that this striking distance corresponds to a lightning current of 10 kA. Statistics show that more than 90% of lightning flashes to ground

---

*In spite of the fact that the "rolling sphere" method is widely accepted and commonly used, it should be noted that the validity and scientific basis of this method has been challenged and is controversial. Alternative approaches for lightning protection criteria have been suggested, for instance, the least time–maximum probability (LT–MP) principle, described in Briët, R., "The Least Time–Maximum Probability Theory of Lighting Propagation", and "The Least Time–Maximum Probability Theory of Lightning Protection Design," both from the *Proceedings of the 2004 IEEE International Symposium on Electromagnetic Compatibility*, Santa Clara, CA, 2003. This approach is more conservative than the "rolling sphere" method.

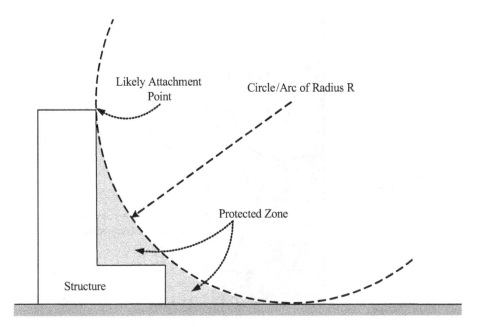

Figure 6.30. Rolling sphere method of defining lightning zones of protection.

have current greater than 10 kA. This implies that a protection system design based on a rolling sphere of a 45 m radius will provide a probability of protection greater than 90%.

For simplicity, some lightning protection codes [12, 20] include the use of a cone of protection characterized by its cone tip protective angle for defining the zone of protection provided by a structure such as a mast. The protection afforded increases as the protective angle (decreases. Apart from assuming that the lightning attachment point is at the top of the structure, the cone of protection method does not have any theoretical basis for its application, but it is very easy to apply. In general, the protective angle method should not be applied to tall structures. BS 6651 [20], for instance, limits this method of protection to structures not exceeding 20 m.* Figure 6.31 provides a comparison of the zones of protection provided by the rolling sphere method and the protective angle method. A tall structure is purposely selected to show that the protective angle method may give a sense of false security.

6.4.3   Components of the Lightning Protection System

A lightning protection system for structures and buildings against direct strikes basically comprises three subsystems:

---

*The SI 1173, *Israel Standard, Lightning Protection Systems for Buildings and Installations,* 2003 (Draft) ICS Code: 91.120.40, specifies a "cone of protection" with an extremely narrow cone tip angle ($\alpha < 20°$). This standard too appears to be significantly more conservative than the common "cone of protection" method used by other standards ($\alpha < 30°$ to $60°$).

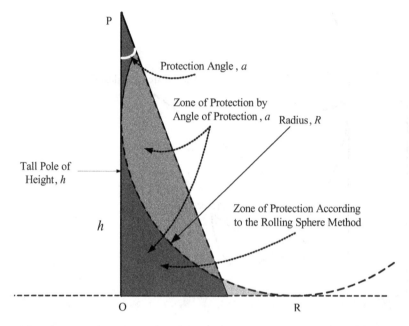

<u>Figure 6.31.</u> Comparison of zones of protection by rolling sphere and cone of protection (protection angle) methods.

1. Air termination subsystem that serves as the attachment point to intercept the lightning current
2. Down-conductor subsystem to bring down the lightning current safely to the earth surface
3. Earth termination subsystem to effectively dissipate the lightning discharge energy into the general mass of the earth

The components of a conventional lightning protection system are depicted in Figure 6.32.

6.4.3.1 *The Air Termination.* The objective of lightning protection is to ensure a high probability that the successful streamer originates from the lightning protection system, and not from a part of the structure or building being protected.

The essence of lightning protection is to provide a preferred path to divert lightning current safely and efficiently to the earth mass without causing hazards to human, animals, and property.

In order to effect lightning interception, air terminations must be designed and so positioned as to ensure a high probability that lightning will attach to the air termination network and not to any part of the protected object that could be damaged by the lightning current. Air terminations should be designed by taking into consideration of protection using existing metal work.

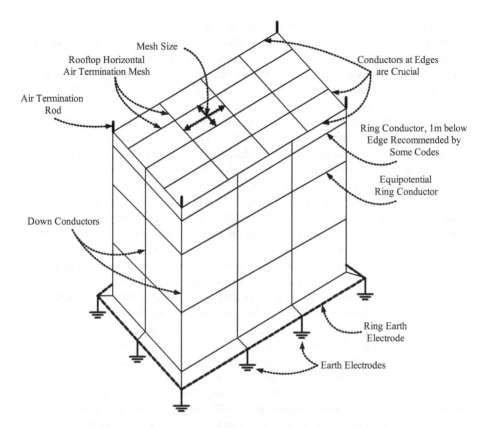

Figure 6.32. Components of a lightning protection system.

Air termination networks may consist of vertical* or horizontal conductors or their combination. In the past, it was thought that pointed objects such as the vertical rods would be able to more readily generate upward streamers to intercept the lightning strikes. Studies in recent years have shown that the vertical rods are not as effective as they were thought to be and, hence, modern air terminations use mainly a mesh of horizontal conductors. IEC 61024-1-2 [12], for instance, does not provide any criteria for the selection of the air-termination systems because it considers rods, stretched wires, and meshed conductors as equally effective. Selection of the appropriate mesh size depends on the degree or level of protection required. In the Israel code for lightning protection [39], a mesh size of 5 m by 5 m is recommended, whereas in BS 6651 [20] and SS CP33 [21] a mesh size of 10 m by 20 m is proposed for normal protection levels.

6.4.3.2 The Down Conductors. Having successfully intercepted the lightning discharge current, the lightning protection system should provide a subsystem to

---

*The "Franklin rod" was invented by Benjamin Franklin in 1752.

effectively direct the large lightning current down the side of the structure or building through a system of down conductors. Each down conductor terminates in an earth electrode. Special precautions should be taken to avoid side flashes from the down conductors to other earthed metallic objects nearby. A good lightning protection system should be one that provides the most direct route to ground in order to keep inductance as small as possible.

It is desirable to have multiple down conductors around the side of the building and at the edges. If the lightning current is shared by more down conductors, the risk of side flash and electromagnetic interference to electronic equipment inside the building is reduced. Normally, the down conductors are uniformly placed with a symmetric configuration along the perimeter of the building. The current sharing is improved by increasing the number of down conductors and by the equipotential interconnecting rings. The building structural steel and reinforcing bars are increasingly being designed to form part of the down-conductor system by appropriate bonding to the air and earth terminations. A steel framed or reinforced concrete structure, if properly implemented with electrical continuity, may not require added down conductors because the framework itself provides an effective natural network of many parallel paths to earth.

The lightning protection system must be designed and installed with an understanding that lightning discharge current is characterized by very short-duration current of high amplitude. The median peak current is about 30 kA but the maximum peak current can exceed 200 kA. The typical rise time is on the order of several microseconds (μsec). The exceedingly high di/dt could cause problematic induced transients in other electrical circuits. Standard voltage and current waveforms are well defined in standards [40] for assessing the susceptibility of equipment to transient overvoltage due to secondary effects of lightning, such as the 1.2/50 μsec open-circuit voltage waveform, the 8/20 μsec discharge current waveform, and the 0.5 μsec/100 kHz ring wave. For protection from direct strike, the 10/350 μsec double exponential waveform is used for evaluation.

When the lightning current flows through the down conductors and discharges into the earth termination network the upper end of the down conductors can have a potential as high as several megavolts with respect to the remote or true ground. This is a result of the inductive voltage drop ($L \cdot di/dt$) in the down conductors plus the ohmic ($I \times R$) voltage drop of the ground electrode resistance. If there is a grounded metallic object close to the down conductors, a side flash may occur. Severe arcing associated with side flashes, as shown in Figure 6.33 may lead to fire and other property damages. Side flashes may be avoided by increasing the distance of separation between the down conductors and the earthed metallic object nearby. Alternatively, bonding of grounded metallic objects nearby the down conductor may prevent the hazards of arc flashes. It is important, therefore, that down conductors and internal circuits be well separated. ITU-T K.17* and K.21† testing requirements against interference call for a testing waveform of 10/700 μsec, which is rather difficult to meet.

---

*ITU-T Recommendation K.17, "Test on Power-Fed Repeaters Using Solid-State Devices in Order to Check the Arrangements For Protection from External Interference."
†ITU-T Recommendation K.21, "Resistibility of Subscriber's Terminal to Overvoltages and Overcurrents."

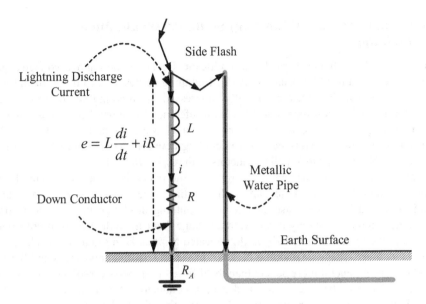

Side Flash

Lightning Discharge
Current

$$e = L\frac{di}{dt} + iR$$

L

i

Down Conductor

R

Metallic
Water Pipe

Earth Surface

R_A

<u>Figure 6.33.</u> Side flash of lightning discharge current to a nearby grounded metal
body.

At the ground level, the lightning current should be dissipated efficiently into the
general mass of the earth via the earth termination network. In order to limit the hazardous ground potential rise and to avoid dangerous step and touch potentials, the
shape and dimensions of the earth-termination system are very important considerations, apart from aiming for a low earth resistance.

6.4.3.3 The Earth Termination Network. The earth-termination network
consists of one or more earth electrodes, and any interconnecting conductors between
earth electrodes serves the purpose of delivering the lightning current into the general
mass of the earth. The footings of large reinforced concrete buildings or structures will
generally provide a better earth connection than can be provided by electrodes driven
around the periphery.

Hazardous potential gradients may be produced at the point where the lightning
current enters the ground. The earth potential gradient is highest at the electrode as the
current density is the highest. As will be discussed in Section 6.6.1, the current density
decreases rapidly with distance away from the electrode. The hazardous potential gradients near a building may be avoided by placing distributed earth electrodes symmetrically around the circumference of a structure and bonding them together.

For a given lightning discharge current, the magnitude of the surge voltage increases
proportionally with earth impedance. For the purpose of minimizing the surge voltage,
the earth impedance should be kept as low as economically practicable. International and
national standards and codes [9, 20, 21, 25] generally recommend 10 Ω as a benchmark
value. The National Electrical Code (NEC) [6] requires a value not exceeding 25 Ω.

6.4.4   Influence of LV Earthing Schemes on Lightning Overvoltage

The type of LV earthing scheme has a significant effect on lightning overvoltage experienced by consumer LV installations [5]. Lightning surges, due to direct or indirect strikes on the supply line, are normally suppressed by lightning surge arresters installed at the substation incoming line. With reference to Figure 6.34, the attenuated lightning surges propagate as common-mode voltage waves to the LV windings of the transformer mainly via stray capacitive coupling between the HV and LV windings. The surge voltage appears on all conductors with respect to earth.

Earthing the transformer neutral suppresses the overvoltage between neutral and supply ground. Overvoltage appears in differential mode between phase and neutral conductors and between phase conductors. Lightning discharge near the substation raises the potential of supply earth, A, to a voltage that can be thousands of volts with respect to the remote consumer's earth, indicated as point B in Figure 6.34. The common-mode overvoltage developing between phase and neutral conductors and the PE conductor is in general more severe than the differential-mode overvoltage that occurs between phases or between phase and neutral conductors. Surge protective devices (SPD) for common-mode protection of TT systems require higher ratings compared to transient protection devices (TPD) used for differential-mode protection as in the case of TN system.

When lightning strikes near the consumer's installation, as shown in Figure 6.35 for a power supply with a TT earthing system, high overvoltage may occur between the supply earth, A, and the consumer installation earth, B, where the installation PE conductor terminates. Similar to the discussion of overvoltage related to Figure 6.34, insulation in the LV installation may break down if no surge protection is implemented [33].

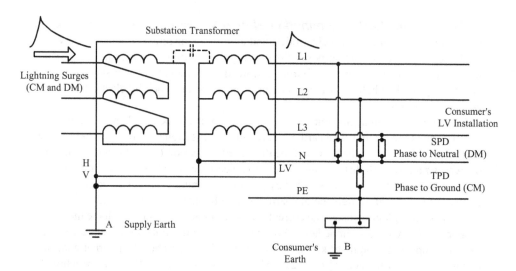

Figure 6.34. Protection of lightning surges in LV supply with TT earthing system.

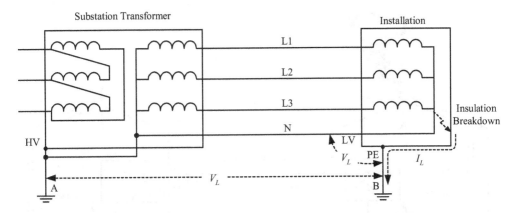

<u>Figure 6.35.</u> TT scheme of power supply earthing: lightning discharge causes high potential difference between the two earth electrodes and may lead to insulation breakdown of LV equipment.

Figure 6.36 shows an example of possible damage that can be caused by propagation of lightning surge current through a communication link between two buildings, each with its own power supply and earthing system. A lightning strike at building A causes a current I_1 to flow down the earth electrode A, raising its potential and that of ground bar A to value of $I_1 \times R_A$, which is some thousands of volts above that of earth electrode B of the other building. If the communication line is grounded at both ends, a surge current I_2 will flow along the cable shield to building B and to earth via earth electrode B. If the shield of the cable is insulated from earth, the potential of the cable shield can become high enough with respect to the earth to cause an insulation breakdown as shown in Figure 6.36. The lightning surge is attenuated as it travels down the cable, but the residual amplitude is still sufficient to cause damage to device B at the other end of the communication link.

One way to overcome the problem of uncontrolled lightning surges is to implement equipotential bonding as shown in Figure 6.37. The two independent earthing systems in the two buildings are linked by an equipotential bonding conductor. To avoid magnetic field coupling that can arise in an above-ground cable installation, the cables should be buried underground and enclosed in ferrous conduits.

Lessons Learned

- Lightning discharge is a high-voltage and high-current transient phenomenon.
- A lightning protection system comprises three components: air termination, down conductor, and earth termination subsystems.
- Lightning overvoltage in electrical distribution systems is influenced by the LV earthing schemes

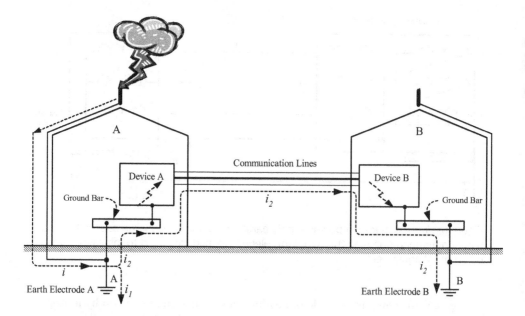

<u>Figure 6.36.</u> Propagation of lightning surges over a communication link between two buildings.

6.5 THE EARTH CONNECTION

The cloud-to-earth lightning flash is essentially a huge electrostatic discharge phenomenon involving the neutralization of charges between two electrodes, the cloud and the earth, producing an electric field on the order of hundreds or thousands of megavolts. Lightning protection earthing is one of the rare examples that require a good connection to the general mass of earth.

Grounding of electrical equipment or a supply system involves a connection to the general mass of the earth. The final connection to earth is normally achieved through buried conductors called the earthing electrodes. The function of an earthing electrode is to provide effective connection to the general mass of the earth. The effectiveness of a earth electrode or a group of electrodes is primarily determined by the impedance between the earthing system and the general mass of the earth.

Safety is a primary consideration of grounding design. IEEE STD 80-2000 [2] specifies the following two major objectives of an earthing design:

1. Provide means to carry electric currents into the earth under normal and fault conditions without exceeding any operating and equipment limits or adversely affecting continuity of service
2. Assure that a person in the vicinity of grounded facilities is not exposed to the danger of critical electric shock

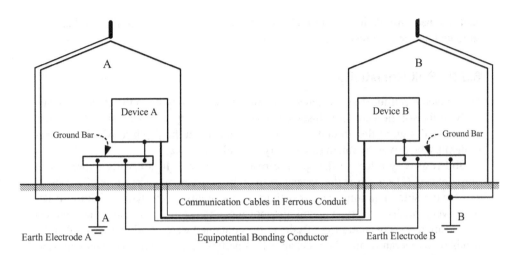

<u>Figure 6.37.</u> Equipotential bonding to minimize lightning surge voltage.

6.5.1 Resistance to Earth

The earth has been used as a conductor of electricity since the earliest days of electricity utilization. In telegraphy, for example, the earth was used as a current return path. The deliberate use of the earth as a conductor is seldom practiced nowadays except for power system protection and lightning protection.

In general, the earth is not a good conductor of electricity. The soil resistivity, defined as the resistance between opposite faces of a unit cube, is very high compared with any metallic conductor. However, the cross-sectional area of the current path in earth is very large and, hence, despite the poor conductivity of the soil, the earth resistance could be quite small. In order to exploit the general mass of the earth as a low-resistance conductor, effective means must be implemented for passage of current into and out of the earth. These means are normally earth electrodes in the forms of rods, pipes, or plates buried in the ground.

At normal power-supply frequency, resistance to current through an earth electrode has three major components:

1. Resistance of the electrode itself and all conductors and leads connected to it
2. Contact resistance between the surface of the electrode and the soil adjacent to it
3. Resistance of the volume of earth immediately surrounding the electrode

The electrodes are usually made of highly conductive materials of sufficient size and cross-sectional area. The electrode resistance is, therefore, negligible. The resistance of the metallic conductors in the earthing system can also be ignored.

The contact resistance between the surface of an electrode and the soil can be reduced to a low value by ensuring that the backfill material is of the appropriate type. Large, dry stones should clearly be avoided. The contact resistance could be significant in a new installation where the soil has not yet stabilized. Once the soil has consolidated, the contact resistance becomes negligibly small.

6.5.2 Soil Resistivity

For a given electrode, the resistance to earth depends mainly upon the electrical resistivity of the soil in which it is installed. The soil resistivity profile on site often determines the design of the ground electrode subsystem and the depth to which the electrode must be driven to obtain satisfactory ground resistance.

Soil resistivity refers to the specific resistance of soil. It is usually expressed as ohm · meters ($\Omega \cdot$ m). It is the resistance in ohms between opposite faces of a cube of soil having sides 1 m long. Soil resistivity is essentially electrolytic in nature. It depends very much on the type of soil, the moisture content of the soil, and the chemical composition and concentration of salts in the contained water. Soil resistivity also depends on temperature and the mechanical properties such as grain size and compactness of the soil. The resistivity of some common materials is shown in Table 6.1. There is a very large variation in resistivity between different types of soils and with different moisture contents. Although dry soil has very high resistivity, measurements have shown that there is no significant reduction in resistivity when the moisture content is above about 15% to 20%. Table 6.2 shows the effect of moisture on the resistivity of soil.*

The resistivity of soil varies widely at different locations as the type of soil changes. At some locations, the soil can be very nonhomogeneous with multilayer soil structure not far away from the earth surface, as illustrated in Figure 6.38. Often, there are several layers of soil made up of loam, sand, clay, gravel, and rocks. The layers may be roughly horizontal to the surface or inclined at an angle to the surface. The resistivity also fluctuates seasonally due to changes in rainfall and temperature. The impact of temperature is only important near and below freezing point.

Soil resistivity is one of the most important factors affecting the performance of the earthing system because the value of the earth resistance is directly proportional to the resistivity of the medium in which the earth electrodes are installed. At locations where multilayer soil exists at rather shallow depths, the average or apparent resistivity depends on the layer structure, the soil resistivity, and the depth of each layer.

6.6 TYPES OF EARTH ELECTRODES

There are many different types of earth electrodes in use. Some are referred to as "natural" [6] because their presence at the facility is due to reasons other than the purpose of providing earthing. Natural electrodes include metal underground water pipes, metal frame of the building or structure, and the reinforcing bars in a concrete foundation. Others are sometime referred to as "made" electrodes because they are specially installed for providing earthing. Made electrodes consist of vertical rods or pipes driven

---

*Values in Table 6.1 and Table 6.2 are from Tagg [41] and Charlton [42].

Table 6.1. Typical values of soil resistivity

| Type of materials | Resistivity, $\Omega \cdot m$ | |
	Typical	Usual range
Sea water	0.2	0.15–0.25
Inland lake water	20	10–500
Damp clay	10	2–12
Loams, garden soil	25	5–50
Alluvium at river banks	25	10–100
Clay, sand, and gravel	100	40–250
Rock	3,000	1,000–50,000
Concrete	100	40–5,000
Porous limestone	50	30–100
Porous sandstone	100	30–300

Table 6.2. Variation of soil resistivity with moisture content

| Moisture content (percent by weight) | Typical resistivity, $\Omega \cdot m$ | |
	Clay mixed with sand	Sand
0	10,000,000	—
2.5	1,500	3,000,000
5	430	50,000
10	185	2,100
15	105	630
20	63	290
30	42	—

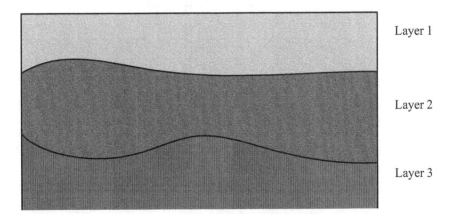

Figure 6.38. Multilayer soil structure.

into the earth, metallic plates or wire mesh buried in the earth, or a horizontal copper wire ring around the installation. Aluminum electrodes or underground gas pipes are not permitted for use as ground electrodes. Aluminum is prone to corrosion when buried in soil, whereas gas pipes may pose an explosion hazard.

The greater part of the fall in potential occurs in the soil within the first two meters of the electrode surface because the current density is the highest within such proximity to the electrode. To obtain a low overall resistance, the current density should be as low as possible in the region surrounding the electrode. A good electrode should be designed to enable current density to decrease rapidly with distance from the electrode. This desirable characteristic is met by making the dimensions in one direction large compared with those in the other two. Thus, a pipe, rod, or strip has a much lower resistance than a plate of equal surface area.

6.6.1 The Earth Rods

The vertical earth rod shown in Figure 6.39 is the most common form of electrode because it can be driven deep into the moisture level below the ground to take advantage of low-resistivity soil. Earth rods are manufactured in a range of lengths, diameters, and materials complying with relevant standards. The diameter of the rod has no significant impact on resistance. For example, doubling the diameter reduces the resistance only by about 10%. Size of rods or pipes is generally determined by mechanical resistance to bending or splitting.

As shown in Figure 6.40, the path of current flowing outward from a rod electrode consists of successive cylindrical and hemispherical shells. The effective resistance of

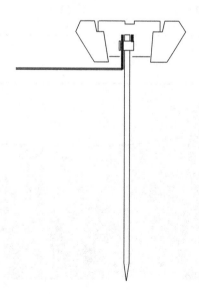

Figure 6.39. Earth rod.

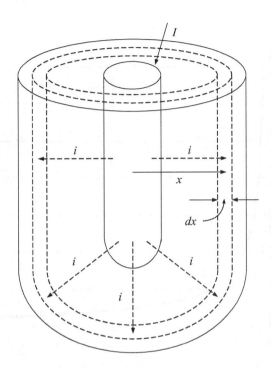

Figure 6.40. Cylindrical shell of the current flow surrounding a rod electrode.

the soil to the current flow from the electrode is the sum of the series resistances of virtual shells of earth, located progressively outward from the rod. The shell nearest the rod has the smallest circumferential area or cross section and, therefore, has the highest resistance. Shells further away have progressively larger areas and, thus, progressively lower resistances. As the distance from the rod increases, the incremental resistance contribution from successive shells decreases to practically zero.

The resistance to earth of a single rod electrode may be calculated using [1, 7, 41]

$$R = \frac{\rho}{2\pi L}\left[\ln\frac{4L}{a}-1\right]\ \Omega \qquad (6.12)$$

where:
R = Resistance of the single rod or pipe, Ω
L = Length of rod, m
a = Radius of rod or pipe, m
ρ = Soil resistivity, $\Omega \cdot$ m

The length of the rod has a major effect on resistance. Figure 6.41 shows the effect of variation of the length of the rod on effective earth resistance. In uniform soil, doubling the length causes a reduction of about 44% of resistance.

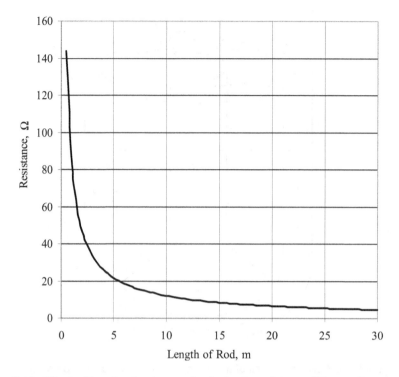

Figure 6.41. Effect of length of rod on earth electrode resistance (16 mm diameter rod, soil resistivity of 100 Ω · m).

The most important parameter influencing the effective earth electrode resistance is the soil resistivity in the immediate surrounding region of the earth electrode. The curve in Figure 6.41 is based on a uniform soil. Often, soil at site may have different resistivity as the depth increases from the ground surface. When the top layers in a multilayer soil have relatively high resistivity, it is highly desirable to extend the rods beyond these high-resistivity layers to get nearer to the water table in order to achieve an acceptably low resistance. Deep-driven rods help to keep a good degree of stability to the overall impedance of the earthing system. This stability is achieved by avoiding seasonal variation in moisture content or temperature of the top soil layers, including freezing problems in winter. In practice, there are physical limitations on how deep a rod may be driven down at the site.

Changing the diameter of the rods has a relatively minor effect, as shown in Figure 6.42. The size of the rod or pipe is generally selected from physical consideration such as resistance to bending or splitting.

If the resistance of a single rod is higher than the required value, two or more rods may be used in parallel. Parallel connection of multiple rods has a similar effect as increasing the rod length. Connecting two rods in parallel does not yield a total resistance half that of a single rod because of the proximal electric field interaction of the

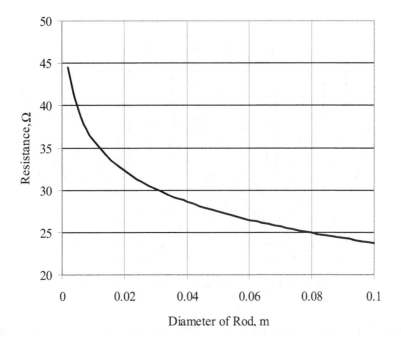

Figure 6.42. Effect of diameter of rod on earth electrode resistance (3 m long/16 mm diameter rod, soil resistivity of 100 $\Omega \cdot$ m).

two electrodes, unless the two rods are placed several rod lengths apart. Figure 6.43 shows the zones of influence of two parallel electrodes.

For two rods in parallel where the spacing, s, is greater than the length of each rod, L, Equation (6.13) applies [1]:

$$R = \frac{\rho}{4\pi L}\left(\ln\frac{4L}{a} - 1\right) + \frac{\rho}{4\pi s}\left(1 - \frac{L^2}{3s^2} + \frac{2L^4}{5s^4}\right) \qquad (6.13)$$

When spacing between the two rods is less than the length of each rod, the equivalent resistance is given by

$$R = \frac{\rho}{4\pi L}\left(\ln\frac{4L}{a} + \ln\frac{4L}{s} - 2 + \frac{s}{2L} - \frac{s^2}{16L^2} + \frac{s^4}{512L^4}\right) \qquad (6.14)$$

where:
R = Resistance of the two rods connected in parallel, Ω
L = Length of each rod, m
s = Separation of the rods, m
a = Radius of rod or pipe, m
ρ = Soil resistivity, $\Omega \cdot$ m

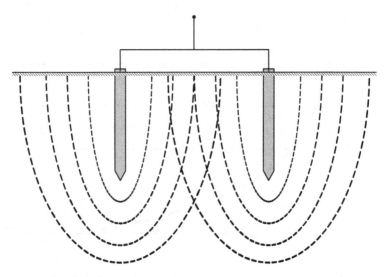

Figure 6.43. Proximity effect of two parallel rods.

The effect of electrode spacing on the effective combined resistance of two similar electrodes is shown in Figure 6.44. For a single 6 m long rod 0.01 m in radius in soil of resistivity 100 $\Omega \cdot$ m, the electrode resistance is 18 Ω. Two similar electrodes at a spacing of a rod length yield a combined resistance of 10.4 Ω, a value which is within 85% of two perfect electrodes in parallel. When the spacing is two rod lengths apart, the effective resistance is better than 90% of two perfect electrodes in parallel.

Crowding multiple vertical rods is not as beneficial in terms of cost per ohm as is achievable with fewer rods properly spaced. The common practice is to place the two parallel rods at a separation greater than one rod length. Connecting a number of rods in parallel is preferred over a single long rod. Deeply driven rods are effective where the soil resistivity decreases with depth.

Earth rods are normally selected based on their resistance to corrosion, mechanical strength, and cost. They are relatively cheap and can be installed with little excavation and backfilling. Where available area is limited at site, vertical rods may be a most effective option. Because of a lack of mechanical strength, copper rods are not suitable for application where the rods have to be driven deep into the earth. Galvanized steel rods, though relatively cheap, may not exhibit as good resistance to corrosion. A common compromise is the application of steel-cored earth rods in copper or stainless steel sheaths.

6.6.2 Earth Plates

The resistance to earth of a plate, R, may be calculated from the following equation [41]:

$$R = \frac{\rho}{4}\sqrt{\frac{\pi}{2A}} \quad \Omega \tag{6.15}$$

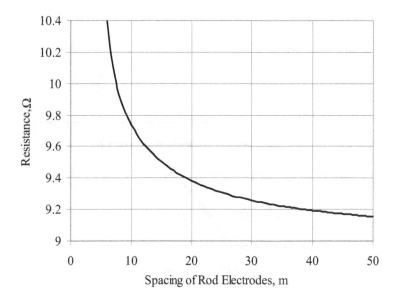

Figure 6.44. Effect of spacing on the effective resistance of two electrodes in parallel (6 m long rod of 0.01 m radius, soil resistivity of 100 Ω · m).

where:
R = Resistance of the single rod or pipe, Ω
A = Area of one face of the plate, m²
ρ = Soil resistivity, Ω · m

For the case of a circular plate of radius r m, Equation (6.4) can be simplified to

$$R = \frac{0.1768\rho}{a} \quad \Omega \tag{6.16}$$

The configuration of an earth plate electrode is shown in Figure 6.45. This type of earth electrode does not have a competitive advantage in term of the cost–performance relationship compared with the more common earth rods. A 40 m² plate with soil resistivity of 100 Ω · m only provides an earth electrode resistance of 5 Ω, as shown in Figure 6.46. Installation of plate-type of earth electrodes is mainly confined to tower footings where design of the structure facilitates the laying of the electrode.

6.6.3 Horizontal Strip or Round Conductor Electrode

Horizontal strip or round conductor electrodes are commonly used in situations in which high-resistivity soil or rock is found below shallow surface layers of low-resistivity soil. They are usually in the form of bare copper conductors. The resistance R for a single strip is approximated by

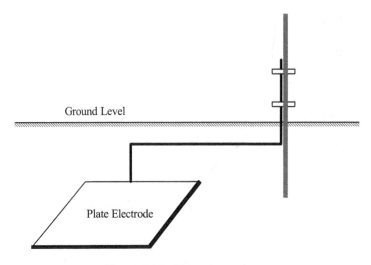

Figure 6.45. Plate electrode.

$$R = \frac{\rho}{2\pi}\frac{1}{L}\left[\ln\frac{2L^2}{wh} - 1\right], \ \Omega \qquad (6.17)$$

where:
R = Resistance of the single rod or pipe, Ω
L = Length of the strip, m^2
h = Depth of electrode, m
w = Width of strip, m
ρ = Soil resistivity, $\Omega \cdot$ m

The thickness of the strip is generally less than one-eight of the width, and within this range the thickness has little effect on the electrode earth resistance. The width, w, and the depth of burial, h, have relatively minor effects. Earthing resistance is determined mainly by the length of the strip. The cross-sectional area of the strip or conductor is selected to provide protection against mechanical damage and corrosion, and to have adequate surge current capacity.

For power frequency considerations, optimum low resistance for a given amount of electrode materials is achieved by installing the electrode in a straight single trench or in several trenches radiating from a point. Four common configurations of earth electrode arrangements are shown in Figure 6.47.

The resistance to earth for the four electrode configurations depicted in Figure 6.47 can be calculated using the expressions in Equation (6.18) through (6.21) [1]:

1. For Configuration (a), single-length electrode:

$$R = \frac{\rho}{2\pi L}\left(\ln\frac{2L}{a} + \ln\frac{2L}{s} - 2 + \frac{s}{L} - \frac{s^2}{4L^2} + \frac{s^4}{32L^4} \cdots\right)$$

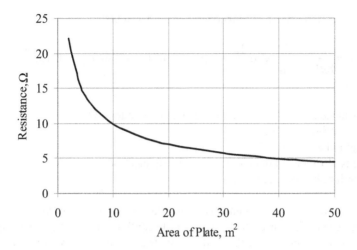

Figure 6.46. Resistance of a buried plate electrode (soil resistivity of 100 Ω · m, with a rod, driven 0.5 m deep).

2. For Configuration (b), two-length electrodes at 90°:

$$R = \frac{\rho}{4\pi L}\left(\ln\frac{2L}{a} + \ln\frac{2L}{s} - 0.2373 + 0.2146\frac{s}{L} + 0.1035\frac{s^2}{L^2} - 0.0424\frac{s^4}{L^4}\cdots\right)$$

3. For Configuration (c), three-length electrodes at 120°:

$$R = \frac{\rho}{6\pi L}\left(\ln\frac{2L}{a} + \ln\frac{2L}{s} + 1.071 - 0.209\frac{s}{L} + 0.238\frac{s^2}{L^2} - 0.054\frac{s^4}{L^4}\cdots\right)$$

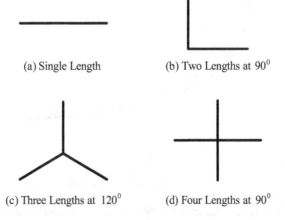

(a) Single Length   (b) Two Lengths at 90⁰

(c) Three Lengths at 120⁰   (d) Four Lengths at 90⁰

Figure 6.47. Strip or round conductor electrode configurations.

4. For configuration (d), four-length electrodes at 90°:

$$R = \frac{\rho}{8\pi L}\left(\ln\frac{2L}{a} + \ln\frac{2L}{s} + 2.912 - 1.071\frac{s}{L} + 0.645\frac{s^2}{L^2} - 0.145\frac{s^4}{L^4}\cdots \right)$$

where:
R = Resistance to earth of the complete electrode system, Ω
ρ = Soil resistivity, $\Omega \cdot$ m
L = Length of each arm of electrode, m
a = Radius of round conductor or width of strip electrode, m
$s/2$ = Depth of buried electrode, m

Figure 6.48 presents the computed earth resistance of the configurations of horizontal strip earth electrodes for each of the electrode configurations (a) through (d) depicted in Figure 6.47, respectively. When the total strip length is short, configurations with more radiating arms tend to have noticeably higher resistance compared with the case of single strip having the same total strip length. This is due to the relative significance of the sphere of influence of the various strips. When each arm length is long, there is little difference in the effectiveness of between the four configurations. In general, the

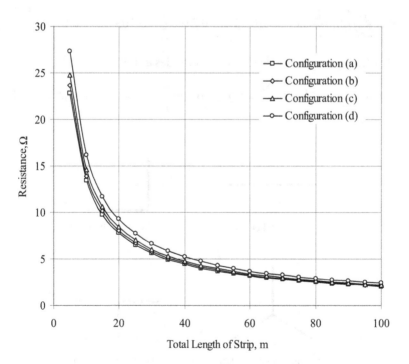

Figure 6.48. Earth resistance of the different electrode configurations (a) through (c) (soil resistivity of 100 $\Omega \cdot$ m, electrodes buried 0.5 m below the earth surface).

electrode resistance decreases inversely proportional to the length of the buried conductor. Similar to other types of electrodes, the resistance of a buried strip increases linearly proportional to the soil resistivity.

The strips are normally buried in trenches at a depth of about 1 m. The effect of buried depth on the earth resistance of a straight strip electrode is shown in Figure 6.49. The earth resistance of the buried strip was computed using both Equation (6.17) and Equation (6.18). The result indicates that there is no appreciable advantage to burying horizontal strip electrodes below a depth of 1 m since the incremental reduction in resistance is insignificant. This resistance versus buried depth relationship applies generally to horizontal electrodes.

Current flows out in a radial direction from a buried conductor as shown in Figure 6.50. As explained in Section 6.5.2, the effective resistance of the soil to the current flow from the electrode is the sum of the series resistances of virtual shells of earth, located progressively outward from the conductor. The shell nearest the rod has the smallest circumferential area or cross section and, therefore, has the highest resistance. Shells further away have progressively larger areas and, thus, progressively lower resistances. As the distance from the rod increases, the incremental resistance contribution from successive shells decreases to zero practically. About 80% of the earth resistance is contributed by the soil within a radius of 1 m from the electrode conductor.

With long strips, reactance increases proportional to the length. Low impedance is desirable for minimizing lightning surge voltages. Hence, several wires, strips, or cables arranged in a star pattern, with the facility at the center, is preferable to one long

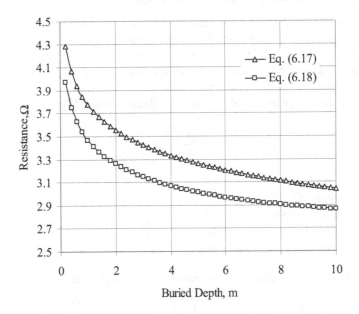

Figure 6.49. Effect of buried depth on earth resistance of strip electrode ($L = 50$ m, $a = 13$ mm, $\rho = 100 \ \Omega \cdot$ m).

Earth Surface

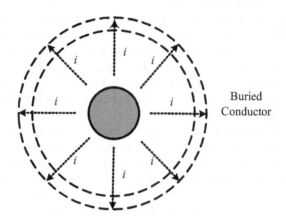

Figure 6.50. Current spreading radially from a buried conductor.

length of conductor. Figure 6.51 provides some examples of such star arrangements [39].

6.6.4 The Mesh or Grid Earth Electrode

A mesh or grid system, as shown in Figure 6.52, consists of copper conductors buried in the earth and forming a network of squares extending over a wide area. Such a system is mainly used to provide equipotential areas in power stations, substations, or high-voltage equipment rooms where very high fault currents are likely to flow into the earth resulting in hazardous step potentials.

The resistance of a buried grid, R_g, may be calculated using [2]

$$R_g = \frac{\rho}{4}\sqrt{\frac{\pi}{A}} + \frac{\rho}{L_T} \qquad (6.22)$$

where:
ρ = Soil resistivity, $\Omega \cdot$ m
L = Length of each arm of electrode, m
A = area occupied by the ground grid, m²
L_T = total buried length of conductors, m

A modified equation [2] taking into account the effect of grid depth is

$$R_g = \rho\left[\frac{1}{L_T} + \frac{1}{\sqrt{20A}}\left(1 + \frac{1}{1 + h\sqrt{20/A}}\right)\right] \qquad (6.23)$$

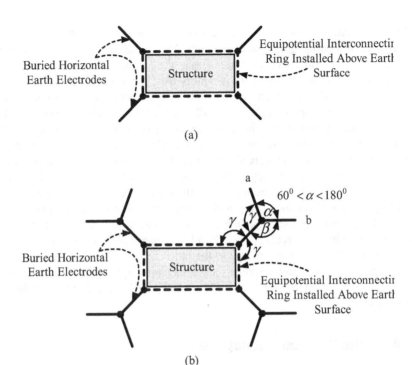

Figure 6.51. Examples of various star arrangements of horizontal earth electrodes around a structure.

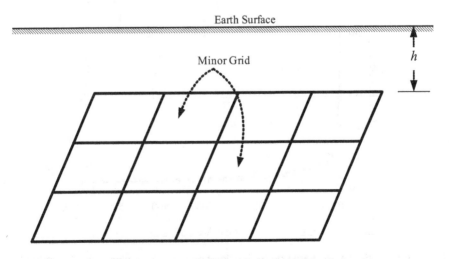

Figure 6.52. A grid electrode buried at depth h.

The two equations yield almost identical values when applied to evaluate the values of resistance of a buried grid at shallow depth. The results, as shown in Figure 6.53, also indicate that the grid resistance diminishes as the grid area becomes larger.

As discussed, the grid network is mainly used for achieving acceptable touch and step potentials. The grid is not a cost-effective design for obtaining low electrode resistance. Due to overlapping spheres of influence, or the so-called proximity effect, the inner conductors that define the minor grids are not effective in dissipating the discharge current from the grid. Some computed values of earth resistance of buried grids of various minor mesh sizes are shown in Table 6.3. Increasing the mesh size from a 1 m square to a 2 m square only increases the grid resistance by about 5%. For a grid area of 400 m^2 with 1m square mesh size, the effective earth resistance is 2.33 Ω, realized by using total conductor length of 840 m. For a large single mesh of 20 m square, the earth resistance is 3.46 Ω, achieved by using total conductor length of 80 m. Thus, the resistance has been reduced by a third by using ten times more conductor.

The last column in Table 6.3 gives the ratio of the grid earth resistance to that of a single buried strip having the same total conductor length. It may again be concluded from the comparison that the mesh grid is not a cost-effective design for achieving low earth resistance.

6.6.5 The Ring Earth Electrode

With reference to the dimensions given in Figure 6.54, a ring earth electrode has resistance to earth given by

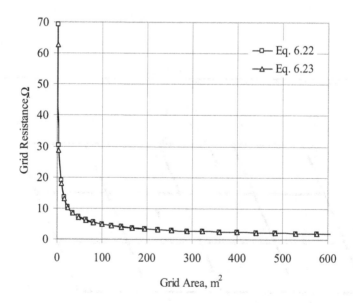

Figure 6.53. Effect of grid area on grid resistance (1 m ⨯ 1 m minor grid size, buried 0.1 m deep, $\rho = 100\ \Omega \cdot$ m).

Table 6.3. Comparison of earth resistance of buried grid of varying minor grid sizes (ρ = 100 $\Omega \cdot$ m, total grid area A = 400 m²)

Size of minor grid	No. of minor grids	Total conductor length L (m)	Grid resistance, R_g (Ω)	Resistance of a long strip of same length L, R_S (Ω)	Ratio of R_g to R_S
1m × 1m	400	840	2.33	0.363	6.42
2m × 2m	100	440	2.44	0.646	3.78
4m × 4 m	25	240	2.63	1.10	2.39
5m × 5 m	16	200	2.71	1.30	2.09
10m × 10m	4	120	3.05	2.02	1.51
20m × 20m	1	80	3.46	2.87	1.21

$$R = \frac{\rho}{2\pi^2 D}\left[\ln\frac{8D}{d} + \ln\frac{4D}{S}\right] \qquad (6.24)$$

where:
ρ = Soil resistivity, $\Omega \cdot$ m
D = Diameter of ring, m
d = Diameter of conductor, m
$S/2$ = Depth of conductor in soil, m

Figure 6.55 is a plot of ring electrode resistance with variation in perimeter length. A comparison with Figure 6.49 shows that for the same soil resistivity and conductor size, the ring electrode has resistance to earth about the same as that of the horizontal

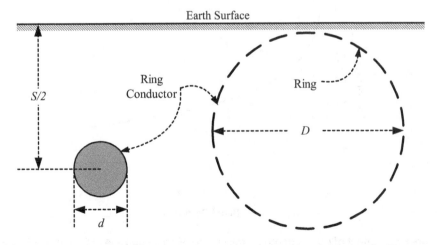

Figure 6.54. Ring electrode buried at depth S/2.

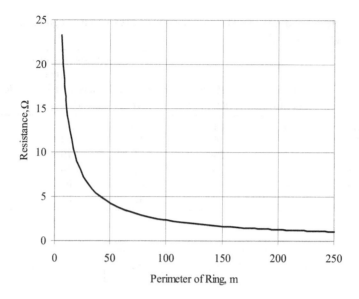

Figure 6.55. Effect of perimeter length on resistance of ring electrode (0.013 m conductor, buried 0.5 m deep, $\rho = 100\ \Omega \cdot m$).

strip electrodes of same conductor length. This is so because the ring electrode is just another form of horizontal strip electrode.

As shown in Figure 6.56, buried depth does not have a significant effect on the resistance of a ring electrode. As a general practice, the ring earth electrode should be installed more than 1 m from the structure and at a depth of 0.5 m or more, and it should

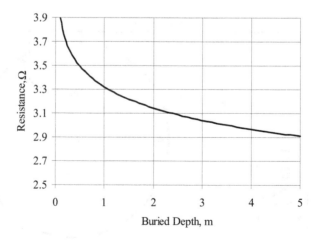

Figure 6.56. Effect of buried depth on resistance of ring electrode (0.013 m conductor, 20 m diameter ring, $\rho = 100\ \Omega \cdot m$).

entirely surround the structure to be protected. This clearance and depth are optimum in normal soil for potential control to protect people in the vicinity of the structure wall.

The ring earth electrodes also perform the function of potential equalization between the vertical electrodes at ground level, since the various electrodes give different potentials due to the unequal distribution of lightning or ground fault currents in them due, for example, to variations in the earth resistance. The different potentials result in a flow of equalizing currents through the ring earth electrode, so that the maximum rise in the structure is brought to approximately the same potential.

In addition to potential equalization, the ring helps to provide potential grading by reducing the touch voltage when a person comes in contact with the down conductor during a lightning discharge.

6.6.6 Foundation Earth Electrode

Conductors that are installed in the structural foundation below ground form the group of earth electrodes commonly called foundation electrodes or concrete-encased electrodes. Concrete below ground level is a semiconducting medium of about 30 Ω-m resistivity at 20°C, or somewhat lower than the average loam soil [1].

Large structures, including tall buildings and industrial plants, are usually constructed with reinforced foundations. Steel reinforcing bars in concrete foundations and footings are excellent ready-made grounding electrodes. Each footing electrode has a resistance substantially lower than that of a driven rod of equal depth. This reduction of effective earth resistance is achieved because the medium, (the concrete), in contact with the primary electrodes (the steel structural bars) has a lower resistivity compared with the surrounding soil. The surface areas between the structural steel bars and the concrete are substantially higher than normal driven rods in contact with surrounding soil. The large number of interconnected rods and wire meshes assures a subdivision of the total lightning current into many parallel discharge paths.

Figure 6.57 shows a telecom tower footing under construction. There are many reinforcing steel bars that can be used to achieve low earth impedance. From the design to the construction phase, special attention should be given to ensure the steel reinforcement elements are electrically continuous in concrete elements. A mechanical clamping device or threaded ferrule can provide a high degree of continuity. For economic reasons, reinforcing rods are frequently tied together by steel tie wire at splice joints.

Figures 6.58 and 6.59 further illustrate the construction details of a tower foundation. With a typical penetration depth of over 10 m, the reinforced concrete piles of the tower foundation can provide a very effective earth electrode system both in terms of cost and electrical performance.

An example of the effectiveness of concrete-encased electrodes is reflected in the setting of the grounding requirements of the National Electrical Code (NEC) [6]. The Code requires that a single electrode consisting of a rod, pipe, or plate that does not have a resistance to ground of 25 Ω or less must be augmented by one additional electrode located not less than 1.8 m away. The grounding requirement can also be satis-

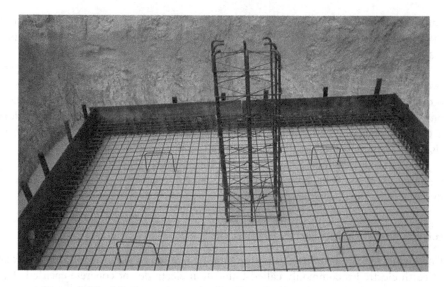

Figure 6.57. A Telecom tower footing/foundation under construction.

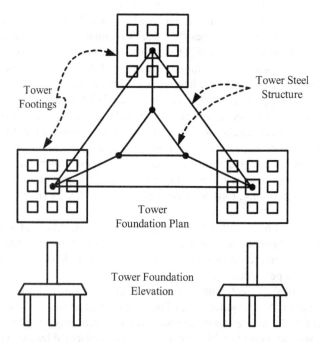

Figure 6.58. Layout of telecommunications tower foundation.

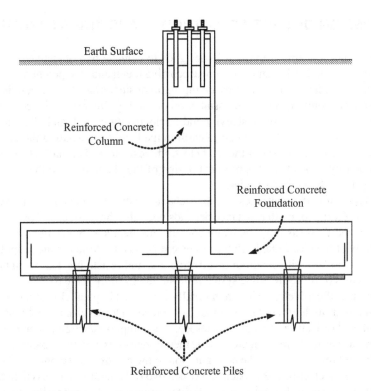

Figure 6.59. Structural details of a foundation of a telecom tower.

fied by an electrode encased by at least 50 mm of concrete, located within and near the bottom of a concrete foundation or footing that is in direct contract with the earth, consisting of at least 6.0 m of one or more bare, zinc galvanized, or other electrically conductive coated steel reinforcing bars or rods of not less than 13 mm (½ in) in diameter, or consisting of at least 6.0 m (20 ft) of bare copper conductor not smaller than 4 AWG. Reinforcing bars are permitted to be bonded together by the usual steel tie wires or other effective means. Further, the NEC also requires bonding of all grounding electrodes present in the building: (1) metal underground water pipes, (2) metal frames of buildings or structures, (3) concrete-encased electrodes, (4) ground rings, (5) rod and pipe electrodes, and (6) plate electrodes.

Lessons Learned

- Three basic earth electrodes are used: vertical rods, plates, and horizontal strips.
- Structural steel and building foundations are effective components of earth termination.
- An effective earth termination is formed by interconnection of the various earth electrode components.

6.7   DESIGN OF EARTH ELECTRODES AND THEIR LAYOUT

Earth electrode systems are primarily designed to achieve two objectives: to obtain a required impedance value and to ensure acceptable touch and step potentials for safety.

Earth electrodes are designed by taking into consideration the site conditions and the particular requirements of the installation or facility. Important site conditions include space availability, the resistivity and homogeneity of soil, and climatic effects. Generally, resistance of the volume of earth immediately surrounding the electrode is the predominant component of the total electrode resistance. The earth electrode resistance primarily depends on the physical design of the electrodes and the resistivity of the soil surrounding the electrodes.

The types of earth electrodes and their characteristics have been discussed in Section 6.6. The value of earth resistance depends upon the shape, size, and position of the electrode, and the resistivity, moisture content, and degree of ionization of the soil in the vicinity of the electrode. Each type of electrode has its particular niche applications with respect to cost, site conditions, soil structure, and resistivity. The advantages and disadvantages of these electrode systems are summarized in Table 6.4.

A number of earth electrodes are normally connected in parallel in situations where the required grounding resistance cannot be achieved. Due to electric field interaction, the combined resistance of parallel electrodes is a complex function of a number of parameters such as the number of electrodes, their dimensions, and the separation between them, the soil resistivity, and the configuration of the electrodes. To optimize the effectiveness of each electrode, the spacing between two rods should be no less than the combined length of the rods. Due to space limitation on-site, it is generally acceptable to have spacing between rods no less than their individual length, obviously with somewhat higher resistance than would be the case if the rods were more widely spaced out.

6.7.1   Selection of Material

Durability of the earthing system throughout the lifetime of the electrical installation is the most important selection criterion for of electrode material. The durability mainly depends on the capability of the electrode system to withstand corrosion.

Special attention is given to the corrosion property of the type of soil it is to be used in and the possible galvanic effects when the earth electrode is bonded to other exposed structural steel. Copper is still one of the best and most popular materials. Copper-plated steel provides enhanced mechanical strength. Steel in concrete-encased electrodes is protected against corrosion by the layer of concrete. Steel in concrete may be connected to copper or copper-covered earth electrodes [7].

Galvanized steel is strongly electronegative to both copper and steel in concrete. Earth electrodes of galvanized steel should be used in soil when no steel parts in the concrete are directly connected to the earth electrode in soil. A recommended method is to connect the galvanized steel earth electrodes in soil to the steel reinforcement in concrete by spark gaps if lightning protection is a major consideration [40]. Bare aluminum is normally not permitted for use underground because it corrodes easily when buried in soil.

Table 6.4. Comparison of the principal types of earth electrodes

Type	Advantages	Disadvantages
Vertical rods	Simple design Easiest to install, particularly where space is limited Low cost Readily available hardware Extended rods can be used to reach water table or low-resistivity soil layers More stable impedance; not subject to too much seasonable variation	High impulse impedance Not useful where soil layer is thin above large rock formations beneath Excessive step voltage on earth surface under high fault current or during direct lightning strike
Earth ring	Simple design Easy to install Readily available hardware	Not useful where large rock formations are near surface
Horizontal bare conductor strips (radials)	Useful where rock formations make vertical rod unsuitable Low impulse impedance Star pattern of conductors provides very good RF counterpoise	Subject to seasonal fluctuations of resistance with soil drying or freezing
Horizontal grid (bare conductor)	Minimum surface potential gradient Easy installation if carried out along with construction work Can achieve satisfactory resistance where rock formations prevent use of vertical rods Often combined with vertical rods to stabilize resistance fluctuations	Subject to seasonal resistance fluctuations with soil drying or freezing if vertical rods not used More open space required
Plates	Can achieve low resistance to earth in limited area	Most difficult to install Not cost-effective
Structural steelwork	Can provide very low resistance to earth Low cost	Corrosion Little control over future alternations

6.7.2 Grounding Requirements of Power Distribution Systems

For a simple installation, such as a domestic home, a couple of ground rods driven into good soil may be sufficient to meet the basic requirements [6]. Within industrial or commercial premises, the fault current level is expected to be high and, hence, the touch voltage is an important design consideration. For HV installations in power stations or substations, the earth electrode system and the associated grounding and bonding of exposed conductive parts are designed with consideration of the significance of step and touch potentials as well as transferred potential [2]. A common design is to lay a mesh of grid with vertical rods and horizontal conductors.

For installation with equipment that demands high EMC performance, such as IT (information technology) centers and C^3 (command, control, and communication) facilities, the best form of earth electrode system is a meshed network buried under and around the installation or building.

For a HV installation in substations or power stations, a typical grid is usually supplemented by a number of vertical earth rods, as shown in Figure 6.60, and may be further connected to auxiliary earth electrodes to lower its resistance with respect to remote earth.

Figure 6.60 is a simplistic illustration of using a combination of grid and vertical rods. The optimum design for a particular site is normally carried out by computer simulation. Minimizing all aspects of hazardous touch and step potentials is a primary objective of such study in addition to achieving low impedance to earth. The grid of horizontally buried conductors is most effective in reducing the danger of high step and touch voltages on the earth's surface when installed at a shallow depth of about 0.5 m below ground surface. The vertical rods, as discussed earlier, are advantageous in stabilizing the seasonal variation of earth resistance. Deep rods are particularly effective in dissipating faults currents in situations in which a two-layer or multilayer soil is encountered and the upper soil layer has higher resistivity than the lower layers.

Figure 6.61 shows the resultant voltage distribution across section X-Y marked in Figure 6.60. The grid tends to transfer the hazardous potential gradient to the periphery of the grid. Designers of power stations and substations where high-voltage faults are likely have to take mitigating measures to deal with such hazards.

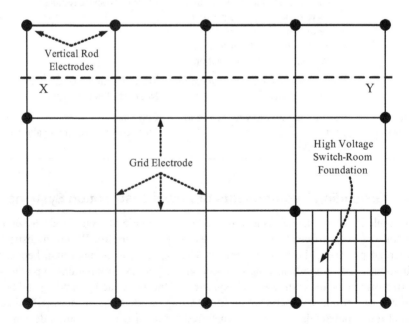

Figure 6.60. A combination of grid and vertical rod electrodes.

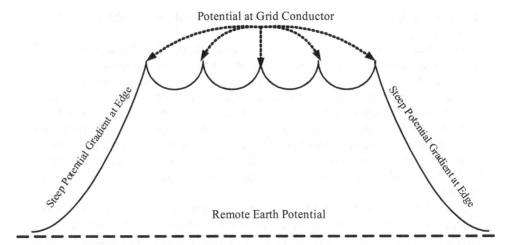

<u>Figure 6.61.</u> Variation of surface potential along section X-Y of a combination of grid and vertical rod electrodes (refer to Figure 6.60).

6.7.3 Measures to Reduce Transient Impedance of Earth Electrodes

Lightning current has rise times on the order of 10^{10} amperes per second and, hence, is most appropriately treated as a high-frequency phenomenon. An effective earthing system should provide low earth impedance as well as low resistance. The equations given in Section 6.6 for calculation of electrode resistances assume perfect conductivity of the conductor and consistent characteristics of the medium surrounding the electrode. Such assumptions introduce little error in the calculation of DC or power-frequency electrode resistance.

When designing an effective electrode system for dissipating the energy of lightning impulses, the variation of the electrode impedance with time and the magnitude of the current should be considered. This transient or surge impedance mainly depends on the following factors [21]:

1. Resistance and surge impedance of the earth electrode system and the connecting conductors
2. Contact resistance between the earth electrode and the surrounding soil
3. Resistivity of the soil surrounding the electrode
4. Degree of soil ionization

The resistance of the earth around an earth electrode contributes the major part of the total earth resistance of the electrode, with the contact resistance between the electrode and the soil making up about 10%. The high voltage and current of a lightning discharge causes ionization and arcing over the soil in contact with the electrode. Due to ionization of soil in the vicinity of the electrode, there is a significant reduction of resistance of earth connection under impulse conditions, as shown in Table 6.5 (Refer to Table B.1 of [21]).

The effective impedance to earth may be lowered by using flat tape instead of circular conductors. This increases surface area, reduces high-frequency resistance due to the skin effect, and increases both capacitive coupling and the ground contact area for a given cross section of conductor. Vertical electrodes, for a given dimension, are generally effective in providing low surge impedance in areas of low-to-moderate soil resistivity. The surge impedance of two or more horizontal conductors laid in parallel will be less than that of a single conductor of equivalent length. A conductor with a center-point feed has the same effect as two parallel connected lines. These methods of lowering the effective earth impedance are illustrated in Figure 6.62.

When lightning current discharges through a single buried conductor earth electrode, the initial value of the effective electrode impedance, the surge impedance, is several times the DC resistance. As the wave of energy propagates through the electrode system, more and more of the wire of the electrode makes effective contact between the propagating energy and the medium that dissipates the energy. For the same total conductor length, a star formation with short multiple lengths is, therefore, more effective than a single length of long conductor in dissipating the lightning energy more efficiently into the earth. Figure 6.63 [9] illustrates the characteristic that lightning surge impedance of the electrode decreases with time and approaches the DC resistance value within a few microseconds.

An example of earthing systems of a telecommunication tower and its associated equipment room are shown in Figure 6.64 [43]. A system of radial horizontal electrodes in combination with driven vertical rods is used to achieve low surge impedance for efficient dissipation of lightning discharge. The ring earth electrode around the equipment room provides potential equalization to reduce the hazard of step potential.

Figure 6.651 provides a conceptual visualization of the manner an effective earth termination system supports the propagation and dissipation of the lightning surge current, protecting the communications facility.

Table 6.5. Example of reduction of resistance of earth connection under impulse conditions

Soil resistance characteristics	Number of rods and arrangement			
	One isolated rod		Four rods at corners of a square with 3.05 m edges	
Soil Resistivity, $\Omega.m$	100	1,000	100	1,000
Resistance at Low Current, Ω	30	300	10.5	105
Resistance at Peak Current, Ω	11.3	54	6.8	37
Ratio of Resistance at Peak Current /Resistance at Low Current	0.38	0.18	0.65	0.35

Note: based on the following parameters: diameter of rods = 10 mm; Depth in earth = 3.05 m, peak current injected = 80 kA, time to current crest = 4 μs.

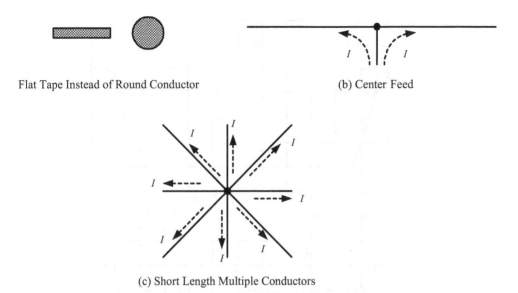

Flat Tape Instead of Round Conductor   (b) Center Feed

(c) Short Length Multiple Conductors

<u>Figure 6.62.</u> Methods of reducing earth impedance.

6.7.4   Earthing Requirements for Lightning Protection

Low earth impedance has the advantage of reducing the potential gradient around the earth electrodes when discharging lightning current. It also reduces the risk of side-flashing to metal in or on a structure. A high potential difference may be hazardous or even fatal to personnel. As this potential difference is a function of the product of the lightning current and the resistance of the earth electrode, the importance of keeping the latter as low as possible, particularly in areas where people are likely to be present, is evident. As discussed in Section 6.4.4, a large potential rise on the earth electrode and associated ground bar reference can also cause insulation breakdown of electrical equipment.

Lightning protection codes require that an earth electrode should be connected to each down conductor to ensure efficient discharge to the ground mass. For practical purposes, some codes specify as a general guide a maximum value for each earth electrode system. Usually, the soil resistivity and other site conditions are such that extended or parallel electrodes are installed or that the individual earth electrodes are interconnected by a ring earth electrode. Buried ring earth electrodes laid in the ring manner are considered to be an integral part of the earth termination network and are taken into account when assessing the overall value of resistance to earth of the installation.

A number of standards require that the whole of the earth termination network should have a combined resistance to earth not exceeding 10 Ω without taking account of any bonding to other services [9, 20, 21, 25]. NEC [6] permits a maximum value of 25 Ω. The maximum acceptable value of lightning earth electrode resistance of 10 Ω

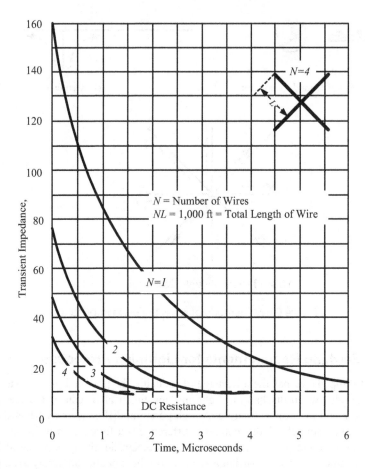

<u>Figure 6.63.</u> Transient impedance of an earth electrode subsystem as a function of the number of radial wires.

or 25 Ω is obviously not intended to constitute a "safe" value. A current of 5 kA flowing down a 10 Ω earth electrode will theoretically raise the earth reference voltage to 50 kV. The specific values should be considered as practically achievable values with reasonable cost. Where safety is concerned, an integrated system of equipotential bonding and surge protection is necessary.

NFPA 780 [19] does not specify a maximum permissible value for resistance or impedance limits of the ground electrode subsystem. It emphasizes that properly made ground connections are essential to the effective functioning of a lightning protection system, and every effort should be made to provide ample contact with the earth, the so-called volumetric efficiency, to permit the dissipation of a lightning strike without damage. The code favors an increased number of bonds of the lightning conductor system to other grounded conductors within the building and use of more down-conductor paths so as to reduce dangerous side flashes.

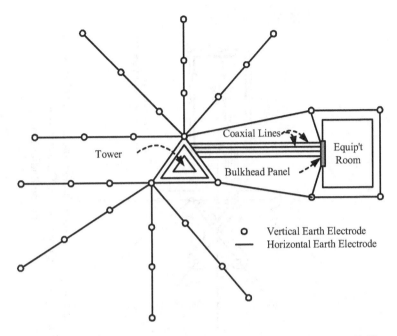

Figure 6.64. Earthing for lightning protection of telecommunications facility.

6.7.5   Earth Potential Rise and Surface Potential Gradients

The earth electrode is raised to a high potential with respect to the general mass of the earth when lightning discharge current or high fault current flows through it. All exposed metalwork connected to the earthing system will experience a rise of earth potential with respect to the true earth. The potential rise is proportional to the magnitude of the current and also proportional to the effective earth impedance. Under severe fault or lightning discharge conditions, ground surface potential may be raised to thousands of volts.

The potential gradient around the ground electrode depends on the soil resistivity, the physical design of the earth termination network, and the interconnection of the electrodes. The step potential is highest in areas immediately surrounding the buried electrodes. Exposed metalwork bonded to the substation grounding network will be raised to a high voltage under fault conditions. The potential at a point on the ground surface some distance away from the electrode will be lower. The potential difference between the hand of a person touching the exposed metalwork and his feet standing on the ground is called the touch potential or contact voltage. The touch potential is larger at distances further away from the buried electrodes.

Codes and standards [29] usually specify a certain low impedance value, such as 1 Ω, in order to limit the ground fault potential rise. New developments, however, are moving towards ensuring touch and step potentials of the earthing system and the associated equipment are below specified safety limits.

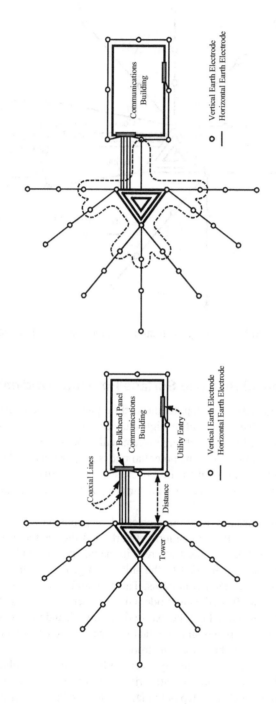

(a) Example of Recommended Site Earth Termination
System about to be Struck by Lightning

(b) Neglecting the Currents through the Coaxial Cables, the
Lighting Surge Current Flows Outwards from the Tower
Base along the Radial Line

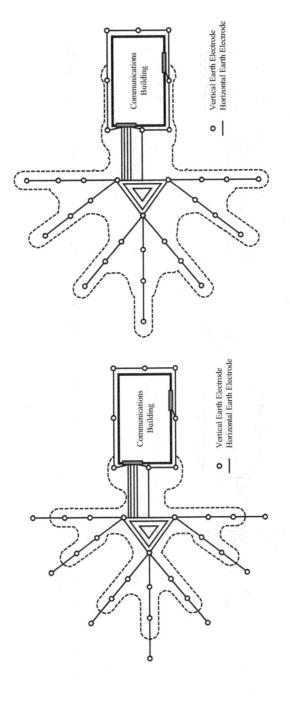

(c) In a Well-Designed Earth termination System, The Surge Current Spreads Out Initially from the Building

(d) As Current Spreads, Energy is Dissipated due to Spreading and I^2R Losses

Vertical Earth Electrode
— Horizontal Earth Electrode

Figure 6.65. Conceptual illustration of lightning surge current spreading/dissipation in the earth termination subsystem of a telecommunications facility. (The images are adapted from *The Grounds for Lighting Protection* and used with permission from PolyPhaser Corp.) (*continues*)

449

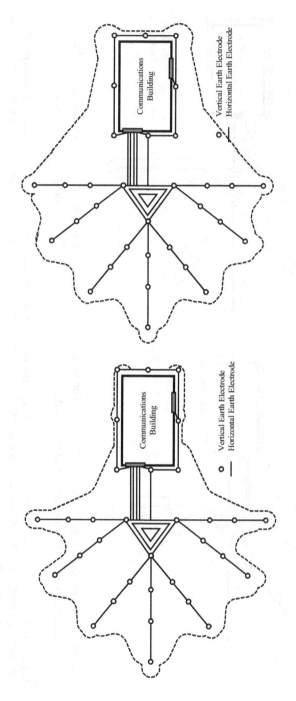

(e) As the Current Reaches and Saturates the Radial System, it Traverses the Building Perimeter

(f) By the Time the Current Surrounds the Building, the Radials have Spread Out much of the Surge Energy

Vertical Earth Electrode
Horizontal Earth Electrode

Figure 6.65. *Continued.*

450

As the earth fault current or the lightning current is discharged through the earth electrode, the surrounding soil is raised to a high potential with respect to the body of the earth. It is recognized that earth surface potential gradients inside or adjacent to an electrode are mainly a function of the topsoil resistivity [2]. In contrast, the earth electrode resistance is primarily a function of deep soil resistivity, especially if the electrode is very large.

Vertical Earth Rod

For a single vertical rod, Equation (6.25) is a simplified expression [7] for calculating the fraction E of the potential rise at a point P on the earth surface:

$$E = \frac{\ln\left(\dfrac{L}{r} + \sqrt{\left(\dfrac{L}{r}\right)^2 + 1}\right)}{\ln\left(\dfrac{4L}{D}\right)} \tag{6.25}$$

where:

L = Buried length of an electrode, meters
r = Distance of point P on the earth surface to the electrode, meters
D = Diameter of the electrode(s), meters

The resulting earth surface potential gradient is shown in Figure 6.66 for the case of a single 16 mm electrode driven 6 m and 12 m into the earth. It is seen that the potential gradient is greatest near the rod and decreases rapidly with distance away from the rod.

The expression given in [9] for the step potential expressed as a fraction of the earth rod potential is

$$\frac{V_0 - V_P}{V_0} = \frac{\ln\left[\dfrac{3r}{d\left(1 + \sqrt{\dfrac{r^2}{L^2} + 1}\right)}\right]}{\ln\left(\dfrac{3L}{d}\right)} \tag{6.26}$$

where:

V_0 = Potential of the earth electrode, volts
V_P = Potential of the earth surface at point P from the electrode, volts
r = Distance of point P on the earth surface to the electrode, meters
L = Length of rod, meters
d = Diameter of rod, meters

The maximum step potential occurs for the case in which a person has one foot directly above the electrode, which is installed flush with the ground surface. The values in

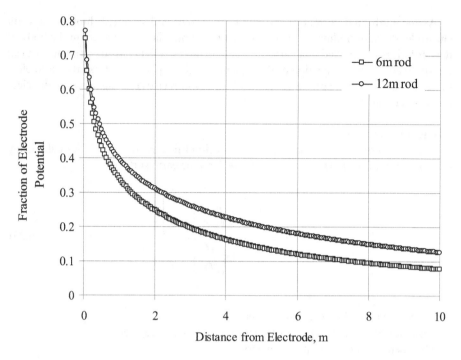

Figure 6.66. Distribution of earth surface potential around a single rod.

Table 6.6 were calculated using Equation (6.26) for a step length of 1 m between the two feet. The small variation of step potential with rod length indicates that increasing the rod length does not significantly reduce the hazard of step potential associated with a vertical rod.

The magnitude of step potential drops rapidly with distance away from the earth electrode. Equation (6.26) may be used to calculate the potential difference between two points located 1 m apart. The step potential calculated for a single vertical rod is shown in Figure 6.67.

Table 6.6. Step potential for a buried vertical earth rod (one foot directly above the earth electrode, rod diameter of 16 mm)

Rod length (m)	Ratio of step potential to electrode potential
2	0.75
5	0.66
10	0.60
20	0.55
40	0.51

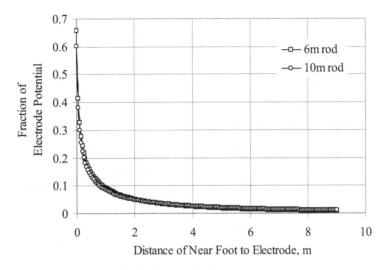

<u>Figure 6.67.</u> Step potential between a person's two feet, one meter apart.

Horizontal Electrodes

For horizontal electrodes, the fraction E of the electrode potential at a point P on a line perpendicular to the line of the electrodes may be calculated with the following expression [7]:

$$E = \frac{2 \ln \left(v + \sqrt{v^2 + 1}\right)}{\ln \left(\dfrac{L^2}{hd}\right)} \tag{6.27}$$

where:
h = Depth of electrodes from the ground surface, meters
r = Distance along the ground surface from point P to a point vertically above the electrode, meters
L = Length of horizontal electrode, meters
d = Diameter of electrode, meters

For a strip electrode, $d = 2w/\pi$, where w is the width of a strip electrode in meters.

The distribution of ground surface potential around a horizontal strip electrode is plotted as shown in Figure 6.68. The ground surface potential drops more gradually with increasing distance from the electrode, as compared with the abrupt changes in the case of a vertical rod (Figure 6.66). As a consequence, the step potential near a horizontal electrode is also lower, as shown in Figure 6.69.

The existence of transient high potential rise and associated large potential gradient may be harmful to communication and control cables that either run past the affected ground area or are in contact with metalwork bonded to the grounding network. The

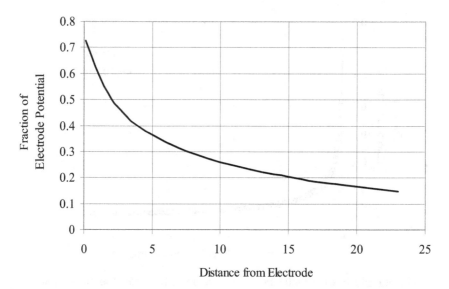

Figure 6.68. Distribution of ground surface potential around a horizontal strip electrode (25 mm wide strip electrode, 50 m long, buried 0.5 m deep).

cores of these cables are substantially at remote earth potential, whereas their sheaths
are raised to the high fault voltage.

Several design approaches may be adopted to reduce the potential hazards of step
potential. The most effective method is to reduce the potential rise of the earth electrode by aiming for electrode resistance as low as possible. Soil resistivity on-site de-

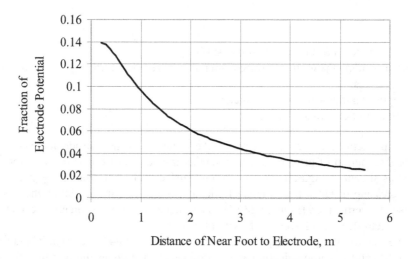

Figure 6.69. Step potential near a horizontal strip electrode.

termines to a large extent how low an electrode resistance may be and still be econom-ically feasible. Burying the earth electrode some distance below the ground surface can reduce substantially the maximum gradient on the ground surface. For this pur-pose, horizontal electrodes are commonly buried about half a meter beneath ground surface. The smaller incremental reduction of surface potential with greater depth may not be justified by the increasing cost of installation.

Excessive ground potential gradients may be mitigated by installing a ground sys-tem of vertical rods and horizontal conductors forming a grounding grid. The grid with interconnected bare conductors buried horizontally near the earth's surface is effective in reducing the danger of high step and touch voltages on the earth's surface. The ver-tical rods are more effective than the horizontal conductors in dissipating high current into the earth. The danger of excessive step potential may be substantially reduced by covering the surface above the ground grid with a high-resistivity surface layer, such as gravel [2]. The current through the body due to the step potential will be consider-ably less because of greater contact resistance between the earth and the feet.

A serious hazard is posed by the transfer of the ground potential rise to locations out-side the ground grid area. This dangerous potential may be transferred via communication circuits, control cables, metallic conduits, pipes, metallic fences, and so on. This risk of transfer potential is usually associated with a person standing at a remote location and touching a conductor connected to the ground grid where a fault occurs. The hazard of transfer potential should be taken into consideration in the design of the grounding system.

Lessons Learned

- Earth electrode systems are designed to obtain a low impedance value and to achieve acceptable touch and step voltages.
- Low transient impedance is desirable for lightning protection.
- The value of earth resistance achievable depends on the shape, size, and position of the electrodes and the resistivity, moisture content, and composition of the soil around the electrodes.

6.8   MEASUREMENT OF SOIL RESISTIVITY AND EARTH ELECTRODE RESISTANCE

The actual resistance value of an earth electrode system depends primarily on the soil resistivity on-site. As a rule of thumb, electrode resistance increases in direct propor-tion to the soil resistivity. Designers of earth electrode system need to know the soil re-sistivity for the particular site in order to produce a design that meets specific resis-tance value as well as satisfies step and touch voltage requirements.

6.8.1   Measurement of Soil Resistivity

It is difficult to determine accurately the resistivity of the soil by laboratory analysis of samples taken from a site. Soil samples cannot represent the actual soil composition on-site. Soil is rather nonhomogeneous, with many types or different soil layers commonly present at a particular location. Additionally, moisture and temperature variations affect the resistivity significantly. Actual on-site measurement is the only way to provide reliable values of soil resistivity.

The most common technique for determining soil resistivity is to inject a known current into a given volume of soil, measure the voltage drop produced by the current through the soil, and then calculate the resistivity from a predetermined equation [41, 44].

An accurate determination of the soil resistivity may be carried out by the four-terminal method of measurement, as shown in Figure 6.70. This is the most accurate method in practice of measuring the average resistivity of large volumes of undisturbed earth [44]. Small electrodes are inserted at depth b and spaced in a straight line at intervals a. A test current I is made to flow between the two outer electrodes and the potential V between the two inner electrodes is measured. The potential terminals should be small relative to the soil sample cross section and located sufficiently distant from the current terminals to assure near-uniform current distribution across the sam-

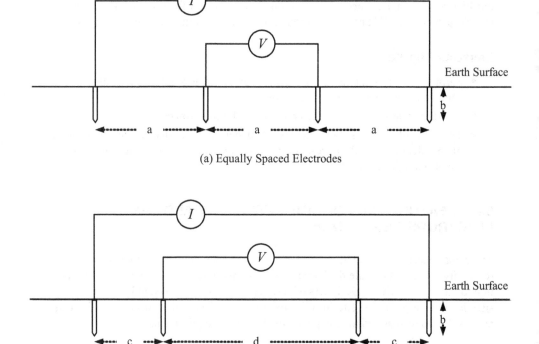

(a) Equally Spaced Electrodes

(b) Unequally Spaced Electrodes

Figure 6.70. Four-point method of measurement.

ple. When all the electrodes are equally spaced and placed in a straight line, the method is called the Wenner arrangement.

The ratio of V/I gives the resistance R in ohms. If the distance between two adjacent electrodes is a, then the resistivity ρ in terms of the length units in which a and b are measured is [44]

$$\rho = \frac{4\pi\, aR}{1 + \dfrac{2a}{\sqrt{a^2 + 4b^2}} - \dfrac{a}{\sqrt{a^2 + b^2}}} \tag{6.28}$$

In practice, the depth of insertion, b, of the four electrodes is kept very much smaller than the distance of separation, a. Then Equation (6.28) reduces to

$$\rho = 2\pi\, aR \tag{6.29}$$

Equation (6.29) gives approximately the average resistivity of the soil to the depth a.

One major shortcoming of the Wenner method is that the potential between the two inner electrodes (V in Figure 6.70) decreases rapidly when the spacing of the potential electrodes is increased to a relatively large value. Such low potential values may not be measured accurately by low-cost commercial instruments. The potential value to be measured may be kept relatively high by the so-called Schlumberger–Palmer arrangement, in which the potential probes are brought nearer the corresponding current probes.

Referring to Figure 6.70(b), if the depth of burial of the electrodes b is small compared to electrode separations d and c, then the measured resistivity can be calculated as follows:

$$\rho = \pi c(c + d)R/d \tag{6.30}$$

With the various dimensions of electrode separation measured, the soil resistivity can be easily determined from Equations (6.28) or (6.30) where appropriate.

6.8.2 Measurement of Earth Resistance

The earth electrode system is typically buried or inaccessible except during construction of the facility or through an earth pit, as shown in Figure 6.71. The resistance of the earth electrode system should be verified at the time of construction because the actual value may vary from the designed value, which normally is calculated based on a set of idealized assumptions. Periodic inspection and testing of the earthing system of a large installation are carried out to check the integrity and efficacy of the grounding system, which may be subject to corrosion deterioration and changes in soil resistivity arising from seasonal and hydrological variations such as changes in the water table.

For new construction, it is common to use an earth tester to measure the resistance of an earth electrode before it is connected or bonded to other earth electrodes. For ex-

Figure 6.71. An earth pit allows inspection and testing of individual electrodes.

isting installations, suitable precautions must be taken in view of the danger of removing the connecting conductor from the earth electrode.

6.8.2.1 Fall-of-Potential Method.

A very popular method is the three-stake measurement shown in Figure 6.72. The earth electrode resistance tester combines a current source and voltage source measurement. A current is passed between the electrode E to be tested and an auxiliary current electrode C. The voltage drop between E and a secondary auxiliary electrode P is measured and the resistance of the test electrode is then the voltage between E and P divided by the current flowing between E and C. Referring to Figure 6.73, it is important that the measuring electrode P be sited on the flat part of the curve. This means that the current probe C must be located at a sufficient distance away from the electrode E, the resistance of which is to be measured so that the regions of influence of E and C are not overlapping.

For a large earthing system, the spacing required to obtain the flat portion of the curve may be too large to be practicable. It can be shown from an analysis of the theory of the fall-of-potential method [44] that the correct spacing, X_0, is obtained when the potential probe to is located at a distance 61.8% of the separation, d, between the electrode under measurement and the current electrode, as indicated on Figure 6.73. The 61.8% rule is applicable provided the soil is fairly uniform. The electrodes should be spaced sufficiently apart such that they are equivalent to the performance of hemispheres, an assumption made in deriving the 61.8% rule.

The resistance of a ground electrode is usually determined with alternating or periodically reversed current to avoid possible polarization effects when using direct current.

6.8.2.2 Two-Point Method.

This method measures the total resistance of the unknown and an auxiliary electrode. The earth resistance value of the electrode under

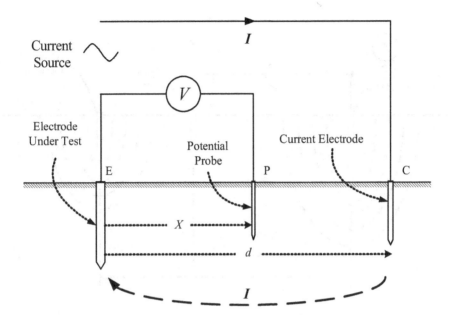

<u>Figure 6.72.</u> Three-stake method of earth electrode resistance measurement.

test is simply given by the measured value in ohms. The resistance of the auxiliary ground has to be negligible in comparison with the resistance of the unknown electrode for an acceptable value to be determined.

One useful application of this method is to determine the resistance of a single vertical rod electrode near a residence that also has a common municipal water supply system that uses metal pipe without insulating joints. The water pipe serves as an excellent low-resistance auxiliary electrode. This method, as illustrated in Figure 6.74, is subject to large errors when a low-resistance auxiliary electrode is not available. It is a very convenient method of providing a good approximate value.

The measured resistance value R is the sum of the electrode resistance R_e and the auxiliary electrode resistance R_a. If R_a is very much smaller than R_e, the measured value R is a good approximation to the electrode resistance under measurement.

6.8.2.3 *Clamp-On Earth Tester.* Figure 6.75 shows a clamp-on earth electrode resistance tester that has both a source transformer and a measurement transformer. The source transformer imposes a voltage on the loop under test and the measurement transformer measures the resulting current [45]. The earth resistance is determined from these voltage and current values. The basic operating principle is similar to that of the two-point method just discussed. In Figure 6.76, the measured loop resistance comprises the electrode resistance R_e in series with the auxiliary resistance R_a, which is a parallel combination of other electrodes. Again, R_e can only be determined accurately if R_a is comparatively much smaller.

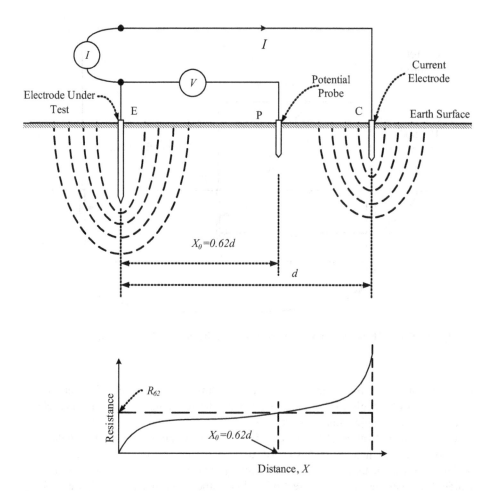

Figure 6.73. Measurement of earth electrode resistance by the fall-of-potential method.

6.8.2.4 Characteristics of Earth Electrode Resistance Testers.
Earth electrode resistance testers have the following characteristics:

- AC test current, because earth does not conduct DC very well
- Test frequency that is close to, but distinguishable from the power frequency and its harmonics. This prevents stray currents from interfering with ground impedance measurements.
- Separate source and measure leads to compensate for the long leads used in this measurement
- Input filtering designed to pick up its particular frequency signal and screen out all other interfering noises

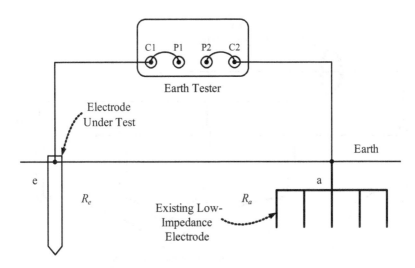

Figure 6.74. Two-point method of earth resistance measurement.

Some manufacturers have introduced models of ground testers that allow electrode resistance to be measured without disconnecting the electrode, thus avoiding the associated safety hazards. This "direct" method is essentially a variation of the fall-of-potential method. Testers with this capability can measure the ground impedance of a specific ground electrode without disconnecting it from an electrode array or from a structure's distribution system.

Both the fall-of-potential method and the "direct" method use stakes to inject current and measure the corresponding voltage drop. In a traditional fall-of-potential

Figure 6.75. Clamp-on earth tester.

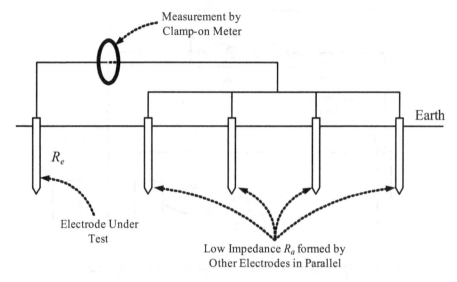

Figure 6.76. Clamp-on tester measures R_e and R_a in series.

method, there is no measurement of current flowing between any particular electrodes. In "direct" testing, the test current flowing in the electrode under test is measured using an integrated, high-sensitivity, clamp-on current transformer. Being able to accurately measure the current in the electrode under test effectively isolates the electrode and allows them to be tested without disconnecting from the system or from other electrodes

A good earth electrode resistance tester invariably incorporates digital filters in the current measurement to minimize the effects of stray currents.

Lessons Learned

- Soil resistivity is a required parameter for design of an earth electrode system.
- Due to variation of soil composition, measurement on-site is the best way to determine soil resistivity.
- Resistance of an installed earth electrode is commonly measured by the fall-of-potential method, the two-point method, or with a clamp-on earth tester.

6.9   REDUCING EARTH RESISTANCE

If value of earth electrode resistance is found to be too high for acceptance, several methods may be considered to improve it. The first method is to increase the effective length of the earth electrodes. The effects of increasing the length of various types of earth electrodes have been discussed in Section 6.6. Doubling the length of a vertical rod will reduce the earth electrode resistance by about 45%. Increasing the length of horizontal strips has a similar effect. At some locations, deeply driven rods may reach close to the water table and, hence, have the benefit of bringing a significant reduction in resistance.

For large earthing systems, multiple vertical rods are used in conjunction with mesh networks to reduce the overall earth resistance and to achieve the specific requirements of step and touch potentials.

An integrated system of earthing and bonding is very cost-effective in bringing down the overall earth resistance. Bonding together the following subsystems in a facility not only lowers resistance of the earthing system, but also creates an effective equipotential zone that minimizes electrical shock hazards:

1. Metal underground water pipe
2. Metal frame of the building or structure
3. Concrete-encased electrode
4. Ground ring
5. Rod-and-pipe electrodes
6. Plate electrodes
7. Other local metal underground systems or structures

Sometimes, difficult site conditions may be encountered where it is not practical to drive earth rods to great depth because of underlying rock. Due to space limitations, it also may not be feasible to implement the multiple rods approach. In such circumstances, chemical treatment or doping of soil to improve its conductivity is a good way of reducing the earth electrode resistance. However, their effects are temporary in nature since the chemicals are soluble and are, therefore, diluted by rainwater. The effect of possible chemical corrosion on electrodes should be considered together with any likely environmental impact.

Instead of chemical treatment, the soil immediately around an electrode may be replaced with material of lower resistivity such as clay-based bentonite, which is formed by the decomposition of volcanic ash. Other commonly available ground-enhancement materials include those based on a form of conductive concrete or cement. Figure 6.77 illustrates the application of ground-enhancement materials.

6.10 BONDING TO BUILDING STRUCTURES

An effective method of preventing the hazardous touch voltage resulting from an earth fault is to create an equipotential zone within which all exposed conductive parts and extraneous conductive parts are essentially at the same potential. This local equipotential zone may be achieved by a proper system of grounding and bonding.*

---

*An extensive discussion of bonding can be found in Chapter 5.

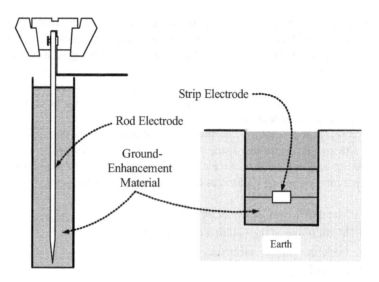

Figure 6.77. Application of ground-enhancement material.

Bonding of extraneous conductive parts in an installation to the exposed conductive parts is vital in order to minimize any potential difference between these parts during an earth fault. In each installation, a main earthing terminal (MET) or bar, as shown in Figure 6.78, is normally provided, either within the switchboard or mounted on insulating supports on the switch room wall. This main earthing terminal or bar, which is directly connected to the earth termination network, provides a common connection

Figure 6.78. Bonding at the main earth terminal (MET).

point for the main equipotential bonding conductors of extraneous conductive parts, including the following:

1. Metallic water service pipes
2. Gas service pipes
3. Other service pipes and ducting
4. Central heating and air conditioning systems
5. Exposed metallic structural parts of the building
6. The lightning protection system

It is a good practice to apply equipotential bonding to metallic sheaths of telecommunication cables at the point of entry to the building. The bonding is carried out at the points of entry of these pipes, cables, or ducts into the building. In order to achieve the best result of bonding for safety as well as EMC performance, the services are usually designed to enter the building at a common point.

It must be noted that codes of practice in some countries may specifically require separate earthing for the lightning protection system and the electrical installation. In any case, the down conductors of any lightning protection system must still be connected to their own earth electrodes, which must be tested before the cross bonding is made.

The term "equipotential" should be taken with the proper perspective. When no current is flowing through the bonded metallic parts, they are at the same potential, that is, they are equipotential. However, significant potential difference can exist between bonded parts when large fault currents or lightning discharge currents flow through bonded components. For example, at selected locations further away from the main earth termination, such as bathrooms, it is necessary to apply supplementary equipotential bonding to simultaneously accessible exposed conductive parts and extraneous conductive parts at the particular locations to reestablish the equipotential reference.

The main purpose of bonding is to equalize potential rather than to carry fault current, although bonding conductors should be sized to carry part of the fault current or lightning discharge current where they form part of a parallel earth return path. Bonding conductors should be short and direct where possible.

A major concern in lightning protection of a building is the occurrence of high potential differences between the conductors of the lightning protection system and other grounded metal bodies and wires belonging to the building. As discussed in Section 6.4.3.2, these potential differences are caused by resistive and inductive effects and can be of such a magnitude that dangerous sparking can occur. In order to avoid damages arising from overvoltage, it is necessary to equalize potentials by bonding grounded metal bodies to the lightning protection system.

In Figure 6.79, all structural steel and metallic reinforcements in a structure are bonded to the lightning protection system. Bonding connections to incoming metallic piped services should be as near as possible to the entry point to the premises and on the consumer's side of any insulating section. This integrated bonding system helps to prevent structural damage or puncture of service pipes arising from electric arc breakdown between these parts through the soil.

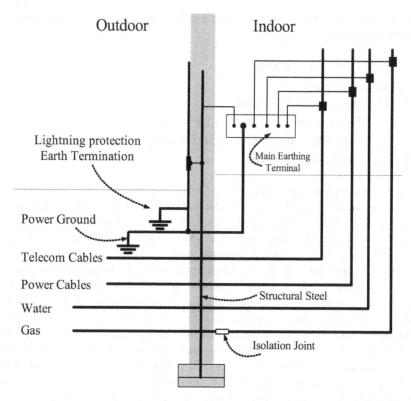

Figure 6.79. Bonding of power, lightning, and facilities grounds.

Lessons Learned

- Effective grounding and bonding create an equipotential zone, thereby preventing hazardous touch voltage resulting from earth faults.
- Grounding and bonding for cables and metallic service pipes, implemented at the building entry point, help to prevent the propagation of lightning surges into the building interior.

BIBLIOGRAPHY

[1]   IEEE STD 142-1991, *Grounding of Industrial and Commercial Power Systems.*

[2]   IEEE STD 80-2000, *Guide for Safety in AC Substation Grounding.*

[3]   IEEE STD 1100-1999, *Powering and Grounding Electronic Equipment.*

[4]   IEC 60364-1, *Electrical Installations of Buildings—Part 1: Fundamental Principles, Assessment of General Characteristics, Definitions,* 2001.

[5]   IEC 60364-4-44, *Electrical Installations of Buildings—Part 4-44: Protection for Safety—Protection Against Voltage Disturbances and Electromagnetic Disturbances,* 2001.

[6]   NFPA 70, *National Electrical Code 2005,* National Fire Protection Association, 2005.

[7]   BS 7430:1998, *Code of Practice for Earthing,* BSI.

[8]   SS CP16:1991, *Singapore Code of Practice for Earthing,* Standard Productivity and Innovation Board, Singapore.

[9]   MIL-HDBK-419A, *Military Handbook, Grounding, Bonding, and Shielding for Electronic Equipments and Facilities,* Washington, DC: U.S. Government Printing Office, 1987.

[10]   IEC 61024-1, *Protection of Structures against Lightning—Part 1: General Principles Edition: 1.0,* 1993.

[11]   IEC 61024-1-1, *Protection of Structures against Lightning—Part 1-1: General Principles—Guide A—Selection of Protection Levels for Lightning Protection Systems,* 1993.

[12]   IEC 61024-1-2, *Protection of Structures against Lightning—Part 1-2: General Principles—Guide B—Design, Installation, Maintenance and Inspection of Lightning Protection Systems,* 1998.

[13]   IEC 61312-1, *Protection against Lightning Electromagnetic Impulse—Part 1: General Principles,* 1995.

[14]   IEC 61312-2, *Protection against Lightning Electromagnetic Impulse—Part 2: Shielding of Structures, Bonding Inside Structure and Earthing,* 1999.

[15]   IEC 61312-3, *Protection against Lightning Electromagnetic Impulse—Part 3: Requirements of Surge Protective Devices,* 2000.

[16]   IEC 61312-4, *Protection against Lightning Electromagnetic Impulse—Part 4: Protection of Equipment in Existing Structures,* 1998.

[17]   IEC 61662, *Assessment of the Risk of Damage Due to Lightning,* 1995.

[18]   IEC 62305-4, *Protection against Lightning—Part 4: Electrical and Electronic Systems within Structures,* 2006.

[19]   NFPA 780, *Standard for the Installation of Lightning Protection Systems,* National Fire Protection Association, 2004.

[20]   BS 6651, *Protection of Structures against Lightning,* 1992.

[21]   SS CP33, *Code of Practice for Lightning Protection,* 1996.

[22]   UL 96A, *Standard for Installation Requirements for Lightning Protection Systems,* 1998.

[23]   MIL-HDBK-1004/6, *Military Handbook, Lightning Protection,* Washington, DC: U.S. Government Printing Office, 1979.

[24]   API Recommended Practice 2003, 6th Edition, *Protection Against Ignitions Arising out of Static, Lightning, and Stray Currents,* American Petroleum Institute, 1998.

[25]   FAA-STD-019d, Department of Transport Federal Aviation Administration Standard, *Lightning and Surge Protection, Grounding, Bonding and Shielding Requirements for Facilities and Electronic Equipment,* 2002.

[26]   M 440.1-1, *DOE Explosives Safety Manual: Electrical Storms and Lightning Protection,* U.S. Department of Energy, 2002.

[27]   DA PAM 385-64, *Ammunition and Explosives Safety Standards,* U.S. Department of the Army, 1999.

[28]   NAVSEA OP 5, Volume 1, *Ammunition and Explosive Ashore—Safety Regulations for Handling, Storing, Production, Renovation and Shipping,* Department of the Navy, 1999.

[29]   Lacroix, B., and Calvas, R., *Earthing Systems Worldwide and Evolutions,* Cahier Technique No. 173, Schneider Electric.

[30] IEEE STD 141-1993, *Recommended Practice for Electric Power Distribution for Industrial Plants.*

[31] Calvas, R., and Lacroix, R., *System Earthings in LV,* Cahier Technique No. 172, Schneider Electric.

[32] IEEE STD 242-1986, *Recommended Practice for Protection and Coordination of Industrial and Commercial Power Systems.*

[33] Kato, J., Asakawa, A., Kuramoto, S., "Experimental Results For Effect Of Common Grounding In TT Power Distribution System At Customer Premises," in *Proceedings of the 19th International Wroclaw Symposium and Exhibition on Electromagnetic Compatibility,* Wroclaw, Poland, June 2008.

[34] IEC 60364-5-548, *Electrical Installations of Buildings—Part 5: Selection and Erection of Electrical Equipment, Section 548: Earthing Arrangements and Equipotential Bonding for Information Technology Installations.*

[35] ITU-T Recommendation K.27, *Bonding Configurations and Earthing Inside a Telecommunication Building,* 1996, International Telecommunication Union.

[36] IEC 60364-4-41, *Low-voltage Electrical Installations—Part 4-41: Protection for Safety—Protection against Electric Shock.*

[37] Uman, M. A., *The Lightning Discharge,* Academic Press, 1987.

[38] Frydenlund, M. M., *Lightning Protection for People and Property,* New York: Van Nostrand Reinhold, 1993.

[39] The Standards Institution of Israel, SI 1173, *Lightning Protection Systems for Buildings and Installations,* ICS Code: 91.120.40, 2003 (Draft, Hebrew).

[40] Hasse, P., *Overvoltage Protection of Low Voltage System,* 2nd Ed., Institution of Electrical Engineers, U.K.

[41] Tagg, G. F.,*Earth Resistances,* Pitman Publishing Corporation, 1964.

[42] Charlton, T., *Earthing Practice,* CDA Publication 119, Copper Development Association, 1997.

[43] Block, R. R., *The "Grounds" for Lightning and EMP Protection,* 2nd Ed., PolyPhaser Corporation, 1993.

[44] IEEE STD 81-1983, *Guide for Measuring Earth Resistivity, Ground Impedance, and Earth Surface Potentials of a Ground System.*

[45] Megger Limited, *Getting Down to Earth—A Practical Guide to Earth Resistance Testing,* 2005.

7

GROUNDING IN WIRING CIRCUITS AND CABLE SHIELDS

The mathematical theory of wave propagation along a conductor with an external coaxial return is very old, going back to the work of Rayleigh, Heaviside and J. J. Thomson.—S. A. Schelkunoff, 1934 [1,2]

7.1 INTRODUCTION: SYSTEM INTERFACE PROBLEMS

Signal interfacing by means of electrical cables and wiring constitutes an inseparable part of any electronic or electrical system. For this reason, the question of how grounding should be implemented extends from the individual assembly (the "black box") up to the level of a complete system.

Most interface problems traceable to improper grounding implementation are caused by lack of sufficient direction or coordination by overall system-cognizant personnel, who often espouse the philosophy of "If my system works, why should I worry about this grounding question?" From a compatibility approach, this philosophy can be disastrous.

Wiring harnesses and cabling are the primary avenues for the intrusion of electromagnetic interference (EMI) into electrical and electronic systems.* The physics of field-to-wire coupling of electromagnetic energy into wiring harnesses and cabling of

---

*Many excellent sources such as [1] discuss field and cable interaction mechanisms.

Grounds for Grounding. By Elya B. Joffe and Kai-Sang Lock
Copyright © 2010 The Institute of Electrical and Electronics Engineers, Inc.

469

susceptible systems has been a concern ever since devices such as radio receivers, microphones, and the like were invented.

For controlling such interference, particularly when little or no control is available on the interference source, techniques such as robust signal interface design, including balancing of power and signal lines; utilizing differential rather than single-ended interfaces; shielding of wiring harnesses, cabling, and equipment enclosures; increasing robustness of signal transmission; implementation of filtering and interference-suppression circuits; or combinations thereof are commonly applied.

Occasionally, a single one of these techniques is sufficient to reduce the undesired noise to an acceptable level but, general, a combination of techniques is necessary. The effectiveness of the above techniques is considerably dependent on proper implementation of grounding in the system and, particularly, on signal interfaces.

This chapter provides reasoning and guidance specific to grounding techniques for wiring harnesses and cabling signal grounding. Grounding of terminal protection devices is addressed in the following chapter.

7.2 TO GROUND OR NOT TO GROUND (CABLE SHIELDS)

"It depends. . . ." Whereas everyone would likely agree with such a reply, this is just about where the consensus ends. One of the all time dilemmas faced by electrical and electronic system designers and application engineers is that of where and how to ground cable shields. How will a grounded or ungrounded shield perform in any particular system configuration? Should the shield be terminated and/or grounded at one enclosure point or the cable end, at both ends, at intervals along the shield, continuously or not at all? How are the grounding and termination* recommendations affected by the shield characteristics? How should shield termination impedances be taken into consideration, if they should?

Literally thousands of scientific and engineering reports, documents, standards, and specifications have addressed these questions, and yet, when a typical application engineer is engaged in a system integration and installation activity, the answer to the question of "where (and how) to ground the shield" is desperately and repeatedly sought. Unfortunately, the number of replies to that question will, no doubt, be equal to or greater than the number of respondents. The primary source of this confusion is the natural tendency of people to relate new situations they encounter to past experience, or adopt solutions well beyond the boundaries of the new application. As an example, approaches that worked well for low-speed, high-level RS-232 links may not be acceptable for high-speed, low-level RS 644 (or LVDS) signaling. No single universal approach exists that will work for every case. Successful wiring harness and cable shield design and performance must correspond to the unique requirements and characteristics of the application, and the interference of concern.

---

*The term "shield termination" is used here to indicate a low-impedance connection of the shield to the reference structure or equipment enclosure, also commonly known as "shield grounding." Both terms are considered interchangeable within the context of this book. From the standpoint of the actual physical manner of connection, the term "termination" is indeed more appropriate.

A glance at Figure 7.1 easily demonstrates that there could literally be dozens of different responses to this question. The complexity of the situation is amplified when the question of the circuit grounding, as it relates to the shield grounding, is added.

When a circuit is grounded to a reference structure at two different points with unequal voltage potentials, an equalizing current must flow between the two points. This notorious "ground loop" often drives designers to the perception that multipoint shield grounding falls short, leading to a disastrous implementation.

Observe Figure 7.2(a), which depicts a situation in which a shielded single-ended conductor interconnects two circuits. In Circuit #1, the signal return plane (Sig. Rtn.) is connected to chassis. In Circuit #2, the signal return plane is floated from chassis. The signal wire is shielded, and the shield is grounded at both ends to the enclosures of the circuits. However, such a shield configuration will be of little avail. The radiating loop formed by the signal and its return is not wholly enclosed by the shield. The return current is allowed to flow outside the shield, thus the shield has little effect on the magnetic field emission from the signal current loop. Emission is controlled, in this case, only by the size of the loop area formed between the signal and the return conductor. Routing the return conductor within the shield, adjacent to the signal wire, would correct this flawed design.

In Figure 7.2(b), a different situation is presented. The signal returns in the two circuits are grounded to the chassis of their respective enclosures. The shield is terminated to the chassis of Circuit #1 only, and is left floating at Circuit #2, based on the misconception that if terminated at both ends, a "ground loop" will result. In this arrangement, the current is forced to return to Circuit #1 via the signal reference, forming an extremely large loop, again resulting in intensified magnetic field emissions. Terminating the shield to the enclosure or chassis of Circuit #2 would allow the higher frequency signal current to return via the shield rather than the signal reference, improving this design. Best practice would route the return inside the shield together with the signal. In this manner, although the return signal would still have a path via the reference structure, the lowest impedance for the return current back to the source would be via the intentional path routed with the signal, and not the reference structure.

In conclusion, without a clear understanding of the function of the shield, a flawed shielded cable configuration may result, counteracting the desired shielding action.

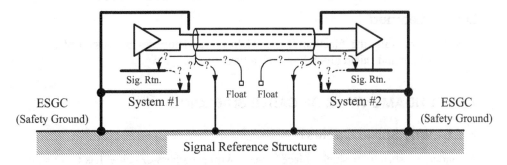

Figure 7.1. The dilemma of shield grounding.

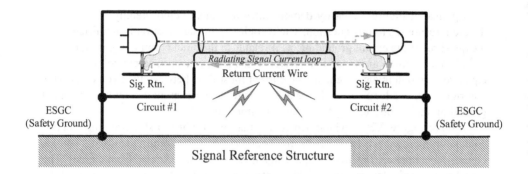

(a) Return Conductor Routed Outside the Shield Degrades Shielding Effectiveness

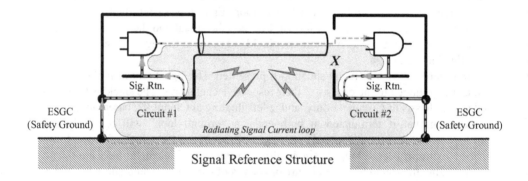

(b) Floated Shield Forced Return Current Flow through the Signal Reference Structure

Figure 7.2. Improper shield grounding results in increased radiation.

The following sections discuss the performance of a cable shield in balanced and unbalanced wiring circuits for preclusion of electric and magnetic field interactions.

Lesson Learned

- Understanding the function of the shield is mandatory for preclusion of cable shielding and shield grounding design flaws.

7.3 FUNDAMENTALS OF CABLE SHIELDING

7.3.1 Why Shield Cables?

A conductive sheath, or shield, placed over a wiring harness or cable, has two main purposes: to confine electromagnetic energy radiating from the enclosed wiring and to preclude external electromagnetic energy from coupling to the enclosed wiring. Shielding is thus the decoupling of radiated electromagnetic energy interactions between system wiring and its surroundings.*

Electromagnetic interference may be transferred from one circuit to another by an interconnecting cable. Interference may be radiated from wiring harnesses or cabling or may be coupled into a cable from ambient external electromagnetic fields. Once interference has been transferred into an electronic or electrical system, it can be conducted throughout the system by interconnecting wiring harnesses or cabling. Owing to the proximity between wires in cable bundles, crosstalk may also occur as a result of capacitive and inductive coupling between the wires. Proper wiring harness or cabling shielding can be used to mitigate all of the above.

7.3.2 Fundamental Shielding Mechanisms

Whenever an electromagnetic wave is incident upon a metal surface, attenuation of electromagnetic fields occurs due to two distinct mechanisms: (1) reflection (due to impedance mismatches) of the interference wave at the air–metal boundary as the incident electromagnetic wave strikes the metal surface and reflection at the metal–air boundary as the interference wave emerges from the metal shield, and (2) absorption of the interference wave in passing through the metal shield between the two boundaries. The remainder propagates beyond the metal barrier (transmission) (Figure 7.3).

This shielding effect can be visualized as being the result of the incident electric and magnetic fields inducing charges and a current flow on the shield surface, respectively. These induced charges and currents are of such a polarity and direction that their associated electric and magnetic fields oppose the incident fields, reducing the EM field intensity beyond the shield.[†]

Using transmission line theory, the shielding effectiveness, SE, of a solid, infinitely large shielding surface can, therefore, be expressed (in dB) as [1,4]:

$$SE_{dB} = \underbrace{20 \cdot \log\left(e^{\gamma \cdot l}\right)}_{A} + \underbrace{20 \cdot \log\left(\frac{1}{\tau}\right)}_{R} + \underbrace{20 \cdot \log\left(1 - \Gamma \cdot e^{2\gamma \cdot l}\right)}_{B} \qquad (7.1)$$

where:
A = Absorption (penetration) loss of the material (dB)
R = Single reflection loss from both surfaces of the sheet (dB)
B = Multiple reflection correction term (dB)
and:
Γ = Reflection coefficient
γ = Propagation coefficient
τ = Transmission coefficient

---

*A discussion of cable shield characteristics and shielding mechanisms is beyond the scope of this book. [1], [3], and [4] provide detailed information on shielding mechanisms.
[†]Although not lending itself to efficient calculation of the degree of shielding provided by a particular shield, this concept does provide a useful understanding of the physical effect of shielding.

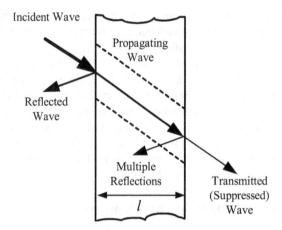

Figure 7.3. Fundamental shielding mechanisms.

Note that the above formulation, as well as the following discussion, assumes a planar incident wave front (i.e., far-field interaction).

The suppression of the field due to reflection is called reflection loss, resulting from the mismatch between the intrinsic impedance of the incident field* and the surface impedance of the shielding barrier [1,4].

It is convenient to make a distinction between the initial reflections from both surfaces of the shield and subsequent reflections that may occur at these surfaces. These effects are called the single-reflection loss, R, and multiple-reflection correction term, B, respectively. For thick shields, when absorption loss is greater than about 15 dB, B is negligible.

Absorption or penetration loss results from energy dissipation within the shield and, at any given frequency, is proportional to the shield thickness. Penetration of higher frequency current induced on the surface into the interior of the metallic barrier is limited due the skin effect,[†] thus it is forced to flow on the surface (or skin) of the shield facing the field source. Since the skin depth is inversely proportional to frequency,[‡] absorption constitutes the chief suppression mechanism at higher frequencies, but will be of little significance at lower frequencies. Effective low-frequency shielding against electric fields (or high-impedance waves) must be primarily achieved, therefore, by reflection [5]. Effective shielding against low-frequency magnetic fields (or low-impedance waves) may be achieved by the use of materials exhibiting relatively high permeability.

---

*The intrinsic wave impedance Z_W is defined as the ratio between the E-field and H-field components of the incident field, $Z_W = E/H$. Dimensions of the wave impedance are in ohms (Ω). Predominantly electric field waves are, therefore, called "high-impedance waves," whereas predominantly magnetic field waves are called "low-impedance waves."
[†]Skin effect is discussed in Chapter 2.
[‡]The skin depth of copper (Cu) at 1 MHz is 0.1 mm, approximately, falling proportionally to f^{-1} thereafter.

The performance of the shield significantly differs between incident E-field (high-Z_W) and H-field (low-Z_W) waves. Any reasonable conductive surface will provide moderately good E-field suppression since large reflection losses are easily obtained. H-field reflection, on the other hand, requires extremely low conductivity of the surface metal in order to have any effect on the incident field. Practical shielding against magnetic fields, relying on classical shielding mechanisms, therefore depends primarily on absorption losses.* Absorption loss is the same for electric and magnetic fields.

Lessons Learned

- The performance of shields significantly differs with predominantly E-fields and H-fields.
- Reasonably conductive surfaces will provide moderately good E-field suppression owing to the effective field reflection.
- Shielding against magnetic fields, relying on classical shielding mechanisms, depends primarily on absorption losses.

7.3.3 Configuration of Shielded Cables

To aid in the understanding of the function of a cable shield, observe the configuration depicted in Figure 7.4 [6,7]. This cable could be considered as a combination of two transmission-line systems with the shield acting as the boundary between the two. Circuit #1 represents a functional circuit, carrying the intended (differential-mode) current and exhibiting controlled designed parameters, in particular, source, load and characteristic impedances, Z_{S1}, Z_{L1}, and Z_{01}, respectively. Circuit #1 can be "floated" or grounded, depending on the circuit configuration.

Circuit #2 makes up a transmission line formed by the shielded cable and the signal reference structure. The characteristics of this circuit are controlled to a significantly lesser extent, in particular with respect to terminating impedances, Z_{S2} and Z_{L2}, and characteristic impedance, Z_{02}, all of which are governed by the system's geometry and layout.

The effectiveness of shielding performance serves as a figure of merit for the degree of energy suppression from the external circuit (Circuit #2) to the internal circuit (Circuit #1) for EMI immunity and emission considerations. Normally, a shield will be expected to accomplish both objectives, perhaps with differing efficiency.

7.3.3.1 Balanced and Unbalanced Shielded Signal-Interface Cables.
Circuit #1 in Figure 7.4 may be balanced or unbalanced, depending on its grounding configuration.[†] The performance of the shield will differ significantly in balanced and

---

*More effective shielding against low-frequency magnetic fields can be achieved by using annealed, high-permeability shields. However, the shielding mechanism in this case is not based on classical absorption.
[†]A discussion of the concept of circuit balancing is presented in Chapter 2.

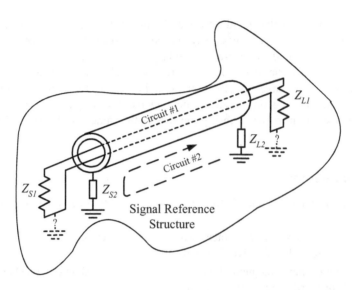

Figure 7.4. A cable shield constitutes a barrier between two transmission-line systems.

unbalanced circuits. In unbalanced shielded cables, such as coaxial, the shield constitutes the intended return path of the transmission line. In coaxial cables, a return conductor surrounds an inner signal conductor (Figure 7.5). The return current path in such a system will depend on the grounding configuration of the shield circuit (Circuit #2):

1. When the shield is grounded at one end, all current is forced to return through the shield. The circuit in this case could be considered to be balanced in the sense that all signal current is forced to return through the shield, which acts as the return conductor [Figure 7.5(a)].
2. When the shield is grounded at both ends, high-frequency (HF) current will still return through the shield, which constitutes the path of least impedance. Some portion of low-frequency (LF) current may flow through the signal reference structure, rather than the shield, making this similar to an unbalanced circuit [Figure 7.5(b)].

Coaxial cables constitute excellent examples of unbalanced shielded cables.

On the other hand, a shielded cable will be considered balanced when the circuit consists of signal and return conductors, independent of the shield, such as that depicted in Figure 7.6(b). Circuit #1 will be truly balanced if the functional (intentional) current flowing in Circuit #1 circulates between the source and load impedance Z_{S1} and Z_{L1}, respectively, through Circuit #1 wiring only. The cable shield does not carry any intentional circuit current. This condition will be fulfilled if:

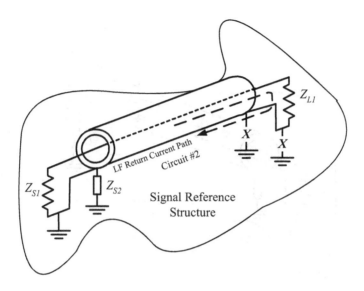

(a) Shield Grounded One End, at Most; Circuit is Still Balanced

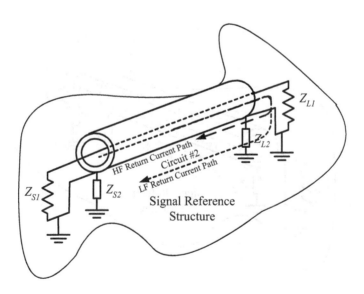

(b) Shield Grounded Both Ends; Unbalanced Circuit

Figure 7.5. In unbalanced shielded cables, the shield constitutes the return signal conductor, forming part of the active circuit.

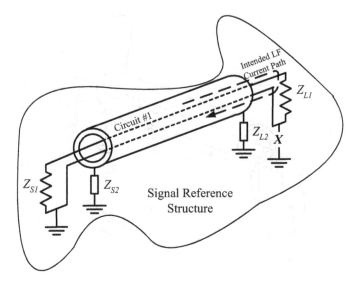

(a) Circuit Grounded One End, at Most

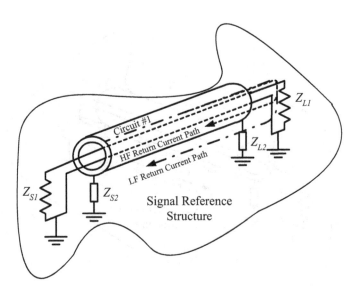

(b) Shield Grounded Both Ends

Figure 7.6. In balanced shielded cables, the shield should not carry intended signal current.

1. Circuit #1 carries low-frequency current and the circuit is grounded at one side, at most, forcing the current to return through the return conductor only [Figure 7.6(a)], regardless of the shield grounding configuration. The return current is forced to flow via the return conductor owing to the "floating" of the load).
2. Circuit #1 carries high-frequency current, so that even if grounded at both ends, the current will return through the intended return conductor because this is the path of least impedance. If Circuit #1 were to be grounded at both ends, some lower frequency current could return through the signal reference structure and a fraction of the current could even find its way through the shield, making Circuit #1 unbalanced [Figure 7.6(b)].

Examples of balanced shielded cables include twisted shielded pairs (TSP) or twisted shielded cables (TSC) comprised of multiple wiring pairs enclosed within a common shield.

The difference between the two configurations is fundamental for understanding the manner in which the cable shield functions for suppression of interference and the effect of shield grounding topology on the performance of cable shields.

Lessons Learned

- In balanced shielded cable circuits, the shield should not carry any intended signal current.
- In unbalanced cable circuits, the shield constitutes part of the signal current path providing a return path for the signal current.

7.3.3.2 *Transmission-Line Model of Shielded Cables.* Consider the circuit in Figure 7.7(a), which illustrates a physical representation of an (unbalanced) coaxial cable above the signal reference. The equivalent circuit is shown in Figure 7.7(b) [9].

Note that in the equivalent circuit, two reference conductors are portrayed: the signal reference structure (commonly called the "ground plane") and the cable shield, which in an unbalanced transmission line also serves as the return conductor. Observe that this configuration constitutes an equivalent multiconductor transmission line (MTL)* consisting of three conductors.

The general expressions for the interactions between the conductors across an infinitesimal section, Δx, in such a MTL system, assuming transverse electromagnetic (TEM) propagation, are in the form [10]

$$\frac{d}{dx}\widehat{\mathbf{V}}(x) = -(\mathbf{R} + j\omega\mathbf{L})\widehat{\mathbf{I}}(x)$$

$$\frac{d}{dx}\widehat{\mathbf{I}}(x) = -j\omega\mathbf{C}\widehat{\mathbf{V}}(x)$$

(7.2)

*General concepts of transmission lines are discussed in Chapter 2.

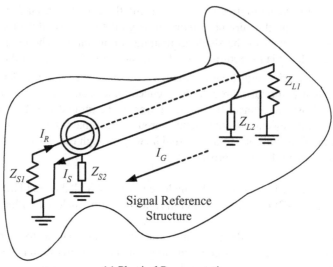

(a) Physical Representation

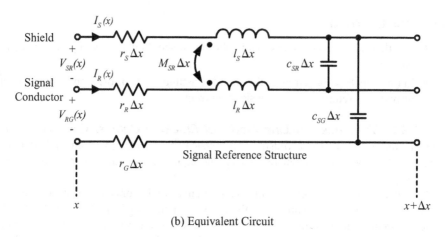

(b) Equivalent Circuit

<u>Figure 7.7.</u> Physical representation and transmission-line equivalent circuit of a shielded wire.

where

$$\widehat{\mathbf{V}}(x) = \begin{bmatrix} V_S(x) \\ V_R(x) \end{bmatrix}; \quad \hat{\mathbf{I}}(x) = \begin{bmatrix} I_S(x) \\ I_R(x) \end{bmatrix} \tag{7.3}$$

and

$$\mathbf{R} = \begin{bmatrix} r_S + r_G & r_G \\ r_G & R_R + r_G \end{bmatrix}; \quad \mathbf{L} = \begin{bmatrix} l_S & l_{RS} \\ l_{RS} & l_R \end{bmatrix}; \quad \mathbf{C} = \begin{bmatrix} c_{SG} + c_{RS} & -c_{RS} \\ -c_{RS} & c_{SG} + c_{RS} \end{bmatrix} \tag{7.4}$$

The losses in the medium between the conductors are neglected in this approximation (i.e., $G = 0$). The voltages are defined with respect to the signal reference and the currents are assumed to return through the reference structure. Note that the capacitance between the shielded conductor and the reference, C_{RG}, is excluded from the model, since the shield effectively eliminates capacitive coupling between the conductor enclosed by the shield and the signal reference.

Expressions for the per-unit-length parameters developed in [10] are as follows:

$$r_S = \frac{r_b}{NC\cos\theta_w}; \ \Omega/m \tag{7.5}$$

where r_b is the resistance of the strand, θ_w is the weave angle, N is the number of strands per carrier, and C is the number of carriers (Section 7.4). Assuming that the current is uniformly distributed over the shield's cross section, the shield's DC resistance is approximated as

$$r_S = \frac{1}{\sigma 2\pi r_{sh} t_{sh}}; \ \Omega/m \tag{7.6}$$

where r_{sh} is the shield interior radius and t_{sh} is the shield thickness.

The per-unit-length self-inductances of the wire and the shield, respectively, both placed at a height, h_G, above the signal reference structure are given by

$$l_R = \frac{\mu_0}{2\pi}\ln\left(\frac{2h_G}{r_{wR}}\right); \ H/m \tag{7.7}$$

and

$$l_S = \frac{\mu_0}{2\pi}\ln\left(\frac{2h_G}{r_{sh}+t_{sh}}\right); \ H/m \tag{7.8}$$

The mutual inductance between the inner wire at a height, h_R, above the signal reference structure and the shield, m_{SR}, is

$$m_{SR} = \frac{\mu_0}{2\pi}\ln\left(\frac{2h_R}{r_{sh}+t_{sh}}\right) = l_S; \ H/m \tag{7.9}$$

The result presented in Equation (7.9) is fundamental and of utmost importance, and merits further explanation [1].

Consider a tubular conductor carrying a uniform axial current (Figure 7.9). If the cavity in the tubular conductor is concentric with the outer surface of the conductor, the magnetic field vanishes within the cavity and exists only outside the conductor. When another conductor is placed inside the cavity, thus forming a coaxial cable, all the magnetic flux Φ_S produced by the current on the shield I_s encloses the inner conductor as well. By definition, the inductance of the shield is equal to [Figure 7.8(a)]

$$L_S = \frac{\Phi_S}{I_S} \qquad (7.10)$$

Since no magnetic flux from the shield current exists within the shield, magnetic flux encircling the shield equally encircles the inner conductor ($\Phi_R = \Phi_S = \Phi$); therefore, the mutual inductance between the shield and the inner conductor is, similarly, equal to [Figure 7.8(b)]

$$M_{SR} = \frac{\Phi_R}{I_S} \qquad (7.11)$$

It follows, therefore, that the mutual inductance between the shield and the inner conductor is equal to the self-inductance of the shield :*

$$M_{SR} = L_S = \frac{\Phi}{I_S} \qquad (7.12)$$

Finally, the capacitance between the inner wire and the shield, C_{SR}, is

$$c_{SR} = \frac{2\pi\varepsilon_0\varepsilon_r}{\ln\left(r_{sh}/r_{wR}\right)} \qquad (7.13)$$

C_{SG} can be obtained using the reciprocal relationship for conductors in a homogenous medium:

$$\begin{bmatrix} c_{SG} + c_{RS} & -c_{RS} \\ -c_{RS} & c_{SG} + c_{RS} \end{bmatrix} = \mu_0\varepsilon_0 \begin{bmatrix} l_S & l_{RS} \\ l_{RS} & l_R \end{bmatrix}^{-1} \qquad (7.14)$$

resulting in

$$\mathbf{LC} = \mu_0\varepsilon_0 \begin{bmatrix} 1 & 0 & 0 \\ 0 & \varepsilon_r & \varepsilon_r - 1 \\ 0 & 0 & 1 \end{bmatrix} \qquad (7.15)$$

The above per-unit-length formulations are applicable to circuits consisting of electrically long conductors. In low-frequency situations, the conductors may be considered electrically short, allowing the application of "quasistatic" approximations,[†] and the per-unit-length parameters can be replaced by equivalent lumped circuit elements.[‡]

---

*Based on reciprocity, the opposite also holds.

[†]An interaction of a system with its surroundings that is carried out so slowly that it remains arbitrarily close to equilibrium at all times is said to be quasistatic. In the context of this book, the term "quasistatic" refers to low-frequency electric/magnetic processes to which static electricity/magnetism models can be reasonably applied.

[‡]A conductor may be considered "electrically long" when the cable length exceeds $\lambda/10$, where λ is the wavelength associated with the frequency of concern; in this case, transmission-line effects must be taken into account and lumped-element circuit analysis cannot be applied. Note that some practitioners would choose a longer wire length than $\lambda/10$; the above definition is, therefore, somewhat arbitrary.

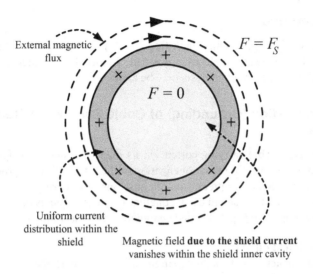

(a) No Inner Conductor: Self-Inductance of the Shield

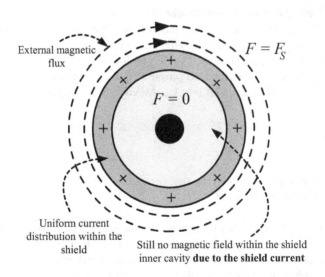

(b) Inner Conductor Present: Mutual Inductance
between Shield and Inner Conductor

Figure 7.8. Magnetic field produced by current in tubular and concentric conductors.

Lessons Learned

- In electrically large circuits (high frequencies), per-unit-length (distributed) circuit analysis is used; in electrically small circuits (low frequencies), equivalent lumped-element circuit analysis may be applied.

7.3.4 Termination (Grounding) of Cable Shields—A Qualitative Discussion

Consider a coaxial cable carrying current on its internal conductor (Figure 7.9). This model is used henceforth for the rationalization of the effect of shield grounding on the performance of the shield for preclusion of emissions from wiring.

The performance of the shield can be qualitatively explained using the simplified illustrations in Figure 7.10 [1]. A detailed quantitative analysis follows in Section 7.3.5.1.

Case (a): No Shield on Wire
Figure 7.10(a) depicts a nonshielded, single-ended long wire the loads of which are connected at their ends to the signal reference through infinitely small termination impedances Z_{S2} and Z_{L2}.* Electric field flux emerges radially to the wire, whereas magnetic flux surrounds the wire.†

Case (b): Floating (Ungrounded) Shield Placed on Wire
In Figure 7.10(b), a shield surrounds the conductor; however, it remains totally floating (i.e., not connected to a common reference at either end) at both ends. This situation offers no improvement compared to the previous one.

Why must the shield be terminated (grounded) at least at one end? The answer lies with Gauss's Law for Electric Fields (a.k.a. Maxwell's first equation):

$$\nabla \cdot \vec{D} = \rho \quad \text{or} \quad \oint_S \vec{D} \cdot d\vec{s} = \int_V \rho \, dv \qquad (7.16)$$

where:
$\vec{D}$ = electric displacement (or electric flux density) (C/m²)
ρ = volume (V) charge density (C/m³)

Gauss's Law simply states that the net electric displacement emerging from a closed surface is equivalent to the net positive charge enclosed by the surface. If there are no electric charges within an arbitrary hollow, perfectly electrical conductor (PEC) metallic enclosure (otherwise called a conductor with a PEC cavity), no electric field can exist within that cavity, even when external electrostatic fields are incident on its external metallic surface. It follows therefore, that if a cavity is completely enclosed by a PEC enclosure, no static distribution of electrical charges (or fields) outside the metallic en-

---

*A "single-ended wire" actually implies that the return current path exists at an infinitely large separation from the current-carrying conductor.
†A uniform field flux distribution exists, in practice, for infinitely long wires only, where the "fringing effect" at the ends of the line can be neglected.

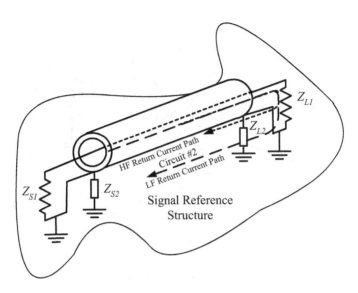

Figure 7.9. Signal and return paths in a coaxial cable.

closure can ever produce any fields inside the enclosed cavity. This, in fact, explains the basic principle of electrostatic shielding provided by a "Faraday cage" [8].

The reverse is not true, however. Any charge that resides inside a completely enclosed PEC surface will induce an equal but opposite charge on the internal surface of the enclosure. This charge will, in turn, induce an electric charge equal in magnitude to that enclosed to appear on the external surface of the enclosure. This is a direct and unavoidable consequence of Gauss's Law. Thus, when the shield is left "floating" [i.e., not connected to a 0 V reference at either end (Figure 7.10(b)], charges contained inside the shield will produce a nonzero electric displacement flux emerging from the enclosure. If the shield is connected to a common reference, the charge on the shield can flow off the shield without causing buildup. The short version of the foregoing is that a PEC enclosure simply does not work as an electrostatic/electric shield for internal charge distributions if not connected at least at one end to a common reference!

Connecting the shield to a 0 V reference, at least at one end, is obviously necessary for a PEC enclosure to effectively shield the external world from flux emerging from the enclosed charge. In addition to providing a means whereby the charge residing on the external surface of the shield can migrate away from the shield, connecting the shield to a common reference also forms a path for an oppositely polarized charge that may exist on the 0 V reference to neutralize that on the external surface of the shield. As a result, the net total charge enclosed by the shield is equal to zero [Figure 7.11(b)].

Case (c): Shield Placed on Wire Connected to a 0 V Reference at One End Only
With a PEC shield placed around the wire, connected to a 0 V reference at one end [Figure 7.10(c)], and applying low-frequency assumptions, no electric field will emerge from the metallic shield boundary enclosing the wire.

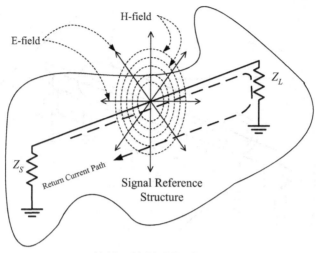

(a) Nonshielded Conductor

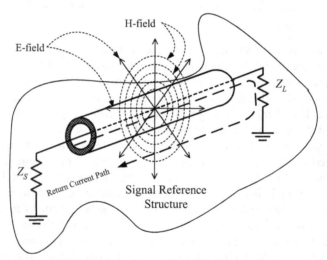

(b) Shielded Conductor; Shield Not Terminated

Figure 7.10. Contribution of the shield for suppression of radiated emissions from a current-carrying wire.

The magnetic field is not affected by the shield and penetrates the shield, experiencing no attenuation [Figure 7.10(b)]. Gauss Law for Magnetic Fields (a.k.a. Maxwell's second equation) [Equation (7.17)] reveals that there can be no magnetic charges to counteract the magnetic fields generated by the current on the wire inside the shield:

$$\nabla \cdot \vec{B} = 0 \qquad (7.17)$$

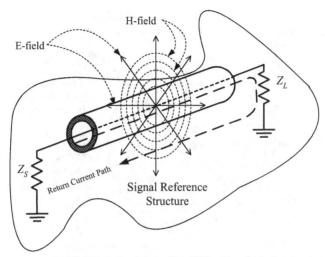

(b) Shielded Conductor; Shield Not Terminated

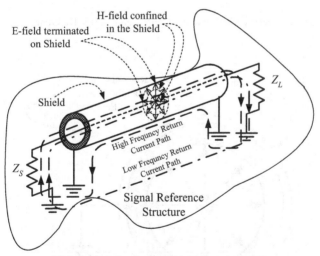

(d) Shielded Conductor; Shield Terminated to
Reference Structure at Both Ends

Figure 7.10. *Continued.*

where $\vec{B}$ = magnetic flux density (Tesla). In 7.3.2 it was demonstrated using shielding theory that the absorption and reflection from a thin conductive metallic surface are of little consequence to low-frequency magnetic fields.

Case (d): Shield Placed on Wire Connected to a 0 V Reference at Both Ends
When the shield is connected to a 0 V reference at both ends [Figure 7.10(d)], return current now flows through the shield. If the signal return current is made to flow en-

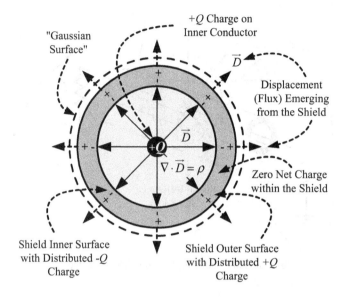

(a) Electrostatic Shield not Connected to a 0 V Reference

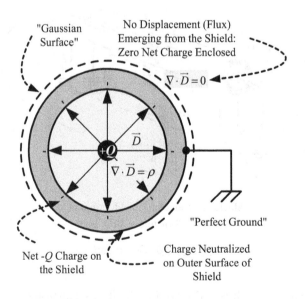

(b) Electrostatic Shield Connected to a 0 V Reference

<u>Figure 7.11.</u> The effect of connecting an electrostatic shield to a common reference at least at one end.

tirely through the shield, equal but opposite magnetic flux is produced by this current, counteracting the magnetic flux emerging from the inner conductor, which is unaffected by the presence of the shield. This flux cancellation effect minimizes magnetic field emissions.

High-frequency current will flow back to its source on the inner surface of the shield, this path being the path of least impedance. Note, however, that at lower frequencies, the majority of the return current will not flow through the shield but rather through the signal reference, which for low-frequency current constitutes the path of least impedance. Consequently, poor suppression of low-frequency magnetic field emission (and high-field pickup) effects may be observed.

When the load, Z_L, as well as the shield at the load end are floated and are connected as depicted in Figure 7.12, this difficulty is overcome. The lower frequency return current is forced to return through the shield, and flux cancellation occurs. Note that the important difference between this and the configuration shown in Figure 7.10(c) is the connection of the return side of the circuit to the shield, and not to the common reference. This is a typical configuration for headset and audio circuits that cannot be connected to a common reference at the headset termination.

Lessons Learned

- Electric fields are much easier to suppress than magnetic fields.
- For the purpose of electric (E) field suppression, the shield acts solely as a boundary ("Faraday cage") between the internal and external zones, and should not carry any intentional current.

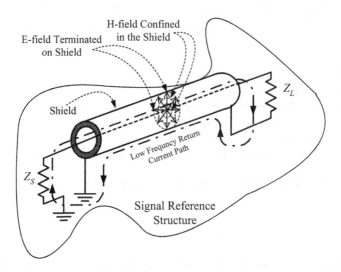

Figure 7.12. Floating both one end of the shield and load, with the load return connected to the shield, ensures flux cancellation for effective H-field suppression.

- Connecting a shield to a 0 V reference at least at one end is mandatory for suppressing low-frequency (quasistatic) E-field coupling.
- The key to shielding against magnetic field emission and pickup is reducing the loop area of the circuit. For shielding against high-frequency magnetic field emission (and pickup), the shield should be connected to a common reference at both ends. The shield's effectiveness in low-frequency magnetic flux cancellation is directly determined by the portion of the return current flowing through the return path.

7.3.5 Termination (Grounding) of Cable Shields—A Quantitative Discussion

Following the discussion presented above, a quantitative analysis of the effect of the shield and its termination to a reference structure at one or both ends on electric and magnetic field coupling now follows. In both cases, a shielded wire is assumed to be immersed in an electromagnetic field (Figure 7.13).

7.3.5.1 Shielding Against Electric Fields Interactions.
Current induced on the surface of a shield, I_S, due to an incident electromagnetic field (Figure 7.13), creates a voltage drop across the shield's series impedance, Z_S, which comprises the shield's series resistance and inductance. Some capacitance will still exist between the shield and the inner conductor or conductors in the case of a multiconductor cable* [1].

This configuration is illustrated and modeled in Figure 7.14(a) and (b), respectively. Using the equivalent circuit shown in Figure 7.14(b) for evaluating the voltage across the wire terminating impedance (noting that it is valid at lower frequencies only) we obtain [4]:

$$\frac{V_L}{V_S} = \frac{Z_G}{Z_S + Z_G} \cdot \frac{R_L}{1/j\omega C + R_L} = \frac{Z_G}{j\omega L_S + R_S + Z_G} \cdot \frac{j\omega C R_L}{1 + j\omega C R_L} \quad (7.18)$$

where:
V_L = voltage across the load (Volts)
V_S = voltage at the source (Volts)
Z_G = impedance from the shield to the reference structure (ohms)
Z_S = impedance of the shield (ohms)
R_L = resistance of the load (ohms)
C = capacitance between the shield and the inner conductor (farads)

Equation (7.18) can be approximated at lower frequencies as

$$\left.\frac{V_L}{V_S}\right|_{Lo-F} \cong Z_G \cdot j\omega C \cdot \frac{R_L}{R_S} \quad (7.19)$$

---

*This analysis assumes electrically small circuits; in electrically long circuits, per-unit-length formulations must be used, whereby the circuit elements will be computed using the per-unit-length values (e.g., c_{SR}, capacitance per unit length) multiplied by the conductor's length, ℓ. For instance, the total wire-to-shield capacitance, C_{SR}, along the cable length, ℓ, would thus be: $C_{SR} = c_{SR}\ell$.

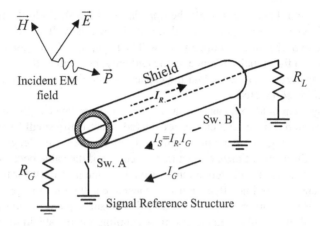

Figure 7.13. A shielded cable immersed in an electromagnetic field.

and at higher frequencies as

$$\left.\frac{V_L}{V_S}\right|_{Hi-F} \cong \frac{Z_G}{\omega L_S} \tag{7.20}$$

The crossover frequency between the high- and low-frequency band occurs at

$$\omega_{CO} \cong \frac{1}{\sqrt{(L_S/R_S)/(C \cdot R_L)}} \tag{7.21}$$

The peak voltage coupled onto the load due to the capacitive mechanism is, therefore,

$$\left.\frac{V_L}{V_S}\right|_{Peak} = Z_G \cdot \sqrt{(L_S/R_S) \cdot (C \cdot R_L)} \tag{7.22}$$

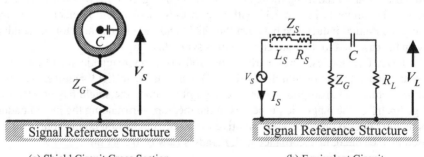

(a) Shield Circuit Cross Section   (b) Equivalent Circuit

Figure 7.14. A model describing the shield performance against electric fields.

Bearing in mind that Z_G represents the impedance of the shield to the reference structure, it follows from Equation (7.22) that if Z_G is low, the voltage, V_{LO}, across the terminating load of the inner conductor, R_L, will be negligible. Consequently, if the shield is connected to the signal reference at one end only, be it either the source or load, the shield voltage reduces to zero, and the capacitive coupling of interference due to the electric field is eliminated.

This argument holds only as long as the intrinsic impedance of the shield, Z_S, is sufficiently low and if the circuit can be considered "electrically small," such that approximately zero voltage is maintained uniformly along the entire length of the shield. However, if these conditions are invalid, particularly if the cable becomes "electrically long," exceeding approximately a tenth of the wavelength ($\lambda/10$) for the frequency of concern, transmission line effects must be considered, and the voltage distribution will no longer be zero uniformly along the shield. The subsequent shield current distribution will result in an enhanced electric field coupling via the shield to inner conductor capacitance (per unit length), c_{SR}, onto the internal wiring.

It follows that for ensuring effectiveness of shields over long cables, termination of the shield at one end only is insufficient. Multiple shield terminations to a reference structure at intervals along the cables' length spaced no more than $\lambda/10$ are necessary.*

Lessons Learned

- Effective shielding against E-fields is achieved by using highly conductive shielding materials, and low-impedance termination of the shield.
- For shielding against E-fields at lower frequencies only ($\ell < \lambda/10$), single-point and low-impedance termination of the shield to a reference structure is sufficient.
- For shielding against E-fields at higher frequencies ($\ell > \lambda/10$), shield terminations at both ends, as well as at multiple points along the cables' length is necessary.

7.3.5.2 Shielding for Control of Magnetic Fields Coupling onto Wiring.
The suppression afforded by the shield for control of magnetic field coupling onto wiring is analyzed using the equivalent circuit in Figure 7.15, which is a representation of the circuit in Figure 7.13 with respect to inductive field interactions. In this circuit, the mutual inductance between the shield and the inner conductor (or conductors, in the case of a multiconductor cable) is symbolized by M_{SR}.

At lower frequencies, particularly below 100 kHz,[†] nonmagnetic shield materials, such as copper braids or aluminum foils offer no magnetic field suppression, since both reflection and absorption losses are negligible. Since the shield has no effect on the geometry or the magnetic properties of the medium surrounding the inner conductors, neither does it preclude electromotive force (emf) induction on the conductor enclosed by the shield, in accordance with Farady's Law of Induction:

---

*Further discussion of low frequency vs. high frequency shield grounding can be found in Section 7.3.6.2.
[†]The skin depth of copper at $f = 100$ kHz is approximately 0.21 mm.

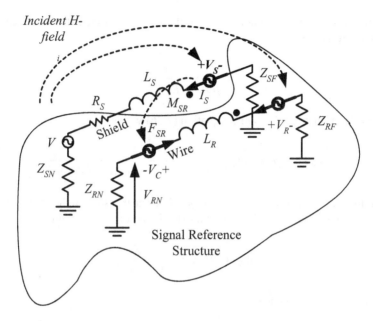

Figure 7.15. The effect of the shield on suppression of inductive/magnetic coupling.

$$emf = \oint_C \vec{E} \cdot d\vec{l} = -\frac{\partial}{\partial t} \int_S \vec{B} \cdot d\vec{s} = -\frac{\partial \Phi_i}{\partial t} \tag{7.23}$$

Assuming the shield is connected to a common reference at both ends, incident magnetic flux, Φ_i, will also induce an emf in the shield-reference circuit, which, in turn, will produce a shield current, I_S. The shield current thus produced will create magnetic flux, which tends to induce a reverse polarity emf on the internal conductors, due to the mutual inductance between them. By virtue of this process, the shield is able to suppress inductive or magnetic field coupling.

From Equation (7.23), the emf coupled onto the inner conductor due to the current on the shield can be expressed as

$$V_C = -j\omega M_{SR} I_S \tag{7.24}$$

The voltage developed across the near-end (Z_{RN}) and far-end (Z_{RF}) loads of the inner conductor due to the induced emf, a consequence of the voltage divider comprising of both loads, is [1,4,10]

$$V_{RN} = -V_C \cdot \frac{Z_{RN}}{Z_{RN} + Z_{RF}} = -j\omega M_{SR} I_S \cdot \frac{Z_{RN}}{Z_{RN} + Z_{RF}} \tag{7.25}$$

$$V_{RF} = -j\omega M_{SR} I_S \cdot \frac{Z_{RF}}{Z_{RN} + Z_{RF}} \text{ and } V_{RN} = -j\omega M_{SR} I_S \cdot \frac{Z_{RN}}{Z_{RN} + Z_{RF}} \tag{7.26}$$

In similar fashion, the current induced on the shield due to the shield-induced emf, V_S, can be expressed as

$$I_S = \frac{V_S}{j\omega L_S + R_S + Z_{SN} + Z_{SF}} \tag{7.27}$$

Substituting Equation (7.27) into Equation (7.26) yields

$$V_{RN} = -j\omega M_{SR} V_S \frac{Z_{RN}}{\left(j\omega L_S + R_S + Z_{SN} + Z_{SF}\right)\cdot\left(Z_{RN} + Z_{RF}\right)} \tag{7.28}$$

Assuming from Equation (7.12) that $M_{SR} = L_S$ or nearly so at higher frequencies ($\omega \rightarrow \infty$), the voltage induced on the near-end load [Equation (7.28)] reduces to

$$\left|V_{RN}\left(HiF\right)\right| = V_S \frac{Z_{RN}}{\left(Z_{RN} + Z_{RF}\right)} \tag{7.29}$$

whereas at lower frequencies

$$\left|V_{RN}\left(LoF\right)\right| = \omega M_{SR} V_S \frac{Z_{RN}}{\left(R_S + Z_{SN} + Z_{SF}\right)\cdot\left(Z_{RN} + Z_{RF}\right)} \tag{7.30}$$

Rearranging Eq. (7.28), again using the assumption that $M_{SR} = L_S$ [Equation (7.12)] or nearly so, yields

$$\left|V_{RN}\right| = V_S \cdot \frac{Z_{RN}}{\left(Z_{RN} + Z_{RF}\right)} \cdot \left[\frac{j\omega}{j\omega + \frac{\left(R_S + Z_{SN} + Z_{SF}\right)}{L_S}}\right] = V_S \cdot \frac{Z_{RN}}{\left(Z_{RN} + Z_{RF}\right)} \cdot \left[\frac{j\omega}{j\omega + \frac{Z'_S}{L_S}}\right] \tag{7.31}$$

The notation Z'_S used in Equation (7.31) represents the sum of the series impedances of the entire shield-reference loop. The shield cutoff frequency (ω_c) is defined as

$$\omega_c = \frac{Z'_S}{L_S} = \frac{R_S + Z_{SN} + Z_{SF}}{L_S}, \text{ or } f_c = \frac{Z'_S}{2\pi L_S} = \frac{R_S + Z_{SN} + Z_{SF}}{2\pi L_S} \tag{7.32}$$

For negligible shield terminating impedances (Z_{SN}, $Z_{SF} \rightarrow 0$ and, thus, $Z'_S \rightarrow R_S$), Equations (7.31) and (7.32) reduce to

$$\left|V_{RN}\right| = V_S \cdot \frac{Z_{RN}}{\left(Z_{RN} + Z_{RF}\right)} \cdot \left[\frac{j\omega}{j\omega + \frac{R_S}{L_S}}\right] = V_S \cdot \frac{Z_{RN}}{\left(Z_{RN} + Z_{RF}\right)} \cdot \left[\frac{j\omega}{j\omega + \frac{R_S}{L_S}}\right] \tag{7.33}$$

and

$$\omega_c = \frac{R_S}{L_S}, \text{ or } f_c = \frac{R_S}{2\pi L_S} \tag{7.34}$$

respectively.

A plot of Equation (7.33) is provided in Figure 7.16, which displays the efficiency of the shield versus frequency, for magnetic interactions.

From this plot, it is evident that the load noise voltage is zero at DC and increases to almost V_S at higher frequencies* (i.e., five times the shield cutoff frequency). Therefore, if current is allowed to flow on the shield, a voltage must be induced onto the inner conductor(s) approaching the shield voltage at frequencies significantly higher than the cutoff frequency.

Considering that the total voltage on the load due to an external incident magnetic field is the difference between directly induced voltage on the inner conductor (V_R in Figure 7.15) and the voltage coupled from the shield onto the inner conductor (V_C), and remembering that they are in reverse polarity to each other,

$$V_{RN}(\text{Total}) = V_R(\text{from field}) + V_C(\text{from shield}) \tag{7.35}$$

Using V_{RN} (Total), Equation (7.31) can be rearranged to produce

$$SE_H = \frac{V_{RN}(\text{Total})}{V_S} = \frac{Z_{RN}}{(Z_{RN} + Z_{RF})}\left\{1 - \left[\frac{1}{1 + \dfrac{Z_S{}'}{\omega L_S}}\right]\right\} \tag{7.36}$$

Equation (7.36) is called the magnetic shielding effectiveness (SE_H).

Observe that when the shield termination (i.e., connection to the reference structure) impedances are small ($Z_{SN}, Z_{SF} \to 0$), the shield impedance is dominated by the shield properties only (R_S, L_S). Equation (7.36) points, therefore, toward two very interesting and important characteristics of this situation:

1. Increasing the shield termination, Z_{SN}, Z_{SF}, decreases the voltage coupled from the shield to the inner conductor, resulting in an increase in the total voltage on the load. This can be easily explained by observing that magnetic coupling between the shield to the inner conductors is a function of the shield current, I_S, [Equation (7.25)]. Therefore, increasing the impedance of the shield and the reference structure by disconnecting one of the shield's ends (Z_{SN} or Z_{SF}) from the ground results in a reduction in the shield current. Clearly, from Equation (7.36), the efficiency of the shield (SE_H) is, thus, appreciably degraded.

---

*The cutoff frequency is a function of the shield characteristics (series resistance, inductance) and the shield termination impedances. For a shield terminated to a reference through low impedance, such that the termination impedance can be ignored, and assuming a shield resistance of 5 mΩ/m and inductance of 1 μH/m, this cutoff frequency will occur at a frequency of approximately 5 kHz.

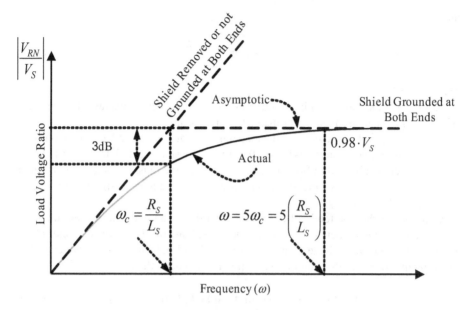

<u>Figure 7.16.</u> Effect of shield termination on magnetic coupling to a shielded wire versus frequency.

2. On the other hand, with higher quality shield terminations (Z_{SN}, $Z_{SF} \rightarrow 0$), the shield performance is dominated by the shield characteristics themselves. Below the shield cutoff frequency, ω_c, little magnetic shielding can be expected; however, at frequencies exceeding $5 \cdot \omega_c$, excellent shielding effectiveness can be expected.

Lessons Learned

- Effective shielding against coupling of H-fields is achieved using highly conductive shielding materials only if the shield is grounded at both ends through a low impedance to the reference.
- Magnetic shielding is effective only at frequencies exceeding the shield cutoff frequency, determined by the shield intrinsic properties (resistance and inductance).

7.3.5.3 Shielding for Control of Magnetic Field Emissions. Similar to shielding against the coupling of external magnetic fields, emission of magnetic fields is also controlled by "flux cancellation."

Figure 7.18 portrays a simplified equivalent circuit of the problem examined in Figure 7.17. E_S is depicted as a signal source, whereas Z_G and Z_L, symbolize source and load impedances, respectively.

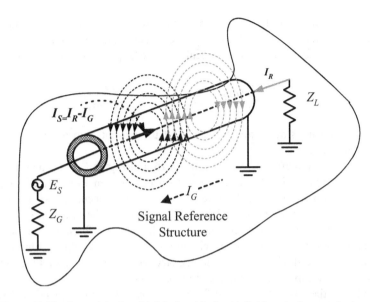

Figure 7.17. A model of a shielded cable for H-field emission analysis.

Inductances L_R and L_S symbolize the partial self-inductance of the signal and the shield, respectively, whereas M_{SR} denotes the partial mutual inductance between the two. Signal current is symbolized by I_R, the current flowing through the shield, I_S, whereas current flowing through the signal reference is I_G. Applying Kirchhoff's Voltage Law to the loop A–B–R_S–L_S, formed by the signal reference structure and the shield, we obtain

$$\left(R_S + j\omega L_S\right)\cdot I_S - j\omega M_{SR}\cdot I_R = 0 \tag{7.37}$$

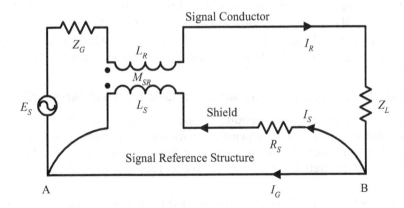

Figure 7.18. Simplified equivalent circuit of the system for H-field emission analysis.

Note that the current in the signal conductor is inductively coupled into this loop via mutual inductance, M_{SR} between the signal conductor and the shield.

Rearranging Equation (7.28) and recalling that $M_{SR} = L_S$ [Equation (7.12)] yields:

$$\frac{I_S}{I_R}(\omega) = \frac{j\omega}{j\omega + R_S/L_S} = \frac{j\omega}{j\omega + \omega_c} \tag{7.38}$$

where w_c constitutes the cutoff frequency of the current ratio in the intended return current conductor.

Observe the similarity between Figure 7.16 and Figure 7.19, which demonstrates that at lower frequencies the return current selects the signal reference structure as its primary return path. As a result, little flux cancellation occurs between the flux produced from the inner conductor and the flux generated by the shield current. Moreover, the loop formed by the inner conductor and the signal reference structure could be quite large, exacerbating magnetic field emissions from the cable. At higher frequencies, however, the shield serves as the primary path for the signal return current,* allowing effective flux cancellation (see Figure 7.17).

Similar to the case of magnetic field coupling onto cables (Section 7.3.5.2), the shield cutoff frequency directly depends on the shield's intrinsic resistance and inductance:

$$\omega_c = \frac{R_S}{L_S}, \text{ or } f_c = \frac{R_S}{2\pi L_S} \tag{7.39}$$

In conclusion, for ensuring magnetic field emission control (effective shielding for H-fields), the return current must be "forced" to flow through the shield in order to allow effective flux cancellation.

Lessons Learned

- Effective shielding against emissions of H-fields is achieved by controlling the path of the return current to encourage effective magnetic flux cancellation.
- Magnetic shielding is effective only at frequencies exceeding the shield cutoff frequency, determined by the shield intrinsic properties (resistance and inductance).

7.3.6   Frequency Considerations in Cable Shield Termination

7.3.6.1   Shielding and Ground Loops. The topic of so-called "ground loops" associated with cable shield termination configurations is probably one of the most confusing, controversial, and difficult to understand concepts in shielding theory. However, when carefully considering the basics of shield termination, it is self-evident

---

*Recall that at lower frequencies the return current follows the path of least resistance, whereas at high frequencies it follows the path of least impedance (dominated by inductance). Current *always* follows the path of least impedance. At low frequencies, the impedance is dominated by resistance.

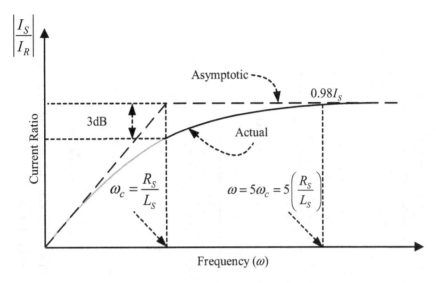

Figure 7.19. Ratio of the current on the shield versus current on the inner conductor.

there is nothing mysterious about "ground loops" [1]. Furthermore, the issue is much less an issue than envisioned and is limited to very distinct situations. The following discussion presents concerns associated with "ground loops" created by cable shield termination to a reference structure and puts them in their proper perspective. Concepts presented herein further support conclusions derived in the previous sections and constitute a basis for the application of "frequency-selective shield termination" to be presented in Section 7.3.6.3.

Observe the circuit presented in Figure 7.20, which presents commonly used configurations of typical single-ended circuits, namely, a nonshielded single-ended circuit and a shielded single-ended circuit, with the shield terminated to a reference structure at one or both ends. The signal reference structure is assumed to carry externally generated interference current, I_{EMI}, due to stray power currents resulting in a voltage drop, V_G, due to the finite (common) impedance of the structure.*

From the standpoint of EMI control, the basic principle governing the termination of the cable shield connected to sensitive electronic circuits is to preclude the flow of interference currents through the shield that could introduce common-mode interference voltage on the internal wires in the shield through the shield surface transfer impedance mechanism.†

In Figure 7.20(a), the signal reference constitutes an essential part of the signal circuit. With no connection to the reference structure, no return path exists for the signal current, I_R, and the circuit will not be functional. The reference is shared by both the

---

*See Chapter 4 for deeper insight into the question of common-impedance coupling through the signal reference structure.

†Shield surface transfer impedance is discussed in Section 7.4.

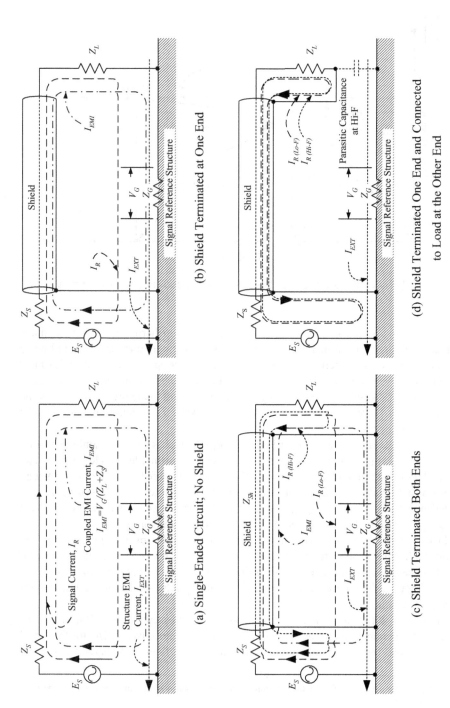

(a) Single-Ended Circuit; No Shield

(b) Shield Terminated at One End

(c) Shield Terminated Both Ends

(d) Shield Terminated One End and Connected to Load at the Other End

Figure 7.20. Ground loops in single-ended circuit interconnects.

signal and the interference current paths. Under this condition, an interference voltage will appear across load Z_L:

$$V_L = V_G \cdot \frac{Z_L}{Z_L + Z_S} = I_{EXT} \cdot Z_G \cdot \frac{Z_L}{Z_L + Z_S}; \text{ volts} \qquad (7.40)$$

Galvanic coupling occurs at all frequencies; however, at lower frequencies, the voltage divider between the source and load impedances [Equation (7.40)] constitutes the primary coupling mechanism, due to the primarily resistive impedance, Z_G at low frequencies. As frequencies increase, inductive reactance dominates ($Z_G \approx j\omega L_G$). The large loop area formed by the signal current path, I_S, also increases magnetic field interactions with the circuit.

In Figure 7.20(b), the shield is terminated at one end only. Similar to the previous case, common-impedance coupling in the reference is not prevented. Also, no suppression of magnetic field interactions is accomplished as no flux cancellation can occur. Low-frequency electric field (electrostatic) suppression is effectively achieved.

With the shield grounded at both source and load [Figure 7.20(c)] two situations may occur, depending on frequency. At lower frequencies, signal return current, $I_{R(Lo-F)}$, still flows through the lower resistance of the signal reference, making the shield ineffective in protecting the circuit from being galvanically coupled and causing common impedance interference. The shield is virtually "ignored" by the signal return current. This situation is the source of "ground loop phobia."

Electric field suppression occurs in a manner similar to case (b). However, a difficulty arises relating to magnetic field suppression: effective magnetic flux cancellation is achieved only if the return current flows through the shield, which is now grounded at both ends. What makes the current flow through the shield at frequencies below the shield cutoff frequency? Nothing, unless the current is forced to flow in that manner!

At higher frequencies, the signal return current, $I_{R(Hi-F)}$, flows through the shield following the path of least impedance, resulting in excellent flux cancellation.* External high-frequency interference current I_{EMI} flowing in the reference will not flow through the shield interference but may still couple into the circuit due to the voltage drop in the reference through the shield surface transfer impedance mechanism, associating the shield current with the common-mode voltage induced on the internal conductors, V_I:

$$V_I \triangleq I_{EMI} \cdot Z_T' = \frac{I_{EXT} \cdot Z_G}{Z_{Sh}} \cdot Z_T'; \text{ volts} \qquad (7.41)$$

where Z_{Sh} is the shield intrinsic impedance and Z_T' is the cumulative shield surface transfer impedance, in ohms.*

---

*The performance of coaxial cables for RF applications is based on this principle, eliminating magnetic field coupling to and from the coaxial cables.
*As will be seen in Section 7.4, the shield transfer impedance is expressed per unit length, in Ω/m. the symbol Z_T' is intended to symbolize the cumulative shield surface transfer impedance along the entire length of the cable.

In an attempt to overcome the problematic situation in Figure 7.20(c), the scheme in Figure 7.20(d) should be considered, whereby the stray power current path through the shield is interrupted. To interrupt the path for stray power currents, the system's signal reference must be connected to structure at one end only. The shield and the load impedance are now interconnected in the manner illustrated in Figure 7.20(d); thus, the return signal current is driven to flow through the shield, resulting in effective flux cancellation. Also, since stray interference current does not flow through the shield, the interference voltage across the signal reference structure, V_G, presumably has now no effect on the circuit's performance.

Necessary isolation can be accomplished either by floating the load equipment or its internal circuitry. This scheme works for lower frequency applications. Either method is generally difficult to implement and maintain at higher frequencies, at which parasitic capacitance virtually grounds the line. Also, when the length of a cable exceeds 0.05 λ at the frequency of interference, transmission-line effects cannot be ignored. Shield termination considerations in such electrically long circuits are discussed in Section 7.3.6.2.

Lessons Learned

- Single-point shield grounding will not eliminate coupling of EMI from structural currents into the circuit, unless the shield and signal conductor are terminated at one end only and interconnected at the other end.

- Ground loops in shielded circuits terminated at both ends will result in noise coupling and ineffective shielding performance at lower frequencies. At high frequencies, coupling will be suppressed through the shield surface transfer impedance coupling mechanism.

- Effective shielding against magnetic field interactions is achieved by controlling the path of the return current to promote effective magnetic flux cancellation.

- Single-point termination is ineffective when the length of the cable is greater than 0.05 λ at the frequency of interference.

- Single-point termination should serve as a last resort when other methods, such as balanced interface, cannot be used.

7.3.6.2 Shield Termination at High Frequencies. At higher frequencies when the length of a cable, ℓ, exceeds 0.05 λ, transmission-line (t-line) effects are observed between the shield and the signal reference.* This t-line is characterized by distributed resistance, R_X, inductance, L_X, capacitance, C_X, and conductance, G_X, per unit length (Figure 7.21). Also, the shield and the inner conductors constitute a set of tightly coupled transmission lines [10].

As a result, the point on the shield connected to the GRP will ideally be at zero potential ("0 V"); however, other points along the shield will exhibit a distance-dependent voltage and current distribution; V_X and I_X, respectively. Consequently, the input

---

*Transmission-line basics and phenomena are discussed in Chapter 2.

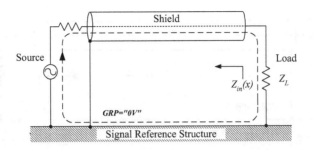

(a) Actual Circuit

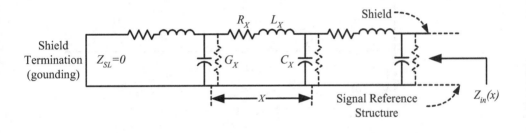

(b) Equivalent Circuit

<u>Figure 7.21.</u> A shielded conductor forming a transmission line with the signal reference structure.

impedance at any point along the line also varies with distance. In particular, for a lossless t-line shorted at one end to the "0 V" point, the input impedance at any point x along the line, at a frequency of concern f, exhibits a periodic variation expressed as*

$$Z_{in} \triangleq jZ_0 \cdot \tan\left(\frac{2\pi}{\lambda} \cdot x\right) = jZ_0 \cdot \tan\left(\frac{2\pi f}{v} \cdot x\right) \qquad (7.42)$$

where Z_0 is the characteristic impedance of the t-line formed between the shield and the reference, and v is the phase velocity of the propagating wave. In the special case where $x = \lambda/4$, the phase difference between voltage and current distribution is $\pi/2$ (or 90°). The short circuit at the load is transformed into an "open circuit" at the input, resulting in infinitely high input impedance at the frequency associated with this wavelength, λ (Figure 7.22):

$$Z_{in} = jZ_0 \cdot \tan\left(\frac{\pi}{2}\right)\bigg|_{x=\frac{\lambda}{4}} \to \infty \qquad (7.43)$$

---

*See Chapter 2 for the full derivation of this expression. Note the similarity between this situation and that of the "daisy chain" grounding topology, described in Chapter 4.

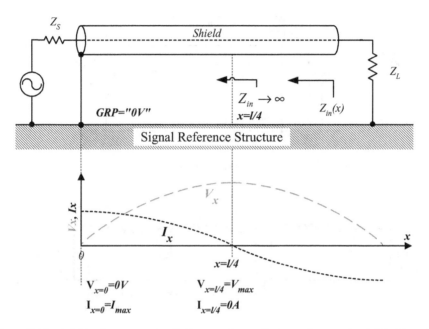

Figure 7.22. A 0 V reference potential is transformed into an "open circuit" (∞ V) at the point $x = \lambda/4$.

The implication of this situation is that at this point $x = \lambda/4$ along the line, the shield exhibits a $\lambda/4$ resonance. Since the voltage at the GRP is set to 0 V, the shield will be experiencing a standing wave with a frequency of

$$f = \frac{v}{4\ell}; \text{ Hz} \tag{7.44}$$

where ℓ is the length of the shielded cable and v is the phase velocity of the propagating wave.

Moving away from the "0 V" GRP along the shield, the input impedance will alternate between $Z_i \to \infty$ and $Z_i = 0$ for parallel and series resonance of the t-line at odd and even multiples of $\lambda/4$, respectively, occurring at (Figure 7.23):

1. $x = (2k+1) \cdot \dfrac{\lambda}{4}$, $k = 0, 1, 2, \dots$, parallel resonances, where $|Z_{in}| \to \infty$

2. $x = 2k \cdot \dfrac{\lambda}{4}$, $k = 1, 2, \dots$, series resonances, where $|Z_{in}| = 0$

Within cables, transmission line resonant effects are superimposed on the AC resistance of the shield. Hence, the input impedance at the series resonances, occurring in case (2), is limited by the line AC resistance, $|Z_{in}| \to R_{AC}$ (Figure 7.23).

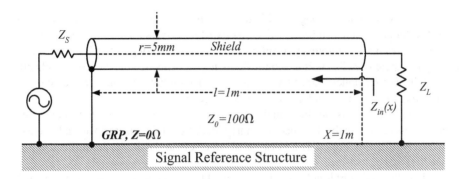

(a) Circuit Configuration

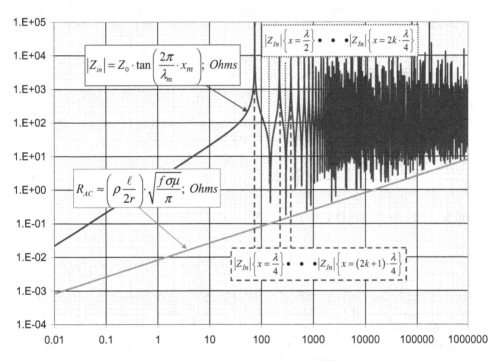

Input Impedance of a Cable Shield, $|Z_{In}|$ [Ω]

f [MHz]

<u>Figure 7.23.</u> Parallel and series resonances along a cable shield, $|Z_{In}|$ (Ω) versus frequency.

As a result, the points along the shield removed by distance $x = (2k + 1) \cdot \lambda/4$ from the 0 V reference point will be effectively isolated from the signal reference at the corresponding resonance frequency. It follows that at an arbitrary point x along the shield, the input impedance, Z_{In}, will exhibit a frequency dependence appearing periodically as an inductive reactance, capacitive reactance, open circuit, or a pure resistance, providing an extremely poor and unreliable 0 V reference.

So, here is the key issue: In a shielded line, when the shield is shorted to the reference structure, is it really because a "short circuit" exhibits an intrinsic inductance? Conversely, if it is left open, is it because an "open circuit" exhibits an intrinsic capacitance?

With an "electrically long" shield, terminated to the reference structure at one end, the high-frequency shield surface current can no longer be assumed to be uniformly distributed along the shield. The standing waves developed along the shield will result in a nonuniform, time-variant current and voltage distribution along the shield.

A severe outcome of this situation, coupled with the actual impedance exhibited by the shield, is increased coupling from the shield to the inner conductors, contrary to the objective of the shield.

Assume an arbitrary point x along the shield, sufficiently far from the single ground reference point, GRP. Due to shield voltage distribution (Figure 7.24), the open-circuit voltage will vary at the open end of the shield between $V_x(\max)$ and $-V_x(\max)$, with a total voltage swing of ΔV_x. There will always be some mutual capacitance, C_M, between the shield and inner conductors. The longitudinal interference current, $I_i(x)$, developing along the inner conductors at that point is*

$$I_i\left(x\right) \cong C_M \frac{\Delta V_x}{\Delta t}; \text{ amperes} \tag{7.45}$$

The per-unit-length shield to inner conductor capacitance can now be approximated as

$$C_M = \frac{2\pi\varepsilon_0\varepsilon_r}{\ln\left(r_{sh}/r_{ic}\right)}; \text{ F/m} \tag{7.46}$$

where r_{sh} and r_{ic} represent the radii of the shield and the inner conductor, respectively and ε_0 and ε_r stand for the permittivity and the relative permittivity of the dielectric between the shield and the inner conductor.

Similarly, owing to the mutual inductance, L_M, also known as the "shield transfer inductance" (discussed in Section 7.4.2) between the shield to the inner conductors, the current will vary along the shield between $I_x(\max)$ and $-I_x(\max)$ with a total current swing of ΔI_x, resulting in an induced voltage along the inner conductors, $V_i(x)$:

$$V_i\left(x\right) = -L_M \frac{\Delta I_x}{\Delta t}; \text{ volts} \tag{7.47}$$

---

*A similar reasoning exists for the voltage waveform distribution along the shield; however, the shield voltage would couple capacitively into the shield.

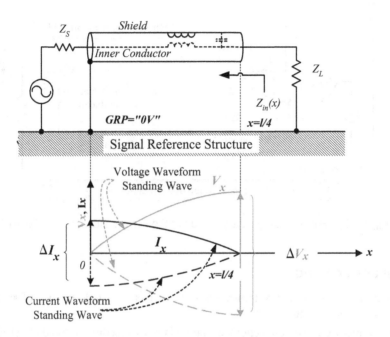

Figure 7.24. Inductive and capacitive coupling between the shield current/voltage and the inner conductor.

The per-unit-length mutual inductance between the shield and the inner conductor, L_M, is

$$L_M = \frac{\mu_0}{2\pi} \ln\left(\frac{2h_R}{r_{sh}+t_{sh}}\right); \ H/m \tag{7.48}$$

where r_{sh} and t_{sh} represent the radius and thickness of the shield, respectively, h_R, stands for the height of the inner conductor above the signal reference, and μ_0 signifies the permeability of the dielectric between the shield and the inner conductor. Figure 7.25 depicts an equivalent circuit of the system in Figure 7.24.

The above line of reasoning indicates that when high-frequency signals couple to an electrically long cable shield, the shield, rather than suppressing the interference, may be efficiently driven as an unintentional radiator while effectively coupling undesired noise to the inner conductors. The effectiveness of the shield in interference suppression in such a case is significantly compromised.

This outcome can be evaded by using multiple ground connections at intervals not exceeding λ/20 (0.05 λ). This ensures that resonance conditions are unlikely to occur. When termination of the shield at such intervals is impractical, shields should at least be terminated at both ends. In addition, a high-quality shield (low transfer impedance, Z_T)* should be used for this application.

---

*See Section 7.4.2 for a discussion on shield surface transfer impedance.

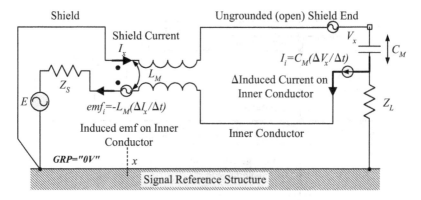

Figure 7.25. Simplified equivalent circuit of the system illustrated in Figure 7.24.

Lessons Learned

- At higher frequencies, transmission-line effects preclude the use of single-end shield grounding.
- Multiple ground connections at intervals not exceeding $\lambda/20$ (0.05 λ) must be utilized when shielding against high-frequency interactions.

7.3.6.3 Frequency-Selective Shield Termination.
In general, the multiple-point shield termination described in Section 7.3.6.1 serves as an adequate suppression strategy in most cases. In lower frequency circuits, such as audio applications, single-point shield termination may be more effective for precluding the reference structure current problem. In an attempt to provide for that situation, the configuration in Figure 7.20(d) was proposed, whereby the shield (and circuit) were terminated at one end only and interconnected at the other end.

In the previous section, it was shown that this scheme suffers from a drawback at higher frequencies, namely, that "floating" of the shield at one end will not be maintained, in practice, due to parasitics, while the electrically long conductors approach resonance. This may intensify coupling of interference between the shield and internal conductors.

A resolution to this problem is found in frequency-selective shield termination.* Observe the depiction in Figure 7.26, where a multiple-shield triaxial† cable configuration is used [5,9]. In this configuration, the inner shield is connected to the signal reference at one end only. The outer shield is connected to the signal reference at the other end. Low-frequency isolation is maintained between the two shields along their entire length. Assume that the inner shield is used as the signal return path. Thanks to this isolation, the low-frequency signal current returns to the source on the inner shield,

---

*Basics of frequency-selective grounding schemes were discussed in Chapter 4.

†A triaxial cable consists of an inner conductor, an internal shield (often serving as the signal return), and an outer shield. See Section 7.4.6.2 for details.

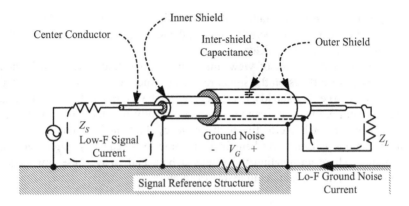

(a) Physical Configuration

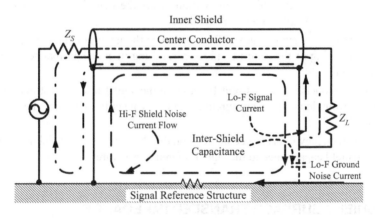

(b) Circuit Representation

<u>Figure 7.26.</u> Using frequency-selective shield termination in high-impedance, low frequency circuits.

which also acts as the return conductor, keeping off the signal reference structure. As a result, excellent flux cancellation is achieved and little radiated emission will result.

High-frequency noise, on the other hand, which may couple to the shield, is effectively shunted to the signal reference structure, thanks to high-frequency coupling between the inner and outer shield afforded by the intershield capacitance. It follows that from knowledge of the interference frequency, the geometry of the triaxial cable can be determined so as to provide necessary low-impedance coupling.

The per-unit-length intershield capacitance is approximated as [9]

$$C_M = \frac{2\pi\varepsilon_0\varepsilon_r}{\ln\left(r_{os}/r_{is}\right)}; \text{ F/m} \qquad (7.49)$$

where r_{sh} and r_{is} represent the radii of the outer shield and the inner shield, respectively, and ε_0 and ε_r stand for the permittivity and the relative permittivity of the dielectric between the outer and inner shields.

From the immunity point of view, this configuration finds use when high-level, low-frequency noise flows on the signal reference structure. As a result of the capacitive coupling between the shields, in effect between the return conductor and the signal reference structure, this noise cannot couple efficiently into the circuit and interfere with its performance while effective high-frequency shielding is maintained.

This technique is especially beneficial in some low-frequency, high-load-impedance circuits, where terminating the shield at both ends may cause low-frequency noise currents on the shield to couple into the circuit.

Lessons Learned

- Frequency-selective shield termination offers the benefits of both single-point and multipoint cable shield termination.
- Low-frequency signal current flows between the inner conductors and the inner shield, acting as the return conductor, resulting in excellent magnetic flux cancellation.
- Lower frequency EMI current flowing on the signal reference structure is kept away from low-frequency circuits through capacitive coupling between both shields.
- High-frequency interference induced on the shield is effectively shunted through intershield capacitance to the signal reference structure.

7.4 SHIELD SURFACE TRANSFER IMPEDANCE

When an external electromagnetic field interacts with an imperfectly shielded cable, cable shield currents and voltages, $I_S(x)$ and $V_S(x)$ in Figure 7.27(a), are produced, subsequently resulting in induced currents and voltages $I_i(x)$ and $V_i(x)$ appearing on any internal conductor(s). Exterior-to-interior field coupling in shielded cables occurs by means of two mechanisms: diffusion through the imperfectly conducting shield material and penetration of shield apertures. Coupling by these mechanisms may occur along the entire length of the cable (i.e., "distributed" coupling) or at isolated points, as a consequence of the presence of connectors or localized shield imperfections or defects.

Voltages and currents induced on the internal conductors can be calculated using transmission-line equations with a distributed-source current. If the shield contains apertures such as those in certain types of braided shield construction, the transmission line will contain both a distributed shunt current and a series voltage source. The distributed-source voltage is a function of properties of the shield and the shield current. Distributed-source current is determined by shield properties and construction, includ-

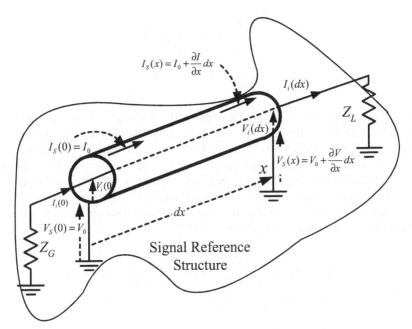

(a) Voltages and Currents Associated with Shielded Cable Analysis

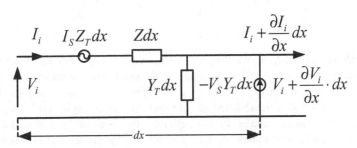

(b) Equivalent Circuit for the Internal Circuit, Considering Both
Transfer Impedance and Transfer Admittance

Figure 7.27. Voltages and currents associated with shielded cable analysis.

ing apertures as well as the voltage between the shield and the signal reference structure, serving as the shield-current return path.

A section of transmission line of infinitesimally short length, dx, contains a distributed voltage source, $E_x(x) = Z_T \times I_0(x)$, where $I_0(x)$ is the current on the shield, and a
distributed current source, $J(x) = -Y_T \times V_0(x)$, where $V_0(x)$ is the external voltage on the
shield, is shown in Figure 7.27(b). The properties of the shield and external structure
are incorporated into the transfer impedance, Z_T, and the transfer admittance, Y_T. The
differential equations for the internal voltage, V_i, and current, I_i, are [10,11]

$$\begin{cases} \dfrac{dV_i}{dx} + ZI_i = E_x(x) \\[2mm] \dfrac{dI_i}{dx} + YV_i = J(x) \end{cases} \tag{7.50}$$

where Z represents the series impedance per unit length and Y is the shunt admittance per unit length of the transmission line formed by the internal conductors and the shield. $E_x(x)$ is the source voltage per unit length and $J(x)$ is the source current per unit length produced by the external current and voltage of the shield. For the case of tubular shields, the current source, $J(x)$, is negligible compared to the voltage source, $E_x(x)$ [10]. For an electrically short cable* that is terminated by impedances Z_G and Z_L at the source and load ends, respectively, the induced currents through, and voltages across, the terminating impedances are [Figure 7.27(b)]

$$\begin{cases} I_G \approx I_S Z_T \ell \cdot \dfrac{1}{Z_G + Z_L} + V_S Y_T \ell \cdot \dfrac{Z_L}{Z_G + Z_L}; \quad (\ell \ll \lambda) \\[3mm] V_G = -I_G Z_G \end{cases} \tag{7.51}$$

$$\begin{cases} I_L \approx I_S Z_T \ell \cdot \dfrac{1}{Z_G + Z_L} - V_S Y_T \ell \cdot \dfrac{Z_G}{Z_G + Z_L}; \quad (\ell \ll \lambda) \\[3mm] V_L = -I_L Z_L \end{cases} \tag{7.52}$$

The shield current, I_S, flows from the source end (G) to the load end (L), and the shield voltage, V_S, is measured with respect to the shield current return path, the signal reference structure [Figure 7.27(a)]. It is clear from Equations (7.51) and (7.52), if $I_S \times Z_T \gg V_S \times Y_T \times Z_{1,2}$, the transfer admittance term can be neglected.[†] This condition is satisfied if the shield voltage, V_S, is negligible (that is to say, low-impedance external circuit with shield termination), if Y_T is small or if the source and load impedances, Z_G and Z_L, are small (low-impedance internal circuit).

7.4.1 Methods for Cable Shielding

Commonly used types of cable shielding include (1) woven braid, (2) flexible conduit, and (3) rigid conduit [8]. Of these, the most widespread form is woven braid. Advantages of braided shields are flexibility, durability, strength, and relatively light weight. However, the percentage of optical coverage by a braided shield constitutes a critical parameter in the design of such cables. The percent of optical shield coverage (OC) may be computed by using

---

*A cable of length ℓ that is much shorter than wavelength λ at the highest frequency of concern. Beyond this frequency, transmission-line effects come into effect.

[†]For the purpose of the discussion in this book, it will indeed be assumed that the above condition prevails and that the transfer admittance can be ignored.

$$OC\ [\%] = \left(2F - F^2\right) \times 100 \tag{7.53}$$

(see Figure 7.28) [4,12], where F is the fill factor, expressed as

$$F = \frac{pNd}{\sin\theta} \tag{7.54}$$

$$\theta = \arctan\left(2\pi\frac{(D+2d)p}{C}\right) \tag{7.55}$$

where:
N = Number of strands per carrier
D = Diameter of core under the shield (inch)
d = Strand diameter (inch)
p = Picks per inch
C = Number of carriers
θ = Weave angle

Typically, braided shields provide an optical coverage ranging from 60% to 98%, and thus experience the disadvantage, compared with solid shields, that for radiated fields, their effectiveness in field suppression decreases with increasing frequency and with the reduction in the density of the weave.

Conduit, whether solid or flexible, may also be used to shield cables and wiring, providing excellent shield suppression. Degradation of the conduit's shielding is usually not the result of insufficient shielding properties of the conduit material but, rather, the result of discontinuities in the conduit, resulting from poor splicing techniques, poorly bonded interconnects, or improper termination of the shield at connectors or plugs.

The principal types of shielded cables that are available include shielded single wire, shielded multiconductor, shielded twisted pair, and coaxial configurations.

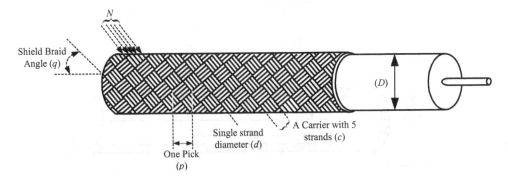

Figure 7.28. Definition of cable shield parameters.

Where a high degree of shielding is needed, cables with multiple shields separated by insulation are commonly used. Note that the improvement in shielding is not linear with respect to the number of shield layers, with two providing the greatest improvement over a single shield. Such cables are available in many different forms and with a variety of physical characteristics.

Much of the previous discussion relates to the shielding of cables against plane wave or high-impedance (predominantly electric) fields. For shielding against low-frequency/low-impedance (predominantly magnetic) fields, the use of annealed high-permeability (e.g. μ-metal) shields can be beneficial, offering high suppression (20 dB, approximately) at frequencies as low as 400 Hz with moderate penalties in weight and flexibility. This has to be traded off against their increased cost and somewhat degraded suppression of electric fields.

7.4.2 Shield Surface Transfer Impedance in Coaxial Lines

First formulated by Schelkunoff for coaxial cables in 1934 [1,2,7], the concept of shield surface transfer impedance, Z_T, provides an unambiguous figure of merit for the effectiveness of a cable shield, and, thus, constitutes a particularly convenient means of shield characterization. Interference currents induced by external electromagnetic interactions generally flow on the outer surface of the shield as a result of the skin effect and an axial voltage gradient is developed along the inner conductor (Figure 7.29). For an electrically short cable, the ratio of induced conductor-to-shield voltage per unit length to the shield surface current is defined as the surface transfer impedance, Z_T:

$$Z_T \triangleq \frac{1}{I_S} \frac{dV_i}{dl} \bigg|_{I_i=0} \; ; \; \Omega/m \qquad (7.56)$$

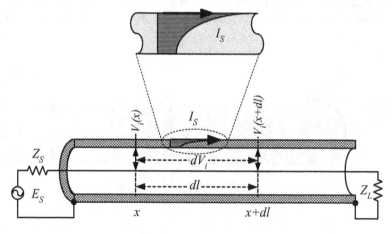

Figure 7.29. Definition of shield surface transfer impedance (Z_T) of a coaxial cable.

where I_S represents the total current flowing along the shield in amperes,* and $dV_i/d\ell$ symbolizes the per-unit-length longitudinal (or common-mode) voltage drop generated by the shield current along the transmission line formed by the shield and the inner conductors (V/m). The surface transfer impedance represents the open-circuit voltage developed between the conductors and the shield, normalized to the shield current, per unit length. Figure 7.29 illustrates the physical significance of shield surface transfer impedance. By observation, it is clear that the lowest achievable surface transfer impedance is desirable [13].

The shield transfer impedance comprises two primary components: a resistive or transfer resistance, R_T, and a primarily inductive component, or transfer inductance, M_T. It can be expressed as

$$Z_T = R_T + j\omega M_T ; \ \Omega/\text{m} \tag{7.57}$$

At lower frequencies, typically below 100 kHz, Z_T is dominated by the transfer resistance, R_T. Transfer resistance is associated with the shield's conductivity and permeability, and the contact resistance between strands in the braid as well as frequency of use. Resistance is also affected by frequency, exhibiting both a DC and AC value.

At very low frequencies, surface current can penetrate the shield through the process of diffusion. At such frequencies, the shield transfer resistance is equal, in practice, to the shield's DC resistance, R_{DC}†:

$$R_T \left(\text{Low} - f\right) \approx R_{DC} \tag{7.58}$$

As frequency increases, skin effects begin to dominate and shield current will tend to concentrate on the outer surface of the shield. Consequently, R_T decreases with the decrease of the skin depth [7]:

$$R_T \left(\text{High} - f\right) = \frac{R_{DC}}{1 - e^{-\delta/t}} = R_{DC} \cdot \left(1 - e^{\sqrt{\pi f \mu \sigma}/t}\right)^{-1} \tag{7.59}$$

where δ represents skin depth (m),‡ t is total thickness of the shield (m), f is frequency (Hz), μ is the permeability (H/m), and σ is the conductivity (S/m).

As frequency increases, transfer inductance, M_T, becomes of greater significance in leaky shields. The transfer inductance is associated with the penetration of magnetic flux through apertures in the weave of braided-wire shields due to current flow on the surface of the shield, inducing electromotive force (emf) on the inner conductors, known as "Faraday induction" (Figure 7.30). Unlike transfer resistance, transfer induc-

---

*Possible sources of the shield current include coupling from external electromagnetic waves or ground loop currents, when the shield is grounded to the reference structure at two points having different voltage potentials.

†DC and AC resistance of conductors is defined in Chapter 2.

‡δ, the skin depth, is the depth of penetration at which the current density is reduced to e^{-1} (~37%) of its initial value on the surface of the shield. Typical values of skin depth at 100 kHz are 0.5 mm for copper and aluminum and 0.1 mm for steel.

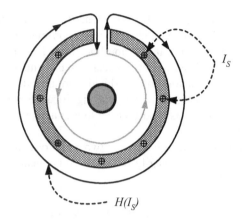

(a) Magnetic (Inductive) Coupling Mechanism

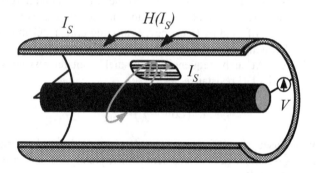

(b) Physical Representation of Magnetic Coupling

Figure 7.30. Mechanisms of braided shield transfer inductance, M_T, due to coupling through an aperture in a coaxial cable shield.

tance is a complex term, related to the type of shield (e.g., braided, foil, or solid conduit) and its construction. For most common braided cable shields, the surface transfer impedance becomes primarily inductive at about 1 MHz.

Because Z_T is normalized to a length of 1 meter, all applications require that the nominal Z_T of the shield in ohms per meter (Ω/m) be multiplied by the physical length of the cable, ℓ, in meters (when the length of the shielded cable is shorter than a quarter of wavelength, $\lambda/4$). Practical cable lengths normally extend over several wavelengths at frequencies of impinging fields. However, the increase of the shield transfer impedance with length does not continue indefinitely but, rather, levels off when the length of the shielded cable approaches a quarter of a wavelength, $\lambda/4$. The shield transfer impedance is thus upper-bound by its value at the frequency corresponding to that wavelength, λ:

$$Z_T = \begin{cases} Z_T\left[\Omega/m\right]\cdot\ell \approx 2\pi\cdot c\cdot\dfrac{\ell}{\lambda}\cdot M_T, \ \ell < \dfrac{\lambda}{4}; \ \text{ohms} \\[4mm] Z_T\left[\Omega/m\right]\cdot\dfrac{\lambda}{4} \approx c\cdot M_T\cdot\dfrac{\pi}{2}, \ \ell \geq \dfrac{\lambda}{4}; \ \text{ohms} \end{cases} \qquad (7.60)$$

where the transfer impedance is represented by the transfer inductance, M_T, and c symbolizes the speed of light [4]. The curve in Figure 7.31 illustrates the frequency dependence of the cable shield surface transfer impedance.

Good EMI control mandates that field pickup by cables be controlled by the cable shields, exhibiting the lowest possible transfer impedance at the frequency of the incident interference field. For meeting that goal, a shield should be terminated at as many points as practically achievable along the cable length in order to reduce the ratio λ/ℓ to the lowest possible value. A small ℓ/λ ratio will result in a better shield performance and avoid the need to revert to more expensive shield construction. Once the above objective is achieved, performance of the shield is dominated by the quality of the shield termination.

7.4.3 Where Should a Shield of a Balanced Line be Terminated?

Unlike coaxial transmission lines, in which the shield serves as a return path for the signal current, in shielded balanced lines the shield serves strictly as a barrier between the electromagnetic environment and the inner volume of the shield (Section 7.3.2). The shield does not serve as a return current path and carries no intentional signal current. It may seem, therefore, that the designer has full freedom in the selection of the shield termination topology of the circuit.

A more careful look at this condition reveals that this is not the situation in practice if a consistent approach for functionality, product safety, and EMI control is to be ap-

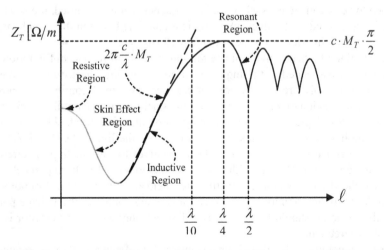

Figure 7.31. Frequency dependence of the cable shield surface transfer impedance.

plied. Often, EMI control requires that a cable, even when balanced, be enclosed within a shielded enclosure for precluding undesired interference with the circuit, as shown in Figure 7.32(a), where a balanced cable is routed between two nonmetallic enclosures. Clearly, nothing precludes field-to-cable interaction between the system wiring and electromagnetic fields, resulting in potential EMI failures. Note also that the internal signal return is not connected to the signal reference structure, which may be unacceptable from the standpoints of functionality, electrical safety, and ESD control.

In an attempt to solve potential problems, Figure 7.32(b) depicts an improved situation in which the equipment enclosures as well as the cable are shielded. In order to ensure adequate EMI control, shield continuity must be maintained by ensuring that the cable and enclosure shields form a continuous metallic barrier with respect to the external electromagnetic fields. Note that both conductors are routed within an "extended" shielded enclosure while the internal signal return remains floated, whether or not the shielded enclosures come into contact with the signal reference structure. The optional floatation or bonding of the enclosures to the reference structure is symbolized by switches [Figure 7.32(b)]. This condition may be unacceptable from a functionality aspect, even if the enclosures are grounded and electrical safety is not a concern.

In Figure 7.32(c), the signal return conductors in both enclosures are connected to the metallic enclosures, which, in turn, are connected to the signal reference structure. This represents a typical system configuration often encountered in practice. It is similar to the condition described in Figure 7.20(c) for single-ended circuits, and suffers from the same drawbacks. Interference current will now circulate between the cable shield terminated at both ends to the signal reference. Lower frequency EMI current along the shield may result from external interference current flowing across the structure impedance, Z_G, or from magnetically coupled EMI into the loop formed between the shield and the reference. Also, higher frequency radiated interference coupling onto the cable shield will introduce high-frequency current flow between the shield and reference structure.

When lower frequencies, such as audio frequencies, are concerned, the situation in Figure 7.32(c) is indeed troublesome. Efficient coupling between the shield and inner conductors could interfere with the circuit's performance.

The question of whether to terminate the shield or not in such a configuration stems from concerns associated with single-ended audio circuits, discussed previously. If low frequencies only are of concern, interrupting the shield continuity is possible. However, the question of where to interrupt the shield requires careful consideration.

Figure 7.33(a) depicts the baseline configuration in which the shield and circuit are floated. The line A–C constitutes the "Hi Line" and the line B–D, the "RTN Line," serves as the signal return conductor. The alternatives for terminating the circuit at a single point on the line B–D are either from point D to point 1 or from point B to point 2, whereas the alternatives for terminating the shield are either to point 1 or point 2 [4].

In order to avoid common-mode noise pickup and coupling, the voltage potential across the capacitor should be zero. This will occur only when the capacitor is effectively short-circuited.

If the circuit and the shield are both terminated (or referenced) at one end only [Figure 7.33(b)], no EMI current will flow through the circuit. The shield-to-conductor ca-

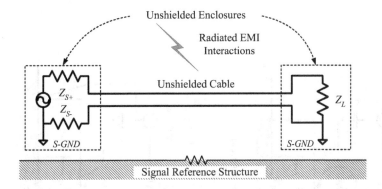

(a) Unshielded Enclosures and Cables: Inadequate EMI Control

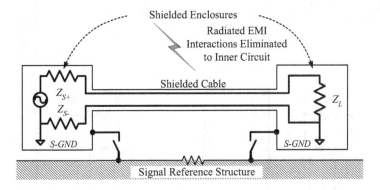

(b) Shielded Enclosures and Cables: Acceptable EMI Control, but Other Concerns Remain

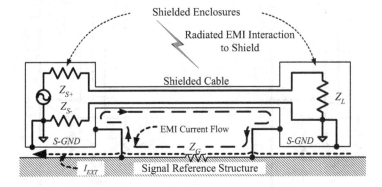

(c) Shielded Enclosures and Cables: Inner Circuits Grounded Both Ends

<u>Figure 7.32.</u> Shielding a balanced cable for preclusion of EMI coupling to the wiring.

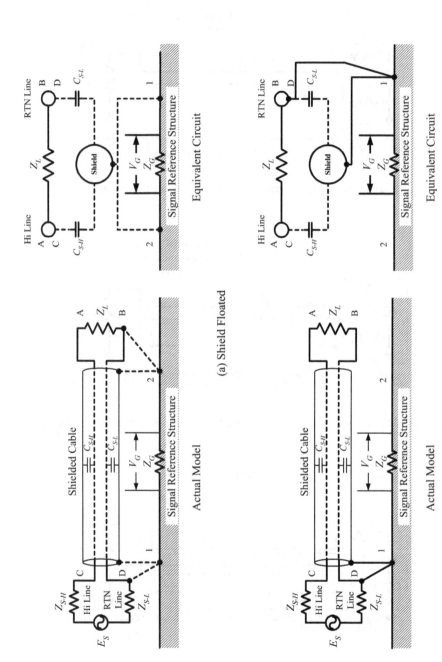

(a) Shield Floated

(b) Shield and Circuit Terminated at the Same End

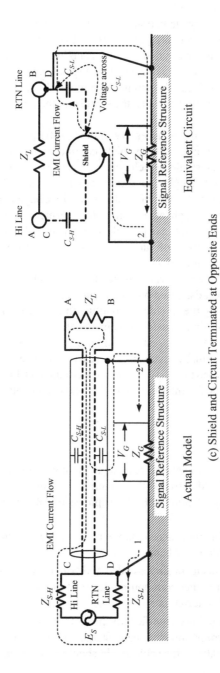

Actual Model

(c) Shield and Circuit Terminated at Opposite Ends

Figure 7.33. Where should the shield of a balanced cable be terminated?

521

pacitance, C_{S-L}, is shorted out by the termination conductor and interference current is diverted away from the load.

If, however, the circuit is terminated at point "1" and the shield is terminated at point "2," a different situation occurs [Figure 7.33(c)]. EMI current is now able to flow through load impedance Z_L as the capacitance C_{S-L} is no longer shorted out.

It follows that even in the case of a balanced wiring pair, attention must be paid to where the referencing point is located when interruption of the shield is attempted. It is also imperative that the connection between the shield and the circuit referencing point be kept as short as possible.

Is interruption of the shield really necessary in the low-frequency case? Reasoning leads to a recommendation to interrupt the shield continuity in order to preclude current from flowing through the shield, which generally acts as a return path for the signal current. This is prevalent in single-ended signaling and does not hold in the case of a balanced wire pair. Fortunately, solutions described in Chapter 4 for interrupting low-frequency ground loops, such as isolation transformers and optical isolators placed at the balanced line interface, can provide an adequate solution to this situation. Utilizing a single-point reference configuration should be avoided if these other solutions are economically and volumetrically feasible.

At higher frequencies, when the cable is considered electrically long, the shield should be grounded at both ends, based on a rationale presented previously. The high-frequency current flows along the shield and coupling into the shield will be governed by the shield surface transfer-impedance mechanisms previously discussed. The good news however, is that coupling occurs equally to both conductors of the balanced circuit, limited only by longitudinal conversion loss (LCL).* Hence, by using a shielded cable with low transfer impedance and a balanced circuit and load exhibiting high common-mode rejection (CMR), interference in the load can be minimized to a tolerable level.

Lessons Learned

- Shield surface transfer impedance represents the coupling mechanism between externally induced shield currents to the internally coupled line voltage. Similarly, due to reciprocity, it represents the suppression of interference emissions from the wiring.

- Shield surface-transfer impedance exhibits resistive low-frequency performance and inductive high-frequency behavior.

- Similar mechanisms occur in single-ended and balanced circuits; however, balanced circuits exhibit higher common-mode rejection, suppressing internally induced common-mode interference.

- Shield termination topology significantly impacts low-frequency performance, when single-point termination is implemented. At high frequencies, when length of the cable exceeds 0.05 λ, the cable shield should be terminated at both ends.

---

*Longitudinal conversion loss (LCL) is discussed in detail in Chapter 2. It represents the extent of common-mode to differential-mode conversion in a balanced system due to imperfections in circuit balancing.

- With the exception of coaxial cables at higher frequencies, cable shields should never intentionally carry signal current in configurations in which the outer shield serves as the return conductor. Coaxial cables should be used only for signals in which the lowest signal frequency spectral component in MHz is above 15/L [m]* approximately.

7.4.4 Shield Termination—The Key to Optimal Shielding Performance

Where and how a shield is terminated is critical to its performance. Performance specifications provided in data sheets of shielded cables normally consider only the intrinsic performance of the shield barrier itself but fail to account for the effect of shield termination on the overall transfer impedance. In an otherwise adequately shielded system, RF currents that are conducted along shields can easily couple to the system wiring from any point of an improper cable termination. This is particularly important in the case of cables exposed to high-strength RF fields comprising the electromagnetic environment the equipment is located within. Furthermore, inappropriate termination may not only degrade shielding performance of the circuit but could even produce secondary affects due to interactions between the shield and the circuit signal conductors. Often, such interactions are attributed to the shield rather than to an unsuitable circuit grounding topology. Designers subsequently often interrupt the shield rather than the path of current flow into the sensitive circuits.[†] If effectiveness of a shield is to be maintained, therefore, a low-impedance shield termination must be provided.

Figure 7.34 illustrates the contribution of the shield termination impedances, $Z_{TC(S)}$ and $Z_{TC(L)}$ to the total shield transfer impedance.

Shield termination is carried out using two main methods: so-called "pigtail" and peripheral (360°). The efficiency of the latter surpasses the former as it provides a lower impedance path. Figure 7.35 illustrates the advantage of implementing peripheral shield termination. Continuity of the shield is achieved through peripheral metal-to-metal contact between the shield of the cable to the enclosure, thus forming an extension of the enclosure shield. As frequency increases, interference current tends to concentrate on the surface of the shield (skin effect), further enhancing the effectiveness of the peripheral shield termination method. Figure 7.36 illustrates an actual manner by which peripheral shield termination can be accomplished.

A pigtail (or drain wire) termination (Figure 7.37) is the least preferred method of shield termination and should be avoided, because at RF frequencies the inductance of the pigtail counteracts the objective of low-impedance shield termination necessary for achieving effective shielding performance [4].

7.4.4.1 Effect of Pigtail Shield Termination. Observe the circuit in Figure 7.38. In Section 7.3.5.3, the conditions for achieving effective shielding performance

---

*This implies that the cable should be considered "electrically short," that is, $L \le 0.05\ \lambda \leftrightarrow f \le 0.05(c/L)$, or $f\,[\text{MHz}] \le 300/(20 - L\ [\text{m}]$ approximately.

[†]Section 7.5 discusses in detail the relationship between signal circuit grounding and cable shield termination.

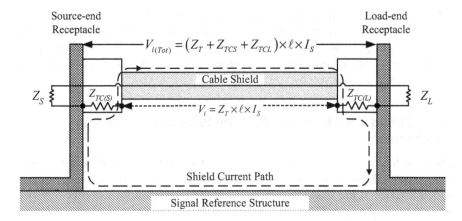

Figure 7.34. Total cable transfer impedance results from the shield transfer impedance and its termination.

in controlling magnetic field emissions by a magnetic flux cancellation mechanism were presented. In that case, perfect (zero impedance) shield termination was assumed, that is, Z_{TS} and Z_{TL} in Figure 7.38 are assumed to be zero. The design goal was shown to minimize the ground return current, I_G, thus facilitating the desired effective flux cancellation.

When using pigtail shield termination, the shield termination impedances Z_{TS} and Z_{TL} now include an inductive component.* These inductances are represented as L_T in the equivalent circuit of Figure 7.38(b).

This situation is similar to one described in Chapter 2, where it was shown that for an inductive return path, a condition for ensuring that the return current flows through the shield is developed:

$$\frac{I_s}{I_i} = \frac{j\omega \cdot L_s}{R_s + j\omega \cdot (L_s + 2L_T)} = \frac{1 + j\omega \cdot \left(\dfrac{L_s}{R_s}\right)}{1 + j\omega \cdot \left(\dfrac{L_s + 2L_T}{R_s}\right)} \tag{7.61}$$

For higher frequencies ($\omega \gg R_s/L_s$), this expression reduces to

$$\left.\frac{I_s}{I_i}\right|_{\text{High } \omega} \approx \frac{L_s}{L_s + 2L_T} \tag{7.62}$$

*For a typical "hookup wire," an inductance of 10 nH/cm or 1 μH/m can be assumed. Therefore, a typical 5 cm (2 inch) pigtail could exhibit an inductance of 50 nH per termination (or 100 nH for termination at both ends).

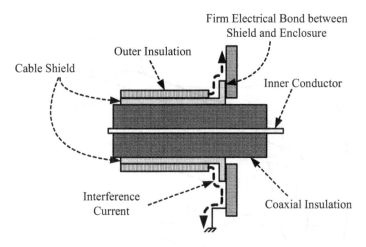

Figure 7.35. Advantage of peripheral shield termination.

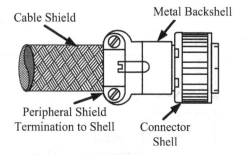

Figure 7.36. Peripheral shield termination.

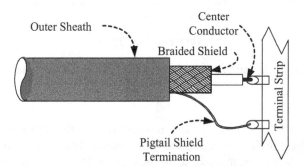

Figure 7.37. Pigtail (drain wire) shield termination.

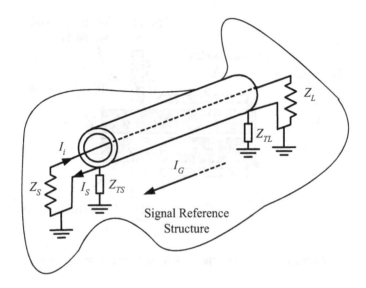

(a) Physical Representation

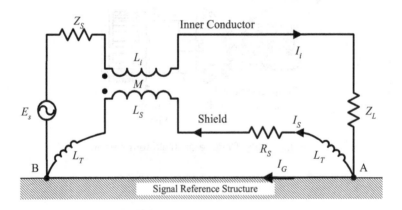

(b) Equivalent Circuit

<u>Figure 7.38.</u> Model for analyzing the effect of pigtail shield termination on shielding performance.

From Equation (7.62), it is obvious that in order to ensure that all return currents flows through the shield or, in other words, no return current flows through the signal reference structure, L_T must be minimized, that is, $L_T \to 0$. This result clearly indicates that use of an inductive pigtail shield termination should be avoided whenever possible if effective high-frequency shield performance is required.

Figure 7.39 presents the performance of a cable shield when terminated using both a peripheral (360°) and pigtail shield termination. The degradation in the shielding efficiency when a pigtail is used is evident, and is similar to that expected with no shield present. Measurements of EMI coupling with 100BaseT cables clearly indicate that pigtail termination is of small benefit compared to a disconnected shield (Figure 7.40).

The effect of the pigtail length on coupling between two shielded cables at frequencies beginning from 40 MHz is examined in Figure 7.41 [5]. The cables were 25 cm (10 inch) long. The aggressor line was excited by an RF signal generator having an output signal of 100 mV (–7 dBm @ 50 Ω).

A pigtail termination is adequate only for low-frequency applications but is ineffective at higher frequencies. The break point between the efficiency of a pigtail at both low and high frequencies cannot be determined in a formal manner. Rather than attempting to set a strict criterion, which at times may appear to be too liberal while at other times too severe, a reasonable rationale would be to select the shield termination methodology such that the degradation of shield performance due to a particular type of termination would be tolerable when evaluated in comparison to the shielding objectives.

For example, if a 2 meter long cable shield has a transfer impedance $Z_T \approx j\omega L_T = $ 100 mΩ/m ($L_T \approx 1.5$ μH/m) at a frequency of 10 MHz of 200 mΩ (–14 dBΩ), the shield termination should not contribute more than –34 dBΩ or 10% of the overall cable shield transfer impedance. This will limit the shield termination impedance to 20 mΩ, equivalent to an inductance of about 0.3 nH at 10 MHz. In such a case, a pigtail could not be used. In practice, if pigtail termination is unavoidable, the pigtail should be kept as short as possible.

For common cable shields, pigtails lengths of approximately 5 cm (or 2 inches) can be tolerated without significant degradation at frequencies below 100 kHz. However,

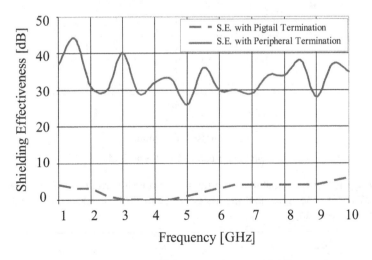

Figure 7.39. Effect of pigtail shield termination on shielding performance.

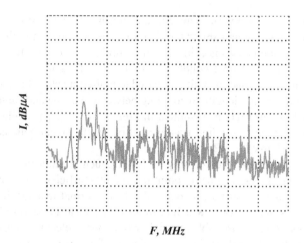

(a) 100Base-T Transmitter:

Pigtail Termination at Near End

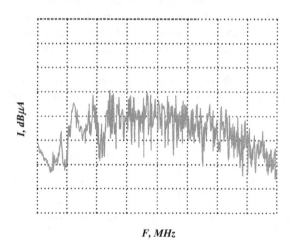

(b) 100BaseT Transmitter:

Shield Disconnected at Near End

Figure 7.40. Effect of pigtail shield termination on EMI current coupling (peak) onto a 100BaseT cable. (Measurement results courtesy of Prof. Dr.-Ing. H. Garbe, University of Hannover, Germany.)

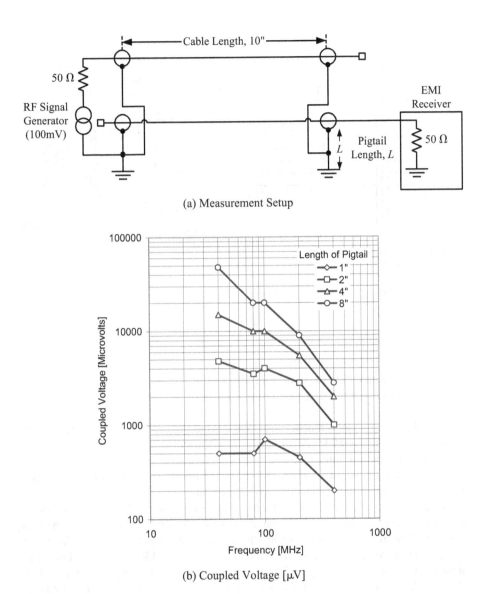

(a) Measurement Setup

(b) Coupled Voltage [μV]

Figure 7.41. Effect of length of pigtail shield termination on shielding performance.

at higher frequencies and when higher shield performance is required, pigtail shield termination can no longer be tolerated and peripheral shield termination must be applied.

7.4.4.2 *High-Performance Shield-Termination Techniques.* In essence, a shield termination consists of a sequence of junctions or joints (Figure 7.42), all of which contribute to its overall quality:

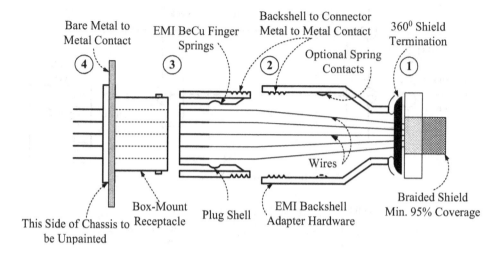

<u>Figure 7.42.</u> Components of a shield termination sequence.

1. Cable shield to backshell
2. Backshell to connector (plug)
3. Connector to bulkhead connector (receptacle)
4. Bulkhead connector to the enclosure (chassis)

Ideally, shield termination transfer impedance should be sufficiently low such that its contribution to the overall transfer impedance of the shield itself plus terminating connector assemblies is minimized. The preferred approach for achieving high-performance shield termination is when the entire periphery of the shield is terminated through a low-impedance connection to the enclosure.

A frequently used method of shield termination is illustrated in Figure 7.43(a) [14]. In this arrangement, the cable shield is flared so that it extends over the rear portion of the sleeve and the crimp ring is slid over the sleeve. The ring is then crimped onto the sleeve.

An alternative method to crimping is shown in Figure 7.43(b). The shield is placed through the ground ring and flared over and around the ring. The ground ring is then slid into the rear of the sleeve that has a tapered base. Tightening the cable clamp onto the end of the sleeve guarantees a 360° peripheral shield termination and provides strain relief. A variation on this termination method includes environmental seals in the backshell construction, which is of particular utility in equipment that is to operate in a corrosive environment such as on board ships with harsh environment effects.

Figure 7.44(a) presents construction details of shield termination using a threaded assembly. The ground ring can serve for terminating both overall [Figure 7.44(a)] and individual [Figure 7.44(b, c)] shields.

Figure 7.45 presents another common technique for overall shield termination by means of "IRIS" compression rings [12]. The "IRIS" is a compressible metal coil spring, conductive gasket, or other compressible material that provides electrical con-

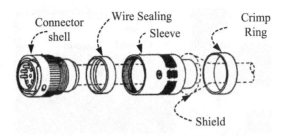

(a) Shield Termination Using Crimping

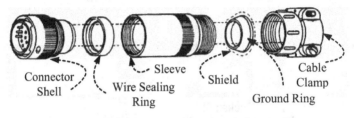

(b) Shield Termination Using Threaded Assembly

Figure 7.43. Two common methods for implementing peripheral shield termination.

tact between the overall shield and the backshell/adapter. The compression ring constitutes part of the connector backshell or adapter and encircles the cable at the shield. The advantages of using "IRIS" compression rings lie in the simplicity of installation, since the shield only has to be trimmed and there is no other disturbance to the shield. Also, the compression ring fits a wide range of cable sizes and creates a good bond to cables that have outlines other than round.

Another technique for effective overall shield termination utilizes EMI shrink boot adapters that feature a unique termination scheme. Shrink boots fit over the junction of the backshell and cable and shrink when heated to provide mechanical strain relief and some environmental protection. The braided shield is locked between the backshell collar and termination nut when the shield is pulled over the collar threads and the termination nut is tightened (Figure 7.46).

When terminating individual shielded wires within a cable assembly, the practice of terminating a number of conductor shields by means of a single wire to one of the connector pins should be avoided. Such a shield termination is highly inductive, which degrades overall shielding performance. Furthermore, this strand of wire constitutes a common impedance element across which interference voltages can be developed and coupled from one shielded circuit to another. It also provides a very good path for interference currents flowing on the shield to penetrate into the enclosure, thus obviating its shielding effectiveness.

Figure 7.47 presents the "halo technique" which is recommended for terminating a large number of individual wires shields [13,15,16]. The exposed unshielded leads

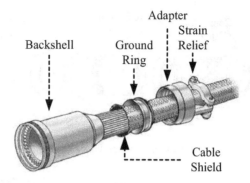

(a) Construction of Conical Ring Style Backshell for Termination of Overall Cable Shield

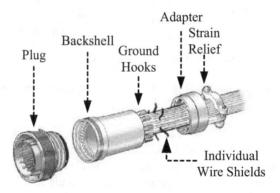

(b) Construction of TAG® Ring Style Backshell for Termination of Individual Wire Shields

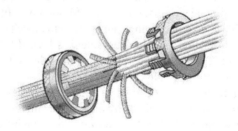

(c) Construction Details of TAG® Ring Style Backshell

Figure 7.44. Construction details for the shield termination using threaded assembly. (Courtesy of Glenair, Inc.)

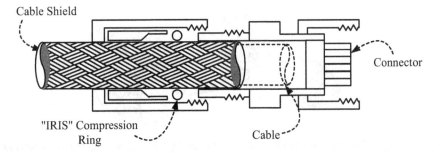

Figure 7.45. Overall shield termination by means of "IRIS" compression rings.

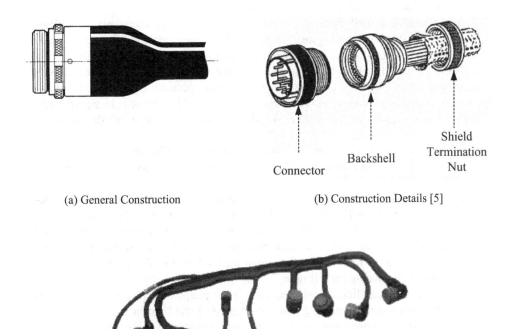

(a) General Construction

(b) Construction Details [5]

(c) Example of Shield Termination Utilizing EMI Shrink
Boot Adapter on a Shielded Cable Assembly

Figure 7.46. Overall cable shield termination utilizing an EMI shrink boot adapter. (Images (a) and (c) Courtesy of Glenair, Inc.)

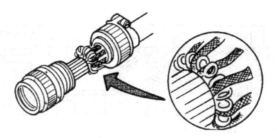

Figure 7.47. Termination of individual conductors' shields using the "halo technique."

should be made as short as physically possible to reduce undesired coupling between conductors. In this method, the individual shields are also eventually connected to the enclosure, which may not always be desirable.*

An alternative method makes use of connectors that provide coaxial pins specifically adapted for this purpose (Figure 7.48) [5]. In this case, isolation of the inner shields from the outer, overall shield is maintained. Multiple shields offer the advantage of a higher degree of shielding, however, galvanic isolation between the different shields must be maintained. This technique provides for that additional benefit when both inner and overall shields are used. The main drawback of this scheme is its significantly higher cost compared to the former. Weight and volume are also primary concerns here.

Shield termination devices and techniques also exist for other than circular connectors, such as the rectangular-shaped D-Subminiature type. Examples are shown in Figure 7.49.

For high-quality mating of the backshell to the connector, threaded contacts are recommended, as shown in Figure 7.42 as Junction #2. For improved shielding integrity, particularly when a high grade of shielding is required, further improvement can be achieved at the mating area of a backshell–connector interface by placing flexible spring contacts on the inner circumference of the backshell so that positive contact is made between the mating parts (Figure 7.51).

One of the weakest links in this sequence is the plug-to-receptacle joint (Junction #3). Some circular connectors (e.g., the 38999, series I and II or the 83723 series connectors) utilize a bayonet-type, quarter-turn latch mechanism. This type of interface provides for only a small number of contact points and provides a poor 360° shield termination, with corresponding higher transfer impedance. Threaded coupling, used in other configurations (e.g., MIL-C-38999, Series III) offer a superior performance.[†]

Flexible spring contacts placed on the inner circumference of the mating parts (Figure 7.51) provide for improved shielding integrity of a connector pair (i.e., two inter-

---

*There are certain situations, particularly where multiple shields consisting of an external overall braid as well as inner individual shields are used, where isolation may be required between the individual and overall shields.

[†]This is the type of shield termination used in the "BNC" RF connector. Compare it to the "TNC" connector, in which a threaded coupling is used.

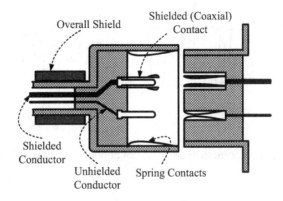

Figure 7.48. Method of terminating individual wire shields within an overall shield using coaxial contacts.

Figure 7.49. Shield termination devices and techniques used with rectangular connectors. (Courtesy of Glenair, Inc.)

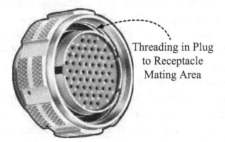

Figure 7.50. Threaded coupling provisions in a 38999 series plug for improved bonding at the plug-to-receptacle joint. (Courtesy of Glenair, Inc.)

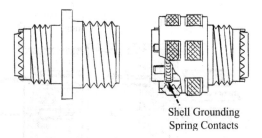

Shell Grounding
Spring Contacts

Figure 7.51. Spring contacts in an RF-proof connector. (Courtesy of Glenair, Inc.)

connecting connectors). The advantage gained in using circumferential spring contacts over bayonet coupling are dramatically illustrated in Figure 7.52. In this figure, the spring contacts were silver-plated beryllium–copper (BeCu) [13].

Finally, it is important that the connector be properly electrically bonded to its mounting surface (Junction #4 in Figure 7.42). Superior bonds at this interface can be achieved by maintaining metal-to-metal contact between the connector and equipment housing. In cases in which a good metal-to-metal contact cannot be achieved, appropriate conductive gaskets may be used for interfacing the connector structure to the metal enclosure surface* (Figure 7.53). Use of conductive gaskets is especially important when D-subminiature and similar rectangular connectors are used or when high shielding integrity is required, due to their inferior bonding performance as a result of their small contact area.

Realistic and achievable cumulative impedance of cable shield termination paths through connector assemblies is on the order of 10 mΩ from the shield to the equipment enclosure for a cadmium-plated aluminum assembly, with a typical impedance of 2.5 mΩ for any particular joint. In most cases, bonding of cable shields is more critical to performance than enclosure bonding. Figure 7.55 shows common methods of shield termination in descending order of preference.

7.4.4.3 Maintaining Cable Shield Continuity. As discussed, adequate performance of cable shields is achieved by maintaining continuity of the shield from source to load. When routing shielded cables through bulkheads using through-connectors, the configuration of the cable shield termination should be maintained along the entire length of the cable, including passage through the bulkhead connectors. Figure 7.56 presents three commonly used configurations for cable shield termination schemes when crossing bulkhead through connectors [16]

In Figure 7.56(a), the cable's shield is wired through a pin in the bulkhead connector. This configuration suffers from two main drawbacks. First and foremost, the shield carries interference current. The shield is intended to enclose the protected inter-

---

*Note that if gaskets are used for this application, metal compatibility must be ensured. Often, silver-filled gaskets are used, which are incompatible with aluminum and steel and may lead to surface corrosion. See Chapter 5 for a discussion on dissimilar metals considerations.

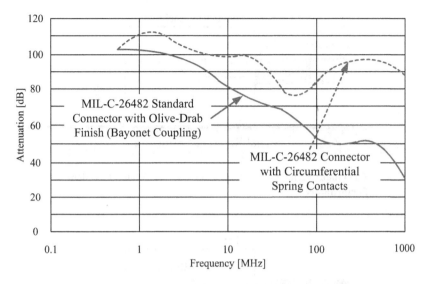

Figure 7.52. Effect of spring contacts in improvement of connector shielding performance.

Figure 7.53. Conductive gaskets for improved bonding of box-mount receptacles. (Courtesy of MAJR Products Corporation.)

Figure 7.54. Bonding provisions of the D-subminiature connector. (Courtesy of Glenair, Inc.)

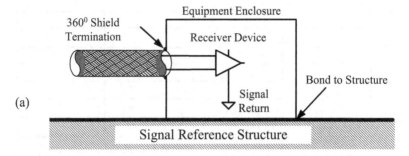

(a)

The Good: Peripheral shield termination to enclosure. **Excellent Performance.**

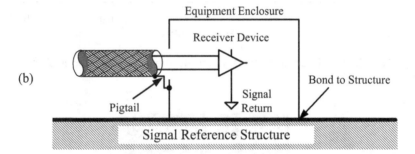

(b)

The Bad: Short pigtail shield termination to outside of enclosure. Inductive shield termination limits performance at higher frequencies and will allow EMI current to couple to inner conductors. **Acceptable for lower frequencies** if pigtail is short

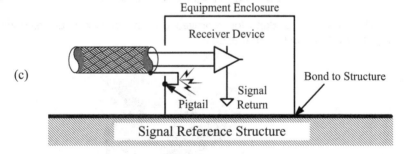

(c)

The Ugly: Short pigtail shield termination to inside of enclosure. Severe violation of enclosure shield integrity. Not recommended. The pigtail must be very short.

Figure 7.55. Methods of shield termination: The good, the bad, the ugly, and the dirty.

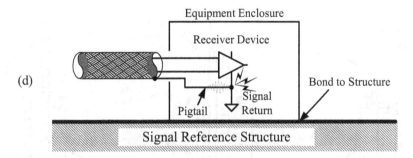

(d)

The Dirty: Short pigtail shield termination to the signal return
(reference) inside enclosure. Poor EMC design; EMI injected into the
circuit. **Highly objectionable.**

Figure 7.55. *Continued.*

nal wiring from the external environment, providing a close-to-perfect shielded enclosure. However, in the long transmission path through the bulkhead connectors, the shield is routed in parallel to the very wires it is supposed to protect. Crosstalk between the shield wire and the internal wire, as well as between the pins and sockets in the connector itself, will result in capacitive and inductive coupling of interference from the shield to the inner conductors.

The mutual inductance between two parallel conductors, enclosed within an overall, cylindrical shield is [9]

$$L_m = \frac{\mu_0}{2\pi} \cdot \ln \left[\frac{d_r}{r_{sh}} \sqrt{\frac{\left(d_g d_r\right)^2 + r_{sh}^4 - 2 d_g d_r r_{sh}^2 \cos\theta_{gr}}{\left(d_g d_r\right)^2 + d_r^4 - 2 d_g d_r^3 r_{sh}^2 \cos\theta_{gr}}} \right]; \; H/m \qquad (7.63)$$

where L_m is the per-unit-length mutual inductance between the conductors (in this case, the wiring and pins run in parallel) in H/m, and μ_0 represents the permeability of the medium between the conductors H/m. The shield radius is r_{sh} and the wires have radii r_{wg} and r_{wr}, and are separated from the shield inner surface by distances d_g and d_r, respectively (see Figure 7.57).

The per-unit-length mutual capacitance, C_m, between the wires can now be approximated from [9]

$$C_m = \frac{L_m}{v^2 \left(L_g L_r - L_m^2\right)}; \; F/m \qquad (7.64)$$

Where L_g and L_r represent the per-unit-length self-inductances of the wires (Figure 7.57):

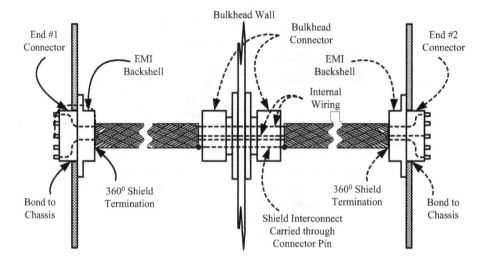

(a) Wire the Shield through a Pin in the Bulkhead Connector

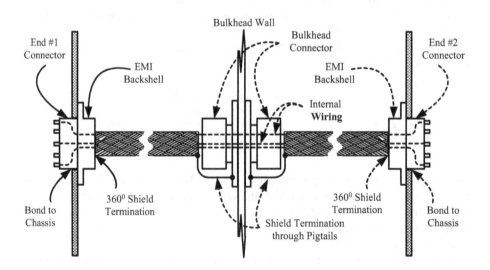

(b) Pigtail Shield Termination

Figure 7.56. Approaches for shield termination at bulkheads for maintaining shield continuity.

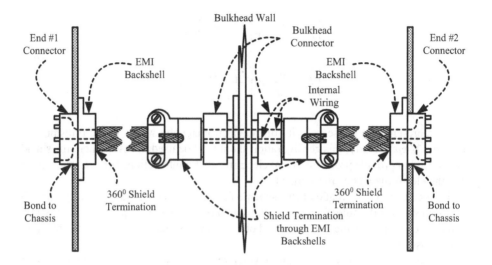

(c) Peripheral (360^0) Shield Termination

<u>Figure 7.56.</u> *Continued.*

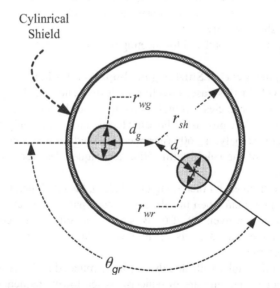

<u>Figure 7.57.</u> Computation of per-unit-length mutual inductance between wires enclosed within an overall cylindrical shield.

$$\begin{cases} L_g = \dfrac{\mu_0}{2\pi} \cdot \ln\left(\dfrac{r_{sh}^2 - d_g^2}{r_{sh} r_{wg}}\right); \ \mathrm{H/m} \\[2em] L_r = \dfrac{\mu_0}{2\pi} \cdot \ln\left(\dfrac{r_{sh}^2 - d_r^2}{r_{sh} r_{wr}}\right); \ \mathrm{H/m} \end{cases} \tag{7.65}$$

and ν represents the phase velocity of the electromagnetic waves in the transmission line, $\nu = \sqrt{\varepsilon\mu}$. Clearly, this configuration should not be used in cases where the shield is intended to provide protection from electromagnetic interactions between the wiring and the electromagnetic environment.

Limited use may be made of this scheme for maintaining shield continuity through the connectors if the sole objective of the shield is to provide suppression of crosstalk coupling contained inside the shield and other, similarly shielded wires in a cable having a gross overbraid.

Figure 7.56(b) depicts a situation in which the cable shields are terminated through pigtails to the bulkhead connector's strain relief or directly to its chassis. Although the shield is no longer routed parallel to the shielded wires in the bulkhead, it still bears the disadvantage of a less than optimal shield termination for reasons previously discussed.

7.4.4.4 *Termination of Multiple Shields.*

Up to now, the discussion of shield termination considered either the overall shielding braid or the individual shields of the wires internal to the shielded cable assembly, but not both simultaneously. Figure 7.58 illustrates a situation consisting of a cable, an overall braid, and an internal balanced, shielded wire pair.

Multiple shields are considered for meeting one or both of the following objectives:

Achieving a Higher Level of Shielding for Improved Radiated EMI Emission and Coupling Control. In this case, effective high-frequency shielding is normally required. At higher frequencies, ground loops are typically not a concern. Therefore, when multiple shields are provided, the same termination rules apply for both the inner and overall shield, namely, a 360° (peripheral) shield termination should be implemented, using one of the shield termination techniques described in Section 7.4.4.2 [Figure 7.58(a)].

When isolated from each other along their length, the combined transfer impedance of the cable is higher than when in galvanic contact, particularly at low frequencies at which a theoretical improvement of 6 dB is achieved, due to the reduced conductive coupling between the shields. (This situation is similar to that of two conductors sharing a common return path.)

At the ends of the cable, both shields may be terminated to the same EMI backshell when higher frequencies only are of concern. When lower frequencies are involved, two situations may arise, requiring possible single-point individual shield grounding, depending on internal wiring configuration and functionality [Figure 7.58(b)]:

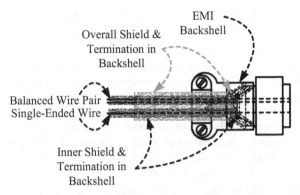

(a) Peripheral Termination of Both Overall and Individual Internal Wire Shields
to a Backshell for High-Level High Frequency Shielding

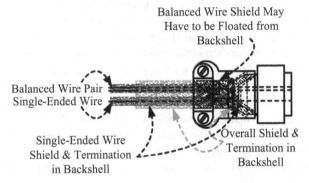

(b) Peripheral Termination of Overall Shield to a Backshell and Floated Individual Internal
Balanced Wire Shield for Both High- and Low-Frequency Shielding

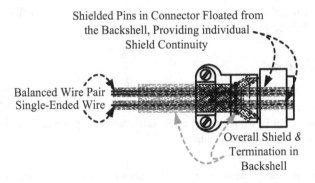

(c) Coaxial Shield Pins for Increased Isolation for Low-Frequency Crosstalk Control

Figure 7.58. Approaches for terminating multiple shields.

1. In a balanced pair,* the inner wire shield may need to be floated from the back-shell at all but one end of the cable, for preclusion of low-frequency ground loops.

2. When a single-ended wire carries lower frequency signals, its individual shield cannot be floated and should be grounded at the backshell or within the enclosure, which is even less advisable, potentially compromising lower frequency functionality.[†]

Achieving Low-frequency Isolation Between the Inner Shielded Conductors in Addition To Shielding For Control of EMI Emission and Coupling. In this case, shielding against external high-frequency EM fields is achieved only by the external shield, offering moderate shielding performance.

Here also, two situations may arise regarding the second objective, namely, low-frequency isolation between conductors for crosstalk control, depending on internal wiring configuration and functionality [Figure 7.58(c)]:

1. An individual single-ended wire shield can be floated with its continuity maintained using coaxial pins internal to the connector, as described in Section 7.4.4.2 (Figure 7.48). The shield should be grounded internally to the load, which should be taken into consideration in the load circuit's "grounding tree" (Figure 7.58(c)).

2. Shielded pins in connectors for balanced wire pairs are not commonly used. Normally, such wires would have to routed through regular pins in the connector, with the individual shields grounded to the backshell, as described above. In extreme cases, twinaxial[‡] pins may be utilized; the shield continuity would be maintained without grounding at the connector backshell itself. Alternatively, "guard/shield" pins connected to the wire shield, but not to the connector shell or to the backshell, may be used around the balanced wire pair (Figure 7.59).

It is important to bear in mind that whenever floating of shields is considered for achieving a single-point shield termination, careful attention should be paid to the concerns presented in Section 7.3.6. Fortunately, as the overall shield is normally terminated at both ends, the inner conductors benefit from the shielding offered by the overall braid and their respective grounding method.

Lessons Learned

- Serious interference problems arise when shielded wires or coaxial cables are not properly terminated at their end connector points. All junctions or joints from the shield to the mating enclosure must be optimized. A shield is just as good as the weakest junction.

---

*Recall that even in shielded balanced wire pairs, difficulties may arise at lower frequencies with two-end shield termination.

[†]In the case in which the shield of the single ended wire is floated in the load circuit, this configuration may be considered a "balanced unshielded pair," where the shield actually acts as the return conductor and not as a high-frequency EMI shield.

[‡]Twinax is an impedance-controlled shielded wire pair configuration.

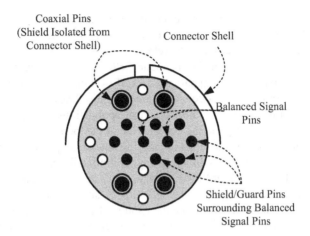

Coaxial Pins
(Shield Isolated from
Connector Shell)

Connector Shell

Balanced Signal
Pins

Shield/Guard Pins
Surrounding Balanced
Signal Pins

Figure 7.59. Connector insert arrangement consisting of coaxial pins and standard pins used a "guard/shield" configuration for a balanced signal pair.

- Pigtail shield termination is highly inductive and should be used only when absolutely unavoidable.
- The best shielding performance is obtained using a low-impedance peripheral (360°) shield termination.
- Individual wire shields should be terminated either through the "halo technique" or similar methods.
- In multiconductor cables, where both individual shields and a common overall braid are used, all individually shielded cables should have insulating sleeves or coverings over the shields within the cable, particularly on lower frequency signal lines, for minimizing conductive cross coupling. Individual shields must be isolated from the overall bundle shield, equipment chassis and enclosures, junction boxes, conduit, cable trays, and so on.
- The overall shield braid should be terminated at both ends for achieving high performance of the shield at higher frequencies. The inner shield termination scheme should follow the circuit configuration and functionality.
- Overall shields of multipair cables should not be used as signal return paths.
- Connectors with coaxial pins can be used for providing continuity of single-ended shielded wires. Alternatively, guard/shield pins may be used for maintaining shielding integrity and isolation of the shields.

7.4.5 Twisted Cables and the Effect of Grounding

Twisted wire pairs are often believed to serve as a "magic solution" to most EMI couplings into or from wiring circuits. In practice, the effect of cable twisting is primarily limited to lower frequency magnetic interactions within the cable, and its performance is highly dependent on the wiring circuit's grounding scheme.

As outlined in Chapter 2, magnetic coupling to and from circuits is dominated, first and foremost, by the geometry of the circuit, the loop area in particular. The larger the physical loop area, the greater the EMF (electromotive force) induced into the circuit (Faraday's Law) and the higher the magnetic field emissions from the circuit (Ampere's Law). The first and most apparent effect of twisting wire pairs is that of minimization of RF current or induction loop area (Figure 7.60). Clearly, this effect is small, maybe even negligible for most practical applications. Actually, the effect of pair twisting extends well beyond the contribution of the loop area reduction.

Figure 7.61 depicts a twisted wire pair carrying equal but opposite signal and return currents, I_+ and I_-, respectively. Observing any arbitrary adjacent wire "twists," j and $j + 1$, it is evident that the current flow in each opposes that in the adjacent twist. These currents generate equal but opposite magnetic flux, $\vec{B}_j$ and $\vec{B}_{j+1}$, at an arbitrary observation point, O, sufficiently far away from the j and $j + 1$ twists by distances r_j and $rj + 1$ respectively, such that $r_j, r_{j+1} \gg s$, and $r_j \approx r_{j+1}$ (s represents the spacing between the centers of any two adjacent twists). Consequently, magnetic flux cancellation occurs at point O, considerably reducing the total radiated field emerging from the wire pair. Clearly, only the two extreme loops cannot produce efficient flux cancellation, ending up as the chief contributors to the emissions. The tighter the twisting, the smaller and more uniform the loops will be, resulting in enhanced magnetic flux cancellation.

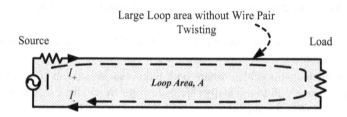

(a) Untwisted Wire Pair: Large Current Loop Area

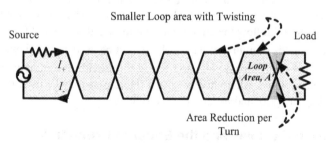

(b) Twisted Wire Pair: Small Current Loop Areas

Figure 7.60. Effect of wire pair twisting on current loop area reduction of magnetic field interactions.

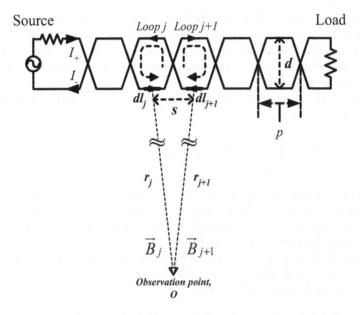

Figure 7.61. Cancellation of magnetic flux due to wire pair twisting.

The twisting efficiency in producing flux cancellation, or "twist factor," R_T, in dB, is expressed by [17]

$$R_T = -20 \cdot \log\left\{\left(\frac{1}{2nl+1}\right) \cdot \left[1 + 2nl \cdot \sin\left(\frac{\pi}{n\lambda}\right)\right]\right\}; \text{ dB} \tag{7.66}$$

$$R_T \leq 60 \text{ dB } @ f \leq 100 \text{ kHz for } 30 \div 40 \text{ twists/m}$$

where n is the number of twists per unit length (twists/m), l represents the total length of the cable (m), and λ, corresponds to the wavelength (Figure 7.62) [17]. In practice,

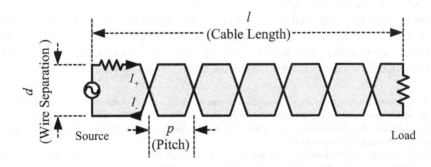

Figure 7.62. Factors affecting twisting effectiveness.

R_T is normally limited to 60 dB, and should not be considered for frequencies exceeding 100 kHz. The number of twists per unit cable length is limited by physical size of the cable's jacket and wire diameter, however, the greater the number of twists per unit length (or the smaller the pitch), the smaller the effective loop area and consequently magnetic emission or pickup.

The resultant maximum magnetic field emission from a twisted wire pair is expressed as [17]

$$\left| \vec{B} \right|_{Max} = \frac{\mu_0 I \cdot d}{p \cdot r} \cdot q \cdot I_0(q) \cdot e^{\frac{-2\pi \cdot r}{p}} \; ; \; q = \frac{\pi \cdot d}{p} \tag{7.67}$$

where $I_0(q)$ represents the zeroth order modified Bessel function of the first order, I is the current flowing through the circuit, d symbolizes separation between wires, and p is the pitch of the twist, both in meters [17]. The distance between the wire and the observation point is represented by r, in meters.

The benefit of tight pair twisting can be easily observed when Equation (7.67) is compared to the expression for a simple wire pair with the same spacing, d, carrying an equal current, I (discussed in Chapter 2):

$$\left| \vec{B} \right|_{Max} = \frac{\mu_0 I \cdot d}{2\pi \cdot r \cdot (r+d)} \tag{7.68}$$

The wire pair twisting is equally effective for reducing magnetic interactions from and to a wire pair.*

The key issue associating cable twisting with grounding theory is related to the misconception that use of twisted cables is a "magic solution" to all low-frequency magnetic interference problems. This is false. Several examples are presented herein proving this concept wrong.

In Figure 7.63(a) the circuit is unbalanced due to the circuit's multipoint grounding connection to a reference ground. At lower frequencies, most of the return current will flow through the reference structure, yielding poor flux cancellation or large magnetic field pickup into the loops. Twisting will not offer any significant flux cancellation when the current flowing through both conductors is not equal and opposite, making the twisting virtually useless.[†]

In Figure 7.63(b) the signal wires of two single-ended circuits are twisted together as a wire pair. The two circuits, being single-ended, are grounded to a reference ground, serving as the return path for each. This scheme cannot produce magnetic flux cancellation. The signal currents flow in the loops formed by each of the signal wires

---

*Use of twisted wires reduces the total inductively coupled noise EMF (Faraday's Law) since the induced EMF in each small twist area is approximately equal and opposite to the EMF induced in the adjacent twist. Superposition of the EMFs induced along the entire cable appears to be destructive, resulting in effective minimal interference across the load.

[†]In high-frequency circuits multipoint grounding does not contradict flux cancellation, since high-frequency return current will tend to flow through the adjacent return conductor. However, at higher frequencies, common-mode electric interactions become dominant, and twisting offers little benefit.

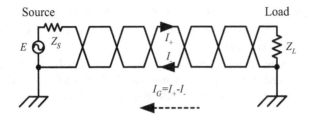

(a) Unbalanced Circuit: Part of the Signal Current Returns

through the Signal Reference Structure (I_G)

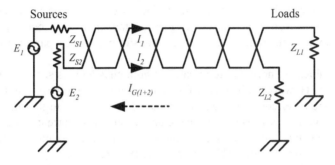

(b) Twisting Separate Single-Ended Signal Wiring: Return

Currents of Both Circuits ($I_{G(1+2)}$) Return through the Signal

Reference Structure

Figure 7.63. Common grounding-related mistakes in twisted circuits resulting in no magnetic flux cancellation.

and the reference ground rather than the loop formed between the wires. The twisting will have no effect on magnetic field suppression.

Note that the same line of reasoning should be applied to balanced cables consisting of more than two conductors, such as three-phase AC power supply circuits, multi-phase motor control lines, and so on. Grounding of both the source and load ends will similarly counteract the circuit balancing in spite of circuit wire twisting. If twisting is used, all current-carrying circuit conductors, such as the three-phase and neutral, must be twisted together or magnetic emissions and field coupling will not be eliminated.

Lessons Learned

- Twisting of wire or cables is only effective in differential, balanced circuits and will be of little benefit in lower frequency unbalanced circuits.

- Circuit grounding topology plays a vital role in the twisting effectiveness.
- Twisting can be used not only on wire pairs but also in any balanced system, such as three-phase power circuits.
- Twisting wiring circuits carrying high-frequency signals has little effect on control of the predominantly common-mode EMI interactions. It does contribute, however, to impedance control of the transmission line.

7.4.6 Strategies for Shield Termination in Common Types of Shielded Cables

7.4.6.1 Coaxial Cables. Coaxial cables are uniform, low-loss, impedance controlled transmission lines constructed as a pair of two conductors with the one conductor enclosed by an outer conductor, separated by a dielectric material. The geometry, construction, and dielectric material determine its characteristic impedance (Figure 7.64). High-performance coaxial cables may contain more than one layer of braid, which contribute to the reduction of surface transfer impedance, particularly at higher frequencies.

Coaxial cables are normally terminated at both ends and, as such, are not considered balanced. However, when used for higher frequencies (above the shield cutoff frequency), the outer conductor, when terminated at both ends to the signal reference structure, carries the vast majority of the return current. This current flows in the outer conductor, constituting the path of least impedance (inductance), minimizing emission and coupling from and to the cable, respectively (Figure 7.65).

Grounding of the outer conductor of a coaxial cable should be accomplished through special coaxial connectors offering both a low-impedance shield termination as well as impedance matching to the characteristic impedance of the transmission line. The same termination schemes apply to single-braid and nonisolated, multiple-braid coaxial cables.

Bayonet-coupling connectors (BNC) should be avoided except when high performance is not required, as it provides for only a small number of contact points. Threaded couplings, used in other configurations (e.g., TNC and N-Type) exhibit significantly lower transfer impedance and thus offer superior performance.

Coaxial cables should be selected based on their characteristic impedance (typically 50 Ω to 93 Ω). Primary applications include RF and high-frequency data and video cir-

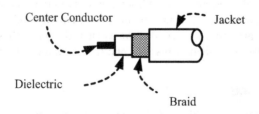

Figure 7.64. Configuration of a coaxial cable.

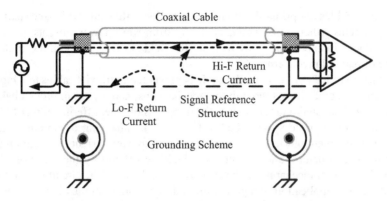

Figure 7.65. Grounding scheme and circuit functionality of coaxial cables.

cuits. However, coaxial cables should not be used for low-frequency applications (audio frequency circuits), with the outer conductor serving as the return path. If a circuit consists of a floating reference, terminating the outer conductor to the reference structure (chassis) at both ends will introduce a severe violation to the overall system grounding topology, resulting in ground loops detrimental to the circuits' performance. Therefore, if used in low-frequency circuits, the outer conductor of the coaxial cable should be connected via insulated (floated) contacts and grounded internally to the equipment chassis according to the grounding topology implemented. This configuration may introduce severe EMI, since the shield could carry interference currents into the equipment enclosure.

7.4.6.2 Triaxial Cables. A coaxial cable with a second braid is commonly called a triaxial cable. Triaxial cables consist of two braids, insulated from each other. The outer braid typically serves as a true shield, and is terminated to the chassis, whereas the inner braid acts solely as a signal return conductor, normally connected to the circuit signal return, thus maintaining a floating electrical reference. As such, triaxial cables function as truly balanced and shielded transmission lines (Figure 7.66).

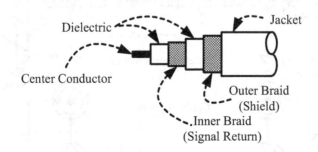

Figure 7.66. Configuration of a triaxial cable.

Triaxial cables are primarily used in low-level sensitive and wideband (particularly video) systems, where noise voltages arising from stray power currents or from currents induced in cable shields by incident RF fields flowing through the cable shield can be of particular concern.

Induction of RF currents into the cable is prevented in triaxial cables to a higher degree than in coaxial cables through two mechanisms, namely (1) improved balance of the triaxial signal and return circuit in the transmission line, offering higher LCL (longitudinal conversion loss, described in Chapter 2), enhancing its immunity to common-mode coupled interference as well as to stray power currents flowing on the signal reference structure; and (2) superior shielding offered by the outer braid to the cable. For the first mechanism to be effective, imbalance of the circuit should be minimized through application of a proper grounding topology in the circuit, whereas, for the second to be effective, the outer braid (shield) of the triaxial cable should be peripherally bonded to the terminating equipment chassis (Figures 7.67 and 7.68).

At higher frequencies, the triaxial cable significantly reduces the shield surface transfer impedance compared to an equivalent coaxial configuration (by up to 30 dB), due to the isolation and decoupling between the two braids.

Terminating triaxial cables should be accomplished through special triaxial connectors, which, in addition to the features of the coaxial connectors described in Section 7.4.6.1, provides isolated shield and return contacts. The shield contacts constitute part of the shell of the connector, thereby terminating to the chassis, whereas the inner braid is connected to isolated contacts (Figure 7.68) [12]. Similar to coaxial connectors, threaded coupling should be preferred over bayonet-coupling connectors.

Triaxial cables should be selected based on their characteristic impedance (typically 50 Ω to 93 Ω). Primary applications of triaxial cables include video and some high-frequency data circuits.

7.4.6.3 Twinaxial Cables.
Twinaxial cables contain two signal lines that are balanced (differential-mode signaling), exhibiting well-controlled characteristic im-

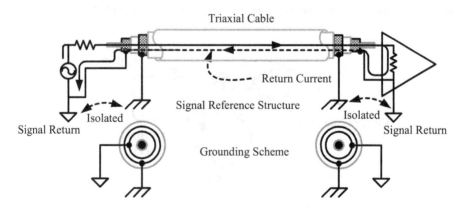

Figure 7.67. Grounding scheme and circuit functionality of triaxial cables.

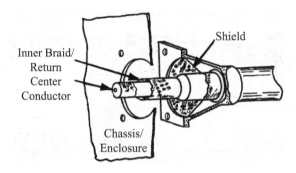

Figure 7.68. Application of triaxial cable shield grounding.

pedance with an overall braid around the wire pair. The braid serves as a shield, and is terminated to the chassis, whereas the signals propagate between the balanced wire pair. One wire serves as a signal conductor, and the second as the return conductor. Both wires may be parallel or twisted in construction. The return conductor is normally connected to the circuit signal return, thus maintaining a floating electrical reference (Figure 7.69). With the exception of the highly controlled characteristic impedance and the high-quality shield, the twinax is equivalent to any twisted and shielded wire pair. The drawback of the twinax is its frequency range of operation: usefulness of twinax cables is limited to an upper frequency of 10 MHz, approximately.

Similar to the triaxial cables, twinaxial cables benefit from the separation between the signal and shield circuits. The outer shield provides a high degree of protection against incident electromagnetic fields, whereas the inner conductors constitute a balanced transmission line, further contributing to the immunity of the circuit to common-mode coupled interference, as well as to stray power currents flowing on the signal reference structure. The shield should be peripherally bonded to the chassis of the terminating equipment, whereas the inner conductors should be connected to the equipment circuit returns in accordance with the system's grounding topology (Figure 7.70).

Terminating triaxial cables must be accomplished through special connectors, which also maintain the characteristic impedance of the line. If the characteristic impedance is of no concern, regular connector pins may be used; however, the need for use of expensive twinaxial cables should be carefully considered. Regular twisted shielded pairs may be used instead. Either way, the shield of the Twinax should be ter-

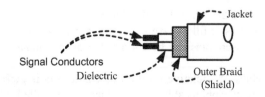

Figure 7.69. Configuration of a twinaxial cable.

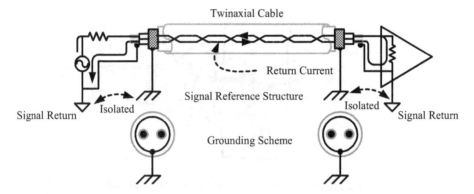

<u>Figure 7.70.</u> Grounding scheme and circuit functionality of twinaxial cables.

minated to the equipment enclosure or chassis, using one of the techniques presented in Section 7.4.4.2.

Twinaxial cables should be selected in consideration of their characteristic imped-ance (typically 78 Ω to 124 Ω). Primary applications of triaxial cables include differ-ential video and some balanced data lines, operating at frequencies typically not ex-ceeding 10 MHz, due to their useful bandwidth of operation.

Lessons Learned

- Coaxial cables are comprised of a center conductor and an outer braid(s). They are intended for high-frequency applications. The outer conductor of the coaxial cable serves as the signal return and should be terminated to the chassis at both ends. Coaxial cables should not be used for low-frequency applications.
- Triaxial cables are shielded coaxial cables, comprised of two isolated braids sur-rounding a center conductor. The inner braid serves as a signal return and, to-gether with the center conductor, serves as a balanced transmission line. The in-ner braid should be connected to the signal return conductors in the equipment. The outer braid serves as a shield and should be terminated at both ends of the cable to the equipment's enclosure.
- Twinaxial cables are similar to a shielded twisted pair. The inner wire pair serves as a balanced transmission line, whereas the outer braid serves as a shield and should be terminated at both ends of the cable to the equipment's enclosure.

7.4.6.4 Ribbon Cables. When flexibility requirements and space constraints are critical, ribbon cables fulfill diverse applications, typically inside equipment enclo-sures. Ribbon cable assemblies can be constructed of individual or twisted pair con-ductors.

With regular cables, the relative position of conductors is generally random to a great extent. In contrast, ribbon cables are considered "controlled cables," maintaining consistent electrical characteristics (e.g., impedance, capacitance, inductance, time delay, cross-talk, and attenuation) due to the fixed position and relative orientation of the conductors (Figure 7.71). The key concern associated with the application of ribbon cable assemblies relates to the manner of individual conductor assignment, particularly corresponding to the relative position of signal ground conductors [3].

Figure 7.72(a) depicts a ribbon cable assembly consisting of multiple conductors sharing a common single return conductor. This configuration suffers from two major disadvantages: (1) large common-impedance coupling between the different transmission line conductors through the single return path, constituting a common-impedance; and (2) the large loop area between the signal(s) and return conductors, particularly those at the extreme end. A large loop results in excessive radiated EMI emission and pickup, as well as increased crosstalk, both capacitive and inductive, between the individual conductors.

Figure 7.72(b) presents an enhanced arrangement in which multiple return conductors are used, and are located adjacent to each signal conductor. Certain ribbon cables can be purchased as twisted pairs and twisted and shielded pairs, which is of particular advantage for balanced circuits. Each signal conductor is routed adjacent to a corresponding return conductor. Common-impedance coupling between different circuits is thus eliminated while the smaller circuit loop area minimizes crosstalk as well as as EMI emission and pickup.

In Figure 7.72(c), an intermediate configuration is shown. It is slightly inferior to (b), as it contains a smaller number of return conductors. Each signal conductor still has an adjacent return conductor. As a result, some circuits must share a common return path. This configuration may be adequate in applications that do not carry higher frequency RF energy.

Ribbon cables are also available with a ground plane placed under the conductors, across the entire width of the cable [Figure 7.72(d)]. Since the spacing between the wires to the ground plane is much smaller than the interconductor separation, the ground plane serves as the primary return path for both single-ended and differential pairs. Consequently, the area of the current return path is minimized resulting in an enhanced EMI performance. However, high-quality termination of the plane at the source and load is mandatory to preclude compromising the performance of the ground plane as an effective return path. This is difficult to achieve as metal connectors are

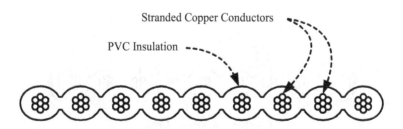

Figure 7.71. Typical construction of a ribbon cable assembly.

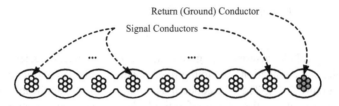

(a) Single Return (Ground) Conductor

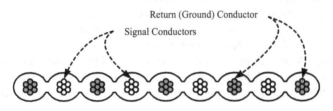

(b) Multiple Return (Ground) Conductors

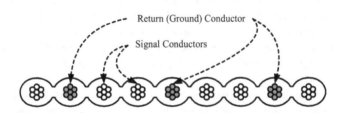

(c) Compromised Number of Return (Ground) Conductors

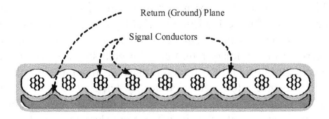

(d) Return (Ground) Plane

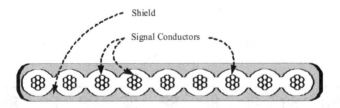

(e) Shielded Ribbon Cables

Figure 7.72. Common grounding configurations in ribbon cable assemblies.

generally large and bulky; hence, their usage is not often desirable, especially for small-size systems.

Shielded ribbon cables are also available [Figure 7.72(e)]. Their performance is significantly dependant on the availability of high-quality peripheral (360°) shield termination; this is not easy to achieve in a ribbon cable assembly. Also, due to the nonuniformity of the shield current, conductors at the center of the cable appear to be better shielded than those located at the edges, requiring careful analysis on how one assigns the pinout of the cable assembly of the conductors' signal assignment [3].

Lessons Learned

- Flat ribbon cables offer the advantage of cost and control of configuration.
- Their primary drawback is the difficulty in providing high-performance signal return paths and adequate cable shielding, due to the difficulty of achieving high-quality return and shield termination.
- Ribbon cables should be used with care, particularly where EMI and signal integrity constitute a major concern.

7.5 GROUNDING CONSIDERATIONS IN SIGNAL INTERFACES

Up to this point, the discussion has focused on grounding of the cable shields. In certain cases, nonetheless, confusion arises, particularly when the shield of a coaxial cable is used as a signal return conductor, rather than a "Faraday cage." This section addresses the question of signal circuit grounding.

One of the most common claims against multipoint shield grounding is that "ground loops will arise." An outcome of this "ground loop" myth is that shields should also be terminated at one end only. As shown previously with few exceptions, ground loops are a concern primarily for lower frequency signal circuits only.

In practice, multipoint grounding of signal interconnects is advantageous, with the exception of lower frequency circuits only, and more so when the shield also serves as the return conductor. For many applications, this constitutes a poor design in the first place. In higher frequency circuits, multipoint shield grounding is mandatory for ensuring acceptable high-frequency performance.

This section is intended to demystify the question of signal grounding versus shield grounding and to provide guidance as to the manner of the signal circuit grounding in various situations.

7.5.1 Interfacing Low-Frequency Unbalanced Signal Circuits

Unbalanced single-ended shielded transmission lines carrying low-frequency signals. The shield serves as the return conductor and may be adversely affected by multipoint circuit and shield grounding. When the shield is single-point grounded to a reference structure (broken line A in the load circuit within Figure 7.73 is not connected), the signal return current is confined to the shield, resulting in effective low-frequency mag-

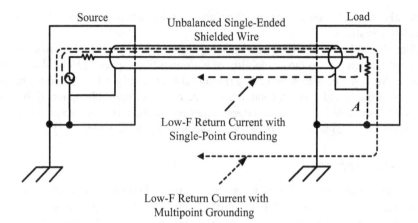

Figure 7.73. Grounding of a single-ended shielded circuit at low frequencies.

netic flux cancellation and no susceptibility to ground loops. However, if multipoint grounding (broken line A in the load circuit) is present, the return current splits between the reference structure and the shield, resulting in poor magnetic field immunity and emission performance. Furthermore, any interference induced onto the shield will directly couple into the signal circuit, as the shield constitutes an inseparable part of the circuit. The skin depth is large at this frequency, resulting in no or insignificant isolation from the outer to inner shield surface (Figure 7.74).

When unbalanced signal interfacing must be used, the signal return must be grounded at one end or the other, but never both. Deciding in advance to ground the signal line at only the source (driver end) or the load (receiver end) generally leads to implementation problems in complex installations, particularly since some equipment may serve as both a load for one signal circuit and a source for another load at the same time (Figure

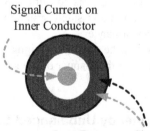

Signal Current on
Inner Conductor

Low-Frequency Signal and Interference
Currents Share a Common Path through
Entire Shield Cross Section

Figure 7.74. Lower-frequency signal return and interference currents are inseparable and share a common return path through the shield.

7.75). In such cases, the single-point circuit ground will be compromised if each line is grounded at its respective source or at the load. Moreover, most modern electronic installations are so complex and widely distributed that even if unbalanced interfaces could be implemented without violating the facility's lower frequency signal grounding network, it is highly probable that future equipment additions and modifications would compromise the single-point ground system. As a result, the equipment must be designed to include isolation from ground at either the source or load. Better still, unbalanced signal interfacing should be avoided at low frequencies whenever possible.

7.5.2 Interfacing High-Frequency Unbalanced Signal Circuits

Contrary to the low-frequency situation discussed in the previous section, higher frequency interfacing signal lines between equipment may be unbalanced or be constant-impedance transmission lines such as coaxial cables. The current return conductor (the shield in the case of a coaxial cable) should be grounded to the equipment enclosure at both ends (Figure 7.75) as well as at intermediate points along the cable run, when possible.

Higher frequency return currents flow primarily through the shield, which constitutes the path of least inductance, even when grounded at both ends. In fact, the shield performs as two virtually isolated current paths, namely, the inner skin carrying the higher frequency signal return current resulting from the proximity effect* and the outer surface carrying high-frequency interference current thanks to the small skin depth at higher frequencies (Figure 7.76).

If the shield is erroneously single-point grounded (line A is not present but line B is connected in the load in Figure 7.77), the path of the signal's high frequency return currents remains virtually unchanged. For electrically long cables, induced noise currents in the cable shield, arising from incident RF field (i.e., the antenna effect) flowing through the cable shield, can be troublesome. Coupling of the shield EMI current to the inner conductor will occur through the parasitic capacitance C and vice versa. From the standpoint of EMC, this situation could be detrimental.

Worse yet will be the outcome of keeping the shield totally afloat at the load circuit (line B is not present, see Figure 7.77). The load impedance is, hence, grounded by line A to the signal reference structure and high-frequency return current is forced to flow through the signal reference structure rather than the shield, resulting in a highly inductive radiating loop, with possible disastrous results from the perspective of EMC.

7.5.3 Interfacing Equipment Containing both Low- and High-Frequency Signals

Many types of equipment process both low- and high-frequency signals based on operational requirements. For instance, a typical VHF or UHF receiver will require both a high-frequency input from the antenna and a low-frequency output to audio amplifiers (Figure 7.78).

---

*The proximity effect is discussed in Chapter 2.

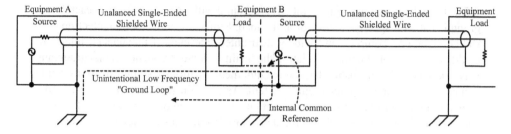

<u>Figure 7.75.</u> Effect of arbitrarily grounding the source end of unbalanced equipment interconnecting cables.

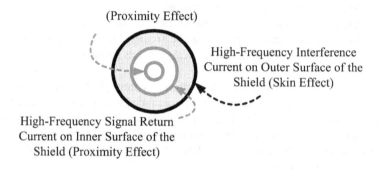

<u>Figure 7.76.</u> Higher frequency signal return and interference currents use separate return paths through the shield.

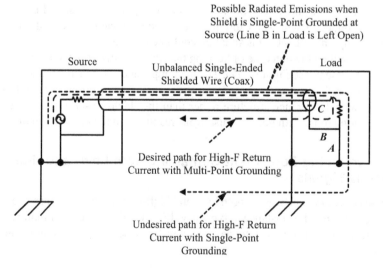

<u>Figure 7.77.</u> Grounding of a single-ended shielded circuit at high frequencies.

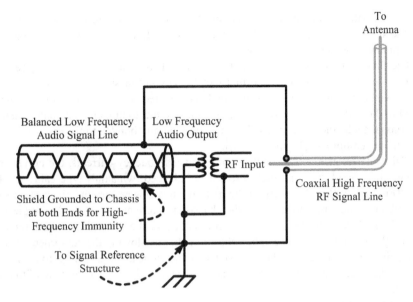

<u>Figure 7.78.</u> Grounding practices in equipment containing both high- and low-frequency signals.

If the low- and high-frequency circuits are functionally independent internally to the equipment and can be separated, a hybrid signal grounding topology can be implemented with the low-frequency and high-frequency circuits utilizing single-point and multipoint topologies, respectively, interconnected in a "hybrid" manner.* However, when both low- and high-frequency circuits share a common signal ground because of design constraints, both signal circuits should be grounded as in high-frequency applications.

The higher frequency interfaces to all transitional-type equipment should exhibit constant-impedance, shielded lines, with the shield grounded around its periphery to the equipment enclosure or chassis. The low-frequency interfaces may be shielded, and should be balanced, twisted-pair lines, as illustrated in Figure 7.78.

7.5.4 Interfacing of Broadband (Video) Signal Circuits

Low-level broadband circuits, particularly video, introduce a special challenge due to the inherent conflicting circuit grounding schemes for both low (single-point) and high (multipoint) frequency content within the video signal spectrum. Noise voltages arising from RF-induced, as well as stray power-frequency currents flowing in the multipoint grounded cable shield may upset sensitive video circuits if unbalanced interfaces are utilized.

*Single-point, multipoint, and hybrid grounding topologies are discussed in Chapter 4.

An alternative approach for combating RF problems is to enclose the shield carrying the signal return current inside another shield, or to make use of a balanced rather than an unbalanced type of transmission line.

The first of these alternatives can be accomplished by using a triaxial* type cable, or the coaxial cable can be routed in a metallic conduit or sleeve. The inner shield of the triaxial cable or the shield of the conduit-protected coaxial cable should be terminated to the signal reference (return) ground inside the equipment. The outer shield of the triaxial cable and the conduit should be peripherally bonded to the enclosure of the terminating equipment [Figure 7.79(a)].

If the interference is the result of stray power frequency currents, the current path through the shield must be interrupted, or a balanced twinaxial[†] type cable must be used [Figure 7.79(b)]. To interrupt the path for stray power currents, the system's signal reference must be connected to the chassis structure at one end only (single-point grounding). Thus, either source or load reference must be isolated from both the chassis and RF ground. Isolation can be achieved either by floating the equipment or its internal circuitry. Generally, either is easier said than done and the best approach is to resort to a balanced interface or relocate the source and reduce the magnitude of the stray current.

7.5.5   Interfacing of Balanced Signal Circuits

The above discussions clearly indicated that balanced signal interfaces benefit from the standpoint of EMI control for circuits.[‡] In particular, balanced shielded lines ensure that the signal return current flows through a designated return conductor rather through the shield; the shield is used strictly for shielding and must not carry any signal (or return) current.

Figure 7.80 demonstrates the application of grounding considerations when interfacing balanced circuits. If the shield is grounded at one end only, be it the source or load, effective shielding against low-frequency E-fields is attained; however, the effectiveness of the shield in precluding low-frequency H-field as well as high-frequency E-field interactions will be compromised. On the other hand, when grounded at both ends, a shield offers effective high-frequency shielding for both E-fields and H-fields.

All signal inputs and outputs should be balanced with respect to the signal reference structure. The signal paths between interfacing circuits should also employ balanced, shielded, twisted-pair lines. For best performance with respect to EMI emissions and pickup, twisting should be as tight as feasible, with a goal of 18 twists per foot or more [13].

Since the shield does not constitute part of the signal circuit, grounding of the shield has no impact on any circuit grounding topologies and the system designer is freed

---

*Triaxial cables and their associated grounding schemes are presented in Section 7.4.6.2.

[†]Twinaxial cables and their associated grounding schemes are presented in Section 7.4.6.3.

[‡]At RF circuits operating at high frequencies only, use of impedance controlled, coaxial transmission lines are still preferred, as long as strictly high-frequency signals are concerned.

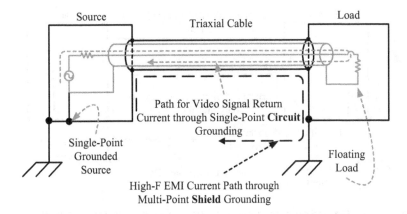

(a) Utilization of Balanced Triaxial Cables against RF-Induced EMI

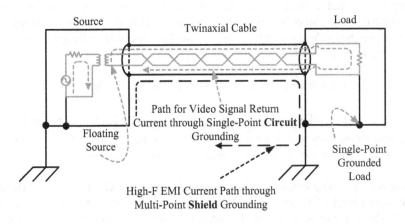

(b) Utilization of Balanced Twinaxial Cables against Stray Power Frequency EMI

<u>Figure 7.79.</u> Preclusion of grounding-induced interference in broadband (video) signal interfaces.

from concerns regarding uncontrolled or undesired grounding system interactions. Coupling between the shield and signal circuits occurs through the transfer impedance mechanism only.

Nevertheless, in some applications such as low-frequency, high-load impedance circuits, grounding the shield of the twisted shielded-pair cable at both ends may allow low-frequency noise currents to circulate through the shield. Although conductive coupling cannot occur in balanced circuits, interference may be introduced into the sensitive, high-impedance circuit through capacitive coupling between the shield and internal wires.

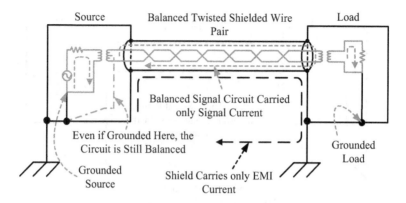

Figure 7.80. Grounding considerations when interfacing balanced signal circuits.

In this case, and in this case only, the shield of lower frequency signal lines should be grounded at one end only. The shield grounding termination to the signal reference structure may be made at either the source or the load end. Generally speaking, the following practices are commonly suggested [13]:

1. Shields of sensitive low-frequency signal lines should be grounded at the load end.
2. Shields of high-level* signal lines should be grounded at the source end.
3. Shields on lines from high-impedance, low-frequency, and DC sources such as sensors and thermocouples, should be grounded at the source end.

If emission or pickup concerns arise because the shield is electrically long (length *l* longer than a tenth of a wavelength, $\lambda/10$ at critical frequencies of concern), the shield may be divided into segments in the manner illustrated in Figure 7.81. Each of the shield segments should be grounded at one end only.

This technique should be applied with caution and only when strictly necessary, owing to its significant inherent drawback. It offers no shielding to higher frequency radiated EMI, which exists in all but very rare cases. Figure 7.82 shows a possible grounding scheme that may be applied in order to alleviate this problem, making use of a double-shielded twinaxial cable [17].

With this method, the twisted-wire pair is enclosed by a double shield similar to a twinaxial cable (Section 7.4.6.3). The outer and inner shields are isolated from each other at DC potential along their entire length and each of the shields is single-point grounded at opposing ends of the cable. At higher frequencies, intershield capacitance (capacitance between the inner and outer shields) is such that the shield acts as if it

---

*High level versus low level is a matter of degree and will depend upon the characteristics of the particular system. A suggested rule of thumb is that if the voltage levels of two signals differ by a factor greater than 10 to 1, then the larger should be treated as high level relative to the smaller.

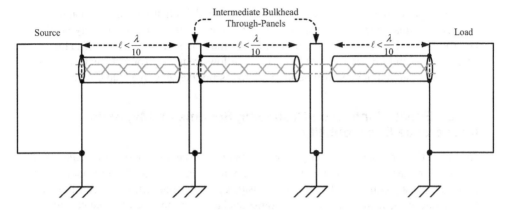

Figure 7.81. Method for grounding individual shields on electrically long low-frequency balanced shielded signal lines. *Note:* Individual shields may be grounded at either end.

were electrically grounded at both ends. This technique is effective against higher frequency radiated fields. The expression for the intershield capacitance is given by [9]:

$$C = \frac{2\pi\varepsilon}{\ln\left(r_{OS}/r_{IS}\right)}; \; \mathrm{F/m} \qquad (7.69)$$

where C is the intershield capacitance, in F/m, r_{OS} is the radius of the outer shield (m), r_{IS} represents the radius of the inner shield (m), and ε symbolizes the permittivity of the dielectric medium between the conductors (ε_0 in free space) (F/m); see Figure 7.82.

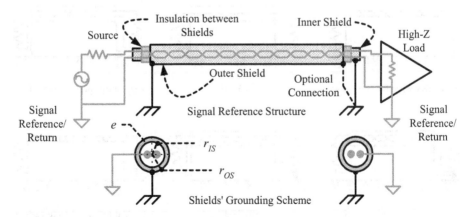

Figure 7.82. Using a triaxial shielding arrangement for low-frequency, high-impedance circuits against high-frequency interference.

If high performance of this arrangement is required, both at low, medium and high frequencies, this arrangement may not be sufficient and the outer shield may need to be grounded at both ends (see the broken line at the load side in Figure 7.82). However, analysis should be carried out to ensure that low-frequency (outer) shield currents do not penetrate it at levels sufficiently high to disrupt the sensitive load circuit.

7.5.6 Effect of Interface Grounding Scheme on Magnetic Interference Susceptibility

An interesting comparison of magnetic interference susceptibility of cable-connected circuitry for various interfacing circuits is shown in Figure 7.83. The evaluation was performed at a frequency of 100 kHz, considered as the boundary between low and high frequencies, with measurement parameters based on the type of cable configuration used, the grounding arrangement of the load and manner of grounding of the cable shield. All circuits were placed 1 inch (2.54 cm) above the signal reference structure [5,14]. The frequency of 100 kHz was well beyond the intrinsic shield cutoff frequency.

The arrangement in Figure 7.83(a) offers virtually no magnetic shielding and is used as a reference, hence it is assigned 0 dB of rejection. The circuit in Figure 7.83(b)

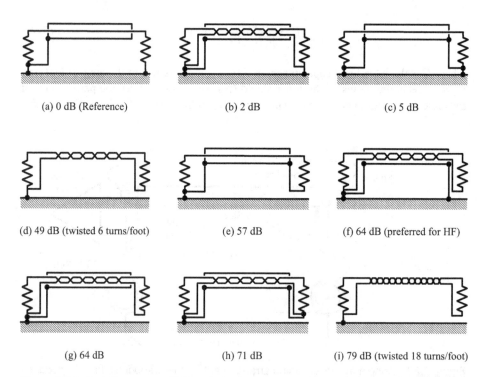

| (a) 0 dB (Reference) | (b) 2 dB | (c) 5 dB |

| (d) 49 dB (twisted 6 turns/foot) | (e) 57 dB | (f) 64 dB (preferred for HF) |

| (g) 64 dB | (h) 71 dB | (i) 79 dB (twisted 18 turns/foot) |

Figure 7.83. Relative susceptibility of circuits to magnetic interference at $F = 100$ kHz. (Note relative suppression for each arrangement.)

is not significantly better. As in the previous case and for the same reasons, the shield provides no attenuation of the incident magnetic fields. As the signal circuit is grounded at both ends, it appears to be unbalanced at 100 kHz, resulting in poor rejection of magnetic field pickup. Although twisting is broadly considered advantageous for reducing magnetic field coupling,* that is true only for balanced circuits. Magnetic fields are coupled into the large loop formed by the twisted pair and the signal reference, resulting in the development of noise voltage across the unbalanced load.

The circuit in Figure 7.83(c) offers some magnetic field suppression, since the field is above the intrinsic cutoff frequency of the shield.† If the circuit were grounded at one end only rather than at both ends, converting it into a balanced circuit, the suppression of magnetic fields would have been even higher [see Figure 7.83(f)].

The circuit in Figure 7.83(d) is an unshielded twisted pair (UTP). As a balanced circuit, it could be anticipated that interference rejection would be higher than 49 dB. That would be correct only as far as magnetic fields are concerned. However, at 100 kHz, the frequency of this example, some common-mode (CM) interference current now shows up as a result of the electric field associated with the magnetic field, flowing through parasitic capacitance to the signal reference structure. This capacitance somewhat degrades the circuit, which is assumed to be "floated" at its right-hand side and thus balanced. If a properly terminated shield were present such as in case (f), E-field coupling would be alleviated by the shield with greater RF suppression.

A significant improvement in magnetic field suppression is achieved in the circuit shown in Figure 7.83(e), representing a balanced coaxial arrangement. The pick-up loop created by the inner conductor and the shield acting as a return conductor is reduced and the ground loop that could potentially defeat the shield performance is nonexistent. In an ideal coaxial cable, this arrangement could yield a very high magnetic rejection but, in practice, coaxial cables exhibit some eccentricity, resulting in a small but nonzero effective pick-up loop compared to a standard (six twists-per-foot UTP cable. Similar to case (d), the performance of this scheme is still limited by E-field (CM) interaction with the cable.

The best performance is achieved by using the circuit in Figure 7.83(f). The twisted, balanced circuit offers high magnetic field rejection. The shield offers high electric field suppression, due to grounding at both ends. As the shield constitutes a separate circuit from the inner signal wiring, they share one common grounding point only. The "ground loop" formed by the shield has a negligible effect on the inner circuit. This circuit is, therefore, preferred for higher frequencies.

Figure 7.83(g) further emphasizes the similarity between the shielded twisted-pair (STP) cable [case (g)] and the balanced coaxial [case (e)] arrangements. The fact that the performance of case (g) is superior to that of case (e) indicates that the specific coaxial cable used presented a larger pick-up area compared to the STP cable. This may not necessarily be valid in certain applications.

Figure 7.83(h) offers a slight enhancement compared to the previous case (g). It constitutes an amalgamation of cases (e) and (g).

---

*The effect produced by twisting of cables and the effect of grounding on twisted cables is discussed in Section 7.4.5.
†Description of the shield cutoff phenomenon is discussed in Section 7.3.5.2.

The circuit in Figure 7.83(i), represents an improvement over case (d), due to the higher twisting rate per unit length, resulting in higher rejection of magnetic fields.

Among other results, the comparison of these arrangements illustrates the disadvantage of returning any load current through the signal reference structure* and the advantages gained using tightly twisted leads.

For low-level signals and low-impedance circuits in which the distance between the interfacing input connector and the actual circuit input is small [up to 5 cm (2 inches)], the use of a twisted-pair cable alone may prove adequate. For long runs, the use of shielded twisted-pair cables becomes mandatory, both in unbalanced and balanced interfaces. Care should be given to ensure the preservation of circuit balancing utilizing a proper grounding scheme of the circuit and shield, or both. Single-point shield grounding is applied for short runs and multipoint grounding for long runs. Particular attention is required when a shielded twisted pair is part of a cable routed through a connector if isolation of the twisted conductors and the cable shield from the connector's (grounded) shell is required.

High-level signals will, in general, not be bothered by a circuits' susceptibility to external incident fields. Rather, they may be a source of emissions and interference coupling to sensitive, lower-level signal lines. For this reason, and dependent on other characteristics of the signal, either a twisted-pair or a shielded lead should be used. Multiple shielding may be required if the signal contains a sufficiently high level of induced RF current. Grounding of the shield should be applied at both ends to prevent electric field radiation from the cable.

Lessons Learned

- Balanced lines, for example, twisted shielded pairs, are preferred for low-frequency or broadband signals containing a major low-frequency spectral content signal interfacing.

- Unbalanced (single-ended) shielded lines should be avoided in low-frequency circuits. The shield forms part of the circuit path of single-ended lines, leading to noise coupling into the circuit.

- At higher frequencies, grounding both ends of the shield and circuit is required for acceptable EMC performance. The shield represents "isolated" current paths for the signal return current (inner surface) and EMI current (outer surface).

- A double-shield arrangement can be used in sensitive low-frequency, high-impedance circuits with each shield being single-point grounded at an opposing end, taking advantage of the intershield capacitance for effective high-frequency shielding.

- Coaxial and similar single-ended interfaces are effective at high frequencies but should be avoided at low frequencies.

---

*Note, however, that at higher frequencies depending on the intrinsic shield and cable cutoff frequency, the circuit wiring may also utilize multi-point grounding, as the high frequency return current tends to flow through the return wire and not the reference structure, in an attempt to minimize the loop inductance. See Chapter 2 on the "Path of Least Inductance" principle.

7.6 GROUNDING OF TRANSDUCERS AND MEASUREMENT INSTRUMENTATION SYSTEMS

An especially challenging situation related to grounding in signal interfaces and wiring is concerned with high-resolution measurement and data acquisition systems. Such systems consist of transducers, measurement instrumentation, and the associated wiring and interconnects.

Many, if not most, transducers and measurement instrumentation systems are concerned with detection and quantification of physical phenomena (or changes in them), requiring periods of observation ranging from a few milliseconds to several minutes or longer. Owing to the relatively slow nature of such events, the fundamental frequency of the (typically analog) transducer's outputs may range from DC to a few hundred hertz. Often, extremely low signals are to be detected with high measurement accuracy.

Sensitive, high-resolution measurement apparatus is often assumed to adequately address this need. However, extraneous interference voltages resulting from such sources as power distribution systems, electromechanical circuits, and atmospheric noise may compromise measurement accuracy. The spectral content of interference from such sources is typically concentrated within the lower frequency region, typically overlapping the spectrum of the measured physical phenomena themselves. Improper grounding connection between a device under test (DUT) and the measurement apparatus often increases interference coupling into the instrumentation, producing measurement errors. Particular techniques are generally required in order to keep the interference voltages or currents from obscuring transducer outputs and producing measurement errors. This section addresses grounding considerations in measurement transducers and instrumentation systems.

7.6.1 Measurement Accuracy Concerns

Consider the circuit shown in Figure 7.84, where measurement using grounded apparatus is illustrated. E_T represents the measured signal at the output of the measurement transducer, Z_H and Z_L represent the impedance of the "high" and "low" leads of the wiring, and Z_M symbolizes the input impedance of the measurement apparatus itself. The transducer and measurement apparatus are referenced to their respective local grounding points, GND_T and GND_M, respectively. As long as no voltage potential difference exists between the signal reference points GND_T and GND_M, no current flows through the impedance of the return lead Z_L. However, if a potential difference does exist between GND_T and GND_M, interference common-mode (CM) current splits between Z_L and Z_H in series with Z_M. Typically, the input impedance of most measurement apparatus is high ($Z_H \gg Z_M$) and Z_H and Z_M appear in parallel with Z_L; therefore, most of the ground common-mode voltage, E_{CM}, also appears across the measurement apparatus terminals input impedance Z_M (Figure 7.85), resulting in measurement errors.

7.6.1.1 Floating Measurements. A common solution to the problem shown above is to "float" the measurement with respect to the reference ground. An ideal

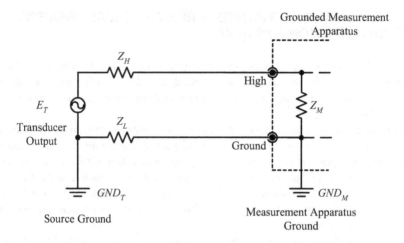

Figure 7.84. Measurements using a grounded apparatus.

floating measurement would be insensitive to the ground-generated common-mode interference and would only measure the differential mode signal appearing between the output terminals of the transducer. Three situations of common practical floating measurements are illustrated in Figure 7.86.

In the first case [Figure 7.86(a)], a measurement apparatus is shown to be connected to the same reference ground as the device being measured. However, the measured voltage is not directly referenced to that ground. Rather, the measurement takes place across the upper resistor, R_U, using the voltage across the bottom resistor, R_B, as refer-

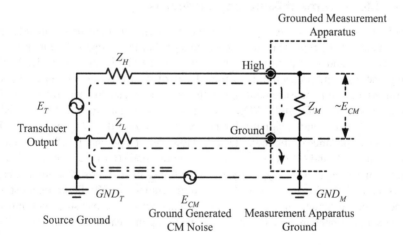

Figure 7.85. Ground-generated interference in measurements using a grounded apparatus.

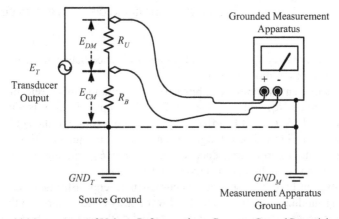

(a) Measurement of Voltage Reference above Common Ground Potential

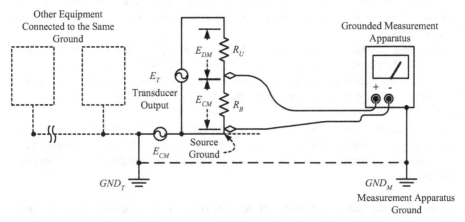

(b) Measurement in the Presence of Common Ground Interference Currents

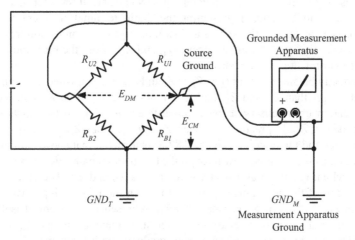

(c) Bridge Measurements

Figure 7.86. Examples of floated measurements.

ence. The voltage across R_B thus constitutes the difference between the two reference potentials, the common-mode voltage.

If the measurement apparatus exhibits sufficient common-mode rejection, accurate measurements across R_U will be achieved. But if its input circuitry is similar to that shown in Figure 7.85, R_B will be effectively shorted out (observe the equivalence between R_B to Z_L in Figure 7.85). Large currents will, consequently, flow through the ground circuits and the entire transducer output voltage, E_T, will appear across R_U. The measured voltage will be higher than the intended value.

The situation in Figure 7.86(b) may occur when several instruments are connected to the same grounding system. Due to the flow of stray currents through the finite impedance of the ground structure, a voltage potential difference exists between different points along the line. Somewhere in the system, the measurement apparatus is grounded to the common grounding system, resulting in common-mode interference across its input terminals.

The bridge circuit shown in Figure 7.86(c) is commonly used in transducer measurements when high resolution is required. Since both sides of the bridge are above ground potential, common-mode noise will be present. If the measurement apparatus does not exhibit high common-mode rejection (CMR), severe measurement errors will result.

In conclusion, none of the techniques illustrated in Figure 7.86 can be carried out using a grounded measurement apparatus, since all require high common-mode rejection in order to provide floated measurements. Floated instrumentation is necessary for that purpose.

7.6.1.2 *Floated Measurement Apparatus.*

Floated measurement apparatus can be made by including a shield between the inner circuits and the apparatus enclosure or chassis (Figure 7.87). The input circuit now consists of three terminals, namely "high," "low," and "ground," representing the "high" terminal in Figure 7.85, the internal (floated) shield, and the apparatus instrument ground connection, respectively. Also shown are the isolation impedances, Z_{HG} and Z_{LG}, from the high and low terminals to the instrument ground, respectively.

In a perfectly balanced circuit (i.e., $Z_H = Z_L$ and $Z_{HG} = Z_{LG}$), where Z_{HG}, and Z_{LG} are also much larger than Z_H and Z_L, high common-mode rejection is achieved and no measurement errors are observed, that is, no differential voltage offset exists across the input impedance, Z_M. In practice, however, $Z_{HG} \gg Z_{LG}$, thus Z_{HG} actually amounts to an open circuit and the circuit in Figure 7.87 transforms to that shown in Figure 7.88. Common-mode currents now flow through the two parallel paths, developing a voltage across Z_L, which results in an offset voltage across Z_H and Z_M, Therefore, common-mode rejection of the apparatus depends entirely on the relationship between Z_L and Z_{LG}. When $Z_L \ll Z_{LG}$, errors are small and Z_{LG} represents the extent of isolation between the inner shield and the chassis of the measurement apparatus. Z_{LG} is typically largest at DC, but due to capacitance it drops gradually with frequency.

Common-mode rejection anywhere between 60 to 120 dB can typically be achieved at lower frequencies using this setup. However, when higher resolution or sensitivity is required, a guarded measurement apparatus is essential.

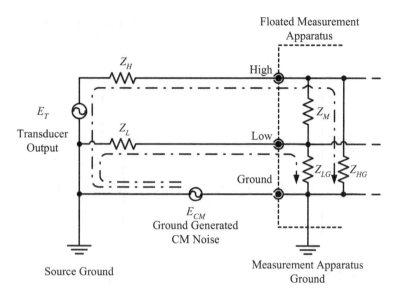

Figure 7.87. Measurement using an ideal floated apparatus.

7.6.1.3 Guarded Measurement Apparatus.

A guarded measurement apparatus includes an additional shield between its low and ground terminals, effectively increasing the leakage impedance between them. The additional shield is called a guard and it may be connected to a circuit or device under test through a dedicated "guard" terminal. This additional shield effectively divides the low-to-ground impedance into two series impedances, Z_{LG1} and Z_{LG2} (Figure 7.89), increasing resistance and reducing overall capacitance. Consequently, a somewhat higher common-mode rejection is obtained, but the greatest advantage of this scheme is attained when a proper connection is made to the measured device or circuit.

Figure 7.89 illustrates how the guard works. In Figure 7.89(a), the guard terminal is left disconnected, resulting in a situation similar to that of a floated measurement apparatus, except that larger low-to-ground isolation impedance is present. Ground-generated common-mode current, ICM, now flows through the two parallel path, again resulting in interference voltage across Z_L, subsequently producing interference across Z_M, the measurement apparatus input impedance. Similar to the floated case, measurement error will be small as long as $Z_{LG1} + Z_{G2} \gg Z_L$. At DC, the guard significantly increases the leakage impedance but at AC only a small increase is observed. As a result, common-mode rejection is considerably improved at DC but only slightly at higher frequencies.

When properly connected, however [Figure 7.89(b)], the guard provides considerable improvement. It shunts the common-mode current away from the two current paths consisting of Z_H and Z_L, practically eliminating them from the common-mode circuit. As virtually no common-mode current now flows through Z_L, almost no error can be produced. Furthermore, since the low and guard terminals are almost at the

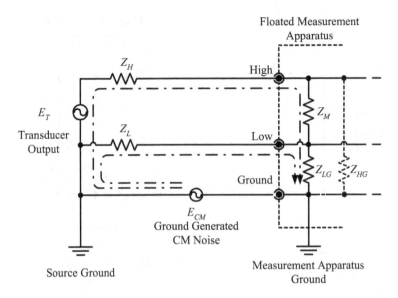

Figure 7.88. Measurement using a practical floated apparatus.

same potential, the voltages on the top and bottom of Z_{LG1} are also almost equal; hence, the voltage difference across Z_{LG1} is maintained very small, yielding higher common-mode rejection across a wider frequency band.

7.6.2 Guard Shields and Instrumentation Wiring Shield Interconnection

One of the most common applications of guard shields is associated with instrumentation wiring circuits. Starting from Figure 7.89(b) and representing the measurement apparatus by an instrumentation amplifier yields the situation shown in Figure 7.90. Now the transducer is connected to the instrumentation amplifier through a shielded cable, which, in turn, is connected to the guard shield of the amplifier, following the reasoning in Section 7.6.1.2 and Figure 7.90(b). Note that the amplifier common is connected to the guard shield, which remains ungrounded (or "floated") to the reference structure, that is, $Z_{LG2} \rightarrow \infty$, represented as capacitance, C_G [3].

This scheme suffers from the drawback that at higher frequencies, capacitance between the guard shield to the local reference structure (ground) contradicts the effect of the guard and cable shields. Current can now flow through the cable shield as well as through the "low" wire to the reference ground through the guard shield, making the guard shield useless at all but very low frequencies.

A second shield, totally enclosing the guard shield and amplifier, however, rectifies this problem (Figure 7.91). The cable shield is connected to the guard shield but does not have any electrical connection with the case shield. The case shield, in turn, is now grounded to the ground reference structure, satisfying functional and safety require-

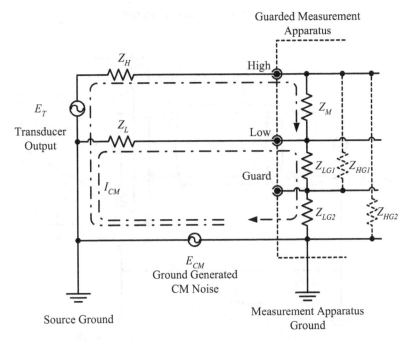

(a) Improper Guard Connection: Common-Mode Interference Produced

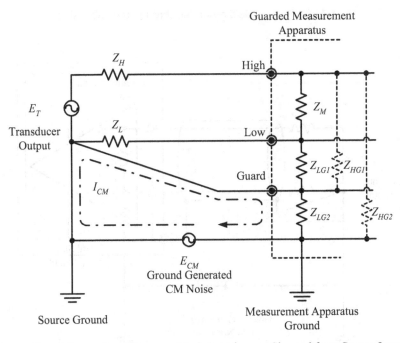

(b) Proper Guard Connection: Common-Mode Interference Shunted from Source Impedance

Figure 7.89. Measurement using a guarded apparatus.

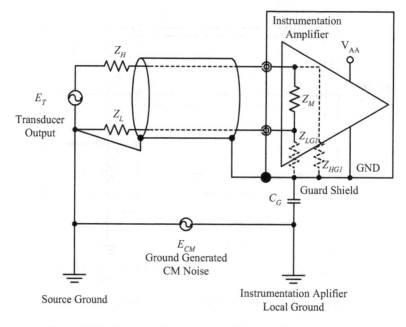

Figure 7.90. Guard shield grounded through the cable shield.

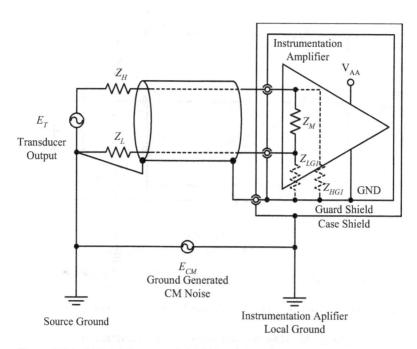

Figure 7.91. Maintaining guard shield integrity with a grounded case shield.

ments with no violation of the guard integrity. Note that in this case, too, the cable shield remains floated from the ground reference at the amplifier end.

This one-end shield termination may result in compromised performance of the shield at higher frequencies (see Section 7.3.6.2). In this case, frequency-selective shield grounding, described in Section 7.3.6.3, or, better still, a second cable shield grounded at both ends and enclosing the inner guarded shield, may be required (Figure 7.92).

7.6.3 Grounding of Wiring Shields in Analog-Data Acquisition Systems

Analog signals such as those detected and processed in measurement-data acquisition systems are primarily lower frequency in nature; thus, single-point grounding architecture should be implemented. The signal circuit must be balanced with the signal return conductor grounded at one end only or totally floated. Similarly, individual cable shields used over signal lines should normally be grounded at one end only, at the signal grounding reference connection point [7]. This grounding scheme is particularly effective when high-impedance, lower frequency interference coupling dominates.* When high-accuracy measurements are required, unique cable shield termination or grounding schemes are commonly applied with a distinction made between measurement equipment guards and enclosure shields (see Section 7.6.2). The discussion is now further extended to consider several particular cases related to transducer types [7,13].

Note that regardless of the type of transducer used, whether grounded or floated, the grounding schemes of the wiring and its associated shield are adequate for lower frequencies only, but ineffective against higher frequency EMI interactions. Following the same reasoning in Section 7.6.2, protection against higher frequency interference coupling (without violation of the guard shield performance) can be achieved by using multiple shields, with the outer shield utilizing both-end shield grounding and the inner shield grounded as described herein. Alternatively, frequency-selective cable shield grounding may be utilized (Figure 7.92).

7.6.3.1 Grounded Transducers. A bonded (grounded) transducer, such as a thermocouple, is a single-ended data amplifier, the output of which drives data acquisition systems or recording devices. Figure 7.93[†] illustrates implementation of several key aspects in single-ended transducer circuits. In particular, the shield surrounding the transducer signal leads is grounded at the same point as the transducer to ensure that the shield and signal return wire are at virtually the same (low-frequency) voltage potential.

---

*When higher frequency interference is present, single-ended shield grounding architecture may be ineffective, when length of the cable, ℓ is greater than $\lambda/10$, where λ is wavelength of the interference signal. Multiple, isolated shields or selective-shield grounding should be considered in this case. Selective-cable-shield grounding is discussed in Section 7.3.6.3.

[†]Note that in all the cases discussed below, the wire pair should be twisted rather than routed simply as a wire pair, for better control of magnetic EMI coupling. This twisting is omitted from the figures below for the purpose of clarity.

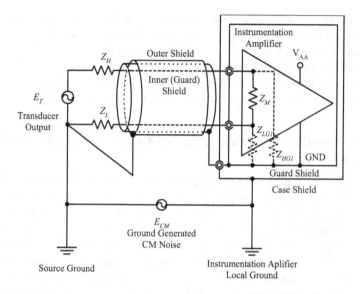

(a) Frequency-Selective Cable Shield Grounding

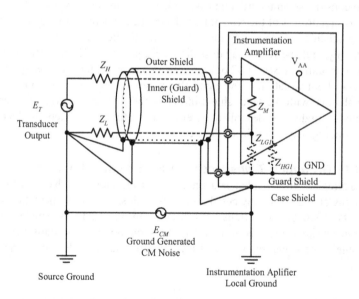

(b) Independently Grounded Dual Cable Shields

Figure 7.92. Solutions for maintaining guard shield integrity at higher frequencies.

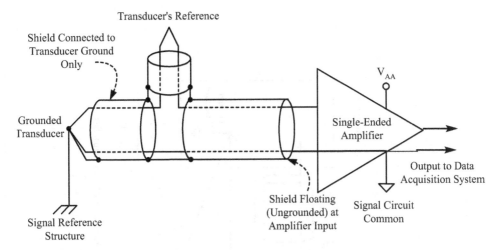

Figure 7.93. Grounding scheme for a grounded transducer and single-ended amplifier.

When the bonded transducer (e.g., thermocouple) is connected to the input of an isolated differential amplifier (Figure 7.94), the shield of the input cable should be connected to the amplifier internal guard shield* in order to maintain continuity of the signal cable shield within the amplifier.

The grounding reference bus interconnects the data acquisition system signal reference and structural chassis ground of the test facility or platform. This ground reference bus is required in any instrumentation system, including isolated differential amplifiers, in order to provide reference for the signal circuitry within the data acquisition system. This common reference precludes high-voltage electrical safety hazards and minimizes high common-mode voltage potentials that may otherwise exist between the amplifier input and output, if the data acquisition system were to be referenced to a separate earth or facility structural ground. Such high-voltage potential difference could result in damage to the data acquisition system if it exceeds the amplifier's input-to-output voltage difference rating. The enclosure of the amplifier and its output shield are connected to the data acquisition system (or load end) ground.

Grounded bridge transducers (Figure 7.95) introduce additional complexity, as they require a DC excitation voltage source. Such transducers should be excited with a balanced DC source as shown in Figure 7.95; thus, the entire bridge is balanced with respect to ground and the unbalanced impedance presented to the single-ended amplifier input will be due only to the leg resistances in the bridge. Although a ground loop does still exist, its consequences are greatly reduced when a balanced excitation supply is utilized [7].

*Guard shields and their grounding schemes are discussed in Section 7.6.1.

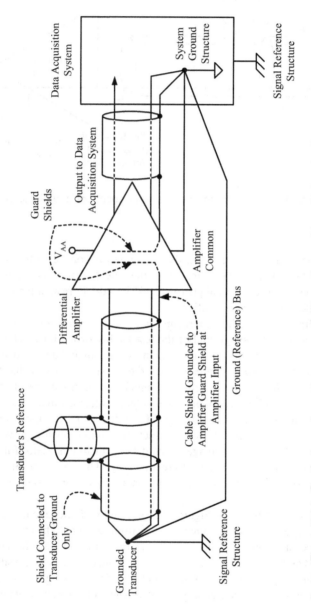

Figure 7.94. Grounding scheme for a grounded transducer with an isolated differential amplifier.

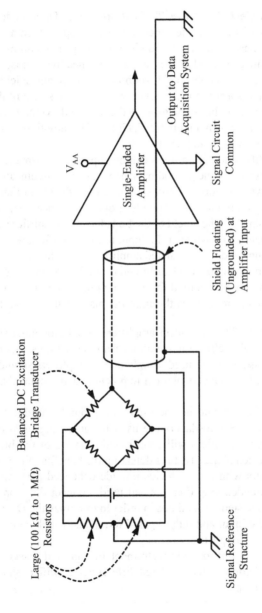

V_{AA}

Single-Ended Amplifier

Output to Data Acquisition System

Signal Circuit Common

Shield Floating (Ungrounded) at Amplifier Input

Balanced DC Excitation Bridge Transducer

Large (100 kΩ to 1 MΩ) Resistors

Signal Reference Structure

Figure 7.95. Grounding scheme for a bridge transducer with a single-ended amplifier

581

Whenever possible, it is best to use isolated differential amplifiers with grounded bridge transducers in the manner illustrated in Figure 7.96, rather than single-ended amplifiers with grounded bridge transducers. In this configuration, both the transducer and the amplifier can be grounded without degrading system performance.

7.6.3.2 *Ungrounded ("Floated") Transducers.* Ungrounded ("floated") transducers remove the difficulty of detrimental ground loops, which may severely degrade high-accuracy measurements. Grounding techniques recommended for ungrounded transducers are presented in Figure 7.97. The metallic enclosure of the transducer is now connected to the cable shield, yet both enclosure and shield are floated at the transducer end with respect to the reference structure. The shield of the input cable to the single-ended amplifier should be grounded at the load, which typically exhibits high-input impedance as shown in Figure 7.97(a).* The amplifier circuit common should also be grounded at the load [7,14].

Here too, isolated differential amplifiers exhibit superior performance compared to single-ended ones. When used, the grounding scheme of such circuits and their respective cable shields, shown in Figure 7.97(b), should be considered. Similar to the single-ended circuit, a single, common, system-wide ground reference connection point is provided for all cable shields. The shield of the transducer-to-amplifier cable should be grounded at the transducer end, while at the input to the amplifier the shield of the input cable should be connected to the isolated amplifier guard shield.

Note that when certain nonisolated differential amplifiers are used, a transducer ground bus connection may be required for ensuring proper amplifier performance. In such cases, the instructions provided by the amplifier manufacturer should be consulted.

7.6.3.3 *Amplifiers.* The transducer amplifier and its grounding scheme in particular can have considerable effect on overall measurement results, especially in cases in which grounded transducers are used. Single-ended amplifiers should not be used, therefore, with grounded transducers in order to preclude detrimental channel-to-channel ground loops.

Single-ended amplifiers should not be used with grounded bridge transducers, in particular. Typically, bridge transducers are used for achieving high-accuracy measurements. Single-ended amplifiers will short-circuit one leg of the bridge, severely corrupting the measurements' quality, thus defeating the benefits of the bridge.

Transducer amplifiers with guard shields provide enhanced performance with respect to simple single-ended amplifiers. When used, connect the amplifier's output guard shield to the data system ground bus in order to minimize the DC voltage potential difference across the input and output guard terminals, which could otherwise inflict severe damage to the amplifier.

Finally, if a permanent unavoidable instrumentation ground exists at test area as well as at the data acquisition system, isolated differential amplifiers should be used to break resultant ground loops.

---

*The circuit grounding to the reference structure is accomplished at the typically high input impedance load, providing higher immunity to stray electrostatic and lower frequency, predominantly high-impedance EM fields.

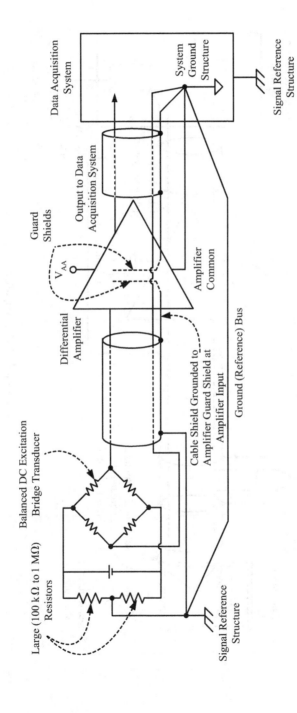

Figure 7.96. Grounding scheme for a bridge transducer with an isolated differential amplifier.

583

584

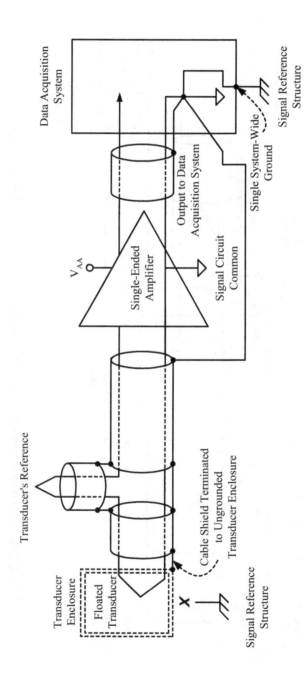

(a) Single-Ended Amplifier

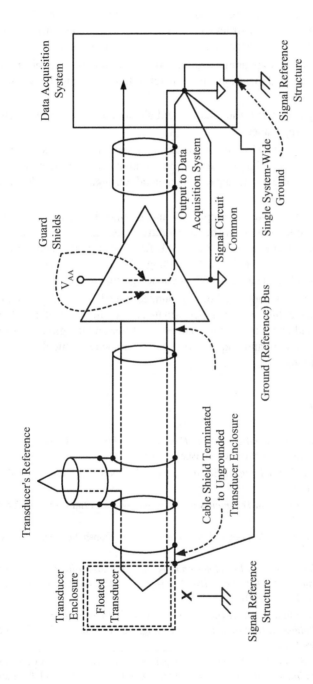

Transducer's Reference

Transducer Enclosure

Floated Transducer

Signal Reference Structure

Cable Shield Terminated to Ungrounded Transducer Enclosure

Guard Shields

V_{AA}

Ground (Reference) Bus

Data Acquisition System

Output to Data Acquisition System

Signal Circuit Common

Single System-Wide Ground

Signal Reference Structure

(b) Isolated Differential Amplifier

Figure 7.97. Grounding scheme for an ungrounded ("floated") transducer.

585

Lessons Learned

- Measurement accuracy can be significantly compromised by improper grounding of the measurement instrumentation (i.e., transducers, amplifiers, and data acquisition systems) and its associated wiring and wire shields.

- Guarded measurement apparatus exhibit superior common-mode rejection compared to simple floated measurements apparatus, thus providing higher measurement accuracy.

- When properly connected, the guard shield shunts common-mode interference from the source impedance.

- The guard and the "low" input signal terminal should be connected such that they are at the same voltage potential; no common-mode current should flow through the low source impedance or, in general, through any impedance affecting the input signal voltage.

- Single-ended instrumentation amplifiers and grounded (particularly grounded bridge) transducers should never be used jointly. When grounded transducers and data acquisition systems are used, isolated differential amplifiers, preferably with guard shields, should be utilized. Grounding of interconnecting cable shields should be accomplished so that lower frequency ground loops are precluded.

- Measurement circuits comprised of ungrounded transducers and isolated differential amplifiers exhibit superior performance; yet, the level of isolation and voltage withstand exhibited by the guarded amplifier and the grounding scheme of the amplifier's input and output wiring and its associated shields must still be carefully considered.

BIBLIOGRAPHY

[1] Tsaliovich, A., *Electromagnetic Shielding Handbook for Wired and Wireless EMC Applications,* Boston: Kluwer Academic Publishers, 2001.

[2] Tsaliovich, A., *Cable Shielding for Electromagnetic Compatibility,* New York: Van Nostrand Reinhold, 1995.

[3] Ott, H. W., *Noise Reduction Techniques in Electronic Systems* (2nd Ed.), New York: Wiley-Interscience, 1988.

[4] Hartal, O., *Electromagnetic Compatibility by Design,* W. Conshohocken, PA: R&B Enterprises, 1994.

[5] NAVSEA OD 30393, *Design Principles and Practices for Controlling Hazards of Electromagnetic Radiation to Ordnance (HERO Design Guide),* Second Revision, Naval Sea Systems Command, April 2001.

[6] White, D., and Mardiguian, M., *A Handbook Series on Electromagnetic Interference and Compatibility, Volume 3, Electromagnetic Shielding,* Gainesville, VA: Interference Control Technologies, Inc., 1988.

[7] Mardiguian, M., *A Handbook Series on Electromagnetic Interference and Compatibility, Volume 2, Grounding and Bonding,* Gainesville, VA: Interference Control Technologies, Inc., 1988.

[8]   Feynman, R. P., Leighton, R. B., and Sands, M., *The Feynman Lectures on Physics, Volume II,* Reading, MA: Addison-Wesley, 1975.

[9]   Paul, C. R., *Introduction to Electromagnetic Compatibility,* New York: Wiley, 1992.

[10] Vance, E. F., *Coupling to Shielded Cables,* New York, Wiley-Interscience, 1978.

[11] Casey, K. F., and Vance, E. F., "EMP Coupling Through Cable Shields", *IEEE Transactions on Antennas and Propagation,* Vol. AP-26, No. 1, January 1978.

[12] MIL-HDBK-1857, *Department of Defense Handbook, Grounding, Bonding and Shielding Design Practices,* Washington, DC: U.S. Government Printing Office, March 1998.

[13] MIL-HDBK-419A, *Military Handbook, Grounding, Bonding, and Shielding for Electronic Equipments and facilities,* Washington, DC: U.S. Government Printing Office, 1987.

[14] Department of the Navy, Naval Air Systems Command, *NAVAIR AD 1115, Electromagnetic Compatibility Design Guide for Avionics and Related Ground Support Equipment,* Third Edition, June 1988.

[15] Recht, E., and Bar-Natan, V., "Methodology for EMC Cable Design," in *Proceedings of the 2002 Wroclaw International Wroclaw Symposium and Exhibition on Electromagnetic Compatibility,* Wroclaw, Poland, June, 2002.

[16] Department of the Air Force, Air Force System Command, *Design Handbook 1-4, Electromagnetic Compatibility,* Fourth Edition, Revision 1, January, 1991.

[17] White, D., *EMC Methodology and Procedures,* Gainesville, VA: Interference Control Technologies, Inc., 1988.

8

GROUNDING OF EMI TERMINAL PROTECTION DEVICES

8.1 FILTERING AND TRANSIENT-VOLTAGE SUPPRESSION— COMPLEMENTARY TECHNIQUES TO SHIELDING

Electrical and electronic equipment cannot always be protected from the electromagnetic environment (EME) by shielding alone. Although shielding prevents both entry and exit of radiated EMI, undesired conducted energy can still couple into the system through power, signal, control, and other wires penetrating the shielded enclosure of the equipment and compromising its electromagnetic shielding integrity. This coupled energy can, thus, degrade equipment performance or even result in malfunction. Filtering and transient-suppression devices installed at the point of entry of cables into the shielded enclosure minimize EMI. Terminal protection devices (TPDs) such as a variety of EMI filters and transient-suppression devices (TSDs) are commonly used to suppress undesired conducted electrical energy to tolerable levels by shunting, bypassing, absorbing, or reflecting the interference energy.

The effectiveness of filtering and transient suppression is dependent on proper implementation of grounding and bonding, particularly for counteracting common-mode interference phenomena. Since many references are available on the design of filter and suppression circuits [1–4], this chapter will provide only a brief overview of filters and transient suppressors but will focus on the role of proper grounding for accomplishing their desired performance.

Grounds for Grounding. By Elya B. Joffe and Kai-Sang Lock
Copyright © 2010 The Institute of Electrical and Electronics Engineers, Inc.

8.2 TYPES OF CONDUCTED NOISE

In order to properly design filters, it is important to understand the different types of conducted noise. The first type, known as differential-mode (DM) noise, propagates out on one conductor (e.g., wire or PCB trace) and returns on another. Noise generated by switching-current waveforms in power supplies is typically DM and its amplitudes are usually minimal above a few MHz. This is because the line-to-line and line-to-ground capacitance inherent in the system, as well as wiring inductance, tend to filter out this type of noise. The other type of conducted noise, known as common-mode (CM), travels in the same direction on both (or all, in the case of multiple-phase power lines, for instance) wires and returns through the chassis or reference structure. In power and signal systems that have a single reference to ground or single-point ground, CM noise is capacitively coupled to the reference structure. Because of this capacitive coupling, CM noise is generally dominant at higher frequencies [2].*

Figure 8.1 illustrates examples of DM and CM noise. Understanding and distinguishing between these modes of conducted interference is important as it results in different design approaches for mitigating each of these two types of noise.

8.3 OVERVIEW OF FILTERING AND TRANSIENT-VOLTAGE SUPPRESSION

8.3.1 Fundamental EMI Filter Devices and Circuits

The purpose of an EMI filter is to prevent the propagation of undesired conducted electromagnetic energy. A filter absorbs noise energy through the use of lossy elements such as resistors and ferrite components, or reflects the noise energy back to the source through use of reactive elements.

An electrical filter can be defined as a two-port network of lumped or distributed passive elements consisting of capacitors, inductors, and, occasionally, resisters or their equivalent, or any combination thereof. Filter networks are designed to attenuate signals at certain frequencies while permitting energy at other frequencies to pass through unchanged.

Reflective filters accomplish this by using combinations of capacitance and inductance to create a high series or low shunt impedance for interfering (EMI) currents, reflecting them back to their source. Lossy filters (e.g., ferrite compounds) do this by absorbing and dissipating the interference energy. The filter's passband is the frequency range in which it exhibits little or no attenuation, whereas the stop band is the frequency range in which high attenuation is desired. The attenuation may vary in the stop band and is usually least near the cutoff frequency, the frequency at which a 3-dB insertion loss is obtained, increasing to higher values at frequencies considerably removed from the cutoff frequency. The characteristic of a filter to pass or suppress signals is called insertion loss, IL, defined as (see Figure 8.2)

---

*Fundamentals of common-mode (CM) and differential-mode (DM) signals were discussed in Chapter 2.

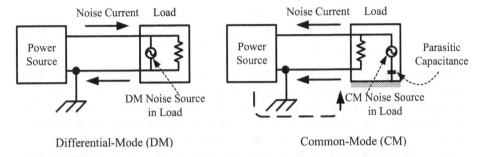

Figure 8.1. DM and CM noise propagation (example for power lines' conducted noise).

$$IL(f) = 20 \cdot \log\left(\frac{E_{L2}(f)}{E_{L1}(f)}\right) = 20 \cdot \log\left(\frac{E_L(f) \text{ filter omitted}}{E_L(f) \text{ filter inserted}}\right); \text{ dB} \qquad (8.1)$$

EMI filters are generally low-pass filters with their effectiveness depending on the impedances of the elements at either end of the filter. The insertion loss of low-pass filters is expressed in general form as

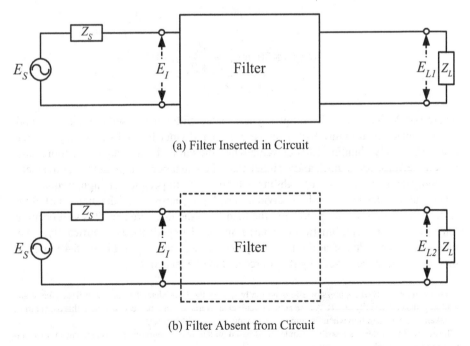

(a) Filter Inserted in Circuit

(b) Filter Absent from Circuit

Figure 8.2. Definition of a filter insertion loss.

$$IL(f) = 10 \cdot \log \left[1 + \sum_{i=1}^{N} k_i \left(\frac{f}{f_0} \right)^{2 \cdot i} \right]; \text{ dB} \qquad (8.2)$$

The index "i" is the "order of the filter" and represents the number of reactive elements in the filter and, subsequently, the number of poles provided by it. The constant k_i is associated with a particular filter topology. f is the frequency of interest and f_0 is the cutoff frequency, or 3-dB knee frequency of the filter. By observation, the higher the order of the filter, the higher the suppression above the cutoff frequency, f_0.

The effectiveness of any EMI filter is greatly influenced by the impedance of both source and load terminations.* For a filter to effectively suppress EMI by reflecting noise, the filter should provide a maximum impedance mismatch to the interference signal, while exhibiting little effect on the desired signal. If the load impedance is low, the impedance of the filter from the load viewpoint, Z_{LF}, should be high. Conversely, if the load impedance is high, the filter should exhibit a low impedance from the load viewpoint.

The most fundamental filter configuration is the discrete filter consisting of a single reactive element—a shunt capacitor or series inductor. The approximate insertion loss of discrete capacitive and inductive filters is expressed in Equations (8.3) and (8.4), respectively [3,4]:

$$IL(f) \cong 20 \cdot \log \left[\omega C \cdot \frac{Z_S \cdot Z_L}{Z_S + Z_L} \right]; \text{ dB} \qquad (8.3)$$

$$1/\omega C \ll Z_S, \ Z_L$$

$$IL(f) \cong 20 \cdot \log \left[\frac{\omega L}{Z_S + Z_L} \right]; \text{ dB} \qquad (8.4)$$

$$\omega L \gg Z_S, \ Z_L$$

where $\omega = 2\pi f$ is the angular frequency associated with f, expressed in radians/second.

The capacitor in Figure 8.4(a) acts as a current divider for high-frequency interference, effectively shunting EMI currents when the parallel combination of source and load impedance are significantly higher than the capacitor's impedance. Conversely, the inductor in Figure 8.4(b) exhibits a high series impedance at high frequencies, serving as an interference voltage divider, developing across itself the major part of the EMI voltage, thus, again, protecting the load from the interference. This will occur as long as the series combination of source and load impedance is significantly lower than that of the inductor at the EMI frequencies.† Z_{SF} and Z_{LF} in Figure 8.4 represent the filter impedance as seen by the source and load, respectively.

---

*Manufacturers of EMI suppression filters normally specify the filter characteristics with fixed source and load impedances (usually 50 Ω). As the actual circuit characteristics may be very different, this aspect must be taken into consideration when designing, specifying, or using EMI filters.

†The effect of the filters can easily by understood based on Kirchhoff's current and voltage laws for capacitive and inductive filters, respectively.

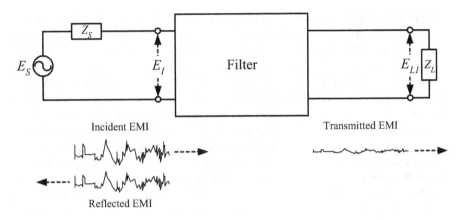

Figure 8.3. Effect of an EMI filter on incident EMI.

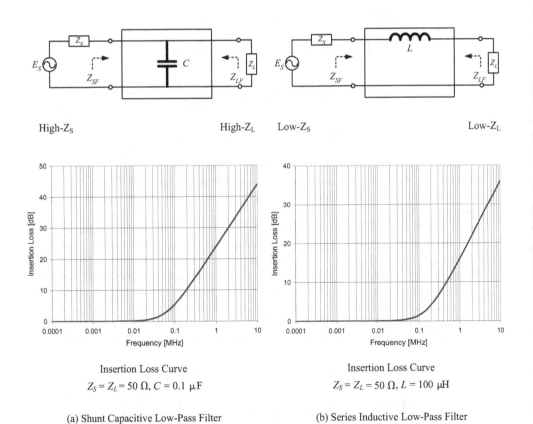

Insertion Loss Curve
$Z_S = Z_L = 50\ \Omega$, $C = 0.1\ \mu F$

Insertion Loss Curve
$Z_S = Z_L = 50\ \Omega$, $L = 100\ \mu H$

(a) Shunt Capacitive Low-Pass Filter

(b) Series Inductive Low-Pass Filter

Figure 8.4. Low-pass EMI filters consisting of a single reactive element.

A primary disadvantage of single-element filters is that their out-of-band theoretical* roll-off rate or slope is limited to 20 dB per frequency decade, or 6 dB per octave.
When a discrete filter does not provide the required suppression, several single-section
filters can be cascaded to obtain more attenuation.

In symmetrically loaded circuits similar to those in Figure 8.4, symmetrical filter sections are used, the most fundamental of which are the "π-section" filter and "T-section"
filter. These filters consist of three reactive elements, resulting in a typical insertion loss
curve with a theoretical slope of 60 dB per decade or 18 dB per octave. In high-impedance circuits, "π-section" filters provide the best results, owing to the intrinsic mismatch
between the filter elements and the source and load impedances. In fact, "π-section" filters are the most commonly used types of EMI suppression networks, whereas in low-
impedance circuits, "T-section" filters are the most commonly used. Both, however,
have a tendency to develop oscillatory ringing when excited by impulsive transients.

The approximate insertion loss of "π" and "T" filters is expressed in Equations
(8.5) and (8.6), respectively [3,4]:

$$IL(f) \cong 20 \cdot \log \left[\omega^2 \cdot LC + \left(LC^2 \cdot \omega^3 + 2\omega C \right) \frac{Z_S \cdot Z_L}{Z_S + Z_L} \right]; \text{ dB} \tag{8.5}$$

$$1/\omega C < Z_S \text{ and } 1/\omega C < Z_L \text{ \& } \omega L > 1/\omega C$$

$$IL(f) \cong 20 \cdot \log \left[\omega^2 \cdot LC + \frac{\left(L^2 C \cdot \omega^3 + 2\omega L \right)}{Z_S + Z_L} \right]; \text{ dB} \tag{8.6}$$

$$\omega L > Z_S \text{ and } \omega L > Z_L \text{ and } \omega L > 1/\omega C$$

In these filter configurations, the capacitors act as low-impedance current dividers and
the inductors act as high-impedance voltage dividers. "π" and "T" filters offer enhanced performance in the same applications in which capacitive and inductive filters
would be used, respectively.

The insertion loss of lossless "π" and "T" filter networks operating with equal source
and load impedances, $R_S = R_L = R$, is defined in Equations (8.7) and (8.8) respectively [5].
These equations are of the same form, but differ with respect to the definition of the equation parameters. The damping factor, d, relates the magnitudes of the filter elements to the
magnitude of the source and load impedances. When the damping factor equals unity, $d =
1$, the response is optimally damped and produces an abrupt transition from the passband
to the stop region. This is known as the ideal Butterworth response curve. When $d > 1$, the
response is in overdamped mode. When $d < 1$, the response is in underdamped mode.

$$IL(f) \cong 10 \cdot \log \left[1 + f^2 D^2 - 2f^4 D + f^6 \right]; \text{ dB} \tag{8.7}$$

$$D = \frac{1-d}{\sqrt[3]{d}}; \ d = \frac{L}{2CR^2} = \text{damping factor}; \ f_0 \cong \frac{1}{2\pi} \left(\frac{2}{RLC^2} \right)^{\frac{1}{3}}; \text{ Hz}$$

---

*In practice, the insertion loss curve can significantly deviate from the theoretical curve, due to parasitics.

$$IL(f) \cong 10 \cdot \log\left[1 + f^2 D^2 - 2f^4 D + f^6\right]; \text{ dB}$$

$$D = \frac{1-d}{\sqrt[3]{d}}; \; d = \frac{CR^2}{2L} = \text{damping factor}; \; f_0 \cong \frac{1}{2\pi}\left(\frac{2R}{L^2 C}\right)^{\frac{1}{3}}; \text{ Hz} \tag{8.8}$$

In asymmetrically loaded circuits (Figure 8.6), asymmetrical filter configurations must be used, the most fundamental of which is the "L-section" filter. Two fundamental representations of low-pass "L-section" filters consist of two reactive elements with a typical insertion loss curve and theoretical slope of 40 dB per decade (12 dB per octave).

The approximate insertion loss of the two "L-section" filters are expressed in Equations (8.9) and (8.10), for the high Z_L [Figure 8.6(a)] and high Z_S [Figure 8.6(b)], respectively [3,4]:

$$IL(f) \cong 20 \cdot \log\left[\frac{\omega L}{Z_L} + LC\omega^2\right]; \text{ dB} \tag{8.9}$$

$$Z_S \ll Z_L$$

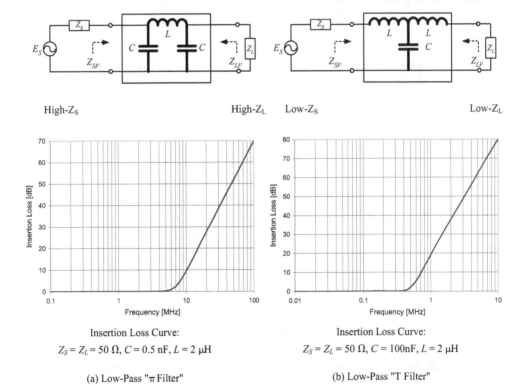

High-Z_S High-Z_L Low-Z_S Low-Z_L

Insertion Loss Curve:

$Z_S = Z_L = 50 \, \Omega, C = 0.5 \text{ nF}, L = 2 \, \mu\text{H}$

(a) Low-Pass "π Filter"

Insertion Loss Curve:

$Z_S = Z_L = 50 \, \Omega, C = 100\text{nF}, L = 2 \, \mu\text{H}$

(b) Low-Pass "T Filter"

Figure 8.5. Low-pass "π" and "T" EMI filters.

$$IL(f) \cong 20 \cdot \log \left[\frac{\omega L}{Z_S} + LC\omega^2 \right]; \text{ dB}$$

$$Z_L \ll Z_S$$

(8.10)

The insertion loss for the "L-section" filter is independent of the direction of inserting the "L-section" into the line, if source and load impedances are equal. When source and load impedance are not equal, the greatest insertion loss will usually be achieved when the capacitor shunts the higher impedance.

The insertion loss of a "L-section" filter is not changed when it is "turned around" so that the source and load terminals are transposed and with equal source and load impedances, $R_S = R_L = R$, expressed by Equation (8.11) [5]. The effect of the damping factor, d, is similar to that for the "π-section" and "T-section" filter networks:

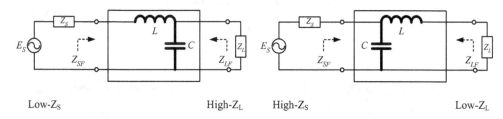

Low-Z_S High-Z_L High-Z_S Low-Z_L

(a) Capacitor Shunts the High Z_L (b) Capacitor Shunts the High Z_S

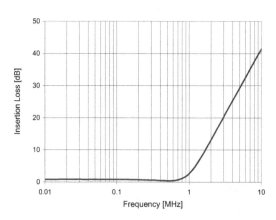

Insertion Loss Curve (Identical for Cases (a) and (b))

(a) $Z_S = 5\ \Omega$, $Z_L = 50\ \Omega$, $C = 3$ nF, $L = 10\ \mu H$ (b) $Z_S = 50\ \Omega$, $Z_L = 5\ \Omega$, $C = 3$ nF, $L = 10\ \mu H$

Figure 8.6. Low-pass "L-section" EMI filters in asymmetrically loaded circuits.

$$IL(f) \cong 10 \cdot \log\left[1 + \frac{f^2 D^2}{2} + f^4\right]; \text{ dB}$$

$$D = \frac{1-d}{\sqrt{d}}; \ d = \frac{L}{CR^2} = \text{damping factor}; \ f_0 \cong \frac{1}{2\pi}\left(\frac{2}{LC}\right)^{\frac{1}{2}}; \text{ Hz}$$

(8.11)

All of the above filter topologies can be used for suppressing both common-mode (CM) and differential-mode (DM) interference signals.* When used for suppression of CM EMI, the shunt capacitors are referenced to the signal reference structure or chassis, as shown in Figure 8.7. For DM interference suppression, the capacitors are connected in a line-to-return configuration. No connection to the signal reference structure is required. Most EMI filters are of common-mode type, with few exceptions, primarily in low-frequency sources within equipment, switch-mode power supplies in particular. Common-mode interference is the most frequently encountered interference mode, both from sources internal to equipment and coupled from the external environment, and is most prevalent at frequencies above 100 kHz [3,4].

The effect of practical grounding and bonding of filter circuits' performance and proper grounding practices for achieving the necessary performance of EMI filters are further discussed in Section 8.4.

8.3.2 Special EMI Filter Applications

Several types of special filter components are commonly used for EMI control, often forming part of the filter topologies discussed above, namely common-mode chokes and feed-through capacitors. These devices are of special interest with respect to grounding considerations in circuit and system design.

8.3.2.1 Common-Mode Chokes. Suppression of common-mode (CM) interference signals can be achieved by either line-to-ground (reference structure) capacitors (CM capacitors) or series inductors. For achieving a high insertion loss, large CM capacitors may be required in power-line filters. A large value of capacitance on the high-voltage input of power supplies is prohibited by electrical safety codes, due to concerns relating to power-frequency leakage currents and the hazard of electrical shock should someone physically touch the pins of the receptacle by accident.† Signal circuits, on the other hand, cannot tolerate large capacitance due to the detrimental effect it may have on the signal waveforms themselves.

Unfortunately, large series inductors cannot typically be used in filters as well. High-permeability cores are easily saturated by excessive magnetic flux induced by large power-supply currents, whereas in signal lines large inductors will be just as harmful to the signal waveforms as large value capacitors. Consequently, the CM filter

---

*Common-mode and differential-mode signals and their generation/propagation are discussed in Chapter 2.
†See Chapters 3 and 6 for a discussion of safety concerns associated with leakage of power-frequency current through CM capacitors to the equipment chassis.

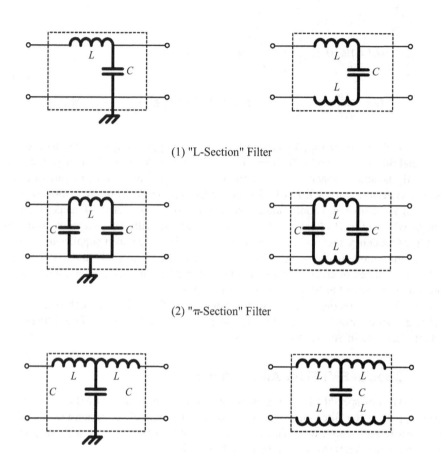

(1) "L-Section" Filter

(2) "π-Section" Filter

(3) "T-Section" Filter

(a) Common-Mode (Line to Chassis) Filters   (b) Differential-Mode (Line to Line) Filters

Figure 8.7. Common-mode and differential-mode filter networks.

configurations shown in Figure 8.7 are rarely used with ordinary high-permeability inductors. This is where common-mode chokes become useful.*

Figure 8.8 illustrates the physical representation of a multiturn CM choke. The wires (signal/power and return) are wound several turns around a high-permeability ferrite core. Magnetic flux (B_{DM}) induced by the differential-mode (DM) current (I_{DM}) in the core is canceled out by an equal but opposite magnetic flux induced by the return (DM) current that encircles the core.

---

*CM chokes were described in Chapter 4 with respect to their ability to "open" ground loops. The same attributes are used here for suppressing CM interference.

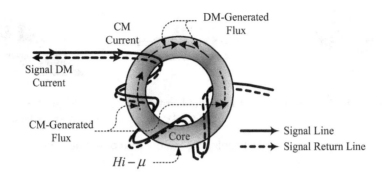

Figure 8.8. Configuration of a CM choke and its effect on CM and DM currents.

The DM current is, therefore, not attenuated if perfectly balanced.* However, magnetic flux (B_{CM}) due to any CM current (I_{CM}) that may be present will cancel itself out with flux from one wire traveling clockwise and the return current traveling counterclockwise, intermingling with each other. Attenuation of the CM noise is, thus, achieved by the series combination of inductive reactance and resistive losses of the core. Figure 8.9 shows typical CM choke configurations.

8.3.2.2 *Power-Line Filters.* Power-line filters are commonly used for compliance with emissions and immunity requirements limits of international EMC standards. Owing to the nature of their application, they warrant particular interest with respect to grounding and bonding considerations. There are two conflicting requirements to conform to: strict electrical safety codes and shunting RF noise into a ground reference structure.

Figure 8.10 presents the configuration of a typical single-phase power-line filter. For three-phase power circuits, the same configuration is duplicated for each phase).[†] Single-series chokes L_1 and L_2 provide suppression of DM interference and are typically on the order of tens to hundreds of μH. Inductors L_3 and L_4, wound on a single toroidal core, constitute a CM choke and normally exhibit high CM impedance due to relatively high inductance (tens of mH). The inherent stray (DM) inductance of CM choke inductors (L_3 and L_4) often provides sufficient DM suppression, eliminating the need for additional DM chokes. The resistor R_1 is required in order to discharge the line-to-line capacitor, C_X, for product safety reasons, in addition to eliminating stray voltage present, thus minimizing oscillations of the filter due to its low-frequency high Q-factor.

---

*Actually, CM chokes exhibit some "leakage (DM) inductance" due to imperfections in their construction. This leakage inductance will introduce some DM attenuation, which, in power-line filters is normally not a major concern.

[†]In three-phase power-line filters, C_Y (CM) capacitors are placed between each (phase and neutral) line to the electrical safety ground conductor (ESGC) and C_X capacitors are included between all phases and between them to the neutral. Also, three-phase lines and the neutral (in "star" power networks) are all wound on a common CM choke core, whereas DM chokes are placed on separate cores.

(a) Multi-Turn CM Choke   (b) Two-Turn Ferrite Bead   (c) Single-Turn CM Choke
                        CM Choke   over a Ribbon Cable

Figure 8.9. Practical configurations of common-mode chokes.

The chassis of the equipment itself may be RF noisy with internally generated EMI. The protective earth (PE) or electrical safety ground conductor (ESGC) lines can carry internally generated interference current, resulting in conducted and, subsequently, radiated emissions. The series inductor L_5, sometimes included in power-line filters, is intended to control these emissions from the ESGC wire. In compliance with electrical safety codes, the impedance of the ESGC must not exceed an approved value, which is typically 100 mΩ (equivalent to approximately 300 μH in a 50 Hz power system). This inductor must be able to handle any and all fault current without damage until the fuse on the power line is activated [3].

A category of capacitor known as "Y" capacitors (C_Y) are used between phases (or neutral) lines and the chassis or ESGC/PE to shunt CM currents to ground. These capacitors are most effective at higher frequencies. The Y capacitors are limited to very low values owing to electrical safety limits set by regulatory bodies with regard to equipment leakage current. Typical capacitive values of Y capacitors are on the order of tens of pF to tens of nF. The X capacitors (C_X) are connected between phase and neutral or phase to phase, producing attenuation of DM interference. Typical values of X capacitors are in the range of tens to hundreds of nF.

Clearly, the key to high-frequency performance of any power-line filter is the effectiveness of the Y (CM) capacitors. With ideal capacitors, which in reality do not exist, higher frequency performance is guaranteed. Unfortunately, use of capacitors is limit-

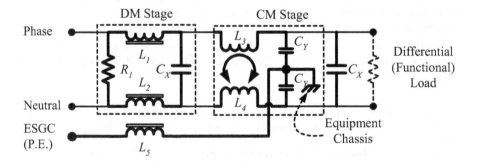

Figure 8.10. Configuration of a typical single-phase power-line filter.

ed by parasitics. The effects of parasitics and the manner of overcoming them are the subject of Section 8.4.

8.3.3 Transient-Voltage Protection Devices and Circuits

Physically long interconnects, such as power lines and cables between equipment, may be subject to very large voltages and current transients due to a wide variety of phenomena. Conducted phenomena include inductive switching transients and induced surges from lightning or nuclear electromagnetic pulses (NEMP) [5,6]. Unlike ordinary EMI, which normally cannot cause permanent damage to equipment, transients are fast events of high magnitude and could significantly harm electrical and electronic apparatus.

8.3.3.1 A Transient-Effects, Grounding-Related Case Study

The following case study illustrates the adverse effect of improper grounding in enhancing vulnerability of a system to lightning-induced surges. Two computer systems, installed in sites separated by several hundreds of meters, were interconnected using a RS-422 data communication link (Figure 8.11). The chassis of each of unit contained a local signal reference to earth ground. Due to lightning surge current, the earth potential increased abruptly between the central and the remote sites, causing a voltage potential, ΔV_E, to be developed between the earthing points of the two facilities. Since each computer was grounded locally to the facility's signal reference network, the same potential difference, ΔV_E, also appeared as common-mode voltage between both

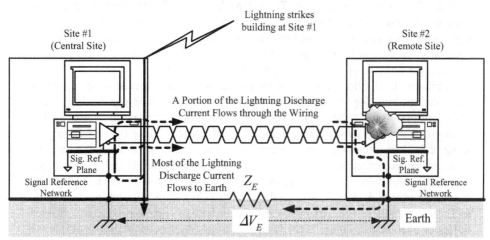

Figure 8.11. A case study: Inadequate grounding as a source of lightning-produced transient effects.

input terminals of the RS-422 line receiver, far exceeding its damage threshold, resulting in damage to the line receiver.

Ideally, preclusion of damage could have been achieved if an equipotential structure were to be provided between the two interconnected computers. This was impractical, however, since regardless of the quality of the equipotential structure construction, inductance of the grounding network conductors would counteract the objective of attaining a true equipotential structure between the two sites. A voltage potential difference of thousands or more volts will ultimately be present between the grounding points of the two computers during a lightning event. To the system designer, this indicates that although a differential interface such as RS-422 uses a low-voltage differential signaling, a remote node may observe the signal voltage superimposed on a transient of hundreds or thousands of volts with respect to that node's local signal reference structure.

How can system nodes be interconnected knowing that these large voltage potential differences between the signal references may be present? The first step toward successful protection is to assure that each device in the system is referenced to only one reference structure, eliminating the path through the device for surge currents searching for a return path.

Isolation of the signal return (reference) plane within one or both of the interfacing computers may constitute a satisfactory solution under certain circumstances. In most cases, however, this approach is impractical or ineffective. The other approach is to tie the circuit signal reference terminal through a low-impedance connection to the signal reference structure. This, however, cannot always be achieved at DC levels, due to functional constraints (e.g., balancing of the interface). However, this goal can be met by means of transient-suppression devices that provide a transitory path for shunting harmful surge currents while limiting the residual voltage transients to tolerable levels.

8.3.3.2 Fundamentals of Transient-Voltage Protection.
Creating one common ground connection at the host device provides a safe place to divert surge energy as well as a voltage reference to attach transient-suppression devices to. Shunting harmful surge current to the signal reference structure before it reaches the protected circuit is the function of transient-suppression devices (TSDs).

Similar to EMI filters, protection against transients is achieved by diverting the surge current by using low-impedance shunting devices such as gas discharge tubes and thyristors, limiting the surge voltage by using nonlinear shunting devices such as metal-oxide varistors (MOVs) and avalanche diodes, and blocking the current surge by using series high-impedance components such as resistors and inductors. Occasionally, transient-protection devices are combined with standard linear EMI filters. Figure 8.12 demonstrates the effect of transient-suppression devices on an incident surge with a series resistor and an avalanche diode, serving as blocking and limiting devices, respectively. Similar to EMI, transients may occur both as common-mode and differential-mode phenomena. Figure 8.12 illustrates a transient-protection scheme of a single-ended circuit in which the return signal is connected to the signal reference structure or chassis. In differential and balanced circuits, TSDs would be required between each line to the reference as well as line to line if protection against both CM and DM transients is necessary.

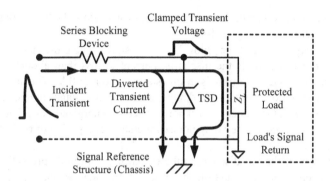

Figure 8.12. A typical transient-protection scheme for a single-ended circuit.

8.3.3.3 Commonly Used Transient-Suppression Devices (TSDs). A
large variety of shunting TSDs is available to choose from. With the exception of linear EMI filters, all designated TSDs are nonlinear in nature. The most commonly used devices are described below.

Spark Gaps and Gas Discharge Tubes (Figure 8.13)
A spark gap is a pair of electrodes, insulated by air or the dielectric material of a printed circuit board, and spaced so that the gap sparks over when the voltage across its contacts exceeds a specified level. The spark-gap firing voltage increases with the rate of rise of the applied surge and ranges from hundreds of volts to hundreds of kilovolts. In their nonconducting state, spark gaps behave as open circuits with extremely low capacitances.

Gas discharge tubes are spark gaps encapsulated in glass or ceramic tubes filled with a low-pressure inert gas and terminated on each end with an electrode providing a lower spark overvoltage. Gas discharge tubes are used primarily for secondary protection of wire pairs entering a facility from a long external shielded cable, or for exposed intrafacility wiring. The gas discharge tube is connected in parallel with the circuit being protected, exhibiting a very high off-state impedance.

Once an incoming transient exceeds the breakdown voltage of the tube, it fires, causing an arc in the tube. This arc ionizes the inert gas that provides a low-impedance

(a) 2-Electrode Gas Tube and its Circuit Representation

(b) 3-Electrode (CM and DM) Gas Tube and its Circuit representation

Figure 8.13. Gas tubes and their circuit representations.

path for the transient. When the transient voltage drops below the hold voltage, the gas discharge tube returns to it high off-state impedance.

Spark gaps may operate in the "glow" or "arc" states. The glow state occurs in circuits whose impedance limits the discharge current to less than about 100 mA. The voltage across the device's terminals in this state is about 100 V. The arc state occurs when large currents flow through the device, resulting in voltages across its terminals of usually 10 to 20 V. Gas tubes should not be used on energized lines that can sustain an arc or glow discharge, as they may not extinguish after a period of time, resulting in their damage.

One advantage of the gas discharge tube is that it can provide protection against a high surge current. For transients on the order of 10 microseconds, a gas discharge tube can withstand up to 20 kA. For longer but lower current pulses, the gas discharge tube can handle many hundreds of transients before the device enters a failure mode.

The greatest disadvantage of spark gaps and gas discharge tubes is their relatively slow response, resulting in large overvoltage transients across their terminals while the inert gas is ionizing. This overvoltage can be on the order of four times the rating of the tube. Spark gaps are rarely used in high-technology products without successive complementary protection devices.

Metal-Oxide Varistors (Figure 8.14)
Metal-oxide varistors (MOVs) are passive, nonlinear devices made from sintered metal oxides, primarily zinc, acting as nonlinear resistances, that is,

$$V = I \times R\left(V\right) \tag{8.12}$$

Where V represents the voltage across the MOV terminals, I represents the surge current through the device, and $R(V)$ is the nonlinear, voltage-dependent resistance of the MOV. When exposed to a voltage greater than the rating of the device, the MOV's resistance changes to a low value, allowing it to dissipate the transient energy as heat.

MOVs are capable of diverting currents up to tens of kiloamperes and are effective for large rate-of-rise transients. A MOV stops conducting and returns to its state of high resistance when the applied voltage decreases below the "knee" of the V-I curve and is ideal for protecting energized lines, since it has no current-extinguishing problems. The MOV typically introduces a shunt parasitic capacitance on the order of nanofarads and shunt resistance of megaohms to the protected circuit, limiting its use-

Figure 8.14. Circuit representation of an MOV.

fulness in high-frequency and high-impedance circuits. The maximum energy dissipation capability for large MOVs is tens of kilojoules.

Avalanche Diodes (Figure 8.15)
Avalanche diodes are passive devices with a high doping content that limit the voltage across a component or signal line by clamping the incoming transient voltage after the stand-off voltage is exceeded. Avalanche diodes are capable of operating at much lower voltages levels than MOVs and gas tubes (1 to 100 V), but are less tolerant to large peak currents than the other devices and may be damaged by full-threat transients. Peak current ratings up to about 100 A are available. The advantages of using an avalanche diode are its speed and repeatability. Since it is made of silicon, the diode is much faster than an MOV or a gas discharge tube, and its capability will not degrade with time, unless it is stressed beyond its power rating. The design restriction for avalanche diodes is that they are both voltage- and current-dependent. Therefore, as the voltage requirement increases, the current capability will decrease. These devices introduce a shunt parasitic capacitance on the order of nanofarads to the protected circuit and may aggravate high-frequency signals.

Of all the TSDs presented here, diodes are unique in that they are unipolar in nature. When forward biased, they exhibit the P–N junction voltage drop across their terminals, whereas, when reverse biased, the voltage across them is limited by their reverse breakdown voltage capabilities. This feature allows the combination of two "back-to-back" avalanche diodes into a "bidirectional" configuration, which, regardless of the surge polarity, will introduce an equal (but reverse) voltage drop across its terminals. The bidirectional devices are typically used in balanced, differential lines.

Linear Filters (Figure 8.16)
Linear filters are often used in combination with transient-suppression devices. On power lines, for example, the line filter usually cannot tolerate peak voltages. For this reason, a supplemental spark-gap surge arrester is used to limit the voltage while the filter isolates the internal circuits from the voltage overshoot resulting from the spark-gap discharge. The shunt input capacitance of the filter may also be used to reduce the rate of rise of the voltage. This allows the surge suppressor to spark over at a lower voltage.

(a) Circuit Representation of a Unidirectional   (b) Circuit Representation of a Bi-Directional

Avalanche Diode   Avalanche Diode

Figure 8.15. Circuit representations of avalanche diodes.

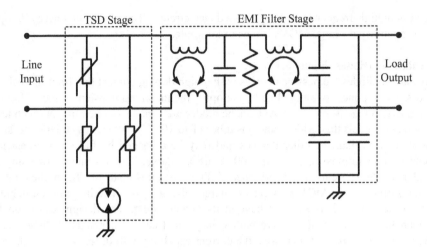

(a) Power-Line Filter Combined with TSDs

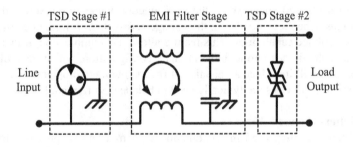

(b) Balanced Signal-Line Filter Combined with TSDs

Figure 8.16. EMI filters combined with TSDs.

8.3.3.4 Hybrid Transient-Suppression Circuits. Often, discrete TSDs cannot provide acceptable protection for the circuits. Devices capable of handling high-energy surges (such as gas discharge tubes) exhibit a slow response, whereas fast devices (e.g., avalanche diodes) cannot handle a high surge current. For this type of application, three-stage hybrid transient-suppression circuits are used (Figure 8.17).

The first stage is typically a gas tube, which can handle high surge currents and diverts most of the surge current once it fires. However, it has a high spark-over voltage and is too slow to protect sensitive solid-state circuits on its own.

An avalanche diode usually serves as the "fast" TSD, and constitutes the third stage of the circuit. It is fast enough to respond "instantly" to the fast front time of the transients, clamping the voltage across the protected circuit to a safe level. However, its energy handling capabilities are low and it cannot operate without damage for a long period of time.

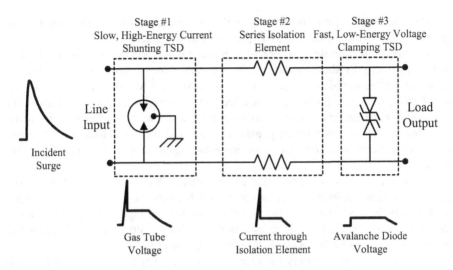

Figure 8.17. Hybrid transient-suppression circuit.

A small series impedance* serves as an isolation element in the second stage of the circuit and is intended to fulfill two primary objectives, namely, limiting the incident surge current flowing into the third, lower energy stage of protection, and creating a sufficiently high voltage drop between the third and the first stage to ensure the firing of the gas tube in a timely manner. For optimal protection, coordination between the different TSD stages and between them and the protected circuit is required [3].

Lessons Learned

- Energetic transients such as those generated by lightning, N-EMP, and inductive load switching, could potentially damage electronic circuits and components.
- Transient suppression is accomplished by transient-suppression devices and circuits, typically comprised of nonlinear devices such as spark gaps, MOVs, and avalanche diodes, as well as linear components such as resistors, inductors, and low-pass filters.

8.4 GROUNDING OF FILTERS AND TRANSIENT-SUPPRESSION DEVICES

8.4.1 When is Ground Not Equal to Ground?

The design and installation of terminal protection devices, such as filters and transient-suppression circuits are an art as well as a science. Much depends on parasitics charac-

---

*Typically, a resistor is used as an isolation element on signal lines, whereas small inductors are used on power lines. Often, the isolation element is implemented as a common-mode filter.

terizing the components and their installation practices, for which bonding and grounding of the filter elements are of particular importance.

Terminal protection devices such as transient-suppression devices and EMI filters operate primarily by shunting the interference energy to a return path (i.e., the return conductor or the signal reference structure for DM and CM protection, respectively) or reflecting it back to the source. For effectively accomplishing these objectives, the return path must present minimal series impedance to the shunted current at all frequencies of interest. Unfortunately, parasitics, particularly inductance, often counteract the opportunity for establishing low-impedance grounding, thus impairing performance.

Common-mode filters, in particular, are intended to shunt undesired high-frequency energy to a signal reference structure or chassis by means of CM capacitors. Figure 8.18 depicts the CM section of the power-line filter from Figure 8.10, containing an ideal CM capacitor, C_Y. Capacitors have intrinsic parasitic attributes: ESR (equivalent series resistance) and ESL (equivalent series inductance).*

As a result of predominantly inductive effects, every capacitor will exhibit capacitive reactance up to its self-resonant frequency. Above this frequency point, the capacitor behaves as an inductive element. Figure 8.19 depicts the intrinsic impedance and the insertion loss (in a 50 Ω/50 Ω source/load resistance circuit) of an ideal and "real-world" 10 nF capacitor.

In addition to intrinsic parasitic attributes, installation of a capacitor cannot be ignored, as the length of its leads adds to the capacitor's total equivalent inductance. The longer the device leads, the higher the parasitic inductance, which will degrade higher-frequency performance.

The significance of bonding terminal protection devices to a reference structure cannot be overemphasized. In particular, when a filter is used at a point of entry for a wire traveling through an enclosure shield, the desired effect is accomplished by directly bonding the filter return, usually the chassis, to the shielding barrier. A poor bond will compromise the filter's insertion loss beyond the effect of the parasitic attributes of the filter itself.

The consequences of poor bonding of a common-mode, π-section filter are shown in Figure 8.20 [3,6,7]. If the return side of the filter, usually the housing, is inadequately bonded to the signal reference structure (the equipment case or rack), bond impedance, Z_B, may be sufficiently high to impair the filter's performance. As shown, the filter constitutes a low-pass filter intended to remove high-frequency interference signals from the line. The filter achieves its goal in part due to the low reactance of the shunt capacitor, X_C, at a frequency of the interference current.

Interfering signals present on input lines should normally be blocked by the inductive reactance, ωL, and shunted to the signal reference structure along the intended Path 1, thereby staying away from the load impedance, Z_L. Interference currents will still follow undesired Path 2 to the load if, on the other hand, the bonding impedance of the filter, Z_B is high relative to the capacitive reactance in series with the load impedance:

*The nonideal nature of passive circuit elements and interconnects was discussed in Chapter 2.

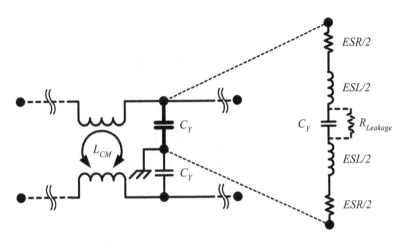

Figure 8.18. Nonideal properties of Y capacitors.

$$\left(\frac{1}{j\omega C}+Z_L\right)<\left(R_B+j\omega L_B\right) \tag{8.13}$$

The effectiveness of the filter at high frequencies could, thus, be severely compromised, even before the condition in Equation (8.13) has been reached (see also Figure 8.22). The circuit schematics depicted in Figure 8.21 were analyzed using Pspice™* for both well-bonded and poorly-bonded π-section low-pass filters (LPFs), respectively, for illustrating the effect of poor bonding on filter performance.

The simulation results (Figure 8.22) clearly reveal dramatic transformation in the performance of the filter circuit as frequency increases. At lower frequencies, well below the filter resonant frequency (50 MHz, approximately), the LPF functions as intended. The bonding impedance is too low to have any significant unfavorable consequence and the interference current flows through Path 1 (see Figure 8.21). At higher frequencies, the impedance of a poor bond connection, Z_B, dominates the filter's performance. The interference current can now take either of two alternate paths, constituting the least of two evils: Path 1, which produces the smallest loop area (the path of least intrinsic inductance), or Path 2, bypassing the blocking inductor through the filter's capacitors and returning through the load impedance, Z_L. The loop impedance of this path may be higher; however, if Equation (8.13) prevails, the poor bond impedance value will be added to the loop impedance, exceeding that of Path 2, which now constitutes the path of least impedance. Consequently, the filter cannot fulfill its role and the load is not protected. In effect, the filter undergoes a "metamorphosis" from a low-pass to a high-pass filter.

A similar situation, with greater potential for harmful impact on a circuit or system's integrity, occurs when transient currents discharge through the above circuit.

---

*"PSPICE" is a trademark of Cadence Design Systems. Evaluation Version 9.1. Web Update 1, Level 000, Build 101 was used for the purpose of the above simulations.

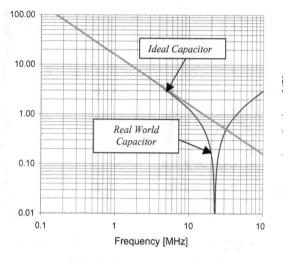

(a) Intrinsic Impedance of Capacitor

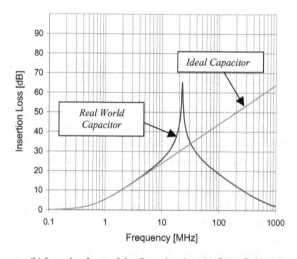

(b) Insertion Loss of the Capacitor in a 50 Ω/50 Ω Circuit

<u>Figure 8.19.</u> Frequency response of an ideal (C = 10 nF) and "real-world" (C = 10 nF, L_S = 5 nH, R_S = 2 mΩ) capacitor.

Fast transients constitute high-frequency, high-energy phenomena. The incident current transient, $I_T(t)$, produces a high-amplitude transient voltage, $V_B(t)$, across an (inadequate) bond impedance, Z_B (Figure 8.23) [6]:

$$V_B(t) = R_B \cdot I_T(t) + L_B \frac{dI_T(t)}{dt} \qquad (8.14)$$

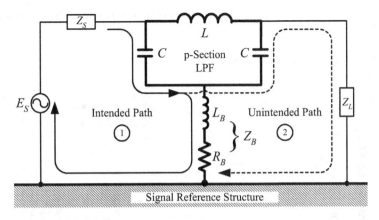

Figure 8.20. Effect of poor bonding on the performance of a π-section low-pass filter.

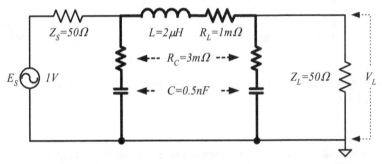

(a) Model of a Basic Filter Circuit with Zero Impedance Bonding

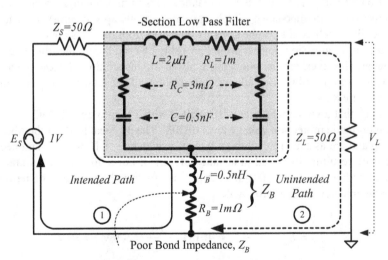

(b) Model of a Poorly Bonded Filter Circuit

Figure 8.21. PSpice™ simulation models for evaluation of the effect of poor bonding on the performance of a π-section low-pass filter.

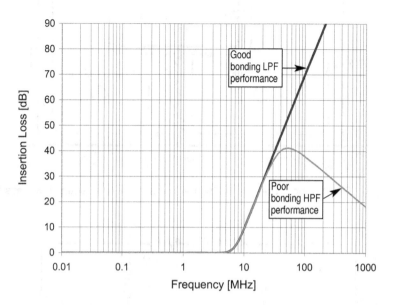

<u>Figure 8.22.</u> PSpiceTM simulation results for good and poor bonding of a π-section low pass filter.

where R_B and L_B represent the bond resistance and inductance, respectively. Not only will the filter not protect the differential load connected between its terminals, but it may also physically damage it due to breakdown resulting from high voltages developing across the bond impedance, which may subsequently appear across the load. Actually, the filter itself could also be severely damaged from the transient event.

Transient suppression devices would probably be used before the filter to eliminate this threat. However, similar concerns apply to the installation of these devices. The following example illustrates adverse effects of lead inductance on grounding of a MOV (Figure 8.24).

Assume that a vulnerable circuit is installed with a line-to-chassis MOV to provide a common-mode protection voltage level of 30V. The physical leads of the MOV are 4 cm long and are assumed to exhibit a typical inductance of 10 nH/cm (and negligible resistance), resulting in a total inductance of 40 nH. For a 50 A peak common-mode transient with a front time of 10 nSec,* the maximum voltage across the terminals of the vulnerable device is, approximately

$$V_L = L_{Leads} \cdot \frac{dI_{Surge}}{dt} \approx 40\,\text{nH} \cdot \frac{50\,\text{A}}{10\,\text{nS}} = 200\,\text{V} \gg 30\,\text{V} \qquad (8.15)$$

*A typical "early time" nuclear EMP waveform based on the IEC (International Electrotechnical Commission) International Standard 61000-2-10, Electromagnetic Compatibility (EMC), Part 2-10: Environment—Description of HEMP Environment—Conducted Disturbance, Geneva, Switzerland, First Edition, November 1998.

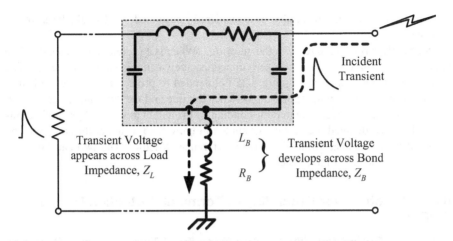

Figure 8.23. Effect of poor bonding on the transient operation of π-section low-pass filter.

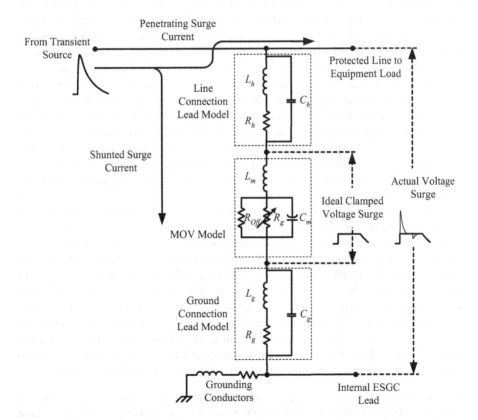

Figure 8.24. Adverse effect of lead inductance on protection level provided by a MOV TSD.

Obviously, the circuit can hardly be considered "protected" and could be severely damaged.

We finally arrive at the reply to the question, "When is a ground not a ground?" The answer is, "Whenever there is non-zero impedance between two circuit elements." The terminal protection device in Figure 8.25 is intended to protect a sensitive load from incident EMI and transients appearing between the line and the "ground" (signal reference structure). However, due to poor bonding of the TPD circuit, the "ground" terminal of the TPD, denoted the "virtual ground" in Figure 8.25, is not at the same voltage potential as the "actual ground" or signal reference structure. The two can be many volts apart in the case of a transient.

8.4.2   Practices for Grounding of Terminal Protection Devices (TPDs)

8.4.2.1   Optimizing Filter Grounding—Feed-Through Capacitors and Filters.   Feed-through capacitors and filters* are three-terminal devices designed to reduce lead inductances. Feed-through filters are shown in Figure 8.26. In (a), a standard capacitor is shown. One electrode is connected to a metal chassis while the other terminal is attached to the protected line. The lead bond wires to the electrode plates contain undesired inductance that limits high-frequency performance. In (b), the inductance of the lead between the protected line and one electrode is eliminated; however, the grounded electrode still has one long connection lead to chassis. In (c), the two electrodes are scrolled around each other. One electrode of the capacitor is now connected to and actually forms the feed-through housing ("ground") while the protected line is directly attached to the other electrode, literally "fed through" the device.

Figure 8.27 shows the actual construction of a feed-through capacitor. A wound foil is made with an extended foil-type construction so that each plate of the capacitor can be soldered to a washer-shaped terminal. One washer is then soldered to the center lead while the other washer is soldered to the case that is identified as the ground terminal.

The theoretical insertion loss of three-terminal capacitors is the same as for an ideal two-terminal capacitor. However, the insertion loss of a practical three-terminal capacitor follows the ideal curve much more closely than does a two-terminal capacitor. The useful frequency range of a feed-through capacitor is improved further by its case construction, enabling a bulkhead or chassis enclosure to isolate the input and output terminals from each other. The lead inductance in a feed-through capacitor is not part of the shunt circuit. Compared with leaded devices, an insertion loss feed-through filter is not degraded as rapidly with an increase in frequency. Consequently, whereas the short-lead construction capacitor is ideally suited for EMI suppression in the frequency range of up to 1 GHz, feed-through capacitors generally exhibit close to ideal performance up to 1 GHz with a self-resonance frequency well above that frequency.

---

*Even though the discussion covers mainly capacitors, it could easily be extended to composite filter circuits, considering the fact that in L–C filter networks grounding considerations apply to the capacitors and, in that context, the discussion in this section applies just as much to the capacitors incorporated in such filter circuits.

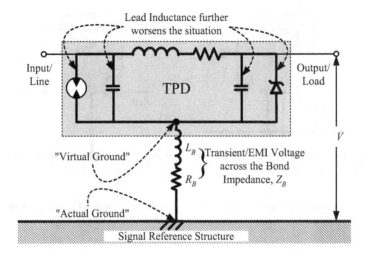

Figure 8.25. When is a ground not a ground? *Note 1:* It is important to note that this will occur even when the leads of the components of the devices comprising the TPD circuit (e.g., the capacitor, gas tube, or avalanche diode leads) exhibit "zero inductance." If this is not the case, the inductance of the components' leads will further exacerbate the situation. *Note 2:* In cases of differential (line-to-line) protection devices, the same concerns apply to the leads connecting the devices to the lines.

The commonly used circuit schematic symbol of the three-terminal and feed-through capacitors are provided in Figure 8.28(a) and 8.28(b), respectively. A practical three-terminal feed-through capacitor schematic is illustrated in Figure 8.28(c). As mentioned above, the lead inductance in a feed-through capacitor is not part of the shunt circuit; however, the parasitic lead inductance of the capacitor allows the three-terminal capacitor to act as a "T"-section filter, further enhancing its intended performance.

Feed-through filters are available in all filter configurations—"π," "T," and "L" —where ferrite beads or inductive elements are threaded on the feed-through bus inside the package of the filter. In addition, certain feed-through devices incorporate transient suppressor devices, such as avalanche diodes or MOVs between the line and the case.

8.4.2.2 *Filter Connectors.*

Another form of filter that extends the concept of feed-through filters is the filter connector. Filtered connectors offer reduced size and improved performance compared to discrete passive elements, and serve as an efficient, compact, and cost-effective solution when multiple lines must be filtered. Capacitors as well as lossy (ferrite-based) filter elements are built directly into a connector assembly, offering low-pass filter performance. Due to their effective feed-through filtering characteristic, connectors with integral capacitors with values as low as 100 pF provide effective bidirectional suppression of unwanted interference traveling into and out of equipment enclosures in frequencies ranging from several MHz up to hundreds of MHz.

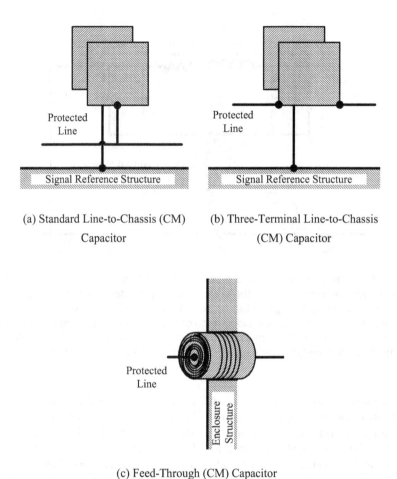

(a) Standard Line-to-Chassis (CM)
Capacitor

(b) Three-Terminal Line-to-Chassis
(CM) Capacitor

(c) Feed-Through (CM) Capacitor

Figure 8.26. "Transformation" from a standard two-terminal capacitor to a three-terminal feed-through capacitor.

Filtered connectors incorporate the popular "π," "T," and "L" filter circuits and are available in D-Type, Micro-D subminiature, cylindrical connectors, and RJ-45 as well as other less common types (e.g., ARINC), both in plug ("male") and receptacle ("female") configurations.

One of the most commonly used filter connector technologies is the filter-pin connector, in which the filter elements are incorporated within the connector contacts themselves (Figure 8.29). Grounding of the feed-through capacitive elements incorporated in the contacts is achieved by means of the grounding plate included in the connector assembly.

Another filter connector technology utilizes planar array filter devices [8]. Planar array filter connectors, widely used, contain ceramic capacitor arrays and ferrites that

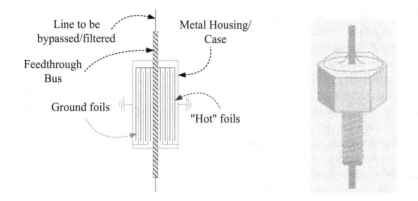

Figure 8.27. Construction of a three-terminal feed-through capacitor.

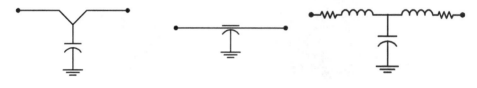

(a) Electrical Symbol of a Short-Lead
Two-Terminal Capacitor

(b) Electrical Symbol of a
Feed-Through Capacitor

(c) Practical Equivalent Circuit of a
Feed-Through Capacitor

Figure 8.28. Electrical schematic symbols of feed-through capacitors.

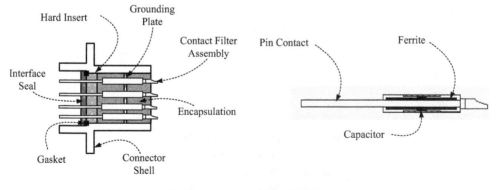

(a) Filter-Pin Connector Assembly

(b) Contact Filter Assembly

Figure 8.29. Filter-pin connector.

externally surround each contact (Figure 8.30). A planar array can be designed with different capacitive values on individual and pin groupings, and can also be selectively equipped with transient suppression devices (e.g., avalanche diodes or MOVs), providing protection against voltage spikes from transient sources such as EMP, lightning, or electrostatic discharge (ESD).

The incorporation of filter assemblies into a standard cylindrical or rectangular connector will often increase the overall length of the shell, typically at the nonmating side of the connector receptacle (inside the enclosure).

Another drawback of filter connectors is their high cost. The decision whether to use filter connectors in lieu of standard feed-through or other filtering techniques should be based on considerations regarding cost of filter connector assemblies versus discrete filters, available space, labor involved in construction, and flexibility of design.

(a) Integrated Cylindrical Connector Incorporating a Planar array Filter Assembly

(b) Planar array Filter Assembly, assembled including its Ferrite Elements and Connector Contacts, Ready for Insertion into the Connector Shell

(c) Cross-sectional view of a Planar Array Filter connector
Note the ferrite elements sandwiched between the ceramic capacitors

Figure 8.30. Multiway planar array filter connector (Courtesy of Glenair, Inc.).

8.4.3 Terminal Protection Devices (TPDs) Installation and Mounting Practices

The principle cause of failure of terminal protection devices is improper mounting and grounding. EMI filters are intended to effectively suppress conducted interference at frequencies ranging from hundreds of kHz to frequencies well beyond 1 GHz. If good performance of the filter is to be achieved, particularly at frequencies above 10 MHz, it is absolutely necessary to follow certain installation and grounding guidelines.

Two mounting faults are often commonly observed:

1. **High-impedance connection to the chassis.** The RF impedance between case and ground must be kept as low as possible. Inductive chassis connections will result in serious degradation to the filter insertion loss or transient suppressor clamping voltage.
2. **Capacitive bypass of the TPD.** Separation of input and output wiring is mandatory in order to minimize capacitive coupling between the protected lines and the nonprotected leads carrying the interference. This is achieved by placing filters at equipment boundaries. This precludes coupling from input wires carrying interference signals directly to the output wires, thus circumventing and nullifying the effects of shielding and filtering.

Both of the above mounting flaws can be readily mitigated by the use of bulkhead-mounted filters such as the one shown in Figure 8.31. The TPD (a power-line filter in this case) is directly mounted on and bonded to the enclosure's boundary.

The preferred method of installing feed-through filters in equipment used in a high-level electromagnetic environment is to mount the filters in a metal enclosure behind the front panel (known as a "doghouse"). This enclosure can be constructed with an access panel on one of the sides. However, the panel should be attached to the dog-

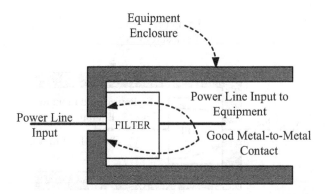

Figure 8.31. Bulkhead-mounted power-line filters for optimal filter performance.

house using a conductive gasket for obtaining an effective seal (Figure 8.32). This configuration is commonly used for power-line filtering.

Signal-line filtering is commonly applied using tubular feed-through filters. The EMI filter is installed on a bulkhead or enclosure case, threaded through a properly machined hole that provides perfect bonding of the filter to the bulkhead. This method is of particular use in RF and microwave assemblies. For many signal lines, brackets containing multiple feed-through filters can be used (Figure 8.33).

The doghouse concept can similarly be implemented with feed-through filters, offering similar advantages as those described for power-line entry. Alternate configurations of this technique are depicted in Figure 8.34.

When using feed-through filters, a bulkhead method of installation must be implemented. Mounting of feed-through filters in the flawed manner shown in Figure 8.35 does not provide the full benefits of the feed-through filters and can even enhance interference and coupling.

The installation of the feed-through filters on a mounting bracket does not provide a solid barrier between the filtered and nonfiltered zones, and, worse still, grounding of the filters is achieved through a bracket that ultimately introduces nonzero impedance between the filter housing and the enclosure of the equipment.

Another effective filtering method makes use of connectors where the EMI filter constitutes an integral part, as explained in Section 8.4.2.2. The filter connector provides high-density, multiple-line filtering, and generally requires less space for mounting than a connector/filter/doghouse assembly. Also, bonding of the filter connector to the enclosure provides the necessary effective low-impedance grounding of the filter elements (Figure 8.36).

When use of either of the above-described feed-through techniques is not desired, either because of cost or long lead time concerns, effective filtering can be accomplished by discrete filter elements installed on the printed circuit board (PCB). A chassis plane, if provided in the PCB, enables low-impedance grounding of the filter capacitors, and must be connected to the equipment enclosure immediately behind the

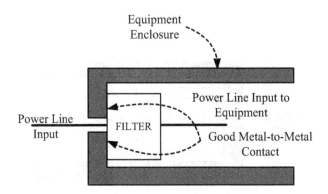

Figure 8.32. Doghouse configuration of power-line filter mounting.

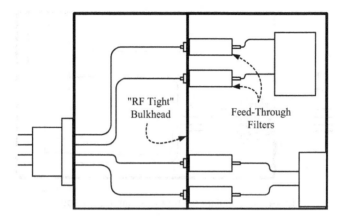

Figure 8.33. Bulkhead-mounted feed-through filters.

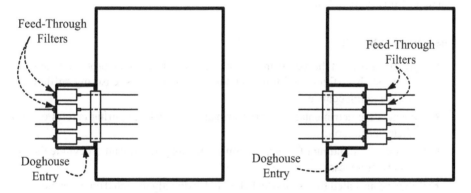

Figure 8.34. Doghouse mounting of feed-through filters.

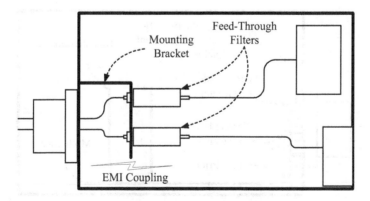

Figure 8.35. A flawed application of feed-through filters.

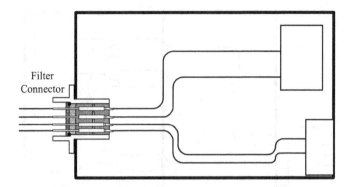

Figure 8.36. Bulkhead-mounted filter connector.

connector for minimizing the length of the wiring leads that could receive or transmit radiated EMI (Figure 8.37).

Lessons Learned

- Special care must be taken in the installation of terminal protection devices (TPDs), such as filters and transient suppression devices, ensuring sufficiently low ground impedance.
- Parasitics are detrimental to the performance of TPDs, requiring careful control of their installation.
- Inadequate grounding of a suppression device may significantly degrade the devices effectiveness.
- Bulkhead-installed filters, in which the TPD is directly mounted on the enclosure's boundary, provide optimal grounding and preclude EMI bypass of the device.

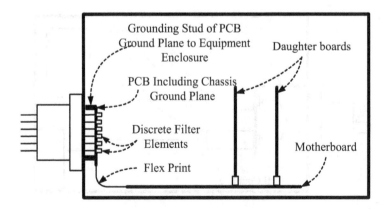

Figure 8.37. Filters mounted on a printed circuit board behind the connector.

BIBLIOGRAPHY

[1]   Clark, T. L., McCollum, M. B., Trout, D. H., and Javor, K., *NASA Reference Publication 1368, Marshall Space Flight Center Electromagnetic Compatibility Design and Interference Control (MEDIC) Handbook, CDDF Final Report, Project No.93-15,* Alabama, June 1995.

[2]   Ott, H. W., *Noise Reduction Techniques in Electronic Systems* (2nd Ed.), New York: Wiley-Interscience, 1988.

[3]   Hartal, O., *Electromagnetic Compatibility by Design,* W. Conshohocken, PA: R&B Enterprises, 1994.

[4]   Nave, M., *Power Line Filter Design for Switched-Mode Power Supplies,* New York: Van Nostrand Reinhold, 2003.

[5]   MIL-HDBK-419A, *Military Handbook, Grounding, Bonding, and Shielding for Electronic Equipments and facilities,* Washington, DC: U.S. Government Printing Office, 1987.

[6]   MIL-HDBK-232A, *Military Handbook, Red/Black Engineering—Installation Guidelines,* Washington, DC: U.S. Government Printing Office, March 1987.

[7]   Department of the Navy, Naval Air Systems Command, *NAVAIR AD 1115, Electromagnetic Compatibility Design Guide for Avionics and Related Ground Support Equipment,* (3rd. Ed.), June 1988.

[8]   *EMI/EMP Filter Connector Designer's Guide Plus Series 801 "Mighty Mouse" Filter Connectors,* Glenair, Inc., Glendale, CA, 2005.

9

GROUNDING ON PRINTED CIRCUIT BOARDS

"What is the primary use of a ground plane and when should it be used?" This question, commonly asked by novices, refers to a key issue in the design of printed circuit boards (PCBs) for achieving EMC and signal integrity. The instinctive (and facetious) reply by experts may be: "A ground plane is where the airplane lands" or "a ground plane is used if there isn't enough room for an airplane." However, the question is a valid one, as it relates to one of the most (if not the most) important measures applied in modern high-speed digital design circuits, that of grounding implementation on printed circuit boards.

9.1   INTERFERENCE SOURCES ON PCBS

It is said that all EMI problems begin and end on the printed circuit board. Interference reduction is a significant design issue in most electronic systems. Along with other performance considerations, interference is an omnipotent factor that must be dealt with for a successful design. Much has been written in previous chapters about system-wide grounding considerations for EMI control. In this chapter, attention will be paid to the design and implementation of grounding on PCBs for meeting EMC and signal integrity objectives.

Grounds for Grounding. By Elya B. Joffe and Kai-Sang Lock
Copyright © 2010 The Institute of Electrical and Electronics Engineers, Inc.

PCBs constitute a compact electronic system encompassing a large number of components on one board, making efficient power distribution and controlled signal propagation possible. PCB technology has not changed significantly in recent decades. An insulating substrate material (usually FR4, an epoxy/glass composite) with copper plating on both sides has portions of copper etched away to form conductive paths. A number of layers of plated and etched substrates are glued together in a stack with additional insulating substrates in between the etched substrates. Holes are drilled through the stack. Conductive plating is applied to these holes, selectively forming conductive connections between the etched copper of different layers. Although there have been advances in the areas of material properties, number of stacked layers, geometries, and drilling techniques (allowing holes that penetrate only a portion of the stack), the basic structures of PCBs have not changed. These structures, formed through the processes outlined above, are abstracted to a set of physical/electrical structures: traces, planes (or planelets), vias, and pads. The evolution of printed circuit board technology has contributed to the increase in circuit performance, speed in particular.

However, the complexity of PCBs also constitutes an increasing concern with respect to EMI: The expectation that the evolution of the microelectronics industry will not deviate from Moore's Law* is driving microprocessors to power levels in the tens of watts and clock frequencies well into the microwave region. The increasing integration of modern high-speed, very large scale integrated circuits (VLSIs) at higher density on PCBs, together with the continuing growth in devices' switching speed and operating frequency, have resulted in increased noise generation. When coupled with the reduction of signal levels, an exponential increase in the potential for interference and susceptibility is to be expected and observed in practice.

Studies on the radiation mechanisms for semiconductor devices and packages [1], the effects of very large scale integrated circuits (VLSI) on radiated EMI from circuits, revealed that inasmuch as most VLSI devices are too small to act as direct sources of radiated EMI, noise coupled from these devices constitute a severe concern. Coupling was found to occur through three principal mechanisms namely: (1) coupling to heatsinks, (2) coupling to traces, and (3) driving reference planes in printed circuit boards. Figure 9.1 presents plots of near-field H-field surface scans of an RDR memory module clocked at 200 MHz.

The major concerns with respect to interference mechanisms associated with modern PCBs are EMI (emission and susceptibility) and crosstalk (interference on the PCB).

---

*Moore's Law describes a long-term trend in the history of computing hardware. Since the invention of the integrated circuit in 1958, the number of transistors that can be placed inexpensively in an integrated circuit has increased exponentially, doubling approximately every two years. Moore's original statement that transistor counts had doubled every year can be found in his publication "Cramming More Components onto Integrated Circuits," *Electronics Magazine* 19, April 1965:

> The complexity for minimum component costs has increased at a rate of roughly a factor of two per year. . . . Certainly over the short term this rate can be expected to continue, if not to increase. Over the longer term, the rate of increase is a bit more uncertain, although there is no reason to believe it will not remain nearly constant for at least 10 years. . . .

Almost every measure of the capabilities of digital electronic devices is linked to Moore's law: processing speed, memory capacity, even the number and size of pixels in digital cameras.

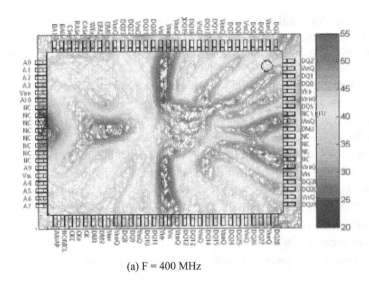

(a) F = 400 MHz

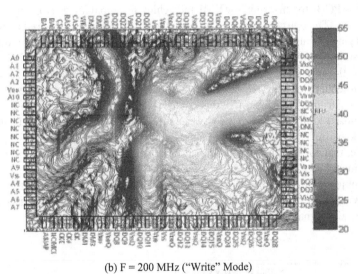

(b) F = 200 MHz ("Write" Mode)

Figure 9.1. Near-field H-field surface scan of RDR memory module clocked at 200 MHz. (Courtesy of Dr. Todd Hubing, Clemson University.)

Crosstalk is also often a major contributor to EMI. Poor PCB layout may increase coupling from internal, noisy circuits and outgoing input/output (I/O) lines that will, consequently, "export" EMI (emission). Conversely, interference "imported" into the PCB through I/O lines will couple onto internal, sensitive circuits resulting in their undesired response (susceptibility). Signal integrity is associated with both EMI and crosstalk, since distortions in signals on the PCB (poor signal integrity) will couple to adjacent signal traces (i.e., crosstalk) and may subsequently contribute to emissions from the PCB due to the increased level of higher frequency harmonics (i.e., EMI) (Figure 9.2). As a result, the following common interference mechanisms are observed on PCBs (Figure 9.3):

1. Common-impedance coupling through power supplies and their associated conductors, shared by multiple circuits
2. Common-impedance coupling through return conductors, shared by multiple circuits
3. Mismatch on high-speed transmission line, resulting in reflections
4. Crosstalk coupling between adjacent conductors of different circuits
5. Radiation crosstalk coupling between high gain, low-level analog amplifiers, resulting in feedback
6. Transients resulting from inductive load switching within the circuit, coupling into adjacent circuits
7. Power-supply-generated noise entering sensitive circuits

Designing PCBs for ensuring low EMI and achieving signal integrity in high-speed systems is easy. The one common driving factor behind the above undesired interactions is time varying or transient current, which exists within power distribution circuits and signal transmission lines on PCBs. The intensity of EMI interactions within PCBs is determined by the amount of current flowing in the circuit and the efficiency of coupling between circuits, particularly associated with their frequency content and geometry.

Where currents are concerned, the existence of round-trip path is implied. Currents follow closed-loop paths, from the power source, through the circuits and components, and back again. This is true for both signal and power-distribution circuits. It follows, therefore, that for both power and signal circuits a low-impedance return path should be provided.

The objective of this chapter is to discuss the design and implementation of signal- and power-return paths (grounding) on PCBs and the role of planes in accomplishing the EMC and signal integrity design goals.

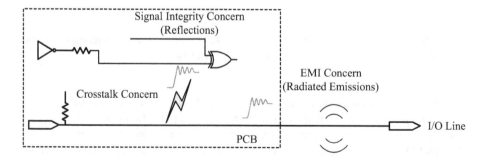

Figure 9.2. Interference interactions on PCBs.

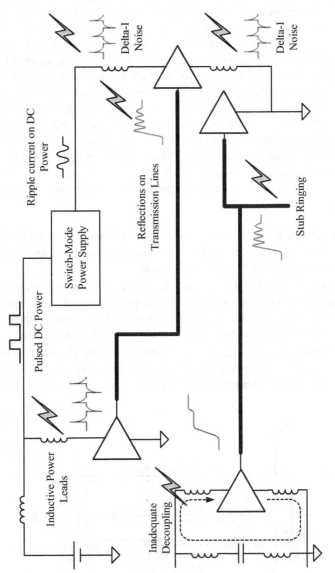

Figure 9.3. Noise sources on PCBs.

9.2 "GROUNDING" ON PCBs

Too often, the overused term "ground" is applied with respect to the function of the "second conductor" associated with signal traces and transmission lines on printed circuit boards. In actual fact, "grounding" on printed circuit boards is associated, first and foremost, with the function of the path for current return. Furthermore, the AC signal return current is not affected by the absolute DC voltage potential of the return conductor. As such, the actual AC current return path may in fact be the plane designated V_{CC} or V_{AA}, namely, the digital or analog supply voltage conductors, respectively. At other times, the GND, or low-voltage conductor, may serve for this purpose (see Section 9.3).

Note: For that purpose, the term "reference plane" or "return plane," rather than "ground plane," is often used throughout this book, iterating the fact that either the ground or the power planes may serve as the AC current return path. From the standpoint of AC, they are totally equivalent. Throughout this chapter, these terms should be considered interchangeable. See Section 9.3.

In practice, "grounding" on PCBs it related for the following distinct functions (Figure 9.4):

1. **Signal Return Path.** Most electronic devices, digital devices with high-speed CMOS logic in particular, utilize single-ended signaling, where the "ground" serves as a return path for the signal currents.

2. **DC Power Return Path.** Electronic devices, whether digital, analog, or mixed, draw current from some DC power source (often denoted "V_{CC}" and "V_{AA}" for digital and analog circuits' supply, respectively). The "ground" conductors serve as the return path for the DC supply current to the DC power source.

3. **Image Plane.** The presence of a solid metal plane beneath and close to a printed circuit board has the effect of reducing radiated emissions as a result of the equivalent "image" currents produced in the plane [2]. As will be shown further

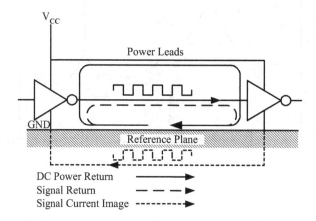

Figure 9.4. Functions of "ground" on a printed circuit board.

in this chapter, the image plane will have similar effects on single-ended and differential signals.

Different conductors may perform the above different functions concurrently. For instance, certain AC return signals may flow through a V_{CC} conductor, while the DC current return associated with the same circuits occurs through the DGND (digital ground conductor).

The manner in which these objectives are met on printed circuit boards is discussed in detail throughout this chapter. It will be shown that the three are inseparable and are tightly interrelated.

9.3   SIGNAL PROPAGATION ON PCBs

Signals on high-speed multilayer PCBs* typically propagate in transmission line structures, which are typically formed by the combination of a signal trace and one or two reference planes.[†] The reference planes serve as a path for the signal return current, particularly for single-ended circuits (e.g., TTL logic). Yet, even in differential circuits (e.g., LVDS, RS-422/RS-485, etc.) the planes provide a path for the greater part of the high-frequency return currents (see Section 9.4.2.2).

Note: In certain logic families, such as ECL (emitter coupled logic), the V_{CC} (power) plane is actually a 0 V supply voltage, whereas the V_{EE} (ground) plane is actually connected to the −5.2 V (low) DC supply.

As transition (rise and fall) times of digital devices continually decrease[‡] [3], usage of power and reference planes have become increasingly indispensable. Digital devices such as the 74HC series that have been around for nearly 20 years and exhibit significant emission only up to approximately 200 MHz, are now found to significantly emit undesired energy up to frequencies as high as 1 GHz, due, among other reasons, to progress made in fabrication techniques of the integrated circuits (ICs), such as die shrinking.

In high-speed circuits, planes are essential for accomplishing numerous EMC and signal integrity (SI) functions, including:

1. Providing for optimal return signal propagation through a solid plane adjacent to the signal traces, serving as the path of lowest impedance (see Section 9.3.5)
2. Guaranteeing a low-impedance power distribution system, offering fast transition time decoupling, thus also serving as a stable reference at high frequencies (due to the impulsive power transients), consequently reducing emissions and increasing immunity of the circuit (see Section 9.5.4)

---

*Discussion of single- and two-layer PCBs is beyond the scope of this book. In two-layer boards, use of planes may not be possible and traces may be used for signal current return.

[†]Coplanar traces placed above the dielectric laminate without a return plane (single- or two-sided PCBs) also constitute transmission lines, but are rarely used in high-speed signal propagation due to their relative inferior performance.

[‡]Effect of signal transition (rise/fall) times on signal spectra is discussed in detail in Chapter 2.

3. Serving as part of a controlled-impedance transmission line, providing for control of reflections due to impedance mismatch (see Section 9.3.4)

4. Reducing ground and power bounce [also known as simultaneous switching noise (SSN)] effects resulting from high-level impulsive currents through the inductance of the power distribution systems by providing a path of minimum inductance for those currents (see Section 9.3.5)

5. Providing control over crosstalk between traces and circuits, by terminating electric fields as well as providing magnetic field flux cancellation (see Section 9.3.7)

6. Offering some degree of shielding for signal traces routed in adjacent and embedded signal planes in the PCB by virtue of their function as image planes (see Section 9.6.1)

7. Reducing emissions on cables attached to the PCB due to common-mode EMI present on the circuits

In this section, the role of the reference planes in signal propagation is presented, with respect to high-frequency (and high-speed) interconnects in particular.

9.3.1   Circuit Representation of Transmission Lines on PCBs

At low frequencies, at which circuit dimensions are electrically small, PCB traces may be considered close to ideal (lossless or low-loss) lumped circuits (Figure 9.5). At higher frequencies, on the other hand, interactions along and between interconnects are best understood by treating them as transmission lines.* Transmission lines constitute interconnects with their respective return conductors, capable of guiding a signal between a signal source (the "transmitter") and a load (the "receiver") through a medium with controlled electrical characteristics. Alternating current (AC) circuit characteristics dominate the performance of interconnects, causing distributed (per-unit-length) resistance (R_X), conductance (G_X), inductance (L_X), and capacitance (C_X) to become prevalent in the conductor (Figure 9.6). Traditionally, transmission lines were considered in telecommunications cables operating over long distances. Yet as transmission rates increase and, furthermore, transition times reduce, even the shortest PCB traces and transmission lines may exhibit undesired transmission line effects such as ringing, crosstalk, and reflections, and may introduce ground-bounce effects, seriously hampering the integrity of the signal. These issues can be overcome primarily by maintaining controlled characteristic impedance along the line and by following adequate design techniques and simple layout guidelines.

Note: Transmission line effects and their control are beyond the scope of this book. EMC and signal integrity issues (including crosstalk) pertaining to the design of transmission lines and the control of transmission line effects (e.g., reflections, propagation delay, and ringing) are not discussed in any detail here, as they are very adequately covered in a number of excellent sources [3, 4]. This section focuses only on the role of reference planes in signal propagation along the transmission lines.

---

*An overview of transmission line fundamentals is provided in detail in Chapter 2.

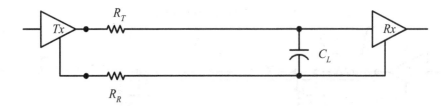

Figure 9.5. Low-frequency (lumped circuit) representation of a trace in a printed circuit board.

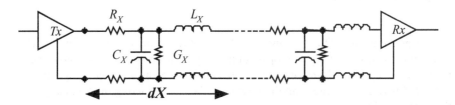

Figure 9.6. High-frequency (transmission line) representation of a trace in a printed circuit board.

For attaining the best comprehension of the role of reference planes in signal propagation on PCBs, it is important to realize that the high-frequency signal constitutes a voltage difference between the signal and the return path, that is, between the trace and the reference plane(s) [5]. This voltage difference propagates as the distributed capacitance along the line charges, generating a current flow forward into the transmission line. This current provides the charge for the capacitance of the next section of the line, between the signal and return conductor (see Figure 9.7).

As a result, significant current flows on the solid reference plane beneath both single-ended and differential signal traces.* When a high-frequency (such as high-speed digital) signal propagates down a single-ended transmission line, displacement currents follow the electric flux and flow through the distributed capacitance of the trace to adjacent metallic objects, and to the large, solid plane underneath the trace in particular. This capacitive effect provides the means for a returning current flow to the solid reference plane. The return current flows underneath the signal trace along the reference plane back to the source.

9.3.2 Electromagnetic Field Representation of Transmission Lines on PCBs

Circuit representation of transmission lines does not provide a full account of the behavior of return current paths in reference planes, particularly when the plane adjacent

---

*The distribution assumed by the current in the return plane is presented in Section 9.3.6.

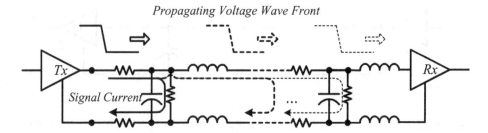

Figure 9.7. Signal voltage and current waveform propagation along a transmission line in a printed circuit board.

to the signal trace is *not* the "ground" or "0 V" plane. A more generalized approach for explaining the nature of signal propagation along a transmission line employs electromagnetic theory. Signal propagation can thus be viewed as an electromagnetic wave conveyed between the trace and the reference plane, considering the boundary conditions imposed by a metal plane.

Whereas DC is conducted in the copper, the signal RF energy propagates throughout the PCB in the form of guided quasi-TEM* (transverse electromagnetic) waves in which the E-field and H-field are transverse to the propagation direction (or Poynting) vector.

Consider a trace carrying a signal above a metallic plane. An electric field exists between the trace and the metallic plane, and magnetic field surrounds the trace (Figure 9.8). A boundary exists between the dielectric substrate of the PCB and the metallic plane. The tangential components of the electric field vector $\vec{E}$ and the magnetic field vector $\vec{H}$ must be continuous across the boundary between the two media, that is $\vec{E}_{td} = \vec{E}_{tm}$ and $\vec{H}_{td} = \vec{H}_{tm}$, where the subscripts "$t_d$" and "$t_m$" represents the tangential component of the fields in the dielectric substrate and the metal surface, respectively.[†] In addition, from the law of continuity, the normal component of the electric flux density vector $\vec{D}$ and the magnetic flux density vector $\vec{B}$ must be continuous across the boundary between the two media, that is $\vec{D}_{nd} = \vec{D}_{nm}$ (assuming that no surface charges exist on the boundary between the two media) and $\vec{B}_{nd} = \vec{B}_{nm}$, where the subscripts "$n_d$" and "$n_m$" represents the normal component of the fields in the dielectric substrate and the metal surface, respectively.[‡]

For simplicity of discussion, the metal plane is assumed to be a perfect electrical conductor (PEC), so that $\sigma_m \to \infty$. Consequently, all fields within the metal plane must vanish, that is, $\vec{E}_m = 0$ and $\vec{D}_m = 0$, as well as $\vec{H}_m = 0$ and $\vec{B}_m = 0$. As a result, the tangential electric field vector and normal magnetic flux density field vector components in the dielectric at the boundary must vanish too, that is, $\vec{E}_{td} = 0$ and $\vec{B}_{nd} = 0$. The tan-

---

*Ideally, propagation would occur in the TEM mode. However, due to the nonhomogenous discontinuous interface between the dielectric substrate and the surrounding air in a microstrip transmission line (discussed in the following section), a quasi-TEM rather than perfect a TEM pattern arises.

[†]Boundary conditions are discussed in detail in Chapter 2.

[‡]A detailed discussion of boundary conditions is presented in Chapter 2.

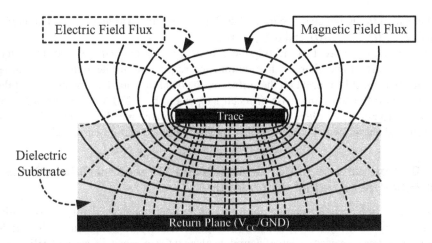

Figure 9.8. Contours of constant H-field (solid lines) encircle a signal conductor microstrip; electric field force lines (broken lines) connect the signal conductor to the reference plane. (8 mil wide trace, 8 mils above plane, 65 Ω.) (Courtesy of Bruce R. Archambeault, IBM Corporation.)

gential magnetic field vector and normal electric flux density vector components in the dielectric substrate do not vanish, however. To satisfy this discontinuity, the tangential magnetic field component creates a surface current distribution K_S (in A/m) along the surface of the boundary that is orthogonal to $\vec{H}_t$, so that $\vec{H}_{td} = K_S$. In a similar manner, the normal electric field flux density component deposits a surface charge distribution ρ_s (in C/m^2) along the surface of the boundary so that $\vec{D}_{nd} = \rho_s$ (refer to Figure 9.9).

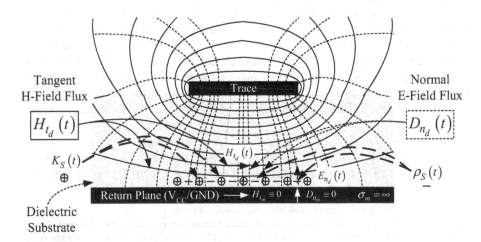

Figure 9.9. Surface current K_S and charge ρ_S distributions on the surface of the reference plane due to signal propagation in the structure.

Note that all fields and charge/current distributions are time-varying values, that is $H = H(t)$, $D = D(t)$, $K_S = K_S(t)$, and $\rho_s = \rho_s = (t)$.

It follows, therefore, that meeting the boundary conditions requires that time-varying return current and charge* distributions, corresponding to the signal propagating on the trace, must be present on the surface of the metallic boundary (or surface) of the plane even if the adjoining mass of metal happens to be a 230 VAC mains, a 5VDC (V_{CC}), a ground "0 V" plane (*GND* or V_{EE}), or an isolated metallic surface [5]. These current and charge distributions are also known as the "AC return currents."

9.3.3 Equivalence of Power and Ground Planes as Return Paths for High-Speed Signal Propagation

Planes on printed circuit boards constitute an uninterrupted solid (or, occasionally, high-density meshed) area of metal covering the entirety of a PCB layer. In certain cases, planelets[†] are used in lieu of complete planes; for instance, in cases of split-power (5 V, 3.3 V, etc.) power planes or split reference planes (*AGND* and *DGND*, for analog and digital ground planelets, respectively). The plane is generally designated as "ground" (or 0 V) plane if it serves as the DC current return conductor and a "power" plane if it is connected to a power supply voltage.

As distinction is often made between power (e.g., V_{CC}) and ground (*GND*, V_{EE}) planes, the distinction should be applied, therefore, for DC signals and power distribution only. AC signals can and do travel on either the power or ground plane (Figure 9.10). This concept should not be disturbing in any way and can be explained as follows.

The E- and H-field flux distribution in a transmission line structure (discussed in the previous section) and the corresponding charge and current distributions are shown by Maxwell's equations to be dependent on permittivity, permeability, and conductivity of the dielectric substrate and the metallic surface, ε, μ, and σ, respectively, only. The spatial distributions of the E ($= D/\varepsilon$) and H ($= B/\mu$) fields are linked in the dielectric substrate between the trace and the metal plane through Ampere's Law, $\nabla \times \vec{H} = \vec{J} + \partial\vec{D}/\partial t$ and Faraday's Law, $\nabla \times \vec{E} = -\partial\vec{B}/\partial t$, whereas the E-field to surface charge distribution is linked through Gauss's Law, $\nabla \times \vec{D} = \rho$. Nowhere in these equations is the DC potential or bias of the metallic plane to be found. The boundary conditions discussed above hold, therefore, from the standpoint of time-varying propagating fields, regardless of the DC potential of the metallic surface; Q.E.D.

As another example, consider a coaxial cable (a transmission line), which guides RF signals between its inner conductor and its outer conductor (i.e., shield), concurrently delivering DC power along the same conductors[‡] (Figure 9.11). Transmission

---

*Note that time-varying charge distribution is actually current: $\partial Q/\partial t \leftrightarrow I$.
[†]A planelet, a variation of a plane, is an uninterrupted area of metal covering only a portion of a PCB layer. Typically a number of planelets exist in one PCB layer. Applications of split power and return planes are discussed in detail in Section 9.7.
[‡]Consider the DC voltage source as an equivalent short circuit.

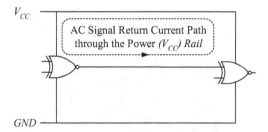

(a) Return RF Current Flow through V_{CC}

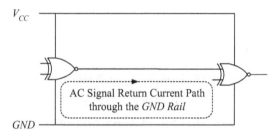

(b) Return RF Current Flow through GND

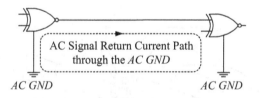

(c) Return RF Current Flow through AC GND

<u>Figure 9.10.</u> Equivalence of ground and power planes as AC current ground planes.

lines on printed circuit boards are no different. The actual DC voltage potential (represented as an ideal voltage source in Figure 9.11) with respect to some other reference plane has no effect on either the DC or RF signal propagation. In fact, note that TEM propagation will occur in the coaxial transmission line even if no connection exists to any reference plane, as long as the transmission path is complete and forms a closed circuit. So why should it be any different in signal transmission lines on printed circuit boards (e.g., microstrip or stripline)?

From this point on, the terms "ground" or "return" will, therefore, apply to either the "power" (V_{CC} or V_{AA}) or "return" (GND) planes with respect to AC energy propagation on PCB transmission lines unless explicitly stated otherwise.

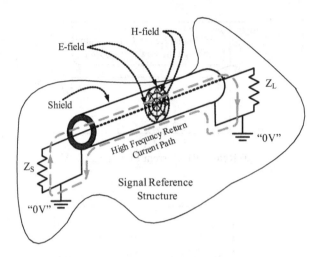

(a) No DC Bias; Shield at Potential of "0V";
Wave Propagates in TEM Mode in
Transmission Line Structure

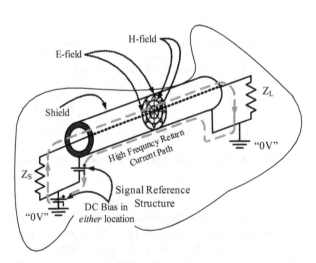

(b) DC Bias Introduced; TEM Wave Propagation in
Transmission Line Structure Prevails

Figure 9.11. DC bias has no effect on high-frequency wave propagation in a coaxial transmission line.

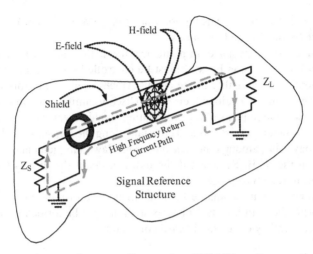

(c) No Reference Structure Connection; TEM Wave Propagation in
Transmission Line Structure Prevails

Figure 9.11. *Continued.*

Lessons Learned

- Use of solid copper planes for power distribution and return is essential for high-frequency (short transition time) circuits.
- Planes minimize noise produced by switching currents through stray inductance, help to mitigate the effects of crosstalk, control impedance of transmission lines, suppress reflections and ringing, and decrease emissions and susceptibility.
- Distinction between "power" and "ground" (or "0 V") planes is valid for DC only; AC signals can and do travel on either type of plane. The term "AC current return" applies to both the "power" and "ground" planes.
- Planes constitute a cost-effective measure for ensuring EMC and signal integrity at the PCB level. Without the use of planes, EMI control would be very difficult if not impossible.

9.3.4   Common Transmission Line Configurations on PCBs

Transmission lines are characterized by two main parameters, namely the characteristic impedance, Z_0, and the signal velocity of propagation, v. Both the characteristic impedance and the signal velocity of propagation are associated with the particular transmission line geometry and depend on the conductor's shape and its geometry with respect to the conductor(s) carrying its return current. They also depend on the relative permeability, μ_r and permittivity, ε_r, of the PCB's dielectric substrate (or laminate) material associated with the transmission line (e.g., FR4 or alumina) and on its proximity of other conductors and insulators.

The characteristic impedance, in particular, forms the one most important feature associated with transmission lines; it must be controlled for ensuring signal integrity and, subsequently, functionality of the circuit. Traces used as transmission lines must maintain a fixed value of Z_0 along their route. Discontinuities in Z_0 produce reflections of high-frequency components of the propagating signal waveform, introducing distortions and, consequently, emissions and crosstalk along the transmission line. Such discontinuities may arise, among other reasons, from flawed implementation of the reference planes on the PCB. Section 9.4 discusses in detail the effects that arise from discontinuities in reference planes.

Several techniques are available for implementing transmission line structures in multi-layer printed circuit boards. Transmission lines can be formed by a single-ended signal conductor and by differential signal conductors.

9.3.4.1 Single-Ended Transmission Line Configurations. In single-ended interconnects, both the signal trace and one or more reference planes are essential for guiding a signal across the PCB. Two fundamental topologies are commonly used: microstrip (surface and embedded) and stripline (single and dual) [3].

In the single-ended transmission line, a single conductor interconnects the source to the load. The reference plane provides the signal return path.

9.3.4.1.1 MICROSTRIP. Microstrip is a common configuration of transmission lines used to implement impedance-controlled interconnects on external layers on a printed circuit board. The two most common configurations of microstrip transmission lines are surface microstrip and embedded microstrip. The geometries of the two microstrip transmission line topologies are illustrated in Figure 9.12.

A surface microstrip transmission line consists of a trace routed in parallel to a solid and continuous reference plane, separated by a dielectric substrate; air constitutes the dielectric material above the conductor (in practice, a thin surface of solder mask is typically present above the trace). The approximate expression for the characteristic impedance Z_0 of a surface microstrip is [3]

$$Z_0 \approx \frac{87}{\sqrt{\varepsilon_r + 1.414}} \cdot \ln\left(\frac{5.98h}{0.8w + t}\right), \Omega \ (15 < w < 25 \text{ mils})$$

$$Z_0 \approx \frac{79}{\sqrt{\varepsilon_r + 1.414}} \cdot \ln\left(\frac{5.98h}{0.8w + t}\right), \Omega \ (5 < w < 15 \text{ mils})$$

(9.1)

Refer to Figure 9.12(a) for a description of the dimension parameters.

An embedded (or buried) microstrip transmission line is similar to the surface microstrip; however, the signal line is embedded in a dielectric and located a known distance h from the reference plane. The approximate expression for the characteristic impedance Z_0 of a surface microstrip is [3]

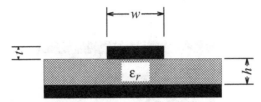

(a) Geometry of a Surface Microstrip Transmission Line

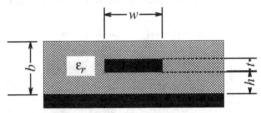

(b) Geometry of a Embedded Microstrip
Transmission Line

Figure 9.12. Geometry of microstrip transmission lines.

$$Z_0 \approx \frac{87}{\sqrt{\varepsilon_r \varepsilon_r \left[1 - e^{\frac{-1.55b}{h}} \right]}} \cdot \ln\left(\frac{5.98h}{0.8w + t} \right), \ \Omega \tag{9.2}$$

Refer to Figure 9.12(b) for a description of the dimension parameters.

9.3.4.1.2 STRIPLINE. Stripline refers to a trace located between two solid and continuous reference planes. The volume between the two reference planes contains a dielectric substrate that totally surrounds the trace. The stripline offers some advantages over the microstrip topology, namely, reduced dispersion of electric fields and enhanced magnetic flux cancellation, thus minimizing emissions from and crosstalk within the PCB.

The two most common configurations of stripline transmission lines are centered stripline and dual stripline. The geometries of the two stripline transmission line topologies are illustrated in Figure 9.13.

A centered stripline transmission line consists of a single trace centered in between two parallel solid and continuous reference planes. The approximate expression for the characteristic impedance Z_0 of a surface microstrip is [3]

$$Z_0 \approx \frac{60}{\sqrt{\varepsilon_r}} \cdot \ln\left(\frac{4h}{0.67\pi \cdot (0.8w + t)} \right), \ \Omega \tag{9.3}$$

Refer to Figure 9.12(a) for a description of the dimension parameters.

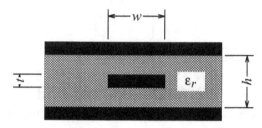

(a) Geometry of a Centered Stripline Transmission Line

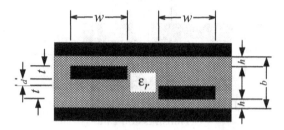

(b) Geometry of a Dual Stripline Transmission Line

<u>Figure 9.13.</u> Geometry of stripline transmission lines.

A dual stripline transmission line is a variation on the centered stripline and consists of a pair of traces asymmetrically placed in between two parallel solid and continuous reference planes. Each of the traces serves as a single-ended interconnect (this configuration is typically found in a stackup, in which two signal traces are placed in between two reference planes).*

Note: This configuration should not be confused with a differential transmission line topology, discussed in Section 9.3.4.2.

The approximate expression for the characteristic impedance Z_0 of a dual stripline is [3]

$$Z_0 \approx \frac{80}{\sqrt{\varepsilon_r}} \cdot \ln\left[\frac{4h \cdot (2h+t)}{0.67\pi \cdot (0.8w+t)}\right] \cdot \left[1 - \frac{2(h-t)}{4 \cdot (h+d+t)}\right], \ \Omega \qquad (9.4)$$

Refer to Figure 9.12(b) for a description of the dimension parameters.

9.3.4.2 Differential Transmission Line Configurations. For differential[†] interconnects (e.g., LVDS), the transmission line is formed by the combination of two

---

*A special subcategory of the dual stripline is the "offset stripline," in which only one signal layer embedded between the reference planes is present but is intentionally off-center.

[†]Although denoted "differential," the term "pseudodifferential" is more appropriate. The so-called differential interconnects are not truly differential since the drivers are typically formed by two complementary single-ended drivers (i.e., each is referenced to the common reference and utilizes a common ground plane for return

traces and one or more reference planes. In the differential transmission line, two conductors interconnect the source to the load. The reference plane provides the signal return path for the largest part of the returning current from each trace, not the adjacent trace. The presence of a reference plane is, therefore, not strictly necessary in the case of differential signals but it is vital for practical implementation of high-speed differential traces in PCBs.

Two fundamental topologies are commonly used: edge-coupled microstrip and edge-coupled stripline [3]. The geometry of the two differential transmission line topologies is illustrated in Figure 9.14.

9.3.4.2.1 DIFFERENTIAL EDGE-COUPLED MICROSTRIP. In a differential edge-coupled microstrip topology, the two conductors are routed in a manner similar to that of a surface microstrip transmission line. However, by virtue of their implementation, the lines are coupled, serving as a paired transmission line. To guarantee their performance, the separation between both traces of each differential pair should be maintained as small as practical and remain constant to avoid discontinuities in differential characteristic impedance and maintain the balance along the line. The approximate expression for the characteristic impedance Z_0 of a differential edge coupled microstrip is [3]

$$Z_{DIFF} = 2 \times Z_0 \cdot \left(1 - 0.48 \cdot e^{-0.96\frac{s}{h}} \right), \ \Omega \tag{9.5}$$

where

$$Z_0 \approx \frac{60}{\sqrt{0.457\varepsilon_r + 0.67}} \cdot \ln \left(\frac{4h}{0.67 \cdot (0.8w + t)} \right), \ \Omega \tag{9.6}$$

Refer to Figure 9.14(a) for a description of the dimension parameters.

9.3.4.2.2 DIFFERENTIAL EDGE-COUPLED STRIPLINE. In a differential edge-coupled stripline topology, the two conductors are routed in a manner similar to that of an embedded stripline transmission line. However, by virtue of their implementation, the lines are coupled, serving as a paired transmission line. Similar to the edge-coupled microstrip, the separation between both traces of each differential pair should be maintained as small as practical and remain constant to avoid discontinuities in differential characteristic impedance and maintain the balance along the line. The approximate expression for the characteristic impedance Z_0 of a differential edge coupled stripline is [3]

$$Z_{DIFF} = 2 \times Z_0 \cdot \left(1 - 0.347 \cdot e^{-2.9\frac{s}{h}} \right), \ \Omega \tag{9.7}$$

currents. The receivers, on the other hand, are differential as they respond to the difference between the two input terminals. A detailed discussion of balanced versus differential interfaces is provided in Chaper 2.

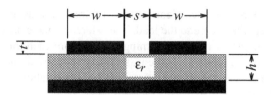

(a) Geometry of a Differential Edge Coupled Microstrip
Transmission Line

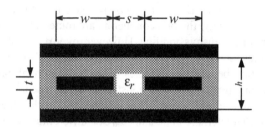

(b) Geometry of a Differential Edge Coupled Stripline
Transmission Line

Figure 9.14. Geometry of differential transmission lines.

where

$$Z_0 \approx \frac{60}{\sqrt{\varepsilon_r}} \cdot \ln\left(\frac{4b}{0.67\pi \cdot (0.8w+t)} \right), \; \Omega \tag{9.8}$$

Refer to Figure 9.14(b) for a description of the dimension parameters.

9.3.4.2.3 BROADSIDE COUPLED STRIPLINE. Contrary to the previous coplanar structures, in broadside coupled stripline transmission lines the two conductors are routed in a stacked-up manner in two adjacent layers, embedded between two return planes (Figure 9.15). Broadside coupled striplines are particularly useful in backplane design as these utilize only one routing channel and may be easier to route through high-density connector pin fields. The approximate expression for the characteristic impedance Z_0 of a differential broadside coupled stripline is

$$Z_{DIFF} \approx \frac{80}{\sqrt{\varepsilon_r}} \cdot \ln\left(\frac{1.9 \cdot (2h+t)}{0.8w+t} \right) \cdot \left(1 - \frac{h}{4(h+s+t)} \right), \; \Omega \tag{9.9}$$

Refer to Figure 9.15 for a description of the dimension parameters.

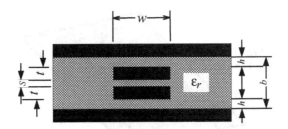

Figure 9.15. Geometry of broadside coupled stripline.

9.3.5   Return Current Path on Printed Circuit Boards

The fact that electrical current flows in a closed loop constitutes an absolutely fundamental physical principle.* It follows that each signal has an equal and opposite complimentary return current associated with it.† A key to ensuring that optimum PCB design is achieved lies in the understanding of how and where signal return currents actually flow. Signal integrity (functionality concerns) and EMC (radiated emissions and immunity concerns) performance of the circuit are directly associated with the inductance of the signal and return current paths, which, in turn, is directly related to the area occupied by them. Careful attention is normally paid to the signal current path, whereas little or no consideration is often given to the path utilized by the return current. It is simply taken for granted that the return signal will sort of take care of itself.

In a detailed discussion in Chapter 2, it was illustrated that electrical current will always flow in the path exhibiting the lowest impedance possible. Rather than repeating that discussion here, a practical example related to signal propagation on PCBs will be reviewed.

Consider a two-layer printed circuit board comprised of a trace above a solid reference (ground) plane, illustrated in Figure 9.16. The circuit includes a signal current source on the top layer, driving a single copper trace winding along its path in a signal layer, and terminated (through a resistive load) at its end through via #2 to the reference plane. The return lead of the signal source is also connected through via #1 to the reference plane. Ideally, impedance between via #1 and via #2 is zero and the voltage appearing across the current source should likewise be zero.

The simple schematic hardly begins to demonstrate the actual subtleties, but an understanding of how the return current flows back from via #2 to via #1 makes the realities apparent and illustrates how interference can be avoided at high frequencies.

A simulation of this configuration was made using Agilent Technologies' "Momentum," a three-dimensional planar EM simulator. The signal propagation path from the

---

*This a direct consequence of Ampere's Law, or Kirchhoff's Current Law, discussed in detail in Chapter 2.
†A detailed generalized discussion regarding return current propagation in electrical systems is included in Chapter 2. This Section focuses on the unique application of those principles to current propagation on printed circuit boards.

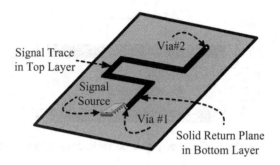

Signal Trace in Top Layer

Via#2

Signal Source

Via #1

Solid Return Plane in Bottom Layer

Figure 9.16. A model for evaluation of the current return path in reference plane interconnects.

source to the reference plane is well defined, as the trace can be considered as a one-dimensional structure. However, the reference plane, being a two-dimensional structure, offers an infinitely large number of alternative paths for the current to flow back to the source. The question is, "what is the actual return path utilized by the signal current in the plane?" The answer is revealed in the simulation results presented in Figure 9.17 (for legend, see Figure 9.18).

At this point, it is apparent that without a-priori knowledge of the frequency of the signal current, coming up with an unambiguous response is not possible. Fundamental physical laws, such as Ohm's and Kirchhoff's Current Laws, maintain that current always flows in the path of the lowest impedance. From the derivation in Chapter 2 it follows that the fraction (ΔI in percent) of the return current following the path other than that of the shortest electrical distance can be expressed as

$$\Delta I = \frac{j\omega}{j\omega + R_C/L_C} \times 100, \ \% \tag{9.10}$$

Based on an intuitive application of Ohm's Law ($V = I \times R$) considering the DC resistance of the current contour formed by the signal and return current paths, one might surmise that the return current follows the shortest electrical distance (having the lowest resistance), that is, a straight line from the end of the trace back to the source (assuming uniform conductivity of the surface). It would be reasonable to expect, under these assumptions, that the bulk of the return current would indeed take this shortest path. Obviously, some (insignificant) current spreading from this main path would be expected, to an extent inversely proportional to the increased length (and, consequently, the resistance) of the alternative spreading current paths. This is clearly supported by Equation (9.10) when applied for DC (and very low frequencies) currents, where $\omega \to 0$ and $\Delta I \to 0$

$$\lim_{\omega \to 0} (\Delta I) = \lim_{\omega \to 0} \left(\frac{j\omega}{j\omega + R_i/L_i} \times 100 \right) = 0\% \tag{9.11}$$

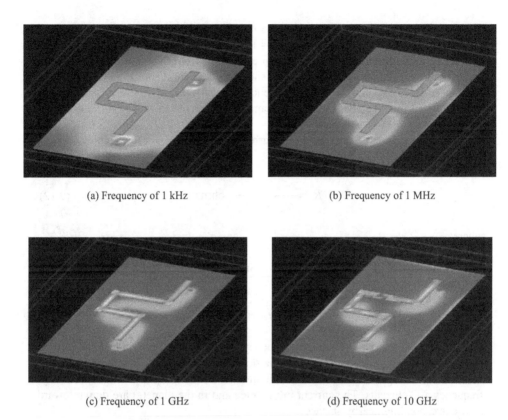

(a) Frequency of 1 kHz   (b) Frequency of 1 MHz

(c) Frequency of 1 GHz   (d) Frequency of 10 GHz

Figure 9.17. Simulation results visualizing the return current pattern in a solid refer-ence plane under a signal trace at various frequencies. (Simulation run on Agilent Technologies "Momentum" three-dimensional planar EM simulator; courtesy of Alexander Perez, Agilent Technologies.)

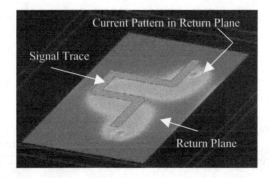

Figure 9.18. Legend for the plots depicted in Figure 9.17.

This is in support of the outcome of the simulation at 1 kHz presented in Figure 9.17(a). Practically all the current follows the shortest electrical distance; therefore, only a negligible fraction of the current spreads away from the main return path. The majority of the return current follows the shortest electrical distance, the path of least resistance. Since DC and very low-frequency (typically below a few kHz) current is uniformly distributed across the ground plane, it spreads out to occupy the entire cross section of the plane.

For a microstrip structure, the return current path resistance at low frequencies can be computed as [6]

$$R_0 = \frac{l}{\sigma_{cu} \cdot w_p \cdot t_{cu}}, \text{ ohms} \tag{9.12}$$

where:

R_0 = Resistance of the ground plane, Ω
σ_{cu} = Conductivity of copper ground plane, Si/m
l = Length of the microstrip trace, meters
w_p = Width of the ground plane, meters
t_{cu} = Thickness of the copper plane, meters

As frequency increases, additional frequency-dependent perturbation of the current distribution in the planes occurs due to the skin and proximity effects.* Figure 9.19 illustrates an example of the current distribution in a microstrip and stripline for high frequency AC currents. The current in the trace and in the plane(s) are drawn toward their corresponding nearby surfaces.

At frequencies at which the skin depth δ is less than t_{cu}, Equation (9.13) should be used in place of Equation (9.12) [6]:

$$R_0 = \frac{l}{\sigma_{cu} \cdot w_p \cdot \delta} \tag{9.13}$$

where the skin depth, δ, is defined as

$$\delta = \frac{1}{\sqrt{\pi \cdot f \cdot \mu \cdot \sigma_{cu}}} \tag{9.14}$$

where:

f = Frequency of interest, Hz
μ = Permeability of the plane material (copper), H/m

Two important features stand out: (1) the signal current flows only on the outer surface of the conductors, owing to the skin effect; and (2) the current distribution in the reference plane(s) is concentrated underneath (or above) the signal trace, due to the proxim-

---

*Skin and proximity effects were discussed in detail in Chapter 2.

(a) Current Distribution in a Surface Microstrip
Transmission Line

(b) Current Distribution in a Centered Stripline
Transmission Line

Figure 9.19. Illustration of high-frequency AC current distribution in the trace and reference planes for a microstrip and stripline due to skin and proximity effects (the lighter the shading, the higher the current density).

ity effect. Above the frequency at which the proximity effect begins to dominate, the current in the conductors attains a minimum-inductance distribution and is invariant with frequency [4].

Quite a different situation occurs at higher frequencies, as depicted in Figure 9.17, parts (b) through (d), (illustrated for the frequencies 1 MHz, 1 GHz, and 10 GHz, respectively). Now the voltage drop across the current path, $V = I \times Z_{AC}$, results from the AC (RF) impedance of the current contour, Z_{AC}, which is more accurately described by resistive and reactive terms:

$$Z_{AC} = R_C + jX_{L(C)} \approx R_C + j\omega L_C, \text{ ohms} \tag{9.15}$$

R_C and $X_{L(C)}$ represent the resistance and inductive reactance of the current contour, respectively, while $X_C = \omega L_C$ is where ω is the angular frequency of the signal and L_C, the net inductance of the current contour. When the resistance of the ground plane is significantly lower than the branch inductive reactance of the ground plane (typically at frequencies exceeding a few MHz), the return current is no longer distributed uniformly over the surface of the plane and does not follow the shortest path.

From Equation (9.10), when applied for high frequencies, where $\omega \to \infty$ and $\omega I \to$ 1 (100%) :

$$\lim_{\omega \to \infty} (\Delta I) = \lim_{\omega \to \infty} \left(\frac{j\omega}{j\omega + R_i / L_i} \times 100 \right) = 100\% \tag{9.16}$$

Figure 9.17, parts (b) through (d), clearly support this outcome at higher frequencies.*
The inductive reactance, $X_C = \omega L_C$, is proportional to frequency and the area of the loop
enclosed by the current contour. As frequency ω rises, the inductance, L_C goes somewhat down, but at a rate lower than the increase in frequency. The influence of the inductive reactance becomes increasingly dominant. For the current to follow the path of
least impedance, it must flow through the contour exhibiting a lower inductance.

The total inductance of the current contour is associated with the net flux per unit of
current flowing through the signal-return contour. By definition, the self-partial inductance of the ith leg of the loop, L_{pii}, is[†] (refer to Figure 9.20)

$$L_{p_{ii}} \triangleq \frac{\phi_{ii}}{I_i} = \frac{\phi_{ii}}{I} \tag{9.17}$$

where the index i relates to the legs $i = 1, 2, 3, 4$ of the loop. Note that in an electrically small circuit or loop $I_1 = I_2 = I_3 = I_4 = I$. Similarly, the mutual partial inductance,
L_{pij}, between ith and the jth legs of the circuit is defined as

$$L_{p_{ij}} \triangleq \frac{\phi_{ij}}{I} \tag{9.18}$$

The total inductance of the current contour (neglecting the effect of legs 2 and 4 of
the circuit[‡]) is associated with the sum of the magnetic flux, ϕ_{11} and ϕ_{33}, generated by
and surrounding the conductors represented by legs 1 and 3, respectively, and the magnetic flux, ϕ_{31} and ϕ_{13}, generated by the one leg, but surrounding the other, relative to
the loop current, I. The magnetic flux fields ϕ_{11} and ϕ_{31} as well as ϕ_{33} and ϕ_{13} are opposite in polarity due to the equal but opposite currents producing them. The resultant
total magnetic flux confined by the contour of the current path is, thus, the difference
between the flux produced by the signal and return conductors:

$$\phi_{Net} \approx \left| \left(\phi_{33} - \phi_{13} \right) + \left(\phi_{11} - \phi_{31} \right) \right| \tag{9.19}$$

Combining Eq. (9.17), (9.18), and (9.19) results in

$$L_{Total} \approx \left| \frac{\left(\phi_{33} - \phi_{13} \right) + \left(\phi_{11} - \phi_{31} \right)}{I} \right| \triangleq \left| \left(L_{33} - L_{13} \right) + \left(L_{11} - L_{31} \right) \right| = \left| \left(L_{33} + L_{11} \right) - \left(L_{31} + L_{13} \right) \right| \tag{9.20}$$

The diminishing inductance (occurring when $L_{11} \rightarrow L_{31}$ and $L_{33} \rightarrow L_{13}$) forces the return current to concentrate even more beneath the signal trace, further reducing the

---

*At the frequencies 1 GHz and 10 GHz [Figure 9.17, parts (c) and (d), respectively], the current distribution
in both the trace and the return plane is not uniform as a result of resonance at the high frequencies on the
electrically large structures.
[†]The concepts of partial, self-, and mutual inductance were discussed in detail in Chapter 2.
[‡]This approximation is valid thanks to the very short length of the vias interconnecting the signal trace and
the return plane.

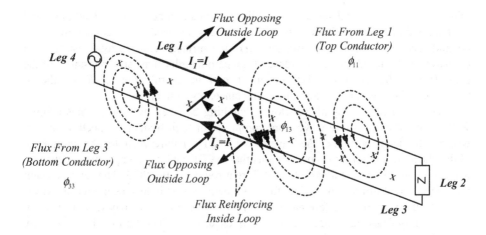

(a) Electromagnetic Representation

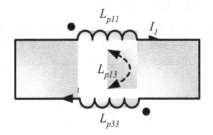

(c) Circuit Representation

Figure 9.20. Self- and mutual inductance of the signal-return contour (flux from legs 2 and 4 neglected).

contour area confined by the signal and return current paths (i.e., decreasing the spacing between the signal and return paths). Equation (9.20) thus reduces to

$$L_{Total} \approx 2\left|L_{33} - L_{13}\right| \tag{9.21}$$

The simple illustration presented in Figure 9.17 visibly demonstrates that the loop with the least area is quite evidently formed by the top trace and the portion of the reference plane directly underneath it. Whereas Figure 9.17(a) illustrates the return current path at frequencies DC (or at very low frequencies), Figures 9.17(b) through 9.17(d) reveal that the path that most of the high-frequency current takes in the reference plane resulting in the smallest loop area directly under the top conductor. Even at frequencies as low as 1 MHz, the return current path is nearly all under the top trace. In practice, the resistance in the reference plane causes the current distribution at low and mid frequencies to spread and attain an ellipse-like pattern between the feed-point and the load end of the trace. This distribution balances the need of the current to spread out and occupy the maximum cross-section of the surface (reducing total resistance of the parth) with the subsequent increase of the path length (increasing resistance of the path). [Eq. (9.10) demonstrated the dependence of the return current on R_C/L_C, the ratio between the contour resistance and inductance, respectively].

Finally, consider a digital waveform switching between two logic states (Figure 9.22). The initial transition wave front comprises of high-frequency spectral content, whereas the steady portion of the waveform is of a DC nature. As a direct consequence of the above discussion, it is now obvious that the DC and AC components of the signal return currents may follow totally different routes on the PCB. The return current during the waveform transition time follows a path of least inductance; after the waveform settling time (e.g., a few nanoseconds) the return current may take a different route—that of least resistance. As a consequence, some distortion of the waveform, particularly during the transition time, could occur.

9.3.6 Return Current Distribution

The reasoning in Section 9.3.5 may lead to the conclusion that all the return current tends to flow directly and only underneath the signal trace. Observation of the plots in Figure 9.17 illustrates that although the highest current density is underneath the signal

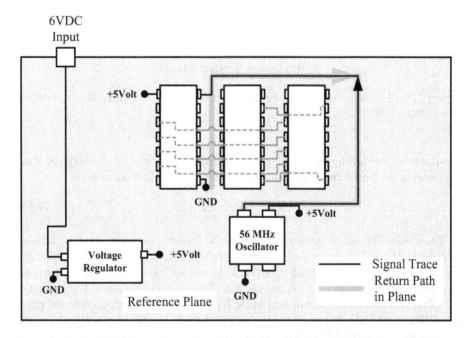

Figure 9.21. 56 MHz clock signal currents follow the path of least inductance. (Courtesy of Dr. Todd Hubing, Clemson University.)

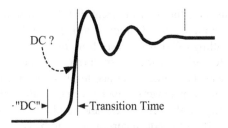

DC ?

·"DC"► ◄Transition Time

<u>Figure 9.22.</u> Digital waveforms represent a transition (high-frequency spectral content) between two "DC" states.

trace, the RF current also spreads out in the plane to either side of the trace's route. The distribution of high-frequency return current in a reference plane adjacent to a signal trace routed in the form of microstrip is more accurately described by [6,7,8]

$$J_{GP}(d) = -\frac{I_0}{\pi} \cdot \left\{ \frac{1}{w} \cdot \left[\tan^{-1}\left(\frac{d + \frac{w}{2}}{h} \right) - \tan^{-1}\left(\frac{d - \frac{w}{2}}{h} \right) \right] + \frac{4h}{w_p \cdot \pi} \cdot \frac{1}{\sqrt{\left(\frac{w_p}{2}\right)^2 - d^2}} \right\}, \text{ A/m}$$

(9.22)

where:
$J_{GP}(d)$ = Current density at a distance, d, from the trace centerline, amperes/meter
I_0 = Magnitude of the total current flowing on the source trace, amperes
d = Distance from source trace centerline, meters
h = Height of trace above the reference plane, meters
w_p = Width of the ground plane, meters
w = Width of trace, meters

An approximate expression for the return current distribution at a distance, d, from the centerline of the signal trace (assuming a narrow trace close to an infinitely wide reference plane) is [9]

$$J_{GP}(d) \approx \frac{I_0}{\pi h} \cdot \frac{1}{1 + (d/h)^2}, \text{ A/m}$$

(9.23)

Using this current distribution, the effective resistance, R_0, of the path occupied by the actual return current can now be computed as [6]

$$R_0(d) = \frac{J_{GP}(d) \cdot l}{I_0 \cdot \delta \cdot \sigma_{cu}}, \text{ ohms}$$

(9.24)

Note how R_0 decreases the farther the coupled trace is from the source trace.

Figure 9.23 [10] illustrates the return current distribution in a reference plane beneath a signal trace in a typical printed circuit board. The peak current density lies directly under the trace, falling off sharply away from the trace. From Figure 9.23, it is evident that approximately 95% of the return current flowing in the reference plane travels in a strip as wide as three times the trace-to-plane spacing (h) or three times as wide as the trace width (w), whichever is smaller, centered along the trace's path.

This current distribution in the reference plane results from the opposite effect of two contradictory forces. If *all* the return current were to be concentrated immediately underneath the trace, it would have exhibited a higher partial self inductance, as it would have, in effect, acted as a narrow trace, disregarding the entire breadth of the plane. The impedance resulting from this inductance is proportional to frequency ($Z_L \approx \omega L_S$, where L_S represents the partial self-inductance of the return path). Conversely, if the return current were to spread farther apart from the trace, taking full advantage of the entire plane, the inductance resulting from the large loop area created by the widely spread return current will increase. The impedance resulting from this inductance is also proportional to frequency ($Z_L \approx \omega L_L$, where L_L represents the loop inductance between the signal trace and the return path). Equilibrium is reached, therefore, when optimum current distribution is attained, balancing the two opposing inductances, resulting in some spreading of the return current in the reference plane.

9.3.7 Crosstalk Mechanisms on PCBs

Consider a case in which two traces are routed adjacent to each other (Figure 9.24). When a signal propagates along an active transmission line (called an "aggressor" or

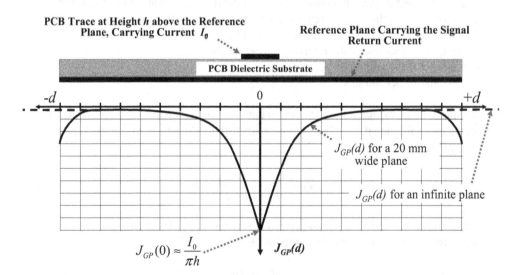

Figure 9.23. Current distribution across the reference plane for a microstrip trace as a function of the distance from the centerline of a microstrip trace centered 1 mm above a 20 mm wide plane. (Courtesy of Keith Armstrong, Cherry Clough.)

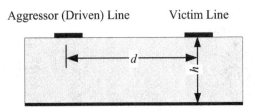

Aggressor (Driven) Line   Victim Line

Figure 9.24. Fundamental configuration for crosstalk between aggressor and victim lines.

"driven" line), electric flux emerges from the line while magnetic flux encircles the line. These flux fields are not confined to the immediate space between the signal and its associated return path. Rather, some of the flux extends into the surrounding area. These fields are called "fringe fields" (Figure 9.25).

Electric field flux emanating from the active trace terminates on any adjacent metallic structure (Gauss's Law of Electric Fields), whereas magnetic fields surrounding the transmission line also partially encircle nearby metallic structures (Gauss's Law of Magnetic Fields). Consequently, if the adjacent metallic structure happens to be a signal trace, stray current and voltage (actually, electromotive force or emf) sources are deposited into that trace. Clearly, as separation increases between the two traces, the fields (and hence the coupling) drop off quickly. However, if they are sufficiently close, noticeable interference may be picked up by the victim trace. The resultant current flowing on the receptor or "victim" trace is now subject to the same behavior as the intended signal flowing on the victim line and may experience reflections and distortions as well as reradiate fields or couple to other closely spaced traces. Likely outcomes of crosstalk are degradation of signal integrity and increase of EMI from the PCB. Many failures in EMC radiated emission tests are due to emissions from "quiet" I/O lines carrying residual interference current induced onto that line by "noisy" signals internal to the PCB, due to poor PCB layout.

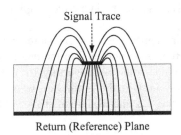

Signal Trace

Return (Reference) Plane

(a) Electric Field Flux Emerging from an Active (Current Conducting) Trace

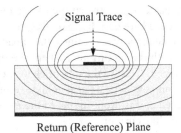

Signal Trace

Return (Reference) Plane

(b) Magnetic Field Flux Surrounding an Active (Current Conducting) Trace

Figure 9.25. fringe electric (a) and magnetic (b) field flux due to current flowing in an "active" transmission line.

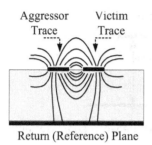

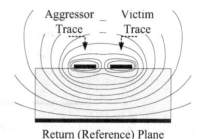

(a) Electric (or Capacitive) Coupling: Electric Flux   (b) Magnetic (or Magnetic) Coupling: Magnetic
Terminating on an Adjacent Victim Trace   Field Flux Encircling an Adjacent Victim Trace

<u>Figure 9.26.</u> Fringe electric (a) and magnetic (b) field flux coupling to adjacent traces.

The electric and magnetic field coupling between conductors is commonly also known (using equivalent circuit models) as "capacitive" and "inductive" coupling, respectively. Through both capacitive and inductive coupling play a role in crosstalk, their respective effects significantly depend on the circuit layout and topology. Figure 9.27 depicts a simplified model for crosstalk, consisting of capacitive and inductive coupling between short* sections of traces on a printed circuit board. Capacitance, C_G, exists between the traces to the return plane, playing a role in the characteristic impedance of the traces and the signal propagation delay along the line. Also, the intertrace capacitance, C_M, forms the path for undesired signal coupling (capacitive crosstalk) between the traces. In addition, the aggressor and victim traces each exhibit self-partial inductances, L_A and L_V, respectively, which are also relevant for determining the characteristic impedance and signal propagation delay (similar to the capacitance to the return plane). Due to the asymmetry of this configuration—a narrow trace above a broad return plane—the flux contribution due to the current flowing on the trace and through the return plane are unequal, yielding partial flux cancellation only, hence resulting in radiated emissions from the circuit. The mutual inductance, L_M, between the two trace circuits gives rise to the inductive coupling (or crosstalk) between the two circuits.

In electrically short lines, the capacitive coupling appears as a current source in parallel with the victim line, where the induced current is proportional to the mutual capacitance, C_M, and the rate of change of the source line voltage. Conversely, the inductive coupling appears as a voltage (emf) source in series with the victim line, where the induced emf is proportional to the mutual inductance, L_M, and the rate of change of the source line current:

*In electrically long lines, this expression is correct for electrically small fractions of the traces only. Coupling in such lines occurs at the position of the aggressor signal wave front (position of transition of the voltage or current waveform) as it propagates along the line. The local properties of the transmission line determine the coupling, whereas the termination impedance comes into play only as soon as the wave front arrives there.

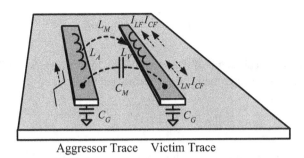

Aggressor Trace   Victim Trace

Figure 9.27. Fundamental model of capacitive and inductive coupling between trace sections.

$$\begin{cases} I_C = C_M \dfrac{dV_S}{dt}, \text{ amperes} \\[2mm] V_L = -L_M \dfrac{dI_S}{dt}, \text{ volts} \end{cases} \tag{9.25}$$

where I_C and V_L are the capacitively induced current and inductively induced voltage in the victim line, respectively, due to a changing source voltage V_S and current I_S in the aggressor circuit.

In practical situations, both capacitive and inductive crosstalk mechanisms are present simultaneously. The capacitively induced currents in the victim line emerging from the capacitance between the traces (represented as a current source in parallel with the line) propagate both forward, toward the far end, as well as backward, toward the near end (I_{CF} and I_{CN}, respectively in Figure 9.27) of the victim trace. The inductively coupled voltage source (represented as an induced emf in series with the line), on the other hand, drives current in the victim line (I_{LF} and I_{LN} in Figure 9.27) in a direction opposite to the current flowing in the aggressor line [note the negative sign in Equation (9.25)]. Consequently, the capacitive and inductive coupled signals reinforce themselves in the backward direction, whereas they tend to cancel in the forward direction, in particular in stripline configurations. The total coupled signals flowing backward are known as "backward crosstalk" or "near-end crosstalk" (NEXT), whereas the total coupled signals flowing forward (and practically cancelling out) are known as "forward crosstalk" or "far-end crosstalk" (FEXT). The features of NEXT and FEXT differ dramatically: NEXT constitutes a constant-magnitude pulse, with a width extending to twice the propagation time down the coupled region between the traces, whereas FEXT is characterized by a narrow pulse, having a width equal to the transition time of the aggressor signal, but an amplitude growing as the coupled region between the traces increases [5].

NEXT reaches its peak magnitude and then remains constant at the so-called "critical length," L_C, the length at which the round-trip propagation time equals the transition (rise or fall) time of the signal, t_r:

$$L_C = \frac{2 \cdot T_{PD}}{t_t}; \text{ meters} \qquad (9.26)$$

where L is the physical length of the transmission line in meters and T_{PD} is the propagation delay in seconds per meter. T_{PD} is estimated using

$$T_{PD} \approx 1.017 \cdot \sqrt{a \cdot \varepsilon_r + b}; \text{ sec/m} \qquad (9.27)$$

where:
ε_r = the relative permittivity of the dielectric substrate on the PCB
$a = 1$ and $b = 0$ for stripline
$a = 0.475$ and $b = 0.67$ for microstrip

Defining the maximum crosstalk coefficient as k_x, we thus obtain

$$NEXT = \begin{cases} k_x \cdot \dfrac{L}{L_C}; & \text{if } \dfrac{L}{L_C} < 1 \\[3mm] k_x; & \text{if } \dfrac{L}{L_C} \geq 1 \end{cases} \qquad (9.28)$$

The crosstalk coefficient, k_x, is estimated as follows (Figure 9.28) [11]:

$$k_x \approx \frac{1}{1 + \left[d^2 / (h_1 \cdot h_2) \right]} \qquad (9.29)$$

$$k_x \approx \frac{1}{1 + \left[d^2 \bigg/ \left(\dfrac{h_{ia} \cdot h_{ib}}{h_{ia} + h_{ib}} \right)^2 \right]} \qquad (9.30)$$

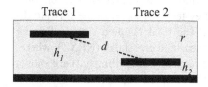

(a) Microstrip Configuration *Note:* For coplanar microstrip traces, use $h_1 = h_2$ in Equation (9.29).

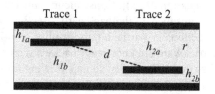

(b) Stripline Configuration[1] *Note:* For co-planar centered stripline traces, use $h_{ia} = h_{ib}$ ($i = 1, 2$) in Equation (9.30).

Figure 9.28. Estimation of the maximum crosstalk coefficient. In stripline, the crosstalk drops with the fourth power of distance from the centerline of the trace; therefore, the quadratic expression provided in Equation 9.30 serves as a conservative estimate of crosstalk.

The proximity of the trace to the return plane significantly reduces the effects of capacitive coupling, owing to the increase of trace-to-return capacitance. On the other hand, if the signal traces are kept closer to the return plane than the mutual separation, the plane becomes the preferred return path, and inductive crosstalk also diminishes considerably.

In digital circuits, inductive coupling is more predominant than capacitive coupling, owing to the low-impedance nature of digital drivers, whereas capacitive coupling is more predominant in high-impedance (often analog) circuits.

9.3.8 Common-Impedance Coupling on PCBs

A third and very important coupling mechanism in PCBs is common-impedance coupling. As was shown in Chapter 4, common-impedance coupling is associated with the interaction between the return current of the aggressor circuit and the current flowing through a return plane in a path common to the victim circuit. This could be of particular concern when noisy, high-current (e.g., digital) circuits share a common return path with sensitive (e.g., analog) circuits.

High-frequency return currents tend to flow in the return plane immediately beneath the signal trace (regardless of the DC potential of that plane). The current slightly spreads out in the plane, but otherwise (if not obstructed) remains under the trace. Equation (9.31) [derived from Equation (9.23)] expresses the approximate extent of spreading of the current distribution under the trace, $S(d)$:

$$S(d) \approx \frac{1}{1+(d/h)^2} \tag{9.31}$$

where d is the horizontal distance from the trace center line in the return plane and h represents the height of the aggressor trace above the return plane. Figure 9.29 illustrates a mechanism by which coupling occurs between an aggressor trace and a victim trace, due to the fraction of the aggressor trace return current, $J_{GP}(d)_A$, flowing under the victim trace. The path of the victim return current distribution is also shown, symbolized by $J_{GP}(d)_V$. The "influence zone" is the area where an overlap occurs between the two current distributions, and it determines the extent of interaction between the two traces. The fraction of the return aggressor current flowing in the victim return

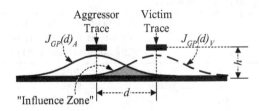

Figure 9.29. Common-impedance crosstalk through a shared return plane.

current path efficiently couples into it, as it adds to the victim's return current flowing through that common path, appearing as noise in the victim circuit.

The significance of common-impedance crosstalk is easily illustrated: Consider a digital processor drawing a current surge of 10 A through the return plane common to an analog circuit. The analog circuit contains a 24-bit A/D converter, with a least significant bit equivalent to 5.9 nV per 1 V of supply voltage. Assuming a practical plane impedance of 40 $\mu\Omega$, this voltage is equivalent to 0.15 mA, approximately, constituting 0.15% of the digital switching current. The necessary separation, d, between the digital and the analog traces, in order to preclude any interference, must be increased so that 99.97% of the digital return current is contained within the distance from the signal trace centerline, d.

Table 9.1 presents the fraction (in percent) of the cumulative return current contained within a distance d (normalized to height of the trace above the return plane) from the centerline of the aggressor signal trace, $\pm d/h$, where d is the horizontal distance and h is the height of the trace above the ground plane.

From Table 9.1, it follows that with typical heights, h, of 10 mil, a separation, d, greater than 200 mil (or 5 mm) would be required. Such separations are almost always impractical in contemporary PCB designs. One of the solutions for eliminating common-impedance crosstalk by means of split return planes is addressed in Section 9.7.

9.3.9 Consequences of Transmission Line Topology on EMI and Crosstalk Control

Figure 9.30 illustrates the effect of transmission line topology on the EMI and crosstalk performance of the circuit. In a coplanar transmission line [Figure 9.30(a)], the traces are exposed and no reference plane is present. The electric flux terminates on adjacent metallic structures. If those happen to be signal traces, significant crosstalk may be the result. Effective magnetic flux cancellation will be limited by the separation between the signal and the corresponding return traces.

Placing a solid metallic reference plane beneath the signal traces (i.e., microstrip) provides a considerable improvement [Figure 9.30(b)]. The greater part of the electric flux now terminates primarily on the large mass of the reference plane while the equal

Table 9.1. Fraction (in percent) of the return current contained within a normalized distance of $\pm d/h$ from the signal trace centerline

d/h	Fraction of cumulative current density at distance, d from aggressor trace centerline (%)
2	70%
5	87%
10	94%
20	97%

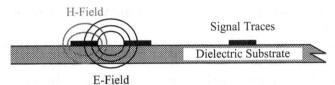

(a) Exposed Traces on a PCB with No Reference Plane

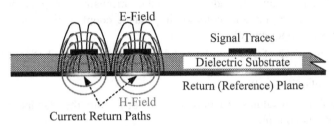

(b) Traces Routed as Surface Microstrip on a PCB with an Adjacent Reference Plane

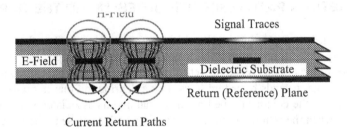

(c) Traces Routed as Centered Stripline on a PCB with Adjacent Reference Planes

<u>Figure 9.30.</u> Effect of reference planes on H- and E-field distribution.

and opposite return current flowing in the plane* provides for substantial magnetic flux cancellation.

Embedding the trace between two solid metallic reference planes (i.e., stripline) enhances the performance of the circuit [Figure 9.30(c)]. Electric flux now terminates on both reference planes while the equal and opposite return current flowing in the planes provides for even superior magnetic flux cancellation. Consequently, stripline constitutes the ultimate topology from the standpoint of EMI and crosstalk control.

Lessons Learned

- Return current does not all flow directly under signal traces; although the highest current density is underneath the traces, the RF current also spreads out in the plane to either side of the trace's route.

*See Section 9.3.5 for a detailed discussion on the path of return current flow in reference planes.

- Return current spread in the planes can affect collocated circuits routed across the board.
- Crosstalk constitutes an "on-board" EMI problem, associated with coupling between aggressor and victim traces.
- Capacitive, inductive, and common impedance mechanisms all contribute to crosstalk.
- In digital circuits, the inductive crosstalk mechanism is predominant; in mixed digital–analog circuits, common-impedance coupling through the return plane plays an important role.
- The presence of a return plane and the proximity of the both the aggressor and victim traces to it (in microstrip or stripline configurations) significantly enhance control of crosstalk.
- Stripline constitutes the ultimate topology from the standpoint of EMI and crosstalk control.

9.4 RETURN PATH DISCONTINUITIES: "MIND THE GAP"

If constrained to traces, the signal and its return currents will flow through the path to which they are confined, ultimately compromising EMC and, possibly, functional performance of the circuit. When the return signal is allowed to flow through a plane, on the other hand, it tends by nature to assume a distribution that minimizes the overall impedance of the contour formed by the signal and its associated return path. This is achieved when the return current converges to a path as close to the trace as possible, namely, directly next to it.

Up to this point, solid, continuous reference planes, serving as return paths, were assumed. Interruptions (or gaps) such as holes, slots, or isolation splits in the reference plane continuity beneath a trace can drastically degrade the performance of the transmission line and any devices sharing the reference plane. Regretfully, reference planes without numerous perforations are almost nonexistent in practice because vias and other through-holes introduce unavoidable holes in the planes. Using high-density PCB fabrication techniques (making use of "blind vias," which are visible only on one surface, or "buried vias," which are visible on neither) can significantly lessen the problem of excessive perforation of the planes. Fortunately, the effects of vias and through-holes, if carefully treated, are usually (but not always!) minor.

Slots and gaps in the planes, on the other hand, can and do introduce considerable EMI and signal integrity concerns. When traces cross or are routed in the near vicinity of gaps in the reference plane, the discontinuities in the plane obstruct the return current path, forcing the return current to flow in larger loops. Such situations should be avoided or carefully controlled.

Figure 9.31 illustrates some common layout mistakes made when traces are routed too close to antipads or edges and crossing gaps in reference planes, thereby increasing the impedance to the return currents associated with the traces. From the standpoint of signal integrity, adding impedance in a return current path is equivalent to adding impedance in the signal trace [10].

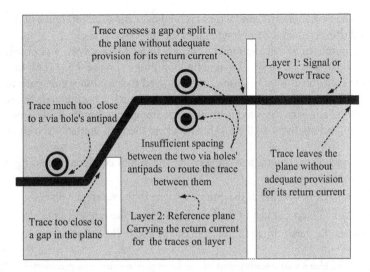

<u>Figure 9.31.</u> Some examples of common trace routing errors. (Courtesy of Keith Armstrong, Cherry Clough.)

The path occupied by the return current in a gapped reference plane and the resultant loop (i.e., antenna) formed by the signal and its return path in the plane can be easily demonstrated experimentally [11]. In this demonstration and in those presented in Section 9.4.1, the test board depicted in Figure 9.32 [12, 13] was used. The board consisted of two 24 gauge wires, approximately 12 cm in length, routed from a BNC connector and taped down to the circuit board to form approximately 50-Ω transmission lines with the reference plane. The conductors were terminated by 47-Ω resistors on

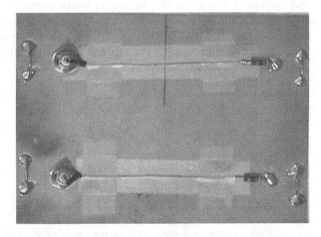

<u>Figure 9.32.</u> Test board consisting of a conductor crossing a gap in the reference plane. (Copyright 2003 Douglas C. Smith, http://www.dsmith.org, used with permission.)

the right. The first (bottom) conductor is routed over the solid reference plane, whereas the second (top) crosses a 5 cm long gap in the reference plane.

For the purposes of this demonstration, a 5 to 50 MHz square-wave oscillator was used as a signal source. A small magnetic loop probe (Figure 9.33) [14], the output of which was connected to an oscilloscope, was used to trace out signal current paths on the board. The oscilloscope was triggered directly from the oscillator so the relative directions of the edges of the waveform could be compared. Figure 9.34 depicts the four positions and orientations on the test board where the loop output was measured:

1. Position A, horizontal and adjacent to the signal conductor crossing the gap in the reference plane
2. Position B, horizontal at the end of the gap in the reference plane
3. Position C, vertical just to the left of the gap in the reference plane
4. Position D, vertical just to the right of the gap in the reference plane

Figure 9.35 shows an oscillogram of the loop probe output at position A. The output voltage of the loop, V_O, is the derivative of the signal current ($V_O = M \cdot di/dt$), but could be used to sense the course and direction of current flow. The positive and negative peaks correspond to the edges of the square wave. The edges of the square wave are somewhat more rounded than expected due to the heavy load presented by the 47 Ω resistor. When the loop probe was moved to position B with the same orientation, the plot in Figure 9.36 was produced. It is the inverse of the one in Figure 9.35, indicating that the current was now flowing in an opposite direction around the edge of the break in the reference plane. The amplitude is a little smaller in Figure 9.36, most probably because the current was not parallel to the loop probe for its entire length (as it bends around the edge of the gap) and the inductance of the ground plane is lower compared to the signal wire.

Similarly, Figure 9.37 and Figure 9.38 show the output of the loop probe at positions C and D respectively.

The data above noticeably illustrates that the signal return current flowing in the reference plane is diverted to the edge the gap and thus forms a substantially larger signal current loop area with the signal conductor. This larger loop area increases the total loop inductance and, consequently, the RF voltage drop across the reference plane

Figure 9.33. The small "paperclip" magnetic loop probe used for the test. (Copyright 1999 Douglas C. Smith, http://www.dsmith.org, used with permission.)

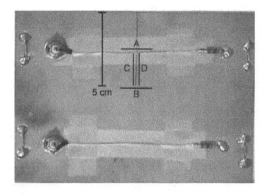

Figure 9.34. Test board with positions of magnetic loop probe labeled. (Copyright 2003 Douglas C. Smith, http://www.dsmith.org, used with permission.)

$(V_{Gap} = L \cdot di/dt)$:

$$V_{GP} \approx L_{Gap} \cdot \frac{dI}{dt}, \text{ volts} \tag{9.32}$$

where:
V_{GP} = Voltage drop across the reference plane, volts
I = Signal current, amperes
t = Time, seconds

The RF voltage drop across the gap in the reference plane results in increased emissions from the PCB and larger susceptibility to externally induced EMI (an EMI con-

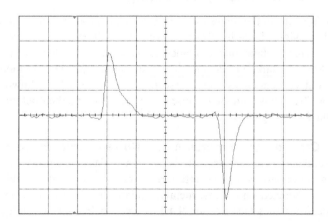

Figure 9.35. Loop voltage (position A), 10 ns/div (horizontal)/10 mV/div (vertical). (Copyright 2003 Douglas C. Smith, http://www.dsmith.org, used with permission.)

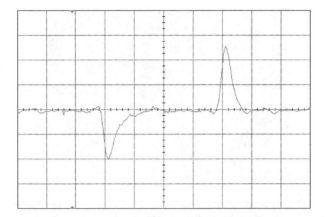

Figure 9.36. Loop voltage (position B), 10 ns/div (horizontal)/10 mV/div (vertical). (Copyright 2003 Douglas C. Smith, http://www.dsmith.org, used with permission.)

cern), as well as higher crosstalk between adjacent traces, introducing waveform distortion (signal integrity and functionality concerns). In addition, gaps introduce significant impedance discontinuities in controlled impedance transmission lines formed by the trace and the reference planes since the return path discontinuity alters its characteristic impedance, resulting in reflections and distortions in the signal waveform and, again, ending up with higher EMI and degraded signal integrity. Gaps also comprise a significant contributor to power/ground bounce and simultaneous switching noise (SSN)* in signal circuits and in power distribution systems on PCBs.

9.4.1 Undesired Effects of Traces Crossing Gaps in the Reference Planes: "Seeing is Believing"

Signals crossing gaps in the reference plane breaks can produce numerous adverse undesired EMI and signal integrity effects in printed circuit boards due to traces crossing a break in a reference plane, including increased emissions of EMI, impairment of immunity to EMI, crosstalk, and degradation of the signal rise time. Demonstrations of these undesired consequences were presented in [13, 15–18]. Following the well-known saying, "seeing is believing" those demonstrations and their outcomes are summarized in the following subsections.

9.4.1.1 Creation of Common-Mode Currents and Emissions. Signals flowing in conductors crossing gaps in the reference plane on a printed circuit boards can result in common-mode (CM) interference currents on system cables, followed by excessive conducted and, consequently, radiated EMI emissions, in addition to any signal quality issues (see Section 9.4.1.4), even at frequencies as low as tens of

---

*Power/ground bounce (also known as ΔI noise) and simultaneous switching noise (SSN) are discussed in detail in Section 9.5.

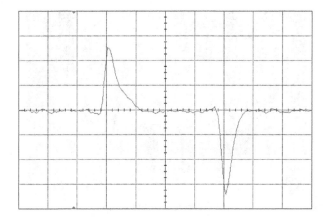

Figure 9.37. Loop voltage (position A), 10 ns/div (Horizontal)/10 mV/div (vertical). (Copyright 2003 Douglas C. Smith, http://www.dsmith.org, used with permission.)

megahertz. The effect on emissions due to CM currents on system cables resulting from a signal crossing a plane break is addressed in the following demonstration [15].

Figure 9.39(a) depicts the setup used for this demonstration. A small oscillator operating at about 30 MHz with signal edge rates of about 2 nsec and an amplitude of 3 V peak to peak, approximately, was coupled to the test board shown in Figure 9.32 through a short BNC coupler. A pair of 2-meter-long wires was attached to the reference plane at both sides of the board. The resultant CM current in these wires was measured by means of a pair of matched Fischer F-33-1 current probes. A close-up of the test board, generator, and current probes is presented in Figure 9.39(b).

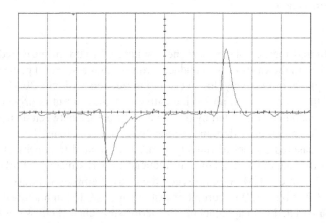

Figure 9.38. Loop voltage (position B), 10 ns/div (Horizontal)/10 mV/div (vertical). (Copyright 2003 Douglas C. Smith, http://www.dsmith.org, used with permission.)

(a) Overview of the setup

(b) Close-Up of the Test Board, Generator,

and Current Probes

<u>Figure 9.39.</u> Test setup for measuring common-mode current emissions due to a gap in the reference plane. (Copyright 2005 Douglas C. Smith, http://www.dsmith.org, used with permission.)

Figure 9.40 and Figure 9.41 present the resultant CM current measured on the 2-meter-long wires in the cases of the conductor routed above the solid plane and across the slot in the plane, respectively, for an excitation frequency of only 30 MHz. Visibly, the output of the probes are in phase, indicating that the current is driven along one wire and onto the other, stimulated by the voltage difference between them on the reference plane. The wires are acting, in practice, like a dipole antenna.

The peak measured current of roughly 1 mA is estimated to consist of about 250 μA of current flowing at the fundamental frequency, approximately 30 MHz,. Radiated emissions due to this CM current exceed the class B emissions limits of standards such as CISPR 22* by 34 dB (a current of just about 5 μA fed into a tuned $\lambda/2$ dipole will produce a radiated field approaching the permissible class B emissions level).

*CISPR 22, *Information Technology Equipment, Radio Disturbance Characteristics, Limits and Methods of Measurement.*

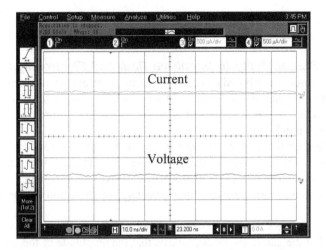

Figure 9.40. Measured common-mode current in the case of the conductor routed above the solid plane, vertical scale of 500 μA/div. (Copyright 2005 Douglas C. Smith, http://www.dsmith.org, used with permission.)

The outcome of this demonstration is a direct consequence of the findings of the experiment described in Section 9.4, which demonstrated that a sizeable loop is formed, composed of the signal conductor and the return path, in the reference plane when the return current is forced to find its way around the edge of the break in the plane. As a result, RF voltage develops across the inductance of the loop in the plane caused by the

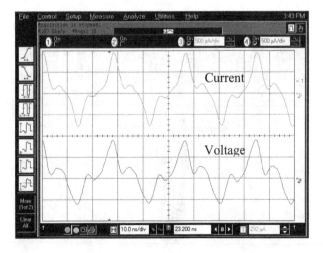

Figure 9.41. Measured common-mode current in the case of the conductor routed above a slot in the plane, vertical scale of 500 μA/div. (Copyright 2005 Douglas C. Smith, http://www.dsmith.org, used with permission.)

signal current. The voltage across the gap in the reference plane drives CM current onto the cables connected to the plane, stimulating them as efficient dipole radiators.

9.4.1.2 Susceptibility to Impulsive Radiated Electromagnetic Fields.
Figure 9.42 depicts a test board placed 20 cm from an impulsive radiated EMI source, emulating the field that may be produced by the indirect effects of a low-voltage electrostatic discharge (ESD), a severe yet common interference source in the workplace. The board shown in Figure 9.32 was used.

An oscilloscope screenshot is shown in Figure 9.43 for the conductor routed over the solid plane. The amplitude of the signal due to the radiated interference is about 100 mV, not enough to disrupt the performance of most digital circuits. Edge rates of the impulsive signals were very fast, approximately 300 ps or faster at the output of the signal source.

Comparing Figure 9.43 to the screenshot shown in Figure 9.44, taken using the conductor crossing the gap in the plane, reveals that the amplitude of the induced impulsive signal is increased from 100 mV to over 3 volts peak, an increase of about 30 dB (note that the scale changes from 100 mV/div to 1 V/div) in Figure 9.44. Such a level constitutes a significant loss of immunity to EMI as it by far exceeds the noise immunity level (NIL) of most logic families.

Yet another adverse effect produced by routing signal conductors over breaks in the reference plane has been demonstrated. The large loop formed by the signal conductor and the return signal current traveling around the edge of the break in the reference plane makes an efficient antenna, effectively picking up the induced radiated interference.

9.4.1.3 Crossing Reference Plane Breaks as a Source of Crosstalk.
An additional problem that often results from traces crossing reference planes is exces-

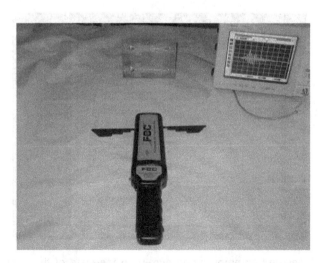

Figure 9.42. Setup for the impulsive radiated EMI demonstration. (Copyright 2003 Douglas C. Smith, http://www.dsmith.org, used with permission.)

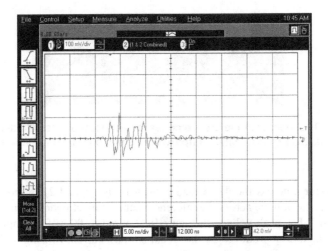

Figure 9.43. Signal from path over the solid plane, 5 ns/div (horizontal)/100 mV/div (vertical). (Copyright 2003 Douglas C. Smith, http://www.dsmith.org, used with permission.)

sive coupling (or crosstalk) between conductors crossing the gap [16]. Figure 9.45 depicts a part of the PCB shown in Figure 9.32 showing a single trace (of approximately 50 Ω), approximately 12 cm in length, crossing a 5 cm long slot in the reference plane. The board's solid plane extends downward from the bottom of the figure for about another 10 cm. A 300 mV signal with a rise time of about 300 ps was launched on the signal conductor.

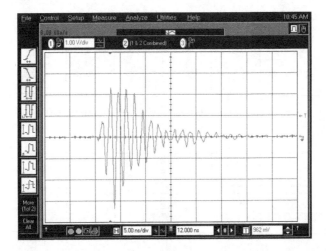

Figure 9.44. Signal from path over slot in the plane, 5 ns/div (horizontal)/1 V/div (vertical). (Copyright 2003 Douglas C. Smith, http://www.dsmith.org, used with permission.)

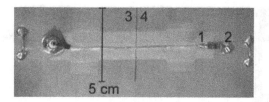

Figure 9.45. Test board with reference nodes labeled for the crosstalk demonstration. (Copyright 2002 Douglas C. Smith, http://www.dsmith.org, used with permission.)

The waveform across the 47 Ω load resistor (nodes 1 and 2) is shown in Figure 9.46, whereas the voltage across the reference plane break at nodes 3 and 4 is depicted in Figure 9.47. The peak voltage in Figure 9.47, almost 100 mV in amplitude, represents the voltage drop produced by the signal's return current flowing in the reference plane through the primarily inductive impedance of the path around the edge of the slot in the plane.

Significant undesired coupling (crosstalk) would result if a second parallel conductor were to cross the gap, even as far as 1 cm away from the signal path (about halfway between the existing signal path and the top edge of the board above nodes 3 and 4). Ordinarily, the level of crosstalk to be expected in such a situation would be insignificant (see Section 9.3.6).

In this case, both conductors must share a common return path (i.e., common impedance) around the edge of the break in the reference plane. The voltage measured across the break in Figure 9.47 couples onto the second (victim) conductor. Considering the typical low output impedance of a gate driving a high-impedance input of a gate in the victim path, the ~80 mV peak voltage of Figure 9.47 will show up as

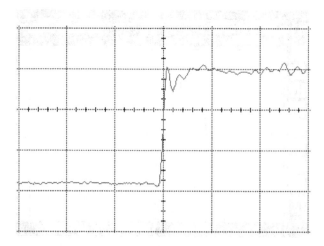

Figure 9.46. Signal voltage ($V_1 - V_2$), 5 ns/div (horizontal)/100 mV/div (vertical). (Copyright 2003 Douglas C. Smith, http://www.dsmith.org, used with permission.)

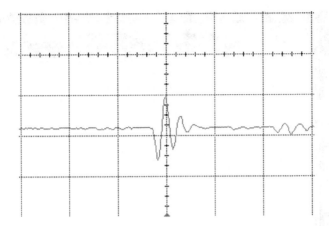

Figure 9.47. Voltage across the slot ($V_3 - V_4$), 5 ns/div (horizontal)/100 mV/div (vertical). (Copyright 2003 Douglas C. Smith, http://www.dsmith.org, used with permission.)

crosstalk between the signal conductor of Figure 9.45 and the fictitious victim path. The amplitude of the peak crosstalk in this case is nearly 30% of (approximately 10 dB below) the source (aggressor) signal, a value that would not remain unobserved from the standpoint EMC and signal integrity. As a general rule, any conductors crossing a gap in a reference plane (whether a "ground" or a power plane) are likely be strongly coupled, resulting in significant crosstalk.

9.4.1.4 *Crossing Reference Plane Breaks, Rise Time Effects on Signals.* Signals on conductors crossing gaps in a reference plane on printed circuit boards may experience significant transition (rise/fall) time distortion (constituting a signal integrity concern) even with relatively small gaps in the reference plane and at rise times on the order of 300 ps [16, 17].

A test signal with a 300 ps rise time, approximately, was applied directly through a BNC coupler from a signal generator to the two conductors on the test board shown in Figure 9.32, with one path routed over a continuous plane while the second crosses a 5 cm gap in the plane.

Figure 9.48 depicts the rising edges of both signals as measured at the terminations. The signal flowing in the conductor routed above the continuous plane exhibits a significantly faster edge rate (about 50% faster) and slight overshoot, whereas the path that passes over the gap in the plane is slower and the overshoot is filtered out.

Once again, the reason for the slower edge rate of the signal crossing the gap is that the signal's return path in the reference plane is forced to circumnavigate the edge of the gap. In addition, the impedance discontinuity in the return signal path when crossing over the gap results in reflections of the higher frequency components of the signal. Consequently, the edge rate of the signal is substantially slowed down by the gap in the plane to about two-thirds of that of the signal routed over a continuous plane. Such distortion could be intolerable from the standpoint of signal integrity and functionality.

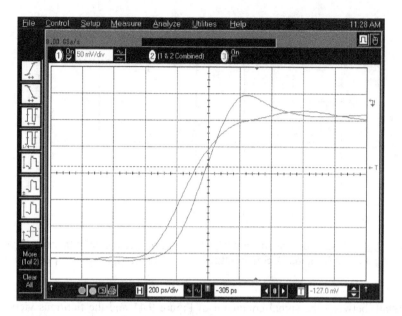

Figure 9.48. Comparison of the waveforms of the two signals measured at the terminations, horizontal scale of 200 ps/div. (Copyright 2005 Douglas C. Smith, http://www.dsmith.org, used with permission.)

9.4.1.5 Radiation from a Split Plane, Verification by Simulation. The effect of traces crossing gaps in return planes in PCBs was also investigated using CST Studio Suite™ 2008.* Simulations used for verification of the effects shown to occur through measurements in the previous sections were carried out on a stripline structure with the objective of investigating a particular case of interest: A stripline with only one of the return planes gapped while the second was kept solid. The purpose of this simulation was to show that even when one of the adjacent planes was solid, the second adjacent plane, which was gapped, nevertheless had a measurable effect on the electromagnetic performance of the PCB. Three cases were investigated for this purpose, namely:

1. Solid plane microstrip—a trace is routed over a solid return plane (reference case #1)

2. Solid planes stripline—a trace is routed between two solid return planes (reference case #2)

3. Gapped plane stripline—a trace is routed across a gap in one of the adjacent planes and adjacent to another solid plane (test case)

In both cases, a PCB having dimensions of 400 × 800 mil was used for the simulation. In reference case #1, a solid return plane was used, with a trace, 5 mils wide, separated

*CST Studio Suite is a product of CST, Computer Simulation Technology.

by a 10 mils thick dielectric substrate from the return plane. In reference case #2, two solid return planes were used. A 10 mil thick dielectric substrate was used both above and below the trace, separating it from the two return planes with a trace, 5 mils wide, routed in the center plane. In the test case, a gap 50 mils wide was cut in the bottom return plane while the upper layer remained solid. The gap was crossed by the trace described above. The relative permittivity of the dielectric substrate, ε_r, was 4.5. The models for the simulations are illustrated in Figure 9.49.

The outcomes of the simulation presented herein include:

- Surface currents (Figure 9.50)
- Magnetic (H-) and electric (E-) fields at the gap in the split plane (Figure 9.51)
- Radiation patterns ("far-field" directivity) (Figure 9.52)

The simulations clearly support the measurement results presented in previous sections. Figure 9.50 demonstrates that distortion of the return surface-current distribution occurs due to the presence of the gap in the return path. Figure 9.51 further shows that at the point of the gap crossing, where surface-current discontinuity occurs, significant enhancement of the H- and E-field distribution occurs. This is indicative of the fact that wave propagation does occur across the gap; however, it is accompanied by discontinuity of the electromagnetic wave structure and, consequently, the RF impedance introduced by the gap.

In effect, the signal-carrying conductor crossing the gap results in slot excitation, in a manner similar to that occurring in a "slot antenna." When a plane is driven as an antenna by a signal trace routed across the gap, the slot launches electromagnetic waves in a manner similar to a dipole radiator. Equivalence between the two is founded on a corollary to Babinet's Theorem (Babinet's Slot-Dipole Reciprocity Theorem), whereby $Z_{Slot} \times Z_{Dipole} = Z_0^2/4$. Z_{Slot} and Z_{Dipole} represent the slot and dipole input impedances, respectively, whereas Z_0 stands for the intrinsic wave impedance of free space (in vacuum). At low frequencies, the slot drives the cables in the "dipole radiator" mode; at higher frequencies, when the PCB dimensions approach a half-wavelength, λ, resonance effects may occur, producing direct excitation of the PCB. Resonance effects on PCBs are discussed in brief in Section 9.9.5

Let it be noted however, that whereas a "slot antenna" consists of a metal surface with a limited-sized slot cut out through it, the common situation on PCBs is somewhat different as the gap often occurs across the *entire* PCB. Slots and gaps in PCBs (where good design practices* are implemented) exist primarily because multiple power (and occasionally return) planelets (i.e., islands) are used for different supplies (e.g., 5 V, 3.3 V, 1.5 V, etc.) and returns (analog/digital) on the same layer of the PCB are used.

Figure 9.52 presents the "far-field" directivity of the antenna formed by the PCB. In Figure 9.52(a), the directivity radiation pattern at a frequency of 1 GHz from a solid plane (serving as a reference case) is shown. Radiation efficiency from this structure is

---

*Slots created by routing signal traces in return planes are often limited in length and do not totally cut the plane. This, however, constitutes a poor design practice and should be avoided.

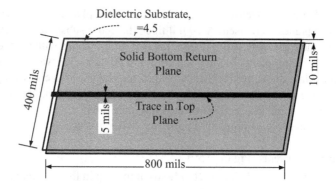

(a) Solid Plane Microstrip Case (Reference Case #1)

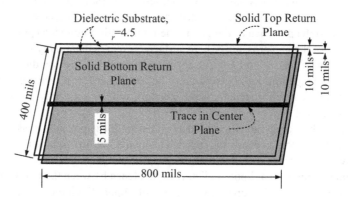

(b) Solid Plane Stripline Case (Reference Case #2)

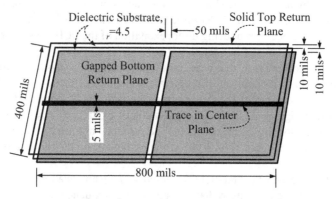

(c) Gapped Plane Stripline Case (Test Case)

Figure 9.49. Models used in the simulation.

low (10^{-11}). When a slot is present, as shown in Figure 9.52(b), it seems at first sight first site that only a minor change in the directivity pattern occurred. Radiation efficiency increased, however, to 3×10^{-2} in this case.

Note: The radiation efficiency referred to in Figure 9.52 is defined as the ratio of gain to directivity, or, equally, the ratio between the radiated to accepted (input) power of the antenna, $G(\theta, \varphi) = e_{rad} \cdot D(\theta, \varphi)$ or $P_{rad} = e_{rad} \cdot P_{in}$.

These simulation outcomes fully support measurements that demonstrated that signal traces crossing gaps in adjacent return planes considerably exacerbate emissions from the PCB, indicating the severity of such designs.

9.4.2 Reference Plane Discontinuities and Mitigation Strategies

Discontinuities in reference planes may be intentional ("by design") or due to unintentional flawed design. The most common return path discontinuities on PCBs are discussed in this section.

9.4.2.1 Traces Crossing Slots and Splits in Reference Planes. In actual PCBs, it is not rare that traces cross a gap in the reference (power or return) plane. Slots and splits in reference planes are among the most commonly observed discontinuities in high-speed PCBs, yet constitute a severely flawed design, often implemented due to misapprehension of fundamental high-speed design principles or, worse yet, intentionally. In many cases, splits and slots are used simply "because we always used to do it." Figure 9.53 illustrates a trace crossing a slot in a reference plane.

A slot in the current return plane results in increased inductance to the path comprising the trace passing perpendicularly over the slot and the return plane, creating signal integrity and EMI problems, due to the increase in the area of the signal and its return current loop. Equation (9.33) provides an expression for this increase in inductance associated with this configuration (note definition of variables in Figure 9.53) [18]:

$$
L_{Slot} \approx \begin{cases} L_{Plane} + \dfrac{\mu I}{2\pi} \ln\left(\dfrac{g}{t}\right); \ g > h \\[2em] L_{Plane} + \dfrac{\mu I}{2\pi} \ln\left(\dfrac{h}{t}\right); \ g \le h \end{cases} \text{, henry} \tag{9.33}
$$

Displacement across the gap (also known as the "capacitance across the gap") allows some current to flow across the gap. This mechanism, however, will be effective only when a relatively large crossover area exists, again requiring current spread and, subsequently, increased inductance across the gap (Figure 9.54).

Some similarity exists between slots and splits in reference planes, but, then again, they differ appreciably. A "slot" refers to a partial gap in an otherwise solid and continuous reference plane (Figure 9.55), whereas split reference planes result in unrelated planelets (Figure 9.56). Both equally constitute EMI and signal integrity horrors.

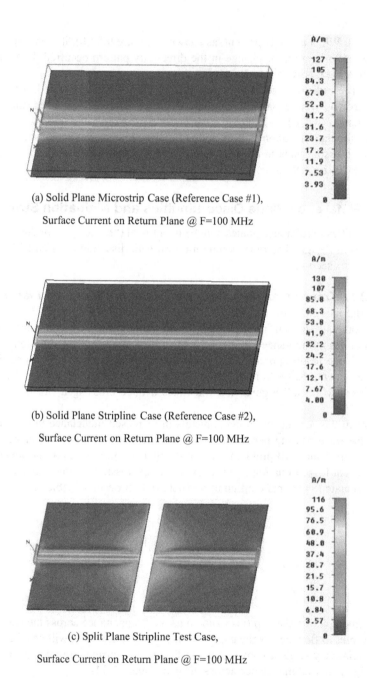

(a) Solid Plane Microstrip Case (Reference Case #1),

Surface Current on Return Plane @ F=100 MHz

(b) Solid Plane Stripline Case (Reference Case #2),

Surface Current on Return Plane @ F=100 MHz

(c) Split Plane Stripline Test Case,

Surface Current on Return Plane @ F=100 MHz

Figure 9.50. Comparison of surface current on solid and split return plane versus frequency. (Simulations run on CST "Studio Suite" 2008 3D EM Simulator; courtesy of Edoardo Genovese, CST.)

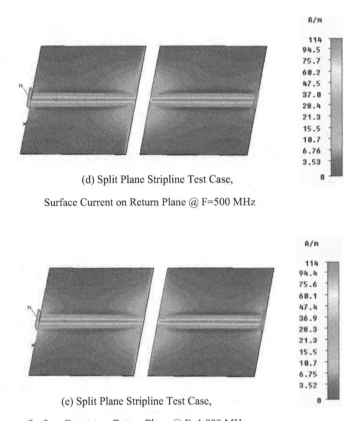

(d) Split Plane Stripline Test Case,

Surface Current on Return Plane @ F=500 MHz

(e) Split Plane Stripline Test Case,

Surface Current on Return Plane @ F=1,000 MHz

Figure 9.50. *Continued*.

A similar yet somewhat different case is depicted in Figure 9.57, where a trace crosses a slot totally enclosed in a return plane. The effective increase in inductance associated with the slot may be considered as being in series with the trace (and current loop), and can be estimated by using the expression of a loop formed by two flat parallel conductors with a center-to-center spacing d, having width W and length ℓ. Assuming that the return current around the gap flows onto two parallel conductors of width $W = 3W_T$, length $\ell = D$, thickness of trace t_T, and spacing $d = W_C + 3W$, the inductance associated with this configuration n can be expressed as [19]

$$L_{Slot} = \frac{1}{2} \cdot \left\{ \frac{\mu_o}{2\pi} \cdot 2D \cdot \ln\left[\left(\frac{3W_T + W_C}{3W_T + t_T} \right) + \frac{3}{2} \right] \right\} = \frac{\mu_o}{2\pi} \cdot D \cdot \ln\left[\left(\frac{3W_T + W_C}{3W_T + t_T} \right) + \frac{3}{2} \right], \text{ henry}$$

(9.34)

The factor of ½ is added in Eq. (9.34) in order to account for the fact that the slot results in two loop inductances in parallel and the factor 2 is included since the objective

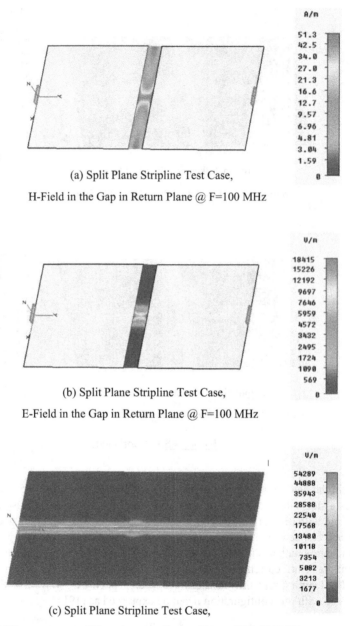

(a) Split Plane Stripline Test Case,

H-Field in the Gap in Return Plane @ F=100 MHz

(b) Split Plane Stripline Test Case,

E-Field in the Gap in Return Plane @ F=100 MHz

(c) Split Plane Stripline Test Case,

E-Field between around Trace between Return Planes @ F=100 MHz

Figure 9.51. Magnetic (H-) and electric (E-) field strength distribution in the structure. (Simulations run on CST "Studio Suite" 2008 3D EM Simulator; courtesy of Edoardo Genovese, CST.)

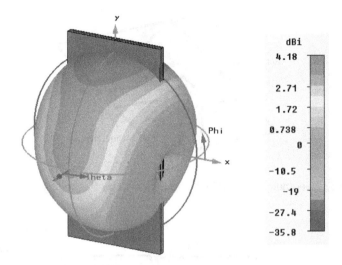

(a) Solid Plane Stripline Case, Directivity @ F=1 GHz

Note: Total Radiation efficiency: 10^{-11}

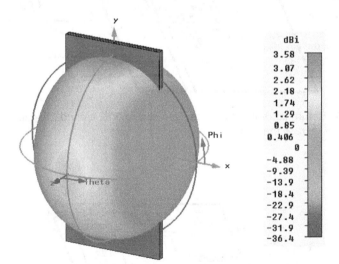

(B) Split Plane Stripline Case, Directivity @ F=1 GHz

Note: Total Radiation efficiency: 2×10^{-2}

<u>Figure 9.52.</u> Comparison of radiation patterns ("far-field" directivity) from solid and split return plane versus frequency. The scaling in both cases is different, as can be observed from the values of the radiation efficiency, implying significantly higher emission (exceeding 8 orders of magnitude) from the split plane configuration. (Simulations run on CST "Studio Suite" 2008 3D EM Simulator; courtesy of Edoardo Genovese, CST.)

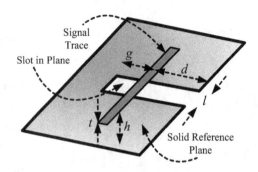

Figure 9.53. Geometry of a slot in the reference plane crossed by a signal trace.

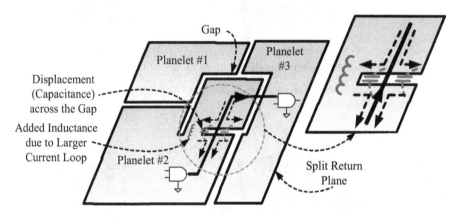

Figure 9.54. Return current in the plane is forced to flow around the gap and, through displacement, across the gap.

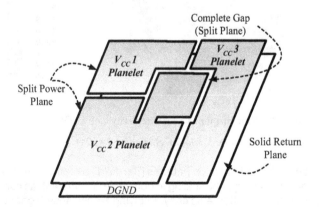

Figure 9.55. A split in a power plane creating coplanar power planelets in a multisupply voltage circuit.

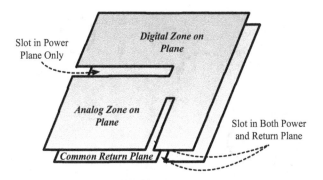

Figure 9.56. Slots in power and ground planes in a mixed analog/digital circuit sharing a common ground plane.

of the calculation is the inductance of one of the loops that is twice the effective inductance associated with one branch of the loop. In this calculation, edge effects were neglected for simplicity.

Equation (9.34) was validated by numerical modeling carried out at a frequency of 1 GHz, using CST Microwave Studio. The test PCB depicted in Figure 9.58 had the dimensions of $20 \times 100 \times 0.77$ mm, dielectric substrate thickness of 0.7 mm, relative permittivity $\varepsilon_r = 4.4$, $W_T = 0.35$ mm, $t_T = 0.035$ mm, length $\ell = 100$ mm, gap created in the middle of the PCB of size $D = 7.5$ mm, and $W_C = 5$ mm [19]. The source was an ideal voltage source with trapezoidal waveform of amplitude 1 V, $t_y = t_f = 1$ ns, $t_{hold} = 5$ ns, and $t_{tot} = 10$ ns. The trace was loaded with its characteristic impedance, $Z_0 = 90.36$ Ω. The resulting contribution of the slot to the loop inductance, L_{Slot}, is 4.82 nH. Such high inductance may result in considerable distortion to the signal waveform and degradation of the rise time of the signal in particular. In addition, common-mode emissions are increased, particularly when a cable is attached to the PCB.

Observe in Figure 9.58(b) the distribution of the surface current around the slot. Note that the assumption that the largest part of the return current concentrates in a

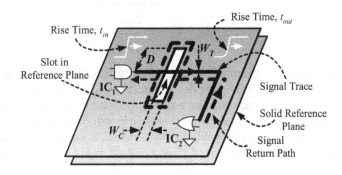

Figure 9.57. A trace routed over a slot in an adjacent reference plane.

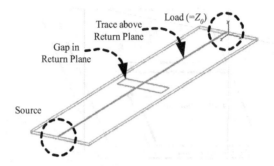

(a) Model of the PCB Structure

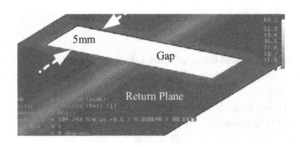

(b) Return Current Distribution in Proximity

to the Gap at F=1GHz

<u>Figure 9.58.</u> Simulation model of a trace crossing a gap in a return plane on a PCB. (Images from CST "Microwave Studio" 3D EM Simulator; courtesy of Spartaco Caniggia.)

space equal to $3W_T$ is valid. Equation (9.35) expresses the increase in rise time of the propagating signal waveform due to the increased inductance caused by the gap [19]:

$$t_{out} = \sqrt{t_S^2 + t_{in}^2} \approx \sqrt{\left(2.2 \cdot \frac{L_{Slot}}{2Z_0}\right)^2 + t_{in}^2} \qquad (9.35)$$

where t_{in} and t_{out} represent the rise time of the input (source) and output (load) current waveforms, respectively. The rise time degradation term, t_S, is explicitly shown.

Figure 9.59 illustrates a trace crossing a slot in an adjacent reference plane. A common application in which such a situation is purposely introduced is illustrated in Figure 9.60. A slot (often even a total split, discussed below) is introduced by the designer in the reference planes of a mixed analog/digital* board with the intention of

*Grounding considerations in mixed analog/digital circuits are discussed in detail in Section 9.8.

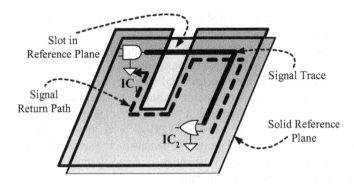

Figure 9.59. A trace routed over a slot in an adjacent reference plane.

reducing crosstalk from the digital to the analog zones of the PCB. To reduce crosstalk between analog and digital circuitry, the EMC literature often recommends gapping the ground plane between analog and digital areas on PCBs. This technique reduces common impedance but not inductive or capacitive coupling interactions [6]. In the case presented here, the opposite is achieved. As the designer failed to notice that several digital signal traces cross the slot, the digital return current from the driver cannot flow immediately beneath the trace and is diverted around the slot. Only a little portion of the return signal current may flow through the capacitance of the gap. The diverted return current forms a large loop with respect to the signal trace and strongly increases

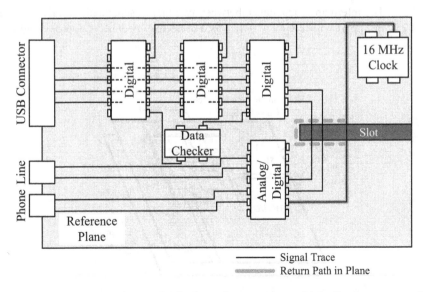

Figure 9.60. Traces crossing a slot in the reference plane. (Note the traces crossing the gap in the circled areas.) (Courtesy of Dr. Todd Hubing, Clemson University.)

the current loop inductance, radiated emissions, and crosstalk that the designer was attempting to eliminate in the first place.

The condition in Figure 9.59 resembles an unintentional "slot antenna." Some undesired adverse outcomes of this situation are presented in detail in Section 9.4.1.

In many well-designed high-speed digital circuits, the power planes are split into several "planelets" (i.e., unrelated partial planes), allowing coplanar distribution of multiple supply voltages from the main power source or monolithic voltage regulators/converters to reside on the same PCB "V_{CC}" layer, making efficient use of the power plane area (Figure 9.61). Figure 9.62 depicts this situation in an actual PCB layout [20]. If the digital return (*DGND*) plane is solid and continuous, it may still provide an unobstructed signal current return path.

Note: This situation is commonly found in CPLDs (e.g., FPGAs, ASICs, etc.) when multiple power supplies for the I/O and the "core" of the device are required. Multiple supplies are often likewise required on "mixed supply logic" PCBs, in which voltages such as 5 V, 3.3 V, 2.5 V. and 1.8 V are utilized, or in high-resolution (>10 bits) mixed analog/digital circuits (briefly discussed below and in more detail in Section 9.7).

Assume that a trace is routed between digital devices IC_1 and IC_2 across a split in the adjacent power plane (Figure 9.63). The return current must ultimately find an alternative path to get across the split. In this case, illustrated for a four-layer board [Figure 9.64(b)], the current is diverted through the closest decoupling capacitor across to the solid DGND plane (common to all power planelets). Beyond the split, the current returns to the power plane through another decoupling capacitor. The equivalent inductance of the capacitors now dominates the loop impedance, in addition to the effect of the larger loop formed by the path of the current between the planes.

Note: The interplane capacitance between two planes separated 10 mils (0.01 inch or 0.25 mm) apart with FR-4 ($\varepsilon_r = 4.7$) as a dielectric medium is approximately 100

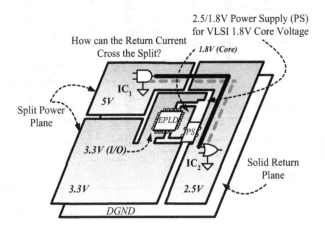

Figure 9.61. Trace crossing a split between coplanar power planelets in a multi-supply voltage circuit.

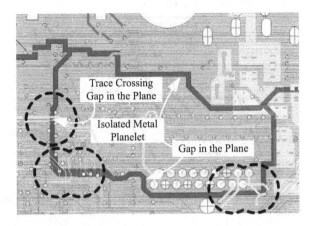

<u>Figure 9.62.</u> Traces crossing a gap in a reference plane. (Note traces crossing the gap in circled areas.) (Courtesy of Dr. Todd Hubing, Clemson University.)

pF/inch^2, much too small to be effective for bypassing the return current across the gap except for frequencies considerably higher than 500 MHz.

A slightly different situation may occur in mixed analog/digital circuits in which separate analog and digital power supplies and return paths are commonly utilized for eliminating undesired crosstalk through the common return path impedance to sensitive analog circuits* (Figure 9.63).

Consider again a trace routed between digital devices IC_1 and IC_2. Ideally, the trace should be routed exclusively over the digital power plane, but if the trace were to be routed over the analog plane (due to some design flaw), where would the return signal current flow? There are two answers to this question, both equally unfavorable.

If the return signal current remains on the digital plane and encircles the analog plane, a large loop is created, resulting in increased EMI. Conversely, if the return signal current somehow finds a path onto the analog plane following the signal trace, the digital signal will probably couple onto analog traces in the area through crosstalk or ground-coupled interference mechanisms.

Some similarity does exist between the case of slots in planes and that of planes split into unrelated planelets, but they differ appreciably. Unlike the case of a slot in a plane, no path is available for the return current (not even surrounding the split) when a trace crosses a split between unrelated coplanar planelets as the gap cuts through the entire board. Furthermore, the situations presented in Figure 9.61 and Figure 9.63 are not equivalent. In Figure 9.61, the power plane is split but an adjacent continuous ground plane is present. By rearranging the layers, this situation may be overcome. In Figure 9.63, on the other hand, both the power and ground planes are split in a similar manner (normally with no overlap between planelets), efficiently obstructing all possible return current paths.*

*Grounding considerations in mixed analog/digital circuits are discussed in detail in Section 9.7.
*In a severe case, the only available current return path may be through the power source feeding both the digital and analog circuits.

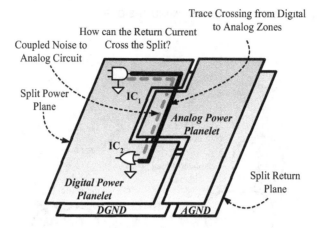

Figure 9.63. Trace crossing a split between coplanar power and ground planelets in a mixed analog/digital circuit.

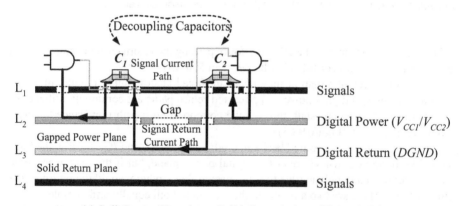

(a) Split Power Plane but a Solid Ground plane (Figure 9.61)

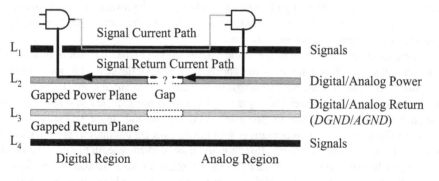

(b) Split Power and Ground planes (Figure 9.63)

Figure 9.64. A trace crossing a split reference plane in a four-layer board.

Common to both situations is the fact that when a trace crosses a slot in the adjacent reference (power or "ground") plane, a sizeable loop is formed, composed of the signal conductor and the return path in the reference plane when the return current is diverted from underneath the trace and forced to find its way around the edge of the slot. The longer the slot, the bigger the loop area becomes. Since emission of and immunity to EMI as well as numerous signal integrity concerns are both related to loop area, a potential EMI and signal integrity situation is now created, where none was present beforehand.

The best approach to solution of problems arising from using slots and split reference planes is to avoid routing traces across slots or splits at the first place, at least with critical high-speed/high-frequency signal traces. While recognizing the significance of this recommendation, it must also be recognized that some design constraint may require this to occur. For instance, in the four-layer stackup depicted in Figure 9.64(a), the ground plane (DGND) is continuous, whereas the power plane is split. Unfortunately, not all traces can be routed in the bottom layer L_4 adjacent to the *DGND* plane, so some traces will ultimately have to be routed in layer L_1 above the (split) power plane.

When crossing slots in the reference plane is unavoidable, the second best strategy is to count on capacitance placed between the reference planes to provide an alternate detour path for the return current. When a solid ground plane is used, adjacent to the split or slotted power plane [Figure 9.64(a)], a large number of decoupling capacitors* is normally present between the boards, accomplishing this task.

The relative position of these capacitors with respect to the position of the gap is of utmost importance. Counting on decoupling capacitors that may be present on the board may appear to be unacceptable as the position of these capacitors may not be optimal. The criteria for placement of decoupling capacitors are not associated with the gap in the plane but, rather, on with the proximity to the electronic device utilizing that capacitor. As a result, they may not be placed adjacent to the slot, in which case the return current is still forced to travel a long distance, producing a large signal-return current loop (see Figure 9.65). Capacitors C_1 and C_2 would provide for the smallest loop area if C_2 were present, but it is not, and the closest decoupling capacitor available on the right hand side of the gap is C_3, resulting in a significantly larger loop. Note that the capacitors used for the "crossover" between the power and ground planes need not necessarily be those associated with the driving and receiving gates, but rather those that minimize the loop impedance.

Note that this solution scheme, based on the decoupling capacitors, can be effective only as long as at least one of the reference planes is continuous under both sides of the gap [Figure 9.64(a)]. If both reference planes are split in the same way [Figure 9.64(b)], the scheme using decoupling capacitors is of little avail, as each decouples to its own reference planelet while the return current must still overcome the gap between both sides of the split in both planes (see Figure 9.66).

An alternate solution comprises of bypass or stitching capacitors, placed across the gap in the reference plane(s) (Figure 9.67). This scheme will work effectively even

---

*Decoupling is briefly discussed in Section 9.5.4.

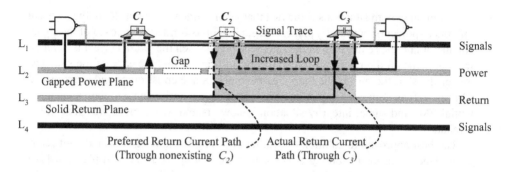

Figure 9.65. You cannot always count on decoupling capacitors for bypassing return current across a split plane because they may be too far away.

with both planes split. The capacitors should exhibit low impedance at the signal frequencies for providing an acceptable return current path. For higher frequency components, the capacitor's impedance is not driven by the capacitance; rather, it is dominated by the inductance, particularly the capacitor's intrinsic equivalent series inductance, ESL, as well as by the inductance resulting from its installation (vias and pads).

In order to ensure the highest possible effective frequency of the capacitors the value, package, and physical size (which primarily determines the ESL) of the capacitors should first be carefully considered. Figure 9.68 illustrates the impedance of two capacitors of equal package size (equal ESL) but differing in their capacitance. Clearly, the larger the capacitance, the lower the capacitor's resonance frequency. Values of 1 nF to 100 nF are normally recommended, depending on the frequency content of the signal, whereas the package sizes of the capacitors should be maintained as small as possible (0805, 0603, or smaller yet, when possible).

Placement and installation of the capacitors play a major role in their overall performance and should next be carefully considered. SMT (surface mounted technology) stitching capacitors should be used and mounted across the gap, preferably one on either side of the signal trace. The capacitors should be placed at near proximity to the

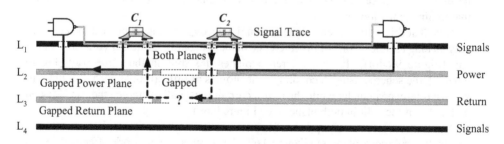

Figure 9.66. You cannot count on decoupling capacitors when both reference planes are split. How can the return current flow, even with C_1 and C_2 in place?

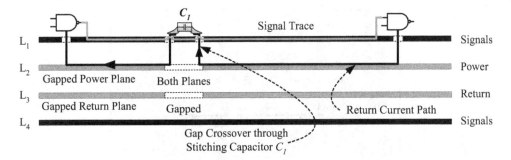

Figure 9.67. Stitching capacitors provide a direct bypass for the return current across a split plane.

trace crossing the gap (preferably within 0.1 inch, or 2.5 mm). For best performance, capacitors should also be placed at regular intervals along the perimeters of all of the plane splits, so that they can be used to interconnect the planelets on either side of the gap. The spacing between the capacitors along a gap should not exceed $\lambda/10$ at the highest frequency of concern, where λ represents the wavelength associated with the highest signal frequency of concern in the dielectric medium of the PCB. Figure 9.69

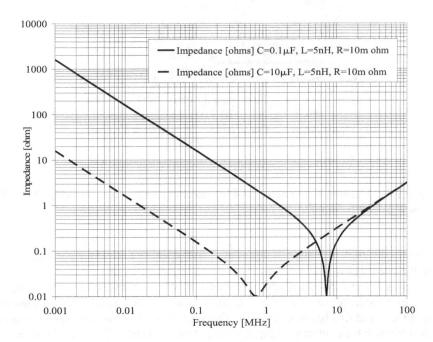

Figure 9.68. Typical impedance curves of capacitors of two different capacitance values.

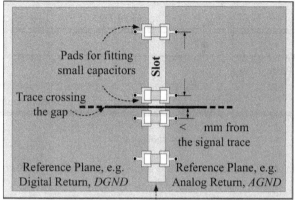

Single Point (Star) Connection between
AGND and *DGND* Reference Planelets

(a) Stitching Capacitors across a Slot in
a Reference Plane

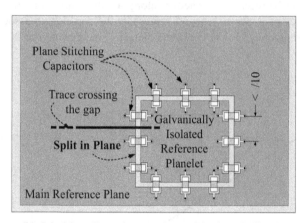

(b) Stitching Capacitors across a Slot in a Split Plane

Figure 9.69. Placement examples of SMT stitching capacitors. (Courtesy of Keith
Armstrong, Cherry Clough.)

illustrates the implementation of this scheme with a slot in a reference plane (a) and in
a split plane (b) [10].

Bearing in mind that the return current must now flow from one planelet, through a
series of "obstacles" namely a via, a trace, a mounting pad, a capacitor, a mounting
pad, a trace, and finally through a via to the other planelet of the split plane, the solution is far from being ideal and considerable inductance is present in series with the return current path (typically no less than 5 nH to 10 nH minimum). Proper design of the
mounting pads is of utmost importance. Figure 9.70 and Figure 9.71 present, in order

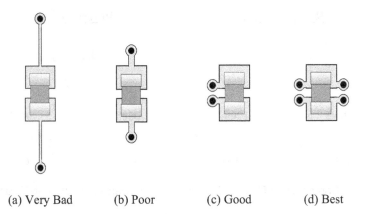

(a) Very Bad   (b) Poor   (c) Good   (d) Best

Figure 9.70. Installation of regular aspect SMT stitching capacitors that minimize inductance (e.g., 0805, 0603). (Courtesy of Keith Armstrong, Cherry Clough.)

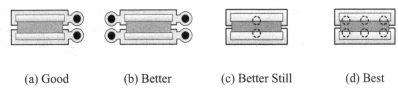

(a) Good   (b) Better   (c) Better Still   (d) Best

Figure 9.71. Installation of aspect reverse SMT stitching capacitors that minimize inductance (e.g., 0508, 0306). (Courtesy of Keith Armstrong, Cherry Clough.)

of improved performance, the manner of installation of the capacitors, shape of the mounting pad, and location of the vias* [10].

A comparison of inductance of equally sized capacitors when applied in regular and reverse aspects revealed, expectedly, that reverse aspect capacitors exhibit significant ESL reduction. Figure 9.72 depicts a comparison of measured ESL associated with several SMT capacitor packages [21].

Although the stitching capacitors do provide an optimal path for the return current (if placed in near proximity to the signal trace at the crossover point), their performance is appreciably limited by their high-frequency impedance[†] [22]. Figure 9.73 depicts a sample plot of the maximum E-field radiated emission from a simple PCB with an exposed microstrip trace. The level of emissions in the frequency band 20 MHz to 1 GHz are illustrated for the trace routed over a solid metal plane and crossing over a split plane. Noticeably, emissions are ten-fold (20 dB) higher for the latter in comparison to the former.

---

*Other capacitor installation schemes, which are less appropriate for plane stitching due width of the gap and the corresponding installation constraints, are useful for decoupling purposes, as discussed in Section 9.5.4.3.
[†]The nonideal behavior of capacitors is extensively discussed in Chapter 2.

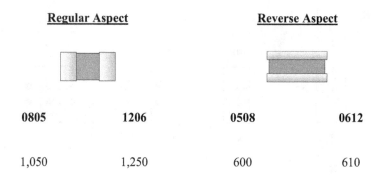

Regular Aspect		**Reverse Aspect**	
0805	1206	0508	0612
1,050	1,250	600	610

Figure 9.72. Comparison of ESL [pH] of regular and reverse aspect of SMT capacitors.

When stitching capacitors are added (Figure 9.74), a significant improvement is observed at frequencies up to 100 MHz but degrade with further increase in frequency. With one stitching capacitor used, a rise of emissions above 100 MHz is visibly observed. When two capacitors are used, emissions drop to the level observed over a solid reference plane.

Up to this point, real-world intrinsic characteristics (ESR and ESL) of capacitors were considered whereas the reactance associated with the *mounting* of the capacitors

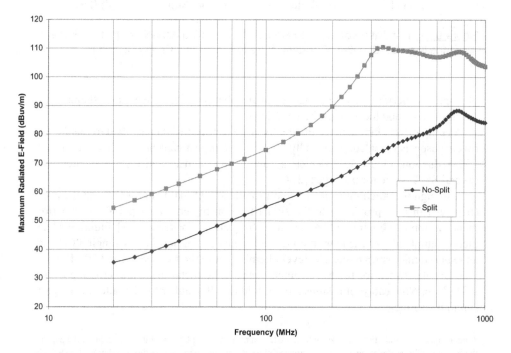

Figure 9.73. Maximum radiated E-field from a microstrip with and without a split in the reference plane. (Courtesy of Bruce R. Archambeault, IBM Corporation.)

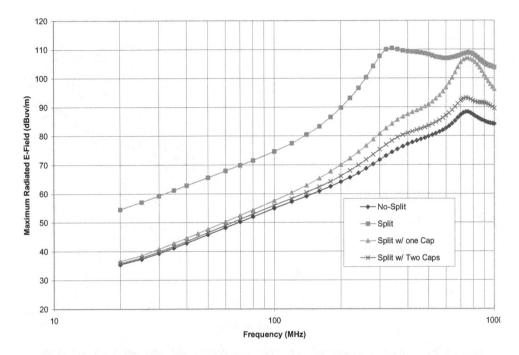

Figure 9.74. Maximum radiated E-field from a microstrip with and without a split in the reference plane and with ideally mounted stitching capacitors. (Courtesy of Bruce R. Archambeault, IBM Corporation.)

(e.g., the vias, mounting pads, and connecting traces) were ignored. If those were to be considered, an additional inductance of approximately 1.5 nH must be added. The adverse effects of this situation are depicted in Figure 9.75. Although far from ideal and definitely inferior to the preceding case, without a doubt stitching capacitors still contribute a significant improvement compared to when not present.

9.4.2.2 Traces Jumping Layers Across Reference Plane(s)

Every valley shall be lifted up, and every mountain and hill shall be made low; and the rugged shall be made level, and the rough places a plain. ISAIAH 40, 4

In the design of multilayer PCBs, it is almost impossible to maintain routing of traces in one layer only. Jumping layers, although undesired from the standpoint of EMC and signal integrity, is often inevitable owing to the nature of routing traces in a crisscross manner in adjacent layers. When a signal trace traverses from one layer to another, the path of the return current following the signal trace is interrupted.

Figure 9.76 illustrates a signal trace going across a single reference plane to another signal layer. In this simple situation, conductive crossover occurs, yet the manner in which the return current flows across both sides of the reference plane is not straightforward.

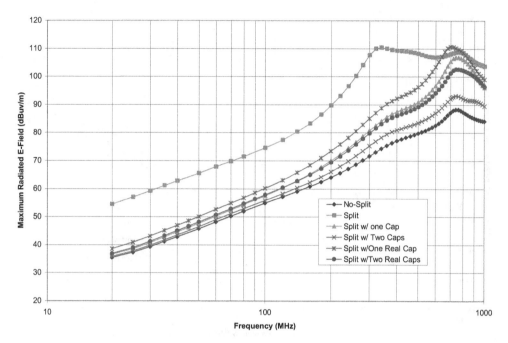

<u>Figure 9.75.</u> Maximum radiated E-field from a microstrip with and without a split in the reference plane and with ideally and real-world mounted stitching capacitors. (Courtesy of Bruce R. Archambeault, IBM Corporation.)

Skin and proximity effects complicate the situation. The high-frequency return current is confined to the surface of the reference plane in proximity to the adjacent to the signal trace and cannot flow (or diffuse) through the plane. In order to cross over to the opposite surface of the reference plane a plated-through via hole must be provided in the plane. The plated-through via hole provides a surface connecting the top and bot-

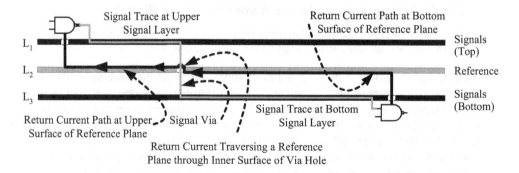

<u>Figure 9.76.</u> Return current path following a signal crossover to the opposite side of the same reference plane.

tom surfaces of the plane and provides a path for the return current to flow from the bottom to the top of the plane (Figure 9.77). When a signal traverses through a via and continues on the opposite side of the same plane, a return current discontinuity does not exist, whereas the impedance of the current return path is strongly influenced by the via geometry.

In many layer stack-up designs, signal layers are grouped in pairs, followed by a pair of reference planes (i.e., power and ground planes). Figure 9.78 depicts a four-layer stackup consisting of two signal layers on the outer layers (L_1 and L_4), whereas the innermost layers L_2 and L_3 comprise two reference planes.* In the simple (yet rare) case of both reference planes being at the same DC potential (e.g., either both are V_{CC} or both GND), the two planes can be short-circuited together using plated-through via(s) immediately adjacent to the signal via, making this situation quite similar to the former, except that the signal now crosses through two reference planes, as seen in Figure 9.78(a). This plated-through via now allows the current to cross over from the bottom surface of the lower reference plane, L_3, to the top surface of the upper reference planes. There may be some nudge (or "bump") in the signal due to the effect of the transition through a longer via.

A considerably mode difficult situation occurs when the signal traverses through entirely different types of reference planes (e.g., V_{CC} and GND). Under these circumstances, interconnecting the layers L_2 (V_{CC}) and L_3 (GND) through vias is obviously impossible. Furthermore, as long as a signal trace is routed above either of the planes, the return current is forced (through skin and proximity effects) to flow on the corresponding plane adjacent to the trace. The current must, therefore, cross the return plane around the signal via. As the two planes are dissimilar in their nature, traversing the layers can only be made possible by passing the two planes through the displacement mechanism (i.e., capacitance) between the planes. However, displacement is confined to the effective area between the two planes occupied by the current distribution, requiring that the current spread around the signal via hole, consequently increasing loop inductance between the signal and return current paths.† An alternate current return path can be provided using stitching capacitors (commonly known as "bypass capacitors") between the two different planes, in a manner similar to that applied for crossing over slots or split planes (Section 9.4.2.1). Figure 9.78(b) illustrates this situation. Once again, the return current must now flow from the receiving gate, through a via to the ground plane, continuing then through a series of "obstacles," namely, a via, a trace, a mounting pad, a bypass capacitor, a mounting pad, a trace, and, finally, through a via to the power plane and from there through yet another via back to the driving gate. Considerable inductance, typically on the order of 5 nH or more, is consequently introduced into the return current path. Then again, without the capacitor the return current would have been forced to take an even longer path. For minimizing the impedance of the return current path near large ASICS, for instance, multiple properly installed parallel bypass capacitors may be required. Refer to Section 9.4.2.1 for capacitor mounting considerations.

---

*Adjacent power and return planes minimize the impedance between the layers, providing tight coupling between the two planes while also decreasing the power distribution system switching noise.

†See the discussion lower in this section regarding the nature of radial current spreading in the vicinity of the signal via.

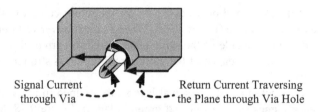

Signal Current   Return Current Traversing
through Via   the Plane through Via Hole

Figure 9.77. Focus on the signal and return currents traversing a via.

With progress of modern CMOS technology, especially the increase in signal bandwidth, now exceeding several gigabits/second (Gbps), time-varying switching currents traversing power and return planes of a printed circuit board (PCB) through vias result in electromagnetic waves being launched across the power distribution network's (PDN) parallel plates' structure [composed of power (V_{CC}) and return (GND) planes], a phenomenon known as "parallel-plate waveguide (PPW) noise" excitation. PPW

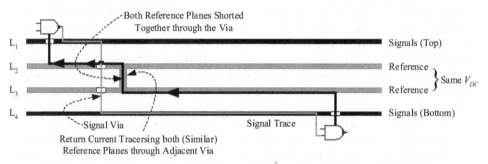

(a) Return Current Traverses through Via between Reference Planes of the Same DC Potential

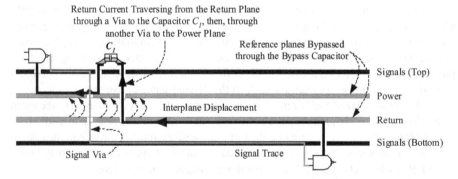

(b) Return Current Traverses through a Bypass Capacitor between Reference Planes of Different DC Potential (e.g., V_{cc} and GND)

Figure 9.78. Return current traversing between reference planes.

noise results from the mode conversion of desired modes into parasitic PPW modes. The PPW noise, sustained in standing wave patterns due to multiple reflections from the edges of PCBs, occurs at power-return parallel-plate-cavity resonance frequencies, governed primarily by structural dimensions.*

An excellent illustration of the effect of a signal trace traversing between two reference planes and its effect on the circuit performance is now presented [23]. Intuitively, for minimizing inductance (and, hence, the loop area) of the switching current path, the most appropriate location for the bypass capacitor would be in proximity to a via carrying the signal traversing the reference planes. It follows that if only one bypass capacitor can be added to suppress the voltage fluctuations between the power and return planes, better performance is attained when the capacitor is installed near the signal via compared to that obtained when placed anywhere else. The complete simulation results shown below validate the intuition.

The test structure containing four metal layers, depicted in Figure 9.79, was used to demonstrate the importance of placement of the bypass capacitor. The top and bottom layers are signal layers, whereas the power and return planes form the internal layers. The size of the planes is 10 cm × 10 cm. The thickness of each dielectric layer is 30 mil† and its relative permittivity, ε_r, is 4.0. Trace a traverses by means of Via 2 from the top signal layer (Signal #1) through the power and return planes to the bottom layer (Signal #2), where it is renamed Trace b. A 50 mA driver, represented as an ideal current source and parallel resistor, is connected between Trace a and Via 1, which interconnects the trace in the top signal layer to the power plane. A termination resister, R_{Term}, is connected to Trace b and to Via 3, connecting the trace in the bottom signal layer and the return plane. The current source output waveform is a pulse of a 300 ps low-to-high transition, 300 ps duration, and 300 ps high-to-low transition.

In this simulation, three cases were examined.

- **Case 1:** No bypass capacitor
- **Case 2:** Bypass capacitor placed near Via 2
- **Case 3:** Bypass capacitor placed near the switching current source

A 100 nF capacitor was used in cases 2 and 3 with a series equivalent inductance (ESL) of 50 pH and a series equivalent resistance (ESR) of 10 mΩ. The spacing between the two leads of the capacitor was 2 mm. The power and return planes were discretized into 2500 cells using a 50 × 50 mesh in order to accommodate the two adjacent vias of the capacitor. The 2500 node voltages on the 50 × 50 mesh, which represent the spatial noise voltage distribution between the power and the return planes, were recorded and are shown in Figure 9.80 and Figure 9.82.

In Case 1 (bypass capacitor not installed), illustrated in Figure 9.80(a), it can clearly be observed that the signal reaches via 2 and the switching current launches an EM wave radially propagating away from the via. When the wave reaches the edges of the board, it reflects, stimulating resonance effects inside the power/return plane structure.

---

*Chassis resonances and their effects and mitigation techniques are discussed in Section 9.9.
†1 inch (25.4 mm) = 1000 mils.

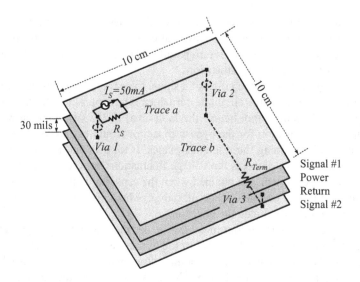

Figure 9.79. Four layer PCB with a signal trace traversing power and return planes through a via. (Simulation performed on Sigrity "Speed97" Suite; courtesy of Sigrity, Inc.)

In Case 2 (bypass capacitor installed near the switching current source), shown in Figure 9.80(b), the spatial voltage distribution suggests that the capacitor is not effective until the peak noise is generated and propagated to the capacitor location. That is why placing the bypass capacitor near the traversing via is more effective (see Case 3).

In Case 3 (bypass capacitor installed near via 2), depicted in Figure 9.80(c), the location of the bypass capacitor can easily be identified. When the plane voltages bounce to positive amplitude, the capacitor's surroundings look like a dent. The capacitor attempts to maintain a "quiet" area in the vicinity of via 2 when fluctuations occurs, effectively suppressing the noise.

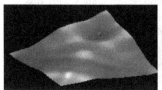  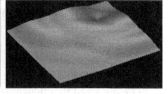

(a) Case 1: No Bypass Capacitor	(b) Case 2: Bypass Capacitor Near Via 1 at Current Source	(c) Case 3: Bypass Capacitor Near Via 2 at Layer Jumping point

Figure 9.80. Simulation results for the three analyzed cases. (Simulation performed on Sigrity "Speed97" Suite; courtesy of Sigrity, Inc.)

The peak noise voltage distribution* level throughout the 2500 meshed circuit nodes across the power and return planes was computed in a manner illustrated in Figure 9.81.

Figure 9.82 presents the distribution plots of the peak noise voltage for each of the three cases discussed above. Curve #1, which represents the peak noise distribution for Case 1 (no bypass capacitor), shows that a noise voltage distribution in the range of 25–45 mV (median of approximately 35 mV) is present across most of the power/return planes (i.e., the largest fraction of the 2500 nodes).

Curve #2, representing Case 2 (bypass capacitor located near Via 1, near the current source), demonstrates that with the 100 nF bypass capacitor placed near the device, some areas on the power/return planes now exhibit a lower noise voltage (approximately 25 mV). However in general, the noise voltage distribution still ranges from 25 mV to 40 mV (but the median is shifted to approximately 25 mV).

Finally, in Curve #3, presenting the results for Case 3 (bypass capacitor located near Via 2), a much better suppression of noise is observed. Now the noise voltage distribution is reduced to the 15 mV to 25 mV range while the greater part of the power/return planes exhibit voltage fluctuations of less than 20 mV.

In conclusion, when signal traces traverse layers, particularly power and return reference planes, a return current path should be provided by placing vias or bypass capacitors in the immediate vicinity of the via of the traversing signal. Any other location could result in significant increase of power- and return-plane noise voltage, PCB resonance, and, subsequently, EMI and signal integrity performance degradation.[†]

It is noted, however, that in practical circuits (unlike the simplified circuit used for the simulation) multiple bypass capacitors rather than a single capacitor should be used across the plane in order to keep the power and return planes resonance-free.[‡]

But why not consider (and benefit from) interplane capacitance? After all, it could be argued, a large capacitance exists between the planes thanks to their large overlapping surface area and the small spacing between the layers. Could we not count on displacement current through the interplane capacitance as a current return path?

For probing into this very valid question, a model for separated power (V_{CC}) and return (GND) planes can be constructed based on the following assumptions (Figure 9.83) [19]:

- The power and return planes exhibit an impedance of a plate capacitor and can be modeled by using the theory of radial transmission lines.
- The field perturbation due to the switching devices travels outward from the component with cylindrical symmetry.
- The effective radius of the plate capacitor is small compared to the distance to the edges of the board, so board-edge reflections can be neglected.

---

*Peak noise voltage distribution is the frequency of occurrence of any peak noise voltage across the planes.
[†]Novel techniques for mitigation of parallel-plate waveguide (PPW) noise, resulting from "layer jumping" and other mechanisms, are presented in Sections 9.5.5 and 9.5.6.
[‡]Resonances in PCBs and their mitigation are further discussed in Section 9.9.

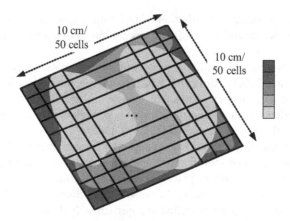

Figure 9.81. Illustration of the manner of peak noise voltage distribution throughout the 2500 meshed circuit nodes across the power and return planes.

- The separation between the power and return planes is very small compared to the wavelength associated with the highest frequency of interest; thus, TEM propagation occurs between the planes in all directions, characterized by the E-field perpendicular to the planes.

Assume that the waves generated by the switching device travel outward with cylindrical symmetry as in a radial transmission line. Inductance and capacitance val-

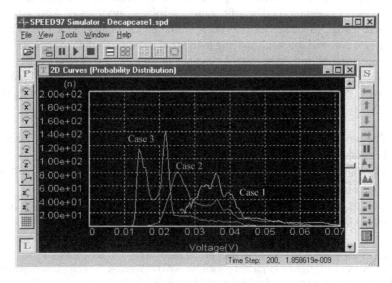

Figure 9.82. Peak noise voltage distributions for the three cases. (Simulation performed on Sigrity "Speed97" Suite; courtesy of Sigrity, Inc.)

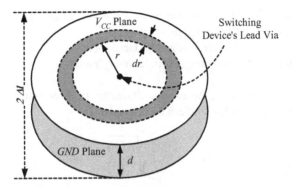

(a) Geometrical Representation

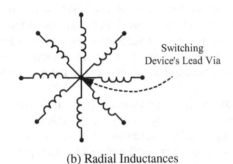

(b) Radial Inductances

Figure 9.83. Representation of nonuniform radial transmission line. (Courtesy of Spartaco Caniggia.)

ues of the radial transmission line exhibit the same behavior of a nonuniform transmission line with line inductance L and line capacitance C parameters dependent of the radial distance r from the origin given by [19]

$$\begin{cases} C_{rad}(r) = \varepsilon \cdot \left(\dfrac{2\pi r}{d} \right), \text{ farad} \\[4mm] L_{rad}(r) = \mu \cdot \left(\dfrac{d}{2\pi r} \right), \text{ henry} \end{cases} \qquad (9.36)$$

where:

$C_{rad}(r)$ = Infinitesimal increment of the radial capacitance the parallel planes for a radius r, farad

$L_{rad}(r)$ = Infinitesimal increment of the radial inductance the parallel planes for a radius r, henry*

---

*The radial inductance is easily obtained using the relationship between the per-unit-length inductance and capacitance for a homogeneous medium: $L_{rad} \cdot C_{rad} = \mu\varepsilon$.

r = Incremental radius of the radial capacitance between the planes
d = Separation between the two planes, meters
ε = Permittivity of the dielectric substrate, farad/meter
μ = Permeability of the dielectric substrate, henry/meter

As illustrated in Figure 9.83, the average characteristic impedance $Z_{0\ell}$ for the segment $\Delta\ell$ of the radial transmission line can be defined as

$$Z_{0\ell} = \frac{1}{\Delta\ell} \cdot \int_{r_{min}}^{\Delta\ell} \sqrt{\frac{L(r)}{C(r)}}\, dr, \text{ ohm} \qquad (9.37)$$

where r_{min} is the radius of the device's connecting via.

In this manner, the power and return leads of every active device on the PCB become local radial transmission lines. The field perturbations travels outward from the device's via in all directions with propagating waves very similar to those occurring in a radial waveguide. The model is valid up to the frequencies at which $\Delta\ell < \lambda_{min}/10$ is valid, where λ_{min} is the wavelength corresponding to the maximum frequency of interest.

Using the expression for the radial capacitance from Equation (9.36), we may now reiterate the above question, namely, could we not consider (and benefit from) displacement current through the interplane capacitance as a current return path?

Consider power (V_{CC}) and return (GND) planes with a separation, d, of 10 mil (0.25 mm).* A dielectric substrate of FR-4 is used. The interplane capacitance per unit area (cm^2) is

$$C_{rad}(r) = \varepsilon_0 \varepsilon_r \cdot \left(\frac{2\pi r}{d}\right), \text{ farad} \qquad (9.38)$$

where:
C_{rad} = Interplane radial capacitance, farad
ε_0 = Permittivity of free space, $\varepsilon_0 \cong (1/36\pi) \times 10^{-9} \approx 8.85 \times 10^{-12}$, F/m
ε_r = Relative permittivity of the dielectric material (assume $\varepsilon_r = 4.5$ for FR-4)
d = Separation between planes, meters

For the dimensions of the discussed case, when substituting in Equation (9.38) and reducing the expression, we obtain

$$C_{rad} = \varepsilon_0 \varepsilon_r \cdot \left(\frac{2\pi r}{d}\right) \approx \frac{1}{36\pi} \times 10^{-9} \cdot 4.5 \cdot \frac{2\pi r}{0.25 \times 10^{-3}} = r \cdot 10^{-6} \text{F} \qquad (9.39)$$

The propagation velocity, v, of the transient waveform through the transmission line formed by the power and return planes is

*10 mils are equivalent to 0.25 mm, approximately.

$$v = \sqrt{\frac{1}{\mu_0 \cdot \varepsilon_0 \varepsilon_r}} = \sqrt{\frac{1}{4\pi \times 10^{-7} \cdot 4.5 \cdot \dfrac{1}{36\pi} \times 10^{-9}}} \approx 1.4 \times 10^8 \, \text{m/sec} = 1.4 \times 10^{-2} \, \text{cm/psec}$$

$$(9.40)$$

Assuming that the digital signal return current waveform has a transition time, t_t, of 100 psec (typical of contemporary high-speed digital devices), the effective upper frequency of the signal spectrum (where the impedance of the interplane capacitance is expected to be most effective) is

$$f_{max} = \frac{1}{\pi t_t} \approx \frac{1}{\pi \cdot 100 \times 10^{-12}} \approx 3.18 \text{ GHz} \qquad (9.41)$$

Since interplane capacitance is effective during the transition time (100 psec in this case) only, where the frequency content of the signal is maximized, the effective radius, r, from which charge is drawn (and TEM propagation occurs) during this period, is

$$r = v \cdot t_t \approx 1.4 \times 10^{-2} \, \frac{\text{cm}}{\text{psec}} \cdot 100 \text{ psec} \approx 1.4 \text{ cm} = 0.014 \text{ m} \qquad (9.42)$$

Substituting this result in Equation (9.39), the total radial capacitance of this "disk" is obtained:

$$C_{rad} = r \cdot 10^{-6} = 0.014 \cdot 10^{-6} = 14 \, \text{nF} \qquad (9.43)$$

At a frequency associated with a transition time of 100 psec, 3.18 GHz [Equation (9.41)], the effective impedance represented by this effective inter-plane capacitance, Z_C, is

$$Z_C = \frac{1}{2\pi \cdot f \cdot C_{rad}} \approx \frac{1}{2\pi \cdot 3.18 \times 10^9 \cdot 14 \times 10^{-9}} \approx 3.5 \text{ m}\Omega \ @ \ 3.18 \text{ GHz} \qquad (9.44)$$

Note that this represents the minimum time-dependent impedance obtained at the end of the state transition time of the device. Initially, higher impedance is present, which decreases during the transition time of the signal thanks to the radial "spreading" of the current wave front across the planes from the transition point. The radial capacitance between the planes can, therefore, be visualizes as a "spreading capacitor," illustrated in Figure 9.84. The capacitor expands out from the device's lead, drawing the current from the plane. This capacitance can be expressed as

$$C(t) = C_{rad}(r) = \varepsilon_0 \varepsilon_r \cdot \left(\frac{2\pi r(t)}{d} \right) \approx \varepsilon_0 \varepsilon_r \times \left(\frac{2\pi v \cdot t}{d} \right), \text{ farad} \qquad (9.45)$$

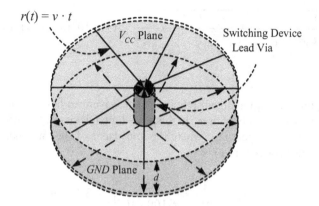

<u>Figure 9.84.</u> Spreading of drawn current (and capacitance) throughout the planes.

This computation does not consider the spreading inductance* in the planes (which appears in series with the effective interplane capacitance). Substituting for L_{rad} [Equation (9.36)], we obtain as beforehand for $d = 10$ mil (0.25 mm) and assuming $\mu_r = 1$

$$L_{rad}\,(r) = \mu \cdot \left(\frac{d}{2\pi r}\right) = 4\pi \times 10^{-7} \cdot \left(\frac{0.25 \times 10^{-3}}{2\pi r}\right) = r \cdot 5 \times 10^{-11},\ \text{henry} \quad (9.46)$$

Substituting $r = 0.014$ m [see Equation (9.42)] in Eq. (9.46) results in

$$L_{rad}\,(r) = r \cdot 5 \times 10^{-11} = 0.014 \cdot 5 \times 10^{-11} = 0.7\ \text{pH} \qquad (9.47)$$

At a frequency associated with a transition time of 100 psec, 3.18 GHz [Equation (9.41)], the effective reactance represented by the effective planar spreading inductance, Z_L, is

$$Z_L = 2\pi \cdot f \cdot L_{rad} \approx 2\pi \cdot 3.18 \times 10^9 \cdot 0.7 \times 10^{-12} \approx 17\ \text{m}\Omega\ @\ 3.18\ \text{GHz} \qquad (9.48)$$

In conclusion, although interplane impedance is dominated by the spreading inductance, the total cumulative interplane impedance, including both the capacitive and in inductive reactance, corresponding to a transition time of 100 psec or a frequency of 3.18 GHz, is approximately 20 mΩ.

If, on the other hand, two bypass capacitors are placed near the layer transition point (1.4 cm away, maximum) and assuming a total effective series inductance, L_{eq}, of the capacitors of 1.5 nH[†] (at higher frequencies the impedance of a capacitor is dominated by its ESL and mounting partial inductance), the effective impedance exhibited by the parallel combination of the capacitors, Z_{Leq} is

---

*Spreading inductance is further discussed in Section 9.5.3.5.
[†]Typical contribution to inductance due to via connections is 1 to 4 nH.

$$Z_{L_{eq}} = 2\pi \cdot f \cdot \frac{L_{eq}}{2} = \frac{2\pi \cdot 3.18 \times 10^9 \cdot 1.5 \times 10^{-9}}{2} \approx 15 \ \Omega \ @ \ 3.18 \ \text{GHz} \quad (9.49)$$

This result indicates that indeed, at the highest frequency component of the return current spectrum, interplane capacitance offers a significant apparent advantage. Regardless, stitching capacitors cannot be eliminated since lower frequency components of the switching current spectrum will find a more effective path through discrete capacitors. At even lower frequencies, the effect of the equivalent inductance of the discrete bypass capacitors diminishes, whereas the impedance of the interplane capacitance increases, resulting in an increasing advantage of the discrete capacitors (at a frequency of 318 MHz, for instance, the effective inter-plane capacitance would be 5 Ω, approximately, whereas the impedance of the two discrete bypass capacitors would drop to 1.5 Ω.

The effectiveness of the interplane capacitance can be appraised by examining the distribution of the displacement current occurring through the interplane capacitance. Effective interplane capacitance occurs when the displacement current is concentrated in a small area around the point of signal via. Higher levels of displacement currents should be found at a smaller radius. Figure 9.85 depicts the distribution of displacement current in the reference planes at two sample frequencies* $f = 500$ MHz (parts a and b) and $f = 800$ MHz (parts c and d) at distances from the location of the traversing signal via of $r = 40$ mils (parts a and c) and $r = 450$ mils (parts b and d). Note the increased amplitude and distribution of the displacement current as separation from the signal via increases.

Figure 9.85 clearly demonstrates the effectiveness of interplane capacitance. For either of the two investigated frequencies, higher displacement current is observed as separation from the current source (i.e., location of the traversing signal) and the subsequent effective area of the interplane capacitance increase, indicative of the increased effectiveness of the interplane capacitance. Compare, for instance, Figure 9.85(a) and (b) for $f = 500$ MHz. Also, as expected, the increase in frequency results in higher displacement current, indicative of the improved effectiveness of the interplane "spreading" capacitance, with frequency, due to reduction of interplane impedance.

It is noteworthy, however, that the results presented above should be carefully interpreted. Since this analysis was carried out in the frequency domain ("steady-state analysis"), propagation time was not considered.

Recalling that interplane capacitance is effective for AC signals only, in digital waveforms this implies that this capacitance can be considered during the pulse transition time only. The duration of the transition time determines, therefore, the time during which the interplane capacitance is effectual and subsequently, the size of the "spreading" capacitance. As transition time shortens, higher frequency spectral components are introduced; however, the effective wave (or current) propagation (spreading) time between the planes shortens. Consequently, this will result in reduction of effective area contributing to interplane capacitance.

---

*This simulation was carried out in the frequency domain; the distribution of the displacement current across the planes was computed at various sample frequencies.

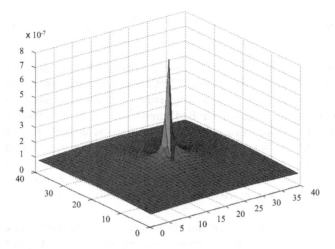

(a) @40 mils from Signal Via, f = 500 MHz

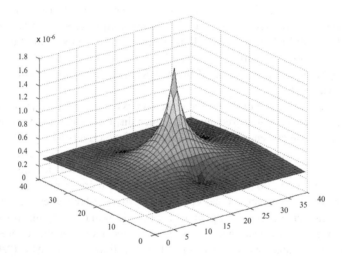

(b) @450 mils from Signal Via, f = 500 MHz

<u>Figure 9.85.</u> Distribution of displacement current around location of signal transition via verus frequency. (Courtesy of Bruce R. Archambeault, IBM Corporation.)

Given that a desired maximum noise voltage (or ripple) across the planes, $V_{max}(t)$, is specified, the required capacitance, C, for achieving this goal is

$$C = \frac{Q}{V_{max}}, \text{ farad} \tag{9.50}$$

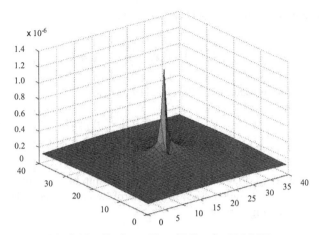

(c) @40 mils from Signal Via, $f = 800$ MHz

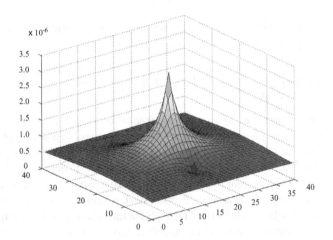

(d) @450 mils from Signal Via, $f = 800$ MHz

Figure 9.85. *Continued*.

The total charge, Q, transferred during transition time, t_r, of the current, $I(t)$ is

$$Q \triangleq \int_{t=0}^{t=t_r} I(t)dt, \text{ coulomb} \tag{9.51}$$

Therefore,

$$C_{\min} = \frac{\int_{t=0}^{t=t_r} I(t)dt}{V_{\max}(t)}, \text{ farad} \tag{9.52}$$

From Equation (9.45), the "spread" capacitance at $t = t_r$ is

$$C(t_r) \approx \varepsilon_0 \varepsilon_r \frac{2\pi v}{d} \cdot t_r, \text{ farad} \qquad (9.53)$$

Therefore, combining Equations (9.52) and (9.53) yields

$$C_{min} = \varepsilon_0 \varepsilon_r \frac{2\pi v}{d} \cdot t_{r\,min} = \frac{\int_{t=0}^{t=t_r} I(t)dt}{V_{max}(t)}, \text{ farad} \qquad (9.54)$$

Rearranging Equation (9.54) we obtain the relationship between the achievable ripple voltage $V(t)$ and the characteristics of the interplane capacitance, the pulse transition time, and the switching current:

$$V(t) = \left(\frac{d}{\varepsilon_0 \varepsilon_r \cdot 2\pi v \cdot t_r} \right) \cdot \int_{t=0}^{t=t_r} I(t)dt, \text{ volts} \qquad (9.55)$$

If the transition current is approximated as a triangular waveform with duration, t_r and peak value, I_p, Equation (9.55) can now be simplified:

$$V(t) = \left(\frac{d}{\varepsilon_0 \varepsilon_r \cdot 2\pi v \cdot t_r} \right) \cdot \left(\frac{1}{2} I_p \cdot t_r \right) = \frac{1}{4\pi \cdot \varepsilon_0 \varepsilon_r} \cdot \frac{d}{v} \cdot I_p, \text{ volts} \qquad (9.56)$$

Note that this approximate expression is independent of the current transition duration, t_r. This expression provides the conditions for effectiveness of the interplane capacitance for reduction of voltage ripple across the planes. When the resultant ripple voltage, $V(t)$, exceeds the desirable level, interplane capacitance is insufficient and it would have to be supplemented by discrete bypass capacitors. Alternatively, the characteristics of the interplane capacitance (separation, d, primarily), would have to be modified.*

In conclusion, displacement current through the interplane capacitance may provide an alternate path for transition of return currents across the two reference planes. However, the relative effectiveness of the interplane capacitance is frequency and geometry dependent. With continued lessening of current waveform transition time, the effectiveness of interplane capacitance, relative to the impedance of the discrete bypass capacitors, increases. That impedance is dominated by the equivalent inductance of the capacitor and its mounting. If the reference planes traversed by the signal are sufficiently close, the interplane capacitance may be adequate.† Conversely, when the signals traverse a relatively large structure, additional interconnecting vias and/or discrete capacitors are still required for handling and bypassing the lower frequency components of the current spectrum.‡

---

*The technique entitled "embedded capacitance" implements this technique for achieving much higher interplane capacitance. Increasing the permittivity, ε_r would be acceptable; larger permittivity results in lower propagation velocities, v.
†Much higher interplane capacitance may be obtained with the use of "embedded capacitance" technology.
‡Refer to Section 9.5.4.3 for further discussion on location and placement of decoupling capacitors.

A concluding experimental demonstration of the effects of signal traces traversing power and return planes is now provided. This illustration demonstrates the effect of signal traces traversing power and return planes on circuit immunity [24]. Figure 9.86 depicts the test setup used for the demonstration. A pulsed field generator was used to produce fast impulse fields. The generator was positioned 30 cm from a test board.

The test board used for this demonstration was similar to the one described in Section 9.4 and Figure 9.32. The board was constructed from a double-sided, copper-clad board with two 50 Ω test conductors placed approximately 30 cm apart on the planes. One was routed from an SMA connector to a 47 Ω resistor over a continuous copper plane. The second was routed over the same copper plane for two-thirds of its length and then traversed through the board and routed over the copper plane on the other side of the board for about one-third of its length. The two copper planes emulate the power and ground planes of a four-layer PCB. The planes were shorted together at the four extremities of the conductors, at the feed points and at the loads. Measurements of the voltage developed on the load of each conductor took place by means of an oscilloscope while the EMI was induced on the test board.

Figure 9.87 shows the voltage waveform induced on the conductor routed over a solid copper plane. The waveform amplitude peaked at approximately 100 mV with very little ringing, thus presenting a low risk of signal corruption. Figure 9.88 depicts the voltage waveform induced on the conductor that traversed the copper planes. A peak voltage of over 200 mV was recorded along with a ringing frequency between 400 and 500 MHz. Such an interference level constitutes a substantial fraction of the noise margin of modern logic families and could potentially result in degradation of the circuit's EMC performance, for example, lowered immunity to EMI and disruption of sensitive circuits. The results presented in Figure 9.87 and Figure 9.88 reinforce the conclusions drawn above.

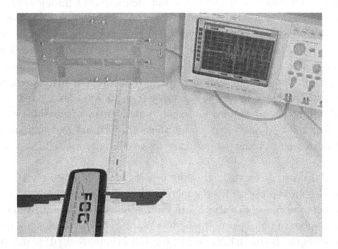

Figure 9.86. Setup for the impulsive radiated EMI immunity demonstration. (Copyright 2003 Douglas C. Smith, http://www.dsmith.org, used with permission.)

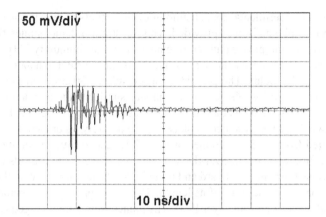

Minimizing the induced voltage when signal traces must traverse reference planes can be achieved by increasing interpane capacitance. This objective can be accomplished by implementing improved PCB stack-ups* with smaller power-return plane spacing.

9.4.2.3 Excessive Pin/Hole Clearance.

In high-density multilayer printed circuit boards and backplanes, traces may have to be routed through regions containing high-density VLSI packages and connectors or multipin sockets. In these regions, the signal traces may be routed in proximity to board plane cutouts, periodically encountering regular patterns of connector pin fields, plated through-holes (vias), antipad cutouts, BGA solderballs, and so on, all resulting in massive perforation (known as "keep-out areas") of all layers, as illustrated in Figure 9.89 [18]. Owing to the clearance holes (antipads) required around each via, a large proportion of the reference planes' area is consumed. Consequently, "solid" reference planes (power and return planes) do not exist as such in reality. High levels of plane perforation significantly increase the plane inductive reactance (i.e., its impedance) and counteract the benefits of solid planes. Moreover, traces routed on opposite sides of a reference plane, passing close to or over periodic plane cutouts and perforations, can line up small resonances and produce significant loss suck-outs and nonzero layer-to-layer crosstalk.

Figure 9.90 illustrates a trace crossing a perforated reference plane, and Equation (9.57) provides an expression for the increase in inductance associated with this configuration (note definition of variables in Figure 9.90) [18]. In Equation (9.57), n represents the number of holes per unit length, λ represents the wavelength, in meters, and L_{Plane} represents the partial inductance of the solid (unperforated) plane in the system

---

*Stack-up considerations are presented in Section 9.6.

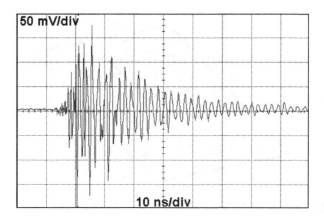

<u>Figure 9.88.</u> Noise induced into the trace traversing through the reference planes, 10 ns/div (horizontal)/50 mV/div (vertical). (Copyright 2003 Douglas C. Smith, http://www.dsmith.org, used with permission.)

shown, whereas ΔL [consisting of all other elements in Equation (9.57)] represents the increase in circuit inductance.

$$L = L_{Plane} + \Delta L \approx L_{Plane} + \left(\frac{\mu a^3}{3\pi^2 h^2} \cdot e^{-\frac{t}{a}} \right); \ a << \frac{\lambda}{n} \qquad (9.57)$$

Figure 9.91 presents two common situations in which such perforation occurs on high-density PCBs. In both cases, the high-density pin-out results in massive perforation

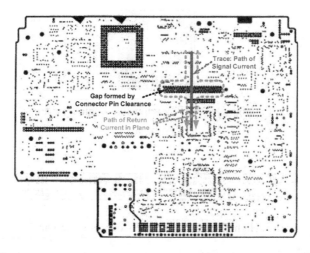

<u>Figure 9.89.</u> Perforated PCBs due to clearance holes (antipads) around vias. (Courtesy of Dr. Anatoly Tsaliovitch.)

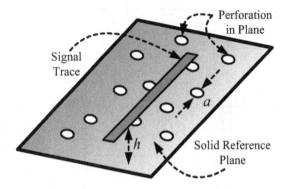

Figure 9.90. Geometry of a perforated reference plane crossed by a signal trace.

due to the antipads around adjacent vias, which may lead to insufficient space for any metallization in between.

Figure 9.92 illustrates traces of a parallel digital bus routed in a backplane interconnecting pins in high-density edge connectors. Excessively large antipads around the pins and vias result in insufficient space for metallization in their midst, leaving no path for return currents to find their way through to a return pin [Figure 9.93(a)]. The clearance in the reference plane constitutes a slot in the reference planes (as discussed in Section 9.4.2.1), forcing the return signal current traveling under the trace to circle around the copper void area to get to the return pin in the other connectors. This situation creates a large current loop that could result in poor EMC performance. When adequate separation is maintained between clearances (antipads) of adjacent vias, a sig-

(a) High-Density Programmable Logic Device

(PLD) Pin-Out

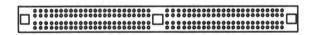

(b) High-Density SMT Edge Connector Pin-Out

Figure 9.91. Massive perforation of reference planes due to clearance holes (antipads) around each via.

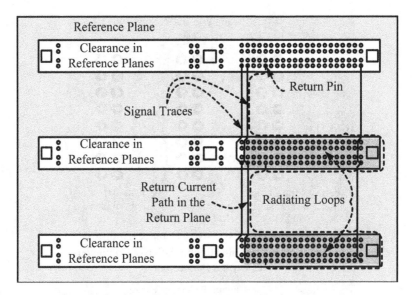

Figure 9.92. Excessive through-hole pin clearance (antipads) leading to excessive current loop area.

nificant improvement is achieved as the return current may take a path following the signal trace. A recommended clearance, S, between adjacent antipads should be no less than one-third of the antipad diameter [Figure 9.93(b)]:

$$\frac{S}{d} > \frac{1}{3} \tag{9.58}$$

A similar situation can occur when a large number of signal traces (particularly those associated with wide buses) must traverse layers ("layer jumping") as a result of layout and routing constraints. As signal traces traverse through reference planes, pads and antipads are automatically placed in all layers, including the reference planes. As a result, a large gap is formed in the reference plane, where continuity of the plane is most needed for providing a path for the return currents. When a PCB has a large number of through-holes and via holes, such as may occur with wide digital buses, care should be taken to prevent their antipads from merging together or encroaching on the current return paths of the traces. Figure 9.94 illustrates a poorly designed reference plane, where the antipads around via holes of a digital bus touch each other.

The above examples are all a result of excessive clearance between pins or vias, which often occurs due to antipads merging into each another, creating larger gaps or slots in a reference plane. This could occur when spacing between via holes is too small or when the computer-aided-design (CAD) system is set up to have too large a minimum radius when drawing planes. Figure 9.95 presents some examples of errors resulting in antipad merger.

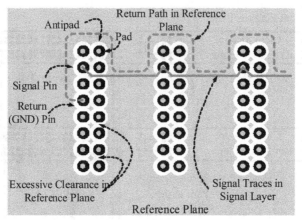

(a) Objectionable: Excessive Clearance (Antipads)
around Field of Via Pads; Return Path Not Continuous
Under the Connectors

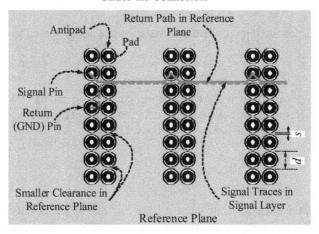

(b) Acceptable: Smaller Clearance (Antipads) around
Field of Via Pads; Continuous Return Path
Under the Connectors

Figure 9.93. Effect of through-hole pin clearance (antipads) leading to excessive current loop area (zoom in on connector pads).

Solutions to this situation include:

1. **Reduction of Antipad Radius.** Antipads should be made as small as possible, thus providing even more space for metallization in the reference planes.
2. **Avoidance of Antipads Merger.** In the unanticipated cases in which pads must be placed in reference planes, the antipads must not merge into one another. This

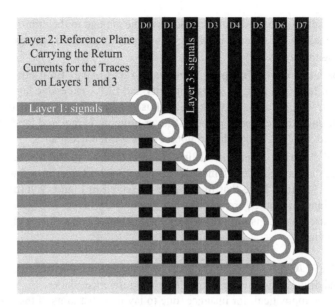

Figure 9.94. A poorly designed reference plane. The antipads around the via holes are touching each other. (Courtesy of Keith Armstrong, Cherry Clough.)

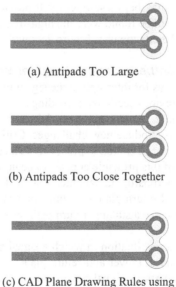

(a) Antipads Too Large

(b) Antipads Too Close Together

(c) CAD Plane Drawing Rules using
a Too Large Minimum Radius

Figure 9.95. Some examples of errors leading to antipad merger. (Courtesy of Keith Armstrong, Cherry Clough.)

can be achieved by reducing the radius of the antipad to the minimum necessary or increasing the spacing between vias (typically possible for signal traces traversing the layers only).

3. **Minimizing Layer Jumping of Signal Traces.** Apart from other adverse effects of layer jumping on signal integrity and EMC, excessive layer jumping of signal traces, particularly wide digital buses, adds a large number of vias crossing through the reference planes. If layer jumping is required [and "blind vias" are used, see (5)], traces should not cross over reference planes.

4. **Use of "Buried Vias" or "Blind Via."** Standard plated-through vias cross through all layers, even if intended to interconnect traces in a few layers only [Figure 9.96(a)]. Note that in the case presented in Figure 9.96(b), both reference planes L_2 and L_4 are not affected by the via, whereas in Figure 9.96(c) only reference plane L_4 is not affected by the via but reference plane L_2 is traversed by the via.

5. **Land Fills/Grids.** When sufficient clearance is available, it is best to provide copper "land fills" between the antipads in the reference planes. If, however, such "land fills" are impractical, for instance, due to layout constraints of the CAD system, it is recommended, alternatively, to crisscross the area between the pins of the connector (or the VLSI device) with traces in the reference planes, forming a grid or mesh, in a manner that interconnects the reference planes across the otherwise cleared area. This solution is not as good as a solid plane but is definitely better than having a slot in the plane. Skinny as they may be, these traces still provide a return path with lower (loop) impedance than would be found were the high-frequency current to encircle the gap formed by the cleared area (compare Figure 9.97 with Figure 9.92).

9.4.2.4 *Improper Motherboard-to-Daughter Board Connections.* Inasmuch as there are many ways for interrupting the signal return paths on printed circuit boards (as illustrated in previous sections), providing a controlled path within the PCB is seemingly a simple task. Interconnections between motherboards to daughter boards, on the other hand, introduce new challenges. Critical signals (from the standpoint of signal integrity) must be routed in proximity to a continuous return path; however, the implementation of this principle appears to be nontrivial when critical signals are routed between a motherboard and daughter boards. The question to be addressed in this case is "how should return planes be interconnected at the interface between two PCBs in order to guarantee a suitable return path for signals crossing the interface between the two boards?"

Figure 9.98 demonstrates a situation in which a signal path exists between a motherboard and a daughter board. Power and return planes are allocated on both boards in addition to the signal layers. A signal trace is routed from driver D_1 in signal layer L_1 to the motherboard-to-daughter board edge connector. This signal layer is adjacent to the return (GND) layer L_2, which subsequently serves as the signal return path for the signal in layer L_2. After the connector, the signal continues in the daughter board in signal layer L_4. This signal layer is adjacent to the power (V_{CC}) layer L_3,

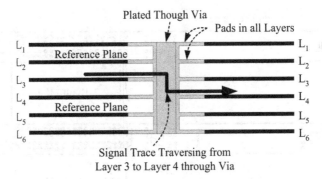

(a) A Standard Via Traverses all Layers

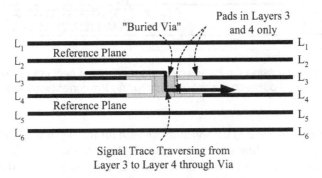

(b) A "Buried Via" Traverses only Interconnected Layers
L_3 and L_4

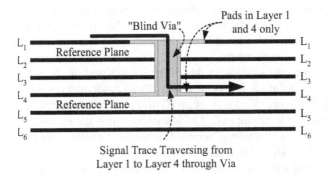

(c) A "Blind Via" Traverses only External Layer
L_1 to Layer L_4

Figure 9.96. Use of "buried vias" and "blind vias" may reduce the number of pads and antipads in reference layers.

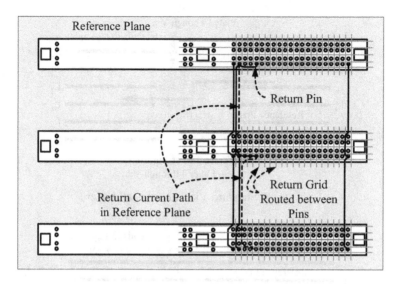

Figure 9.97. Crisscrossing the clearance under backplane connectors by means of a grid of traces routed between the antipads, reducing the current loop area.

which is expected to serve as the signal return path for the signal in layer L_4. The signal return current, traveling from the receiver R_1 back to the driver D_1 must, somehow, cross over from layer L_3 to layer L_2. These layers, however, are of different DC potentials, and cannot be directly interconnected, thus creating a return current path discontinuity.

Indeed, some "leakage" of return current will occur through displacement mechanism thanks to the interplane capacitance between layers L_3 and L_2, as illustrated in Figure 9.98. For this mechanism to effectively happen, however, spreading of the return current across the two planes will have to take place to provide higher effective interplane capacitance.* This spreading, however, could result in increased inductance of the signal and return currents.

One often proposed solution to this difficulty is based on the presence of decoupling capacitors.† After all, it may be argued, these capacitors provide an AC connection between the power (V_{CC}) and return (GND) planes and, therefore, should be able to complete the return current path. Unfortunately, this is not necessarily the case. Figure 9.99 illustrates a situation in which decoupling capacitors are provided near the driver and receiver, D_1 and R_1, respectively.

The decoupling capacitors provide a low-impedance AC connection between the power and return planes in layers L_3 and L_2, respectively. Indeed, the capacitors may effectively provide switching power current for the neighboring switching devices. However, they are of little use in providing a lower impedance return path for the signal current traveling from D_1 to R_1: Since high-frequency return current will always at-

---

*Current spreading in planes and "spreading capacitance" are discussed in Section 9.4.2.2.
†Decoupling and the role of decoupling capacitors are discussed in Section 9.5.4.2.

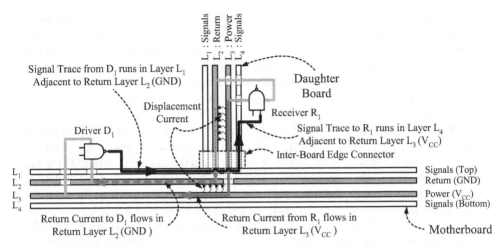

Figure 9.98. Signals crossing the interface between motherboard and daughter board. How is the return path completed? where does the return current cross over from Layers L_3 to L_2?

tempt to flow in the reference plane closest to the signal trace, the favored layer for the return current leaving the receiver R_1 is indeed layer L_3, the power (V_{CC}) layer. The decoupling capacitor in the vicinity of R_1 actually provides a connection to the return (GND) layer L_2, which is a longer way away from the signal trace. The return current will, therefore, remain in layer L_3.

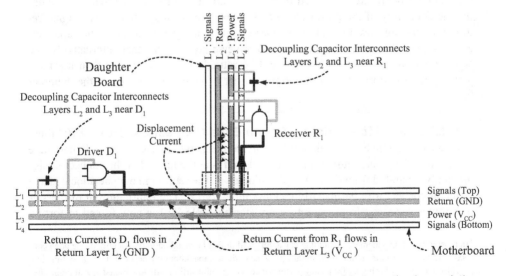

Figure 9.99. Decoupling capacitors do not necessarily provide an optimal completion of the return current path from layers L_3 to L_2.

Similarly, the capacitor placed near the driver, D_1, offers little improvement, since the current must travel the entire path from the edge connector to the driver in layer L_3 until it can traverse from layer L_3 to L_2, used as an optimal return path for the current flowing from the driver to the edge connector. Again, the return current must rely on the (inefficient) displacement mechanism for allowing the return current to traverse from layer L_3 to L_2 and back to D_1. Obviously, the decoupling capacitors adjacent to the switching devices do not provide an adequate solution.

Relocating the capacitors to the near vicinity of the edge connectors yields a totally different situation (Figure 9.100). This capacitor interconnects layers L_3 to L_2, providing a path for the return current traveling in layer L_3 in the daughter board and entering the motherboard at the edge connector to traverse from layer L_3 to L_2 immediately at the entry point to the motherboard. After the capacitor, the return current flows in layer L_2, serving as the optimal return path for the signal in the motherboard, between the edge connector back to the driver D_1. Note that displacement still occurs between the planes in the daughter board and motherboard. Note also that bypass capacitors can (and in most cases should) be installed both in the motherboard and the daughter board.*

This approach was, in fact, proposed and discussed in Section 9.4.2.1, and the effect of the bypass capacitors was presented in Figure 9.73 through Figure 9.75 in the context of traces crossing slots and splits in reference planes. As shown there, the effectiveness of a bypass capacitor is limited by the capacitor's ESL and the inductance associated with its installation, diminishing its effectiveness at higher frequencies.

The best solution is illustrated in Figure 9.101. Note that in the daughter board, the layers L_2 and L_3 exchanged their function, to power (V_{CC}) and return (GND) respectively. Consequently, the signal trace routed in layer L_4 is now routed adjacent to the return rather than the power plane. In the edge connector, layers L_2 and L_3 in the daughter board are connected to layers L_3 and L_2 in the motherboard, maintaining the continuity of the power (V_{CC}) and return (GND), respectively. The signal return currents can now flow along its entire path in a layer adjacent to the same reference plane, eliminating the necessity to traverse layers. Note that, alternatively, the signal trace in the daughter board could be routed in layer L_1 rather than layer L_4, yielding the same result with the original order of layers maintained in the daughter board.

9.4.2.5 Insufficient Return Pins in Edge Connectors.

Figure 9.102 illustrates a case in which a wide digital bus interconnects two PCBs by through their edge connectors. Only one return pin is provided, and is at the far end of the connector, significantly separated from the signal pin. Worse yet, the return pin may be allocated at a different connector (a particularly reckless thing to do). The return current of the signals routed far away from the return pin (for example, D_0) is now forced to form a

---

*In most decoupling schemes, decoupling capacitors will in fact be installed in the vicinity of the edge connector, at the power entry point to the circuit. If present, these capacitors may perform the bypass function, provided they are fit for higher frequency performance (i.e., exhibit sufficiently low impedance at higher frequencies). Correctly selected ceramic capacitors will fulfill this task.

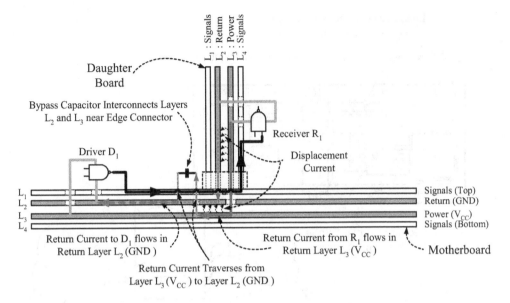

Figure 9.100. Bypass capacitors near an edge connector provide a path for the return current to effectively traverse from layers L_3 to L_2.

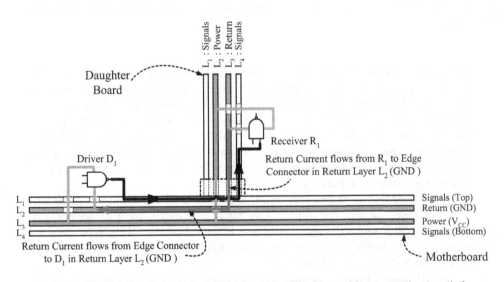

Figure 9.101. Rearranging layers in the daughter board provides an optimal path for the return current all the way from R_1 to D_1 through the edge connector.

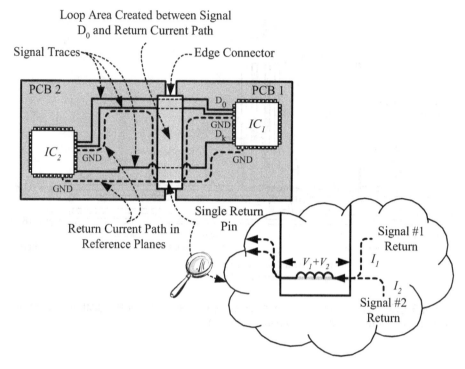

Figure 9.102. Insufficient return pins in a connector increase the interboard current loop area.

large loop. Consequently, any sensitive trace passing through adjacent pins in this connector will pick up the excessive emissions generated by the current loop, resulting in excessive EMI and possibly in potential functionality problems as well. An additional major shortcoming of using a single return pin is that it serves as the current return path for multiple signals, which may significantly differ in their amplitude and sensitivity levels. The consequent common-impedance coupling between the signals may result in degradation of system performance and in increased EMI (see the cloud at the bottom-right corner of Figure 9.102).

For avoiding this undesired situation, a large number of return pins should be allocated in edge connectors, particularly when high-speed signals are used. Ideally, a ratio of signal (S) to return (R) pins should be S:R = 1:1. Normally, a compromise, using 3:1 or 5:1 is acceptable for this purpose. Figure 9.103 illustrates an optimal pin assignment where S:R = 1:1 is implemented.

Note: High-speed return current will always have a tendency to return through the closest return pin, even if multiple pins are available, as that path serves as the path of least inductance (and impedance).

The return leads should be uniformly interspersed throughout the connector. Figure 9.104 demonstrates that for an equal S:R ratio of 2:1, a good arrangement, in which the

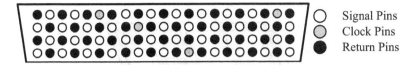

Figure 9.103. Optimal pin assignment of signal and return leads in an edge connector (S:R = 1:1).

pins are uniformly distributed throughout the connector (a), minimizes the loop area between the signal and its associated return pins, whereas in the poor arrangement the return pins are all allocated at the one end of the connector, forming a large loop between the leftmost signal pins and the return pins allocated to the rightmost edge of the connector.

9.4.2.6 Reference Plane Edge Effects. Yet another design error is associated with routing traces above reference planes in near proximity to the edge of the board or, just as common, near (and along) a gap in a reference plane. The former case occurs in the attempt to make the best use of the "real estate" on the board, whereas the latter arises in the belief that such traces are isolated by virtue of the gap in the reference plane placed between the circuits. Both cases are illustrated in Figure 9.105. Figure 9.106 illustrates a practical case of a trace routed at the very edge of the PCB.

Although not a discontinuity in the reference plane in the sense described in the previous sections, the effect of routing traces in near proximity to edges of reference planes (in both above cases) has a similar consequences owing to the distortion of the

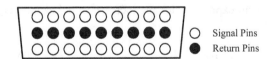

(a) Good Pin Assignment of Return Pins: Small Loop
between Signal and Return Pins

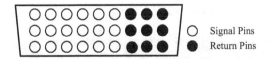

(b) Poor Pin Assignment of Return Pins: Large Loop
between Signal and Return Pins

Figure 9.104. Good (a) and poor (b) pin assignment of signal and return leads in an edge connector for a ratio of signal (S) to return (R) Pins of S:R = 2:1.

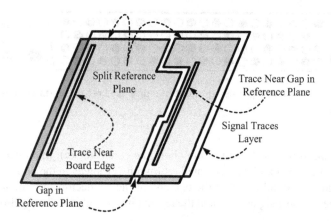

Figure 9.105. Signal traces routed along the edge of the board (left) and along a gap in the split reference plane (right).

return current distribution in the reference plane (see Figure 9.23). Figure 9.107 depicts the effect of fringe electric field radiation (or "edge-fired" emissions) caused by the distortion of the return current distribution in the reference plane resulting from the "truncation" of the normal current distribution at the edge of the board. The consequent increase of current distribution results in distorted E-field flux. In addition, since magnetic flux is also distorted, a similar effect occurs in the magnetic field emissions. Gaps in the ground plane could lead to an increase in radiated EMI and be rather ineffective in most configurations at reducing (and, in certain cases, even increasing) inductive and capacitive crosstalk on the PCB [6].

Figure 9.108 illustrates a trace routed along the edge of a reference plane and Equation (9.59) provides an expression for the increase in inductance associated with this configuration [18]:

$$L \approx \frac{\mu I}{2\pi w} \int_{h'=t}^{h} \left(\arctan\left(\frac{d}{h'}\right) + \arctan\left(\frac{w-d}{h'}\right) \right) dh' \qquad (9.59)$$

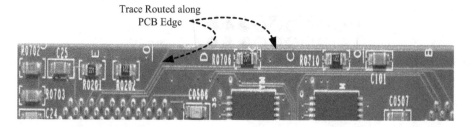

Figure 9.106. Trace routed at near proximity to edge of reference plane. (Courtesy of Dr. Todd Hubing, Clemson University.)

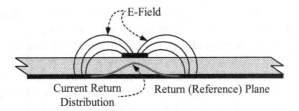

(a) Acceptable: Trace Far from Edge of Reference Plane

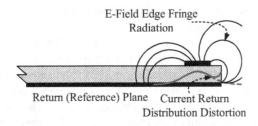

(b) Objectionable: Trace Adjacent to Edge of
Reference Plane

Figure 9.107. Fringe radiation ("edge-fired" emissions) from a trace routed along the edge of a reference plane.

where:
w = Width of the PCB, meters
d = Distance of the signal trace from the edge of PCB, meters
h = Height of the trace above the reference plane

Results of a simulations of the current distribution in the reference plane beneath the trace placed at near proximity to the edge of the PCB are presented for separations of 0

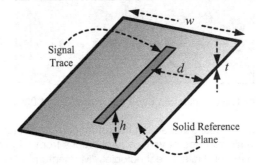

Figure 9.108. Geometry of a trace close to edge of a reference plane.

mil and 30 mil in Figure 9.109(a) and (b), respectively, both at a frequency of 100 MHz. These simulations support the illustrations in Figure 9.107.

As may be observed, currents along edges of PCBs exhibit an unbalanced distribution, producing characteristics similar to those along a wire antenna. When placed near seams in the enclosure, significant radiated emissions may penetrate through the enclosure. Expressions for current distribution along the edge of a return plane for common transmission line configurations are presented in Figure 9.110.

The dimensions shown in Figure 9.110 are the distance from the signal trace to the edge of the return plane, d; the total width of the return plane, w_p; the width of the signal trace, w; and the heights of the trace from the nearest and farthest return planes, h_2 and h_1, respectively (note that in a symmetrical stripline $h_1 = h_2 = h$, whereas in microstrip $h_2 = h$ and h_1 is not used).

In order to evaluate the adverse effects of proximity of a trace to a parallel gap in the ground plane and to better understand the fundamental mechanism at work, simulations were carried out using two-dimensional FEM (finite element method) numerical analysis [6]. The test board geometry used for the simulation is illustrated in Figure 9.111. The test board had a pair of parallel microstrip traces. The properties of the test board were as follows. The width of the traces, w, was 0.114 cm (45 mils); the permeability of the dielectric substrate, ε_r, was 4.3; the height of the trace above the reference plane, h, was 0.114 cm (45 mils); and the coupled length of the traces, l, was 19 cm. Separation between the traces, s, was 0.381 cm (150 mils) and the width of the gap, w_{gap}, was 0.0254 cm (10 mils).

(a) Trace Separation of 0 mil from Edge of Plane

(b) Trace Separation of 30 mil from Edge of Plane

Figure 9.109. Simulation results visualizing the return current distribution in a solid reference plane under a signal trace at various separations from edge of the PCB at $F =$ 100 MHz. (Simulation run on Agilent Technologies "Momentum" 3D Planar EM Simulator; courtesy of Alexander Perez, Agilent Technologies.)

Return (Reference) Plane

Return (Reference) Plane

$$I_{edge} = \frac{I_S}{\pi}\left[\frac{\pi}{2} - \tan^{-1}\left(\frac{w_p - 2\left[\left(w_p/2\right) - d\right]}{2h}\right)\right]$$

$$I_{edge} = \frac{I_S}{2\pi}\left[\frac{\pi}{2} - \tan^{-1}\left(\frac{w_p - 2\left[\left(w_p/2\right) - d\right]}{2h}\right)\right]$$

(a) Microstrip

(b) Symmetrical Stripline

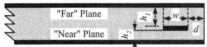

Return (Reference) Plane

$$I_{edge-near} = \frac{I_S}{\pi}\cdot\left(\frac{h_2}{h_1}\right)\cdot\left[\frac{\pi}{2} - \tan^{-1}\left(\frac{w_p - 2\left[\left(w_p/2\right) - d\right]}{2h_2}\right)\right]$$

$$I_{edge-far} = \frac{I_S}{\pi}\cdot\left(\frac{h_1}{h_2}\right)\cdot\left[\frac{\pi}{2} - \tan^{-1}\left(\frac{w_p - 2\left[\left(w_p/2\right) - d\right]}{2h_1}\right)\right]$$

(c) Asymmetrical Stripline

Figure 9.110. Distribution of return current in the return plane with a signal trace routed along edge of a reference plane.

The resulting contour maps of the magnetic flux for the test configuration are shown in Figure 9.112. The highest valued contour line in each of the contour plots is the line that is closest to the source trace, which is the leftmost trace in each figure. Contour lines nearest the source trace were omitted for ease of viewing. Each contour plot uses the same scale [6].

As shown in the contour maps of the flux in Figure 9.112, a partial narrow gap in the ground plane has little or no effect on the magnetic flux. The magnetic flux does

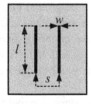

(a) Traces above Solid Plane, No Gap

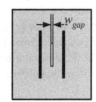

(b) Partial Gap in Ground Plane between Traces

(c) Complete Gap in Ground Plane between Traces

Figure 9.111. Geometry of the simulated models. (Courtesy of Dr. Todd Hubing, Clemson University.)

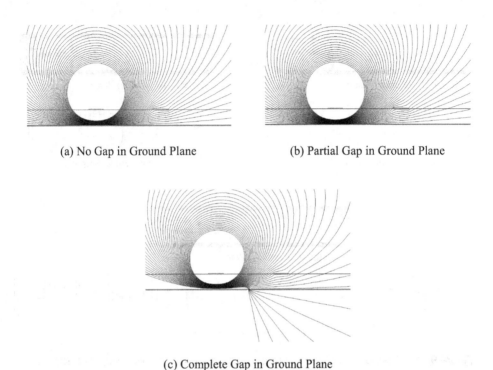

(a) No Gap in Ground Plane   (b) Partial Gap in Ground Plane

(c) Complete Gap in Ground Plane

Figure 9.112. Contour plots of the magnetic flux for a test board, s = 0.381 cm (150 mils). (Courtesy of Dr. Todd Hubing, Clemson University.)

not change because the return current is allowed to spread out across the ground plane just as it was able to do when there was a solid ground plane. With a narrow gap, the return current is not disrupted and the mutual inductance between the two traces is not affected. With a complete narrow gap, the return current is forced to flow only on the side of the ground plane directly beneath the source trace and the magnetic flux near the victim trace is changed. Figure 9.112(c) demonstrates that the complete gap allows some flux to wrap the ground plane of the susceptible circuit, constituting the fringe radiation. The magnetic flux passing between the victim trace and the ground plane in this case is slightly less (~2 dB) than it was with no gap in the ground plane and the magnetic flux coupling the victim circuit decreases. For capacitive coupling, experiments have shown that adding a gap in the ground plane actually increases the mutual capacitance. However, the change is rather insignificant unless the width of the gap is quite large [6].

Reducing edge-fired emissions can be achieved by reducing the spacing, h, between the signal traces and the reference planes, as well as by keeping traces away from the edge of the reference plane or from a parallel gap in it. A recommended separation of high-speed traces from the edge of the PCB is at least 10 times the trace height above the reference plane, that is,

$$d \geq 10 \cdot h \tag{9.60}$$

where d and h relate to the distance from the edge of the reference plane and the height of the trace above the reference plane, respectively [25].

9.4.3   Differential Lines Crossing Gaps in Reference Planes

Up to this point, only single-ended signal traces were considered. A common misconception is that no restriction exists with respect to differential pair traces crossing gaps in the reference plane, in view of the fact that the return current flows on the return trace. That may not always be the case. As a matter of fact, gaps may pose a threat to EMC and signal integrity performance of differential pairs in certain cases almost as severe as in single-ended circuits [4].

9.4.3.1   Differential Return Current Concerns: Return Current "U-Turn." Consider a single microstrip transmission line illustrated in Figure 9.113. As discussed in Section 9.3.2, boundary conditions on the metallic reference plane result in surface current distribution K_S (in A/m) due to the tangential magnetic field $[\vec{H}_{td}(t) = K_S(t)]$ and surface (in C/m²) due to the normal electric field flux density $[\vec{D}_{nd}(t) = \rho_S(t)]$ along the surface of the boundary.

When two transmission lines carry "coincidentally" equal but opposite currents (and guide equal but opposite quasi-TEM electromagnetic waves), the same boundary conditions apply (Figure 9.114) As a result, there are equal but opposite surface current distribution K_S^+ and K_S^- (in A/m) due to the tangential magnetic fields $[\vec{H}_{td}^+(t) = K_S^+(t)]$ and $\vec{H}_{td}^-(t) = K_S^-(t)]$ and surface charge distribution ρ_S^+ and ρ_S^- (in C/m²) due to the normal electric field flux density $[\vec{D}_{nd}^+(t) = \rho_S^+(t)$ and $\vec{D}_{nd}^-(t) = \rho_S^-(t)]$ along the sur-

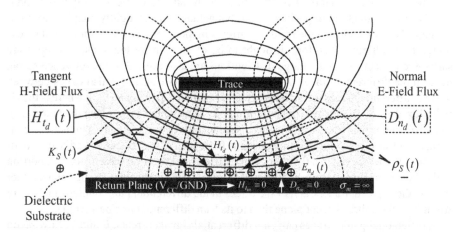

Figure 9.113. Contours of constant H-field (solid lines) and electric field force lines (broken lines) in a microstrip transmission line (8 mil wide trace, 8 mils above plane, 65 Δ).

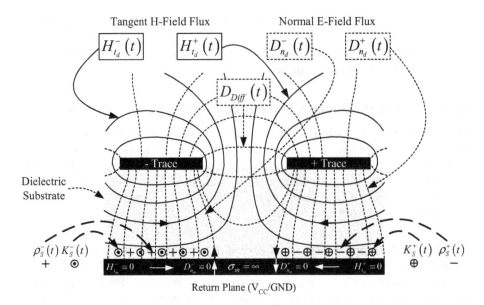

Figure 9.114. Contours of constant H-field (solid lines) and electric field force lines (broken lines) in a differential transmission line (8 mil wide trace, 8 mils above plane, 65/115 Ω).

face of the boundary. The "+" and "−" notations represent the fields and surface charges/current associated with the "+ trace" and the "− trace" of the differential pair. Correspondingly, the "+" and "−" surface currents constitute the return current associated with the signals flowing on the "+" and "−" traces, respectively.

If we are now told that the two transmission line structures are not only "coincidentally" carrying equal but opposite current, but actually form a tightly coupled differential pair, the distribution of the (return) currents in the reference plane should not be expected to be altered. Therefore, it ought to be of no surprise that equal and opposite surface return currents and charge distributions are also present in the reference plane, beneath the differential traces. An E-field (or displacement, D_{Diff}) also exists between the two differential traces (often represented as mutual capacitance, C_M), whereas the H-field flux also couples the two traces (commonly represented as mutual inductance, L_M). The differential pair trace-to-trace electric (capacitive) and magnetic (inductive) coupling results in some induced return currents from each of the traces to its adjacent "buddy." Nevertheless, even in a tightly coupled differential pair of traces, a large proportion of the AC return current from each trace still flows on the adjacent reference plane rather than on the adjacent trace. Each of the traces in the differential pair couples much more strongly to the solid adjacent plane than to its lean differential line partner [4].

A differential pair of traces carrying differential signal currents, I^+ and I^-, flowing on traces 1 and 2, respectively, is illustrated in Figure 9.115. The traces are routed across a gap in the adjacent reference plane. It was shown above that high-speed return currents

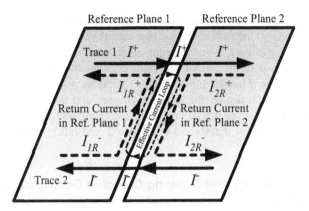

Figure 9.115. A gap in a reference plane separates the return currents from the differential signal traces 1 and 2, resulting in increased current contour impedance and an effective loop antenna.

in differential circuits, follow the path of the respective signal traces, flowing immediately under the signal traces while routed above the solid reference metal plane. However, when the traces cross the gap, the return currents' paths on reference plane 1, I_{1R}^{+} and I_{1R}^{-} and on reference plane 2, I_{2R}^{+} and I_{2R}^{-} are obstructed. Since continuity of the current path must be maintained (Ampere's Law), the return currents are forced to turn away from their "natural" return path, and carry out a "U turn" maneuver as illustrated in Figure 9.115, returning under the second trace of the differential pair [4].

Figure 9.116 demonstrates results of a simulation investigating the current distribution of a differential traces' pair crossing a gap in the reference plane at a fairly low frequency of 100 MHz. The "U-turn" zone is clearly visible.

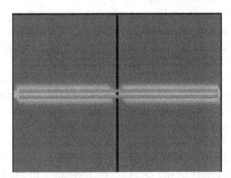

Figure 9.116. Simulation results visualizing the return current distribution in a solid reference plane under a differential pair of signal traces at $F = 100$ MHz. (Simulation run on Agilent Technologies "Momentum" 3D Planar EM Simulator; courtesy of Alexander Perez, Agilent Technologies.)

This separation of the return currents from their respective signal traces results in increased inductance of the differential current contour due to the increased area of the current paths. The gap now forms an equivalent loop antenna, the effectiveness of which as a radiator is determined by the spacing between the differential traces and the width of the gap in the reference plane. With contemporary high-speed digital devices with current transition times on the order of 10 psec and when large effective loop areas are created, the effect of this gap may be devastating both from the signal integrity and EMC points of view, manifested as increased crosstalk, emission, and susceptibility.

9.4.3.2 Differential Lines Crossing Defected Ground Structure (DGS): Putting Slots to Work.

Differential signals play an important role in high-speed digital circuits thanks to their high immunity to noise, low crosstalk, and low electromagnetic interference (EMI). Several high-speed serial link formats, such as PCI Express II or Gigabit Ethernet, have differential signal transmission data rates exceeding 5 Gbps.

In practical circuits, common-mode noise due to the timing skew or amplitude imbalance along differential signal transmission paths is unavoidable (Figure 9.117). In a practical layout scenario, the interconnect topology is unbalanced; for instance, due to bends or the presence of adjacent traces. Consequently, common-mode noise is produced and is superimposed on the (ideally balanced) differential interconnects. Microwave frequency common-mode noise may degrade signal or power integrity of the high-speed circuit and result in excessive emissions from input/output (I/O) cables connected to the differential signals. Suppression of common-mode noise on differential transmission lines, however, normally achieved by utilizing common-mode chokes using high-permeability ferrite cores, may be impractical at microwave frequencies.

A way out of this tricky situation is surprisingly introduced through intentional periodical etching of the return plane beneath the differential pair. This etching perturbs the common-mode return current, providing suppression of common-mode noise superimposed on high-speed differential signal pairs.*

Figure 9.118 illustrates an application of this technique, using etching of dumbbell-shaped structure in the ground plane beneath the differential pair (known as defected ground structure, or DGS) [25, 26]. By carefully selecting a suitable geometry design of the dumbbell-shaped structure, the structure can be designed to exhibit broadband and efficient suppression of the common-mode noise while maintaining a high differential signal quality. This technique could thus be most beneficial when applied to high-speed differential I/O ports in which long cables are generally attached. Radiated EMI emissions produced by common-mode currents present on the cables can thus be significantly reduced. In order to avoid excitation of common-mode noise, the dumbbell-shaped DGS is maintained balanced with respect to the centerline of the two signal traces.

As an example, the effect of the DGS on performance of the line was evaluated by considering a differential transmission line routed above a dielectric RF-4 substrate

---

*All material in this Section is derived from [25],[26]. Images provided courtesy of Tzong-Lin Wu, National Taiwan University, Taipei, Taiwan.

(a) Multi-Gb/s Digital Circuit Driving a Serial Differential High-Speed Data Link

(b) Circuit Layout

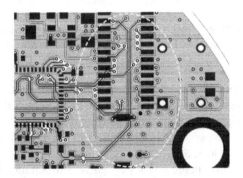

(c) Zoom on Circuit showing Bends in Differential

Lines

Figure 9.117. Unbalanced interconnect topology due to bends or presence of adjacent traces results in common-mode interference. (Images courtesy of Tzong-Lin Wu, National Taiwan University, Taipei, Taiwan.)

with relative permittivity, ε_r, of 4.4. The geometric parameters of the circuit were: $h = 1.6$ mm, $a = 40$ mm, $b = 60$ mm, $W = 2.483$ mm, $S = 1.763$ mm, and $W_d = 7$ mm (see Figure 9.118 for definition of geometrical parameters).

Figure 9.119 demonstrates the effect of dimensions of the DGS, S_d, and, K_d (see Figure 9.119(d) for definition of geometrical parameters) on common-mode and differential-mode insertion loss S_{cc12} and S_{dd12}, respectively, along the differential transmission line. Since the differential signal propagates in odd mode, relatively low return current flows through the ground plane; thus, degradation of the differential signals

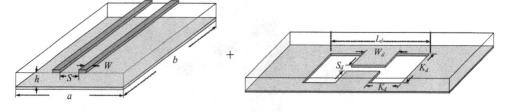

(a) Common Solid Differential Microstrip Transmission Line

(b) Dumbbell-Shape Defected Ground Structure (DGS)

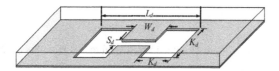

(c) Fundamental Section of the Dumbbell-Shape Defected Ground Structure (DGS)

Figure 9.118. Common-mode filter (CMF) using defected ground structure. (Images courtesy of Tzong-Lin Wu, National Taiwan University, Taipei, Taiwan.)

caused by the DGS will be relatively small. Common-mode (or even-mode) return current, on the other hand, must flow through the ground plane and the DGS may have significant impact on this type of signal.

Observation of Figure 9.119(b) and (c) clearly reveals that the common-mode insertion loss, S_{cc12}, exhibits a resonant behavior. Common-mode noise can thus be significantly suppressed at frequencies surrounding the resonance frequency, f_0:

$$f_0 = \frac{1}{2\pi\sqrt{L_{DGS} \cdot C_{DGS}}} \tag{9.61}$$

L_{DGS} and C_{DGS} represent the capacitive and inductive elements associated with the dumbbell-shape DGS, corresponding to the capacitance between two sides of the center slot and the equivalent inductance of the return signal flowing through the DGS, respectively.

The resonant frequency observed in the insertion loss, S_{cc12}, in Figure 9.119(b) suggests a positive frequency dependence, such that an increase of the center slot width, S_d, results in a corresponding increase in the resonance frequency, f_0. The center slot width exhibits, therefore, a capacitive behavior. Figure 9.119(c), on the other hand, implies negative frequency dependence, such that an increase of the dumbbell square slot width, K_d, results in a corresponding decrease in the resonance frequency, f_0. The dumbbell square slot width exhibits, therefore, an inductive behavior.

As shown in Figure 9.120, the DGS unit cell can be modeled as an ideal transmission line with even-mode (common-mode) characteristic impedance, Z_{even}, and a res-

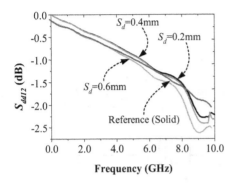

(a) Differential-Mode Insertion Loss (S_{dd12} for Varying S_d, (K_d fixed at 4 mm)

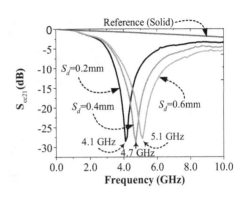

(b) Common-Mode Insertion Loss (S_{cc12} for Varying S_d (K_d fixed at 4 mm)

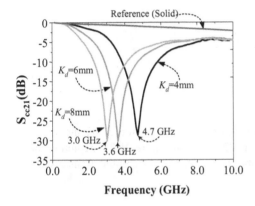

(c) Common-Mode Insertion Loss (S_{cc12} for Varying K_d, (S_d fixed at 0.4 mm)

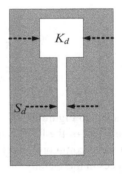

(d) Definition of DGS Parameters K_d and S_d

Figure 9.119. Effect of varying DGS dimensions on even-mode and odd-mode propagation. (Images courtesy of Tzong-Lin Wu, National Taiwan University, Taipei, Taiwan.)

onator cascaded on the ground plane. It is evident that this scheme corresponds to a band-reject common-mode filter.

Figure 9.121 presents the simulated* and measured insertion loss (S_{cc12}) for the differential transmission line. The equivalent inductance, $L_{DGS} = 2.5$ nH, was extracted from TDR (time-domain reflectometer) measurements while the slot capacitance, $C_{DGS} = 0.47$ pF was approximated by [25, 26]:

*Simulations were carried out using Ansoft HFSS, 3D full-wave electromagnetic field simulation software.

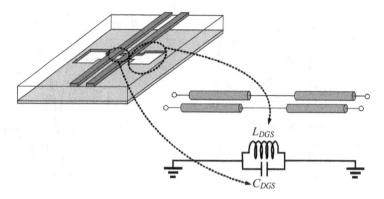

<u>Figure 9.120.</u> Correspondence between DGS physical structure and resonator's equivalent circuit. (Images courtesy of Tzong-Lin Wu, National Taiwan University, Taipei, Taiwan.)

$$C_{DGS} = \frac{\varepsilon_0 \varepsilon_r}{\pi} \cdot \ln\left[\coth\left(\frac{\pi}{4} \cdot \frac{S}{h}\right)\right] + \frac{\varepsilon_0}{\pi} \cdot \ln\left[2\frac{1+\sqrt{k'}}{1-\sqrt{k'}}\right]$$

$$K' = \sqrt{1-K^2}; \ K = \frac{S}{S+2W}$$

$$(9.62)$$

Refer again to Figure 9.118 for definition of the geometry parameters S, W, and h.

The effect of the DGS can be observed qualitatively in Figure 9.122, depicting the even-mode (common-mode) current distribution under the differential transmission line traces in the vicinity of the DGS.

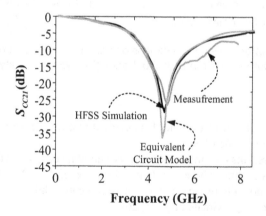

<u>Figure 9.121.</u> Even-mode (common-mode) insertion loss (S_{cc12}) for the DGS-structure. $S_d = 0.4$ mm ($C_{DGS} \approx 0.47$ pF) and $K_d = 4$ mm ($L_{DGS} \approx 2.5$ nH).

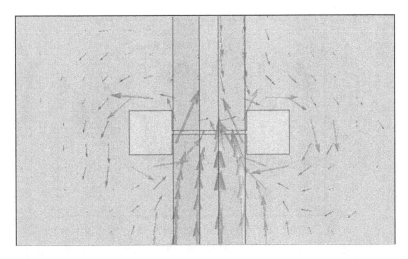

Figure 9.122. Even-mode current distribution on the DGS. (Image courtesy of Tzong-Lin Wu, National Taiwan University, Taipei, Taiwan.)

By periodically etching the dumbbell-shaped structure on the ground plane underneath the differential transmission line, increased bandwidth of the structure stop band can be achieved. Figure 9.123 illustrates a cascaded three unit-cell DGS structure with period $P_d = 10$ mm and the geometrical parameters for each DGS are $W_d = 7$ mm, $S_d = 0.4$ mm, and $K_d = 4$ mm. Over 20 dB of common-mode suppression ($S_{cc12} > 20$ dB) is achieved for this configuration over a wide frequency range between 3.3 GHz and 5.7 GHz (Figure 9.124).

At first glance, this technique seems to be in substantial conflict with the conclusions of the previous section. It is to be noted, however, that the suppression provided by the CMF-DGS technique has substantial effects only on conducted common-mode EMI currents, which are present on both traces of the transmission line (ultimately re-

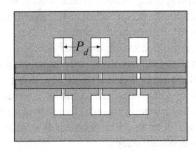

Figure 9.123. Implementation of a common-mode filter using periodic defected ground structure (CMF-PDGS). (Image courtesy of Tzong-Lin Wu, National Taiwan University, Taipei, Taiwan.)

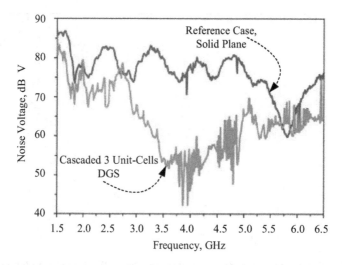

Figure 9.124. Effect of periodic defected ground structure (PDGS) on near-field radiated emissions. (Image courtesy of Tzong-Lin Wu, National Taiwan University, Taipei, Taiwan.)

turning through the ground plane, beneath the traces). The common-mode noise superimposed on the traces is rejected and does not flow into the cables, precluding resultant excessive radiated emissions.

On the other hand, a slot in the ground plane can result in increased emissions due to the "U-turn" effect across the slot. This effect was shown to be strictly associated with differential-mode currents, due to the increased loop area between the traces and across the slot. With cables attached to the circuit, radiated emissions due to common-mode interference dominate, but, with no cables attached, this structure may indeed result in somewhat increased emissions from due to the "differential U-turn" effect across the slot. This effect, however, can be minimized through careful layout and routing of the transmission line and shrinking of the loop area in particular (i.e., by reducing width of the slot and spacing between traces). Radiated EMI emissions and signal integrity compromises due to perturbation of the intentional differential-mode signal may, consequently, be of little or no concern.

Furthermore, if properly implemented, the CMF-DGS technique has only minor effects on signal integrity and the differential-mode signals maintain excellent signal integrity. If the differential transmission line is designed with weak coupling between the traces, that is, $Z_{even} \cong Z_{odd}$, the slot effect could be considerable and the integrity of the differential signal could be compromised. However, with tighter coupling between the traces, the effect of the slot on the SI will be much smaller since much of the return current could couple between both traces.

This can easily be validated through observation of the "eye diagram." The measured differential-mode signal eye pattern is presented in Figure 9.125 with 1 V (com-

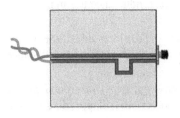
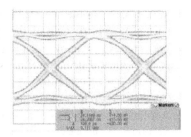

(a) Reference Case, Solid Plane, Maximum Eye Opening of 830 mV and Width of 109.8ps

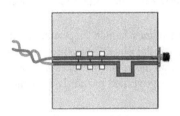
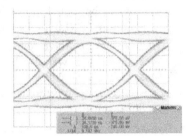

(b) 3 Cell-Unit CMF-PDGS Case, Maximum Eye Opening of 746 mV and Width of 108.9ps

Figure 9.125. Measured differential eye pattern for (a) reference case and (b) cascaded 3-unit-cell CMF-DGS. (Image courtesy of Tzong-Lin Wu, National Taiwan University, Taipei, Taiwan.)

plementary ±0.5 V) signal amplitude at a data rate of 8 Gbps (note the imbalance introduced by the unequal length of traces resulting in differential-mode to common-mode interference conversion). The maximum eye opening obtained in the reference case (solid plane) was 830 mV, whereas in the CMF-PDGS case, the eye opening reduced by a mere 10% approximately, to 746 mV. An even smaller difference was observed in the eye width, 109.8 ps versus 108.9 ps, respectively. This is primarily due to the small length of the slot [$L_d = W_d + 2K_d$, see Figure 9.118(b)], less than 15 mm. Of course, with a much greater length, L_d, the impact would be greater, resulting in the conclusions presented in Section 9.4.3.1.

Lessons Learned

- Planes are necessary but insufficient for control of EMI and ensuring signal integrity.
- Current at high frequencies is always inclined to assume a distribution that minimizes the net inductance of the contour formed by the signal and return path.

- In high-speed printed circuit boards, the path of least inductance lies in the reference plane directly beneath the signal trace.
- The return signal current also spreads out in the plane to either side of the trace's route, with a distribution inversely proportional to the square of the distance from the trace's centerline.
- Uncontrolled return paths due to discontinuities (e.g., gaps) in ground planes constitute a major source of ground bounce and power rail collapse (discussed in Section 9.5), increased crosstalk, and EMI.
- Traversing reference planes can exhibit similar effects, particularly when dissimilar (e.g., V_{CC} and GND).
- Signal traces traveling across an edge connector from motherboard to daughter board should be routed in proximity to the same reference plane [e.g., power (V_{CC}) or return (GND)] in both boards.
- Switching currents traversing power and return planes launch electromagnetic waves across the power distribution network's parallel plates' structure ["parallel-plate waveguide (PPW) noise"]; PPW noise is sustained in standing wave patterns occurring at cavity resonance frequencies governed primarily by structural dimensions.
- Bypass capacitors can be used to circumvent the gap and reference planes traversed by signal traces, providing an alternate path for the return current. However, this solution is frequency limited!
- Low ESL and the capacitor mounting inductance dominate the high-frequency impedance of the capacitors and are far more important than their actual value of capacitance.
- Interplane capacitance provides an alternate path for displacement return current at very high frequencies; at lower frequencies, discrete bypass capacitors are more effective.
- Gaps in reference planes may also be harmful to high-speed differential transmission lines if not carefully implemented.
- Dumbbell-shape defected ground structures (DGS) can be used to provide high common-mode (even-mode) EMI suppression on GHz differential I/O transmission lines while maintaining differential-mode (odd-mode) signal integrity; periodic DGS (PDGS) exhibits enhanced common-mode suppression.
- Perforation in reference planes effectively creates gaps in the planes if size of antipads and their mutual separation are not maintained. "Blind vias" are recommended and, when not necessary, pads of signal traces should not be placed in reference planes.
- Signal traces routed along the edge reference planes or along and in near proximity to gaps in the reference planes may create fringe radiation (an EMI concern) and could increase crosstalk (EMI and signal integrity concerns). High-speed traces should be routed at least 10 trace heights away from the edge of the board or from the gap.

9.5 DELTA-*I* (Δ*I*) AND SIMULTANEOUS SWITCHING NOISE (SSN) IN PCBs

This section provides a detailed description of the generation and control of delta-*I* (Δ*I*) noise and the subsequent phenomenon, commonly known as "ground bounce." This phenomenon, however, is shown to occur in the power and return rails alike; hence, it is often called "rail bounce" to account for both.

Switching of digital devices, especially when driving electrically long transmission lines or heavy capacitive loads, is accompanied by an impulsive variation of current, Δ*I*, having approximately a triangular waveform, flowing in power and return connections in a (short) time, Δ*t*. The fast transition time of the switching current results in high-frequency interference throughout the power distribution network (PDN) of the PCB. Due to the nonideal nature of the PDN, the switching current, Δ*I*, propagating through the nonzero impedance of the PDN's power-return planes structure introduces an interference voltage drop, Δ*V*, across the inductance-dominated impedance or across the characteristic impedance of the PDN,* symbolized as L_{PDN} and $Z_{0(PDN)}$, respectively, and also known as "delta-*I* noise" or "Δ*I* noise." The Δ*I* noise voltage may be expressed as

$$\Delta V \approx Z_{PDN} \cdot \Delta I \qquad (9.63)$$

At higher frequencies, inductive effects dominate the PDN impedance (i.e., $Z_{PDN} \approx j\omega L_{PDN}$), thereby Equation (9.63) becomes

$$\Delta V \approx L_{PDN} \cdot \frac{\Delta I}{\Delta t} \qquad (9.64)$$

where Δ*t* is the transition (rise/fall) time of the switching current waveform.

Since this section is focused on I/O current, a typical CMOS I/O driver "totem pole" output driver configuration is shown for the following discussion (Figure 9.126). When the output of the IC switches from "high" to "low" state, the upper P-MOS transistor exhibits high impedance to the power (V_{CC}) rail, whereas while the lower N-MOS transistor exhibits low impedance to the return (GND) rail. Conversely, when the output of the device switches from "low" to "high" state, the upper P-MOS transistor exhibits low impedance to the power (V_{CC}) rail and the lower N-MOS transistor exhibits high impedance to the GND rail.

Δ*I* noise current is produced by the three primary mechanisms depicted in Figure 9.126 [27]:

- Discharge of the loads' input stray capacitances observed on the driving gate's outputs

---

*Bear in mind that the power distribution network comprises the loop consisting of the power and return conductors.

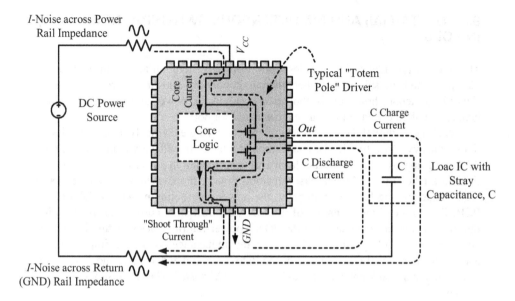

I-Noise across Power
Rail Impedance

DC Power
Source

Core
Current

Core
Logic

Typical "Totem
Pole" Driver

V_{CC}

Out

C Charge
Current

C Discharge
Current

C

Loac IC with
Stray
Capacitance, C

"Shoot Through"
Current

GND

I-Noise across Return
(GND) Rail Impedance

Figure 9.126. Fundamental "ΔI noise" generation mechanisms. (Courtesy of Keith Armstrong, Cherry Clough.)

- "Shoot-through" currents in "totem-pole" drivers
- Core (processing) noise

"Stray capacitance" is unavoidably associated with all signal loads (e.g., an input to a CMOS IC or an operational amplifier) due to device terminals, lead frames, bond wires, and the capacitance resulting from the internal device features. In addition, stray capacitance of the traces and pads of the PCB itself should be considered. Typically, the capacitances involved are quite small, often in the order of a few tens of picofarads in total. As transition time of digital signals continues to decrease the transient current created by alternately charging and discharging the load capacitances increases, resulting in interference voltage across the power (V_{CC}) and return (GND) rails. This mechanism is discussed in further detail in Section 9.5.1.

"Shoot-through" currents are created by "totem-pole" type driver circuits at the instant of their switching. For a very brief time as the two transistors switch states, both the upper and lower transistors in the totem pole are "on" together and exhibit relatively low source–drain (or emitter–collector) impedance, which also appears between the power and return rails. This low impedance momentarily enables current to flow directly from the device's power supply lead to the power return. The rate of rise of the consequential "shoot-through" current is limited only by the impedance (resistance and inductance) in the current path. This phenomenon occurs during both the logic "low" to "high" and "high" to "low" state transitions. The effects of this mechanism are discussed in further detail in Section 9.5.2.

"Core noise" is a feature of VLSIs such as large FPGAs, which have a core of semi-conductors operating at very high speeds. These internal circuits drive internal capacitive loads and may experience "shoot-through" currents.* The core resultant transient current adds to the VLSI's overall current fluctuations. The frequencies associated with the core noise currents are usually much higher than those of the other two mechanisms. For example, the core of some FPGAs can have power current demands with transition times in the order of 50 ps, equivalent to a noise frequency spectrum extending up to 13 GHz, at least.

In summary, there are two primary categories of current drawn from the power distribution system during any switching cycle: a load current demand, which occurs once during each clock cycle; and the "shoot-through" current, which may occur twice during each clock cycle. These currents produce Δ*I* noise on power rails ("power bounce/collapse") and power return rails ("ground bounce"). Furthermore, the Δ*I noise* interference transients may also introduce interference to signal output leads and circuits sharing the same current return path. Since either the power (V_{CC}) or the return (GND) rails may serve as RF signal current return paths, this phenomenon can occur in both rails. A secondary effect of the transient switching current is increased radiated EMI emissions from the printed circuit board (predominantly magnetic field) and, subsequently, from cables connected to it.

Integrated circuit manufacturers have traditionally specified simultaneous switching output (SSO) guidelines for each driver type, based on the number of drivers allowed per power/ground pair. In low-voltage TTL (LVTTL) drivers, ground bounce can be as high as 800 mV before tripping the input-low threshold. Such a level of ground bounce noise might no longer be adequate. Although lead inductance in integrated circuit (IC) packages has been whittled down to less than 100 pH per return pin, PCB-level inductance has more than doubled in recent years to a range of 2 nH to 3 nH due to miniaturized vias, smaller trace widths, and increased board thickness (due to increased stack-ups). Consequently, substantially higher ground bounce noise voltage now develops across the PCB. Actual ground bounce voltage on PCBs using 8 mil vias on 125 mil-thick boards including an FR4 embedded dielectric material could by far exceed 500 mV.

For ease of PCB routing and debugging, it is common practice to cluster wide buses at one edge of very large scale (VLSI) devices. This results in nonuniform current distribution and consequent ground bounce within the device's package. As supply voltage decreases, components tolerate even less noise. Some new devices tolerate as little as a 300 mV undershoot at their input. Also, as load capacitance increases (due to increased fan-out and the intrinsic input capacitance of digital devices), ground bounce further worsens. All in all, these design issues now necessitate more critical treatment of ground bounce than ever before.

9.5.1   Δ*I* Noise Generation in Signal I/O Circuits

I/O circuits constitute a significant source of Δ*I* noise. Some detailed insight into the mechanism by which Δ*I* noise is produced can be gained by observing Figure 9.127

---

*Since "core noise" is a special case of the above other two mechanisms, it will not be further discussed separately.

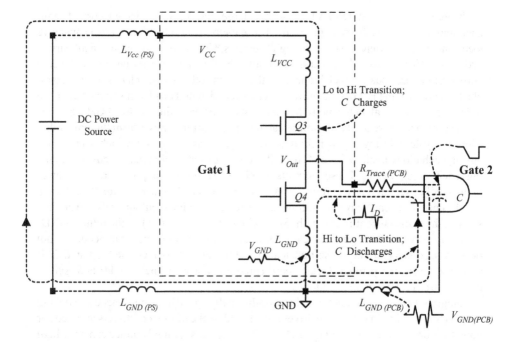

Figure 9.127. Detailed model of ΔI noise generation mechanism within I/O signal circuits in a logic gate.

and Figure 9.128, depicting a simplified CMOS driver totem pole I/O output configuration and its schematic in its two switching states, respectively.

When the output of Gate 1, initially at the "high" state (logic "1") switches to the "low" state (logic "0"), as depicted in Figure 9.128(a) (i.e., the upper P-MOS transistor $Q3$ turns off and exhibits high impedance transistor and the lower N-MOS transistor $Q4$ turns on and exhibits low impedance in Gate 1 in Figure 9.127), the gate's output is strapped to GND plus the source-drain (V_{SD}) voltage drop across $Q4$. Transistor $Q4$, now in saturation, exhibits a low resistance path for discharge current transient, $I_D(t)$, to flow from the load capacitance* to ground. The discharge current transient flowing out of the capacitance, C, holding a voltage, V_C, during a transition time, t_t, is related to the time derivative of the momentary voltage across the capacitance, $V_C(t)$:

$$I_C(t) = C \frac{dV_C(t)}{dt_t} \tag{9.65}$$

But some inductance, L_{GND}, is introduced by the very small GND lead within the device's package (see Figure 9.127). Due to the transient current flowing through that lead, a transient voltage drop develops across this inductance [Equation (9.66)]:

---

*Digital devices commonly exhibit some input capacitance.

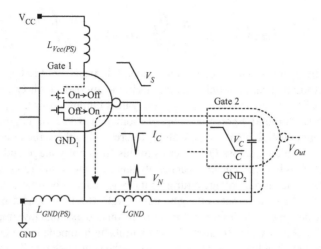

(a) Gate Switching from "Hi" ("1") to "Lo" ("0"),

Producing "Ground Bounce"

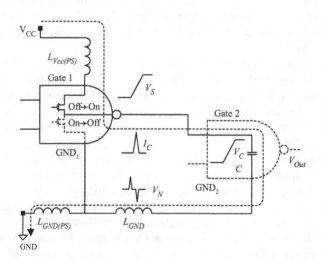

(b) Gate Switching from "Lo" ("0") to "Hi" ("1"),

Producing "Power Bounce"

Figure 9.128. Simplified model of power/ground bounce generation mechanism within a logic gate.

$$V_{GND}\left(t\right) = L_{GND} \cdot \frac{dI_C\left(t\right)}{dt_t} = \left(L_{GND}\,C\right) \cdot \frac{d^2 V_C\left(t\right)}{dt_t^2}\qquad(9.66)$$

This voltage transient is associated with the second derivative of the load capacitance voltage waveform and could initiate oscillations in the low-loss, high-Q circuit. Consequent to this voltage transient across L_{GND}, the gate's output voltage, V_{Out}, "bounces" above the GND ("ground") potential. This phenomenon is known as ground bounce. Any downstream receiving gate (e.g., Gate 2) referenced to the same circuit GND rail will observe the voltage across Q4 in series to this transient voltage and may not respond correctly. When the current reaches its steady state value (and subsequently dI_D/dt_t vanishes) the output voltage returns to its expected steady-state value.

Conversely, when the output of the gate goes "high" (logic "1"), as depicted in Figure 9.128(b) (the upper P-MOS transistor Q3 turns on and exhibits low impedance and the lower N-MOS transistor Q4 turns off and exhibits high impedance in Gate 1 in Figure 9.127), the gate's output is strapped to V_{CC} less the source–drain (V_{SD}) voltage drop across Q3. Current now flows from the V_{CC} rail through Q3 toward the output of Gate 1 to downstream loads (represented by Gate 2), charging its input capacitance, C, to the output voltage of Gate 1. The current that earlier flowed through Q4 and the GND lead inductance is terminated, resulting in a negative dI_D/dt_t voltage drop across this inductance. In addition, a voltage drop between the collector of Q3 and V_{CC} due to the V_{CC} lead inductance, L_{Vcc} results in similar effects. V_{Out} is limited by the voltage drop across the lead inductance. In this case, both power collapse/bounce and ground bounce are produced.

In practice, integrated circuits are intrinsically designed with some level of immunity to this effect thanks to the noise margin of the devices, thus the ground bounce internal to the device is of little functional concern. The inductance associated with the power (V_{CC}) and return (GND) rails on the printed circuit board, on the other hand, yields circuit-level power and ground bounce at much higher magnitudes than those observed within the device.

Now, when Q4 turns on (and Q3 turns off), the output of the gate drops and the load's discharge transient current flows from the load capacitance through both the inductance of the bond wire in the chip, L_{GND}, and the return (ground) rail inductance on the PCB, $L_{GND(PCB)}$. The resulting total gate output voltage, L_{Out}, constitutes a sum of the nominal output voltage of the chip itself and the transient voltage drop across these inductances. This "bounce" can be detrimental to circuit performance. A similar effect occurs when transistor Q3 turns on. Current flows from the power supply through the inductance of both the package bond wires and the power and return rails on the PCB to the transistor, charging the load capacitance.

But power and ground bounce can have even wider-ranging consequence. Consider the circuit in Figure 9.129. Gates 3 and 4 were added to Gates 1 and 2 and share a common return path. The input of Gate 4 is assumed to be driven "high" by Gate 3. At the same time, the output of Gate 1 drops to "low," resulting in the discharge of the capacitance in Gate 2 through Gate 1 and the return inductance, L_{GND}, producing a voltage bounce on the return path inductance, L_{GND}.

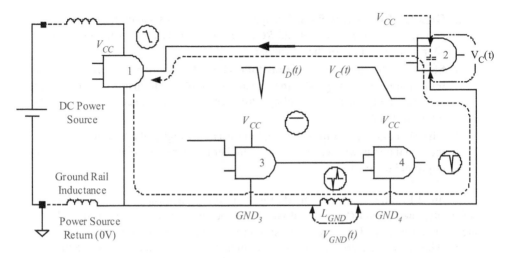

Figure 9.129. Illustration of "Δ*I noise*" interference propagation in a circuit.

Many engineers are lead to believe that logic gates possess the capability to sense the absolute level of incident signals regardless of circuit configuration and layout. In practice, logic gates respond only to the difference between the voltage on the input leads and whatever voltage appears to be present on its designated reference (e.g., GND). Any interference present at the reference leads of the device may consequentially result in erroneous response of the device to the signal at its input port [9].

Due to the ground bounce voltage transient, $V_{GND}(t)$, between points GND_3 (the "GND" reference of Gate 3) and GND_4 (the "GND" reference of Gate 4), Gate 4 may thus misinterpret the transient ground potential and responds to the momentary difference between the voltage on the input signal voltage and the ground voltage, resulting in a corresponding transient on its output line. This voltage bounce may propagate for some distance on the PCB [note that Equation (9.66) represents the equation of a wave propagating in a lossless medium] and with the ever-increasing density of the PCB may affect a large number of adjacent devices.

Power and ground bounce thus constitutes a form of common-impedance coupling* or crosstalk and is a special case of the more general phenomenon known as simultaneous switching noise (SSN) or simultaneous switching outputs (SSO) noise, attributed to the fact that it often occurs while multiple devices switch simultaneously. Clearly, ground bounce in signal circuits is exacerbated by a combination of several causes:

1. **Load Capacitance.** The larger the number of parallel loads (or "fan out"), the higher the capacitive loading of the switching gate and the larger the transient charge and discharge current.

---

*Common impedance coupling was extensively discussed in Chapter 4.

2. **High-Q of Discharge Path.** The gate exhibits low resistance for the discharge current, forming a high-Q (small RC time constant) discharge path, resulting in a short transition time of the discharge current, t_t.

3. **Short Discharge Current Transition Time.** The shorter the current impulse transition time, t_t, the greater the transition rate of the discharge current, dI_D/dt, resulting in higher transient voltage drop across the inductance of the power and return conductors.

4. **Circuit Total Inductance.** Last but not least, the higher the total inductance of the signal and power/return loops, the higher the transient voltage drop (or power/ground bounce).

Figure 9.130 illustrates an example for quantifying the ground bounce* in a large-scale integrated circuit (LSI). I/O driver gates located in two different LSI packages are interconnected and share a common reference and current return path in the GND plane. The devices are powered from a common $V_{CC} = 5$ V voltage. Assume that $n = 32$ gates switch simultaneously in one of the packages, at a (moderate) transition time of $\Delta t_t = 5$ nsec. If the gates are initially at the "high" state, the load capacitance would be charged to 5 V. When switching to the "low" state the gates will hence experience a transition voltage of $\Delta V_C = 5$ V. The maximum switching current can be estimated from the voltage change across the load capacitance [9]:

$$\max\left(\frac{dI_{Sw}}{dt}\right) \approx \frac{1.52 \cdot \Delta V_C}{\Delta t_t^2} \cdot C \tag{9.67}$$

where:
ΔV_C = voltage transition across the total load (fan out) capacitance, volts
I_{Sw} = fanout capacitance switching discharge current, amperes
Δt_t = transition time (10%–90%) of the current waveform, nsec

The resultant maximum discharge current from a load capacitance of $C = 50$ pF is obtained by substitution in Equation (9.67), resulting in

$$\max\left(\frac{dI_{Sw}}{dt}\right) \approx \frac{1.52 \cdot \Delta V_C}{\Delta t_t^2} \cdot C = \frac{1.52 \cdot 5V}{(5 \text{ nsec})^2} \cdot 50\,\text{pF} = 15.2 \times 10^6\,\text{A/sec} \tag{9.68}$$

Although the two devices are assumed to be referenced to the same "absolute" reference plane, in practice the extra inductance introduced by the connections (i.e., pins and bond wires) between the gates to the common reference plane causes a voltage drop between the "virtual grounds" $G_{\#1}$ and $G_{\#2}$, the reference point of each of the gates. Assuming that the partial inductance of the LSI's leads is $L_L = 2$ nH, the reference potentials are no longer equal and the potential difference (ground bounce), ΔV_{GB}, between these two points can be roughly approximated as

---

*This discussion is focused on the return (GND) leads of the devices. Similar phenomena occur in the power (V_{CC}) rail, hence, similar arguments apply to the "V_{CC}" leads as well.

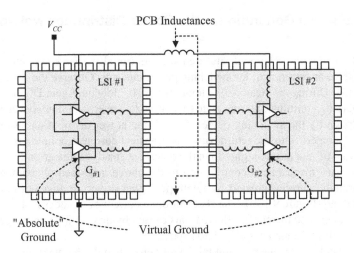

<u>Figure 9.130.</u> Quantitative illustration of "ground bounce" in a large-scale integrated circuit.

$$\Delta V_{GB} = n \cdot L_L \cdot \max\left(\frac{dI_{Sw}}{dt_t}\right) = 32 \cdot 2 \times 10^{-9} \cdot 15.2 \times 10^{6} \approx 0.97 \text{ V} \qquad (9.69)$$

where:
ΔV_{GB} = the ground bounce noise voltage, volts
I_{Sw} = the signal switching current [derived from Equation (9.68)], amperes
L_L = the LSI pin partial inductance, henry
n = the number of I/O signal lines switching simultaneously
$\Delta t_{t(min)}$ = the transition time of the current waveform, nsec

Obviously, such a high level of noise voltage exceeds the noise margin of the device and will not permit acceptable performance of the circuit.

Lessons Learned

- Δ*I* noise is produced during gate switching by two mechanisms of transient current demands from the power bus: load current demand, occurring once per switching cycle; and shoot-through current, occurring twice per cycle.
- Δ*I* noise in I/O signal circuits on PCBs arises whenever transient current flows through inductance of the signal and return paths on a printed circuit board.
- Performance of devices can be severely compromised if adequate design rules are not appropriately applied to mitigate the Δ*I* noise effects.
- Δ*I* noise also constitutes a significant source of crosstalk between signal paths sharing a common return path as well as of common-mode currents driving external interfacing cables, resulting in increased radiated emissions.

9.5.2 Δ*I* Noise Generation in the Power Distribution Network (PDN)

Transient voltage on power and return conductors can also arise since return conductors also serve as return path for switching power currents. Observe the NAND gate in Figure 9.131. During the steady-state logic "1" or "0," low quiescent DC current flows into the gate V_{CC} terminal. During state transition, however, an impulsive current transient is drawn by the switching device through the power distribution system on the PCB and "pumped" into the return path. The transient current amplitude is often more than an order of magnitude higher than the quiescent steady-state current.

The primary mechanisms contributing to the above described phenomenon is the "shoot-through" current, created by totem-pole type driver circuits at the instant of their switching (Figure 9.132). The shoot-through current of a modern 74HC glue logic device can exceed the current fluctuations caused by driving its output signals by as much as 15 dB at frequencies above 30 MHz.

Unfortunately, data sheets of digital devices provide only the maximum (or at best, the typical) transition time of the device (in terms of voltage swing) but not the minimum current transition time, which significantly influences the spectral content of the waveform. Furthermore, owing to "die shrinking," devices, which two decades ago emitted signals acquiring an effective spectrum of 200 MHz to 300 MHz, now produce emissions at frequencies reaching or even exceeding 1 GHz. As this trend continues, broader spectrum will be occupied by future devices. Figure 9.133 depicts the spectral content of the power bus spectrum of a clock driver IDT74FCT807. Note that the sep-

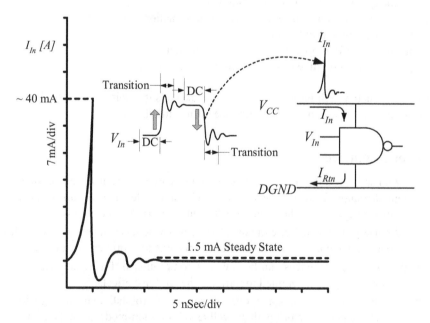

Figure 9.131. Typical time pattern of switching current of a 74LSXXX NAND gate.

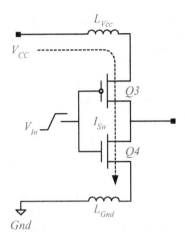

<u>Figure 9.132.</u> "Shoot-through" currents in a "totem-pole" type gate.

aration between consecutive harmonics that corresponds to the fundamental switching frequency (or pulse repetition frequency, PRF) of the clock is 100 MHz and, particularly, that harmonics at significant levels by far exceed 1.8 GHz!

As digital IC technology is scaled to form smaller and faster transistors, the power supply voltage decreases. Conversely, as clock rates rise and more functions are inte-

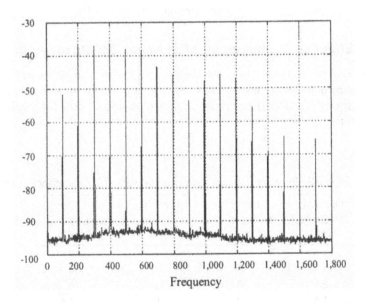

<u>Figure 9.133.</u> Power bus spectrum (dBm) of a clock driver IDT74FCT807. (Courtesy of Dr. Bruce Archambeault.)

grated into VLSIs, such as ASICs (application-specific integrated circuits) the power consumed (and, therefore, current drawn) increases. These trends are summarized for the 1990s in Table 9.2 [28].

Assuming that the amplitude of ripple voltage (noise) is limited to a small fraction of the power supply voltage (i.e., 5%), target impedance for the PDN can be derived [Equation (9.70)]. The target impedance, Z_{TGT}, is falling at an alarming rate ×5 per technology generation:

$$Z_{TGT}\,(\sim 2002) = \frac{V_{CC}\,[\text{V}] \cdot \text{ripple}\,[\%]}{I_{CC}\,[\text{A}]} = \frac{1.2\,\text{V} \cdot 5\%}{0.15\,[\text{A}]} = 0.4\,\text{m}\Omega \qquad (9.70)$$

Figure 9.134 illustrates a situation where Gate A is switched, driven by a clock source. When Gate A is switched to "high" at its output, the input capacitance, C, of the load is charged by the output transient current, I_T. This current must flow through the inductance associated with the power supply wiring, L_{PSW}, producing noise voltage across it. Concurrently, the current returns through the inductance associated with the return wiring, L_{GNDW}, resulting in a similar noise voltage drop across the return conductor. This noise too appears as ΔI noise or "ground/power bounce" (also known as "rail bounce" or "rail collapse").

The transient noise voltage drop across practical (highly inductive) power and return conductors would be intolerable and the circuit would be nonfunctional. In practice, many thousands of gates may switch simultaneously, which could be detrimental to performance of any digital circuit sharing common power and return conductors. Impedance of the power distribution network plays, therefore, an important role in the generation or mitigation of power and ground bounce in PCBs.

9.5.3 Effective Management of ΔI Noise Effects in Signal Circuits

ΔI noise voltage was shown to constitute a serious problem in high-speed digital systems in particular, becoming a greater concern as transition time decreases and bus width increases. The following sections describe techniques used for preclusion of ΔI noise effects.

Table 9.2. Evolution of power distribution network (PDN) impedance goals

Year	Operating voltage, V_{CC} (volts)	Power dissipated, P (watts)	Current consumption, I_{CC} (amperes)	Target impedance, Z_{TGT} (mΩ)	Upper frequency, F (MHz)
1990	5.0	5	1	250	16
1993	3.3	10	3	54	66
1996	2.5	30	12	10	200
1999	1.8	90	50	1.8	600
2002	1.2	180	150	0.4	1,200

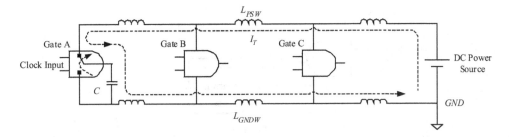

<u>Figure 9.134.</u> A current transient flowing through power and return inductance, L_{PSW} and P_{GNDW}, every 100 nsec, occurs when gate a drives the load capacitance to "high."

9.5.3.1 Reducing Load Capacitance. Δ*I* noise occurs in signal circuits due to load stray capacitance charging and discharging current through inductance of the power distribution network. Large fan-out of a gate will exhibit a larger overall capacitance, thus higher charge and discharge current during logic state transition. Limiting fan-out will reduce the load capacitance, thereby limiting the discharge charge and current through the circuit.

9.5.3.2 Reducing Common Impedance in Connectors and Device Packages. Several signals may be sharing common return pins or leads in device packages and connectors. Often, a single return pin is allocated in PCB edge connectors per three to five signal pins. The Δ*I* noise voltage across the return pin will result from the cumulative effect of the transient switching current from each signal line.

Minimizing noise voltage requires that the number of signals sharing a common return path be significantly reduced, lessening common impedance coupling between signal circuits. Consequently, the pin or lead count inevitably increases along with cost of the connector or package. Ideally, each signal has its dedicated return pin.

9.5.3.3 Increasing Transition Times of the Switching Signals. The voltage drop across return conductors is inversely proportional to the discharge current transition time during state changeover [see Equation (9.69)]. Increasing the transition time of the transient current waveform decreases the Δ*I* noise voltage. Unfortunately, this cannot always be applied, since timing budgets may be compromised, resulting in functionality concerns.

Certain PLDs enable the programming of logic interface slew rates. When practical, increasing output driver transition times in this manner may constitute an adequate solution. In other cases, transition times can be increased by inserting a damping resistor, R_D (Figure 9.135), effectively increasing the R–C time constant and suppressing the impulsive discharge current from the stray load capacitance. Typical values of resistors range from 10 Ω to 51 Ω, and are limited by circuit functionality (e.g., timing budget) constraints.

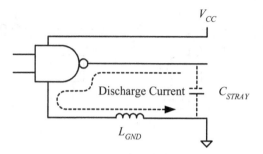

(a) Path of Impulsive Discharge Current

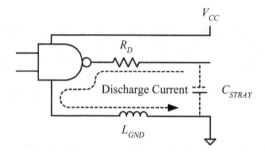

(b) Discharge Current "Damped" by the
Increased Damping Resistor, R_D

<u>Figure 9.135.</u> Insertion of a damping resistor at the driver output increases the circuit R–C time constant.

9.5.3.4 Reducing Circuit Net Inductance ("It's Really All about Inductance").

Without a doubt, the biggest "bang for the buck" in reducing ΔI noise lies in minimizing the total or net inductance of the signal and return circuits. Particular attention must be paid to minimizing board-level inductance during PCB layout. Equation (9.20) has shown that

$$L_{Total} \approx \left|(L_{33} + L_{11}) - (L_{31} + L_{13})\right| \to 2\left|L_{33} - L_{13}\right| \tag{9.71}$$

Since the partial self-inductance of the return path, L_{p33}, is proportional to the length of the path and the partial mutual inductance, L_{p13}, is proportional to the separation between the signal or power and its corresponding return conductors, Equation (9.71) points to the extremely important golden rules for reducing the total loop inductance of the current path [4]:

Rule #1. Decrease the circuit's partial self inductance, by using solid power and return planes rather than traces, minimize PCB thickness and break-out return trace length, and maximize via diameter and break-out trace width.

Planes may also be used instead of single leads within device packages. A two-layer BGA with a ground plane as a second layer will produce less noise voltage than a two-layer BGA that has only return traces. Multilayer BGAs with internal return planes typically have much lower Δ*I* switching noise than leaded packages.

At the integrated circuit (IC) level (particularly in VLSIs), the partial self-inductance is further minimized by using appropriate packages, particularly surface mount devices (SMD) (Figure 9.136 and Figure 9.137) [29, 30], which have shorter leads than through-hole devices. Whenever possible, chip-scale packages (CSPs) should be used, as their leads are shorter than those of other leaded packages. Ball grid array (BGA) leads in particular exhibit low inductance and, therefore, have far superior electrical performance compared to leaded devices. VLSI pin-out assignments should be carefully planned to spread out fast/strong drivers across multiple I/O banks (all of the above should be applied, if possible).

In right-angle PCB edge connectors (such as the D-subminiature PCB-mount connectors), which have multiple rows of pins (Figure 9.138 and Figure 9.139) [29], use of the inner row (with shorter lead lengths) for the return conductors will minimize the circuit's partial series inductance in the connector (note that signal pins should be allocated adjacent to the return pins for minimizing total loop inductance).

Using multiple, parallel return leads in VLSI packages and connectors further reduces the equivalent partial self-inductance of the leads. In older logic devices, only a small number of power and return leads were provided (often only one of each). Similarly, a small number of power and return leads is still often allocated in edge connectors interconnecting printed circuit boards to the motherboard or backplane (again, often only one of each). Multiple signals as well as high transient switching power current must therefore share a small number of DC power (V_{CC}) and return (GND) pins or leads, resulting in common impedance Δ*I* noise coupling across the power and return pins or leads.

By providing multiple pairs of power (V_{CC}) and return (GND) pins in edge connectors (often 2 to 5 signal pins per each return pin) in modern logic devices, this interference coupling can be controlled.

Figure 9.136. Surface mount devices (SMD) provide shorter leads. (Source: [29]; reproduced under the GNU free documentation license.)

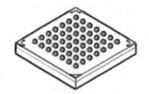

(a) Chip-Scale Package (CSP)   (b) Ball Grid Array (BGA) Package   (c) Thin Small-Outline Package

                                                                        (TSOP)

Figure 9.137. Some configurations of surface mount packages. (Source: http://en. wikipedia.org/wiki/, reproduced under the GNU free documentation license.)

   One application of this technique is demonstrated in Figure 9.140, where the pin-out of an older 74AC00 quadruple NAND device is compared with its contemporary equivalent 74AC11000. In the latter (b), two center V_{CC} and GND pins were allocated rather than one of each on the former (a) at the extreme edges of the device. Thanks to the multiple V_{CC} and GND leads, the total partial self-inductance of the device's leads is approximately halved. Furthermore, the near proximity the power and return pins significantly reduces the loop inductance, resulting in a diminished magnetic field.

   The positive effect of multiple leads on the total partial inductance, L_{pTotal}, can be easily observed. For instance, the total partial inductance of two closely spaced parallel conductors (denoted conductors 1 and 3, exhibiting self-partial inductances L_{p11} and L_{p33}, respectively), both carrying current in the same direction (assuming that the conductors are equal, i.e., $L_{p11} = L_{p33}$, is equal to [30]

$$L_{pTotal} = \frac{L_{p11} \cdot L_{p33} - L_{p13}^2}{L_{p11} + L_{p33} - 2L_{p13}} \approx \frac{L_{p11} + L_{p13}}{2} \tag{9.72}$$

The partial mutual inductance between the parallel conductors, L_{pij}, ($i \neq j$) now adds to the equivalent partial self-inductance of the parallel conductors ($i = j$). The total inductance thus never vanishes and is bound by the partial mutual inductance between the current-carrying conductors. Minimizing the total partial inductance, L_{pTotal}, is

Figure 9.138. Multirow D-subminiature PCB-mount connector. (Source: http://en. wikipedia.org/wiki/, reproduced under the GNU free documentation license.)

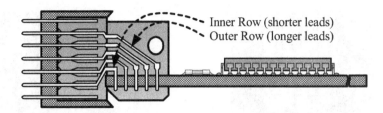

Figure 9.139. Multirow PCB-mount connector. (Observe length difference between inner and outer rows.)

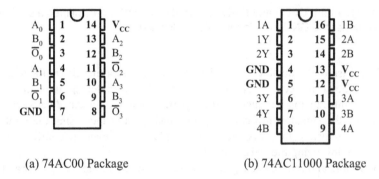

(a) 74AC00 Package (b) 74AC11000 Package

Figure 9.140. Advantage of multiple center return pin allocation. (Example of quadruple 2-Input positive NAND gate.)

achieved by keeping similar parallel leads (e.g., GND or V_{CC} pins) as far apart as practical, at least as much as their lengths [4].

Many more V_{CC} and GND leads are allocated in VLSI packages such as high-capacity ASICs (application-specific integrated circuits), complex programmable logic devices (CPLDs), and field-programmable gate arrays (FPGAs) (Figure 9.141 and Figure 9.142). Ideally, each signal lead would have its associated return pin or lead. Δ*I* noise in pins or leads is thus minimized at the price of the inevitable increase in the pin or contact count.

Rule #2. Increasing the partial mutual inductance between the signal (and power) and return paths (L_{p13}) is accomplished by minimizing the loop area enclosed between the signal (or power) and return paths. The tight coupling between the two current paths diminishes total loop inductance. This goal is best achieved in printed circuit boards by providing for each signal trace an immediately adjacent, wide, and continuous return conductor, ideally in the form of a solid and continuous plane.

When such a return plane is provided immediately beneath the signal trace,* high-frequency return currents can follow the path of the signal trace in the plane serving as

*The role of return planes in PCB stack-ups is discussed in Section 9.6.

(a) An ALTERA MAX 7000 CPLD with 2,500 Gates   (b) An ALTERA FPGA with 20,000 Cells

<u>Figure 9.141.</u> Altera CPLD and FPGA VLSIs. (Source:http://en.wikipedia.org/wiki/, reproduced under the GNU free documentation license.)

the path of least total inductance.* On the other hand, discontinuities in the return path, described in detail in Section 9.4 crossed by signal traces interrupt the path of the return current, introducing a reduction of the partial mutual inductance between the signal trace and the signal return current path, consequently increasing the total inductance of the return path. This increase in inductance severely conflicts with the objective of minimizing the effects of ΔI noise.

In connectors and device packages, once the maximum number of power and return pins or leads is allocated (as discussed above), the pins or leads should be evenly distributed among the signal leads.

9.5.3.5 Spreading Inductance.
Power and return planes still possess some inductance determined by their respective geometry. Although it is much lower than the inductance associated with traces, it is not insignificant. Section 9.4.2.2 introduced, however, an interesting concept associated with inductance of planes that is of great implication in the control of total loop inductance in power distribution networks, namely, the concept of spreading inductance.

Consider two points in the planes interconnecting two devices, such as one connecting an active device to its respective decoupling capacitor(s). As the planes constitute planar structures, current does not just flow through them in one direction but rather tends to spread out as it travels from one point to another in the planes, resulting in a lower magnetic flux density, $\vec{B}$, between the planes and consequently, lower partial inductance (Figure 9.143). For this reason, inductance of planes is commonly called spreading inductance and is specified in units of henrys per square (H/□).[†]

Spreading inductance should be minimized as it retards the ability of decoupling capacitors to respond to transient currents during device state transition. It is primarily determined by the spacing between power and its associated ground planes. The closer

---

*A detailed discussion of the pattern of return current flow in return planes is provided in Sections 9.3.5 and 9.3.6.

[†]Compare the concept of spreading inductance to that of spreading resistance for a sheet of conductive material, for which the surface resistance is expressed in ohms per square (Ω/□) (See Chapter 2).

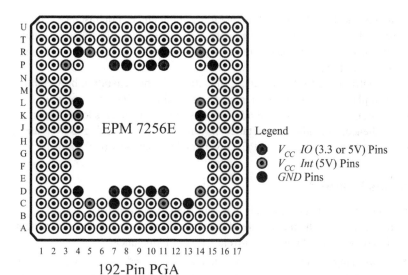

Figure 9.142. Multiple power (*V*_{CC}) and return (GND) leads in an ALTERA® MAX 7000 192-pin CPLD. (Source: ALTERA MAX 7000 Programmable Logic Device Family, September 2005, ver. 6.7 Data Sheet.)

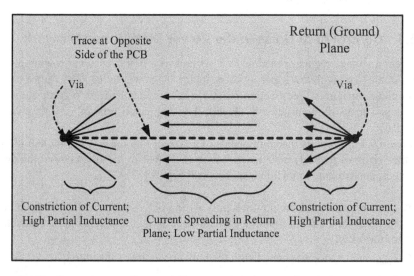

Figure 9.143. Current spreading and the associated spreading inductance in the return (ground) planes.

the spacing, the lower will be the spreading inductance. For instance, considering a separation of 6 inches (or 15.24 cm) between the two vias illustrated in Figure 9.143, the inductance versus distance from the right-hand via was computed and is presented in Figure 9.144 [35].

Also contributing to the spreading inductance is the separation between the power and return planes. Figure 9.145 depicts the spreading inductance obtained for various plane separations at the center of a PCB.* Visibly, inductance is lower for smaller separations, hence, spreading inductance is minimized by keeping power and ground planes close to each other in the PCB layer stackup (see Section 9.6) [36].

Lessons Learned

- Control of ΔI noise effects is accomplished by minimizing the load (fan-out) capacitance (and the magnitude of the switching current), control of current transition time, and reduction of current path net inductance.
- Minimizing the circuit total inductance is achieved by reducing self-inductance and increasing mutual inductance between signal and return conductors, VLSI devices leads, and edge connectors pins.
- A solid and continuous return plane placed immediately adjacent to the signal trace minimizes the total signal and return loop inductance. Inductance is further reduced thanks to current's spreading in the power and return planes.
- Discontinuities in return planes crossed by signal traces disrupt the return current path and increase noise.

9.5.4   Control of ΔI Noise in the Power Distribution Network

Whereas a voltage regulator module (VRM) such as a switch-mode power supply provides stable (e.g., with low superimposed ripple) DC power to the PCB,† aided by bulk decoupling capacitors (see Section 9.5.4.2), ΔI *noise* voltage is present across the PCB's power distribution network (PDN), ΔV, dominated primarily by its inductive impedance.

Figure 9.146 presents the components of a typical PDN in a PCB, while highlighting the "chain of (partial) inductances" from the VRM to the chips within the device packages, contributing to its impedance, consisting of [19]:

- L_{psw}, inductance of the power supply wiring between the VMR and the input to the PCB (and bulk decoupling capacitors), also including the internal inductance of the VRM
- L_{bulk}, inductance associated with the bulk decoupling capacitor(s)

---

*Spreading inductance is dependent on location of the vias with reference to the edge of the PCB, due to its effect on current distribution around the via.

†In practice, a switch mode power supply provides regulated DC power up to a fraction, e.g., 10%, of its switching frequency, due to the response time of the control loop of the power supply. For instance, a 200 kHz switch mode power supply could probably provide effective regulation up to about 20 kHz.

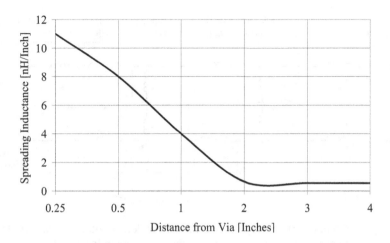

Figure 9.144. Spreading inductance related to distance from via.

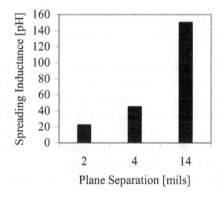

Figure 9.145. Spreading inductance related to separation between power and return planes (at center of the PCB).

- L_{spread}, spreading inductance in the power and return layers of the PCB between the IC leads and the nearest decoupling capacitor(s)*
- L_{lead}, inductance of the connection mechanism between the device's power and return leads and the PCB (this inductance is larger when the device is connected to the PCB through a socket)
- L_{bond}, inductance of the bond wire from the device's die to its package lead and of the lead itself
- L_{cap}, total inductance associated with the IC-related decoupling capacitors[†]

---

*Spreading inductance was further explained in Section 9.5.3.5.
[†]A decoupling capacitor installation for minimizing Δ*I* noise is further discussed in Section 9.5.4.3.

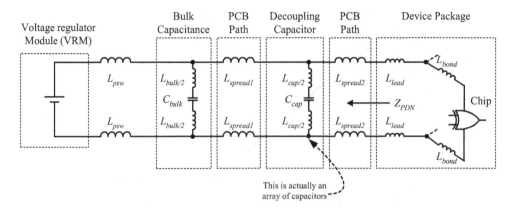

<u>Figure 9.146.</u> Components of a typical power distribution network (PDN). Note that although several capacitors may be present, only one is illustrated for clarity. (Courtesy of Dr. Spartaco Caniggia.)

Note that although only one decoupling capacitor is present, nevertheless, in a PCB populated by ICs there are several capacitors of different categories (see Figure 9.146).

9.5.4.1 Impedance Control in the Power Distribution Network.

The key to the control of ΔI noise produced in the power distribution system (PDS) lies in reduction of the inductive impedance of the power distribution network. As such, it is accomplished in a manner similar to that applied in signal circuits, taking into consideration that, unlike signal circuits, in which current flows between signal and return [power, V_{CC} or DC return (GND) conductors], currents in this case circulate between the power (V_{CC}) and the DC return (GND) conductors.

Addressing and mitigating ΔI noise in the PDN requires that the following measures be implemented:

- Rule #1: Use low impedance return (GND) connections between gates.
- Rule #2: The impedance between power (V_{CC}) leads on any two gates should be just as low as the impedance between return leads.
- Rule #3: A low-impedance path must be provided between the power and return conductors.
- Rule #4: Provide each logic device an adjacent transient current source.

Rule #1 and Rule #2 refer to the need to provide low-impedance connections between the return and the power leads of all interconnected gates. Reduction of the switching transient noise across the power distribution system's per-unit-length inductance, accomplished by (a) using as wide and as short as possible conductors (in other words, using wide and solid planes rather than skinny traces, with as few cuts and perforations as possible) in the power distribution system while (b) maintaining the smallest possible separation between the power and return paths (e.g., by allocating the power and return planes as adjacent layers), and (c) reducing of the loop area between conductors. Utilization of noise-tolerant circuits (e.g., differential signaling) increases the immunity of the circuits to the residual adverse effects of Δ*I* noise in the PDN.

Even through the use of power and return planes (Rules #1 and #2), total loop inductance in many high-speed circuits is just too large. Rule #3 is concerned with minimization of loop inductance by providing a low-impedance connection between the power and return planes, supplementing the preceding two rules. Rule #3 is a direct consequence of the fact that from the standpoint of AC (and transient) currents, the power and the return planes are equally useful for the purpose of AC current return. Clearly, DC connection between the power and return planes is not possible (nor necessary). On the other hand, the essential low-impedance AC connection is achievable through the use of low-inductance decoupling capacitors. Rule #3 and Rule #4 are both concerned with decoupling and are discussed in detail in the following section. Figure 9.147 illustrates the application of the above rules in the power distribution system on a PCB.

9.5.4.2 Decoupling Principles.

The effect of a decoupling capacitor is qualitatively illustrated in Figure 9.148. It is assumed that power (V_{CC}) and return (GND) planes have been implemented and are kept close together. When properly selected and installed in the circuit, decoupling capacitors accomplish the objectives of both Rules #3 and #4 whereby they provide a low-impedance discharge path for transient Δ*I* currents between power and return planes, acting as a readily available source of electrical charge and providing the high-frequency Δ*I* transient current to the active device during abrupt logic state transition. The active device contains this low-impedance source in parallel with the highly inductive power distribution system consisting of the large loop formed by the power (V_{CC}) and return (GND) planes back to the power source. During the quiescent period between state transitions, charge is drawn from the bulk capacitors on the PCB (see below) recharging this capacitor. The current required to replenish the charge in the decoupling capacitors is drawn through the power

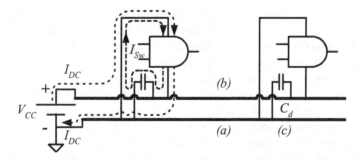

Figure 9.147. Management of Δ*I* noise: (a) return planes, (b) power planes, and (c) decoupling capacitors.

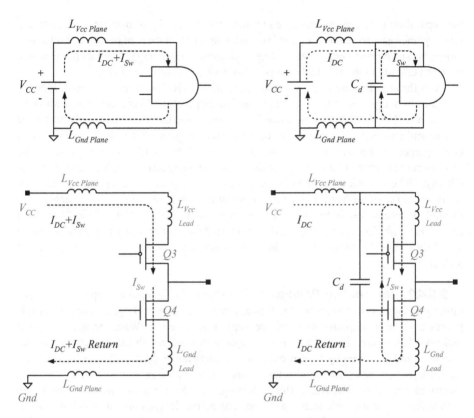

(a) No Decoupling Capacitor: Large Loop Inductance and Large EMI

(b) Decoupling Capacitor Present: Small (High-Frequency) Loop Inductance and Low EMI

<u>Figure 9.148.</u> Effect of decoupling on reduction of high-frequency current loop (and on radiated EMI), a schematic (top) and detailed (bottom) representation.

and return planes at a slower rate than it is removed from them during state transition. As a result, the capacitor reduces ΔI noise as it provides a somewhat regulated power (V_{CC}) and return (GND) at the immediate vicinity of the IC package for the short duration of the switching current transient [32, 33].

Implementation of the above four rules also plays a significant role in reduction of radiated magnetic field EMI emissions from the printed circuit board [34], particularly as a result of the reduced noise voltage stimulus and the corresponding switching current loop area. Electric field emissions are also decreased by minimizing the impedance of the power bus at the point of connection of the gate or IC device drawing fluctuating currents across the entire frequency bandwidth of concern. Noise voltage developing across the GND rail may further excite conductors (such as cable shields) connected to it at a point removed from the actual chassis ("true 0 V") connection, resulting in radiated electric field EMI emissions (Figure 9.149).

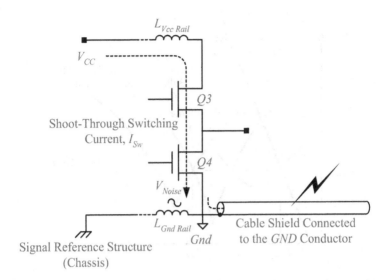

Figure 9.149. Radiated emissions produced by Δ*I* noise through an excited cable shield.

The suitable value of the decoupling capacitor is a subject of much contemporary debate. Traditionally, the value of decoupling capacitor is selected based on the maximum tolerable power supply noise voltage ripple, Δ*V*, at the V_{CC} leads of the device. A common selection methodology is based on defining the target reactive impedance across the frequency range of interest* and selecting decoupling capacitors such that the impedance goal is maintained. This frequency-dependent impedance of the PDN is dominated by capacitors installed between the power and return planes. Excellent references are available for discussing the selection of the value of the decoupling capacitors [3,9]; hence, only a brief discussion is presented here for completeness of the topic.

Mitigating Δ*I* noise in the PDN begins from the inductive power supply conductors powering the PCB. The target (dynamic) impedance, Z_{MAX},[†] is determined by the ratio of the maximum step change in the supply current, Δ*I*, assuming that all gates switch simultaneously as a worst case and the maximum tolerable power supply noise voltage ripple, Δ*V*:[‡]

$$Z_{MAX} = \frac{\Delta V}{\Delta I}, \text{ ohm} \tag{9.73}$$

The frequency beyond which the PDN wiring is deficient due to the high inductance of the power and return wiring from the VRM, L_{PSW}, is (Figure 9.150):

---

*Evolution of typical PDN target impedances was presented in Table 9.2 in Section 9.5.2.
[†]This impedance Z_{MAX} is predominantly reactive (i.e., inductive); therefore, in reality, $|Z_{MAX}| \cong X_{MAX}$.
[‡]A default value of 20% of the logic noise immunity level (NIL) may be used.

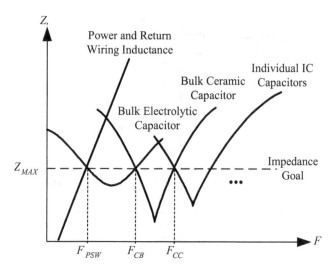

Figure 9.150. Illustration of the power distribution system impedance with the "decoupling capacitor brigade" in place.

$$F_{PSW} = \frac{Z_{MAX}}{2\pi L_{PSW}}, \text{ Hz} \qquad (9.74)$$

Above this frequency, a bulk capacitor (typically electrolytic or tantalum), C_B, must be placed in parallel to the power input leads to compensate for the increasing reactance, maintaining the impedance of the PDN below the target value. The electrolytic or tantalum* bulk capacitors serve as the second-largest source of charges after the VRM. When properly selected, the bulk capacitor should be able to supply charge with sufficient speed to meet the demands by circuits characterized by time constants as low as several hundreds of a nanosecond or less. The minimum value of capacitor required is

$$C_B = \frac{1}{2\pi F_{PSW} Z_{MAX}}, \text{ farad} \qquad (9.75)$$

Unfortunately, no capacitor is ideal. "real-world" capacitors exhibit parasitic inductance (ESL) and resistance (ESR), which, when combined with the capacitance, form a series resonant circuit, which reaches minimum impedance at the frequency at which the inductive and capacitive reactance cancel out.[†] Beyond some frequency, F_{CE}, the

---

*Large ceramic capacitors are currently often used rather than electrolytic and tantalum bulk capacitors due to various considerations associated with high-reliability applications. However, attention must be paid to their higher Q-factor and the subsequent potential resonances they may introduce into the circuit.
[†]Nonideal behavior of capacitors was discussed in detail in Chapter 2.

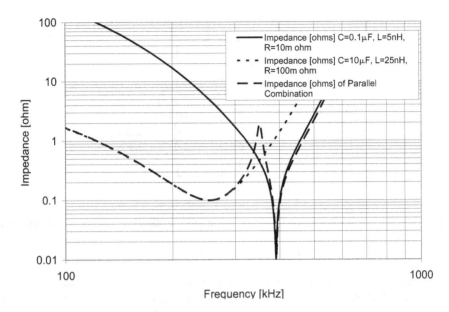

__Figure 9.151.__ Reactance of typical individual electrolytic and ceramic bulk capacitors and their parallel combination.

inductance associated with the capacitor, L_{CB}* dominates (Figure 9.151) and the bulk capacitor becomes ineffective as its reactance exceeds the target impedance, Z_{MAX}:

$$F_{CB} = \frac{Z_{MAX}}{2\pi L_{CB}}, \text{ Hz} \tag{9.76}$$

If the capacitor's resonance is shifted to higher frequencies by lowering its parasitic inductance, effective higher frequency decoupling can be achieved. Further lowering of the PDN impedance is accomplished by installing a ceramic capacitor, which is more suitable for higher frequencies, in parallel with the electrolytic or tantalum capacitor. Its value, C_C, is selected following the same rationale and using Equation (9.75), substituting the power distribution system inductance, L_{PDN} with the capacitor's inductance, L_{DB}, and the frequency, F_{PSW}, with F_{CB}. Properly selected ceramic capacitors are typically able to support charge demands from circuits with time constants as low as a few tens of nanoseconds.

A pair of the bulk electrolytic and ceramic capacitors should be placed at every power supply input to the PCB. Keep in mind, though, that the parallel combination of the two capacitors will result in resonance but will still be capacitive (Figure 9.152),

---

*This inductance results from the internal equivalent series inductance, the ESL, as well as the partial inductance of its leads and interconnects on the PCB. The former is determined by the package of the capacitor, whereas the latter is determined by the installation of the capacitor (to be discussed below).

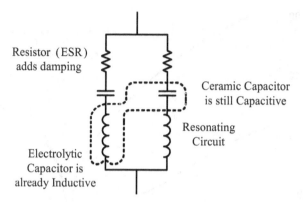

Resistor (ESR) adds damping

Ceramic Capacitor is still Capacitive

Resonating Circuit

Electrolytic Capacitor is already Inductive

Figure 9.152. Resonance mechanism in a parallel combination of bulk electrolytic and ceramic capacitors.

resulting in the curve for the parallel combination in Figure 9.151 (see the footnote on page 768).* If the reactance of the parallel combination of the capacitors exceeds Z_{MAX}, performance of the capacitors is compromised. Typical values of these capacitors are on the order of 1 to 10 μF and 10 to 100 nF for the bulk electrolytic and ceramic capacitors, respectively.

At higher frequencies, effective decoupling is best carried out at the near vicinity of the individual ICs. Selection of these capacitors is carried out based on their low- and high-frequency performance, dominated by their capacitance and inductance, respectively. The maximum tolerable inductance, L_T, associated with the capacitor is determined by the frequency related to the minimum transition (rise/fall) time (t_t) of the switching current waveform:

$$L_T = \frac{Z_{MAX}}{2\pi F_{MAX}} \overset{F_{MAX} \approx \frac{1}{\pi t_t}}{=} \frac{Z_{MAX}}{2} \cdot t_t \, , \text{ henry} \tag{9.77}$$

L_T is typically too low to be provided by a practical single discrete capacitor, considering its ESL (and its installation-associated inductance), that is, $ESL \gg L_T$. Therefore, an array of N parallel equal capacitors may be required, such that

$$N \geq \frac{ESL}{L_T} \tag{9.78}$$

The total array of capacitors must have a capacitive reactance lower than Z_{MAX} at frequencies down to the frequency beyond which the ceramic capacitor, C_C, ceases to be effective, F_{CC}. The total array capacitance, C_N, is thus computed:

*Note that when large ceramic capacitors are used as bulk capacitors, their associated ESR is relatively small, whereas their ESL is relatively higher due to their larger package size; thus, higher Q-factor and more intensive resonance effects may occur.

$$C_N = \frac{1}{2\pi F_{CC} Z_{MAX}}, \text{ farad} \tag{9.79}$$

Finally, the resultant capacitance of each element, C_I, in the capacitance array is computed using

$$C_I = \frac{C_N}{N}, \text{ farad} \tag{9.80}$$

Typical values of these decoupling capacitors are on the order of 0.47 to 10 nF. Present technology allows the use of surface-mount technology (SMD) components, which can be placed adjacent to or immediately beneath the device on the opposite side of the PCB. A very large range of decoupling capacitors is available, differing in package sizes, dielectric materials (electrolytic, tantalum, or ceramic X7R, X5R, Y5V, etc.), and manufacturing technologies [multilayer configuration (MLC), low-inductance chip capacitor (LLIC), interdigitated configuration (IDC), and low-inductance chip array configuration (LICA)]. Ceramic capacitors typically exhibit much lower ESL than electrolytic and tantalum capacitors, even when available in the same package size.

In addition to the individual IC capacitors mounted on the PCB, extremely effective decoupling is achieved by creating capacitors on the silicon die of an IC itself. The next most effective implementation of decoupling capacitors is mounting them in an IC's package ("hybrid circuit" construction) and on multichip modules (MCMs) (Figure 9.153). When choosing or designing devices such as FPGAs, ASICs, or MCMs, design (and cost) of the PCB will be significantly simplified if as much decoupling as possible is embedded on the chip die or on the MCMs themselves.

Beyond the frequency at which the capacitance array is effective, the distributed interplane capacitance formed between the power and return planes dominates and provides low impedance for the power distribution system. The interplane capacitance is able to deliver charge to circuits whose constants are shorter than a few tens of nonoseconds, that is, a charge demand frequency above several hundreds of mega-

Figure 9.153. "On-chip" decoupling capacitors. Note that both decoupling capacitors and termination resistors are present. (Courtesy of Mark I. Montrose.)

hertz. The effect and efficiency of the interplane capacitance was discussed in detail in Section 9.4.2.2.

Figure 9.154 illustrates the resultant "decoupling brigade" consisting of the entire decoupling scheme described above. Figure 9.155 illustrates the effect of the "decoupling brigade" on the PDN impedance and, finally, when decoupling is properly performed, the resultant switching noise suppression is observed as illustrated in Figure 9.155.

Interconnects between the different capacitors comprising the "decoupling brigade" should be accomplished by direct connection to the power and return planes rather than through traces. The contribution of the planes to the total interconnecting inductance is typically negligible compared to the inductance of the vias and leads of the capacitors and active device, primarily thanks to the "spreading inductance" of the planes. The concept of "spreading inductance" constitutes the key for the optimal installation of the capacitors and is discussed in the next section.

9.5.4.3 Placement and Installation of Decoupling Capacitors. The
VRM and the bulk capacitors are commonly few in number* and are positioned in specific areas of the PDN due to their dimensions and other functional constraints. High-frequency ceramic decoupling capacitors, on the other hand, are usually large in number (as shown in the Section 9.5.4.2) and are typically placed with a greater flexibility.

Once a reasonable number of decoupling capacitors has been distributed on the PCB, their usefulness appears is restricted by the inductance associated with their mounting, dominated by the inductance of the vias connecting them to the internal power and return layers. To be effective, the inductance associated with the mounting of the decoupling capacitor, L_{cap}, must be an as low as possible. In addition, the capacitors must be positioned in a manner that minimizes inductance of their installation.

The adverse effects of poor installation of decoupling capacitors are demonstrated in Figure 9.156. With the large series inductance of the capacitor installation, a large proportion of the switching current is still drawn from the power source rather than from the capacitor, contradictory to the objective of the capacitor in the first place.

A preferred installation is illustrated in Figure 9.157. The capacitor is placed near the power (V_{CC}) lead of the active device and connected directly to the power and ground planes. The power and return leads of the device are likewise connected to their respective planes in multilayer PCBs. The total inductance is now minimized and proportional to the length of the vias and the loop area produced between the capacitor and the device.

The scheme illustrated in Figure 9.157 is further expanded in Figure 9.158(a) with emphasis on the inductances associated with the installation of the capacitor. The decoupling capacitor exhibits parasitic inductance and resistance in addition to its capac-

---

*Manufacturers of VLSI (e.g., FPGA) devices often recommend that a bulk electrolytic capacitor be allocated for each device. However, since the area influenced by these capacitors (i.e., the area from which they draw the charge during their performance) is relatively wide due to their lower operational frequency and longer time of response, such capacitors may not necessarily be used in such large numbers and may be shared by several adjacent devices.

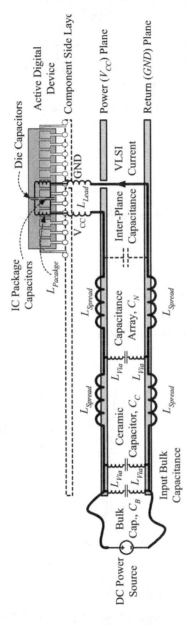

Figure 9.154. The "Decoupling Brigade." (Courtesy of Mark I. Montrose.) 1. First current surge comes from the on-die capacitance, due to the device switching. 2. On-die capacitance is recharged from the internal package capacitance. 3. Internal package capacitance is recharged through the interplane capacitance. 4. Interplane capacitance is recharged by the array of decoupling capacitors. 5. Array of decoupling capacitors recharge through the interplane capacitance. 6. Interplane capacitance is recharged by the input bulk capacitor pair. 7. Input bulk capacitor is recharged through the interplane capacitance. 8. Interplane capacitance is recharged by the power source.

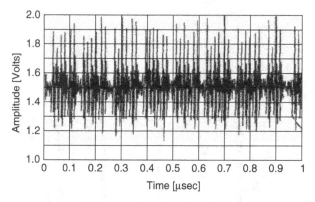

(a) Decoupling Capacitors Removed

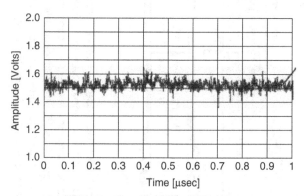

(b) Decoupling Capacitor In Place

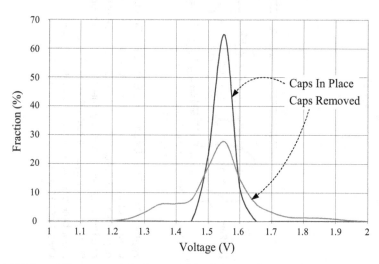

Caps In Place
Caps Removed

(c) Voltage Histogram of the Power Noise Measured across the 1.5V Power Bus

Figure 9.155. Power noise measured across a digital device's supply (1.5 V) with de-coupling capacitors removed and inplace. (Courtesy of Dr. Bruce Archambeault.)

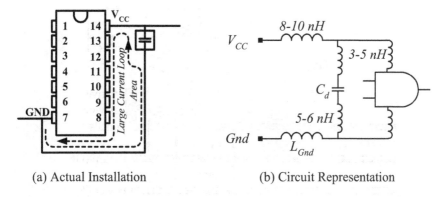

(a) Actual Installation   (b) Circuit Representation

Figure 9.156. Poor installation of a decoupling capacitor: loop inductance dominates.

itance. The resistance is known as the equivalent series resistance or *ESR*. The inductance consists of two factors, the first being L_{ESL}, the equivalent series inductance (*ESL*), which is associated with the capacitor itself, and L_{Loop}, an inductance associated with the means of mounting and connection of the capacitor between the power and return planes. These means consist of the solder pads used to secure the capacitor to the PCB and any traces or vias that establish the electrical connections.

The switching current provided to the digital device by the decoupling capacitor is indicated in Figure 9.158(a) as ΔI. The loop inductance L_{Loop} represents the sum of the effective inductances associated with Loop #1 and Loop #2:

$$L_{Loop} = L_{Loop\ \#1} + L_{Loop\ \#2} \qquad (9.81)$$

Loop #1 is formed by the pad and the trace of the capacitor, the return (GND) plane, and the two via paths between the signal layer and the return plane, whereas Loop #2 is formed by the return and power planes and two vias. Note that the effective induc-

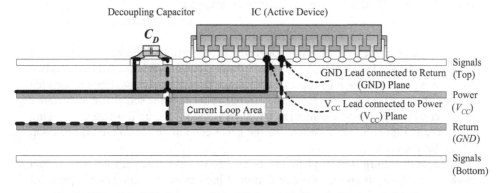

Figure 9.157. Improved installation of a decoupling capacitor.

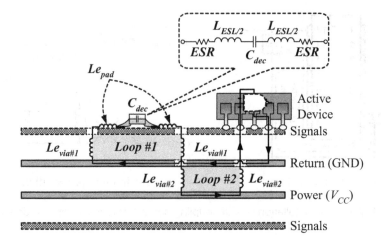

(a) Illustration of Inductances Associated with the Installation of a Decoupling Capacitor

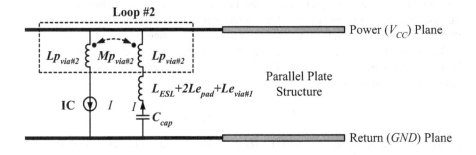

(b) Equivalent Circuit Illustrating the Mutual Inductance in Loop #2

Figure 9.158. Inductances associated with decoupling capacitors and their installation in a multilayer PCB. (Courtesy of Dr. Spartaco Caniggia.)

tance, L_{evia2}, associated with Loop #2 depends on the separation between the two vias incorporated in the loop. Since the current (I flows in the parallel coupled-pair conductors sharing the same loop in opposite directions, the effective inductance associated with the trace and via is computed as the difference between the self-partial inductance of the conductor, L_p, and the mutual-partial inductance, M_p, as illustrated in Figure 9.158(a) and Figure 9.158(b), where the inductive coupling between the vias of Loop #2 can be observed.

Because the return (GND) plane serves as the current return path for the pad and trace, image theory is applied for calculation of inductance. Equation (9.82) provides the resultant expression for the total inductance associated with the decoupling capacitor and its installation (i.e., the total equivalent inductance seen from the return and power planes toward the decoupling capacitor), L_{cap} [Figure 9.158(a)] [19]:

$$L_{cap} = L_{ESL} + L_{Loop} = L_{ESL} + 2Le_{via\,\#1} + 2Le_{pad} + Le_{via\,\#2} \qquad (9.82)$$

where $Le_{via\#1}$ and $Le_{via\#2}$ represent the partial inductance associated with the two mounting vias of the capacitor ($j = 1, 2$), and Le_{pad} represents the partial inductance associated with each of the mounting pads and associated interconnecting traces. These can be expressed in general as

$$
\begin{aligned}
\text{(a)} \quad & Le_{via\,\#j} = Lp_{via\,\#j} - Mp_{via\,\#j} \\
\text{(b)} \quad & Le_{pad} = Lp_{pad} - Mp_{pad}
\end{aligned}
\qquad (9.83)
$$

The expression (b) in Equation (9.83) is derived through the application of image theory, as illustrated in Figure 9.158(b). From Figure 9.158(b), it is also evident that in order to minimize the effective inductance $Le_{via\#2}$ ($Le_{via\#2} = Lp_{via\#2} - Mp_{via\#2}$) the portion of the via of the IC in Loop #2 should be very close to that of the capacitor in order to maximize $Mp_{via\#2}$, keeping in mind that the self-partial inductance $Lp_{via\#2}$ remains constant.* This conclusion implies that the position of the decoupling capacitor with respect to the IC is important for minimizing total loop inductance. The inductance $L_{IC} + Le_{via2}$ in Figure 9.158(b) represents the total inductance associated with the IC and its connection to the PDN, and can be computed in a manner analogous to that of the decoupling capacitor inductance. Furthermore, when placement of the capacitors is optimized, the residual inductance associated with a decoupling capacitor installation is dominated by the mounting pad shape and layout. For any given length of mounting vias, inductance is minimized by moving V_{CC} and GND vias close to each other. Compare, for instance, the examples presented in Figure 9.159.† A detailed discussion of capacitor pad layout was discussed in detail in Section 9.4.2.1 with respect to stitching or bypassing capacitors and is not repeated here.

When multiple power and return leads are available, which is typically he case in large-scale ICs and VLSIs, the same rule applies and a decoupling capacitor should be placed in the vicinity of each V_{CC}–GND pair of leads (Figure 9.160). The connection of the devices and capacitors is thus accomplished through the power and return planes in a manner similar to that illustrated in Figure 9.157. Whether the capacitor is placed on the same side of the PCB or on the side opposite to the device is not critical for thin PCBs (e.g., 12 to 16 layers). However, in much thicker PCBs the separation between the device and the capacitor may lead to a large loop area due to the long vias if placed on opposite sides of the PCB. In this case, placing both components on the same side of the board is preferable.

---

*It should be pointed out that this derivation is a first-order approximation. In an exact analysis, all mutual inductances to adjacent vias should be taken into consideration and not only those associated with the vias physically belonging to Loop #1 and Loop #2.
†Certain schemes [such as the most aggressive capacitor mounting schemes, (e) and (f) in Figure 9.159] may be in conflict with PCB manufacturing restrictions (e.g., IPC manufacturing rules) or be difficult or almost impossible to implement using current technology.

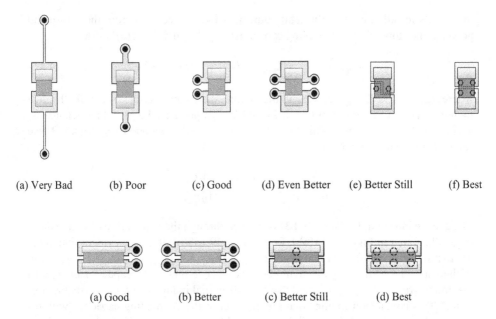

(a) Very Bad   (b) Poor   (c) Good   (d) Even Better   (e) Better Still   (f) Best

(a) Good   (b) Better   (c) Better Still   (d) Best

Figure 9.159. Installation of regular aspect SMT stitching capacitors that minimize inductance (e.g., 0805, 0603). (Courtesy of Keith Armstrong, Cherry Clough.)

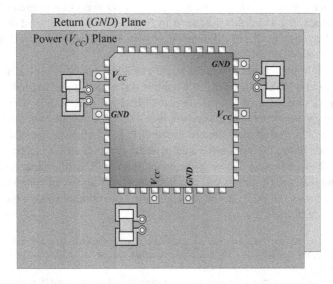

Figure 9.160. Placement of SMT decoupling capacitors in VLSI applications. (Courtesy of Keith Armstrong, Cherry Clough.)

In the case of ICs in which the V_{CC} and GND leads are not in close proximity (such as in Figure 9.160), the key to the location of the decoupling capacitor is inductance rather than physical spacing.

In multilayer boards with closely spaced power and return planes [e.g., less than 30 mils (0.75 mm)], the location of the decoupling capacitors is not critical and they may be mounted in the most convenient locations, for example, within approximately 0.5 inch (1.25 cm) of a typical FPGA perimeter, thanks to the spreading inductance phenomenon associated with the planes, L_{pcb} (see Section 9.5.4). This connection scheme is exceptionally beneficial in ball grid array (BGA) applications, particularly when the decoupling capacitors cannot be placed directly underneath or in between the core power and return ball contacts but rather at the perimeter (placing the capacitors in this dense location might interrupt or block the routing of breakout traces from the device).

On PCBs with widely spaced power and return planes (e.g., larger than 30 mils), on the other hand, the inductance contributed by the planes can no longer be neglected. In particular, mutual inductance between the vias of the active device and the vias of the decoupling capacitor discussed above now becomes of considerable significance. This mutual inductance tends to cause the majority of the current to be drawn from the nearest decoupling capacitor rather than from the planes. The local decoupling capacitors should, therefore, be located as close to the active device as possible (preferably near the pin attached to most distant plane).

In the latter case, the capacitor should be located near the device's lead connected to the most distant plane. If the device is closer to the GND plane layer, the capacitor should be placed closest to its V_{CC} lead. Conversely, if the device is closer to the V_{CC} plane layer, the capacitor should be placed closest to its GND lead. In either case, both the device and capacitor should be connected directly to the power and ground planes using minimum length pin escapes and break-out traces. Measurements carried out on a four-layer PCB demonstrated the negligible difference in the S_{12} parameter associated with the power distribution network when all decoupling capacitors were in place or with capacitors adjacent to the ports removed, compared to the very different situation in which all capacitors were removed (Figure 9.161). A relatively small difference is observed at higher frequencies (and practically no difference at lower frequencies) when the capacitors were not placed immediately near the device ports [37, 38]. The same principle for placement and installation of the capacitor should be applied with even greater weight when power (and often return) traces or "bus bars" are used as the PDN rather than planes* [19].

Sharing a plane via can further improve decoupling effectiveness, as long as no traces are extended in length for this purpose [Figure 9.162(a)]. It is recommended, however, to carefully compute the inductance associated with the capacitor's installation prior to making the decision regarding the use of this configuration. Further improvement is achieved in packages in which adjacent leads are allocated for power and return connections [Figure 9.162(b)]. Observe in Figure 9.163 the installation of ca-

---

*The inductance, L_{pcb}, of a typical bus bar is on the order of 100 nH or greater, whereas in a typical multilayer PCB in which power and return planes are used, this inductance is typically on the order of 0.05 nH/cm and can, therefore, be neglected.

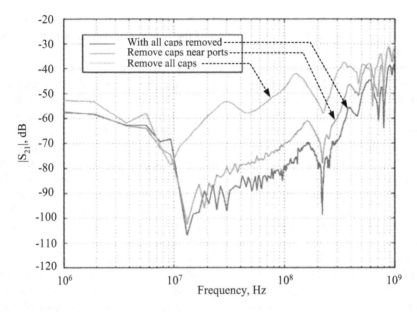

Figure 9.161. Effect of decoupling capacitor location in a four-layer PCB. (Courtesy of Dr. Todd Hubing, Clemson University.)

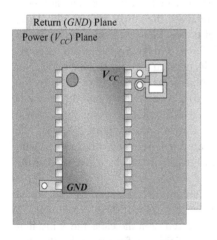

(a) Power (V_{CC}) and Return (GND)

Leads Far Apart

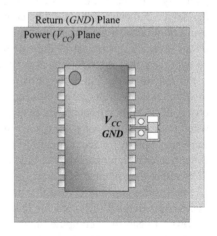

(b) Adjacent Power (V_{CC}) and Return

(GND) Leads

Figure 9.162. Sharing a plane via can improve decoupling if trace length is not extended. (Courtesy of Keith Armstrong, Cherry Clough.)

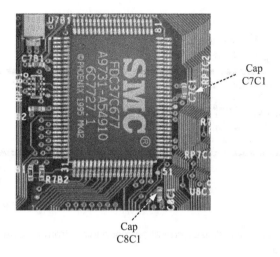

Cap
C7C1

Cap
C8C1

<u>Figure 9.163.</u> Decoupling capacitor location on a multilayer PCB. (Courtesy of Dr. Todd Hubing, Clemson University.)

pacitor C7C1 (to the right of the device) placed in near proximity to and sharing a common mounting pad with the active device, whereas capacitor C8C1 (below and to the right of the device) is connected directly to the planes.

9.5.5 Parallel-Plate Waveguide (PPW) Noise Mitigation Using Electromagnetic Band Gap (EBG) High-Impedance Structures (HIS)

Parallel-plate waveguide (PPW) noise resulting from Δ*I* noise* (or simultaneous switching noise, SSN) in power and return planes as well as from time-varying currents traversing power or return planes of a printed circuit board (PCB) through vias (see Sections 9.4 and 9.5) produce electromagnetic waves propagating throughout the power distribution network's (PDN) parallel plates' structure (Figure 9.164), a phenomenon also known as "parallel-plate waveguide (PPW) noise" excitation. PPW noise results from the mode conversion of desired modes into parasitic PPW modes. When such structures are thus excited, the resultant PPW noise, sustained in standing wave patterns due to multiple reflections from the edges of PCBs and coupling between signal vias, occurs at power-return parallel plate cavity resonance frequencies, governed primarily by structural dimensions.† Since the structure is not a perfect cavi-

---

*Sometimes, the term "parallel-plate waveguide (PPW) noise" is used instead of Δ*I* noise. Δ*I* noise refers to the voltage transients developing across the power distribution network (PDN) conductors or planes, whereas PPW noise refers to the electromagnetic wave propagation between the PDN planes, enhanced due to structure resonances (see Section 9.5.6). Δ*I* noise could, therefore, be considered as the source of excitation of parallel-plate waveguide (PPW) noise. In general, the two terms could be considered synonymous.
†Chassis resonances and their effects and mitigation techniques are discussed in Section 9.9.

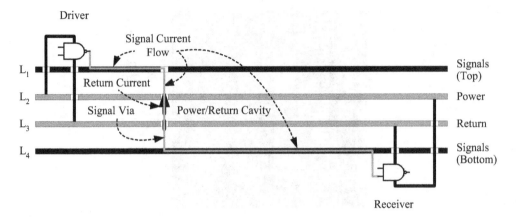

<u>Figure 9.164.</u> Switching noise generation mechanism from a power bus in a four-layer multilayer PCB.

ty, some of the energy is eventually transmitted from the edges of the PCB, resulting in excessive radiated emissions, in addition to the power and ground bounce effects. Figure 9.165 illustrates this phenomenon [39].

Several conventional approaches for mitigating the effects of ΔI noise were discussed in previous sections.* The most prominent of these involve the use of discrete bypass or decoupling capacitors and embedded capacitance [39]. However, decoupling capacitors will fail at high frequency due to their inherent lead inductance, and embedded capacitance is still an expensive technique and reliability considerations limit its practical use. Furthermore, the embedded capacitance does not eliminate higher-order resonant modes. Recent years have witnessed the introduction and development of a novel structure known as the electromagnetic band gap (EBG) on the power or return planes to form a high-impedance surface (HIS)[†] for the purpose of noise mitigation on printed circuit boards (PCBs) [39].

9.5.5.1 Electromagnetic Band Gap (EBG) Structures as Surface Wave Filters.
EBG structures have been previously shown to provide excellent suppression of noise at frequency above several hundred megahertz, at which the abovementioned methods cease to be effective. These novel high-impedance structures (HIS)[‡] essentially consist of electrically small distributed resonators that are characterized by periodic or quasiperiodic metallic patches connected to a common power (or return/ ground) plane by short stubs or plated through-holes (or vias). Electromagnetically, EBG structures represent the region in which electromagnetic fields are in decaying

---

*Other known techniques include use of resistive terminations along the board edges, employing lossy components throughout the board, via stitching and shielding.
[†]EBG structures have also been referred to as high-impedance surfaces (HIS).
[‡]Square patches are shown here; however, other patterns of similar size can be used in optimized designs.

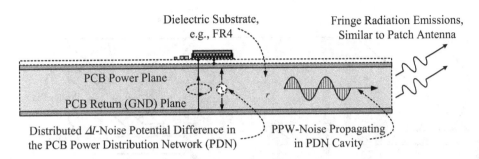

Figure 9.165. Propagation and radiation mechanism of electromagnetic waves caused by Δ*I* noise in a PCB PDN structure.

mode. The wave equation in a source-free region may be expressed in terms of E-field,*

$$\nabla^2 \overline{E} + k^2 \overline{E} = 0 \tag{9.84}$$

where the E-field intensity, E, is expressed as function of distance from source, r, frequency, ω, and time, t:

$$E = E_0 e^{-jk \cdot r} \cdot e^{j\omega t} \tag{9.85}$$

The wave number factor, k, is expressed as

$$k = \omega \sqrt{\mu_{eff}(\omega)\varepsilon_{eff}(\omega)} \tag{9.86}$$

where ω represents the angular frequency in radians/second and μ_{eff} and ε_{eff} symbolize the effective permeability and permittivity of the structure. For the EBG structure to be lossy, the wave number, k, inside the EBG structure is imaginary, implying either μ_{eff} < 0 or ε_{eff} < 0 (representing an H- and E-excitation of the structure, respectively).

A conventional EBG design consists of structures in which the patches are embedded in a plane within the dielectric substrate (e.g., FR4), embedded between the power and return planes of the power distribution network (PDN) of a multilayer printed circuit board (PCB) and are usually equally spaced from each, hence it is known as "embedded EBG" or EEBG[†] (Figure 9.166) [39]. In essence, the EBG structures constitute a frequency-selective surface (FSS) positioned on the top of a conductor-backed material with vias positioned between the FSS and the metallic backing. When such structures are excited, the fields are sustained in standing wave patterns (at resonance) that occur within the structure's band-stop frequency range, consequently suppressing radi-

---

*This presentation is courtesy of Prof. Yahya Rahmat-Samii, Electrical Engineering Department, University of California, Los Angeles.
[†]Owing to its shape, this BG structure is also known as the "mushroom type" EBG.

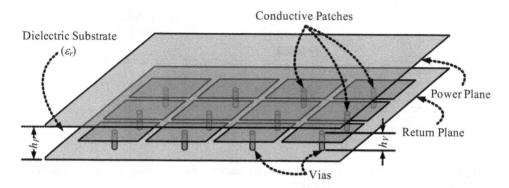

Figure 9.166. A typical embedded EBG (EEBG) structure.

ation. Since the structure is not a perfect cavity, some of the energy eventually radiates from the edges of the PCB. The presence of these surfaces prevents the propagation and, therefore, radiation by creating an inadequate and unsuitable environment for such propagation. The material between the patches and the ground plane can be dielectric or magnetic material, or a combination thereof.

Note that there is no unique choice for the shape of the patches. The square patches described herein represent a generic type of structure; however, this should not diminish the generality of this concept as other topologies of similar size can be designed with increased degrees of freedom to allow optimization for specific design needs. Figure 9.167 presents some alternative shapes of EBG structure patches.

Figure 9.168 demonstrates the advantage, with respect to the frequency ranges for effective switching noise suppression, of simple EBG structures over conventional noise suppression methods (which are efficient up to several hundred megahertz only). The extension of the frequency range for effective suppression thanks to the EBG structures up to frequencies in the 0.5–10 GHz range is clearly noticeable [41].

9.5.5.2 EMI Suppression Using Embedded EBG (E-EBG) Structures.

EBGs have been used in different electromagnetic applications but primarily in im-

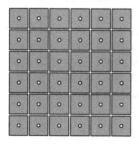 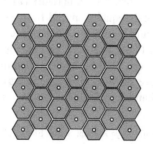 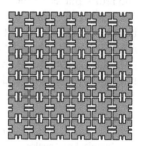

Figure 9.167. Alternate types of EBG (electromagnetic band gap) structures.

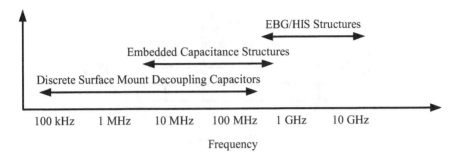

Figure 9.168. Frequency ranges for effective PDN switching noise suppression using conventional and novel EBG methods.

proving the radiation characteristics of antennas. Since EBGs are composed of distributed, electrically small resonators, they mimic, in essence, the behavior of band-stop filters, suppressing the propagation of surface waves over a particular range of frequencies, consequently eliminating interfering surface currents arising from collocated radiators on board the PCB (Figure 9.169). For this particular reason, EBGs have an important role to play in EMC and signal integrity applications and are shown to be of particular use in PPW-noise suppression in PCBs and packages in a manner that cannot be achieved by lumped circuit elements [39, 41, 42].

A commonly employed method for EMI mitigation in PCBs using EBG structures consists of encircling the culprit circuit producing the PPW noise with a high-impedance structure (HIS) strip that is connected to one of the planes of the power distribution network and designed to be effective about a frequency range of interest. Radiated electromagnetic emission from the boundary of the board at frequencies within the band-stop region of the HIS ribbon are minimized thanks to the presence of the EGB structure. This effective frequency range may be narrowband, broadband, or could even consist of multiple bands. Figure 9.170 illustrates a lateral view of a board employing this concept [43].

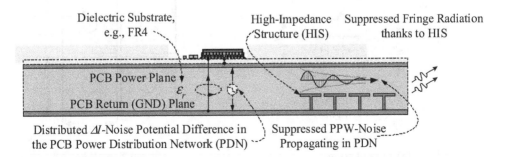

Figure 9.169. Propagation and radiation mechanism of electromagnetic waves caused by Δ*I* noise in a PCB PDN structure in the presence of HIS/EBG structures.

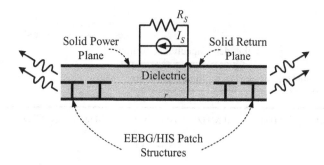

Figure 9.170. Lateral view of a ΔI noise source model enclosed by a HIS/EBG ribbon structure on the PCB perimeter. (Only two rows of patches are shown.)

The stop band of the EBG structures directly depends on the geometrical features of the structure (such as periodicity, patch size, gap size, via diameter, via length, and also board thickness) as well as on the dielectric material used in the PCB as substrate. However, no formulas are available in the literature at this point that provide a good relationship between the geometrical features of the EBG structure and the center frequency and band-stop region. Until such highly accurate closed-form analytic expressions are developed, the stop band of an EBG structure must be derived using numerical computational tools. The critical parameters of an EBG structure are illustrated in Figure 9.171, showing a typical (and the simplest) EBG structure. These parameters include patch size, a, diameter of the stubs (vias), d, patch gap or distance between adjacent patches, g, length of via, h_V, and height between the two planes, h_P. The structure is periodic with period $a + g$ [43].

The effectiveness of this structure in EMI suppression can be easily validated by means of S-parameters analysis or measurement. Figure 9.172 illustrates a model used

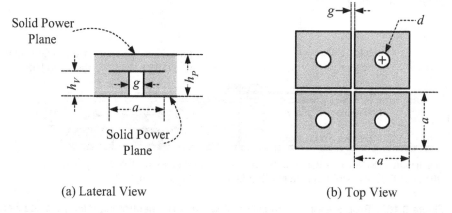

(a) Lateral View   (b) Top View

Figure 9.171. Geometrical design features of an embedded EBG (EEBG) structure.

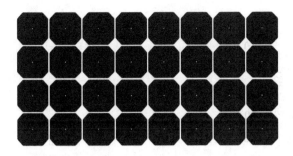

(a) Top View of the Mid (Patch) Layer

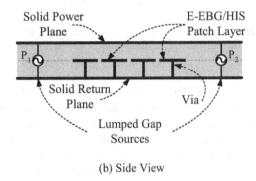

(b) Side View

Figure 9.172. Embedded EBG (EEBG) structure model used for s-parameter validation.

for S-parameter simulations carried out using HFSS.* The model consisted of two ports (P_1 and P_2) and a two-dimensional (2-D), periodic, high-impedance structure located between them [43]. Since the presence of the high-impedance structure between the ports attenuates the wave propagating between the two ports within a given frequency range, the S_{21} parameter,[†] representative of the transmitted power, reflects the effect of the structure.

Sample simulation results for various E-EBG structures are presented in Figure 9.173 and tabulated in Table 9.3 [43]. Considering that the bandwidth of the stop band of the simulated structures is approximately two-thirds of the center frequency and that the Δ*I* and consequent PPW noise power spectrum is mostly concentrated in frequencies below 6 GHz, E-EBG structures can be designed so that their stop band region overlaps the desired suppression frequency. HIS structures with larger patches are required for effective suppression at lower frequency bands; however, patch sizes larger than 5 mm may appear to be impractical. Extension of the technique to lower frequen-

---

*HFSS, High Frequency Structure Simulator, Ansoft Co.
[†]S_{21} represents the forward voltage gain or "transmission" of the system.

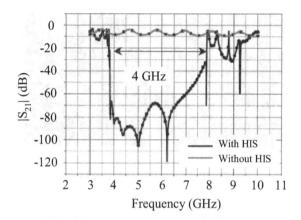

<u>Figure 9.173.</u> Simulation results (S_{21}) for various patch sizes (g = 0.4 mm, d = 0.8 mm, h_V = 1.54 mm, h_P = 3.08 mm, ε_r = 4.1). (Courtesy of Omar M. Ramahi.)

cies can be achieved in reality through the use of more complex HIS structures (see Section 9.5.5.3) in order to achieve the same bandwidth or even multiple bands.

It is also noted that suppression of radiation from PCBs using HIS structures is effective regardless of the power and return planes stack-up (e.g., G-P-G-P, G-P-P-G,* etc.) in a multilayer PCB as pertains to every pair of the PDN V_{CC}–GND layers rather than to GND–GND or V_{CC}–V_{CC} pairs. In particular, HIS structures embedded between V_{CC}–GND planes result in suppressed radiation caused by SSN as well as ΔI noise generated by vias traversing the planes. This suppression does not occur, however, when HIS structures are embedded in GND–GND or V_{CC}–V_{CC} pairs of planes.

Figure 9.174 illustrates a 6.5 cm × 10 cm PCB used to validate the radiation suppression efficiency. The PCB was fabricated on a FR4 dielectric substrate. A ribbon of four rows of HIS patches (with a = 5 mm) surrounded the board, occupying approximately a 2 cm wide ribbon on the perimeter of the PCB (all other dimensions were as defined in Figure 9.173).[†] Excitation of the parallel-plate structure of the PCB took place in the middle of the PCB using a vector network analyzer (VNA). The scattering parameter S_{21} was measured at several test points at a distance of 5 cm from the perimeter of the board using the VNA and used as a representation of the radiated power. Results of the measurements for one of the test points (TP1 in [43]) are shown in Figure 9.175. A wide stop band extending from 4 GHz to 10 GHz demonstrates an average suppression of 30 dB compared to the case without the HIS. These results support the simulations' outcome presented in Figure 9.173 [43].

9.5.5.3 Ultra-Wideband (UWB) EMI Suppression Using EBG/HIS Structures. Extension of the EMI suppression bandwidth of EBG/HIS structures

*G = return (GND); P = Power (V_{CC}).
[†]For a practical 20 cm ×10 cm PCB, this constitutes 28% of its total area, which may introduce some PCB design constraints.

Table 9.3. Bandwidth of EBG/HIS stop band obtained by simulation for various patch sizes*

Patch size, a (mm)	Low frequency (GHz)	High frequency (GHz)	Stop band bandwidth (GHz)
4	5	11	6
5	4	8.1	4.1
6	3.5	7.5	4
8	2.5	5.3	2.8
10	2	4	2
12	1.7	3.3	1.6
15	1.35	2.45	1.1
17	1.2	2.3	1.1
20	1	1.8	0.8
25	0.66	1.46	0.8
30	0.63	1.18	0.55
40	0.45	0.87	0.42

*$g = 0.4$ mm, $d = 0.8$ mm, $h_V = 1.54$ mm, $h_P = 3.08$ mm, $\varepsilon_r = 4.1$.

can be accomplished by employing cascaded high-impedance surfaces (CHIS) between the source of noise and the potential targets. In such arrangements, various ribbons are cascaded around the board, each having different pattern (e.g., different patch sizes, a, or patch gap, g) thus providing noise suppression in a specific frequency range, resulting in a UWB stop band for the propagating wave. Cascading HIS structures has, in fact, the same effect as cascading filter stages so as to achieve larger bandwidths, multiple bands, or increased attenuation. The implication of UWB suppression lies in the fact that CHIS structures provide a practical manner of suppressing radiation over frequency bands that encompass not only the fundamental

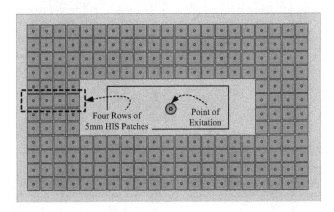

Figure 9.174. Test PCB with ribbon of HIS patch structures surrounding the PCB.

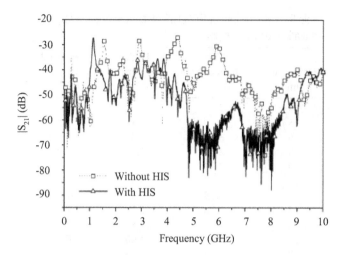

Figure 9.175. S_{21} measurement results with and without the HIS structure.

frequency of interference signals (e.g. a clock sources) but also their immediate harmonics.

As an example, consider the configuration shown in Figure 9.176 with a board size of 20.6 cm × 16.6 cm, excited at its center. In this configuration, an HIS structure with a patch size of $a_2 = 10$ mm (HIS Type 2) is added to a structure with a patch size of $a_1 = 5$ mm (HIS Type 1). All other dimensions of the patches remain identical to the patches in Figure 9.174 [39, 43].

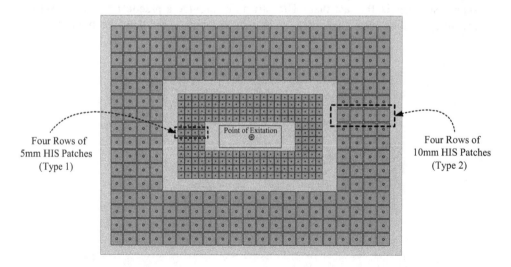

Figure 9.176. Cascading two HIS structures with different configurations (5 and 10 mm patches).

Results of the measurements for one of the test points (TP1 in [43]) are shown in Figure 9.177. Noticeably, this scheme is capable of suppressing not only a noise at 3 GHz but also its second and third harmonics at 6 and 9 GHz, creating UWB radiation suppression in PCBs employing such a design [43].

9.5.5.4 Planar Meander-Line EBG (ML-EBG) Structures.

Earlier EBG structures described above utilized three layers; the EBG pattern layer is embedded between the power and return planes. This structure requires an extra plane for the HIS as well as specially designed vias, resulting in increased complexity and cost of the entire PCB. An alternative technique—low-period coplanar EBG (LPC-EBG) structures with meander-line bridges—also known as planar meander-line EBG (ML-EBG) structures, consists of a two-layer power distribution network with one of the layers patterned in a periodic fashion, producing an effect similar to that of the conventional EBG. These novel structures, in sharp contrast to embedded EBG (EEBG) structures used for power plane noise mitigation, do not utilize vias [45]. The underlying principle behind the planar meander-line EBG structure is to provide suppression of switching and other noise propagating between the power and return planes of the PDN while providing a low-impedance path for DC current on each layer.*

Figure 9.178 shows the geometrical design features of a two-layer power/return stack-up with a planar meander-line EBG (ML-EBG) structure. A unit cell is illustrated in Figure 9.178(c) [45, 46].

The unit cell of the ML-EBG can be conceptually viewed as an electrical filter of a parallel LC resonator, illustrated in Figure 9.178(e) [45]. This qualitative model is inspired by the physical behavior of the fields in the patch. The propagation characteristics between the EBG patch (power or V_{CC}) and the continuous return (GND) plane are represented by equivalent inductance, L_p, and capacitance, C_p. The gap between any two neighboring unit cells introduces fringe capacitance, C_b, shown in [47] to be approximately

$$C_b \approx \left(\varepsilon + \varepsilon_0 \right) \frac{(h)}{\pi} \cdot \cosh^{-1} \left(\frac{w}{d} \right), \text{ farad} \qquad (9.87)$$

In a similar manner, the capacitance between each patch and its adjacent meander line can be obtained.

The bridge connecting the neighboring unit cells is represented by inductor, L_b. Approximate expressions for the inductance of the meander line have been shown to be

$$L_b \approx 0.00266 \cdot a^{0.0603} \cdot h^{0.4429} \cdot N^{0.954} \cdot d^{0.606} \cdot w^{-0.173}, \text{ henry} \qquad (9.88)$$

approximated to an accuracy of 12% [48]. The dimensions a, h, d, and w are given in Figure 9.178(d), and N represents the number of segments of the greatest length, h.

---

*Note that the behavior of this arrangement is independent of the DC potential of each of the planes forming the PDN and, in particular, whether it is the "V_{CC}" or "GND" plane that forms the ML-EBG structure.

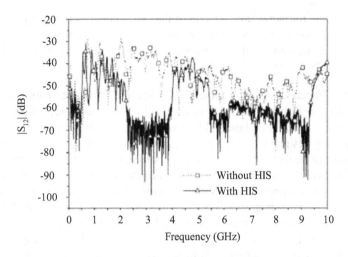

Figure 9.177. S_{21} measurement results with and without a HIS structure.

The center frequency of the EBG unit cell is thus given by [47]

$$f_0 = \frac{1}{2\pi\sqrt{L_b \cdot C_b}}, \text{ hertz} \tag{9.89}$$

Consequently, one observes that for the ML-EBG structure to become effective in the lower frequency region, an increase of either the capacitance, C_b, or inductance, L_b, is necessary. A longer bridge length corresponds to an increase in inductance; thus, a meander line as the connecting bridge (DC link) between adjacent patches is introduced. Capacitance, on the other hand, can be increased by increasing the dimensions of the metallic pattern or increasing the permittivity of the dielectric substrate. Higher permittivity materials possess the distinct advantage of allowing miniaturization of EBG patterns in addition to shifting the band gap to lower frequencies [composite low-loss dielectrics with very high permittivity ($\varepsilon_r \geq 100$) are commercially available] [47].

Figure 9.179(a) depicts dimensions of the ML-EBG structure used in a simulation intended to demonstrate the structure's suppression performance. The overall dimension of this structure constructed of 3 × 5 unit cells of size 30 mm × 30 mm with total dimensions of 90 mm × 150 mm. The solid layer can be used for one DC voltage potential level (e.g., "GND" or "0 V" plane), whereas the EBG-patterned layer may be utilized for the second level (e.g., V_{CC}). Between these two layers, a FR-4 ($\varepsilon_r = 4.4$) uniform dielectric substrate, 1.54 mm in thickness, was included. The distance between the centers of adjacent patches is 30 mm, the patch width is 28 mm, and the gap between the neighboring patches is 2 mm. The width of the meandered line is 0.2 mm [45].

Performance of the ML-EBG structures was evaluated by computation (using HFSS simulations) and measurements of the its S_{21} parameters. The test ports used for the simulation are located at the center of the excited patches, as depicted in Figure

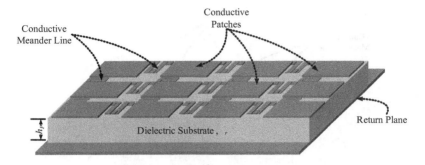

(a) 3-D View of the Entire ML-EBG Structure

(b) Top View of the Entire ML-EBG Structure

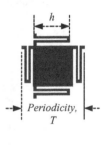

(c) Top View of the Unit Cell

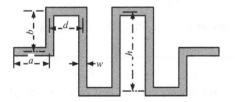

(d) Top View of the Meander Line Bridge (Inductive Part)

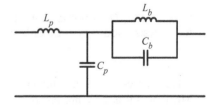

(e) Qualitative Equivalent Circuit Model for Cell of the ML-EBG Structure

Figure 9.178. Geometrical design features of a planar meander-line EBG (ML-EBG) structure. (Courtesy of Omar M. Ramahi.)

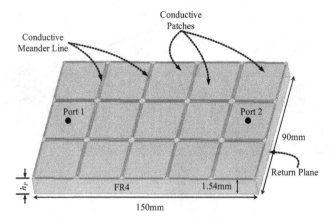

(a) Planar EBG Structure with Meander-Line Bridge

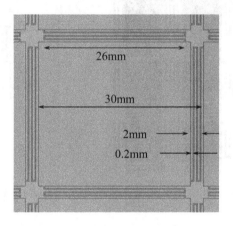

(b) Unit Cell Dimensions

Figure 9.179. Dimensions of the planar meander-line EBG (ML-EBG) structure used for analysis. (Courtesy of Omar M. Ramahi.)

9.179(a). Results of the simulations and measurements suppression performance of ML-EBG surface are presented in Figure 9.180 [45]. The data demonstrates that an ultrawide band gap is obtained using this new structure with strong agreement between simulation and measurements. The stop band frequency range is observed to extend from 450 MHz, approximately, to 12 GHz and beyond.*

---

*The definition of the bandwidth adopted here is the continuous frequency range over which the magnitude of the S_{21} ($|S_{21}|$) is maintained below –28 dB. Note that no standard definition exists for effective switching noise suppression bandwidth. The –28 dB threshold was selected as it represents significant suppression compared to the reference case.

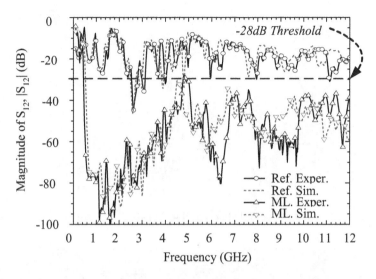

Figure 9.180. Simulation and measurement results for |S₂₁| of the planar meander-line EBG (ML) structure; comparison made with a two-layer power plane (Ref.). (Courtesy of Omar M. Ramahi.)

By introducing the concept of "supercell" EBG structure, even wider band Δ*I* noise suppression can be achieved. The idea behind the "supercell" is to create a cell comprising two patches with different topology. The "supercell" is then cascaded, resulting in a new structure that is expected to embody the band gaps arising from the use of each of the two topologies if they were used separately, in addition to the band gap arising from the periodicity formed by this new "supercell." Figure 9.181(a) shows the basic unit of this supercell structure. It shows adjacent EBG patches with two different connecting bridge topologies: a straight line and a meander line. The whole dimension of this structure, with 3 × 5 unit cells, was 90 mm × 150 mm, whereas the patch was kept at the same size as before, 30 mm × 30 mm. A uniform substrate of FR-4 with dielectric constant 4.4 and layer thickness of 1.54 mm was placed between the two layers of the structure. The fabricated PCB is shown in Figure 9.181(b) [46, 49].

In a manner similar to that above, performance of EBG structures was evaluated using the S_{21} parameter. The simulation results of "supercell" EBG surface (obtained using HFSS) are compared with measured data, as shown in Figure 9.182, and a good consistency is seen. Figure 9.182 also illustrates that not only is an appreciably wider band gap achieved but, more importantly, also a shift of the lower edge of the stop band downward to approximately 250 MHz. The total bandwidth covers from 250 MHz to 5 GHz. This bandwidth improvement obtained using a "supercell" EBG can thus be useful for implementing low-cost ultra wideband suppression of switching noise that is considered a fundamental bottleneck in high-speed printed circuit board design. In particular, this approach could eliminate the need for decoupling capacitors typically used for minimizing noise in the sub-500 MHz range.

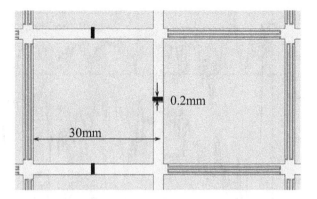

(a) Schematic Top View of a "Supercell" Structure

(b) Photograph of Cascaded "Supercell" EBG
Structure Showing Location of the Ports Used
for S Parameter Measurements

Figure 9.181. "Super cell" EBG structure. (Courtesy of Omar M. Ramahi.)

The supercell structure was conceived based on the principle of increasing the effective inductance to lower the center frequency of the band gap while not affecting the overall capacitance in order to maintain wide bandwidth. In addition, it is founded on the principle of increasing the spatial periodicity in order to decrease the lower edge of the band gap. For instance, if a single square patch of 60 mm × 60 mm was used in conjunction with the meander lines instead of the supercell, the lower edge of the band gap would be close to that of the supercell structure; however, the band gap would be much narrower. Finally, it is to be noted that there is no unique configuration for the connecting bridges between patches. These connecting meander

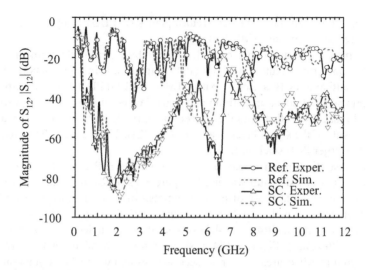

<u>Figure 9.182.</u> Simulation and measurement Results for |S₂₁| of the planar "supercell" EBG structure (SC); comparison made with a two-layer power plane. (Courtesy of Omar M. Ramahi.)

and straight lines can be modified to more complex structures, further enhancing the tunability of the band gap. The structures presented here and in [41] through [49] represent generic structures with increased degrees of freedom for further performance optimization.

9.5.5.5 Partial EBG (P-EBG) PDN Structures Using Remnants of Signal Layers.

A particularly efficient and effective approach for implementing high-impedance surfaces (HIS) using the EBG technique makes use of remnants of signal layers, which would normally be left unexploited (such unused areas are also known as signal layer "desert areas"). In high-speed designs, such areas are rarely used as local ground structures (known as "planelets," "micro islands," or "copper fills") due to the adverse effects they may have on high-speed signal propagation* and their minimal apparent benefits for high-speed propagation. On the other hand, optimizing the performance of the PCB power distribution network could be achieved by making use of such "desert areas" for the implementation of electromagnetic band-gap (EBG) structures made to function as planar band-stop filters for suppression of RF and Δ*I* noise throughout the PDN.†

Previously presented EBG structures such as the embedded EBG (E-EBG) (Section 9.5.5.2) and planar meander-line EBG (ML-EBG) (Section 9.5.5.4) make use of full and dedicated planes for the purpose of producing high-impedance structures, whether

---

*A detailed discussion of local ground structures can be found in Section 9.6.3.
†All material in this section is derived from [50]. Images provided courtesy of Junho Lee, Hynix Semiconductor Inc., South Korea.

by embeding between the power and return planes (E-EBG) or by employing a patterned power or return plane (ML-EBG).

An alternate technique is presented here whereby a partial electromagnetic bandgap (EBG) power distribution network (PDN) structure using remnants of the signal layer in a multilayer PCB is used. A partial EBG (P-EBG) PDN can be incorporated in a conventional PCB stack-up with no additional layers, thus maintaining low cost of the PCB. Figure 9.183 illustrates the concept of partial EBG PDN implementation utilizing remnants of the signal layer in a four-layer PCB. This concept, however, may be extended to larger PCB stack-ups.

As an illustration of performance of the partial EBG (P-EBG) PDN, a four-layer printed circuit board stack-up was used (Figure 1.184). Results for the P-EBG technique were compared to those of the conventional method of solid-copper-filled signal layer remnants as well as with a common four (0.5 oz) layer PCB stack-up, with no copper filling at all. FR4 ($\varepsilon_r = 4.3$) was used as the dielectric substrate, with dimensions as illustrated. Figure 9.184(a) shows a top view of the PCB. Note the area marked "ground filled area." A cross-sectional structure and layer stack-up of this area is shown in Figure 9.184(b)–(d) for the various configurations. Figure 9.184(b) represents a typical four-layer stack-up with no ground fills on the signal layer

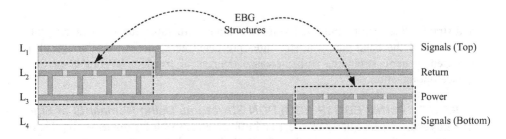

(a) Partial EBG Stricture using Top and Bottom Signal Layers

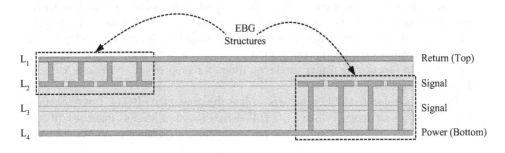

(b) Partial EBG Stricture using Inner Signal Layers

Figure 9.183. Concept of partial EBG PDN using remnants of the signal layer.

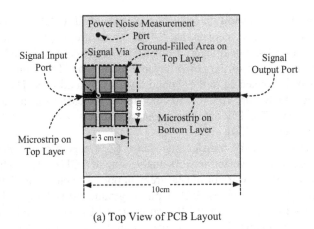

(a) Top View of PCB Layout

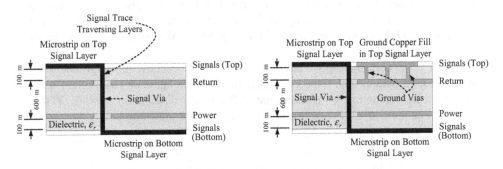

(b) No Ground Copper-Fill (NGF)

(c) Conventional Ground Copper Fill (CGF)

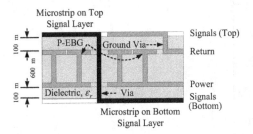

(d) Partial EBG PDN (P-EBG)

<u>Figure 9.184.</u> Cross-sectional view of the test PCB with a partial EBG PDN in the signal entry area versus conventional techniques.

(NGF). Figure 9.184(c) shows a conventional ground fill (CGF) on the signal layer, whereas Figure 9.184(d) depicts an P-EBG structure implementation using the ground fill area on the signal layer. The EBG structures are patterned on the ground layer and are connected to the power layer through vias. The ground filled area on the signal layer is connected to the ground layer by ground vias. The overall dimensions of the test PCB were 10 cm × 10 cm, and the dimensions of the ground fill area were 3 cm × 4 cm. The EBG structure consisted of a 3 × 4 array of 1 cm cells surrounding the signal via [50].

Figure 9.185 shows the transfer impedance (Z_{21}) curves of the PDN of the test PCBs measured with the vector network analyzer up to 5 GHz between the locations of the signal via and the power noise measurement port. The curves in Figure 9.185 represent the impedance of the test PCB with no ground fill (NGF), with conventional ground fill (CGF), and with a P-EBG structure (P-EBG). Several high-impedance peaks, which are generated by the power-to-ground cavity modes, are clearly observed. The effective bandwidth of the band gap appears to be from 1.3 GHz to 4 GHz. Figure 9.185 thus demonstrates that even partial EBG (P-EBG) structures can provide enhanced signal transmission quality and higher signal noise margin than conventional methods. In addition, the method provides suppression of PDN ΔI noise and better noise isolation [50].

Figure 9.186 demonstrates the advantage of P-EBG over conventional ground filled (CGF) PDN structures with respect to propagation of PPW-noise propagation (in terms of E-field propagation in V/m) across the PCB's power distribution network (PDN). Several time-domain stages of waveform propagation are shown for both.

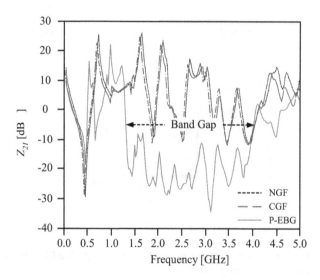

Figure 9.185. Transfer impedance (Z_{21}) on test PCBs with no ground fill (NGF), conventional ground fill (CGF), and partial EBG (P-EBG) between the signal via and power noise measurement port. (Courtesy of Junho Lee, Hynix Semiconductor Inc.)

In conclusion, even PDNs based on partial EBG structures exhibit enhanced EMC and signal integrity performance compared to that offered by conventional ground-fill structures. Furthermore, a particular benefit associated with the P-EBG structure is its usage of otherwise remnant "desert areas" on the signal layer, overcoming a disadvantage of conventional EBG-based PDNs, which typically require additional layers for this purpose. The partial EBG concept can also be extended to multilayer PCBs with stack-ups of more than four layers since inner signal layers can also serve for realization of partial EBG PDNs if "desert areas" are available in such planes.

9.5.5.6 Impact of EBG Structures on SI and Radiated EMI Emissions.
Although low-period coplanar EBG (LPC-EBG) structures offer clear advantages in comparison to conventional multilayer structures in minimizing the effects of ground bounce Δ*I* and PPW noise, the effect of these structures on signal integrity and radiated EMI emissions in high-speed digital circuits should be carefully considered. Since the coplanar EBG (LPC-EBG) structure is used as one of the power planes, high-speed signal traces will inevitably be routed in layers adjacent to the perforated embedded EBG or LPC-EBG power plane and may encounter degradation of signal quality (e.g., reflections and signal distortion) due to the inherent discontinuity arising from this imperfect plane, consequently resulting in possible system malfunction and increased radiated emissions from the PCB [51].

This section addresses aspects of signal integrity associated with power planes patterned with a planar EBG structure by focusing on the change in the eye diagram. In addition, the effect of EBG structure on stimulation of radiated emissions from the PCB is also discussed [43, 45].

The investigated planar EBG structure was identical to that presented in Figure 9.179, consisting of a two-layer power/ground plane with meander-line EBG structure, and its corresponding unit cell is shown in Figure 9.179(b) and overall dimension described in Figure 9.179(a). A uniform FR-4 substrate with dielectric constant 4.4 and layer thickness of 1.54 mm was used between the two layers. A meander line, 0.2 mm wide, provided a connecting bridge (i.e., "DC link") between adjacent patches.

For analysis of the effect of the EBG structure on signal integrity, a four-layer PCB structure (each layer filled with FR-4 of thickness 1.54 mm) is investigated. The signal is propagated along a 50 Ω equivalent microstrip line (with respect to an equivalent solid reference plane) traversing the EBG structure through a via down to the bottom layer and back to the top layer again (Figure 9.187).

A 1 Gbps nonreturn to zero (NRZ), pseudorandom binary sequence (PRBs) 27-1 with 1 V swing and nominal rise/fall time 200 ps was launched at input (Port 1), and the signal propagation properties were monitored at output (Port 2).

The simulated eye patterns obtained for the signal stream at the output of the structure are shown in Figure 9.188(a). The eye is significantly closed for this configuration, primarily due to the strong reflection arising from this nonuniform EBG patterned power plane [43, 45].

In order to improve the signal quality, a differential signaling approach was adopted. In the example considered here, the traces are designed with a differential impedance of 100 Ω-equivalent differential microstrip line (with respect to an equivalent sol-

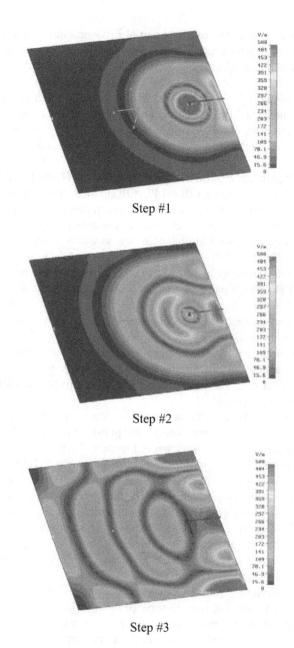

Step #1

Step #2

Step #3

(a) Conventional (Solid) Ground Fill (CGF) Structure

Figure 9.186. Time domain PPW-noise propagation (in terms of E-field, V/m). (Courtesy of Junho Lee, Hynix Semiconductor Inc.)

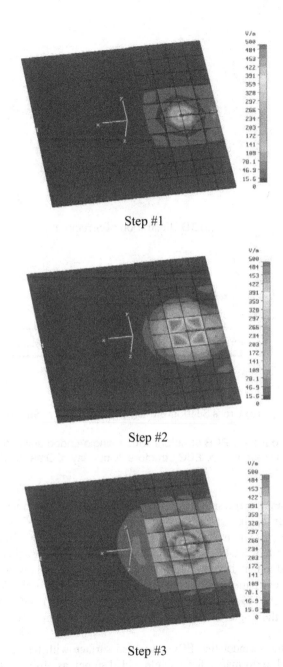

Step #1

Step #2

Step #3

(b) Partial EBG (P-EBG) Structure

Figure 9.186.   *Continued.*

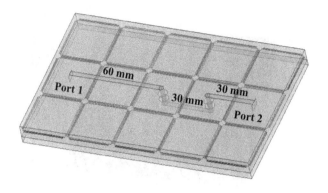

(a) 3D Illustration of Structure

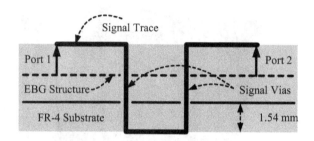

(b) Cross Section through the Four-layers Structure

Figure 9.187. Four-layer PCB structure with a single-ended signal trace routed along and traversing a meander-line EBG structure. (Courtesy of Omar M. Ramahi.)

id reference plane), traversing the EBG structure through two adjacent vias down to the bottom layer and back to the top layer again, as shown in Figure 9.189 [note that the cross section shown in Figure 9.187(b) also applies for each of the two differential traces]. The width of the lines is 2 mm, and the spacing between them is 3 mm. A test pattern identical to that described earlier was applied differentially at the two input terminals, and the eye pattern was monitored at the remote ends of the differential pair, as shown in Figure 9.189. It is evident that the signal quality is greatly improved compared to that of the single-ended configuration as observed in the eye pattern in Figure 9.188(b) [45].

Replacing the meander-line EBG-patterned surface with the "supercell" structure yields enhanced performance for a single-ended signal, as observed from the eye pattern shown in Figure 9.190. The test pattern is the same as that described above. Comparing the results of Figure 9.190 and Figure 9.188(b), we observe that the eye pattern for the "supercell" EBG-based structure is more open than the case of meander line for the single-ended configuration. This is possibly due to the fact that the meander line structure introduces extra phase shift per unit length (due to increased inductance)

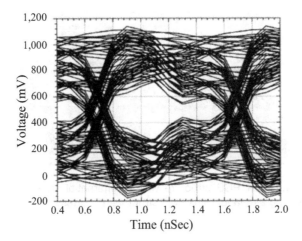

(a) Single-Ended Trace ML-EBG Patterned PCB

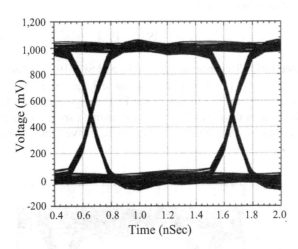

(b) Differential Pair ML-EBG Patterned PCB

Figure 9.188. Simulated eye patterns for signal traces traversing a meander-line EBG (ML-EBG) structure. (Courtesy of Omar M. Ramahi.)

when the signal propagates along the traces, resulting in degradation of signal quality [45].

With respect to radiated emissions (or leakage radiation) through LPC-EBG-patterned PCBs, previous studies demonstrated that radiation from solid power planes emerges at the board edges in a manner similar to radiation from a patch antenna [45]. If a signal is fed at one port of the PCB and received at another, a fraction of the energy radiates at the edge of the board, thus never reaching any of the other ports. The ra-

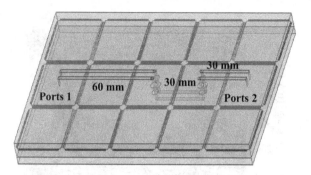

Figure 9.189. Four-layer PCB structure with a differential signal traces pair routed along and traversing a meander-line EBG structure. (Courtesy of Omar M. Ramahi.)

diation loss, L_R, can thus be expressed using S_{11} and S_{12}, representing the reflection and transmission coefficients of the structure, respectively (note that S_{11} and S_{12} are expressed in their absolute, non-dB, values):

$$L_R = 1 - |S_{11}|^2 - |S_{21}|^2 \qquad (9.90)$$

For the perforated LPC-EBG structures (i.e., ML and SC EBG structures), the gaps between the cells could result in increased radiation leakage, especially at frequencies at which the patches resonate, in addition to the PCB's edge radiation. In three-layer or multilayer PCBs with embedded EBG structures using a solid power plane, the effect of LPC-EBG perforation of one of the planes on excessive radiation is expected to be

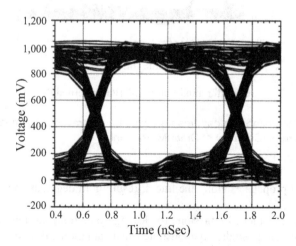

Figure 9.190. Simulated eye pattern for a single-ended signal trace traversing a "supercell" EBG structure. (Courtesy of Omar M. Ramahi.)

much smaller. Strongly coupled differential pairs provide superior EMI performance except at resonant frequencies of LPC-EBG. Further, the larger the coupling coefficient for differential lines is, the less common-mode noise is present [51].

Results of computation of the loss factor, $L_R = -|S_{11}|^2 - |S_{21}|^2$, are examined for EBG-based power planes and compared to the case with a solid power plane (Ref.). Figure 9.191 demonstrates that the radiation peaks at 480 and 960 MHz, which arise from the cavity modes of the 90 mm × 150 mm reference board, are significantly suppressed for the EBG-patterned structures. The strong radiation peaks at 4.7, 6.6, 9.4, and 10.6 GHz for the EBG-patterned power planes are due to the patch resonance [45].

The distinct features of EBG-patterned structures, such as reducing the total radiation, suppressing the resonance modes and surface current propagation, and localizing the field leakage around the excitation region, suggest the use of the EBG-patterned surface as an electromagnetic shield to reduce the radiation from power buses of PCBs. The shielding performance of perforated power planes is demonstrated using near-field emission measurements, whereby the performance of meander-line (ML) and supercell (SC) EBG structures is evaluated. Comparison is made to a reference structure in which both top and bottom layers are made of solid metal [45].

The shielding effectiveness of the EBG planar structure was evaluated by measuring the total amount of radiation from both the edges of the board and open slots in terms of the S_{21} parameter, representative of the radiated power, at various points located at normalized distances of 0.15 (source), 0.44, 0.73, and 1 (far end of board edge) along a "test line" on the PCB (Figure 9.192).

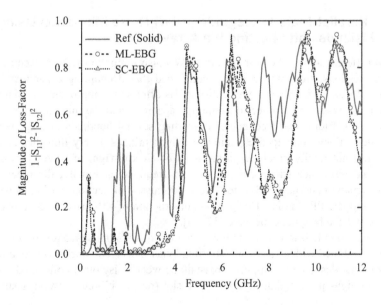

Figure 9.191. Radiation loss for planar "supercell" (SC) and meander-line (ML) EBG structures compared to a two-layer power plane. (Courtesy of Omar M. Ramahi.)

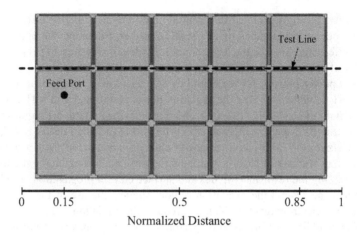

Figure 9.192. S_{21} Near-field measurement test points. (Courtesy of Omar M. Ramahi.)

Shielding effectiveness measurement results are shown in Figure 9.193, presented in terms of the S_{21} parameter. It is clearly demonstrated that for planar EBG-based power planes at a distance spanning a few patches away from the excitation point, wider and deeper band gaps may be achieved. Average suppression of 20 dB is achieved across a wide frequency range from 0.5 to 4.6 GHz and 7 to 12 GHz in comparison to the reference solid-planes case.

9.5.6 Parallel-Plate Waveguide (PPW) Noise Mitigation Using Virtual Islands and Shorting Via Arrays

Several different approaches for ΔI and PPW noise mitigation were presented in the previous sections, from which it was concluded that decoupling capacitors located near signal vias or distributed over entire PCBs offer some improvement. Many vias, however, are needed to connect decoupling capacitors and the mitigation level of ΔI noise is not sufficient for frequencies exceeding several hundreds of megahertz, approximately. Even embedded capacitance, which utilizes a very thin dielectric substrate, does not totally preclude propagation of electromagnetic waves between the planes, and resonance still occurs at specific frequencies. Finally, the recent novel electromagnetic band gap (EBG) technique (discussed in Section 9.5.5) has been applied to mitigate PPW noise. Design and construction of EBG structures, however, is complex and can be quite expensive [52, 53].

An alternative hybrid technique for mitigation of PPW noise makes use of "virtual islands" combined with an array of "shorting" vias. This technique is used in a manner similar to that designed to suppress crosstalk between noisy other collocated sensitive circuits in high-speed digital designs.* The slot (or "moat") cut between islands and

*See Sections 9.6.3 and 9.7 for a discussion of implementation of slots in PCB power and return planes for control of crosstalk.

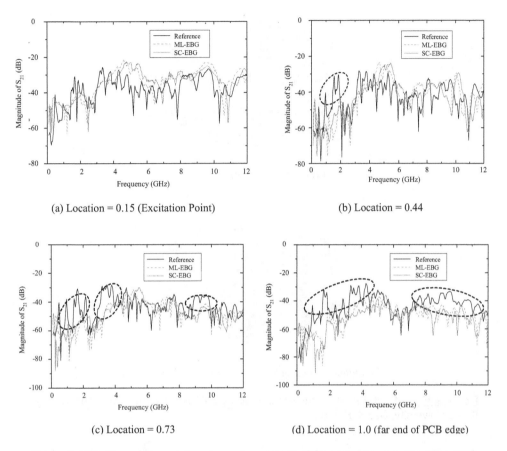

(a) Location = 0.15 (Excitation Point)   (b) Location = 0.44

(c) Location = 0.73   (d) Location = 1.0 (far end of PCB edge)

Figure 9.193. S_{21} measured for planar "supercell" (SC) and meander-line (ML) EBG structures compared to a two-layer power plane (reference) at various test points. (Courtesy of Omar M. Ramahi.)

the rest of the plane is intended to prevent the propagation of the PPW noise from the islands to the larger plane area of the PCB at higher frequencies. The shorting vias are used to provide a low-impedance signal return current path.*

Figure 9.194(a) depicts a three-dimensional structure of conventional via traces traversing layers in a multilayer PCB. Virtual islands with shorting vias are introduced to mitigate the PPW mode excited from signal vias. The three-dimensional structure of signal vias applied to the virtual islands with shorting vias is illustrated in Figure 9.194(b). The virtual islands centered on signal vias are formed by slots on the top and bottom planes. The rectangular conductor on the midplane is striped (note the complementary orientation of the slots in the planes).

---

*Although the terms "Δ*I* noise" and "parallel-plate waveguide (PPW) noise" are considered synonymous, the term PPW noise is used here for consistency with the referenced sources.

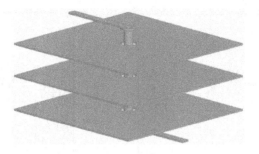

(a) Conventional Signal Vias

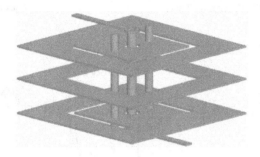

(b) Signal Vias using Virtual Islands with Shorting Vias

Figure 9.194. Structures of signal vias in multilayer PCBs. (Courtesy of Jichai Jeong, Korea University, Seoul, Korea.)

Two shorting vias are used to interconnect the top and bottom planes. These shorting vias in near proximity to the signal vias provide a return current path with low impedance and mitigate the propagation of the PPW noise at low frequencies. The slots on the top and bottom planes block the propagation of the PPW noise from the virtual islands to the remainder of the PDN at high frequencies. Transmission lines should not be allowed to be routed over the slots as this may become another source of mode conversion.*

Figure 9.194 depicts the top and cross-sectional views of a four-layer test PCB model used to illustrate the characteristics of PPW noise propagation in a structure formed by the power and return planes and the effect of the virtual island with shorting vias on PPW noise mitigation. The size and thickness of the PPW formed by the PCB's power and return planes are 10 cm × 10 cm and 0.787 mm, correspondingly. The dielectric material of the dielectric substrate of the PCB is Taconic TLC with relative dielectric constant, ε_r, of 3.2.

The source and probe ports were located at (50, 50) and (50, 15), respectively (see Figure 9.195). The virtual island with the size of 10 mm × 10 mm and gap width of 2

*The adverse effects of current return path discontinuities are discussed in detail in Section 9.4.

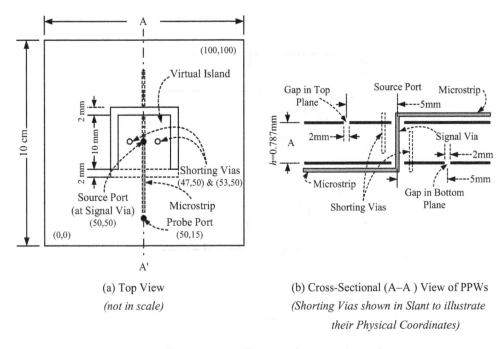

(a) Top View
(not in scale)

(b) Cross-Sectional (A–A) View of PPWs
(Shorting Vias shown in Slant to illustrate their Physical Coordinates)

(All units are in millimeters; drawing not to scale)

<u>Figure 9.195.</u> Basic configuration of virtual island and shorting vias in a four-layer PCB. (Courtesy of Jichai Jeong, Korea University, Seoul, Korea, and Antonio C. Scogna, CST of America Inc.)

mm was used to mitigate the PPW noise excited from the source port. A signal via is located in the middle of the virtual island and two shorting vias are positioned at two sides of the signal via, separated 3 mm away from the source port.

For evaluating the contribution of the virtual island and shorting vias to PPW noise mitigation, the three configurations illustrated in Figure 9.196 were analyzed: (a) conventional source port, (b) source port with two shorting vias, and (c) source port applied to the virtual island with shorting vias. Shorting vias, instead of bypass capacitors, were used to provide a RF return current path between the top and bottom layers in order to simplify the analysis.

The transmission response (in terms of the S_{21} parameter) of PPW noise propagation from the source port to the probe port, located at (50, 50) and (50, 15) respectively, was computed using the FDTD method for the above three cases and presented in Figure 9.197. Figure 9.198 illustrates graphically the propagation of power-to-return-layer PPW noise across the entire PCB, very clearly demonstrating the suppression of noise [53].

Analysis results (see Figure 9.197) demonstrate that with a conventional source port, most of the excited PPW noise propagates efficiently to the probe port through PPWs at some frequencies (top plot). Noise propagation is suppressed by approximately 10 dB thanks to the lower impedance of return current path provided by the

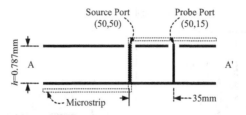

(a) Conventional Source Port

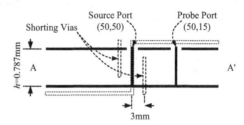

(b) Source Port with Two
Shorting Vias

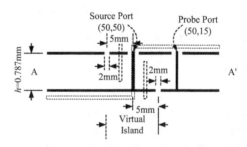

(c) Source Port applied to Virtual
Island with Two Shorting Vias

Figure 9.196. Three configurations of virtual island and shorting vias for investigation of PPW noise propagation. (Courtesy of Jichai Jeong, Korea University, Seoul, Korea.)

shorting vias (middle plot). When the source port is applied to the virtual island with shorting vias, however, transmission of PPW noise from the source to the probe ports is reduced by more than 25 dB while extending the frequency band for effective PPW noise mitigation up to approximately 5 GHz. Computed values were also supported by measurement for all cases [53].

Further improvement of PPW noise suppression is achieved by reducing the separation between the shorting vias and the signal via as illustrated through results of simu-

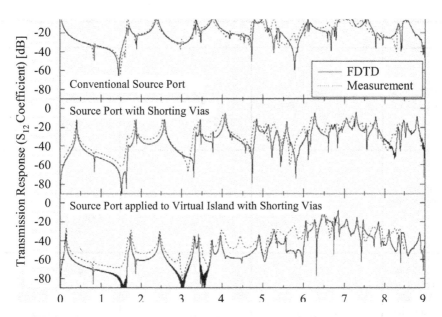

Figure 9.197. Calculated and measured transmission response (S_{21} coefficient) of PPW noise from the source to probe ports in a four-layer PCB. (Courtesy of Jichai Jeong, Korea University, Seoul, Korea.)

lations* for three different distances between the 2 shorting vias and the signal via, $d = 3$ mm, $d = 2$ mm, and $d = 1$ mm (Figure 9.199). A comparison is also made between the PPW noise suppression obtained for near ($d = 1$ mm) shorting vias and a virtual island with shorting vias ($d = 3$ mm) schemes (Figure 9.200).

A significant improvement is observed with reduction of separation, d, thanks to the diminishing loop produced between the signal via and the shorting vias acting as current return paths. Furthermore, suppression obtained in the case of near shorting vias ($d = 1$ mm) is comparable to that obtained by using two shorting vias and a virtual island up to 5 GHz. Above 5 GHz, the performance of the design with near shorting vias is superior to that obtained with the virtual island and shorting vias. This is attributed to the fact that electromagnetic energy can couple through splits, particularly with increase of frequency. Observation of Figure 9.200 reveals, though, that in both schemes a few consistent resonances are still present at several frequencies (1.9 GHz, 2.4 GHz, 3.6 GHz, and 5.4 GHz). These resonant frequencies can actually be evaluated by means of the following equation [52]:

$$f_{mn} = \frac{1}{2a\sqrt{\mu_0\varepsilon_0\varepsilon_r}}\sqrt{m^2 + n^2} \qquad (9.91)$$

*Simulations were run using CST Microwave Studio (CST MWS).

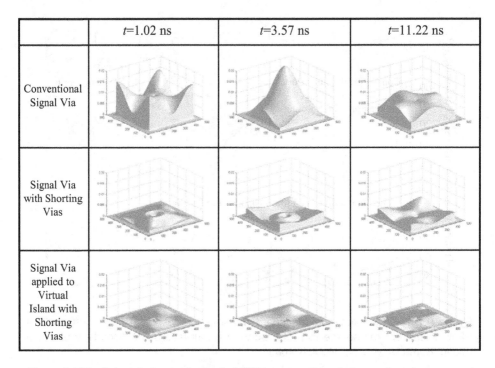

	t=1.02 ns	*t*=3.57 ns	*t*=11.22 ns
Conventional Signal Via			
Signal Via with Shorting Vias			
Signal Via applied to Virtual Island with Shorting Vias			

<u>Figure 9.198.</u> Calculated magnitude of of PPW noise voltage between (inner) power and return layers across the PCBs. (Courtesy of Jichai Jeong, Korea University, Seoul, Korea.)

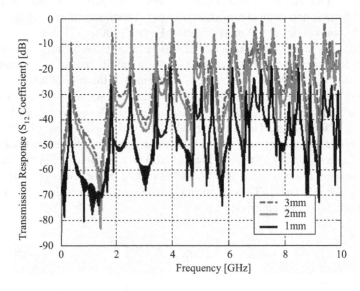

<u>Figure 9.199.</u> S_{21} coefficient for various distances, *d*, between shorting vias and signal via. (Courtesy of Antonio C. Scogna, CST of America Inc.)

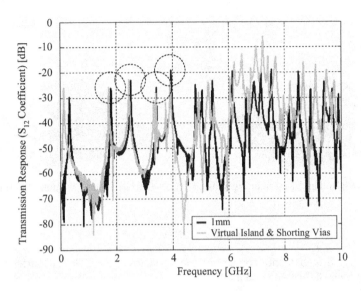

<u>Figure 9.200.</u> S_{21} coefficient for near (d = 1 mm) shorting vias and virtual island with shorting vias (d = 3 mm) schemes. (Courtesy of Antonio C. Scogna, CST of America Inc.)

where m and n are integers, a = 10 cm, and ε_0 and ε_r are the permittivity of free space and relative permittivity of the PCB dielectric substrate (ε_r = 3.2). For achieving even higher PPW noise isolation, other techniques (e.g., arrays of shorting vias or electromagnetic band gap structure) should be used in conjunction with plane segmentation.

Figure 9.201 shows the E-field distribution around the signal via at 5 GHz for the following two cases: (a) virtual island with shorting vias and (b) near (d = 1 mm) shorting vias. It is evident that the amplitude of the field at resonance on the signal via in case (b) is significantly smaller, although in both cases standing waves propagating on the planes are still observable [52].

Finally, the effect of an array of shorting vias is presented. In particular, three different cases are analyzed: (a) virtual island with four shorting vias, (b) array of shorting vias and, (c) virtual island with array of shorting vias. The horizontal dimensions of the metal planes are the same as in Figure 9.195 and the shorting vias array is placed at a spacing of 3 mm from the signal via (1) with a pitch of 1.5 mm (Figure 9.202).

The effectiveness of an array of shorting vias is presented in Figure 9.203 in terms of the S_{21} parameter for all configurations, clearly pointing out the benefit of the virtual island and array of shorting vias scheme, which offers noise suppression of –60 dB across the entire frequency range, whereas removing the virtual island from the same configuration shows noise suppression of –40 dB and only up to 6 GHz. In this case, the splitting on the power and return planes (virtual island) is of little avail for achieving better isolation [50].

Figure 9.203 demonstrates that a virtual island with two near (d = 1 mm) shorting vias provides moderate noise suppression, on the order of approximately 20 dB.

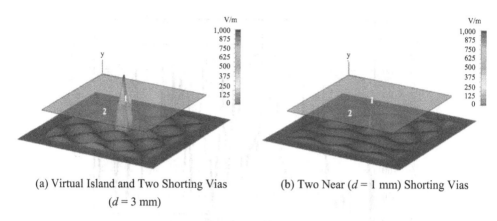

(a) Virtual Island and Two Shorting Vias   (b) Two Near ($d = 1$ mm) Shorting Vias

($d = 3$ mm)

(Port 1: Source; Port 2: Probe)

Figure 9.201. E-field distribution around the signal via at $F = 5$ GHz. (Courtesy of Antonio C. Scogna, CST of America Inc.)

Replacing the two shorting vias with an array of shorting vias, as shown in Figure 9.202, or, alternatively, using four shorting vias together with a virtual island, results in enhanced noise suppression to a total of 40 dB. An additional 20 dB of noise mitigation is gained from the combination of the shorting vias array with a virtual island.

Although the technique represented by the combination of virtual island and array of shorting vias promises effective PPW noise mitigation in the PDN, the potential detrimental effect of the etched slots in the power plane on signal integrity in signal traces routed adjacent to the imperfect power plane is of concern. For evaluating the

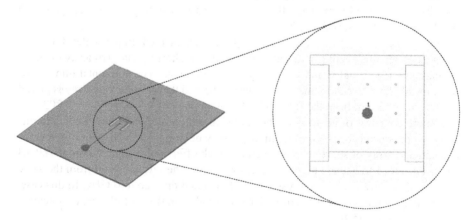

Figure 9.202. Configuration of a virtual island with an array of shorting vias. (Courtesy of Antonio C. Scogna, CST of America Inc.)

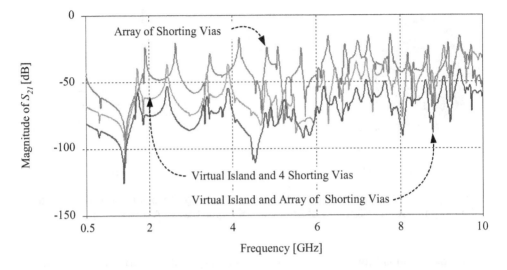

(a) S_{21} Coefficient (Insertion Loss) for Virtual Island and Various Configurations of Shorting Vias

(b) Configuration of a Two Shorting Via Array   (c) Configuration of Four Shorting Via Array

within Virtual Island   within Vrtual Island

Figure 9.203. Effectiveness of an array of shorting vias, in terms of the S_{21} coefficient. (Courtesy of Antonio C. Scogna, CST of America Inc.)

effect of the virtual island on signal integrity, a 35 mm long single-ended microstrip trace is routed such that it traverses the power and return layers from top to bottom by means of a via with radius of 0.1 mm (or 4 mils) (Figure 9.204).

The signal integrity of two configurations is studied: signal traces with virtual island on PDN power and return planes, and with continuous planes. Only a small variation between the two computed TDR (time domain reflectometer) waveforms is observed [Figure 9.205(a)]. Figure 9.205(b) demonstrates the effect of the virtual island by illustrating the S_{21} coefficient obtained with a virtual island compared to that found with continuous power and return planes. When virtual islands are employed, a small resonance appears at approximately 9.5 GHz. Nevertheless, across the entire frequency

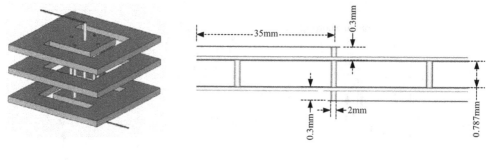

(a) 3D View of Model   (b) Stackup and Dimensions of Model

Figure 9.204. Single-ended microstrip trace traversing the power and return layers from top to bottom. (Courtesy of Antonio C. Scogna, CST of America Inc.)

band the variation between the two waveforms is very small (< 3 dB) and can certainly be considered acceptable [52, 53].

Lessons Learned

- ΔI noise arises whenever transient current flows through the PCB's power distribution network (PDN).

- Electromagnetic waves launched across the PDN [known as "parallel-plate waveguide (PPW) noise"] produce standing wave patterns at the PDN cavity resonance frequencies.

- Transmission of some of the energy from the edges of the PCB produce excessive radiated emissions in addition to power and ground bounce.

- Mitigation of the adverse effects of ΔI noise on the PDN can be achieved by using low-impedance power and return connections as well as a maintaining low interplane impedance.

- Power (V_{CC}) and return (GND) planes should be closely spaced and allocated as close to the top surfaces of the PCB stack-up when high transient current demand exists (e.g., for FPGA core supply).

- A well-designed decoupling scheme, comprised of a "brigade" of bulk and decoupling capacitors, serves as a low-impedance, high-frequency shunt between the power and return planes.

- Interplane capacitance supplements discrete decoupling capacitors and is particularly effective at higher frequencies.

- Electromagnetic band gap (EBG) high-impedance structures (HIS) provide efficient suppression of radiated electromagnetic emissions from PCB power distribution networks (PDNs) at frequencies at which conventional methods are no longer effective (i.e., 0.5 GHz and up).

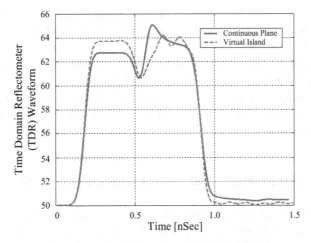

(a) Time Domain Reflectometer (TDR) Waveform

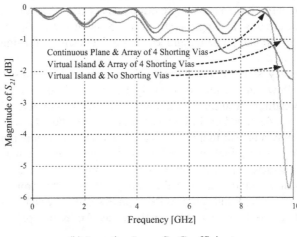

(b) Insertion Loss, S_{21} Coefficient

Figure 9.205. Signal integrity analysis for a single-ended microstrip trace traversing the power and return layers. (Courtesy of Antonio C. Scogna, CST of America Inc.)

- Ultrawideband (UWB) suppression of radiated emissions is achieved by low-period coplanar EBG (LPC-EBG) structures or by cascaded high-impedance surfaces (C-HIS) structures.
- Partial EBG (P-EBG) structures, utilizing otherwise remnant "desert areas" on the signal layer, exhibit enhanced EMC and signal integrity performance compared to that offered by conventional ground-fill structures.

- Shorting vias near the signal via or a combination of a virtual island with an array of shorting vias constitutes an alternate and efficient technique for mitigation of PPW noise.
- The performance of a virtual island with two shorting vias is comparable to that obtained with four shorting vias and no virtual island.
- For high-efficiency of PPW noise mitigation (on the order of 60 dB up to 10 GHz, approximately) the combination of a virtual island and an array of shorting vias is necessary.

9.6 RETURN PLANES AND PCB LAYER STACK-UP

Multilayer printed circuit boards provide some significant advantages associated with effective signal propagation and power distribution, following the principles discussed in previous sections. From the grounding perspective, particular attention is paid to allocation of return ("ground") layers for providing signal and power current return paths and for shielding against EMI radiated emissions. The importance of maintaining large low-impedance return planes is critical to all contemporary analog and digital circuits. Placing the signal or power distribution conductors in close proximity to the ground plane minimizes the net inductance of a signal or power circuit.

High-frequency signal return current, if not interrupted by discontinuities in the return plane, flows in a path immediately beneath and following the signal trace (with little spreading), resulting in the lowest impedance possible. High-speed (digital or analog, single-ended or differential) signal traces should always be routed in a layer immediately adjacent to a return plane layer using transmission line techniques (e.g., microstrip or stripline) for which controlled impedance is required (recall that the required characteristic impedance of the transmission line is dominated by its geometry). Bear in mind that from the standpoint of signal propagation, solid and continuous power (V_{CC}) and return (GND) planes can be equally effective for supporting signal propagation, and may both act as AC return paths in transmission lines.*

Allocation of adjacent power and return planes in the PCB stack-up also has a significant impact on the inductance (and characteristic impedance) of the power distribution system of the PCB. Unlike signal circuits in which relatively high characteristic impedance (typically in the range of 50 to 100 Ω) is desirable, power distribution systems should exhibit the lowest impedance attainable for minimizing the transient voltage and SSN/ground bounce effects.

Finally, return (particularly "ground") planes provide some isolation between signal traces routing layers, reduce EMI emissions from the signal traces, and increase the immunity of the circuit to external radiated EMI, thanks to the improved magnetic flux cancellation and the shielding provided by solid planes.

---

*The rationale for this phenomenon was discussed in detail in Section 9.3.3.

To facilitate the above objectives, the following design principles should be maintained:

1. Power (V_{CC}) and return (GND) planes should be paired in adjacent, closely spaced layers to maximize interplane capacitance and to reduce spreading inductance.
2. Power (V_{CC}) layers with high transient current (e.g., FPGA core supply) should be allocated as close to the top surface (FPGA side) of the PCB stack-up to decrease the vertical length of the transient current path through V_{CC} and GND vias. Such layers should have a paired adjacent ground plane.
3. The signal layer should be adjacent to at least one AC return (V_{CC} or GND) plane layer.
4. High-speed signal layers should be assigned an appropriate transmission line topology with ground plane layer(s).
5. Power (V_{CC}) and return (GND) plane pairs should be interspersed with signal layer pairs. No more than two signal layers should be alternated with a power–return pair.
6. Adjacent signal layers should have traces routed perpendicularly to each other for minimizing coupling (crosstalk) between traces.

9.6.1 Image Planes

Experimental and theoretical practice has pointed to the conversion of desired differential-mode (DM) signals into undesired common-mode (CM) interference, which is consequently radiated from traces or planes, primarily when common-mode currents are allowed to flow on attached cables and other peripheral conductors acting as accidental antennae. Likewise, owing to reciprocity, the reverse process (CM-to-DM conversion) is a significant cause of interference coupling to the PCB and may interfere with circuit's performance. This situation is illustrated in Figure 9.206 (compare this situation to that shown in Figure 9.149).

Consider a pair of identical PCB signal traces routed at a height h above and in parallel to an adjacent conductive plane (but with no electrical connection to that plane), as depicted in Figure 9.207(a). The equivalent image currents* on this conductive plane produce radiated fields that tend to neutralize the fields from the currents (common-mode or differential-mode alike) flowing through the PCB traces. By electromagnetic image theory, the PCB trace currents and the conductive plane are equivalent to a four-trace system illustrated in Figure 9.207(b), in which the image currents are equal but opposite to the original trace currents. The voltages produced across the original traces now become

$$V_1 = L_{p11}\frac{dI_1}{dt} - L_{p12}\frac{dI_2}{dt} - L_{p13}\frac{dI_3}{dt} + L_{p14}\frac{dI_4}{dt}, \text{ volts} \qquad (9.92)$$

---

*From the standpoint of electromagnetic field theory, the "image currents" are a mere circuit-equivalent representation of the boundary conditions of tangential electromagnetic fields on the surface of a perfectly conductive surface. Boundary conditions are discussed in Chapter 2.

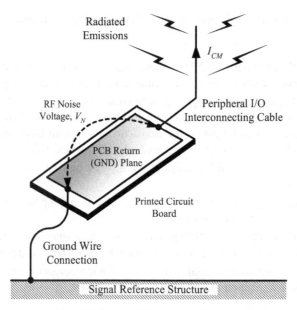

(a) Physical Representation

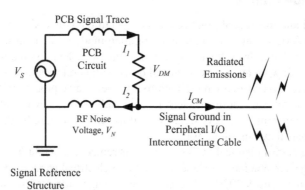

(b) Circuit Equivalent Representation

<u>Figure 9.206.</u> Radiated emissions from ground noise on a PCB with I/O cable attached.

$$V_2 = L_{p22} \frac{dI_2}{dt} - L_{p21} \frac{dI_1}{dt} - L_{p24} \frac{dI_4}{dt} + L_{p23} \frac{dI_3}{dt}, \text{ volts} \qquad (9.93)$$

where:
L_{Pij} = partial self ($i = j$) and mutual ($i \neq j$) inductances of the PCB traces, henrys
I_i = current flowing on the ith trace ($i = 1, \ldots, 4$), amperes
V_i = voltage produced on the ith PCB trace ($i = 1, \ldots, 2$), volts

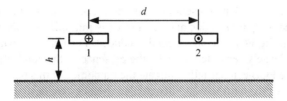

(a) Trace Pair above Conducting ("Image") Plane

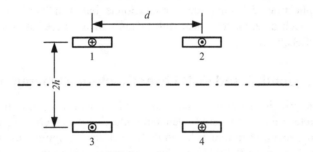

(b) Four-Trace Equivalent System

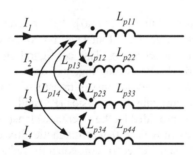

(c) Equivalent Partial Inductance Model

Figure 9.207. Image plane placed beneath PCB signal traces and their equivalent images.

Since $I_1 = I_2 = I_3 = I_4 = I$, Equations (9.92) and (9.93) can be expressed as

$$L_{p1} = L_{p11} - L_{p12} - L_{p13} + L_{p14}, \text{ henry} \tag{9.94}$$

$$L_{p2} = L_{p22} - L_{p21} - L_{p24} + L_{p23}, \text{ henry} \tag{9.95}$$

where L_{Pi} is the net partial self inductances of the PCB ith trace ($i = 1, \ldots, 2$) in henrys.

With a conductive plane present beneath the PCB, the ground noise voltage developed across the net partial inductance of a trace is minimized. Subsequently, radiated emissions from the PCB are significantly reduced. Furthermore, when signal and return traces are largely separated ($d \gg h$), the presence of the conducting plane beneath and close to the PCB can partially mitigate the outcomes of this poor design. This conducting plane is referred to as an "image plane" since this desirable effect is entirely produced by its equivalent image currents [2].

The most important point to note here is that radiated emissions are suppressed dramatically even though the plane is not electrically connected to the PCB circuits. The practical application of this important conclusion is that even "nonfunctional" isolated PCB planes such as embedded heat-sink layers will nevertheless contribute to EMI control on PCBs.*

9.6.2   Frequently Used PCB Layer Stack-up Configurations

Based on the principles outlined above, particularly with respect to the allocation of power and return planes (generally denoted collectively as "reference planes") in the PCB stack-up, several commonly found PCB stack-up configurations are listed below and illustrated in Figure 9.208. Particular attention is paid to the role of the reference planes with respect to their EMC and signal integrity objectives, as well as the desire to maintain pairing of the solid plane layers about the PCB stack-up centerline.†

1. **Four-Layer Stack-up** [Figure 9.208(a)]. A four-layer PCB is the most fundamental multilayer stack-up. On a four-layer board, power and return planes are usually placed on the inner layers. This configuration provides optimal power distribution network (PDN) decoupling, but does not provide for inner-layer routing of signal traces. Only microstrip transmission line topologies are available in this stack-up, utilizing the print side (PS) and component side (CS) of the PCB. No shielding is provided by the reference plane(s) but the proximity to the reference plane does provide efficient magnetic flux cancellation and electric flux termination ("image" effects), consequently reducing radiated EMI emissions from the exposed signal traces. In this stack-up, a yield (fraction of signal layers out of the entire stackup) of 50% is achieved.

2. **Six-Layer Stack-up** [Figure 9.208(b)]. A six-layer PCB provides an increased yield (67%) compared to the four-layer plane. Four of the six layers constitute signal layers, two of which are embedded between the power and return layers; the remaining two are routed as microstrip transmission lines in the external (PS and CS) layers of the PCB. Somewhat less interplane decoupling exists due to the increased spacing between the power and return layers but, on the other hand, two signal layers of the four (Signal #2 and Signal #3) are embedded be-

---

*Notwithstanding the above, due consideration should be given to the grounding of such planes. Isolated planes may introduce other difficulties such as ESD control or safety concerns.
†Solid plane pairing is not an EMC- or SI-related requirement but rather stems from thermal considerations, in order to preclude warping of the PCB as temperature varies.

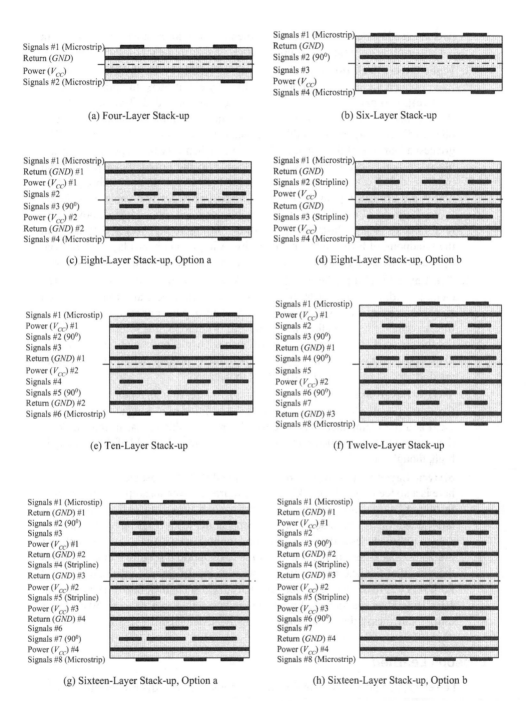

Signals #1 (Microstrip)
Return (*GND*)
Power (*V*$_{CC}$)
Signals #2 (Microstrip)

(a) Four-Layer Stack-up

Signals #1 (Microstrip)
Return (*GND*)
Signals #2 (90°)
Signals #3
Power (*V*$_{CC}$)
Signals #4 (Microstrip)

(b) Six-Layer Stack-up

Signals #1 (Microstrip)
Return (*GND*) #1
Power (*V*$_{CC}$) #1
Signals #2
Signals #3 (90°)
Power (*V*$_{CC}$) #2
Return (*GND*) #2
Signals #4 (Microstrip)

(c) Eight-Layer Stack-up, Option a

Signals #1 (Microstrip)
Return (*GND*)
Signals #2 (Stripline)
Power (*V*$_{CC}$)
Return (*GND*)
Signals #3 (Stripline)
Power (*V*$_{CC}$)
Signals #4 (Microstrip)

(d) Eight-Layer Stack-up, Option b

Signals #1 (Microstip)
Power (*V*$_{CC}$) #1
Signals #2 (90°)
Signals #3
Return (*GND*) #1
Power (*V*$_{CC}$) #2
Signals #4
Signals #5 (90°)
Return (*GND*) #2
Signals #6 (Microstrip)

(e) Ten-Layer Stack-up

Signals #1 (Microstip)
Power (*V*$_{CC}$) #1
Signals #2
Signals #3 (90°)
Return (*GND*) #1
Signals #4 (90°)
Signals #5
Power (*V*$_{CC}$) #2
Signals #6 (90°)
Signals #7
Return (*GND*) #3
Signals #8 (Microstrip)

(f) Twelve-Layer Stack-up

Signals #1 (Microstip)
Return (*GND*) #1
Signals #2 (90°)
Signals #3
Power (*V*$_{CC}$) #1
Return (*GND*) #2
Signals #4 (Stripline)
Return (*GND*) #3
Power (*V*$_{CC}$) #2
Signals #5 (Stripline)
Power (*V*$_{CC}$) #3
Return (*GND*) #4
Signals #6
Signals #7 (90°)
Power (*V*$_{CC}$) #4
Signals #8 (Microstrip)

(g) Sixteen-Layer Stack-up, Option a

Signals #1 (Microstip)
Return (*GND*) #1
Power (*V*$_{CC}$) #1
Signals #2
Signals #3 (90°)
Return (*GND*) #2
Signals #4 (Stripline)
Return (*GND*) #3
Power (*V*$_{CC}$) #2
Signals #5 (Stripline)
Power (*V*$_{CC}$) #3
Signals #6 (90°)
Signals #7
Return (*GND*) #4
Power (*V*$_{CC}$) #4
Signals #8 (Microstrip)

(h) Sixteen-Layer Stack-up, Option b

Figure 9.208. Traditional multilayer PCB stack-up configurations.

tween these layers, normally routed perpendicularly in the adjacent layers, or in a dual-stripline or broadside-coupled differential stripline topologies.

3. **Eight-Layer Stack-up.** Several options are available in the eight-layer stackup, two of which are presented, both providing a yield of 50%. In option a [Figure 9.208(c)], two paired power and return layers are provided. Two signal layers (#1 and #4) carry traces in microstrip configurations in the outer layers and two others (#3 and #4) are embedded between the two power (V_{CC}) planes. The greatest advantage of this stack-up is observed when very high-speed PFGAs are placed in both the CS and PS layers. Thanks to the proximity of paired (and tightly coupled) power and return layers, the PDN impedance and subsequent ΔI noise* are critical. Option b [Figure 9.208(d)] provides two centered stripline layers (Signals #2 and #3), as well as tightly coupled (but centered) power and return planes. The eight-layer design shown in Figure 9.208(c) is optimal from the standpoint of EMC and signal integrity. A larger number of layers are driven by routing needs. The subsequent examples present several such configurations.

4. **Ten-Layer Stack-up** [Figure 9.208(e)]. A ten-layer stack-up is especially efficient, providing a yield of 60% with two microstrip layers and four embedded signal layer pairs. Each signal layer has an adjacent power or return plane; therefore, excellent flux cancellation is achieved. In addition, this stack-up includes a tightly coupled power–return pair of planes, while additional power–return planes (spaced by two signal layers), when interconnected with the central layers, further reduce their respective PDN impedance.

5. **Twelve-Layer Stack-up** [Figure 9.208(f)]. Increasing the number of layers from ten to twelve layers only allows for the addition of two signal layers if high yield is to be maintained. The undesired impact of this is that in this stack-up there are no tightly coupled power–return layer pairs. The yield of this stackup is high, though—67%.

6. **Sixteen-Layer Stack-up** [Figure 9.208(g) and (h)]. The last stack-up discussed here is a sixteen-layer PCB, skipping the fourteen-layer stackup. The interesting feature of this stack-up, which again offers only a 50% yield, is that it is particularly suited for very high-speed signal propagation, while offering tightly coupled power–return planes and possible stripline routing layers. Again, two examples of preferable stack-ups are presented. Of the two, option b [Figure 9.208(h)] offers the benefit of the near proximity of power and return planes to the outer CS and PS layers where the active devices are placed. On such thick boards [with typical 10 mil interlayer spacing and 1 mil layer thickness, they may exceed 150 mils (3.8 mm) in thickness], this arrangement is highly advantageous.

Lessons Learned

- Power (V_{CC}) and return (GND) planes should be paired in adjacent, closely spaced layers.

*ΔI noise and simultaneous switching noise (SSN) are discussed in Section 9.5.

- Power (V_{CC}) and return layers should be allocated so as to minimize high transient current loops.
- The signal layer should be adjacent to at least one AC return (V_{CC} or GND) plane layer; high-speed signal layers should be routed using transmission line techniques.

9.6.3 Local Ground Structures

Manufacturers sometimes recommend making use of partial or local ground* structures either immediately beneath or adjacent to their noisy components, better known as the micro island or copper-fill schemes. They are all similar in their objectives but differ somewhat in the manner of their application [3]. In fact, this technique may be applied to both power and ground structures, as both power and ground equally serve as RF return paths or references. The main purpose of these arrangements, sometimes known as shielding structures (although not always recognized as such), is to reduce radiated emissions and, particularly, crosstalk on the PCB. When placed in a layer underneath a device or signal trace, the structure is called a micro island, whereas when installed in a coplanar fashion, the term copper fill is more commonly used. The following sections discuss in detail the application, advantages, and shortcomings of these techniques.

Maintaining isolation of the local ground structures mandates that power to any device placed over that structure also be isolated. For best performance, keeping in mind that DC isolation is required, monolithic DC/DC switch-mode power supplies (SMPS) are commonly used, with DC isolation between their input-to-output terminals. These devices should be placed on the boundary between the main and the local ground structures.

9.6.3.1 Micro Islands. Micro islands constitute isolated power or return planelets placed immediately underneath noisy components or traces. The notion behind this scheme is that RF flux generated internal to noisy devices (e.g., oscillators) is captured locally and the RF current they draw can be sunk locally, precluding the migration of the effects to the PCB at large.

Consider, for instance, a microcontroller (μC) device connected through a data bus to its peripheral devices such as memory devices. As long as the device is placed above the ground structure along with all its peripherals, a continuous path for the return currents exists in the micro island. For all practical aspects, the micro island serves as a partial yet solid ground plane.

If, conversely, the μC is isolated on a ground micro island, separated from its peripheral devices, through what path will the return currents flow from the peripheral devices back to their source? Ultimately, some return path connection, albeit imperfect, must exist or the circuit will not function at all. But the return current encounters

---

*The term "ground structure" is commonly used, although it may also be applied to power (V_{CC}) or return (*GND*) islands. However, consistent with common terminology, the term "ground structures" will be retained.

an abrupt impedance discontinuity in its path, resulting in generation of excessive EMI. An example of an application of a microcontroller consisting of both low-speed and high-speed digital modules is presented in Figure 9.209. Inasmuch as low-frequency digital return currents may utilize traces as their return paths (even if far from ideal from the immunity viewpoint), that configuration would be totally unacceptable for high-speed digital circuits.

One of the most frequent applications of this technique is related to crystal oscillators and clock-supporting circuitry, generally (e.g., buffers and drivers), considered particularly "noisy" devices. Local ground micro islands immediately beneath oscillators and clock-generation circuits (often even in the same layer as the device itself) are often considered for the following main reasons [3]:

- Circuitry contained in the oscillator generates a high-frequency current spectrum (see, for instance, Figure 9.133). This RF current is sunk to ground through the inductive DC power supply leads of the device. In oscillator devices packaged within a metallic can, the device's case is excited by this RF current flowing in its leads, efficiently emitting interference. In the case of a particularly irregular stack-up, an adjacent return (GND) plane may not be available immediately under the oscillator, contrary to common (and desired) practice. A local ground micro island placed immediately under the device and tied to the main ground plane provides an efficient ground "counterpoise" to the device. Figure 9.210 illustrates a situation in a six-layer PCB in which two signal layers separate the component side (CS, where the oscillator is placed) from the closest ground plane.

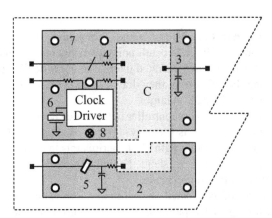

Figure 9.209. Implementation of micro islands in a microcontroller (μC) circuit. 1. High-speed circuits' digital ground micro island. 2. Quiet I/O digital ground micro island. 3. Power input decoupled to digital ground (1). 4. High-speed data bus with source termination. 5. Low-speed I/O bypassed with ferrite and capacitor. 6. Grounded crystal or oscillator. 7. Plated-through via to main internal ground plane. 8. Chassis ground stitch.

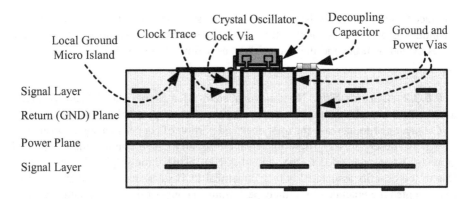

Figure 9.210. Implementation of a micro island with a crystal oscillator.

- If the oscillator is a surface-mount device (SMD) contained in a nonmetallic SMT package, direct emission from the device may occur. Although small, high-frequency spectral components may be efficiently radiated from the device, particularly with the help of the traces it drives.

Placing a local ground micro island under the oscillator and the clock-generation circuits provides a local return and image plane (the "counterpoise"), which traps common-mode RF currents generated internally to the device and its associated circuitry, consequently lowering radiated emissions. The micro island, by virtue of its installation, carries some differential-mode, high-frequency current sunk by the oscillator. In order to preclude it from becoming an efficient radiator itself, and for ensuring its effectiveness, the micro island is tied through multiple vias (depending on its size) to the main, solid internal ground plane of the PCB (fourth layer in the example in Figure 9.210), lowering its overall RF potential. To further enhance the performance of the micro island, clock generation circuits should be positioned and tied to the chassis ground stitch location* whenever possible. For achieving the lowest possible impedance of these interconnects, plated-through vias connected circumferentially rather than through "wagon wheel" connections should be used.

The power for the oscillator is decoupled to the same micro island (note that the power connections are not illustrated for simplicity of the figure, but it should be assumed that they exist and connect to the power plane in the fourth layer).

It is important to extend the micro island to include also the high-frequency/high-current clock oscillator supporting circuitry, such as the clock buffers. These clock buffers are extremely fast and powerful devices. When extended under these circuits, the micro island provides a nearby return path for the high-frequency currents flowing between the devices. Regardless, the major flaw in this technique lies in the routing of the clock trace. The good intentions of the circuit designer in providing a local reference plane are futile. As the local micro island is not placed under the entire span of

*Chassis ground stitches are discussed in detail in Section 9.9.

the trace (even if under all the clock-generation associated circuitry) it cannot provide a continuous path for the clock return current, nor an acceptable image plane, as the traces extend well beyond the micro island as it flows to its loads (see Figure 9.209) [2].

9.6.3.2 Copper Fills.

Nonfunctional solid-filled copper planelets are generally used by PCB manufacturers to produce a more uniform copper balance over the area of a PCB layer and to increase the total unetched portion of copper in a signal layer, also known as "desert areas." These planelets are usually called copper fills or "poured grounds" when connected to the ground (GND) plane through plated-through vias. Due to the benefits associated with symmetrical PCB stack-ups, some manufacturers have adopted the practice of routinely performing a copper fill on every signal layer so that they all maintain an overall percentage of copper similar to that of the solid power and return plane layers. Then copper balance is automatically achieved without anyone having to think about it.*

The benefit of copper fills from the standpoint of EMC in two-layer PCBs is quite apparent. Consider an active signal trace routed on a printed circuit board, as depicted in Figure 9.211. Electric field flux emerging from any current-carrying conductor terminates on the closest metal structures, inducing current on that structure.† If the closest structure happens to be an adjacent signal trace, so be it. The consequent interference current induced on the traces is known as "capacitive coupling crosstalk."

In Figure 9.211(a), two adjacent traces are depicted. As the metallic structure closest to the culprit (aggressor) trace is the second trace, the electric-field lines terminate on that trace, resulting in crosstalk.‡ When the area between the culprit and victim traces is filled with a large mass of copper [Figure 9.211(b)], and that surface is grounded, the vast majority of the electric flux tends to terminate more effectively on the grounded metal surface rather than on the victim trace. Consequently, capacitive crosstalk to the victim trace diminishes [54].

On a multilayer board that consists of solid power and ground planes [Figure 9.211(c)], the copper fill serves no significant function. With the presence of continuous, solid-copper planes extending all over the place around the traces, the electric flux effectively terminates to that mass of metal. The effect of the copper fill would then be marginal. It simply lifts the ground plane one layer higher, to the top layer of the PCB, rather in the layer immediately beneath it, on the order of a mere 10 mils (0.254 mm or 0.01 inch) or less for common stack-up structures. This change will have negligible effect on crosstalk.

---

*For copper-balancing purposes, PCB vendors may put on a so-called "thieving pattern," which differs from ground pour in that it is floating and must be composed of a series of very small islands instead of a solid planelet. This thieving pattern must be applied in most multilayer PCBs even if solid power and return planes are present.

†Note that this current is the "displacement current," $\partial \vec{D}/\partial t$, a concept appended by Maxwell to Ampere's Law, described in detail in Chapter 2.

‡If the "victim" line happens to be an input/output (I/O) line, common-mode interference current on the order of few microampère's is sufficient to result in excessive radiated emissions and, consequently, failure in the EMC tests.

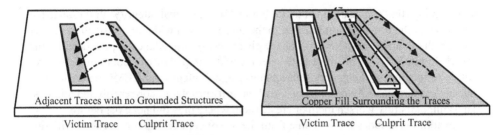

(a) No Copper Fill: Electric Flux Terminates on
Adjacent Signal Traces

(b) Copper Fill Around the Trace: Most Electric Flux
Terminates on the Copper Fill

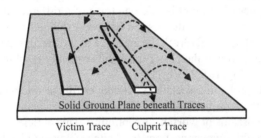

(c) Ground Plane Underneath a Trace: Most Electric Flux Terminates on the Ground Plane
making the Copper Fill Redundant

Figure 9.211. Electric flux terminates on the closest grounded metal surfaces.

Copper fill is exceptionally effective in high-impedance analog circuits, where a solid ground plane may not be present. In high-speed digital circuits, typically structured as multilayer printed circuit boards, the effect of copper fills is marginal. Digital circuits are typically considered as low-impedance circuits (i.e., current sources). Consequently, magnetic (or inductive), rather than electrical (or capacitive), interactions dominate crosstalk on digital PCBs. Magnetic fields do not terminate on any metal surface but rather constitute closed contours. Minimizing magnetic flux coupling can be achieved only through the allocation of a parallel path for the signal return current immediately adjacent to or, better still, beneath the signal trace. A solid and continuous return plane at the layer closest to the signal layer accomplishes this function. Isolated, discontinuous copper fill surfaces do not provide an unbroken path for flow of the signal return current and, hence, will have no true effect on reduction of magnetic-field coupling between traces or on radiated emissions from the PCB. Moreover, copper fills may be detrimental to both signal integrity and EMC if not applied with extreme care. The shortcomings of this technique are now presented in Section 9.6.3.3.

9.6.3.3 Shortcomings of Local Ground Structures. The very few advantages offered by copper fills and local micro islands are tied in with several distinct

shortcomings that can be detrimental for both EMC and signal integrity. Variation in the geometry of the transmission line due to the presence of a wide micro island or copper fills results in an abrupt alteration of the characteristic impedance experienced by the trace* (Figure 9.212). The immediate outcome of this impedance discontinuity could result in reflections, ringing, and, subsequently, signal integrity and EMC concerns. In a nutshell, rather than solving EMC or signal integrity problems, copper fills and micro islands might merely worsen them or create them where they did not exist before.

An additional undesired outcome from the use of copper fills is associated with differential transmission lines, such as those used in LVDS.[†] In such transmission lines, balance constitutes a primary performance concern for ensuring performance. Differing amounts of copper fill on each side of a differential transmission line causes imbalance in the line, predominantly affecting signal integrity. Figure 9.213 demonstrates the effects of a copper fill present beside one trace of a differential transmission line [Figure 9.213(a)]. In contrast to the desired balance exhibited in a differential transmission line system [Figure 9.213(c)], the presence of the metallic structure results in a distorted field pattern around the closer trace, subsequently resulting in imbalance in the transmission line [Figure 9.213(b)] [55].

Yet a third EMC problem associated with copper fills or micro islands is the resonance that can occur on isolated metal structures. Such resonances can subsequently increase radiated emissions or worsen susceptibility. This deficiency can be corrected by making use of multipoint connections (or "stitches") using vias from each copper-filled area directly to the solid return plane (or power plane, as appropriate) on the PCB.[‡] For effectively eliminating the resonance, stitching should be carried out by using at least one via every $\lambda/10$, where λ is the wavelength of the highest frequency of concern (considering the relative permittivity, ε_r, of the PCB's dielectric). But, following the reasoning previously discussed, if a solid plane is present, why is a copper fill or micro island used in the first place?

Lessons Learned

- Copper fills are particularly useful in reducing electric (capacitive) coupling in high-impedance analog circuits on two layer PCBs; they have marginal effects in high-speed digital circuits on multilayer PCBs consisting of solid ground and power planes, where magnetic (inductive) coupling dominates.

- Maintaining isolation between the main and isolated ground structures requires that isolated power sources (e.g., switch-mode power supplies) be incorporated for maintaining isolation through the power systems of the in-between circuits operating across the isolation boundary.

- Local ground micro islands and copper fills may create more problems then they solve and should not be used in high-speed multilayer PCBs.

---

*Reduction of about 30% has been observed in the characteristic impedance, Z_0, as a result of such situations.

[†]LVDS is low voltage differential signaling, defined in the ANSI/TIA/EIA-644 standard.

[‡]"Chassis stitches" are discussed in Section 9.9.

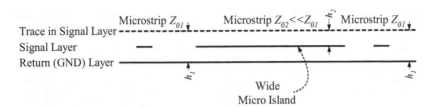

Trace in Signal Layer
Signal Layer
Return (GND) Layer

Figure 9.212. Variation of transmission line characteristic impedance due to copper fills or micro islands.

9.6.4 Shield Traces

9.6.4.1 Guard and Shunt Traces. Guard traces are grounded traces that surround noisy lines (e.g., clocks) or sensitive (e.g., sensitive, high-impedance analog lines) that are to be shielded. As such, they constitute a special case of coplanar ground fills. Shunt traces, on the other hand, are wide traces located directly above or below a

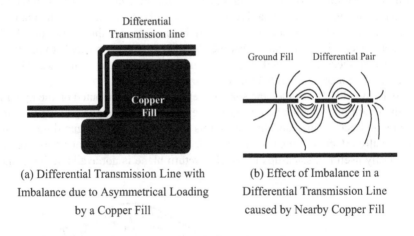

(a) Differential Transmission Line with
Imbalance due to Asymmetrical Loading
by a Copper Fill

(b) Effect of Imbalance in a
Differential Transmission Line
caused by Nearby Copper Fill

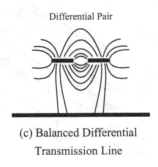

(c) Balanced Differential
Transmission Line

Figure 9.213. Copper fill on one side of a differential transmission line introduces imbalance in the line.

high-threat transmission line along its entire length and constitute a special case of a ground micro island. Both guard and shunt traces have unique applications, advantages, and shortcomings [3]. Figure 9.214 illustrates the guard and shunt trace structures on a PCB.

Guard and shunt traces adjacent to or immediately beneath high-speed lines are often proposed in order to minimize coupling to other traces that may result in EMI problems, for example, coupling to I/O lines, heat sinks, or other "floating" metallic structures. Minimizing crosstalk for signal integrity is yet another motivation. The guard and shunt traces provide an alternate return path for the RF currents flowing in the trace to return to their source. By virtue of their near proximity to the trace, enhanced magnetic flux cancellation is achieved, thereby reducing emissions and crosstalk from the trace (hence, the "shielding effect"). In high-impedance circuits, the guard and shunt traces also serve to contain (or intercept) electric flux that otherwise would terminate on a sensitive trace.

Since magnetic field emissions are directly proportional to loop area, effectiveness of guard and shunt traces is dependent on the spacing between the guard traces (horizontal spacing) and shunt trace (vertical spacing) to the signal trace. Considering the distribution of return current, the width of the shunt trace must be at least three times the separation between the signal layer (containing the said trace) and the plane containing the shunt trace, $3h$ (see Section 9.3.6), but no less than three times the width of the signal trace, $3W$ (to guarantee that the electric and magnetic flux between the signal and shunt traces is restricted to the area encompassed by the shunt trace. The same applies for the purpose of terminating electric flux.

In two-layer PCBs in which the spacing between the traces routed on the upper layer to the return plane on the bottom layer is excessively large, a large loop may form, preventing effective flux cancellation. In this case, closely spaced guard traces can be exceptionally effective. Shunt traces provide an adjacent return plane and are, therefore, mostly useful in stack-ups in which a return plane is not available immediately

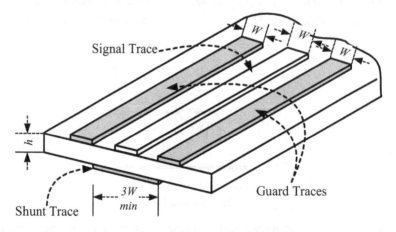

Figure 9.214. Guard and shunt traces.

above or below a noisy transmission line or trace. To be effective, they must be placed immediately above or underneath the signal trace.

In properly stacked multilayer digital circuits, the solid ground planes provide most of the benefits of grounded guard traces. In purely digital circuits, decreasing crosstalk on PCBs is mostly accomplished by spacing traces farther apart, circumventing the effectiveness of guard traces. Consider the example in Figure 9.215. Assuming a trace width (W) of 10 mils (0.01 inch or 0.25 mm) placed above a return plane with a separation, h, of 10 mils, the crosstalk between the traces will have an upper bound of

$$\frac{1}{1+(d/h)^2} = \frac{1}{1+(40 \text{ mil}/10 \text{ mil})^2} \approx 6\% \Leftrightarrow -25 \text{ dB} \qquad (9.96)$$

The fundamental question is how much crosstalk is acceptable. The value obtained in Equation (9.96) is typically not enough to cause upset to digital circuits, but it is large enough to cause excessive common-mode emissions from cables attached to this trace or to interfere with sensitive analog signals. Such levels may also be devastating to the performance of highly sensitive analog instrumentation or audio circuits. In such cases, where much higher isolation (for instance, 80 dB or more) is required, a pair of grounded guard traces running parallel to a sensitive input circuit could reduce crosstalk by orders of magnitude.

The key to achieving effective isolation when using guard traces is to use enough ground vias to "nail the shield down" to the ground plane. When left floating (i.e., not grounded), the guard trace acts just as another signal line. Noise from the aggressor trace will couple onto the guard trace, which, if not grounded, will effectively recouple over to adjacent sensitive lines, circumventing the purpose of the guard trace.

When using guard traces, therefore, ground traces must be stitched to the underlying ground plane using ground vias at least at both source and destination ends of the trace. If the routing of the signal and the guard traces is electrically long, multiple connections to the ground planes by vias along the edges of the guard trace are also required. As a minimum, the vias should be distributed along the guard trace so that there are at least three vias within the spatial equivalent distance of the signal-rise time. The spacing between the vias on the guard trace affects the amount of interference coupling: the closer the spacing, the higher will be the isolation. In order to inhibit the formation of periodic resonance patterns on the guard trace, a judicious use of

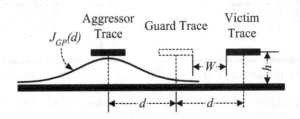

Figure 9.215. Effect of increased spacing due to a guard trace on crosstalk.

these vias is necessary. In particular, an irregular spacing of the "stitch" vias should be maintained.

Many designs have failed because of misapplied guard traces bringing about untold crosstalk and EMI problems. Like any tool, guard traces or shields can be used correctly or they can be abused. In particular, application of this technique requires a good understanding of the signal characteristics.

Particular care should be paid to the application of ground or guard traces to differential and balanced transmission line structures. Contrary to single-ended signals, for which balance is of no concern and flux cancellation is of utmost importance, in differential circuits magnetic flux cancellation occurs primarily between the signal traces and balance is of utmost importance. In LVDS, for instance, use of guard traces is highly encouraged. Attention must be paid to maintaining the balance of the pair. For that purpose, guard traces must be added at both sides of the pair. Running the shield guard trace on one side only creates an imbalance that can actually increase EMI. Furthermore, for reducing the undesired effects of secondary radiation by the guard trace, the separation between the signal traces to the guard traces should be at least $2s$, where s is the spacing between the differential pair traces [Figure 9.216(b)] [55].

In conclusion, in digital circuits implemented on multilayer PCBs, isolation is predominantly achieved by the presence of the ground plane at an adjacent layer. Guard and shunt traces offer some marginal reduction in coupling but, if implemented carelessly, may actually increase crosstalk and EMI. Stripline configuration is considered superior for reducing EMI and crosstalk when compared to guard traces, and should provide greater containment of the magnetic and electric fields associated with the high-speed traces [56].

9.6.4.2 "Picket (Via) Fence." Multilayer electronic circuits commonly utilize stripline transmission lines as electrical interconnects between various high-frequency devices. Even with the high levels of isolation offered by the stripline configuration, designers are required to decrease the spacing between stripline traces to meet new

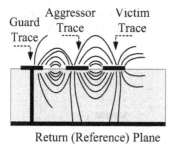

(a) Imbalance Caused by Single Guard Trace

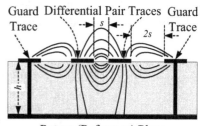

(b) Balanced Guarding with Proper Spacing

Figure 9.216. Negative (a) and positive (b) effect of guard trace on EMI.

size and cost goals. In doing so, coupling between adjacent stripline traces severely limits the overall circuit performance.

Improved isolation between critical stripline interconnects is most typically accomplished by placing metal-filled via fences (also called "picket fences") adjacent to the stripline as electromagnetic barriers to confine the electromagnetic fields around the center strip and decrease electrical couplings between signal paths. These vias are tied to the top and bottom ground planes of the stripline structure. Figure 9.217 illustrates the geometry of a stripline with a continuous plated-through via fence ("picket fence").

With respect to near-end crosstalk (NEXT), it has been shown that via fences with increasingly tighter via pitches, s [see Figure 9.217(a)], increasingly improve isolation as compared to having no via fence. However, for the far-end crosstalk, the introduction of a via fence was actually shown to significantly increase coupling and, thus, degrade isolation between stripline traces. On the other hand, via fences with coplanar guard traces tied to the ground planes have been demonstrated to exhibit high isolation for both near-end and far-end crosstalk. This configuration is commonly known as "boxed stripline" (see dashed line in Figure 9.217).

Figure 9.218 depicts the magnitude of the radiation loss in terms of s-parameters, $1 - |S_{11}|^2 - |S_{21}|^2$, as a function of the transmission line geometry. Note that the radiation loss is indicative of the leakage through the picket fence. It can likewise be observed

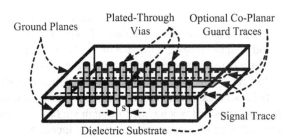

(a) Geometry of the "Picket Fence" Structure

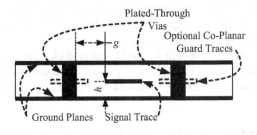

(b) Cross Section

Figure 9.217. Stripline with a continuous plated-through via fence ("picket fence").

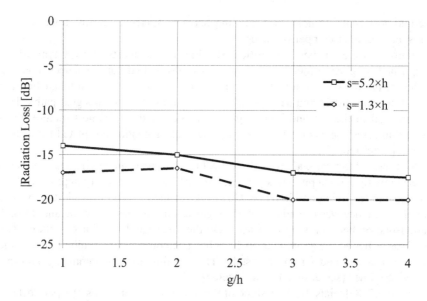

Figure 9.218. Radiation loss, $1 - |S_{11}|^2 - |S_{21}|^2$, of stripline with a continuous via fence as a function of s/h and g/h.

that the radiation loss decreases as separation between the trace and via holes, g, increases. Reduction of the parasitic effects levels off as g/h (see Figure 9.217 for identification of g, h, and s) approaches 2. Furthermore, it is seen that the radiation loss increases as the distance between the via holes, s, increases [57].

Figure 9.218 indicates that the closely spaced via fence completely confines the electromagnetic field, whereas wider spaced via holes results in significant leakage of power. It also demonstrates that when plated-through via fences are placed too close to the stripline trace, radiation loss increases, resulting in increased coupling between adjacent stripline traces. Therefore, in order to minimize radiation losses, the via fence should be kept at least four times the separation of the ground plane from the trace, or $s/h \geq 4$. It was further shown in [57] that use of via fences locally only (that is, not along the entire span of the trace) where isolation is required can actually degrade the stripline characteristics by causing a large perturbation in the electric field that normally extends on either side of the trace.

Finally, careful attention should be paid to antipad clearances in ground planes, as they can be a major source of coupling, depending on their proximity. For achieving high isolation in the gigahertz range, use of "blind vias" or "buried vias" should be considered [58].

9.6.4.3 Guard Rings. Unlike digital circuits, which are typically of a current-driven, low-impedance nature, analog circuits are mostly characterized by their high input impedance, making them sensitive to external noise, predominantly to stray electric fields. Maintaining accuracy in analog circuits, such as in instrumentation that processes signals from high-impedance sensors, presents unique challenges, requiring careful implementation of isolation techniques.

Guard rings are a common type of shielding used with operational amplifiers. They are intended to isolate the noisy environment outside the ring, preventing stray RF currents from coupling into sensitive nodes. The principle is straightforward. Completely surrounding the sensitive node with a guard conductor that is kept at or driven to (at low impedance) the same potential as the sensitive node sinks stray currents away from the sensitive node. Figure 9.219 depicts the guard ring schematics and implementation for inverting and noninverting op-amp configurations.

Lessons Learned

- Guard traces between signal traces constitute a good solution for achieving extremely high isolation and significant reductions in EMI from digital circuits.
- To be of significant effect, the guard traces must serve as the primary current return path.
- The guard trace must be "stitched" at multiple yet irregular points along its span to the main ground plane(s) to preclude resonances and reradiation.
- Special care must be paid to the use of guard traces in balanced transmission lines, to ensure that the balancing of the line is not violated.

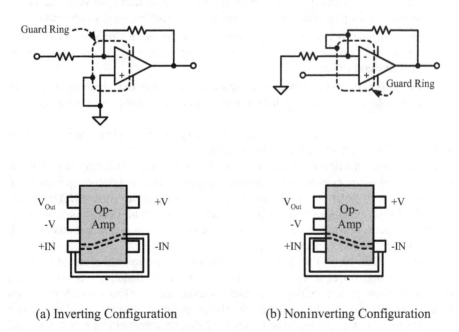

(a) Inverting Configuration   (b) Noninverting Configuration

Figure 9.219. Schematics (top) and implementation (bottom) of guard rings applied in (a) inverting and (b) noninverting configurations.

- A continuous plated-through via fence, or "picket fence," may significantly reduce near-end crosstalk (NEXT), with tight via pitches and large trace-to-via fence separation.
- Via fences with coplanar guard traces tied to the ground planes ("boxed stripline") exhibit high isolation for both near-end and far-end crosstalk.
- Guard rings are useful in sensitive high-impedance analog circuits for intercepting and shunting high-impedance field and stray RF current from the sensitive input ports of the circuit.

9.7   CUTS AND SPLITS IN RETURN PLANES

Most of the time, cutting the return planes to create islands or peninsulas is not recommended. The undesired effects of gaps in return planes, and the complexity they add to circuit design, were discussed in Section 9.4 in detail. And yet, if there are other overriding considerations, splitting return planes may be required. Such situations might arise particularly when vastly different signal levels are present or processed in different zones of the circuit. Splitting return planes would be appropriate and possibly inevitable at least in three instances:

1. Some medical devices, particularly those known as "patient-coupled equipment," must meet extremely strict current leakage requirements between the power mains and the part of the system connected to the patient (safety concerns).
2. In industrial process control equipment, where the outputs are connected to very noisy, high-power electromechanical loads, which may inject noise back into the sensitive controller.
3. Probably the most common of all, though, is found in mixed analog/digital circuits, where high-speed digital circuits are placed on low-noise analog boards.

Of course, poor circuit layout may also mandate splitting the planes in situations in which the split would otherwise not be necessary.

Consider the situation depicted in Figure 9.220. In Figure 9.220(a), the high-frequency digital return current (I_D), following the path of least impedance (primarily inductance), is well isolated from the low-frequency analog return current (I_A) following the path of least impedance (primarily resistance). This path also constitutes primarily the shortest path. The return current slightly spreads out in the plane, but otherwise (if not obstructed) remains under the trace.

In Figure 9.220(b), on the other hand, a conflicting situation exists due to poor layout. Now the higher frequency digital and lower frequency analog return current paths share a common path, resulting in common-impedance coupling through the return plane. This condition can be resolved either by proper routing of the signal traces, in a manner similar to that presented in Figure 9.220(a), or introducing a gap, forcing the analog return current to follow a separate return path in the return plane. This solution is illustrated in Figure 9.220(c).

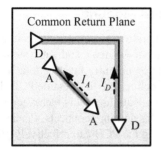

(a) Acceptable Situation: Digital (I_D) and Analog (I_A) Return Currents Paths Isolated

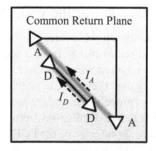

(b) Objectionable Situation: Digital (I_D) and Analog (I_A) Return Currents Paths Coupled

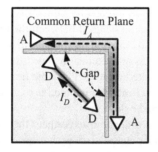

(c) Solution to the Situation in (b): The Analog Return Current (I_A) is Forced to Flow away from the Digital (I_D) Return Current Path

Figure 9.220. Erroneous routing of traces leads to undesired coupling in a common ground plane.

Moats, islands, cuts in the return planes, isolated power planes, floating return regions, and a host of other intricate layout techniques are routinely used by digital circuit designers in order to minimize RF noise current flow across the boundary between two zones. Consequently, crosstalk is reduced, EMI performance is bolstered, and overall system functionality is otherwise improved.

It must be kept in mind that splitting the return plane in this manner constitutes an effective technique only for reducing common-impedance coupling, which generally does not constitute a significant coupling mechanism at frequencies above 1 MHz. On the other hand, it does not preclude inductive or capacitive coupling between adjacent traces and in some cases increases the mutual coupling between printed circuit board traces [6].

9.7.1   Circuit Partitioning, Floating, and Moating

Circuit partitioning [3] refers to the physical separation of components, circuits, and power and return planes from other functional devices, areas, and subsystems. Partitioning is required when sensitive, or "quiet" analog circuitry must be totally isolated from noisy digital circuits, even though both are collocated on the same PCB. I/O partitioning is probably one of the most common applications of this technique intended to prevent radiated emissions from off-board cables and interconnects. Placing all I/O connectors at one edge of the PCB and partitioning the PCB can effectively provide a "quiet" I/O return planelet in that area.

Partitioning is achieved by totally removing the copper from all layers and planes on the PCB through the use of an intentional gap, also known as a "moat," between the zones, including the power and return planes. As a result, the circuits are totally isolated galvanically from each other and the circuits are practically floated from each other.

For ensuring that conductive as well as inductive and capacitive coupling between the zones is minimized, a wide gap, typically 80 to 120 mils (0.08 to 0.12 inch or 2.03 to 3.05 mm, approximately) should be used between the two zones. In other words, an isolated area is a "copper island" created on the board, similar to a castle with a moat (Figure 9.221). The moat creates an "exclusion zone" for all but the signals and power required for the use of the isolated circuits. Only the traces and power supply required in the excluded zone may travel into the isolated area, and even then in an extremely controlled manner, as described below.

Two techniques are used in order to interconnect the isolated zone to the main circuit, without violation of rules associated with signal traces crossing the gap (described in Section 9.4). The difference between the two is associated, essentially, with the design of the power and return planes in the PCB. The first method, called isolation, makes use of isolation devices such as isolation transformers or optocouplers/op-

(a) Photo of the Castle and Moat   (b) Map of the Castle and Moat

Figure 9.221. Photo and map of the Himeji-Jo Castle, Japan, illustrating the moat and bridge.

tical isolators. For lower frequency applications, differential signaling, exhibiting high common-mode rejection, can also be useful. The second, bridging, utilizes a thin strip or "drawbridge" across the gap between the two zones. These two techniques are discussed below with particular emphasis on the implementation of grounding.

Regardless of the technique used, partitioning must be implemented throughout all layers of the PCB, and no overlap should occur between isolated circuits and their associated return planes in any layer. The overlap area between any two planes or planelets exhibits some capacitance, which, small as it is, is sufficient to provide a path through which high-frequency noise may travel from one regulated power supply to another, or from a noisy return plane to a sensitive one. Ultimately, such a flawed design is self-defeating, and counteracts the benefits of separation for which the isolation was implemented in the first place.

9.7.2 Circuit Isolation

In the isolation method, total galvanic isolation is maintained between the excluded zones on the PCB, and no copper interconnects the two in any layer, be it signal (with some exceptions), power, or return. The three most common applications for this technique are the input/output (I/O) area of the circuit, the "core" of FPGAs and other CPLDs (power isolation only), and analog applications, particularly audio frequency circuits contained in a primarily digital circuit. Figure 9.222 illustrates the application of isolation in a mixed digital/analog and I/O circuit.

For maximum effectiveness in I/O applications, the I/O area must be completely isolated from the remainder of the PCB. Only at the metal edge connector is RF bonding of the local return plane accomplished to the chassis, through a high-quality, low-

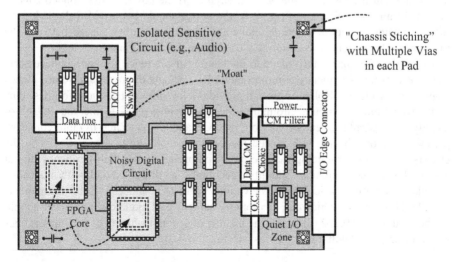

Figure 9.222. Circuit partitioning through isolation. (Note the transformer (XFMR) and optocoupler (O.C. Interfaces).

impedance bonding path. In effect, this creates a chassis "peninsula," rather than a true island, serving as an extension of the equipment chassis or signal reference structure. Often, if shielded wires are used, the shield is also terminated to the chassis of the circuit at this surface.

Only few interconnects between the isolated circuits are shown in Figure 9.222, however, in most cases many more exist, particularly at I/O zones. Due to the lack of return planes common to the isolated circuits, along with the need to avoid such discontinuities as those described in Section 9.4, any interface between the two must be implemented using some interface not requiring a common return plane, such as optical, transformer, or common-mode choke coupling. In the first technique, the signal is transmitted in the form of light pulses, whereas in the second it is transmitted by means of magnetic fields. Contrary to the previous two, common-mode chokes actually do provide a galvanic connection between the circuits while exhibiting high common-mode rejection beyond a certain frequency. Either way, galvanic connection between the planes is unnecessary for effective signal propagation. Alternatively, differential interfaces (e.g., RS-422) may be used for relatively slow circuits (i.e., circuits with lower frequency content). All data lines crossing between the isolated zones must be filtered using R-C or L-C filters (on differential lines, common-mode chokes are often used) using bypass capacitors with typical values in the range of 470 pF to 1000 pF. These filters will be normally located as close as possible to the boundary between the zones or the edge connector (in the case of I/O partitioning), in order to preclude the crossover of noise from the noisy circuit and its return plane to their corresponding sensitive circuits and planes.

For maintaining isolation of the power and return circuits, particularly for internal circuits (e.g., audio or sensing circuits), switch-mode power supplies (SMPS) providing DC-isolated power outputs are commonly used to produce the necessary floating power and return outputs for the isolated circuit.* If the power to the isolated zone is provided from the main part of the PCB, the power should be provided through a "π filter" containing a common-mode choke and bypass capacitors, placed between the two zones. This technique is commonly used at the power I/O area.

Grounding the quiet return plane to the chassis or reference structure is necessary, as it prevents high-frequency common-mode RF currents in power and return planes from coupling into the isolated area and provides a low-impedance shunt for these currents away from the circuit. Grounding the quiet return plane to the chassis subsequently reduces radiated emissions from the PCB and its associated I/O interconnects, particularly those driving unbalanced loads. Often, both zones are grounded to the chassis (multipoint grounding may be most appropriate in the high-frequency digital zone). This "chassis stitching" should be accomplished through a path with the lowest impedance possible, particularly in the vicinity of the I/O connector. Extra components (e.g., "0 Ω resistors," which can be properly named "0 Ω inductors") or traces should not be used. Figure 9.222 illustrates the manner of accomplishment of this chassis connection using a pad with multiple vias (see Section 9.9).

---

*A discussion on grounding considerations and isolation in switch-mode power supply circuits can be found in Chapter 4.

9.7.3 Bridging the Gap

The second method utilizes a "bridge" between the isolated zones on the PCB. Similar to a castle in which a drawbridge serves as a single access to or from the castle interior, the bridge is a break in the moat at one location only, where signal traces, power, and return cross the moat. Contrary to the previous method, single-ended signaling is possible thanks to the continuous (albeit relatively narrow) return path across the bridge, precluding the need for isolation devices. The bridge, therefore, provides an equipotential return plane (at DC), limiting the flow of high-frequency noise between the internal return planes to the restricted I/O zone. The width of the bridge should be just as wide as necessary for the signal conductors traveling across it. It should extend beyond the traces, by at least three times the height of the trace above the return plane, in order to allow unobstructed return current distribution and preclusion of edge fringing effects at the moat).

Figure 9.223 illustrates the application of the "drawbridge" technique in a circuit similar to that in Figure 9.222. The solid return and power planes in the circuit are gapped (with a moat) in two zones, the isolated sensitive circuit zone and the I/O zone. The return plane is interconnected across the moat using the "bridge."

Occasionally, the power plane is totally isolated by the moat, as shown at the I/O area in Figure 9.223. The power is filtered before it is fed into the circuit, typically by means of a π-section filter, where each capacitor is placed at an opposite side of the bridge and an inductor (or ferrite bead) is used to interconnect the two. A DC/DC switch-mode power supply could also be used between the zones in order to produce isolated power for the circuit if necessary. If an isolated power return is not required, a nonisolated switch-mode power supply (e.g., a "buck regulator") or linear regulators (for lower power applications) may be used, providing voltage-level conversion only. Such a con-

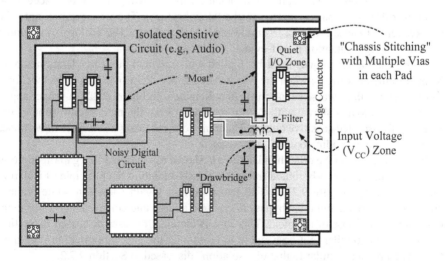

Figure 9.223. Circuit partitioning: bridging the gap using a "drawbridge." Note that isolation devices are not required, unless crossing over the moat.

figuration ends up with a much simpler circuit design. A common application of this technique can be found in internally isolated zones, for instance, in the power supply to the sensitive, low-voltage "core" of VLSIs, such as FPGAs and CPLDs.

Grounding of the return plane to the chassis or reference structure in this configuration should follow the same principles outlined in Section 9.7.2, as illustrated in Figure 9.222 as well as in subsequent figures, which present variants of this configuration.

The greatest advantage of this technique over the preceding one is the fact that no interruption of the signal traces is necessary. Consequently, the traces can be routed above the return plane, avoiding the need for isolation devices (e.g., transformers and optocouplers).

The narrow bridge represents a high-impedance path, precluding high-frequency noise present on the digital return plane in the inner zone of the PCB from efficiently coupling into the sensitive I/O zone. The only exception is the return current associated with those traces intentionally crossing the bridge toward the I/O zone of the PCB.

When the I/O lines carry lower frequency signals (e.g., slow serial data buses, etc.) only, the "drawbridge" may be replaced with ferrite chips installed across the "moat," immediately adjacent to the signal trace crossing the gap (Figure 9.224). The ferrites bridge the two return planes across the gap, allowing lower frequency signal return currents to flow with no interruption, while exhibiting high impedance at high frequencies, suppressing higher frequency noise.*

Typically, ferrites cannot be used when high-frequency signals cross over the "moat" as the ferrite will equally suppress the desired and the interference signals. High-speed I/O signals are, therefore, commonly routed directly to the I/O connector and do not cross through the isolated area (Figure 9.225).

To better illustrate the effectiveness of this technique, observe Figure 9.226. For protecting ships entering the harbor, a breakwater is constructed. The breakwater (the moat) keeps the open sea rough waters outside while allowing the ships to access the inner port through a narrow entrance (the "drawbridge"). The waves smash onto the breakwater but cause little ripple on the inner harbor water. The analogy to the moat and "drawbridge" situation is self-evident.

Both isolation techniques must be applied with care and due consideration should be paid to the crossing between the two isolated zones. Poor planning or lack of foresight might result in worse situations than anticipated. In Figure 9.227, a proper application of the moat is shown with a trace routed between the main digital zone of the PCB and the isolated sensitive circuit. The return current (shown in white) flows under the signal trace, providing optimal performance.

In the I/O area, on the other hand, a trace is shown to cross over the gap (in addition to the two former properly routed traces), seriously violating the principle of isolation through moating (Figure 9.227). The return current (again shown in white) cannot cross over the moat, and must bypass it, creating a large and consequently noisy loop, defeating the purpose of the moat. This case is identical to cases discussed in Section 9.4. If there is no other routing way but to cross over the gap, this situation should be treated in a manner similar to that of "isolation" discussed in Section 9.7.2.

---

*The impedance of ferrite materials is material dependent and is well defined by ferrite bead manufacturers.

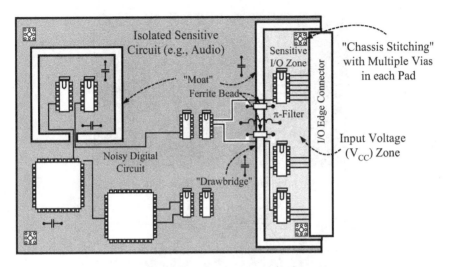

Figure 9.224. Circuit partitioning: bridging the gap using ferrites.

Lessons Learned

- Splitting power and return planes using moats is often beneficial for isolation and preclusion of crosstalk between noisy and sensitive circuits on the PCB.
- Isolation is used to provide 100% galvanic separation between noisy power and return planes and other functional areas from clean or sensitive zones on the PCB.

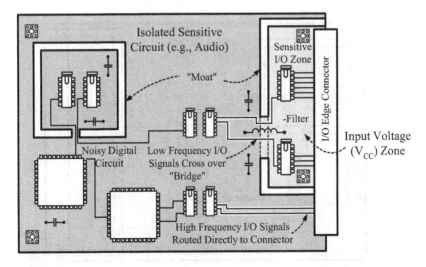

Figure 9.225. Circuit partitioning: higher frequency I/O signals interface directly with edge connector.

(a) Photograph of the Haifa Port

(b) Official Map of the Haifa Port

Figure 9.226. The breakwater in Haifa Port, northern Israel.

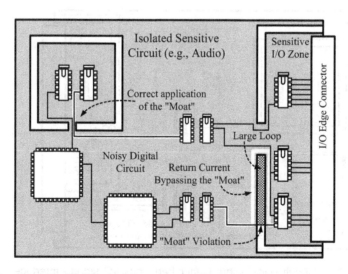

Figure 9.227. Violation of the concept of isolation through moating.

- All data interconnects across the gap should employ balanced interfaces or galvanic isolation devices, for example, transformers, data line common-mode filters, and differential signaling (for low-frequency applications).
- Power to isolated circuits zones should be provided through isolated DC/DC switch-mode power supplies when full isolation is required; when common return connections are required, voltage-level conversion may be accomplished by nonisolated switch-mode power supplies or linear regulators (considering power consumption).
- Bridging the return planes across the moat by means of a narrow copper connection in the otherwise isolated zones simplifies the circuit while controlling high-frequency noise coupling between the zones; only traces intentionally routed between the isolated zones may cross over the bridge.
- Signal traces should never cross the gap between two isolated zones except over the "bridge" or by means of isolation devices; violation of the moat is likely to result in severe interference and crosstalk.
- Grounding the quiet return plane to the chassis precludes coupling of high-frequency common-mode RF currents into the isolated area and reduces radiated emissions.
- Multipoint grounding is commonly recommended in the high-frequency digital zone in addition to the grounding of the quiet zone.

9.8 GROUNDING IN MIXED-SIGNAL SYSTEMS

Designing a high speed mixed (analog/digital) signal system without using a proper ground is like trying to play basketball on a huge trampoline. No matter how well you mount the baskets to the court, the whole court will bounce and wobble as the players jump and try to shoot. To play the game, you must have a solid floor. Similarly, to give a solid ground for your data converter circuit, you must use an analog ground plane. This will put your circuit on a solid foundation. [59]

The analog nature of our physical world and the growing need for processing continuous signals and functions in the digital domain have resulted in the need to design circuits and systems that process both analog and digital signals. Requirements for precision analog signal processing having wide dynamic ranges has further added to the complexity of the problem, particularly when high-resolution conversion of analog-to-digital signals (and vice versa) is necessary. Consequent evolution of mixed-signal processing systems places ever-stringent performance demands on contemporary mixed-signal devices such as analog-to-digital (A/D) converters (ADCs) and digital-to-analog (D/A) converters (DACs) as well as fast digital signal processors (DSPs). With the increase in resolution and drop in the signal voltage scale, these sensitive yet accurate devices have become extremely vulnerable to noise.

There are many ways (and just as many opinions as to what is the best method) to ground ADCs, DACs, and other mixed-signal circuits, keeping in mind that both analog and digital returns should remain at the same DC potential at all times. Unfortunately, most A/D converter manufacturers' data sheets and application notes provide little if any useful and practical information regarding solutions to this difficult situation. If they do provide any information, it is usually applicable only to simple system configurations containing only one A/D converter.

The purpose of this section is to address the question of grounding design in mixed-signal circuits and data converters (ADCs and DACs). It discusses the nature of the currents flowing in these devices, the vulnerabilities of precision data converters to internal and external noise, and the effects of various grounding schemes on their performance. It also develops and provides justification for a general grounding philosophy that should be suitable for most mixed-signal devices in the vast majority of cases.

9.8.1   Origins of Noise in Mixed Digital–Analog Circuits

Noise coupling between digital-to-analog circuits placed on the same PCBs may severely degrade the performance of the sensitive analog circuits, particularly when high-precision signal processing is required. Generally speaking, coupling between such circuits or traces may occur as a result of one or more of the following three coupling mechanisms: common-impedance, inductive, or capacitive coupling [6].

At low frequencies (e.g. below 100 kHz), coupling tends to be dominated by the common-impedance mechanism (at higher frequencies, e.g., above 1 MHz, inductive and capacitive coupling generally dominate). Impulsive changes in digital signal return currents flowing through a common return plane, when impressed across the impedance of the return path, result in voltage change (i.e., ΔI noise or SSN; see Section 9.5). This dynamic switching noise appears as noise in sensitive circuits using the same common return path.

One of the chief challenges in the design of mixed-signal circuit is keeping the switching noise from the digital circuits away from the collocated sensitive analog circuits. The key question is "how"?

This section will begin by presenting the common practices for grounding unconnected analog and digital circuits, followed by a discussion on grounding practices in mixed analog–digital circuits. Finally, we examine the difficulties arising particularly when multiple analog-to-digital data converters (e.g., ADCs and DACs) are included in the circuit.

9.8.2   Grounding Analog Circuits

Analog return plane noise voltages should be kept at least as low as the minimum analog signal level of concern if an acceptable signal-to-noise ratio is to be achieved. The level of tolerable interference in the analog return path depends on the sensitivity of the analog input signal. In the case of ADCs (or DACs), the smallest resolvable signal voltage level, or the least significant bit* (LSB), is a function of the number of the bits

---

*In a binary number, the least significant bit (LSB) is the least weighted bit in the group. For an ADC, the weight of an LSB equals the full-scale voltage range of the converter divided by 2^N, where N is the converter's resolution. For instance, in a 12-bit ADC with a unipolar full-scale voltage of 2.5 V, 1 LSB = 2.5 V/2^{12} = 610 μV.

and the full-scale reference voltage of the ADC. The smaller the reference voltage and the larger the number of bits, the smaller the minimum resolvable signal voltage will be.

Table 9.4 shows the resolution* versus the number of bits for an ADC using a one volt reference. These resolution levels can be scaled for other reference voltages by multiplying the resolution by the appropriate factor. For example, if the converter uses a 3.3 volt reference, the resolution numbers in the table should be multiplied by 3.3.

A large proportion of analog circuits are low-frequency in nature. Single-point grounding† is best suited for these sensitive circuits as it prevents large return currents of collocated noisy components (e.g., digital circuits, motor drives, or inductive loads) from sharing the sensitive analog return path. Also, single-point grounding precludes the creation of ground loops. With proper grounding schemes put into practice in low-frequency analog circuits, control of both intended and unintended return current paths is easily accomplished.

9.8.3 Grounding Digital Circuits

Digital devices exhibit noise margins on the order of several hundreds of millivolts and can typically withstand noise voltage in their return path ranging from tens to hundreds of millivolts. "Image planes" within the multilayer PCB and multipoint "chassis stitching" (see Section 9.9) further improve performance and noise immunity of the digital circuit.

In high-speed digital circuits, multipoint grounding is preferred because of the high-frequency switching currents that create large RF voltage drops across the return conductors if single-point grounding is used. Single-point grounding does not work well for this reason, as parasitics will alter the return paths. Ground loops usually do not constitute a concern within digital circuits, as long as a solid and continuous return path is provided and proper routing of traces is maintained.

9.8.4 Grounding in Mixed-Signal PCBs: "To Split or Not to Split (the Ground Plane)?"

Digital circuits sharing the same return path with the analog circuit on mixed-signal PCBs may severely degrade performance of sensitive analog circuits owing to the transient current flowing through the grounding system during the digital devices' abrupt state transitions. This is of particular concern when high-precision signal processing is required. A frequently asked question regarding systems embedded with data converters is how to ground them for optimal analog performance.

Figure 9.228(a) depicts the classic illustration of a situation in which the digital return current modulates the analog return current. The return path impedance is shared by the analog and digital circuits, causing coupling interaction and the resultant error.

---

*ADC resolution is the number of bits used to represent the analog input signal. To more accurately replicate the analog signal, a higher resolution is required, which also reduces the quantization error.
†Grounding topologies and their application are discussed in detail in Chapter 4.

Table 9.4. Resolution of ADCs versus number of bits
(normalized to a 1 V reference supply level)

Number of bits	Resolution (LSB)
8	4 mV
10	1 mV
12	240 μV
14	60 μV
16	15 μV
20	1 μV
24	59 nV

Consider a PCB hosting an analog circuit including a 24-bit ADC as well as a digital processor comprising large DSPs, ASICs, and peripheral digital logic, drawing a current surge of 10 A through the common return plane during logic state transition. The necessary impedance required to maintain the stray return path voltage at less than 59 nV with 10 A of switching current would be 5.9 nΩ. The common impedance shared between the digital processor and analog area must be less than this value in order not to degrade the desired analog circuit's performance. Even with only 16-bit converters

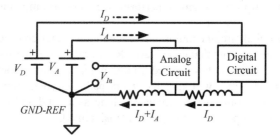

(a) **Incorrect**: Common Impedance in the Return Path
Creates Error Voltage in the Input of the Analog Circuit

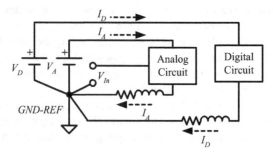

(b) **Compromising**: Split Return Plane Creates
Gaps and Large Current Loops

Figure 9.228. Digital return currents flowing in analog return path create error voltages.

(considered inadequate for high-fidelity audio) a common impedance of less than 0.152 $\mu\Omega$ would still be required, which is practically unattainable. With 8-bit industrial measurement ADCs, an impedance of 39 $\mu\Omega$ would be acceptable and could be achieved with adequate circuit layout and power conditioning.

There are a variety of well-known techniques, and just as many opinions as to which is best, to configure the return paths for mixed analog/digital signal circuits so as to mitigate the coupling between analog and digital circuitry on PCBs while keeping in mind that both analog and digital returns should typically be maintained at the same potential at all times.

In order to reduce undesired coupling between analog and digital circuitry in mixed-signal circuits, containing both analog and digital signal processing, sensitive analog circuitry must be provided a "quiet" return path. The EMC literature often recommends isolating the circuitry in different areas of the printed circuit board by strategically placing a gap (or moat*) between analog and digital areas on printed circuit boards.

The PCB thus consists of separate analog and digital power and return planes in order to prevent noise current from the digital part of the system from corrupting the sensitive analog signals through a common galvanic path common to both circuits. These return paths meet at a single system common reference point (GND-REF), which is usually adjacent to the power supplies or to the PCB's power and return entry point. This, in fact, constitutes the classical concept of a "star" or single-point ground system. The digital and analog return currents are now forced to flow directly to the system common reference point in otherwise galvanically isolated paths, as depicted in Figure 9.228(b). Keep in mind, though, that physically long return conductors exhibiting parasitic resistance and inductance will counteract the objective of low-impedance, high-frequency return current paths. In practice, the current return paths must consist of large planes exhibiting low impedance to high-frequency currents.

When separate, isolated analog and digital return planes are created on the PCB, as shown in Figure 9.229, the grounding scheme of each to the metallic signal reference structure, if present, should be implemented according to its characteristics. Accordingly, the digital return plane typically utilizes a multipoint grounding scheme, whereas the analog return plane is connected to the signal reference structure at a single point only, that point also serving as the system "star ground" (GND-REF).

The back-to-back Schottky diodes placed between the digital and the analog return planes are inserted to prevent accidental DC potential differences from developing between the two return systems, particularly when the PCB is plugged and unplugged. DC or AC potential difference between the planes should typically be maintained at a level lower than 300 mV to prevent possible damage to ICs having connections to both the analog and digital return planes and preclude false triggering of logic gates or possible device latch-up. Schottky diodes are preferable because of their low capacitance and low forward voltage drop. The low capacitance prevents AC coupling between the analog and digital planes. Schottky diodes begin to conduct at about 300 mV, and several parallel diodes in parallel may be required if high currents are expected. In some

---

*Moats and bridges in PCBs and their applications are discussed in Section 9.7. The above discussion on partitioned mixed analog/digital circuits constitutes a special application of this technique.

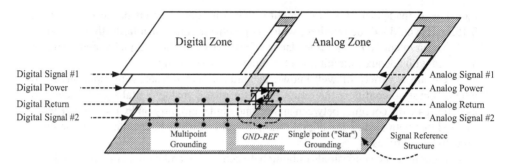

Figure 9.229. Implementation of split analog and digital returns in a mixed-signal PCB.

cases, ferrite beads can be used instead of Schottky diodes; however, they may introduce DC ground loops, which can be troublesome in precision systems.

As shown in general in Section 9.7.1, interconnections between completely isolated analog and digital zones are best accomplished by use of differential, optically coupled, or transformer-coupled interconnects, especially when extremely sensitive analog circuits are used alongside digital components.

Although the split plane approach can be made to work for obtaining good noise performance in low-resolution circuits, it also poses many potential problems with respect to the EMC performance, especially in large and complex systems. One of the major drawbacks is that no traces can be routed across the split in the plane. If traces are to run across a gap in the return plane [Figure 9.230(a)], no immediate return path

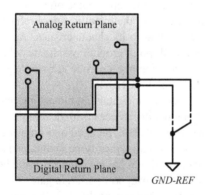

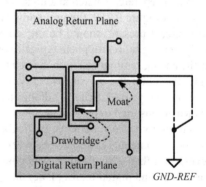

(a) **Incorrect**: Traces Crossing the Gap between Analog and Digital Return: Common Point only at Power Supply

(b) **Correct**: Traces Cross the moat above a "Drawbridge"

Figure 9.230. Signal traces crossing a gap in a return plane in a mixed analog/digital circuit.

is available near the trace, forcing the current to flow through a large loop. Large digital current return paths will results in increased interference emissions, whereas large analog return paths result in increased vulnerability.

A better approach for splitting a mixed analog/digital return plane is depicted in Figure 9.230(b). By interconnecting the two return planes at one point (commonly known as a "drawbridge") and routing all the traces only above the bridge, an immediate return path is provided for the current directly underneath each of the traces (resulting in a very small loop area).

But before continuing, it is important to recall the design objective. It is not that the interference from the analog circuits might interfere with the digital logic. Rather, it is the likelihood that the high-speed digital circuits might interfere with the analog circuits processing low-level signals. Splitting the plane is often proposed in an attempt to prevent the digital currents from flowing in the analog section of the return plane.

High-frequency digital return currents tend to flow in the return plane immediately beneath the signal trace (regardless of the DC potential of that plane), which constitutes the path of least impedance.* The current slightly spreads out in the plane, but otherwise (if not obstructed) remains under the trace. Equation (9.23) [repeated here as Equation (9.97)] expresses the approximate extent of spreading of the current distribution under the trace[†] [59]:

$$J_{GP}(d) \approx \frac{I_0}{\pi h} \cdot \frac{1}{1 + (d/h)^2}, \ \text{A/m} \tag{9.97}$$

Table 9.5 presents the fraction (in percent) of the return current contained within a distance (normalized to height of the trace) $\pm d/h$ from the signal trace centerline, where d is the horizontal distance and h is the height of the trace above the return plane. The data presented in Table 9.5 visibly indicates that the key to a successful layout of mixed-signal PCBs lies in the proper partitioning and routing discipline, rather than in the splitting of the return plane. By physically separating analog and digital components and providing an imaginary boundary on the board while implementing disciplined routing of the PCB traces, digital return currents are confined to the digital zone of the PCB and do not corrupt the analog signals through the plane. Analog signal traces should now be routed only in the analog zone, whereas the digital traces should be routed only in the digital zone and, hence, will not interfere with the analog signals. Problems typically come about when digital traces are routed in the analog zone or vice versa. Extreme care must be taken to ensure that these routing restrictions are categorically adhered to. Why is it necessary, then, to physically split the return plane at the first place, if it is not necessary to preclude the digital return current from flowing in a path it would not follow anyhow?

Carefully study the situation of a PCB hosting an analog circuit including the 24-bit ADC as well as a digital processor. Assume that the digital processor draws a current

---

*See Section 9.4 for a detailed discussion of the effects of gaps and other discontinuities in return planes on EMC and signal integrity performance of the circuit.
[†]This expression assumes a microstrip configuration. In stripline structures, current density drops quadratically with distance.

Table 9.5. Fraction (in percent) of the return current
contained within a normalized distance of $\pm d/h$ from the
signal trace centerline

d/h	Fraction of current density (%)
2	70%
5	87%
10	94%
20	97%

surge of 10 A through the common return plane during state transitions, as discussed in Section 9.8.4. The LSB of this ADC is equivalent to 59 nV. Considering a practical plane impedance of 40 $\mu\Omega$, this voltage is equivalent to 1.5 mA, approximately, equivalent to a fraction of 0.015% of the digital switching current. The separation, d, between the digital and the analog traces must be increased so that 99.97% of the digital return current is contained within that distance. Such separations are impractical in contemporary PCB designs.

9.8.5   The Mystery of A/D and D/A Converters Solved

Maintaining wide dynamic range with low noise in hostile digital environments is dependent upon implementation of sound high-speed circuit design techniques and grounding schemes in particular. A new generation of ADCs has brought enhanced dynamic performance with high-frequency input signals. Consequently, more system designers are facing the challenge of using high-performance ADCs.

All sampling ADCs (i.e., ADCs with an internal sample-and-hold circuit) suitable for signal processing applications operate with relatively high-speed clocks with fast rise and fall times and must be treated, therefore, as high-speed devices, even though throughput rates may seem low. Sigma–delta (Σ–Δ) ADCs also require high-speed clocks because of their high oversampling ratios. Certain ADCs are known to oversample the input signal, sometimes at rates as high as two orders of magnitude higher than the output data rate (effective sampling rate). Even high-resolution, alleged "low-frequency" industrial measurement ADCs (having very low throughputs, on the order of several kHz at most) operate on megahertz-range or higher clocks and offer resolutions as high as 24 bits [60].

To further complicate the issue, mixed-signal devices have both analog and digital ports and, most importantly, separate analog and digital return leads, commonly labeled AGND (or analog return) and DGND (or digital return), respectively. Much confusion has resulted regarding the proper grounding schemes to be implemented in circuit design. In addition, some mixed-signal devices have relatively low digital currents, whereas others have high digital currents, often requiring different treatment with respect to optimum grounding. Too many engineers and, unfortunately, too many data-sheet writers as well, are uncertain as to the manner in which the two pins should be connected.

In spite of the dual interfaces, ADCs and DACs (and often other mixed-signal ICs) should be treated strictly as analog devices and accordingly be grounded to the analog return plane. At first glance, this may seem somewhat contradictory but a simplified model of a converter depicted in Figure 9.231 helps clarify this seeming dilemma [60].

Inside mixed analog/digital devices such as ADCs or DACs, the designers of the device often internally partition the return net into isolated analog and digital return nets in order to preclude coupling of digital signals into the analog circuits. The sensitive analog circuits and analog reference generators connect to the analog return net, whereas the device's high-power I/O drivers connect to the internal digital return net. Physically separate *AGND* and *DGND* leads individually interface between the internal analog and digital nets, respectively, to the host circuit. Power to the analog zone is also provided from dedicated filtered and decoupled power-input leads.

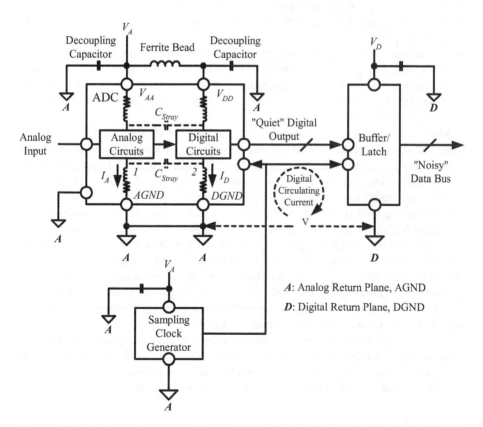

Figure 9.231. Schematic diagram of an ADC illustrating the rational for the grounding scheme in ADCs and DACs. Note that the ADC is powered in this example from a single, analog power source; the Buffer/Latch is supplied from a digital power source, V_D, and decoupled to the digital return.

During state transition of the digital outputs of the ADC, the DGND lead carries large, high-frequency currents, resulting in transient voltages across the inductance of the DGND wire bonds and package leads. The consequent switching noise (or ΔI noise) produces a transient voltage difference between the device's internal digital return net and the PCB ground plane.*

Note: If a single return lead were to be shared by both the analog and the digital circuitry in the converter, the resultant noise voltage could degrade the analog performance of the device. In some high-frequency devices, in fact, multiple analog and digital return leads, connected in parallel, are necessary to minimize the effects of lead inductance (a number of low-power ADCs and DACs, though, for which digital return currents are reasonably low, actually do have just one common return lead).

The resultant interference may not affect the digital logic, thanks to its high tolerance to noise, but it makes the digital return net useless as an analog reference voltage. Furthermore, the resultant transient voltage at point "2" inevitably couples noise into point "1" of the analog circuits through the stray capacitance, C_{Stray}. In addition, some small stray capacitance (typically 0.2 pF) is unavoidable between leads of the converter's package [60].

This problem is resolved by assigning a separate AGND lead in the device. As long as the internal analog circuits do not drive high current through this lead, only a small voltage drop will exist across its impedance. The internal analog circuits connected to the AGND lead are, therefore, referenced to a level close to the actual PCB return voltage at the AGND contact.

Moreover, in order to prevent further coupling, the AGND and DGND leads should be tied together at the device package and to the analog ground plane externally to the device through minimum lead length (and inductance). Any extra impedance in the DGND connection will result in more digital noise at point "2," which, in turn, will couple more digital noise into the analog circuit through the stray capacitance, C_{Stray}, thus interfering with the converter's ability to communicate across its internal analog/digital boundary. It could be expected that this scheme will inject a small amount of digital noise into the analog ground plane. These currents should be quite small, and can be minimized by ensuring that the converter digital input/output (I/O) interfaces do not drive a large fan-out. As a matter of fact, even if the device's application notes suggest that the AGND and DGND leads be separately connected to the circuit's analog and digital returns, respectively, it is generally better to ignore this guidance and connect them together and to the system analog ground.[†]

This raises the question of why the digital return (DGND) lead should be connected to the analog ground plane of the circuit. The answer to this lies in what is referred to as a "category error," whereby it is assumed that the same terms have a similar import in different contexts. The DGND lead of the converter is so named because it is pro-

---

*ΔI noise, or SSN, is discussed in detail in Section 9.5.
†Exceptions to this approach do exist. Large DSPs containing CoDecs draw large transient current through their DGND lead and are typically designed internally so that noise on this lead does not couple to their analog circuitry. In this case, their AGND and DGND leads should be connected to the analog and digital return planes, respectively, unless specifically suggested otherwise.

vides the path for the converter's digital circuitry return current flow, but this is not to say that it is to be connected to the circuit digital ground plane.

Phase noise in the sampling and reconstruction clock generation circuitry can severely degrade the performance of a sampled data system. In order to minimize phase jitter, the sampling clock generation circuitry should ideally be isolated from noisy digital circuits and grounded and decoupled to the analog ground plane with the intention that the converter and the clock share the same ground plane (Figure 9.231). Yet, due to system constraints, this is not always possible. In many cases, the sampling clock must be derived from a higher frequency multipurpose system clock generated in the digital zone of the PCB, thereafter transferred to the ADC on the analog ground plane. Noise between the two planes adds directly to the clock signal and will produce excess jitter. This situation can be somewhat remedied by transmitting the sampling clock signal as a differential signal using either a small RF transformer or a high-speed differential driver and receiver device. Many high-speed ADCs have differential sampling clock inputs to facilitate this approach.

Separate power supplies for analog and digital circuits are also highly desirable. In some of the newer high-speed converters, for instance, the analog circuits are powered by +5 V, whereas the digital interface is powered by +3.3 V to interface to 3.3 V logic.* In this case, the +3.3 V lead of the converter should be decoupled directly to the analog ground plane. If the analog system supply is used for powering the converter's digital circuitry, the digital power supply input lead (V_{DD}) can be isolated from the analog supply by the insertion of a small ferrite bead, as shown in Figure 9.231, in order to minimize the flow of digital currents in the analog return path. The converter's internal digital currents will flow to the ground plane through the V_{DD} lead and the low inductance decoupling capacitor (mounted as close to the converter as possible) and will not spread out in the external analog ground plane.

With appropriate decoupling, the only digital currents to flow between the circuit's analog and digital ground planes are the return currents associated with the converter's logic interfaces. These should be maintained at the lowest level possible. To achieve this objective, parallel noisy data buses should never be driven directly from ADCs' and DACs' parallel I/O port because the digital return currents are likely to be too large and may corrupt the analog signal.

By placing a buffer/latch between a converter and a data bus (as illustrated Figure 9.231), the converter's digital lines are isolated from any noise that may be present on the data bus. Even though a few high-speed converters have three-state outputs/inputs, this isolation latch represents good design practice. This buffer/latch and other digital circuits should be grounded to the digital ground plane of the PCB. Any noise appearing between the analog and digital ground planes reduces the noise margin at the converter digital interface. Since digital noise immunity is on the order of hundreds of millivolts, system performance is unlikely to be affected. Under no circumstances, however, should the voltage between the two ground planes exceed 300 mV, approximately, or the devices may be damaged.

---

*Note, however, that when supplied from two independent sources (or independent outputs of a switch-mode power supply), power sequencing problems should be considered and precluded.

9.8.6   Grounding Scheme for a Single ADC/DAC on a Single PCB

Many ADC and DAC manufacturers, while suggesting the use of split ground planes, state something like the following in their application notes: "The AGND and DGND pins must be connected together externally to the same low-impedance ground plane with minimum lead length. Any extra external impedance in the DGND connection will couple more digital noise into the analog circuit through the stray capacitance internal to the IC."

Yet, some connection must be present between the analog and digital ground planes in order to facilitate the digital bus interface between the converter and the digital part of the PCB and to provide an effective return path for the digital current. The key question is, how should this connection be implemented without violating any of the principles discussed up to now?

In circuits containing low-to-moderate resolution (8 to 10 bits) ADCs or DACs, which typically require only about 60 dB of noise suppression, the use of a large massive and solid ground plane common to both the analog and digital circuits is usually preferred. Nevertheless, the analog and digital circuits (and, consequently, their respective return current paths) should still be confined in well-defined zones, effectively partitioning the board (even if not physically splitting it) in order to control undesired coupling, or crosstalk, between the two. A good practice is often to put digital components and signals on one side of the board and analog ones on the other [Figure 9.232(a)]. ADCs or DACs that process both analog and digital signals straddle the analog and digital zones on the plane. In this fashion, this architecture satisfies the requirement of tying together the AGND and DGND.

Higher resolution circuits (12 bits and more) have resolution voltages in the range of tens of microvolts or less; hence, higher ground noise voltage isolation is required for ensuring their adequate performance. With a solid and common ground plane, noise voltage resulting from even a small fraction of stray digital currents flowing across the analog zone of the PCB's return plane may exceed the LSB of the converter and disrupt its performance (recall from Section 9.8.4 that in 24-bit ADCs, the impedance required to maintain the stray digital voltage at a sufficiently low level is in the range of 100 pΩ to 1 nΩ, which is practically unattainable).

The PCB, including its ground plane, should be partitioned into two physically isolated analog and digital planelets with a gap placed between them in the ground plane (as well as in all other layers) [Figure 9.232(b)]. This gap provides additional noise isolation for the high-resolution ADCs and DACs, by precluding common-impedance coupling between noisy and sensitive return current paths while maintaining an electrically continuous (but not solid) ground plane between both circuits. The boundary between these planelets should pass beneath the converter and separate the device's leads associated with the analog functions from those associated with digital functions. The two ground planelets should be connected at one common point (circuit "star ground") directly beneath the converter and the AGND and DGND leads of a converter should be tied together (to the analog return plane) near that point. A typical width of the splitting gap in the ground plane is 80 to 120 mils (0.08 to 0.12 inch, or 2.03 to 3.05 mm, approximately) for practical PCB constructions

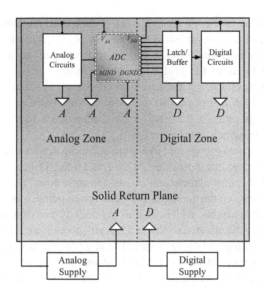

(a) Low Resolution ADC: Solid Ground plane, common
to both Analog and Digital Circuitry

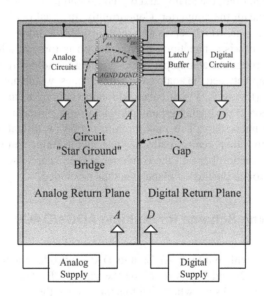

(b) High Resolution ADC: Split Plane Connection

Figure 9.232. Grounding ADCs/ADCs: Single converter on a single PCB.

(e.g., plane thickness of 1-oz copper and FR-4 dielectric medium). The connection between the planelets is accomplished by a narrow bridge, sufficiently short and wide so that little voltage potential difference is present between the two, taking into consideration the thickness of the bridge, the analog return current frequency, and the amount of analog current going through this connection. This narrow connection provides a relatively high impedance to the flow of high-frequency digital return currents with high edge rates, but a relatively low impedance to the analog return currents with lower frequency components [61].

The topological concept depicted in Figure 9.232(b) evades dual-connection situations by tying together the converter AGND and DGND leads, the analog return planelet, and digital return planelet, all at one common point. Since there is only one connection point between the analog to the other return paths, it is observed as a "dead end" for stray digital currents, which are thus precluded from flowing into it. The common connection must be situated so that it does not encourage circulation of return currents between the analog and digital zones. The key to the effectiveness of this scheme is, thus, having separate DGND and AGND planelets. If noise currents are not precluded from passing from one plane to the other, separating the two ground planelets is of little avail.

One often overlooked manner in which such a situation could occur is associated with the analog interface of the converter, when single-ended, chassis-referenced analog input signals are used; the analog and digital ground planelets on the PCB must be interconnected beneath the converter. Concurrently, the analog planelet must also be connected to the chassis, which serves as a reference for the single-ended analog input. Through the two chassis connections, a ground loop is formed, allowing stray noise currents to flow through the analog planelet, defeating the sole purpose of the isolation achieved by the gap between the analog and digital zones, which now no longer constitute a "dead end" (Figure 9.233). If balanced inputs (e.g., by means of a transformer) are used, connection of the analog ground planelet to any particular system-wide reference structure is unnecessary. The analog return region for a floating-input system is thus connected to other return paths at one point only, through the common analog–digital connection beneath the converter and nowhere else. The flow of stray digital currents through the analog planelet is thus precluded.

9.8.7. Grounding Scheme for Multiple ADCs/DACs on a Single PCB

Circuits embedded with multiple ADCs and DACs introduce a further complexity; when numerous ADCs and DACs reside on the same PCB, their various return leads must all be somehow and somewhere tied together. When the analog and digital return planelets are tied together under each converter, numerous interconnection points exist between the two and, thus, they can no longer be considered isolated. Furthermore, interfering return current may flow through the analog zone between two interconnecting points when AC potential difference is present. If on the other hand, the digital and analog return planelets are maintained in total isolation (that is, with no bridge between them under any converter), device manufacturers often recommend tying to-

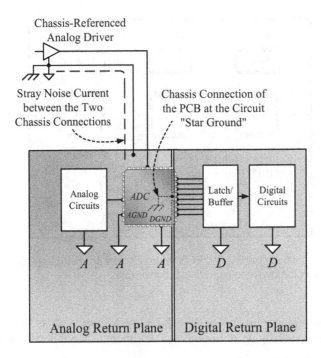

Figure 9.233. Effect of chassis-referenced analog driver connection to an ADC on a split return plane.

gether the analog and the digital return pins to the analog return planelet. In most cases this is highly undesirable.

In low-resolution 8-bit circuits (requiring only about 60 dB of noise rejection), a much better approach is to make use of one large, solid return plane common to all the analog channels and the digital logic. A mindful layout physically separates the analog circuits from the digital circuits to control mutual interference through crosstalk between the different channels (Figure 9.234). This architecture satisfies the requirement of interconnecting the converters' analog and the digital return leads (AGND and DGND, respectively) [4]. The return plane may then be partitioned into analog and digital planelets as illustrated in Figure 9.235. This arrangement, which may be desirable in moderate-resolution (i.e., 10-to-12-bit) converters, (requiring up to 73 dB of isolation, approximately) satisfies the requirement of connecting the analog and the digital return leads together through a low impedance while not compromising EMC concerns associated with the preclusion of creating unintentional loops.

In higher resolution systems (12 bits or more), requiring higher noise isolation, stray digital currents flowing across the analog return planelet of the PCB constitute a serious concern. These currents can interfere with some extremely sensitive analog circuitry (for instance, from Table 9.4 it is easily observed that with a 5 V reference level, the LSB in a 24-bit ADC is equivalent to 0.3 μV, approximately).

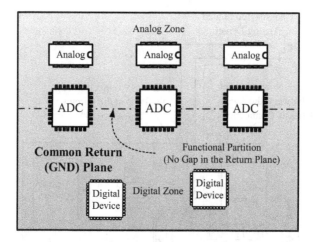

Figure 9.234. A properly partitioned PCB return plane with multiple ADCs provides acceptable isolation for low-resolution (8-bit) converters.

Problems emerging from unacceptably high interference currents flowing in a particular path can generally be addressed in one or more of the following three lines of attack:

1. **Minimize the level of the interfering signal.** High-speed logic is inherently aggressive from the standpoint of EMI, owing to the high edge rate (or short state transition times). When this feature is not necessary for system performance, it

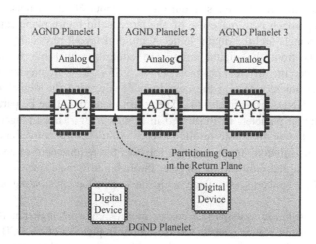

Figure 9.235. A properly partitioned PCB return plane with multiple ADCs provides supplementary noise isolation for moderate-resolution (10-to-12-bit) converters.

is highly advisable to utilize the lowest speed logic possible, commensurate with the performance requirements of the system. In consequence of the distribution of high-speed signal return currents in the return plane, increasing separation between the aggressor signal to sensitive analog signal traces further trims down the coupling between the traces (see Figure 9.23 and Table 9.5) [6].

2. **Interrupt the interference pathway.** In a manner similar to that described in Section 9.8.6, the return plane of the PCB should be split into $N + 1$ planelets, one for each of the total number of A/D and D/A converters combined, N, and an additional planelet common for all remaining digital circuits.* Each of the analog planelets should be tied to the digital return planelet at one point by a narrow copper strip, directly beneath its respective converter. The slots between the planelets disrupt the flow of stray interference currents in the return plane, while the narrow strips across the slot under each of the converters are characterized by high RF impedance. In view of the fact that each analog return planelet is linked to the rest of the system at one point only, it acts like a dead end for stray digital noise currents, prohibiting their entry into that zone. No traces of any kind are allowed cross the gap in the return plane because interrupting the return current path will worsen interference and may result in severe disruption of the circuit performance.

   Use of single-ended, chassis-referenced analog input signals results in a tricky situation; each of the analog-return planelets must come into contact with two grounded objects, namely, the converter's return and the chassis ground for its input signal. The consequent "ground loops" resulting from this scheme will allow interfering stray currents to flow through the analog planelet, similar to the case of a single converter on the PCB, defeating the sole purpose of the gaps between the various analog zones. Use of differential and balanced ("floating") analog signal inputs, for instance, by transformer-coupled lines, provides a satisfactory solution to this situation.

3. **Bypass the affected pathway with a low-impedance shunt element.** When the first two practices cannot be applied, the only remaining strategy lies in the third approach. Shunting stray noise current is achieved by placing a massive solid metal sheet immediately underneath the PCB and interconnecting the edges of all PCB return planelets through "stitches" between the return planes and the shunt sheet (Figure 9.236). The metal shunt sheet, which should contain no gaps or slots, provides the necessary low-impedance shunt around each analog-return planelet, somewhat reducing the impact of stray digital currents in that zone.

   This suggested approach, however, could lead to some further complication, particularly with respect to the analog parts of the circuit. Multipoint grounding of low-frequency, highly sensitive planes (such as found in high-resolution ADC-embedded circuits) could result in significant common-impedance cou-

---

*Note that the analog planelets are also isolated through the gaps, since at such high resolutions, and considering the return current spread at low-frequency analog signals, crosstalk could also occur between the analog circuits. For preclusion of such analog-to-analog common-impedance coupling, gaps must be placed between the analog planelets.

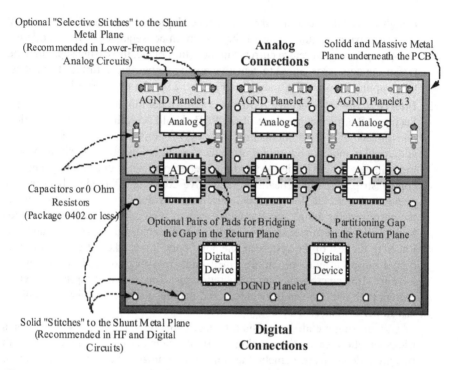

Optional "Selective Stitches" to the Shunt
Metal Plane
(Recommended in Lower-Frequency
Analog Circuits)

Figure 9.236. Stitching all AGND and DGND planelets to an adjacent solid metal plane as a shunt for stray noise currents.

pling in the analog zone of the PCB since low-frequency return currents tend to spread out in the shunt metal plane (now serving as an inadvertent, albeit effective, common return path for the low-frequency analog return currents when multipoint "stitched" to this plane). Consequently, interference coupling to low-frequency analog circuits may even increase.

For avoiding that problem, the shunt metal plane should be truly massive and of low impedance, thus minimizing potential differences across the plane. In addition, selective "stitching" of the analog return planelets through capacitors rather than "hard connections" provides for "floating" connections at the lower frequencies (thus, the connections occur only at the boundary with the digital return planelet) and multipoint connections at higher frequencies, at which shunting of stray interference current is desirable (see "Optional 'Selective Stitches' to the Shunt Metal Plane" in Figure 9.236).

Many variables may take part in this scheme, and small changes may make a big difference. In order to avoid a situation whereby the "cure is worse than the disease," it is important, when applying the third technique (listed above), to keep all options open.

Whenever high confidence does not exist as to the optimal grounding architecture, provisions should be made to interconnect the AGND and DGND planelets, virtually transforming the circuit back into a RF-continuous (even if not solid) plane. Bridging options allow both methods to be tried throughout the design to verify which offers the best overall performance in the system. Here again, the best strategy for putting this into practice is to allocate multiple mounting pads at both sides of the gap, providing the means for interconnecting (or stitching) the analog and the digital planelets together through short jumpers or 0 Ω resistors. The mounting pads should be close together (approximately 1 to 1.5 cm apart). Keeping in mind that this connection should be at the lowest impedance as possible, the size of the jumpers should be kept to the absolute minimum in order to minimize inductance. Therefore, they should be placed as near as possible to the gap, and the distance between any two facing mounting pads should allow the mounting of the smallest possible SMT 0 Ω resistor package (e.g., 0402) across the gap (a width of 80 to 120 mils or 2 to 3 mm was earlier suggested). When necessary, the mounting pads may locally protrude into the gap area on both sides so as to facilitate the necessary connection across the gap without using larger jumpers or traces. Figure 9.237 depicts the recommended manner of stitching the AGND planelet 1 and the DGND planelet of the circuit in Figure 9.236, focusing on that boundary.

Unfortunately, it is difficult to predict whether the "multipoint" (solid return plane) or the "star" ground (separate analog and digital return planelets) method will produce the best overall system performance, and which stitching configuration is best suited for the circuit. Some experimentation with the final PCB through a "trial and error" process using SMT jumpers may be required. The optimum stitching configuration is thus optimized by quantitatively measuring the difference in performance at the different stitching configurations. It may take a few trials in order to determine the most favorable arrangement that befits the particular geometry (if performance improves the

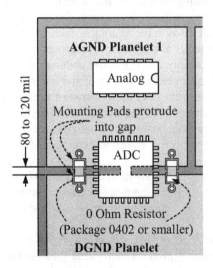

Figure 9.237. Recommended manner of stitching the gap between AGND Planelet 1 and the DGND planelet.

more stitching is carried out). It could be that a solid rather than a split plane constitutes the optimal architecture for this circuit.

It is important to emphasize that extreme care that must be taken to ensure that under no condition will any signal trace, particularly high-speed digital ones (but also analog), cross the gap in any layer of the PCB, whether stitching is implemented or not, except over the bridge between the two planelets (and, even then, only where absolutely necessary for the performance of the circuit). If the planelets are not interconnected, the path of the signal return current will be interrupted by the gap, thereby producing severe EMI; conversely, if the gap is stitched across, particularly at multiple points, return current of signal traces crossing the gap might flow through the stitched points, counteracting the original purpose of the gap and making it useless.

9.8.8   Grounding Scheme for ADCs/DACs on Multiple PCBs

The dilemma increases with respect to the grounding architecture in multicard, mixed-signal systems, in which ADCs or DACs are present in multiple PCBs. Although the approach described in Sections 9.8.6 and 9.8.7 generally works in systems with a single PCB embedded with one or more ADCs/DACs, it is usually not optimal for multicard, mixed-signal systems. In such multicard systems, multiple connections between the analog and digital ground planes at several distributed locations are formed with higher impedance (and, subsequently, higher noise voltage drop) between them, increasing the possibility of harmful ground loops while invalidating the single-point "star" ground strategy. Figure 9.238 illustrates this situation.

The best means of minimizing return path impedance (and, therefore, interference voltage drop between the PCBs) in a multicard, mixed-signal system is to make use of a motherboard or backplane. The motherboard should consist of separate analog and digital planelets, thus providing continuous return plane interconnections between respective connector return pins in the PCBs. A total of at least 30–40% of the PCB edge connector pins should be devoted to the return interfaces. The two return planes in the motherboard are joined together at the system star ground or single-point ground (the system-wide GND-*REF*), usually located at the common ground reference point for the power supplies. The back-to-back Schottky diodes are inserted to prevent accidental DC potential differences from developing between the two ground systems. The Schottky diodes may be located in the backplane, immediately near the edge connector attached to each of the daughter boards. Alternatively, they may be installed on the daughter boards, near their respective edge connectors used for ground interconnection to the backplane.

At this point of the discussion, two distinct cases need to be considered with respect to the actual implementation of the connection between the analog and digital return planelets on the multiple PCBs, according to the level of the digital current flowing in the circuit.

In systems in which the ADC or DAC draws low digital currents, the converters on the different PCBs are treated as predominantly analog components. They are powered from the analog power (V_{AA}), decoupled and referenced to the analog return planelet, AGND (if split rather than solid return planes are used). The analog return plane is not

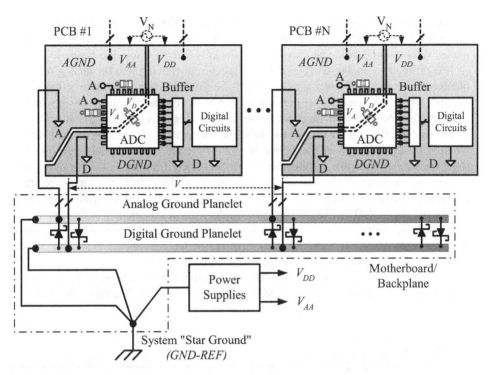

Figure 9.238. Grounding architecture in a ADCs/DACs on multiple PCBs drawing low digital current. Note the power supply from the AGND planelet.

corrupted because the small digital transient currents flow in the small loop between V_{DD}, the decoupling capacitor, and the DGND lead of the device, both placed in the analog zone of the PCB. The noise V_N between the return planes reduces the noise margin at the digital interface but is generally not harmful to performance if maintained at a level of less than 300 mV by using a low-impedance digital return plane all the way back to the system "star ground" (GND-REF) through the motherboard.

In this architecture, a large return current loop is formed at the digital interface between the buffer/latch and the converter as the return current path is completed only through the system "star ground." If it features a high-speed interface, this large loop may not be acceptable due to EMI concerns (even if the 300 mV DC voltage difference is maintained). In this case, bridging the gap under the devices may be necessary. Therefore, optional pads for bridging over the gap with a narrow jumper may be required, in a manner similar to that discussed for multiple ADCs and DACs on the same PCB.

Mixed-signal devices such as sigma–delta (Σ–Δ) ADCs, CoDecs, and DSPs with on-chip analog functions are becoming more and more digitally demanding. As a result of the additional digital circuitry, more digital current is drawn, consequently resulting in higher ΔI noise voltage across the ground system that could no longer be effectively handled by the decoupling capacitor. Any digital current that flows outside the loop between the V_N and DGND leads of the device will then flow through the analog return plane, which could degrade performance, particularly in high-resolution systems.

It is difficult to predict what level of digital current flowing into the analog return plane will become unacceptable to a system's performance. An alternative grounding method, which may yield better performance, is proposed for circumstances in which systems are embedded with mixed-signal devices drawing high levels of digital current.

In this case, the AGND lead of the device is connected to the analog return plane, whereas the DGND lead of the device is now connected to the digital return plane (note the similarity of this architecture to that presented in Figure 9.236 for multiple devices on the same PCB). The digital currents are isolated from the analog return plane; thus, the noise voltage drop between the two return planes develops now directly between the AGND and DGND leads of the device. For this method to be effective, the analog and digital circuits within the device must be isolated to a great degree. The noise between AGND and DGND leads must not be large enough to reduce internal noise margins or cause corruption of the internal analog circuits.

Figure 9.239 shows a ferrite bead (F.B.) interconnecting the analog and digital return planes beneath (or immediately adjacent to) the converter. The ferrite bead pro-

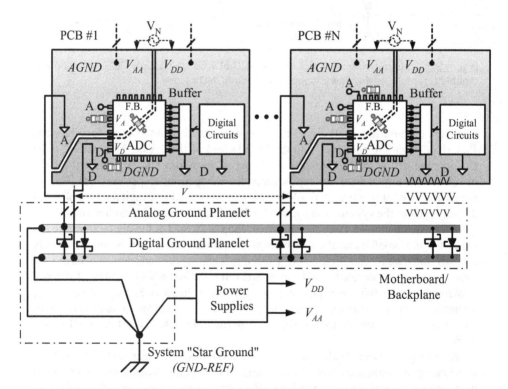

Figure 9.239. Grounding architecture in ADCs/DACs on multiple PCBs drawing high digital current. Note that the digital current is supplied from the DGND planelet.

vides a DC tie between the two planes but isolates them at frequencies above a few megahertz, at which the ferrite bead becomes resistive. This protects the device from excessive DC voltage potential differences between AGND and DGND, which could potentially damage the mixed-signal device if they exceed 300 mV. On the other hand, the DC connection provided by the ferrite bead can introduce unwanted low-frequency ground loops and may not be suitable for high-resolution systems. Alternatively, back-to-back Schottky diodes may be used instead of the ferrite. The Schottky diodes prevent large DC voltages or low-frequency voltage spikes from developing across the two planes. Unlike the ferrites, the diodes provide no noise isolation at high frequency.

Regardless of the scheme employed, it must be ensured that the connection to the return planelets in the motherboard is accomplished directly from both the analog and the digital planelets on the daughter PCBs, as illustrated in both Figure 9.238 and Figure 9.239. Any other configuration, such as connection to the motherboard through either the analog or the digital planelets and corresponding connector pins, may result in significant crosstalk between the digital and the analog circuits. Figure 9.240 illustrates an acceptable (a) and objectionable (b) ground connection between a daughter PCB and the motherboard.

If such a connection is required due to circuit design constraints, it may be best to use a solid return plane and pay careful attention to layout of the traces on the PCB. Even when solid and common return planes are used on the daughter PCBs, the trace layout on the motherboard should be such that the virtual separation between signal return paths is still maintained through disciplined component placement and trace routing in the motherboard, consistent with the daughter PCB layout.

Lessons Learned

- Common-impedance coupling between noisy digital and sensitive analog return currents in mixed A/D systems may be detrimental to the performance of the analog circuits.
- There is no single grounding scheme that will guarantee optimum performance at all times and under all circumstance in mixed analog–digital circuits. A number of possible alternatives are available and their application depends upon the characteristics of the particular mixed-signal device in question.
- A single common solid return plane under both analog and digital zones of the PCB is preferred in lower resolution (e.g., 8-bit) mixed circuits; in higher resolution (10-bit or more) systems, splitting the planes may be inevitable for achieving the best overall system performance.
- The key to a successful layout of a mixed analog/digital PCB lies in disciplined component placement and routing, considering the path of return current flow; the path of digital return currents in the return plane can be easily controlled by careful layout.
- *Never* connect a major data bus directly to an ADC or DAC. Always place a buffer/latch between a data bus and a converter (even when one is shown to be incorporated in the converter).

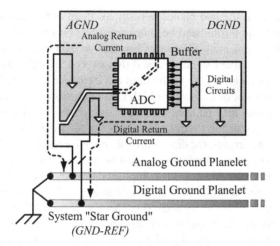

(a) **Acceptable**: The Digital and Analog Return Current
Paths Flow through Separate Return Path

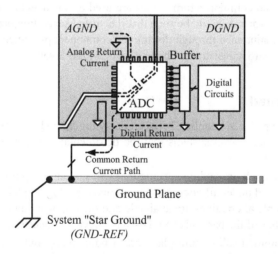

(b) **Objectionable**: The Digital and Analog Return
Currents share a Common Return Path

Figure 9.240. Poor layout of ADC-embedded circuits may result in common-impedance coupling through the common ground connection. Example shown for a single
ADC on the PCB for simplicity; when multiple devices and PCBs are used, the bridge
in the gap would typically be replaced by a ferrite bead, as described in the text.

- As a general rule, AGND and DGND should both be connected to the analog return plane of the split-ground system (with the exception of high-digital-current devices). The lead description DGND does *not* imply that this pin should go to the system digital return.
- When in doubt, it is best to begin with a split analog and digital return plane and later jumper across them (e.g., through 0 Ω resistors or ferrite beads), rather than starting out with a solid plane and later trying to split it.
- As many options as possible should be provided for during initial PCB layout; pads and vias should be allocated at several points for installation of back-to-back Schottky diodes or ferrite beads or for interconnecting the analog and digital return planelets, if necessary.
- Evaluation boards and grounding layout application notes are offered by manufacturers of high-performance mixed-signal devices to assist customers in their initial evaluations and layout; keep in mind, however, that they are optimized for a single device on a single PCB only.

9.9 CHASSIS CONNECTIONS ("CHASSIS STITCHING")

Electronic equipment is frequently constructed using PCBs attached to a metallic chassis, commonly also serving as a signal reference. Since most modern PCBs contain one or more solid return planes in their multilayer stack-up, one or more electrical connections ("stitches") are usually made between the return plane(s) and the chassis, normally near the I/O connectors if nowhere else (Figure 9.241).

The term "chassis" relates to a metallic structure supporting the PCB, for example, any member of the metal enclosure, such as the wall or floor. A shielded metal enclosure makes an excellent chassis for a PCB, but any massive piece of metal (such as a heat sink) could perform just as well. A totally shielded enclosure makes the best RF reference plane, but shielding is not essential for this purpose. Where plastic housings or other structures are used to support a PCB, they can be used as a "chassis" if they are made to be sufficiently conductive (e.g., by metallization).

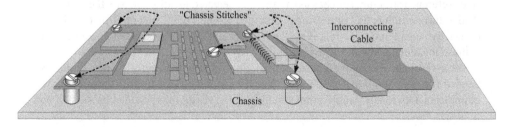

Figure 9.241. Example of a PCB bonded to its local chassis. (Courtesy of Keith Armstrong, Cherry Clough.)

9.9.1   Purpose of Stitching PCB Return Planes to Chassis

Contrary to common belief, the signal return planes are far from being ideal, utopian "0 V" planes. As high-level switching return currents flow through the finite impedance of the digital return plane ("digital ground" or DGND) they produce excessive RF voltage potential differences across the plane, appearing as a common-mode (CM) EMI source. Figure 9.242 depicts the RF voltage distribution across a digital return plane of a PCB.

RF currents that exist within the power and return plane structures produce RF fields, subsequently emitted from the circuit components, interconnect wiring, and the like. These fields tend to couple efficiently to other collocated metallic structures, through distributed impedance (often on the order of a few tens of ohms), generating RF eddy currents* that circulate within the structure [3, 63].

One of the most significant ramifications of this phenomenon is the development of common-mode voltage potentials between the PCB and adjacent metallic members of the equipment chassis. These RF voltage differences give rise to radiated emissions formed at the edges of the planes, sometimes called fringing fields (Figure 9.243). The current to common-mode voltage conversion factor in a PCB return structure is denoted the "return structure transfer impedance".

Common-mode impedance between the PCB return planes and the chassis significantly higher than that associated with the distributed "driving force" (of the eddy currents) will subsequently couple to collocated circuits (e.g., motherboard/backplane and daughter boards) within the same assembly as well as to cables and interconnects attached to the assembly. The length of these conductors often makes them efficient "accidental antennas," resulting in increased radiated EMI emissions and possible degradation of circuit performance.

One of the numerous EMC-related benefits of solid return planes in a PCB is their lower RF impedance. Lowering the planar transfer impedance is highly desirable and may be achieved by interconnecting the thin copper return plane(s) in the PCB to the massive metal chassis [63]. By bonding a PCB's return plane to adjacent conductive structural members or chassis through multiple low-impedance connections, RF noise voltage potential differences are effectively shorted out, resulting in lower fringing fields' radiated emissions and enhanced immunity (by a reciprocal argument).

Additionally, by physically connecting (or "stitching") the return plane to a "massive" reference structure (or chassis) at multiple selected locations across the plane, the PCB return structure is "brute force" strapped to the "0 V" potential of the chassis (normally considered to be relatively "quiet") through low-impedance RF connections at those locations and their near vicinity.†

As a result of the above considerations, high-frequency return planes should be bonded (or "stitched") to chassis whenever not precluded due to other conflicting requirements. Obviously, single-point stitching of the return plane to the chassis will not

---

*Eddy currents (or Foucault currents) are circulating currents made to flow within the conductor, caused by a magnetic field intersecting a conductor or vice versa. Eddy currents are the root cause of the skin effect in conductors carrying AC current.

†See Section 9.9.5 for a more elaborate discussion of techniques for breaking up chassis cavities.

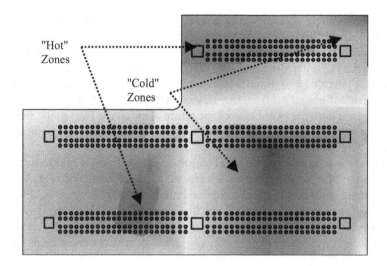

Figure 9.242. Illustration of RF noise voltage distribution across a digital return plane.

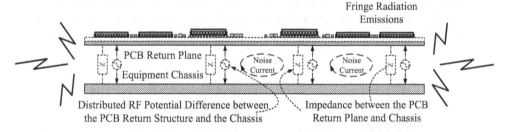

Figure 9.243. Radiated emissions due to PCB and chassis fringing fields. (Courtesy of Keith Armstrong, Cherry Clough.)

do; even if one point on the circuit is maintained at "0 V" potential, other parts of the PCB will experience RF voltage fluctuations.* High-frequency designs generally necessitate the use of multiple low-impedance connections between the signal return plane to the equipment chassis or to a "massive" common reference structure in order to minimize the impedance of the return plane and impose "0 V" potentials throughout the entire return plane. Typically, such connections would be made at least at each of the corners of the PCBs' return plane as well as near the I/O zone interfacing with off-board wiring. Another typical location for a PCB return-plane-to-chassis connection is at the boundary between digital and analog/video/RF return circuits. Such a connection prohibits digital noise currents from flowing into the potentially susceptible signal return of the PCB (Figure 9.237). Additional stitching is commonly positioned near es-

*The rationale for single-point versus multipoint grounding is discussed in Chapter 4.

pecially noisy sources on the PCB (e.g., clock oscillators, exceptionally noisy VLSIs such as powerful DSPs, CPLDs, etc.), as illustrated in Figure 9.241.

Noise on the return planes may also result in high-frequency emissions from cables attached to the PCB. In particular, when driving an unbalanced load, a low-impedance path must be provided for return currents between the PCB and chassis. With a chassis stitching standoff placed far from the edge connector of the PCB, the necessary low-impedance path does not exist between the return plane on the PCB to the chassis, due to the large loop formed between the stitching standoff and the entry point to the enclosure (Figure 9.244).

Noise on the return planes may also result in radiated emissions from I/O cables attached to the PCB. In particular, when driving an unbalanced load (e.g., RS-232), a low-impedance path must be provided for return currents between the PCB return plane to the chassis. With a chassis stitching standoff placed a long way from the edge connector of the PCB, the necessary low-impedance path does not exist, due to the large loop formed between the stitching standoff and the entry point to the enclosure.

Furthermore, the control of radiated emissions from I/O signal lines often requires the installation of EMI filters designed for common-mode suppression. (Note that filters may be required both on single-ended and differential lines.) Signal line EMI filters generally consist of ferrite beads and (typically) small line-to-chassis capacitors. Ideally, the capacitors are connected directly to the chassis. Layout constraints might mandate, however, that the capacitors actually be tied to the return ("ground") plane.

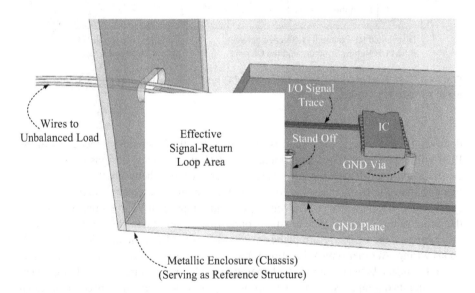

Figure 9.244. With a stitching standoff placed at far from the edge connector, a low-impedance path does not exist between the PCB's ground plane to the chassis. (Courtesy of Dr. Bruce Archambeault.)

In such cases, minimizing the loop area formed between the I/O interface, the capacitor, and the chassis connection point is of utmost importance. Since filters are to be installed on the boundary of the circuit (i.e., near the I/O port), reducing the size of the abovementioned loop requires that the chassis stitching standoff be likewise placed in near proximity to the I/O port of the PCB. An adequate installation is illustrated in Figure 9.245. Note that for control of emissions, only chassis stitching standoffs placed at the I/O port near edge connectors have significant effect. Chassis stitching at points located remotely from the I/O area of the PCB contribute little to reduction of emissions.

In addition to reduction of emissions from the PCB, chassis "stitching" also contributes to increase of immunity, particularly with respect to electrostatic discharge (ESD). Allocation of a proper number of chassis connections and their correct placement help to control the propagation of ESD pulses that may couple into the circuit, throughout the PCB return planes (Figure 9.246). The importance of these factors is illustrated in Figure 9.247 through Figure 9.251.

Consider an ESD pulse injected into a PCB at the left-hand corner of a PCB. With only one chassis stitching post placed in the vicinity of the I/O port of the PCB, the ESD pulse propagates throughout the PCB's ground plane, reflects back from the edges, and produces multiple "hot spots" (Figure 9.247).

With the number of stitching posts at the left-hand I/O port increased to four, little improvement is achieved. Although some suppression of the ESD pulse propagating

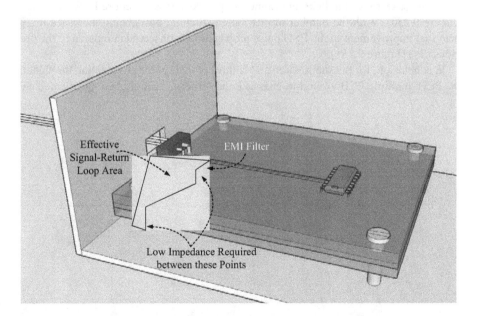

Effective Signal-Return Loop Area

EMI Filter

Low Impedance Required between these Points

Figure 9.245. A stitching standoff placed near the edge connector provides a low-impedance path between the PCB's ground plane to the chassis. (Courtesy of Dr. Bruce Archambeault.)

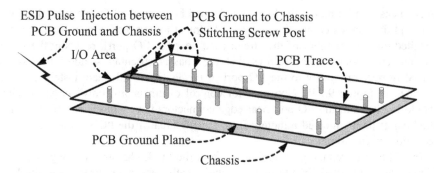

Figure 9.246. Distributed PCB ground plane to chassis stitching posts is important for control of ESD pulse propagation.

across the plane is observed, it is not contained and significant interference to the circuit would still occur (Figure 9.248).

Further increasing the number of stitching posts to eight, placing four at each end of the PCB, offers some improvement, as the ESD pulse reaching the end is partially shorted out, but some residual pulse is still reflected across the board (Figure 9.249).

Rearranging the eight stitching posts, evenly distributing them across the PCB, results in significant reduction of the ESD pulse propagation. The pulse is contained in the leftmost sector of the PCB and cannot propagate across the entire PCB any longer (Figure 9.250). With the number of distributed stitching posts being increased to 20, marginal improvement in the ESD pulse propagation is observed compared to the previous case (Figure 9.251).

In conclusion, for the suppression and control of ESD pulse propagation throughout the PCB, multiple PCB ground-to-chassis bond "stitches" should be implemented and

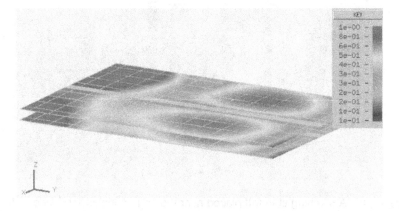

Figure 9.247. With only one stitching post at the left-hand I/O area, high levels of ESD pulse reflections are observed. (Courtesy of Dr. Bruce Archambeault.)

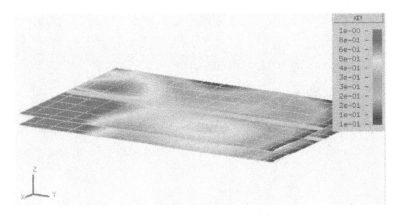

Figure 9.248. With four stitching posts at the left-hand I/O area, high levels of ESD pulse reflections are still observed. (Courtesy of Dr. Bruce Archambeault.)

distributed across the PCB. Considering the high spectral content of the ESD pulse, attention should be paid to ensure as low impedance a connection as achievable.

When a metallic chassis is not available (for instance when the package is not metallic), a "floating chassis plane" internal to the PCB can be used as an alternative. This plane could be positioned immediately adjacent to a digital return plane. Thanks to the large mass of this metallic structure, low distributed transfer impedance will thus be established between both the floating chassis and the return plane. Common construction of "floating chassis planes" is of 12 to 20 oz/sq. ft. When used, all signal traces should be routed in internal layers of the PCB, with both top and bottom layers

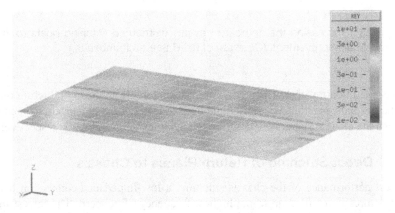

Figure 9.249. With four stitching posts placed at both ends of the PCB, lower levels of ESD pulse reflections are observed but propagation across the PCB still occurs. (Courtesy of Dr. Bruce Archambeault.)

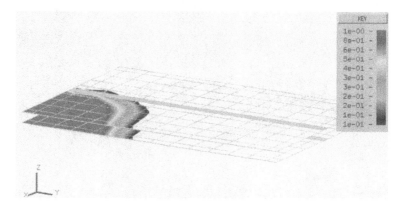

<u>Figure 9.250.</u> Evenly distributing the eight stitching posts throughout the PCB suppresses the ESD pulse, containing it near the injection area. (Courtesy of Dr. Bruce Archambeault.)

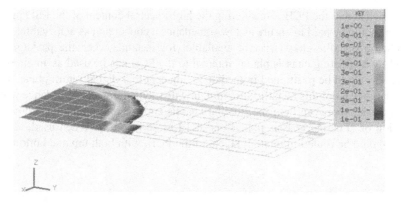

<u>Figure 9.251.</u> Increasing the number of evenly distributed stitching posts to 20 provides marginal improvement. (Courtesy of Dr. Bruce Archambeault.)

used as solid floating chassis planes.* Chassis stitching from the digital return plane to the chassis plane can thus be easily implemented using direct or capacitive connection between the planes (elaborately described in Sections 9.9.2 and 9.9.3, respectively).

9.9.2 Direct Stitching of Return Planes to Chassis

For best performance of the chassis stitching, a low-impedance connection between the structures is required up to the highest frequency of concern. The lowest imped-

---

*When applied to PCBs other than backplanes, where components are mounted on the top (component side, CS) and bottom (print side, PS) layers, the massive solid planes should be placed immediately below and above the CS and PS layers, respectively.

ance bonds are achieved at high frequencies by directly forcing down one metal surface against another. Point contacts are likely to exhibit higher impedance than those achieved through surface contacts at high frequencies.

Large mounting pads present on the bottom of the PCB, at least 3 mm in diameter, are made to overlap and securely press against the metal standoff walls, which in turn press against the chassis. As such, this bond assembly provides the desired low-impedance connection to the chassis. It is quite common to ensure a good bond between these pads on the outer layers of the PCB and an inner return plane by using a ring of via holes, as shown in Figure 9.252.

Figure 9.252(b) and (c) show in further detail the manner of implementation of the bond between the PCB (typically inner) return planes and the metallic chassis on which the PCB is mounted. Note the multiple vias between the bonding ring and the return plane in the PCB, providing a low-impedance connection, thanks to the parallel connection of the vias' inductances. The screw subsequently connects to the metallic chassis in a manner similar to that shown in part (a).

Good surface conductivity is required for all the metal-to-metal contact areas. For ensuring that the contact areas of the mating surfaces are firmly pressed together, screws are commonly used (Figure 9.252). However, screws contain a helical thread, and the edge of the screw thread is the part that mates with the standoff. The standoff must be physically larger in diameter than the screw diameter in order to allow the screw to be properly inserted. As a result, solid and continuous bonding contact between some of the threads to the standoff along the length of the screw is incidental and cannot be guaranteed. Figure 9.253 illustrates this situation [63].

When different metals are joined, galvanic corrosion may occur between the mating metals, counteracting the low-impedance goal. For ensuring the lifelong high performance of the bond, deterioration of the bond due to galvanic corrosion must be prevented by applying compatible protective plating (e.g., tin) on each of the mating surfaces.*

When rubbing occurs between the screw and standoff, the plating can get scraped off, exposing the screw to humidity and corrosive atmosphere. Galvanic corrosion can now occur, degrading the conductivity of the screw contact area. As such, screws cannot be relied upon to provide low-impedance bonds and must be used only for compression between the PCB and chassis.

9.9.3 Hybrid Techniques for Stitching of Return Planes to Chassis

The question of the exact location, manner, and number of PCB-to-chassis connections ("stitches") does not always have a straightforward and precise answer. Flexibility is one of the golden rules that should be maintained when such uncertainty exists, with the objective of allowing a high level of flexibility in implementing chassis stitching during system development and integration. This technique utilizes a variety of types of bonds between the PCB's return plane and the chassis:

---

*Bonding and electrochemical compatibility of metals are discussed in Chapter 5.

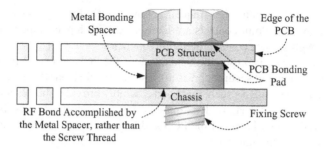

(a) Side View of the Bond Assembly

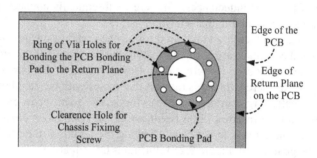

(b) Bottom View of the Bond Assembly

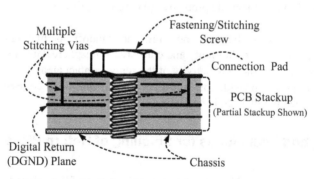

(c) Detailed Side View of the Inner Circuit Connection
made to Upper Layer of PCB

<u>Figure 9.252.</u> Implementation of a PCB to chassis stitching bond. (Parts a and b courtesy of Keith Armstrong, Cherry Clough.)

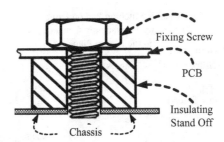

Figure 9.253. Problems arising from bonding the PCB by screws to a standoff. (Courtesy of Mark Montrose.)

- Direct connection (see Section 9.9.2)
- Capacitive connection (see Section 9.9.4)
- Resistive connection, for damping structural resonances (see Section 9.9.5)
- Open (no connection), for leaving the point without chassis stitching at this location

The application of each of the above techniques and "fallback" alternatives should be considered on a case-by-case basis. Hybrid chassis connections may sometimes be the optimal approach despite the fact that capacitive connections are generally less than ideal for broadband RF bonding. It is often difficult to predict what type of bond will exhibit the best overall EMC performance for each location, so prototype PCBs may be required for selection of the best bonding pad pattern.

Figure 9.254(a) depicts one technique for implementing hybrid chassis stitching. By placing a capacitor, resistor, or 0 Ω link across the gap between the circular isolated chassis stitching pad and the pad connected to the inner return plane, the inner return plane can either be connected to the chassis directly (0 Ω), capacitively, or resistively through the fixing screw. If no connection is made across the pads, no bonding is made between the return plane and the chassis at that point.

Figure 9.254(b) depicts a second technique. A circular stitching pad with a ring of via holes on the upper layer of the PCB is permanently connected to the inner return plane. The inner radius of this ring is made to be larger than the thread and head of the fixing screw, and a small washer used to hold the PCB down to the metal standoff and make electrical contact with it.

When the return plane must be connected to the chassis through the standoff (contact stitching) a conducting washer can be used with a diameter large enough to come into contact with the stitching pad connected to the return plane through the ring of vias. When the return plane must be isolated, the washer is left out or a washer with sufficiently smaller diameter is used instead. An insulating washer can also be used for this purpose, as long as it is guaranteed that no inadvertent connection is made to the stitching ring from the metallic screw.

Figure 9.254(c) and (d) provide a detailed side view illustration of the physical implementation of the chassis stitching in both techniques shown in Figure 9.254(a) and (b), respectively. Note that although the standoffs are not shown (for simplicity of the

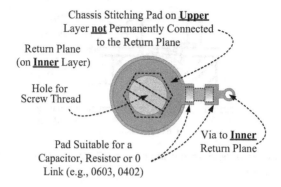

(a) Interconnecting the Return Plane and Chassis Stitching
Pad through a SMT Device. (Courtesy of *Keith
Armstrong, Cherry Clough*)

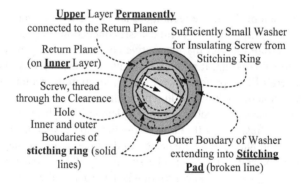

(b) Interconnecting the Return Plane and Chassis Stitching
Pad through Conductive Fixing Washers

Figure 9.254. Two layout techniques for implementing hybrid PCB-to-cassis bond
(shown with two devices in parallel, see Figure 9.255). Insulating standoffs not sown
for simplicity of drawing but should be included as in Figure 9.253.

drawing), it is good practice to insert them between the PCB and the chassis for mechanical mounting purposes.

Chassis stitching must be accomplished through as low impedance as possible, particularly in the I/O area. The primary pitfall in the application of hybrid chassis stitching through an SMT device (e.g., capacitor or resistor) is the series inductance of the device and its connection vias. Zero ohm resistance is not equal to zero impedance, owing to the intrinsic inductance of the SMT devices and, particularly, the inductance

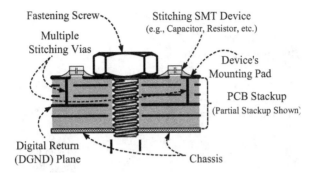

(c) Detailed Side View of the Return Plane and Chassis
Stitching Pad through a SMT Device

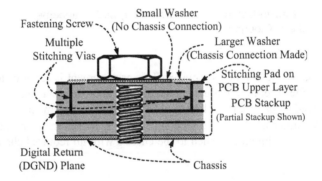

(d) Interconnecting the Return Plane and Chassis Stitching
Pad through Conductive Fixing Washers

Figure 9.254. *Continued.*

associated with their installation. Excessive series inductance due both to the ESL of the device and to its installation can seriously degrade the effectiveness of the chassis connection and should thus be minimized. This inductance can be reduced by extending the technique of Figure 9.254(a) and using multiple devices arranged radially around each bonding location, as shown in Figure 9.255. The physical implementation in Figure 9.254(c) applies to this technique as well.

In this configuration, illustrated with three devices (but could be generally used with two or more), the inductances associated with each device and its pads, traces, and via holes all appear in parallel. The radial arrangement, such as that as shown in Figure 9.255, ensures the cancellation of the mutual inductance between the devices, so that the overall equivalent inductance of this arrangement is simply the value of one

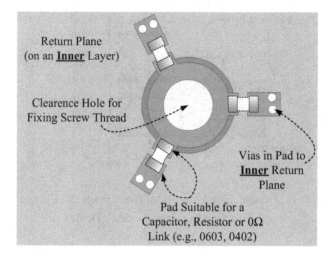

Figure 9.255. Low-inductance chassis stitching pad arrangement. (Courtesy of Keith Armstrong, Cherry Clough.)

bond divided by the number of devices. This arrangement is superior to a parallel array, since in the latter their mutual inductances do not cancel out and the overall inductance achieved would not be as low as with the radial arrangement.

By using either of the above techniques, the PCB return plane may be isolated, single point, or multipoint stitched to the chassis with no required change to the PCB layout.

9.9.4   Capacitive Stitching of Return Planes to Chassis

There are certain circumstances in which a direct conductive connection between the PCB return plane and the chassis may be undesirable, whether due to functional, safety, or other requirements. When DC or low-frequency galvanic isolation is required between a PCB and its chassis, and higher frequency stitching is necessary, capacitive stitching is the only feasible solution.

Consider, for instance, applications in which high-level lower frequency (e.g., power frequency and its corresponding harmonics) interference may be present on the facility or vehicle structures.* For preclusion of such low-frequency structural currents from entering and flowing through a circuit's return plane, multipoint conductive stitching is not advisable as it precisely allows such situations to occur [Figure 9.256(a)].† With all but one conductive connection being replaced by capacitors of properly chosen value, single-point grounding is maintained at lower frequencies, whereas at higher frequencies multipoint grounding is effectively present. Subsequent-

---

*Often, 50/60 Hz power frequency is observed on metallic structural members of facilities. A similar situation exists on aircraft, where 400 Hz power frequency is frequently observed on the aircraft structure, as it commonly serves as the power current return path rather than utilizing dedicated wiring.
†This situation is also known as a "ground loops," discussed in detail in Chapter 4.

ly, the lower frequency interference current remains off the PCB return plane [Figure 9.256(b)] as long as the capacitors are sufficiently small.

A third common application is associated with grounding of lower frequency (e.g., audio) circuits. Such circuits typically utilize single-point grounding architecture in order to evade the adverse effects of low-frequency interference coupling into the circuit through ground loops and common impedance [Figure 9.256(c)]. On the other hand, it is well known that multipoint bonding is required for meeting high-frequency radiated emissions and immunity EMC requirements when higher frequency interference is present. Again, multipoint capacitive chassis stitching, but with only a single-point galvanic connection, keeps the intentional lower frequency signal return current on the PCB's return plane while prohibiting any flow of lower frequency return current

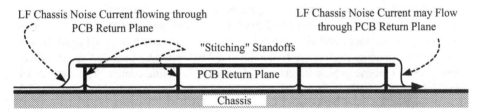

(a) Conductive Chassis Connection: Lower Frequency (LF) Chassis Noise can Flow through the PCB Return Plane

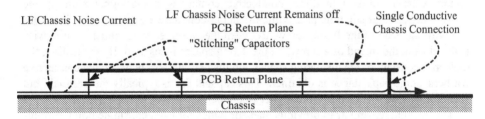

(b) Capacitive Chassis Connection: Low Frequency Chassis Noise Remains off the PCB Return Plane

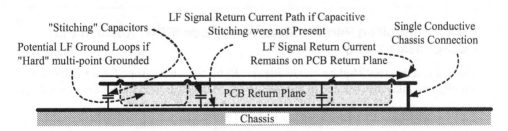

(c) Capacitive Chassis Connection: Low Frequency Signal Return Current Remains on the PCB Return Plane

Figure 9.256. Conductive versus capacitive chassis stitching application. Note: LF = lower frequency.

through the chassis (which would otherwise produce larger loops and increase radiated EMI coupling to or from the circuit). Concurrently, the capacitive chassis connection provides a controlled drain path for high-frequency, common-mode interference currents induced into the circuit and the return plane.

Typical values of stitching capacitors are in the range of 47 pF to 4.7 nF, considering the frequencies up to which effective isolation is required. However, safety requirements may impose more stringent restrictions on the maximum total capacitance allowed between the PCB and the local chassis. For example, galvanically isolated circuits, such as those found in patient-coupled medical equipment, may have to meet very stringent leakage current specifications. Comparable constraints may be imposed in circuits handling hazardous voltages.* In such situations, using suitably rated and, if necessary, safety-approved capacitors may be required.

A particularly important consideration associated with the use of stitching capacitors is the inductance that inevitably appears within (the capacitors' intrinsic inductance, or ESL, typically between 1 to 2 nH for small multilayer ceramics) and in series with (e.g., due to traces and vias, typical increasing with length by 1 nH/mm) the capacitor. The capacitors' effective frequency band is limited by their self-resonant frequency (SRF), above which inductance dominates their impedance. The manner of physical implementation of capacitive chassis stitching was described in Section 9.9.3 and illustrated in Figure 9.254 and Figure 9.255.

A unique implementation of capacitive stitching, using a concept similar to that depicted for hybrid stitching using fixing washers, is illustrated in Figure 9.257. In this technique, a set of two conductive washers and an insulating spacer are utilized. The bottom washer comes into direct conductive contact with the signal return plane through the circular stitching pad, while the upper conductive washer comes into contact with the head of the fixing screw. Between the washers, which must be otherwise isolated from the thread of a screw, an insulating spacer is inserted. By controlling the dielectric constant of the insulating spacer and the overlap area between the conductive washers, some capacitance is obtained. This capacitance is typically rather small, but will probably be sufficient for the higher frequencies.

This technique, similar to previous ones, provides for isolated, single point, or multipoint PCB return plane stitching to the chassis with no required change to the PCB layout.

9.9.5 Controlling Parallel-Plate Waveguide (PPW) Noise in the PCB-Chassis Cavity

The fact that the parallel-plate waveguide (PPW) noise[†] can be excited as a result of ΔI noise[‡] in multilayer PCBs due to the mode conversion of desired modes into parasitic

---

*Voltages up to 60 V RMS or 42.4 V peak, AC or DC, are not generally regarded as unsafe.

[†]In spite of its deterministic nature (associated with the structure resonance frequencies), this mode is referred to as noise.

[‡]ΔI noise is the source of excitation of parallel-plate waveguide (PPW) noise. It refers to the current and voltage transients across the power distribution network (PDN), whereas the PPW noise refers to the electromagnetic wave propagation between the PDN planes. In general, the two terms could be considered synonymous.

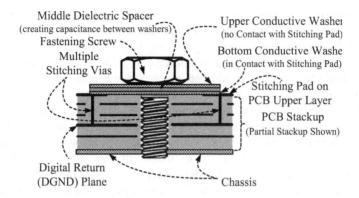

Figure 9.257. Achieving capacitive PCB-to-cassis bond with the conductive washers technique. Insulating standoffs not shown for simplicity of drawing but should be included as in Figure 9.253.

PPW modes is well known. The excited PPW noise can propagate through a PPW composed of power and return planes in multilayer PCBs. The propagation of the PPW noise throughout the entire PCB structure results in excitation of resonances in the circuits, affecting signal transmission characteristics due to multiple reflections from the edge of PCBs and coupling between signal vias. Therefore, the mitigation of the PPW noise excited from signal vias due to mode conversion is very important to improve transmission and coupling characteristics of signal vias in high-speed applications [52, 53].

As operating frequencies of digital circuits further increase and, correspondingly, state transition times keep shrinking, new signal integrity and EMI problems arise due to resonances. Resonances occur at frequencies at which dimensions of the structure correspond to whole numbers of the half-wavelengths ($\lambda/2$) corresponding to the frequencies of concern. They can be calculated (in GHz) by [63]

$$F_{Res} = 150 \cdot \sqrt{\left\{\left(\frac{\ell}{L}\right)^2 + \left(\frac{m}{M}\right)^2 + \left(\frac{n}{N}\right)^2\right\}}, \text{ GHz} \qquad (9.98)$$

where ℓ, m, and n are integers (0, 1, 2, 3, etc.), and L, M, and N are the length, width, and height of the resonating cavity (in mm). As a general rule, the lowest resonant frequencies associated with the longest dimensions of the cavity are of most concern, approximated by setting $\ell = 1$, $m = 0$, and $n = 0$, or when $\ell = 0$, $m = 1$, and $n = 0$. For example, a 150 mm × 75 mm box will correspondingly exhibit its lowest resonant frequencies at 1 GHz and 2 GHz in its length (L) and width (W) directions, respectively.

At resonance, standing waves develop across the structure, making it an efficient ("tuned") radiator (or "accidental antenna") that couples very well indeed with its external electromagnetic environment. As a result, radiated emissions are exacerbated and vulnerability to externally induced interference is increased. At resonance, emissions from physical structures can be significantly higher (often by more than an order of magnitude, or 20 dB) than at nearby nonresonant frequencies. By reciprocity, immunity is similarly degraded. In addition, at resonance coupling between circuits on the PCB (i.e., crosstalk) is increased.

The broader spectral content of present-day digital signals, resulting from their decreasing edge rates, mandates the application of multipoint grounding topologies, including multipoint "chassis stitching," for preclusion of low-order resonances of the structure, which could cause severe excitation of the structure as a radiator. Consequently, multiple cavity resonances between the return plane on the PCB and the chassis may evolve amid the chassis stitch locations, depending on the number of and mutual spacing between the chassis stitches. With PCB-to-chassis connections limited to the four corners of a rectangular PCB, only resonances of one large cavity need to be considered. In RF and high-speed digital circuits, however, more connections are commonly used, resulting in more (and smaller) cavities. Generally, the first (i.e., lowest) resonant frequency of the PCB-to-chassis bonded structure is most dominant and, thus, of greatest concern. This frequency is associated with the longest diagonal of its cavities.

To better illustrate the benefits of a multipoint "chassis stitching," an analogy is presented in Figure 9.258. When two children rotate the rope, a certain height is reached for any given distance between the children. When the rope is held (and fixed) in its middle, it can no longer reach the same height as before, and creates two isolated wave patterns rather than the original one. The analogy should be self-evident. When the separation between the "chassis stitching" points is large, higher level voltage standing waves will be present between the nodes, from relatively low frequencies (large wavelength). When multiple chassis connections are made, standing waves are shifted to higher frequencies, while at lower frequencies only lower amplitude interference voltage may remain.

The effect of multiple PCB-to-chassis stitching is illustrated in Figure 9.259. In this case, the stitching was implemented through capacitors rather than by direct connection (see Section 9.9.4). With no connections [Figure 9.259(a)], high-level noise voltage is present across the entire plane. When stitching was implemented at 10 properly selected locations [Figure 9.259(b)], reduction of the noise voltage distribution occurred (note the "dips" at the stitching locations). With a larger number of "stitches," reaching a total of 23 [Figure 9.259(c)], a considerable noise voltage suppression across the entire plane is further observed.

Equation (9.99) provides a rough estimation* of the likely lowest resonant frequency for a structure that has PCB-to-chassis bonds at its four corners only, assuming a small spacing between the PCB and the chassis structures (as is commonly the case) [62]:

---

*Equation (9.99) is based on the resonances inside a complete metal-sided cavity, whereas the PCB and chassis structure are constructed of only two metal plates with open edges interconnected at several locations through a nonzero impedance. Exact analysis of such complex structures can only be accomplished numerically by means of full-wave, three-dimensional field solvers.

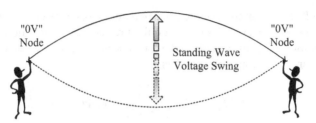

(a) Large Separation between Nodes Allows a Higher
Swing of the Rope

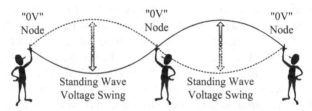

(b) A Center Node yields Limits the Swing of the Rope
to a Lower Height

Figure 9.258. Benefits of chassis ground "stitches"; analogy to a rope jumping game.

$$F_{Low} \approx 150 \cdot \sqrt{\frac{1}{L^2 + M^2}}, \text{ GHz} \tag{9.99}$$

where L and M represent the length and width of the PCB return plane (in mm), respectively. For example, the lowest resonant frequency corresponding to a 150 mm × 75 mm box spaced 5 mm above the chassis structure and bonded to it at its four corners is 0.89 GHz, approximately.

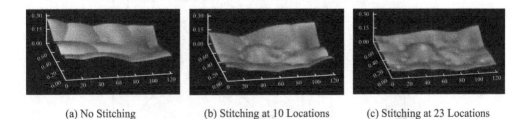

(a) No Stitching   (b) Stitching at 10 Locations   (c) Stitching at 23 Locations

Figure 9.259. Effect of multiple plane stitching on planar noise distribution. In this example, stitching through bypass capacitors was implemented. (Simulation run on Sigrity "SpeedXP" Suite; courtesy of Sigrity, Inc.)

When the PCB-to-chassis bonding is carried out on one side only (e.g., at only two corners) or on two adjacent sides (e.g., at only three corners), excitation of resonances at whole multiples of quarter-wavelengths ($\lambda/4$) occur. The lowest resonant frequency of such structures can be approximated as

$$F_{Low} \approx 75 \cdot \sqrt{\frac{1}{L^2 + M^2}}, \text{ GHz} \qquad (9.100)$$

Figure 9.259 clearly demonstrates that the level of the ripple on the PCB due to resonances is significantly decreased when spacing between chassis connections is reduced.

In order to prevent resonances from occurring in the PCB-to-chassis structure, any two adjacent chassis stitching bonds should be spaced no further than $\lambda/10$ from each other (and preferably closer), where the λ used corresponds to the highest frequency of concern (Figure 9.260). If properly implemented, the resonance frequency of the system can be shifted to frequencies beyond those at which significant spectral content exists. Smaller spacing, for example, $\lambda/20$, will yield better performance [3, 63] but will quadruple the required number of connections. In most PCBs, meeting such a requirement is not realistic.

Multiple chassis connections typically add to assembly time (and cost). A number of techniques exist that allow PCB-to-chassis connections to be added if necessary without increasing assembly time; however, provision for these is required during production of the PCB. One example is illustrated in Figure 9.261. In this technique, PCB pads are provided on the bottom layer of the PCB so that spring-finger (or alternative conductive and sufficiently compressible) gaskets can make contact between the pad and the chassis. The connection between the PCB pads and the PCB return plane may be accomplished conductively or capacitively, using techniques such as those de-

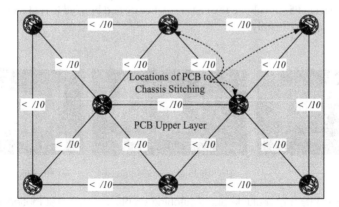

Figure 9.260. Maximum recommended spacing between multiple chassis stitching for mitigating resonance.

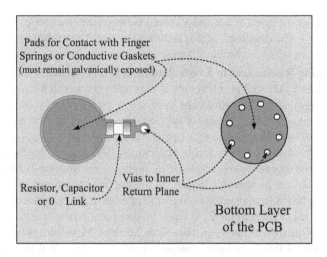

Figure 9.261. Examples of provisions for PCB-to-chassis bonds not requiring screws. (Courtesy of Keith Armstrong, Cherry Clough.)

scribed in Figure 9.252 and Figure 9.254. The spring fingers or conductive gaskets could either be fitted to the PCB or to the chassis. Spring fingers are available as surface-mounted devices that can be soldered to the PCB pads. The key to the effectiveness and durability of this technique is the resistance of the contact to electrochemical corrosion due to metal incompatibility. As the contact pads must remain galvanically exposed, they should be plated with highly conductive metallization, compatible electrochemically with the gasket material.*

With the increasing frequency levels of modern circuits, achieving the goal of a maximum of $\lambda/10$ spacing between any two adjacent stitches may not be feasible. For instance, emerging technologies provide state transition times of 100 pS. The resultant significant spectrum span of such signals exceeds 3 GHz. In order to achieve a minimum resonant frequency of 3 GHz, the separation between PCB-to-chassis stitches, based on the above criterion, should not exceed 10 mm, approximately. In most cases, implementation of such a requirement is not practical. In order to control PCB-to-chassis cavity resonance noise above 500 MHz, approximately, other techniques should be considered. Several such alternative techniques are described below [64]:

1. **Avoid Irregular Spacing of Chassis Stitches.** Using an irregular pattern of chassis connections will produce diverse resonance frequencies in the different cavities. Consequently, the resonant frequencies of the multiple smaller cavities created by the irregular bond arrangement will not coincide in frequency. Breaking up the resonances in this manner will reduce the peak amplitudes of the res-

*Refer to Chapter 5 for more details on galvanic protection and metal compatibility.

onances at the price of a broader spread between the resonances (yielding lower Q values). The drawback of this technique is that it may shift resonances to frequencies at which no problem was seen before.

2. **Designing Resonances to Evade Problematic Frequencies.** Cavity resonances constitute a potential problem only if excited by signals in that frequency range. As clocks typically constitute the most severe interference sources on the PCBs and produce it at their fundamental frequency and harmonics, careful design of the chassis stitches may be able to ensure that these frequencies do not coincide with any of the resonant frequency ranges. The higher the frequency of the clocks, the easier is this technique to apply, as the spacing between their harmonics is thus increased. Contrary to the previous technique (1), it is advantageous in this case to maintain a higher Q of each of the resonances by attempting to make all cavities in the PCB equal in shape and size whenever possible. As a result, avoidance of the resonances with the clock harmonics is made easier. The drawback of this technique is its sensitivity to changes that may affect critical design parameters.

3. **Using Resistors to Dampen Cavity Resonance.** At the resonant frequencies of a PCB-chassis cavity, significantly higher currents flow through the bonding contacts. Inserting a lossy device (i.e., resistor) in series with the high-Q circuit at the resonant frequencies will result in increased losses, resulting in reduction of the cavity's Q). The primary drawback of this technique is the possible adverse influence it has on the EMC performance of the circuit at other than the resonant frequencies due to the higher PCB-to-chassis bonding impedance. The physical implementation of this technique is similar to that used for hybrid stitching (described in Section 9.9.3), using resistors instead of capacitors.

9.9.6 Benefits of Reduced Spacing between PCB and Chassis

The fact that circuit operation causes RF voltage differences between the PCB and chassis, causing fringing fields, was discussed earlier in this chapter. The discussion of image planes in Section 9.6.1 has also indicated that the closer the PCB is placed to an image plane (e.g., the chassis) the lower would be the interaction of the PCB with the electromagnetic environment, both from the standpoint of emissions and immunity. This effect is depicted qualitatively in Figure 9.262.

The closer the PCB is to its chassis, the more effective is the "image" in the chassis in suppression of emissions. When the PCB return plane is also directly bonded to the chassis, whether galvanically or capacitively (as discussed earlier), the closer spacing ultimately lowers the bond (primarily inductive) impedance between them. Since inductance scales linearly with length, halving the PCB's spacing from the chassis will halve the length (and, subsequently, the partial inductance) of the bonds. Reduced inductance in PCB-to-chassis bonds has benefits for the overall noise transfer impedance, and helps return CM currents more effectively to the PCB, thus reducing emissions and improving immunity. Finally, closer spacing will somewhat increase the resonant frequencies of the cavities between the PCB and the chassis.

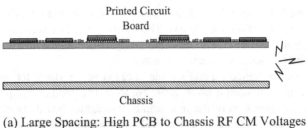

(a) Large Spacing: High PCB to Chassis RF CM Voltages
Produce Fringing Fields Emissions

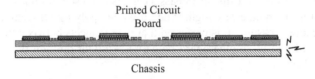

(b) Small Spacing: Low Fringing Fields

Figure 9.262. Effect of PCB-to-chassis image plane spacing on radiated emissions due to fringing fields. (Courtesy of Keith Armstrong, Cherry Clough.)

9.9.7 Daughter and Mezzanine Boards Ground Stitching

Mezzanine and daughter boards suffer from additional complexity, as they typically do not come into direct contact with the chassis, except through the motherboard. The complexity stems from the fact that low-impedance chassis interconnection is limited by the multiple interconnects between the daughter or mezzanine board and the main board, and from it to the chassis or equipment enclosure (Figure 9.263).

All design principles and their respective benefits discussed so far for PCB-to-chassis "stitching" apply to mezzanine- or daughter-board-to-motherboard structure con-

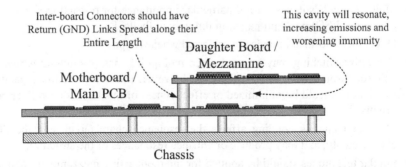

Figure 9.263. Mezzanine and daughter boards chassis stitching. (Courtesy of Keith Armstrong, Cherry Clough.)

nections as well. For such boards, the return plane in their motherboard can provide some of the benefits of a local chassis. The cavity created between them can resonate, creating problems for emissions and immunity, especially where the mezzanine or daughter board is connected to external cables.

In some circumstances, bonding the mezzanine or daughter board frequently enough to the motherboard is often sufficient to ensure good circuit performance, particularly when dimensions of the mezzanine or daughter board are electrically small. Interboard connectors can be used to provide further interconnects between the circuits when numerous return pins are allocated for bonding the two boards' return planes all along the length of the connector.

A local ("floating") yet massive chassis can also be used to improve the EMC performance of a mezzanine or daughter board while applying all techniques described above. Whenever possible, it is preferred that interconnects between the mezzanine or daughter board and the motherboard be carried straight through the motherboard to bond to the chassis.

Lessons Learned

- Bond stitching of the PCB return planes to the chassis is used for reducing RF noise voltage distribution across the PCB return structure due to the return switching current flowing through the plane.
- Lower surface noise reduces radiated emissions from the PCB and its interconnects.
- Low-impedance chassis connection in the I/O area of the PCB is important for control of emissions; chassis stitching at points located remotely from the I/O area of the PCB contribute little to reduction of emissions.
- Multiple bond "stitches" of the PCB return plane to the chassis are recommended for controlling parallel-plate waveguide (PPW) noise in the PCB-to-chassis cavity; bond "stitches" should be allocated at least at each corner of the return plane, a minimum of one connection near each I/O cable port, and at least one near each high-speed device.
- Multiple bond "stitches" are of particular importance for restricting electrostatic discharge (ESD) pulse propagation throughout the PCB.
- Direct metal-to-metal bonds are preferred whenever possible.
- Capacitive stitching may be used when multipoint chassis stitching is undesirable (or forbidden) at low frequencies; however, inductance of the capacitors' installation should be minimized or effectiveness of this technique will be compromised.
- Multipoint chassis bonding affects the resonance characteristics of the PCB-chassis cavity, and may produce or enhance resonances of the structure.
- Similar techniques should be applied for interconnecting mezzanine or daughter boards through the motherboard to the chassis structure while paying attention to the cumulative impedance of the entire chassis stitching path.

9.9.8 PCB Heat Sink Grounding Considerations

9.9.8.1 The Role of Heat Sinks in Generation of EMI.
The increasing density of modern high-speed, very large-scale integrated circuits (VLSI) such as high-density ASICs (application-specific integrated circuits), incorporated on contemporary (and presumably future) PCBs, together with the continuing growth in switching speed, operating frequency, and power density, have resulted in increased heat generation. This heat needs to be dissipated through a heat sink to keep the devices within safe operating conditions, producing close interaction between electromagnetic, thermal, and mechanical considerations, and a significant increase in unwanted parasitic effects (Figures 9.264 and 9.265).

On the other hand, the rapid switching capability of modern digital and power semiconductor devices (e.g., CMOS and IGBT) results in very fast voltage and current variations that act on structures such as heat sinks, generating parasitic transient common-mode currents and, subsequently, increased radiated emissions. Noise coupled from a digital device to the heat sink may cause it to act as an efficient antenna and effectively radiate the noise at one or more harmonics of the switching frequency. Due to increased operating frequencies of microelectronic circuits, extending well into the gigahertz region, heat sink dimensions can no longer be considered electrically small as signal wavelengths are now comparable to heat sink dimensions. In fact, heat sinks used at present tend to strongly resonate around the frequency of 1 GHz, thus becoming efficient radiators of electromagnetic energy.* Containing emissions from heat sinks in high-power microelectronics is rapidly becoming a necessity due to higher power levels and faster clock rates of digital circuits.

Common-mode currents can be controlled by reducing the parasitic capacitance between the switching device and the heat sink. This, however, may lower the cooling efficiency of the heat sink. Thus, the connection of a heat sink to the device and the packaging technology is a design issue involving EMI and thermal performance.

In order to effectively lessen the total amount of radiated emissions, the heat sink must maintain small physical dimensions compared to the wavelength, λ, associated with the highest frequency produced by the switching device. If any dimension of the heat sink is in the range of $\lambda/20$ to λ, significant levels of radiated EMI can be expected. Simulations have revealed that two main dimensions may be excited, yielding two fundamental frequencies and subsequent harmonics that can be enhanced or suppressed by changing the position of the excitation. The addition of fins on the heat sink could reduce the resonant frequency but also increase the amplitude of the electric far-field radiation if they run across the width of the sheet rather than the length. To keep things simple, the heat sink can be considered as a rectangular block of copper on top of an infinite metallic plane that represents the power (V_{CC}) or return (GND or V_{SS}) plane on the motherboard.

A quick estimate of the resonant frequencies and the amplitude of the electric field radiated by an ungrounded heat sink can be investigated by studying the electromag-

---

*The resonances observed are believed to be caused by the parallel combination of the capacitance between the heat sink and the ground reference plane, and the inductance of the grounding posts. As more posts are added, the total inductance decreases and the resonant frequencies increase.

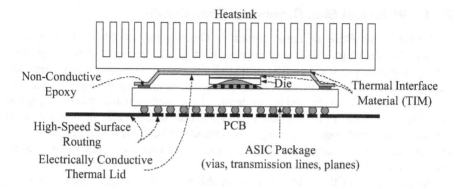

Figure 9.264. Illustration of a heat sink placed on an ASIC package. (Courtesy of David P. Johns, CST.)

netic field within the substrate between the heat sink and the ground plane. The electromagnetic field within the substrate can be derived by treating that region as a cavity bounded by perfect electric conductors at the top and bottom surfaces (tangential electric fields vanish along these two surfaces) and by perfect magnetic walls along the perimeter of the heat sink (tangential magnetic fields vanish along those four walls). When the height, h, of the substrate is very small ($h \ll \lambda$, where λ is the wavelength within the substrate), the field along the height, h, can be considered uniform. In addition, fringing of the fields along the edges of the heat sink is also very small thanks to the small height, and the electric field is practically normal to the bottom surface of the heat sink. Therefore, only TM field patterns can be considered within the cavity. Resonances due to the geometrical dimensions of the cavity are expected at the frequencies derived below.

Figure 9.265. CPU heat sink with fan attached and a smaller heat sink on a Northbridge computing device. (Source:http://en.wikipedia.org/wiki/, reproduced under the GNU free documentation license.)

Using X and Y as the dimensions of the heat sink footprint along the x and y axis, respectively, and h as the height (or separation between the heatsink and the device), the resonant frequency for the general (m, n, p) mode, where m, n, and p represent the number of half-wavelengths along the $x, y,$ and z axis, respectively, is given by [65, 66]

$$(f_r)_{m,n,p} \approx \frac{c}{2\pi\sqrt{\varepsilon_r}} \cdot \sqrt{\left(\frac{m\pi}{X}\right)^2 + \left(\frac{n\pi}{Y}\right)^2 + \left(\frac{p\pi}{h}\right)^2} , \text{ Hz} \qquad (9.101)$$

where c and ε_r represent the speed of light in free space and the relative permittivity, respectively.*

The mode with the lowest resonance frequency is generally the dominant mode. If $X > Y \gg h$, the dominant mode is the TM_{100}, whose resonant frequency is given by

$$(f_r)_{100} \approx \frac{c}{2X\sqrt{\varepsilon_r}}, \text{ Hz} \qquad (9.102)$$

If also $X > Y > Y/2 > h$, the second-order mode is the TM_{010}, whose resonant frequency is given by

$$(f_r)_{010} \approx \frac{c}{2Y\sqrt{\varepsilon_r}}, \text{ Hz} \qquad (9.103)$$

Note that in practice, for very small value of h, the mode number p is always zero at low frequency and need not be considered [65, 66].

9.9.8.2 The Role of Heat Sink Grounding for Mitigation of EMI.
In order to control the emissions from heat sinks mounted on VLSIs or power devices, proper heat sink grounding schemes effectively reduce radiated emissions. Determining the "proper" heatsink grounding scheme is becoming a challenging issue since simple lumped-circuit representations of the heat sinks cannot be used due to their large electrical dimensions. A distributed grounding scheme with multiple grounding points must be utilized. The designers are no longer expected to determine whether heat sinks must be grounded but are challenged by the question of determining how many ground points must be used and where they should be located.

For demonstrating the effect of heat sink grounding on radiated emissions, simulations were carried out using CST Microstripes† to model the ungrounded heat sink structure. Two grounding schemes were investigated to study their relative efficiency in suppressing the emissions and both were compared to the reference ungrounded configuration. The first grounding scheme uses four grounding posts at the four cor-

---

*Note that the ratio $c/\sqrt{\varepsilon_r}$ represents the speed of wave propagation in a medium with relative permittivity ε_r.

†CST Microstripes™ is a three-dimensional electromagnetic simulation tool based on a multigrid formulation of the time-domain transmission-line matrix (TLM) method.

ners of the heat sink. The second grounding scheme utilizes, in addition to the four corner grounding posts, eight additional EMI ground pins.

The structure illustrated in Figure 9.266 was used for the purpose of the demonstration. The heat sink, comprising of 45 fins, was placed over a PPC970 CPU, serving as the EMI source. Thermal contact between the CPU package and the heat sink was made through a thermal interface material (TIM).

With an impulsive excitation applied at the center of the contact area between the CPU and the heat sink, a "natural" or impulse response depicted in Figure 9.267 in terms of the radiated E-field emissions was obtained. This plot will be used as a basis for comparison to impulse responses of the structure with different grounding schemes applied. Figure 9.268 visualizes the electric near field in terms of the surface current density distribution on the heat sink and surrounding ground plane structures with the heat sink floating. Observe the resonance distribution patterns across the structure.

The maximum E-field emission from the heat sink can be computed, as shown above, directly through the use of three-dimensional simulations or, alternatively, may be estimated by means of the reflection coefficient, S_{11}, if known. From the conservation of energy, it follows that

$$P_{rad} \approx P_{in} - \left(P_{ref} + P_{loss}\right) \qquad (9.104)$$

where P_{in} is the input power injected to the heat sink, P_{ref} is the reflected power, P_{rad} is the radiated power, and P_{loss} represents the losses. All power components are expressed in terms of dBm. If the contribution from the loss term, P_{loss}, is negligible, the radiated power, P_{rad}, can be expressed in terms of the input power, P_{in} and the S_{11} coefficient, as

$$P_{rad} \approx \left(1 - |S_{11}|^2\right) \cdot P_{in} \qquad (9.105)$$

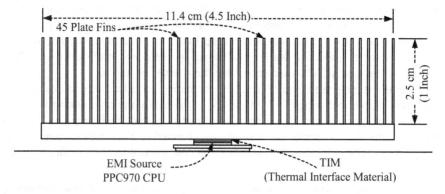

Figure 9.266. Model of heat sink placed on a PPC970 CPU EMI source. (Courtesy of David P. Johns, CST.)

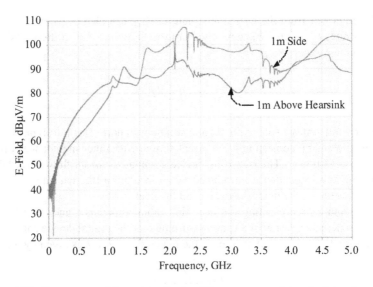

Figure 9.267. Spectrum of impulse ("natural") response of heat sink placed on a PPC970 CPU EMI source. (Courtesy of David P. Johns, CST.)

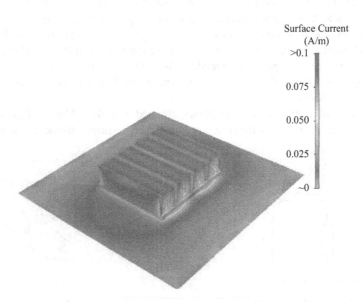

Figure 9.268. Visualization of surface current on heat sink and surrounding ground plane at F = 2.23 GHz. (Courtesy of David P. Johns, CST.)

The radiated power from the heat sink structure, P_{rad}, can also be calculated from the injected current, I (coupled capacitively from the digital device to the heatsink), and the radiation resistance, R_{rad}:

$$P_{rad} \approx \frac{1}{2} R_{rad} \cdot |I|^2 \tag{9.106}$$

Figure 9.269 illustrates a source model and stimulation port. This port type is modeled by a lumped element, consisting of a current source with inner impedance, which excites and absorbs power. The switching device stimulates current flow toward the heat sink through stray capacitance introduced by an insulating thermal interface material layer placed between the device package and the heat sink.

With Z_{in} and Z_S representing the heat sink input impedance and the EMI source (stimulus) impedance, respectively, and assuming that the digital device can be considered as a discrete EMI current source, I_S, the injected current, I, is given by [65, 66]:

$$I \approx \frac{Z_S}{Z_{in} + Z_S} I_S \tag{9.107}$$

The consequent near-field distribution surrounding the heat sink and the far-field directivity pattern are both illustrated in Figure 9.270 for the structure illustrated in Figure 9.266.

Figure 9.271 presents the impulse response of the heat sink structure obtained for the different grounding schemes of the heat sink in terms of spectral distribution of the radiated E-field. Observe the resonance distribution patterns across the structure.

Grounding the heat sink at four corners through metallic supports provides an apparent advantage in reduction of emissions over the floated heat sink scheme, across the larger part of the spectrum extending from DC to 5 GHz. Note, however, that grounding four corners of the heat sink seemingly introduced a new problem at the frequency of 750 MHz.

An increased number of ground points, attained with the addition of eight ground pins, shifts the resonance of the structure to a higher frequency thanks to the reduction

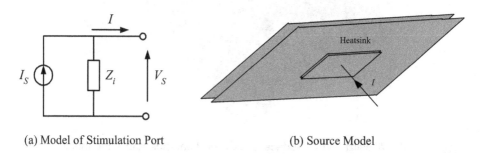

(a) Model of Stimulation Port   (b) Source Model

Figure 9.269. Model of source and stimulus port.

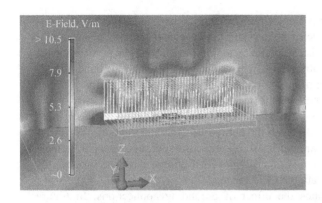
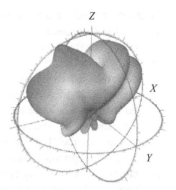

(a) Near Fields, V/m

(b) Far Field Directivity, dB

(Linear Radial Scale from - to 0dBi)

Figure 9.270. Visualization of near and far fields from a heat sink on a surrounding ground plane at F = 2.23 GHz. (Courtesy of David P. Johns, CST.)

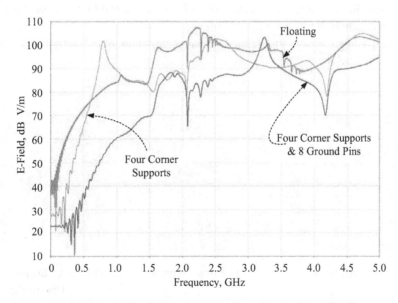

Figure 9.271. Spectrum of impulse ("natural") response of a heat sink placed on a PPC970 CPU EMI source versus grounding scheme of the heat sink. (Courtesy of David P. Johns Ph.D., CST.)

in grounding inductance as more connection points are added. Suppression of emissions at lower frequencies may generate resonant frequencies that could enhance radiation at higher frequencies. In this case, an increase of emissions at the new resonance frequency, 3.25 GHz, is clearly observable, whereas across the rest of the band, the advantage of the increased number of grounding points is evident.

Figure 9.272 visualizes the above results through the surface current density on the ground plane beneath the heat sink. It is apparent that with a floating heat sink, high surface currents are observed underneath the entire heat sink and beyond. With four-corner supports the "hot" areas are confined to a much smaller area beneath the heat sink (and the supports). With the addition of eight EMI ground pins, the surface current concentrates primarily beneath the CPU package.

Figure 9.273 further illustrates the effect of grounding connections. In Figure 9.273(a), four ground pins are used in addition to the four heat sink supports, whereas in Figure 9.273(b) eight ground pins are used in addition to the four heat sink supports. The improvement in surface current distribution, manifested in the lower amplitude of the surface current all over the ground plane, is evident. Clearly, adding more connections from the heat sink to ground improves the EMI performance relating to the surface current distribution and amplitude, ultimately resulting in reduced radiated emissions. The larger number of ground connections also shifts resonances associated with the structure to higher frequencies.

Lessons Learned

- Heat sinks, required for functionality of high-speed and high-current digital devices, can be a major source of EMI.
- Grounding the heat sink reduces radiated emissions from the heat sink and the PCB structure.

 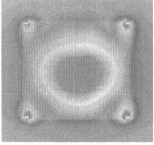 

(a) Floating Heatsink   (b) 4 Corner Supports   (c) 4 Corner Supports + 8 EMI Ground Pins

Figure 9.272. Visualization of surface current on the ground plane beneath the heat sink versus grounding scheme at $F = 2.23$GHz. (Courtesy of David P. Johns, CST.)

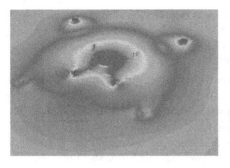 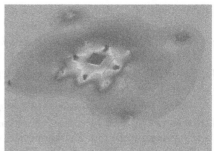

(a) 4 Corner Supports + 4 Ground Pins   (b) 4 Corner Supports + 8 Ground Pins

(c) Surface Current Scale

Figure 9.273. Visualization of surface current on the ground plane beneath the heat sink and surrounding ground plane versus grounding scheme at $F = 2.12$ GHz. (Courtesy of David P. Johns, CST.)

- Multiple ground connections of heat sinks significantly improve the EMI performance of the circuit, resulting in higher resonance frequencies; heat sink corner grounding only may prove insufficient and could produce lower frequency resonances while providing suppression at other frequencies.

BIBLIOGRAPHY

[1] Hubing, T., Beetner, D., Deng, S., and Dong, X., "Radiation Mechanisms for Semiconductor Devices and Packages," in *Proceedings of the 2004 International Symposium on Electromagnetic Compatibility, EMC'04, Sendai,* Sendai, Japan, June 2004.

[2] German, R. F., Ott, H. W., and Paul, C. R., "Effect of an Image Plane on Printed Circuit Board Radiation," in *Symposium Record of the 2000 IEEE International Symposium on Electromagnetic Compatibility,* Washington DC, August 2000.

[3] Montrose, M. I., *EMC and the Printed Circuit Board: Design, Theory and Layout Made Simple,* New York: IEEE Press/Wiley, 1998.

[4] Johnson, H., and Graham, M., *High Speed Signal Propagation: Advanced Black Magic,* Upper Saddle River, NJ: Prentice-Hall PTR, 2003.

[5] Bogatin, E., *Signal Integrity Simplified,* Upper Saddle River, NJ: Prentice-Hall PTR, 2004.

[6] Zeeff, T. M., Hubing, T. H., and Van Doren, T. P., "Traces in Proximity to Gaps in Return Planes," *IEEE Transactions on Electromagnetic Compatibility,* Vol. 47, No. 2, May 2005

[7] Holloway, C. L., and Kuester, E. F., "Closed-Form Expression for the Current Density on the Ground Plane of a Microstrip Line, with Application to Ground Plane Loss," *IEEE Transactions on Microwave Theory and Techniques,* Vol. 43, No. 5, May 1995.

[8] Holloway, C. L., and Hufford, G. A., "Internal Inductance and Conductor Loss Associated with the Ground Plane of a Microstrip Line," *IEEE Transactions on Electromagnetic Compatibility,* Vol. 39, No. 2, May 1997.

[9] Johnson, H., and Graham, M., *High Speed Digital Design, A Handbook of Black Magic,* Upper Saddle River, NJ: Prentice-Hall PTR, 1993

[10] Armstrong, K., "Advanced PCB Design and Layout for EMC: Part 4, Reference Planes for return and Power," *EMC & Compliance Journal, Issue 54,* published online at http://www.cherryclough.com.

[11] Brooks, D., *Signal Integrity Issues and Printed Circuit Board Design,* Upper Saddle River, NJ: Prentice Hall-PTR, 2003

[12] Smith, D. C., "Crossing Ground Plane Breaks—Part 2, Tracing Current Paths, Technical Tidbit—January 2003," *High Frequency Measurements Web Page,* published online at http://emcesd.com/tt2003/tt010103.htm.

[13] Smith, D. C., "Crossing Ground Plane Breaks—Part 3, Immunity to Radiated EMI, Technical Tidbit—February 2003," *High Frequency Measurements Web Page,* published online at http://emcesd.com/tt2003/tt020103.htm.

[14] Smith, D. C., "Using a Paper Clip to Measure Signals and Noise," *High Frequency Measurements Web Page,* August 1999, published online at http://emcesd.com/tt2002/tt120102.htm.

[15] Smith, D. C., "Crossing Ground Plane Breaks—Part 5, Common-Mode Currents and Emissions, Technical Tidbit—December 2005," *High Frequency Measurements Web Page* (published online at http://emcesd.com/tt2005/tt020105.htm).

[16] Smith, D. C., "Crossing Ground Plane Breaks, A Source of Crosstalk, Technical Tidbit - December 2002," *High Frequency Measurements Web Page* (published online at http://emcesd.com/tt2002/tt120102.htm).

[17] Smith, D. C., "Crossing Ground Plane Breaks—Part 4, Risetime Effects on Signals, Technical Tidbit—January 2005," *High Frequency Measurements Web Page,* (published on line at http://emcesd.com/tt2005/tt010105.htm).

[18] Tsaliovitch, A., *Electromagnetic Shielding Handbook for Wired and Wireless Applications,* Norwell, MA, Kluwer Academic Publishers, 2001.

[19] Caniggia, S., and Maradei, F., *Signal Integrity and Radiated Emissions of High-Speed Digital Systems,* New York: Wiley, 2008.

[20] Hockanson, D. M., Drewniak, J. L., Hubing, T. H., Van Doren, T. P., Sha, F., Lam, C. W., and Rubin, L., "Quantifying EMI Resulting from Finite-Impedance Reference Planes," *IEEE Transactions on Electromagnetic Compatibility,* Vol. 39, No. 4, November 1997.

[21] Cain, J., "Parasitic Inductance of Multilayer Ceramic Capacitors," *AVX Technical Information,* (published online at http://www.avxcorp.com/techinfo_catlisting.asp).

[22] Archambeault, B. R., *PCB Design for Real-World EMI Control,* Norwell, MA: Kluwer Academic Publishers, 2002.

[23] Yuan, F., "Where to Place Decoupling Capacitors?," published on line at http://www.sigrity.com/support/techpapers/support_tech_doc.htm.

[24] Smith, D. C., "Signal Paths Passing Through Ground and Power Planes, Effects on Immunity, Technical Tidbit—May 2003," *High Frequency Measurements Web Page,* published online at http://emcesd.com/tt2003/tt050103.htm.

[25] Liu, W. T., Tsai, C. H., Han, T. W., and Wu, T. L., "An Embedded Common-Mode Suppression Filter For Ghz Differential Signals Using Periodic Defected Ground Plane," *IEEE Microwave And Wireless Components Letters,* Vol. 18, No. 4, April 2008.

[26] Liu, W. T., Han, T. W., and Wu, T. L., "A Novel Broadband Common-Mode Filter for High-Speed Differential Signals," in *Proceedings of 2008 Asia-Pacific Symposium on Electromagnetic Compatibility & 19th International Zurich Symposium on Electromagnetic Compatibility,* May 2008, Singapore.

[27] Armstrong, K., "Advanced PCB Design and Layout for EMC Part 5, Decoupling, Including Buried Capacitance Technology," *EMC Compliance Journal, Issue 55,* published online at http://www.compliance-club.com.

[28] Smith, L. D., Anderson, R. E., Forehand, D. W, Pelc, T. J., and Roy, T., "Power Distribution System Design Methodology and Capacitor Selection for Modern CMOS Technology," *IEEE Transactions on Advanced Packaging,* Vol. 22, No. 3, p. 284, August 1999..

[29] *Wikipedia, the Free Encyclopedia,* published online at http://en.wikipedia.org/wiki/Surface_mount_technology, October, 2009.

[30] *Wikipedia, the Free Encyclopedia,* published online at http://en.wikipedia.org/wiki/D-subminiature, September, 2009.

[31] Ott, H. W., *Noise Reduction Techniques in Electronic Systems* (2nd Ed.), New York: Wiley-Interscience, 1988.

[32] Novak, I., *Power Distribution Network Design Methodologies,* Chicago, International Engineering Consortium, 2008.

[33] Novak, I., *Power Integrity Modeling and Design for Semiconductors and Systems,* Upper Saddle River, NJ: Prentice-Hall, 2007.

[34] Montrose, M. I., "Analysis on Loop Area Trace Radiated Emissions from Decoupling Capacitor Placement on Printed Circuit Boards," in *Proceedings of the 1999 IEEE International Symposium on Electromagnetic Compatibility,* Seattle, 1999.

[35] Ott, H. W., "EMC Aspects of Future High Speed Digital Designs," Keynote Address in the PCB Design Conference—East, Worcester, MA, September 2000.

[36] Roy, T., Smith, L., and Prymak, J., "ESR and ESL of Ceramic Capacitor Applied to Decoupling Applications," in *Proceedings of the 1998 Electrical Performance of Electronic Packaging Conference,* West Point, NY, 1998.

[37] Fan, J., Drewniak, J., Knighten, J., Smith, N., Orlandi, A., Van Doren, T., Hubing, T., and DuBroff, R., "Quantifying SMT Decoupling Capacitor Placement in DC Power-Bus Design for Multilayer PCBs," *IEEE Transactions on Electromagnetic Compatibility,* Vol. 43, No. 4, November 2001.

[38] Chen, J., Xu, M., Hubing, T., Drewniak, J., Van Doren, T., and DuBroff, R., "Experimental Evaluation of Power Bus Decoupling on a 4-Layer Printed Circuit Board," in *Proceedings of the 2000 IEEE International Symposium on Electromagnetic Compatibility,* Washington D.C., August 2000.

[39] Shahpamia, S., Iravani, B. M., and Ramahi, O. M., "Electromagnetic Noise Mitigation in

High-speed Printed Circuit Boards and Packaging Using Electromagnetic Band Gap Structures," in *Proceedings of the 54th IEEE Conference on Electronic Components and Technology,* Las Vegas, June 2004.

[40] Xu, M., Hubing, T. H., Chen, J., Van Doren, T. P., Drewniak, J. L., and DuBroff, R. E., "Power-Bus Decoupling with Embedded Capacitance in Printed Circuit Board Design," *IEEE Transactions on Electromagnetic Compatibility,* Vol. 45, No. 1, February 2003.

[41] Qin, J., and Ramahi, O. M.; "Power Plane with Planar Electromagnetic Bandgap Structures for EMI Reduction in High Speed Circuits," in *Proceedings of the 2006 IEEE International Symposium on Antennas and Propagation,* Albuquerque, July 2006.

[42] Shahparnia, S., and Ramahi, O. M., "Electromagnetic Interference (EMI) Reduction from Printed Circuit Boards (PCBs) Using Electromagnetic Bandgap Structures," *IEEE Transactions on Electromagnetic Compatibility,* Vol. 46, No. 4, November 2004.

[43] Ramahi, O. M., Mohajer-Iravani, B., Qin, J., Shahparnia, S., and Kamgaing, T., "EMI Suppression and Switching Noise Mitigation in Packages and Boards using Electromagnetic Band Gap Structures," in *Proceedings of the 2007 International Symposium on Signals, Systems and Electronics (ISSSE '07),* Montreal, July/Aug, 2007.

[44] Mohajer-Iravani, B., Shahparnia, S., and Ramahi, O. M., "Coupling Reduction in Enclosures and Cavities Using Electromagnetic Band Gap Structures," *IEEE Transactions on Electromagnetic Compatibility,* Vol. 48, No. 2, May 2006.

[45] Qin, J., Ramahi, O. M., and Granatstein, V., "Novel Planar Electromagnetic Bandgap Structures for Mitigation of Switching Noise and EMI Reduction in High-Speed Circuits," *IEEE Transactions on Electromagnetic Compatibility,* Vol. 49, No. 3, August 2007.

[46] Qin, J., and Ramahi, O. M., "Ultra-Wideband Mitigation of Simultaneous Switching Noise Using Novel Planar Electromagnetic Bandgap Structures," *IEEE Microwave and Wireless Components Letters,* Vol. 16, No. 9, September 2006.

[47] Mohajer-Iravani1, B., and Ramahi, O. M., "EMI Suppression in Microprocessor Packages Using Miniaturized Electromagnetic Bandgap Structures with High-k Dielectrics," in *Proceedings of the 2007 IEEE International Symposium on Electromagnetic Compatibility,* Honolulu, July 2007.

[48] Stojanovic, G., Živanov, L., and Damjanovic, M., "Compact Form of Expressions for Inductance Calculation of Meander Inductors," *Serbian Journal of Electrical Engineering,* Vol. 1, No. 3, November 2004.

[49] Qin, J., and Ramahi, O. M., "Wideband SSN Suppression and EMI Reduction from Printed Circuit Boards Using Novel Planar Electromagnetic Bandgap Structure," in *Proceedings of the 2006 IEEE International Symposium on Electromagnetic Compatibility,* Portland, OR, August 2006.

[50] Lee, J., Lee, H., Park, K., Chung, B., and Kim, J., Kim, J., "Impact of Partial EBG PDN on PI, SI and Lumped Model-Based Correlation," in *Proceedings of 2008 Asia-Pacific Symposium on Electromagnetic Compatibility & 19th International Zurich Symposium on Electromagnetic Compatibility,* May 2008, Singapore.

[51] Shih C. H., Shiue, G. H., Wu, T. L., and Wu, R. B., "The Effects on SI and EMI for Differential Coupled Microstrip Lines over LPC-EBG Power/Ground Planes," in *Proceedings of 2008 Asia-Pacific Symposium on Electromagnetic Compatibility & 19th International Zurich Symposium on Electromagnetic Compatibility,* May 2008, Singapore.

[52] Scogna, A. C., "PPW Noise Mitigation in Multilayer PCBs by Means of Virtual Island

and/or Array of Shorting Vias," in *Proceedings of the 2007 IEEE International Symposium on Electromagnetic Compatibility,* Honolulu, HI, July 2007.

[53] Nam, S., Kim, Y., Kim Y., Jang H, Hur, S., Song, B., Lee, J., and Jeong, J., "Performance Analysis of Signal Vias using Virtual Islands with Shorting Vias in Multilayer PCBs," *IEEE Transactions on Microwave Theory and Techniques,* Vol. 54, No. 4, April 2006.

[54] Johnson, H., "Ground Fill," *Electronic Design News (EDN),* published online at http://www..edn.com/contents/images/601835.pdf, May 2005.

[55] National Semiconductor, *LVDS Owner's Manual (3rd Ed.), Chapter 3, High-Speed Design,* published on line at http://www.lvds.national.com, Spring, 2004.

[56] Britt, D. S., Hockanson, D. M., Sha, F., Drewniak, J. L., Hubing, T. H., and Van Doren, T. P., "Effects of Gapped Groundplanes and Guard Traces on Radiated EMI," in *Proceedings of the 1997 IEEE International Symposium on Electromagnetic Compatibility,* Austin, TX, August 1997.

[57] Ponchak, G. E., Chen, D., Yook, J., and Katehi, L. P. B., "Characterization of Plated Via Hole Fences for Isolation Between Stripline Circuits in LTCC Packages," *Microwave Symposium Digest, 1998 IEEE MTT-S International,* Volume 3, June 1998.

[58] Johnson, H., and King, M., "Picket Fences," *High Speed Digital Design Online Newsletter,* Vol. 2, Issue 16.

[59] Kester, W., Bryant, J., and Byrne, M., "Grounding Data Converters and Solving the Mystery of 'AGND' and 'DGND'," *Analog Devices Application Note MT-031,* February 2006, published online at http://www.analog.com/zh/content/0,2886,760%255F788%255F97529,00. html.

[60] Ott, H. W., "Partitioning and Layout of a Mixed-Signal PCB," *Printed Circuit Board Design,* June, 2001.

[61] Rempfer, W. C., "The Care and Feeding of High Performance ADCs: Get All the Bits You Paid For," available online at http://cds.linear.com/docs/Application%20Note/an71.pdf.

[62] Gray, N. C., "Attack The Noise Gremlins that Plague High-Speed ADCs," *Electronic Design,* December, 1999.

[63] Montrose, M. I., *EMC and the Printed Circuit Board, Design Techniques,* New York: IEEE Press/Wiley-Interscience, 1999.

[64] Armstrong, K., "Advanced PCB Design and Layout for EMC Part 3, PCB-to-Chassis Bonding," *EMC Compliance Journal, Issue 53,* published on line at http://www.compliance-club.com.

[65] Dolente A., *Analysis of the Heatsink Influence on Conducted and Radiated Electromagnetic Interference in Power Electronic Converters,* Ph.D Thesis in Electrical Engineering, ING-IND/31 XIX Cycle, Alma Mater Studiorum, University Of Bologna, Department Of Electrical Engineering, March 2007.

[66] Dolente, A., Reggiani, U., and Sandrolini, L., "Comparison of Radiated Emissions from Different Heatsink Configurations," in *Proceedings of the VI International Symposium on Electromagnetic Compatibility and Electromagnetic Ecology,* St. Petersburg, Russia, June 2005.

[67] Radhakrishnan, K., Wittwer, D., and Yuan-Liang L., "Study of Heatsink Grounding Schemes for GHz Microprocessors," in *Proceedings of the 2000 IEEE Conference on Electrical Performance of Electronic Packaging,* October, 2000, Scottsdale, AZ.

10

INTEGRATED FACILITY AND PLATFORM GROUNDING SYSTEMS

A facility is a building or other structure, either fixed or transportable in nature, with its power supply system, wiring, grounding network, and all installed electrical and electronic equipment. The facility grounding system is designed and installed with regard to the purpose and physical structure of the facility together with the particular requirements of the electrical and electronic equipment. It is intended primarily to provide protection for personnel and equipment, and to ensure suitable level of signal integrity and electromagnetic compatibility for electronic equipment. Protection of personnel from electrical shock hazards is achieved by a combination of automatic disconnection of faulty circuits and equipotential bonding to the earth termination system of all equipment, noncurrent-carrying metallic objects, piping, and other extraneous electrically conductive surfaces with which personnel may come in contact. Low-impedance ground connection ensures speedy disconnection of faulty circuits by operation of overcurrent or ground-fault protection devices. An effective facility grounding system provides a low-impedance path for discharge of lightning current to the earth mass, precludes fire and ignition hazards to flammable vapors, ordnance, and explosive substances, and protects electronic hardware from damage. Grounding provisions are also necessary under certain operations to provide a current path for preventing accumulation of static electricity charges and minimizing voltage potential buildup, particularly during maintenance of electrostatic-discharge-sensitive (ESDS) electronic equipment, handling of ordnance, and refueling or other flammable vapor operations.

Grounds for Grounding. By Elya B. Joffe and Kai-Sang Lock
Copyright © 2010 The Institute of Electrical and Electronics Engineers, Inc.

For the above reasons, electrical safety codes require that electrical power systems and equipment be intentionally grounded. Consequently, the facility ground system is directly influenced by the proper design, installation, and maintenance of the power distribution systems. Large voltages can develop across grounding conductors as a result of lightning discharges or power faults, and may pose a shock hazard because of potential differences between equipment that may be grounded at different points within the facility. Ground currents routed close to electronic equipment may disrupt equipment operation; thus, a lightning protection system incorporating surge-protective devices is also necessary where a risk of transient over-voltage damages to structure and equipment exists. A well-integrated earth termination system and bonding network will, therefore, ensure personnel safety and minimize equipment malfunction and damage.

EMC and signal integrity for electronic and communication equipment require that minimum voltage potential differences exist between signal reference grounds of circuits. This is achieved by ensuring that the impedance between signal ground points throughout the facility is minimal and that interference from noise sources is minimized.

The grounding system consists of the earth termination system in contact with the soil and the bonding network not in contact with the soil. The main function of the earth termination network is to effectively discharge lightning current into the soil without causing dangerous potential differences in the earth termination system. The bonding network is a low-impedance network formed by multiple interconnections of noncurrent-carrying metallic components in the facility. The main functions of the bonding network are to prevent the occurrence of dangerous potential differences between all equipment in the facility and to provide equipotential reference for electronic equipment.

This chapter provides some examples for implementation of effective earth termination systems and bonding networks in different types of facilities. Although a wide variety of examples is selected, the ones presented in this chapter are not intended to be exhaustive. Within each facility, numerous ways of implementing grounding and bonding exist to achieve the same objective, so long as the fundamental electrical principles are followed. In addition, this chapter covers fundamental grounding concepts and architectures utilized in mobile platforms (e.g., mobile tactical shelters, aircraft, spacecraft, and ships).

10.1 FACILITY GROUNDING SUBSYSTEMS

All terrestrial integrated electrical and electronic facilities are inherently referenced to earth by capacitive coupling, accidental contact, and intentional connections. "Ground" must therefore be considered from an overall system perspective, with various subsystems comprising the total facility ground system. The facility grounding network is composed of four major electrically interconnected subsystems [1–4]:

1. The earth electrode subsystem (EESS)
2. The fault protection subsystem (FPSS)

3. The lightning protection subsystem (LPSS)
4. The signal reference subsystem (SRSS)

The various requirements, principles and techniques for implementing the grounding subsystem have been well discussed in preceding chapters and are summarized as within the context of this chapter.

10.1.1 Earth Electrode Subsystem (EESS)

The earth electrode subsystem is a network of electrically interconnected rods, mats, or grids installed in the earth for the purpose of establishing the facility ground reference for lightning and electrical shock hazard. It is designed to provide a low-impedance contact with the general mass of the earth to ensure that lightning and power fault currents are effectively discharged without producing harmful voltages within the facility. The system should be designed in accordance to the characteristics of the site and the requirements of the facility. Although it is desirable to have an earth electrode resistance as low as possible, compromises must typically be considered with regard to the cost required to achieve such performance.

Ground mats or grids are normally installed to limit step and touch potentials in areas with high-voltage apparatus, such as generating plants, substations, and switching stations. All nearby buried metal objects are interconnected to the earth electrode subsystem. The earth electrode subsystem alone is insufficient for controlling shock hazards arising from power system ground faults. A system of grounding and equipotential bonding must be implemented to ensure that hazardous voltages are not transferred to exposed conductive parts of equipment and all extraneous conductive parts.

10.1.2 Fault Protection Subsystem (FPSS)

As discussed in Chapter 6, one point of an electrical power supply system, normally the neutral, is connected to earth. This connection point to earth, called the "system ground," sets system reference potential to ground. The fault protection subsystem consists of grounding or circuit protective conductors (CPC) that provide a low-impedance connection between the system ground and all noncurrent-carrying metal parts of an electrical wiring system and equipment connected to the system. The equipment electrical safety ground conductor (ESGC) provides an intentional low-impedance path for fault currents so that protective devices can operate properly and speedily to remove the faulty circuit or equipment, thereby minimizing the risk of personal injury or fire.

A main earthing terminal (MET) is normally provided in a facility and is located close to the AC power and telecommunications cable entry to the facility. The MET is connected to the earth electrode subsystem via an earthing conductor of shortest length. Extraneous conductive objects such as piping, ducts, and structural steel are bonded to the MET. A ground bar is provided in all panel boards. The grounding and bonding conductors are sized to carry the maximum fault currents expected.

10.1.3   Lightning Protection Subsystem (LPSS)

The purpose of a lightning protection system in a facility is to protect the building, its occupants, and contents from the thermal, electrical, and mechanical effects of lightning. The lightning grounding system, along with the air terminals and down conductors, is intended to provide a low-impedance path for discharging lightning energy effectively to the earth electrode subsystem. Requirements of lighting protection subsystem are frequently specified by relevant international standards and national codes. Protection against lightning side-flashes is provided by bonding all major adjacent noncurrent-carrying metal objects, including main structural steel support members to the lightning down conductors or grounding system.

10.1.4   Signal Reference Subsystem (SRSS)

Ideally, different points on the ground system should have the same or equal potential relative to an arbitrary reference point. In practice, this equal potential is not achieved. Due to the finite impedance of each element of the grounding and bonding networks, lightning or transient fault current could result in a potential difference between two points on the grounding system. The signal reference subsystem should be designed to prevent the occurrence of such potential differences that exceed the transient immunity of equipment in the facility.

The signal reference subsystem is intended to accomplish three major functions [5]:

1. Enhance reliability of signal transfer between interconnected equipment items by reducing interunit common-mode electrical noise over a broad frequency band.
2. damage to interunit signal circuits by providing a low-inductance and, hence, effective ground reference for all of the externally installed AC and DC power, telecommunications, or other signal-level, line-to-ground/chassis-connected, surge-protective device (SPD) that may be used with the associated equipment.
3. Prevent or minimize damage to interunit signal-level circuits and equipment power supplies when power system ground fault events occur.

A signal reference subsystem that provides an effective means for a broadband common-ground reference for equipment is referred to as the signal reference structure (SRS). A properly designed SRS equalizes ground potential over a broad range of frequencies from DC through the MHz range. An SRS constructed using solid sheet metal surfaces forms an ideal signal reference plane but is expensive. Typically, an SRS is in the form of a signal reference grid (SRG)* made up of a grid of copper straps or copper or aluminum wires. Often, an SRS is also formed by an interconnected raised floor substructure.

All equipment in the contiguous area is connected to the SRS with short bonding straps. Flat foil strips are recommended for providing low-impedance connections. An effective SRS makes use of a bonding network with multiple interconnections of metallic components. These components include structural steel and reinforcing rods,

---

*The SRS/SRG is also known as an MCBN (mesh common bonding network) [5].

metallic pipes, cabinets, racks, conduits, and protective and bonding conductors. All metallic objects crossing or intersecting the SRS are effectively bonded to it. Depending on space and accessibility, a wire mesh grid may be installed at floor level or overhead to function as a signal reference grid. The mesh is bonded to the facility metallic structure at points where structural members are accessible. Equipment enclosures and racks are also bonded to this SRG.

The SRS is an effective technique for mitigation of intersystem ground noise, in particular:

- When logic AC-to-DC power supplies used in the associated electronic equipment are installed with one of the terminals connected to the equipment metal case/enclosure, which is a practice typical of ITC equipment.
- When signal circuits and logic AC-to-DC power supply common terminals are specified by the manufacturer to be connected to "clean" ground or to an externally provided signal ground reference.
- When actual performance problems occur with the equipment, identified as common-mode electrical noise interference related to the existing grounding system or to the signal inter-unit wiring system.

It should be noted, however, that a practical SRS/SRG is poor, at best, for use as an effective signal reference as it exhibits higher impedance than tolerable for most modern communication equipment, analog (due to their higher sensitivity to noise) or digital (due to their higher signaling rates). Equipment individually passing successfully EMC compliance tests may, in fact, encounter problems when integrated in a facility and interfacing with other collocated equipment if it relies on the signal reference structure alone. The SRS/SRG serves therefore, at most, as a supplemental grounding measure at the facility level.

Consequently, a proper grounding scheme should be applied at the facility level considering the practical characteristics of the SRS/SRG. For equipment operating at lower frequencies, a single-point grounding (SPG) scheme may be applied. Higher frequency grounding networks commonly use multipoint grounding (MPG) schemes in the form of an equipotential ground plane or grid. Consideration to use SPG or MPG generally depends on the frequency range of the equipment. Analog circuits with signal frequencies below 300 kHz may use SPG satisfactorily, whereas digital circuits with signal frequencies in the MHz range and beyond would require MPG. In either case, normal practices do not conventionally make use of the SRS/SRG as a signal return path, not even for the "single-ended" RS-232. Practical single-ended signal interfaces normally utilize "dedicated" signal return conductors, rather than the SRS/SRG, whereas high-speed digital communications typically use balanced interfaces, making them virtually independent of the SRS/SRG (Figure 10.1).

In Figure 10.2, one terminal of the switch-mode power supply is used as the common or reference for the electronic circuits of equipment #1. As a common practice, the electronic power supply reference is bonded to the equipment chassis, which, in turn, is also connected to the safety ground as required by safety regulations. Likewise, equipment #2 also has its electronic circuit reference connected to its safety ground.

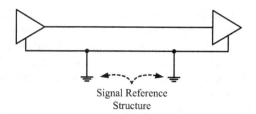

Signal Reference
Structure

(a) "Single-Ended" RS-232C Interface

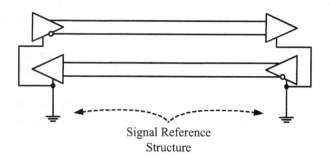

Signal Reference
Structure

(b) Balanced-to-Balanced Interface

Figure 10.1. Practical single-ended and balanced signal interfaces.

Without the SRS, even a small voltage potential difference between the two safety ground connection points could affect the integrity of the signal communication between the two equipment units. This voltage potential difference, ΔV could occur due to transients, power system faults, or electromagnetic interference. The signal reference grid exhibits low impedance between any two points on it over the whole frequency range from DC to tens of megahertz and, hence, helps to maintain an equipotential signal reference for equipment connected to it.

The concept of SRS is rather similar to that of the bonding network used in the telecommunication and IT industries [6–14]. The common bonding network (CBN) is the principal means for effecting bonding and earthing inside a telecommunication or IT building. It is the set of metallic components that are intentionally or incidentally interconnected to form the principal bonding network in a building.

Lessons Learned

- A facility grounding system should incorporate the requirements of electrical safety, lightning protection, and signal integrity of electronic equipment.
- An integrated facility grounding system composes of four interconnected subsystems, namely, the earth electrode subsystem, the fault protection subsystem, the lightning protection subsystem, and the signal reference subsystem.

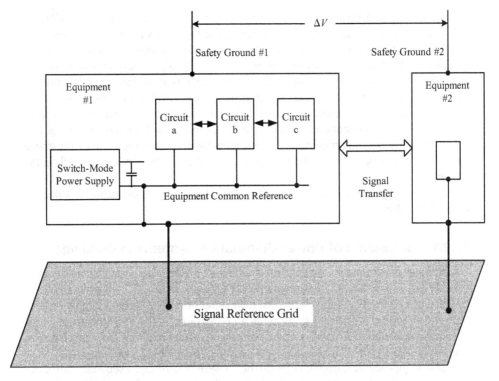

Figure 10.2. Signal reference ground for electronic circuits.

10.2 GROUNDING REQUIREMENTS IN BUILDINGS OR FACILITIES

All grounding subsystems should be bonded together with normally noncurrent-carrying metal objects, including structural steel elements, to form an equipotential facility grounding system. The extent of grounding and bonding requires a careful consideration of personnel safety, equipment protection, and the optimum performance of electronic equipment under both normal and abnormal conditions. Grounding requirements vary greatly depending on the structure of the building and the interconnection with power and signal cables external to the building. The relevant electrical safety code pertaining to power grounding should always be followed in addition to sound engineering design principles. Where electrical wiring is not totally enclosed in a protected building environment, lightning and surge protection requirements are incorporated in grounding implementation.

For a small installation, such as a dwelling house, grounding requirements are often considered solely from the standpoint of electrical safety against electrical shock and lightning hazards. In the case of a critical facility or platform, the challenge is often that of achieving EMC requirements.

The facility's structural steel is ideal as a supplementary grounding element [15–18] as it offers many parallel conducting paths between various points within the facility and between these points and earth. Many of the structural members are in direct electrical contact with earth in the form of foundations, piles, or concrete-encased elements. Structural steel elements can provide a very low-impedance path to earth because of their large mass and volume. Ensuring proper bonding of building structural steel and reinforcement elements at the design and construction phases can help to achieve low earth impedance. Shock hazards arising from lightning discharge and electrical faults are thus minimized by equipotential bonding. Figure 10.3 illustrates the application of reinforcement steel in creating the common bonding network in the building.

The following sections present the principles of grounding application in different types of facilities.

10.2.1   Grounding of Power Distribution Systems in Buildings

Grounding requirements for power distribution systems in buildings are primarily concerned with personal safety and equipment fault protection [15]. National electrical codes such as [16] invariably require that the neutral of the low-voltage supply be connected to an earth electrode at a single point, thus establishing the zero voltage reference of the power supply. Different methods of power system earthing were presented in Chapter 6. In the case of the TN-S system earthing, the facility grounding conductor is traceable to this neutral earthing point. In the TT system earthing, a separate earth electrode system has to be established at the facility. The electrical safety ground conductors (ESGC) are routed along with the supply cables throughout the building in a tree configuration, as shown in Figure 10.4. The ESGC and bonding conductors should have sufficient cross-sectional area to safely conduct the expected fault currents. The ESGC conductor is sized to provide sufficiently low ground fault loop impedance in relation to the settings of the protective device in the circuit. In the event of a ground fault, the resulting fault current is sufficiently large to rapidly activate the overcurrent or the ground fault protective device, thereby disconnecting the power to the faulty circuit.

All metallic parts of equipment such as enclosures, racks, cable trays, equipment grounding conductors, and all extraneous metallic parts are bonded together to form a continuous, electrically conductive system. All earthing electrodes used for grounding of the power system, grounding of lightning protection systems, and grounding of communications systems are effectively and permanently bonded to each other. The main earthing terminal (MET) is located at the service entrance and is connected to the earth electrode subsystem via a short earthing conductor. All incoming metallic pipes, cable shields, and building structural steel are securely bonded to the MET.

An integrated facility grounding and bonding network is illustrated in Figure 10.5. Apart from ordinary electrical equipment, the facility also hosts sensitive electronic equipment operating at both low and high frequencies. In this example, the EMC requirements of the low-frequency equipment are taken care of by the single-point grounding scheme with the signals referenced to a dedicated ground bus, which is also bonded to the main earth terminal (MET). EMC requirements of the high-frequency

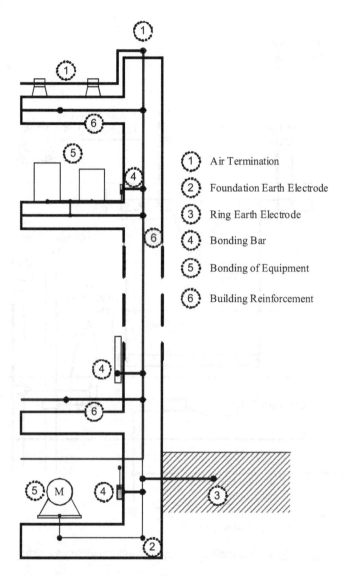

Figure 10.3. Building steel reinforcement structure used to form common-bonding network.

equipment are provided by the multipoint grounding scheme implemented using a signal reference structure which has multiple connections to elements of the common bonding network. Personnel and equipment protection against power fault currents and lightning flashover are provided both by protective conductors and by bonding all normally noncurrent carrying metal objects, including structural steel elements, to the facility earth termination system. To prevent lightning surges from propagating into the

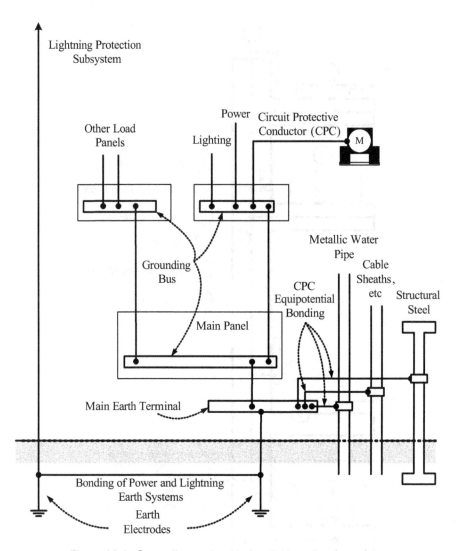

Figure 10.4. Grounding and equipotential bonding for safety.

building, power and communication cables are brought into the facility at a common entry point along with other metallic service pipes so that all noncurrent carrying conductive parts are bonded to the main earthing terminal, which is normally located inside the building wall at the service entry. Surge protective devices* are installed at the point of entry for both power and communication wiring to suppress lightning surges or switching transients.

*Surge and terminal protection devices (SPDs/TPDs) and the grounding considerations associated with their application are discussed in Chapter 8.

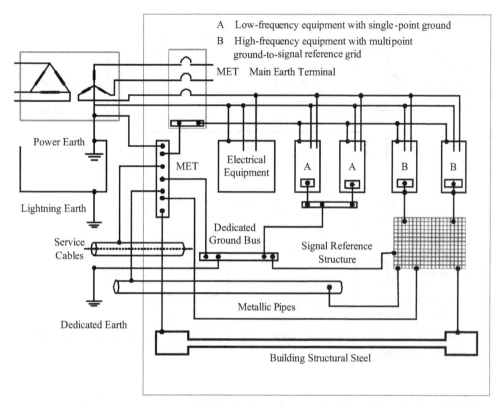

Figure 10.5. Grounding and bonding in a mixed facility.

It is common to have several distribution transformers in a facility to supply the to-tal load demand. Some transformers may be installed at different locations within the facility and are connected to separate earth electrode systems. Intersystem ground noise may arise when interconnecting equipment that are powered from supplies with separate system earths. This ground loop problem is illustrated in Figure 10.6. Mainly due to stray currents that flow in the ground system during normal operation, there ex-ists a potential difference ΔV between the two grounding references. Ground termina-tion at both end of the data cable completes the ground loop, thus permitting interfer-ing current to flow in the loop indicated. In reality, several ground loops may be created due to extensive bonding in a facility. Where possible, it is advisable to supply a group of interconnecting equipment with power derived from the same source. An alterative solution is to bond all the earthing systems of the transformers.

10.2.2 Grounding in Industrial Facilities

Grounding requirements in industrial facilities varies widely depending on the type of facility and the sensitivity of the production or process equipment. The nature and con-

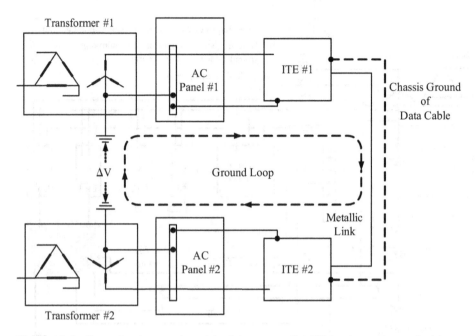

Figure 10.6. Ground noise produced by interconnection of equipment supplied from separated sources.

struction of the building also determine the grounding design. A major challenge often arises from the coexistence of EMC-sensitive control and process equipment along with high-power, highly disturbing loads such as variable-speed motor drives.

Figure 10.7 shows the grounding scheme of an industry facility housing high-voltage and high-current power apparatus along with data center and sensitive electronic equipment. The high-voltage transformers and switch room are provided with a low-impedance earth grid in order to minimize touch and step voltage in the event of electrical power faults. The hazards of power system faults have been presented in Chapter 6. The data center with its associated interconnecting equipment has a signal reference grid to ensure EMC and signal integrity. An effective equipotential ground plane is achieved based on a low-impedance earth electrode subsystem that is extensively bonded to other subsystems. The requirements for sensitive electronic equipment and the implementation of bonding networks are further discussed in the following sections.

Where the facility comprises distributed buildings located in area with high levels of lightning activity, lightning and surge protection are important considerations [18]. Figure 10.8 shows an industrial facility with two main buildings: a tower structure and an outside equipment structure. A mesh earthing termination network is installed over the facility complex to serve as an integrated power and lightning earth. This mesh earthing network provides a very efficient low-impedance path for dissipating lightning discharge current to the earth mass, preventing shock hazard and minimizing

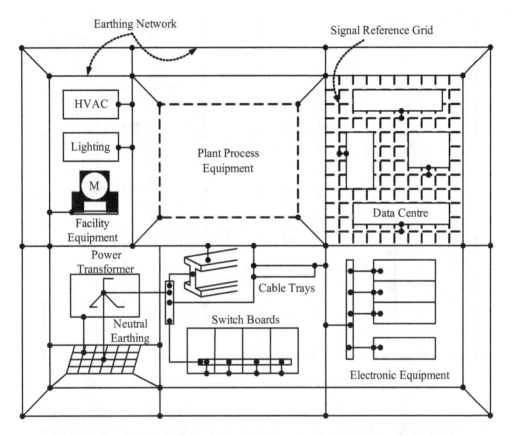

Figure 10.7. Grounding and bonding in a complex industrial facility.

damage or disruption to sensitive equipment. Each building is also surrounded by a ring-electrode subsystem, bonded to the facility earthing network. Cables running between the two buildings and cables between the outside equipment and the building are either armored cables or are carried in ferrous conduits that are bonded to the earthing network at both ends. All incoming facility metallic pipes and cable sheaths are bonded and earthed at the point of entry to the building to prevent propagation of external surges into the building interior.

10.2.3 Grounding for Information Technology Equipment

Information technology (IT) equipment may be used as stand-alone personal or network computers but are also increasingly being used in clusters, hosted in racks, cabinets, or other forms of structures. For stand-alone IT equipment, grounding requirement can usually be met without much difficulty. For large installations, different levels of grounding and bonding are required depending on size, complexity, and technology of

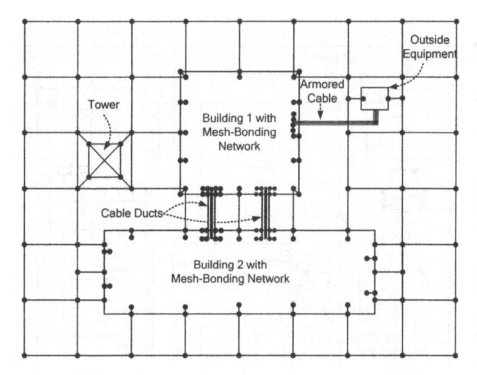

Figure 10.8. Mesh-earth termination system of a multibuilding plant.

the equipment [6–8]. From the standpoint of safety, the ESGC and equipotential bonding conductors should have high current-carrying capacity and low impedance in order to avoid electric shock, risk of fire, or damage to the equipment under normal or faulty operating conditions within equipment or the electricity supply system.

In large IT installations, special emphasis is placed on proper equipotential bonding on both the information technology equipment and the electrical power supply system in order to maintain a good signal reference without compromise on electrical safety. Metallic components included in the building, such as the main earthing terminal or bar, protective conductor, structural steel, metallic plumbing, and reinforcement rods, are bonded to form a basic common bonding network (CBN). The performance of such a basic CBN may be further improved by additional bonding of cable ducts, equipotential bonding conductors, and bonding ring conductors to form a very low-impedance network capable of conducting high current. The screens of all cables entering the building are bonded to the MET at the entry point via a low-impedance path to prevent propagation of surge currents and voltages inside the building.

In Figure 10.9, the small IT installation employs the single-point grounding scheme in which the ESGCs are radially distributed from the common ground bar. This single-point- grounding (SPG) scheme is cost-effective and simple to implement. The major disadvantage of this SPG scheme is the likely intersystem ground noises arising from

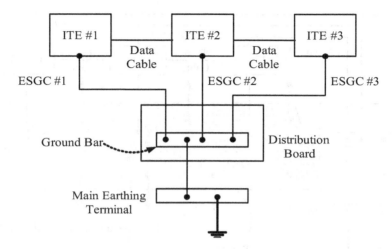

<u>Figure 10.9.</u> A simple IT installation with radially connected grounding conductors (SPG).

ground loops and equipment leakage current that flows through the ESGC under normal operation conditions.

If the electrical safety ground conductors in Figure 10.9 are relatively long, the associated conductor impedance may cause difficulty in maintaining signal integrity for data transfer between the IT equipment. A superior grounding scheme, albeit one with higher installation cost, is the multipoint grounding system depicted in Figure 10.10, where the IT equipment is bonded to the signal reference structure in addition to being connected to the ESGC. The low impedance provided by the signal reference structure over a broad frequency range minimizes intersystem ground noises for data transfer between the equipment.

When signal reference structures (SRS) were first introduced more than three decades ago, they were intended to provide a low-impedance signal reference for interconnected electronic equipment against electromagnetic interference arising from unwanted ground noise on copper wire communication channels that are ground referenced. Early data transfer extensively used single-ended communication channels, such as the RS-232, which are susceptible to intersystem ground noise created by such events as ground faults, lightning, and noise injection from filtering circuits. The use of SRS was highly recommended for applications involving interconnection of sensitive electronic, IT, and communication equipment.

The need for an SRS for interconnection of ITC equipment in modern data communications is no longer as compelling as before. Most interunit signal and telecommunication circuits interface with associated equipment using balanced or isolated interfaces, for instance, through optically or via isolation transformer coupling (e.g., the Ethernet 10/100/1,000BaseT). These interfaces exhibit high common-mode rejection and excellent breakdown-withstand characteristics.

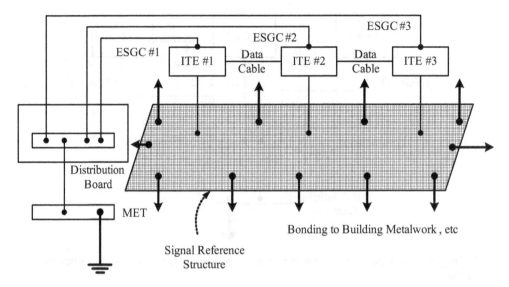

<u>Figure 10.10.</u> Multipoint grounding for IT installation using a mesh signal reference structure.

In a large IT installation, an extensive equipotential bonding network may be realized through the utilization of horizontal and vertical conductors in addition to equipotential meshes or equipotential rings on each floor. Multiple bonds to building metalwork as well as exposed-conductive parts of the electrical installation and metalwork of other services may be implemented, effectively resulting in a three-dimensional mesh structure. Increasing the number of bonding conductors and their interconnections increases the upper frequency limit of the CBN. A complex IT installation with extensive equipotential bonding networks is shown in Figure 10.11 [11].

Excessive voltages between buildings or different areas of equipotential bonding may occur due to lightning and faults on high-voltage distribution networks. These overvoltages can cause hazardous conditions to conductive signal connections. To overcome these problems, metal-free fiber-optic cable or other nonconducting systems are used for signal connections between different areas of equipotential bonding.

10.2.4 Grounding in Telecommunication and C³I (Command, Control, Communications, and Intelligence) Facilities

Grounding and bonding practices differ significantly according to type and application of telecommunication [12–15] and C³I (command, control, communications, and intelligence) facilities [4]. A typical telecommunication/C³I facility is illustrated in Figure 10.12 [4]. The basic requirements of personal safety, equipment protection, and operational performance remain the same. At the user's installation, the increasing use and interconnection of complex electronic telecommunications equipment, such as Inte-

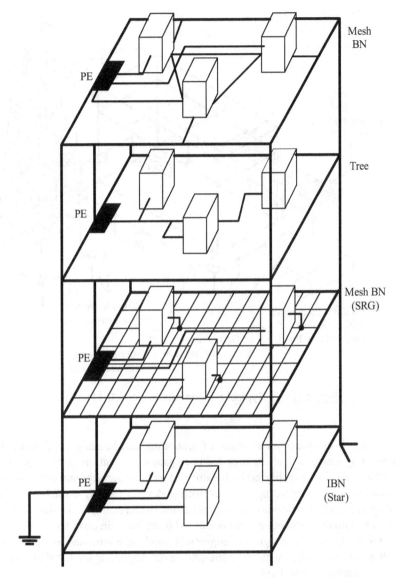

Figure 10.11. Extensive bonding networks in a large IT installation.

grated Services Digital Network (ISDN) terminals, modems, and computers, requires special attention to protect against electrical voltage and current overstresses. Such overstresses are mainly due to exposure of the telecommunications cables and power lines to lightning and the coupling of AC voltages onto the telecommunication cables due to faults on the external power system. Ground potential rise may occur during a power fault when the fault current returns to the power neutral through the ground.

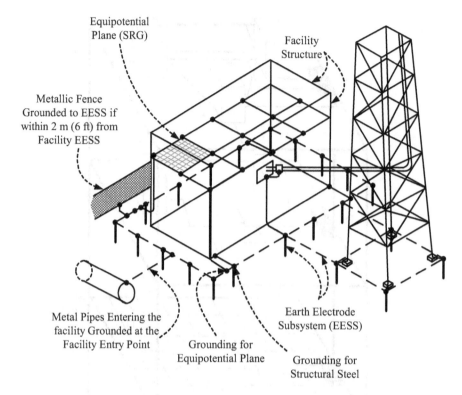

Figure 10.12. Typical telecommunications/C³I facility.

Figure 10.13 illustrates an example of grounding and bonding configuration in a user's building where special attention has been paid to prevent surges from penetrating into the building. This is achieved by bonding all metallic services and cable screens to the building earth and installing surge protective devices (SPDs) at the building entry point on power and telecommunication conductors. In this example, the functional earth for the telecommunication equipment is derived from the main earth terminal.

It is relatively easy to design and implement good earth termination and a bonding network in new buildings. In existing ones, however, upgrading the installation may be technically challenging and expensive.

Telecommunication/C³I facilities located on open terrain [13–15], particularly antenna towers or base stations sited on hilltops, are likely to be subject to severe lightning strikes. A tall tower, such as that shown in Figure 10.14, has a relatively high probability of receiving direct lightning strikes. Equipment and persons at the associated building are exposed to severe lightning effects because conductive services such as telecommunications and power lines, antenna leads, waveguides, earthing conductors, and metallic pipes may carry hazardous lightning surges into the building. The penetration of conducted lightning surge current is mitigated principally by an external earth termination network which effectively disperses the surge current to the earth

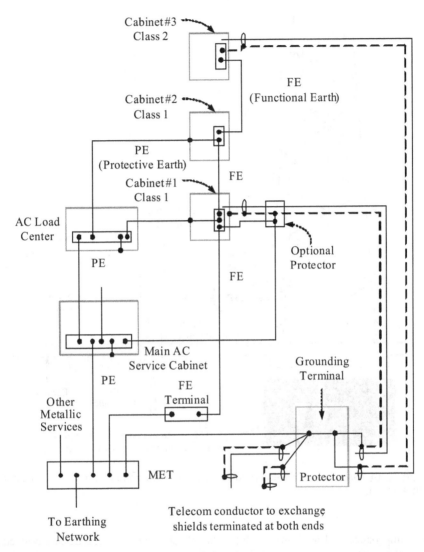

Figure 10.13. Grounding and bonding configuration in a telecommunications/C³I user building.

mass. It is desirable that the earth termination network provide low surge impedance in order to limit the ground potential rise caused by direct lightning strikes. The surge protection devices should be sized for the lightning surge current and voltage they are expected to be subject to.

The tower structure is directly connected to the tower earth electrode system. The tower earth electrode system and the facility building earth electrode system are bonded at more than one point to obtain an integrated earth termination network with low

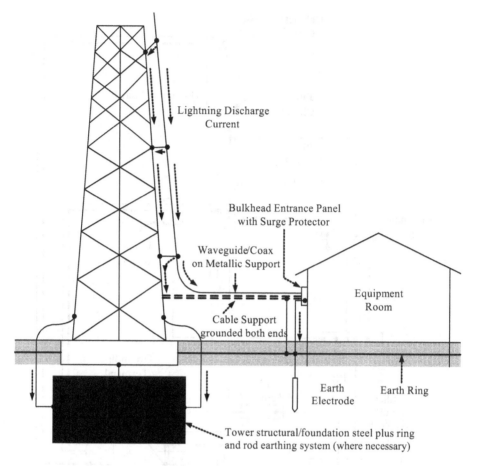

Figure 10.14. Diversion of lightning surges from a telecommunications/C³I facility antennae tower.

surge impedance. All feed lines such as coaxial cables and waveguides are bonded to either the metal tower or to a vertical ground wire. It is desirable that feed lines be bonded to the structure at the point of first contact (near the antenna) and at the bottom where the feed lines bend away from the structure. For a tall structure, intermediate bonding is implemented. The feed line metallic support frames from the tower to the building are bonded to the tower structure and to the exterior earthing electrode system of the building. The multipoint bonding to the tower metallic structure is intended to effectively divert lightning current to the earth termination network of the tower. The remaining lightning current traveling toward the equipment room along the communication cable sheaths and metallic cable support is diverted to the earth termination network outside the equipment room. Figure 10.15 shows the extensive earthing network necessary for lightning protection of a telecommunications facility located at an open

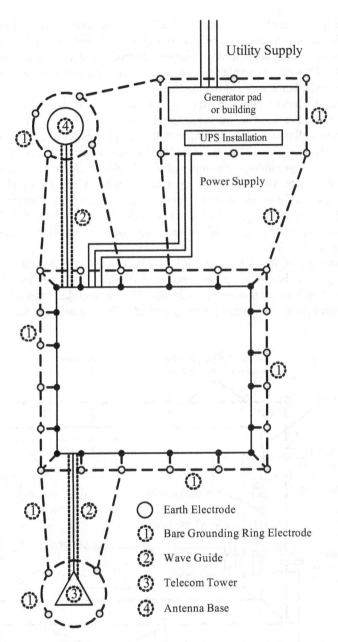

Figure 10.15. Earthing network for lightning protection of an outdoor telecommunications/C³I facility.

site. The facility earth termination network comprises two interconnected earth electrode subsystems: the tower earth termination network and the building earth termination network. A combination of vertical rod electrodes and horizontal ring electrodes is
installed and they are bonded to the structural steel of the tower foundations to achieve
low surge impedance for lightning discharge.

In a large telecommunication facility, a system of equipotential bonding is implemented to obtain a mesh common bonding network or CBN, as in Figure 10.15, by interconnecting all noncurrent-carrying metallic parts by means of low-impedance bonding conductors to the MET and the building structural steel. Figure 10.16 shows the
configuration of grounding, bonding, and signal reference for a small equipment building at a tower site. Inside the building, a ring conductor above ground, also called a
"halo," provides the signal reference for the telecom equipment. The ring conductor is
mounted on insulation support and is only connected at one point to the MET. Noncurrent conducting metallic parts of power cables and service metallic pipes, if any, are
bonded to the MET.

The earth pit of the local generator is bonded to the building earth termination network to prevent hazardous ground potential difference arising from lightning discharge. Where a utility power supply is utilized, lightning discharge to the local earth
termination network can create a large potential difference between the local earth and

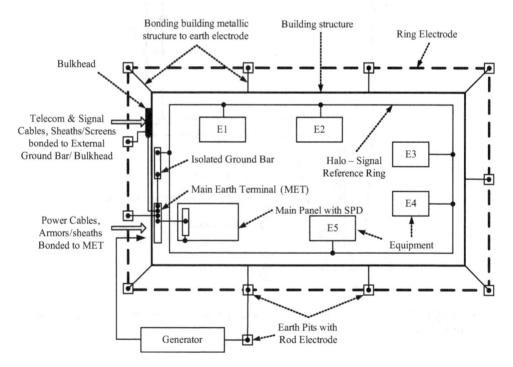

Figure 10.16. Grounding and bonding at a small telecommunications base station with
an adjacent antenna tower.

the remote power earth, leading to flashover at the earthed AC supply panel board. Adequately rated surge protection devices must be installed at the entry point for protection of both power and telecommunications cables.

The telecommunications cables have their shields and screens bonded to an external ground bar, commonly called the bulkhead, at the point of entry to the building. The bulkhead is bonded to the MET and is also connected directly to the earth termination network. A picture of an actual bulkhead bonding is shown in Figure 10.17, in which the arrow points to the cables' entry into the building. The metallic cable supporting structure is also bonded to the bulkhead in order to avoid a possible lightning flashover to the building metallic structure. The bulkhead is directly connected to the earth electrode system and is bonded to the MET. The power cables and telecommunications cables have been arranged to enter the building at the same location to facilitate the required equipotential bonding.

<u>Figure 10.17.</u> Bonding and grounding of shields/screens of telecommunications cables/waveguides at the bulkhead before entry to building. Arrow points to the cables' entry into the building.

10.2.5   Grounding in HEMP-Protected and Secure C³I Facilities

Grounding of equipment, conduit, cabinets, and consoles for safety protection in defense-related command, control, communications, and intelligence (C³I) facilities is typically no different from any other commercial telecommunications/C³I facilities and the discussions in Sections 10.2.3 and 10.2.4 equally apply to these facilities. It is essential to pay special attention to grounding in DoD facilities, however, when survivability of the facility against high-altitude electromagnetic pulse (HEMP)* and the control of compromising emanations from facilities processing classified national security related information are of concern.

For meeting EMP survivability requirements and for controlling compromising emanations, facilities require more elaborate and more expensive grounding, bonding, and shielding (GBS) measures than other types of C³I facilities.

This chapter discusses, in brief, fundamental criteria for grounding and bonding as pertaining to EMP survivability and "red/black" concepts. Detailed discussions of EMP and "red/black" concepts will not be addressed in this book. For detailed information pertaining to "red/black" applications, the information contained in this book should be supplemented with guidance contained in MIL-HDBK-232 and TEMPEST documentation.

10.2.5.1   Grounding for Facility HEMP Survivability. Electromagnetic pulse (EMP) constitutes a significant threat to the survivability of electronic systems and facilities used by defense authorities. In addition to the destructive effects of nuclear detonations—blast, thermal effects, and radioactive fallout—which are restricted to the region near the detonation, a nuclear explosion produces an intense electromagnetic effect. Under proper circumstances, a nuclear detonation generates a high-intensity electromagnetic pulse (EMP) whose frequency spectrum may extend from below 1 Hz to above 300 MHz. This high-intensity EMP can couple to and disrupt or damage critical electronic facilities, unless protective measures are taken in the facilities. The development of such protective measures involves grounding, bonding, and shielding (GBS), and requires an understanding of the EMP itself.

When a nuclear detonation occurs, gamma (γ) rays are produced that propagate outward from the burst location at the speed of light. As the gamma rays collide with air molecules, electrons are dislocated, creating Compton electrons. These electrons are affected by the magnetic fields of the earth, creating an electromagnetic wave that propagates toward the earth (Figure 10.18). High-altitude EMP (HEMP) refers to the high-amplitude EMP produced by an exoatmospheric nuclear burst at an altitude typically exceeding 30 km (19 miles), whereas surface-burst EMP is produced by a nuclear burst within 0.2 km (650 ft) of the earth's surface, and air-burst EMP is produced by a nuclear burst at intermediate altitudes from 2 to 20 km (1.2 to 12 miles). HEMP produced by a nuclear explosion is the form of EMP commonly of most interest because of the large area covered by a single detonation, and will thus be considered in the following discussion only. A burst taking place 560 km (350 miles) above the geo-

---

*HEMP is often known as NEMP (nuclear electromagnetic pulse) or simply as EMP.

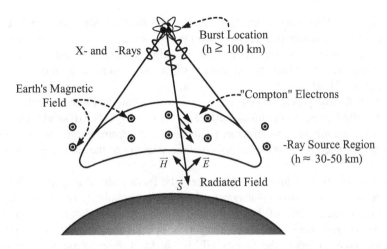

<u>Figure 10.18.</u> Schematic illustration of high-altitude electromagnetic pulse (HEMP).

graphic center of the continental United States would completely blanket the country [1,20].

Unlike lightning, the nuclear HEMP arrives at the earth's surface as a plane wave rather than as a strike channel, and the melting, charring, and splintering effects associated with direct lightning strikes do not usually occur with HEMP. The HEMP exerts its influence through induced effects; that is, the very large electromagnetic fields of the HEMP induce large voltages or currents in antenna-like elements of equipment or structures.

The primary effect of the HEMP is, therefore, the production of large voltages or currents in large structures and conductors such as power lines, buried cables, and antennas, as well as in facility grounding systems. These induced currents and voltages may then result in secondary effects such as insulation flashover and electronic component damage or malfunction. Logic circuits are particularly susceptible to EMP-induced transients. Even small transients in these circuits can cause a false count or status indication that will lead to an error in the logic output, and large transients can destroy the junctions of solid-state components used in these circuits.

The main objective of all HEMP hardening is to prevent the transient produced by a nuclear detonation from causing a system malfunction that degrades mission performance. The only completely effective method uses a barrier or topological schemes comprising electromagnetic shields, transient suppression devices, and isolation elements in harmony to ensure that the mission will not be degraded during HEMP conditions. However, these principles are of little or no consequence if proper grounding and bonding techniques are not applied. The goal of all grounding and bonding techniques is to redirect the HEMP-induced currents to the earth. Furthermore, any transient voltage potential difference across the grounding system, particularly the signal reference structure, subsequent to an EMP event, will result in ground-induced currents with similar damaging outcomes to the equipment installed in the facility. Therefore, extremely effective earth electrode and signal reference subsystems are required, positively bonded to the elements of the facility's shield and to terminal protective devices coupled to the shielded facility points of entry (POE). The facility HEMP shield would then form a major portion of the equipotential ground plane. [1,20–22].

Grounds for equipment and structures enclosed within the protected volume should be electrically bonded to the inside surface of the facility's HEMP shield. Conversely, grounds for equipment and structures outside the electromagnetic barrier should be electrically bonded to the outside surface of the shield or to the earth electrode subsystem.

Bonding straps or cables used to connect the facility shield (equipotential ground plane) to the earth electrode subsystem should correspondingly be electrically bonded to the outside surface of the shield, and at least one such bonding strap or cable should be located at the penetration entry area. Special attention should be paid to all grounding connections to the facility HEMP shield in order to preclude the creation of unintentional points of entry (POEs) into the shielded enclosure (Figure 10.19) [21,22].

For the fault protection subsystem, the NEC [16], for instance, states that a single electrode consisting of a rod, pipe, or plate that does not have a resistance to ground of 25 Ω or less shall be augmented by one additional electrode. Although the language of the NEC clearly implies that electrodes with resistances as high as 25 Ω are to be used only as a last resort, this 25 Ω limit has tended to set the norm for grounding resistance regardless of the specific system needs. The 25 Ω limit is reasonable or adequate for application to private homes and other lower powered type facilities [1].

The above criterion however, is not acceptable for C^3I facilities when consideration is given to the large investments in personnel and equipment. A compromise of cost versus protection against lightning, power faults, or EMP has led to establishment of a design goal of 10 Ω for the earth electrode subsystem (EESS) for defense-related C^3I facilities. For such facilities, the EESS specifies a ring ground around the periphery of the facility to be protected [4]. With proper design and installation of the EESS, the de-

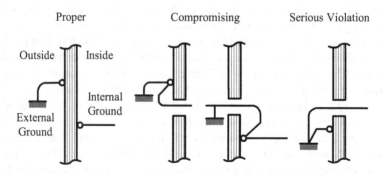

Figure 10.19. Installation of grounding conductor for preclusion of unintentional points of entry (POEs) at facility's HEMP protection enclosure.

sign goal of 10 Ω should be attained at reasonable cost. At locations where the 10 Ω has not been attained due to high soil resistivity, rock formations, or other terrain features, methods for reducing the resistance to earth should be considered.*

A radial, or star, configuration is preferred to other types of earth electrode subsystems because of its lower impulse impedance [1]. Where HEMP protection is to be provided in addition to conventional signal and safety protection, supplemental radials may be added to the conventional system.

A single low-impulse impedance radial should be placed at each location where there are over voltage surge protection devices (SPDs) on incoming external lines or conductors. An example of such a location is the point where commercial power lines enter the first step-down transformer. Another location is at the point where external conductors enter the shelter itself and where protectors or arresters are located. Water pipes or conduit should be connected to the earth electrode subsystem to prevent ground currents from entering the structure. Further, at the first service disconnect the AC neutral should be grounded at only one point (to the EESS) to prevent the possibility of damage to transformers. This does not negate the NEC requirement to ground the neutral at the transformer.

10.2.5.2 Grounding for Facility Emanation Security (EMSEC). Two interrelated concepts apply to secure C³I facilities with respect to emanation security (EMSEC), namely, the "red/black" concept and "TEMPEST." The "red/black" concept provides that electrical and electronic components, equipment, and systems processing national security related, classified, plain text ("red"[†]) information be kept separate from those that process encrypted or unclassified ("black"[‡]) information, whereas "TEMPEST" consists of the measures applied for controlling compromising emanations. Although these terms are often used interchangeably, the concepts are significantly separate and different [20].

A properly installed grounding system is a critical measure for reducing compromising emanations. This is because faulty grounds can radiate signals that may be picked up by equipment and transmission lines, thus negating all other EMSEC installation precautions. The same degree of criticality also applies to bonding. Bonding serves to eliminate differences in potential between metallic structures, such as wire ways and equipment items. Differences in potential in the earthing and grounding system between circuits or assemblies could result in a transfer of compromising emanations between them, particularly with "red" signals. Proper equipment, wire way, and ground bonding can eliminate differences in potential and greatly reduce radiation of compromising emanations.

A "red" and a "black" signal ground is typically established by direct connection totally within a controlled space to an equipotential ground plane and earth electrode subsystem (Figure 10.20) [20].

---

*Techniques for reducing the resistance to earth were discussed in Chapter 6.
†"Red" equipment is any equipment that processes classified information before encryption and after decryption, and should, therefore, be TEMPEST and physically protected.
‡"Black" equipment is any equipment processing unclassified information or encrypted information.

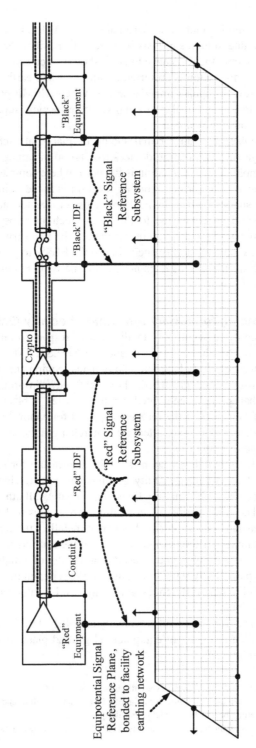

Figure 10.20. Typical "red/black" signal reference subsystem (high-level signals).

938

Cable shields from the "red" equipment to the "red" side of the crypto equipment are terminated at both ends at least. Cable shields from the "black" side of the crypto equipment through the "black" intermediate distribution frame (IDF) to the "black" equipment are also typically terminated at both ends. For unbalanced signaling, signal ground is usually established by direct connection from an isolated signal ground bus in the "red" distribution frame to an equipotential ground plane and in turn to the earth electrode subsystem (EESS).

In secure C^3I facilities, the maximum acceptable impedance of the earth electrode subsystem (EESS) and fault protection subsystem (FPSS) is 10 Ω, similar to the impedance required for EMP. No unique emanation security (EMSEC) requirements associated with the lightning protection subsystems are installed in secure facilities beyond those applicable for commercial telecommunications and C^3I facilities.

The signal reference subsystem (SRSS), however, is crucial for maintaining the EMSEC performance of the facility as it serves as the common signal reference throughout the facility. It provides a path to the EESS for induced static and noise and serves as a ground plane for high-frequency signals between equipment collocated in the facility.

Radiation of compromising emanations is greatly reduced by this subsystem. In modern secure facilities, an equipotential signal plane comprised of a metallic sheet or grid is frequently installed under, over, or beside all equipment in the technical area in order to reduce the effects of compromising emanations. This ground plane is most effective when it extends under or above all "red" and "black" equipment. This includes distribution frames, patch panels, and "red/black" processing equipment. In some cases, physical facility construction may dictate installation of a vertical plane; however, a horizontal plane is more effective than a vertical plane in coupling unwanted signals to earth. All signal ground runs are bonded to the equipotential ground plane (Figure 10.21) [20].

The plane is banded (welded or brazed) to the main metallic structure of the building and to the EESS at multiple points. Older facilities utilize "red/black" single-point signal grounding systems instead of an equipotential plane. "Red" and "black" signal grounds are established by direct connections to the equipotential ground plane, which is bonded to the EESS. For unbalanced signaling, the signal ground is established by a direct connection from an isolated signal ground bus in the "red" distribution frame to the equipotential ground plane and the earth electrode subsystem. A "black" signal ground is used to provide a signal ground reference in the "black" distribution frame [1,20,21,23].

In addition, the following particular grounding applications should be observed [23]:

1. Cable shields for both "red" and "black" signal lines are circumferentially bonded to the equipotential ground plane. Cable shields surrounding individually shielded lower frequency signal lines should be grounded at one end.
2. Cable ducts should be grounded at one end. The duct is then bonded to the equipotential ground plane at the shortest path between the duct and the ground

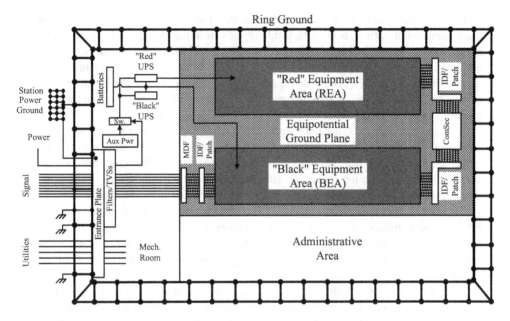

Figure 10.21. Large secure facility grounding scheme.

plane. This is done by bonding the cabinets to the plane since the duct is tied to the cabinets. Cable ducts carrying AC power will be grounded to the AC protective ground bus in the power panel.

3. Correct hookup of FPSS electrical safety ground conductors to equipment housings, cabinets, racks, conduit, ducts, distribution boxes, junction boxes, and other hardware is essential for the protection of personnel and equipment, and suppression of compromising emanations. FPSS ground conductors are normally green. The neutral and FPSS ground conductors are bonded together at the first service disconnect or service transformer and further bonded to the EESS. This is the only intentional grounding of the neutral conductor that is permitted according to standards and national electric codes. If an end item of equipment does not have a ground terminal, provisions for one should be made on the equipment case near the power entry point. In shielded facilities, the FPSS conductor does not penetrate the shield; rather, it should be bonded to the shield on the inside and outside.

10.2.5.3 EMSEC versus HEMP Grounding in Secure C³I Facilities. In EMSEC ("red/black") installation standards (e.g., [1,20]), reference to HEMP is often found. One might conclude that the guidelines for HEMP and EMSEC are the same. Where HEMP protection has been applied, a benign TEMPEST environment external to a secure facility may exist. Within a facility, however, a hostile TEMPEST environment may exist if proper attention had not been paid to EMSEC or "red/black" principles. One may also be led to consider a TEMPEST facility to be immune to HEMP. Although the facility may be EMSEC safe, it may be vulnerable to the high magnitude of the HEMP environment.

Proper installation of the various grounding subsystems—the earth electrode subsystem (EESS), lightning protection subsystem (LPSS), fault protection subsystem (FPSS), and, particularly, the signal reference subsystem (SRSS)—is critical to ensure the containment of compromising emanations and to provide HEMP survivability. An improperly installed grounding system in a classified environment renders "red/black" (or EMSEC) and HEMP installation criteria useless.

Both HEMP and EMSEC provide protection for any ground paths required between the outside and inside of a structure, carefully applying "zoning" concepts.* However, HEMP and EMSEC grounding methods differ. The design of such facilities should account first for EMSEC and then for HEMP protection. In this manner, these attributes that are mutually satisfactory may be successfully integrated. When conflicts exist, trade-offs can be identified, thus achieving the most cost-effective design [20].

10.2.6 Grounding of Instrumentation and Control Equipment Collocated with High-Voltage Power Apparatus

Grounding methods for instrumentation and control (I&C) equipment in areas where high-voltage power apparatus are collocated, such as in a generating station or a substation, require special attention to safety and EMC issues† [24]. The main safety concern in grounding is to ensure that high-voltage faults do not produce shock hazard to personnel working on the equipment. During a power fault that occurs outside the station, some portion of the fault current will return to the station through the station earthing system. A large increase in potential may develop in and around a power station with respect to the remote earth locations. Requirements for safety in substation grounding are covered in relevant standards such as [25]. A comprehensive earthing network with interconnected components minimizes transient voltages from electrical faults or lightning strikes, preventing these transients from affecting I&C equipment.

The typical electromagnetic environment is characterized by many sources of electrical noise such as switching surges, high fault currents, and high-frequency interference from electronic drives. Grounding methods for equipment chassis, cable shields, and signal pairs are well covered in other chapters of this book, and the general principles and good practices are also applicable for I&C equipment.

I&C equipment commonly uses solid-state technology and microprocessor-based subsystems. I&C components and circuits interface with each other through cables and end devices. A cluster or hub is formed with interconnected I&C components and circuits supported in racks, cabinets, or isolated enclosures. The grounding scheme is normally based on two separate grounding systems: the equipment or electrical safety

---

*The concept of "zoning" as it applies to grounding is discussed in Chapter 4.
†Grounding consideration of measurement and instrumentation cables and cable shields are discussed in Chapter 7.

ground and the signal reference ground. For each type, a common ground or reference plane is provided that functions as a return path for the power fault and signal currents, respectively. It is desirable that the ground reference plane offer minimal impedance to all the signals it serves and that the individual signal currents return to their respective sources without creating unwanted coupling and interference to other circuits. Proper grounding and bonding can prevent problems arising from the flow of electrical noise currents in these common signal returns.

Microprocessors, controllers, and solid-state electronics operate at higher frequencies within an equipment enclosure. The majority of signals that exit the cabinet and interface with other systems are either DC or low-frequency signals. A combination of single-point grounding and multipoint grounding may be adopted depending on the particular system requirements. Single-point grounding is more commonly encountered in the substation environment where equipment is located in close proximity and interfacing signals operate at lower frequencies. As shown in Figure 10.22, a separate signal reference ground for I&C circuits is provided within each cabinet. The signal

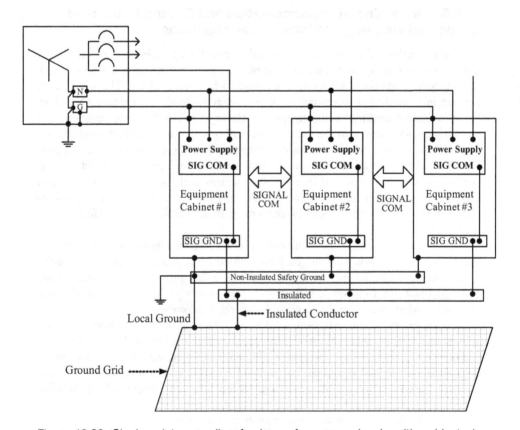

Figure 10.22. Single-point grounding for lower frequency signals with cabinets in close proximity.

reference is intentionally isolated from the equipment or chassis ground. The electrical safety ground conductor (ESGC) that is part of the power supply circuit is not suitable for the purpose of signal reference because it is subject to conducted noise interference from large leakage or fault currents and is also highly inductive. The cluster of equipment has a common signal reference in the form of an insulated ground bar which is connected to the facility ground grid at one point. In addition to the electrical safety ground conductor, the equipment is also connected to a local safety ground, which may be established by local equipotential bonding, including the station earth grid. The signal reference ground and the equipment electrical safety ground conductor are thus tied to the common earth grid.

The single-point grounding method prevents electrical noise current from entering I&C equipment and cables via conductive means. It is very effective and sufficient for equipment operating at frequencies typically not exceeding 300 kHz [24]. At higher frequencies, multipoint grounding should be adopted. An example is illustrated in Figure 10.23. Each cabinet is connected to the local signal reference structure by short conductors. At the equipment, the power supply, I&C, and equipment grounds are tied

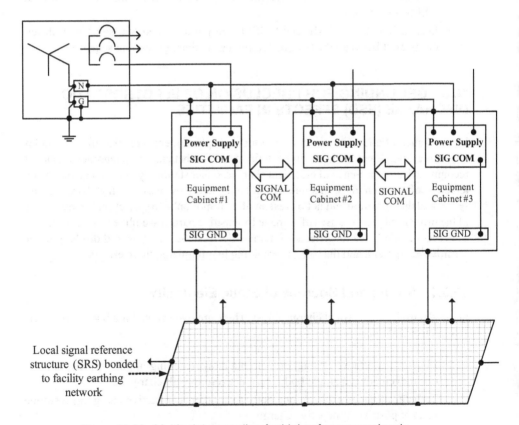

Figure 10.23. Multipoint grounding for higher frequency signals.

together, often at the equipment frame. A disadvantage of the MPG is the possible creation of multiple ground loops that may cause common-mode noise in low-frequency circuits.

Lessons Learned

- Facility grounding systems vary according to the purpose and physical structure of the facility, and the particular requirements of sensitive electronic equipment.
- Extensive equipotential bonding networks are commonly found in information technology (IT), telecommunications, and command, control, communications, and intelligence (C^3I) facilities.
- For instrumentation and control equipment, the single-point grounding method is normally used for low-frequency signals, whereas the multiplepoint grounding method is adopted for high-frequency signals.
- Proper installation of the facility signal reference subsystem (SRSS), earth electrode subsystem (EESS), and fault protection subsystem (FPSS) is critical to ensure the containment of compromising emanations (EMSEC) and to provide HEMP survivability.
- Inasmuch as both HEMP and EMSEC are jointly called out, significant differences exist between the two, including the resultant grounding schemes.

10.3 GROUNDING FOR PRECLUSION OF ELECTROSTATIC DISCHARGE (ESD) EFFECTS IN FACILITIES

Electrostatics (static electricity) and associated phenomena are extremely complex physical events. Electrostatics can be defined as ". . . that class of phenomena which is recognized by the presence of electrical charges, either stationary or moving, and the interaction of these charges, this interaction being solely by reason of the charges themselves and their position and not by reason of their motion" [26]. Electrical charge is one of the primary physical causes of damage to sensitive parts, assemblies, and equipment. This section addresses the effects of human electrostatic charging and discharging on sensitive equipment, and the role of grounding in minimizing these effects.

10.3.1 Nature and Sources of Static Electricity

Static electricity is electrical charge at rest. There are two events for a body that can result in electrical charge:

1. Electrons can move or migrate within a body, resulting in polarization; this can occur even when a single body has a net zero overall charge.
2. Transfer of electrons from one body to another (conductive charging) resulting in a net positive or negative charge.

The movement or transfer of electrons is due to the interaction of charged bodies or charged and uncharged bodies. The magnitude of the charge is primarily dependent on the size, shape, composition, and electrical properties of the substances that make up the respective bodies. Some substances readily give up electrons, whereas others tend to accumulate electrons. A body with an excess of electrons is charged negatively, whereas a body with an electron deficit is charged positively. During friction between two surfaces of different substances, one substance gains electrons and the other loses electrons. This results in each substance becoming charged. When the two materials are subsequently separated, the net positive or negative charge on each substance can be measured. These charges are equal and of opposite polarity. In the case of nonconductors, the charges tend to remain in the localized area of contact. Charges on a conductor are rapidly distributed over its surface and the surfaces of other conductive objects that it contacts (Figure 10.24).

Charged bodies are surrounded by an electrostatic field. Conductive,* dissipative,† and insulative‡ bodies that enter this field may be polarized by induction (that is, without contacting the charged body). In a conductive or dissipative body, electrons closest to the more negative part of the field are repelled, leaving that area relatively positively charged. These electrons are attracted to the more positive part of the field, creating negatively and positively charged areas. The net charge on the body remains zero, though. If a conductive polarized body is subsequently grounded, electrons will flow to or from the polarized surface near the ground, and upon removal of the ground the body retains a net nonzero charge due to the excess or deficit of electrons. In a nonconductive body, electrons are less mobile, but dipoles tend to align with the field, creating apparent surface charges.

The generation of static electricity caused by contacting or rubbing two substances is called the triboelectric (or frictional) charging effect. In the "triboelectric series," substances are listed in an order of positive to negative charging as a result of the triboelectric effect. A substance higher on the list is positively charged (loses electrons) when contacted with a substance lower on the list (which gains electrons). The order of ranking in a triboelectric series is not always consistent or repetitive. Furthermore, the degree of separation of two substances in the triboelectric series does not necessarily indicate the magnitude of the charges created by triboelectric effect. Order in the series and magnitude of the charges are dependent upon the properties of the substance, such as purity. Other factors affecting the magnitude of the charges include ambient conditions, pressure of contact, speed of rubbing or separation, and the contact area over which the rubbing occurs. A sample triboelectric series is provided in Table 10.1 [26]. Substantial electrostatic charges can also be generated through friction when two pieces of the same material are separated.

In many facilities, people are one of the prime generators of static electricity. Typi-

---

*For the purpose of ESD protection, conductive material is material with a surface resistivity less than 10^5 ohms/square (for surface conductive type materials) or with a volume resistivity less than 10^4 ohm(centimeter (for volume conductive type materials).

†For the purpose of ESD protection, dissipative material is material with a surface resistivity equal to or greater than 10^5 but less than 10^{12} ohms/square (for surface conductive type materials) or volume resistivity equal to or greater than 10^4 but less than 10^{11} ohm(cm (for volume conductive type materials).

‡For the purpose of ESD protection, materials not defined as conductive or dissipative are considered to be insulative.

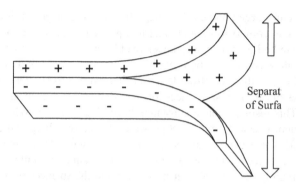

Figure 10.24. Static charge buildup due to friction between two substances and their subsequent separation.

cal prime charge sources commonly encountered in facilities are listed in Table 10.2 [26]. These prime sources are essentially insulators and are typically synthetic materials. Electrostatic voltage levels generated with these insulators can be extremely high since they are not readily distributed over the entire surface of the substance or conducted to another contacting substance. The conductivity of some insulative materials is increased by absorption of moisture under high-humidity conditions onto the otherwise insulating surface, creating a slightly conductive sweat layer that tends to dissipate static charges over the material surface. The generation of 15,000 volts from common plastics in a typical facility is not unusual. In fact, the simple act of a person walking around or repairing a printed circuit board (PCB) can produce a potential of several thousand volts on the human body. Similarly, the movement of carts or other wheeled equipment through the facility can generate static charges that may be transported to the products being conveyed on this equipment. Table 10.3 [26] shows typical electrostatic voltages generated in a facility. These electrostatic voltage levels are indicative of the relative charge (Q) on the object (in coulombs) in accordance with the fundamental relationship $Q = C \cdot V$, where C is the object's capacitance (in farads) and V symbolizes the electrostatic voltage of the charged object (in volts). If not properly controlled, this static charge can easily discharge into ESD-sensitive (ESDS) devices, resulting in their subsequent failure.

10.3.2 Susceptibility to ESD

The susceptibility of assemblies and equipment to ESD is directly related to the susceptibility of the parts used in the assembly and equipment. Numerous parts are susceptible to damage or malfunction when an ESD event occurs, whether directly subject to static discharge or when these parts are exposed to electrostatic fields. ESD-sensitive (ESDS) parts can be destroyed by an ESD event regardless of their electrical and ground connections. Ordnance may be particularly susceptible to inadvertent ignition from electrostatic discharge, often resulting in a catastrophic outcome. The primary

Table 10.1. Sample triboelectric series*

Positive (+)	Human hands
	Rabbit fur
	Glass
	Mica
	Human hair
	Nylon
	Wool
	Fur
	Lead
	Silk
	Aluminum
	Paper
	Cotton
	Steel
	Wood
	Amber
	Sealing wax
	Hard rubber
	Nickel, copper
	Brass, silver
	Gold, platinum
	Sulfur
	Acetate rayon
	Polyester
	Celluloid
	Orlon®
	Polyurethane
	Polyethylene
	Polypropylene
	PVC (vinyl)
	KEL F®
	Silicone
Negative (−)	Teflon®

*This list serves as an example only and is not exhaustive. The precise order of the materials in any triboelectric series is dependent upon many variable factors, and may not be repeatable. It is also noteworthy that charge magnitude is not a function of separation on this list.

concern in ordnance circuits is the bridge wire of electrically initiated explosive devices used to initiate the explosive

Assemblies and equipment containing ESDS parts are often as sensitive as the most sensitive ESDS part they contain. Incorporation of protective circuitry in assemblies and equipment provides varying degrees of protection from ESD applied to their terminals. Such assemblies and equipment are still vulnerable from induced ESD caused by strong electrostatic fields or by direct part, assembly, or equipment contact with a

Table 10.2. Typical prime electrostatic charge sources in facilities

Object or process	Material or activity
Work surfaces	Waxed, painted, or varnished surfaces
	Common vinyl or plastics
Floor	Sealed concrete
	Waxed, finished wood
	Common vinyl tile or sheeting
Clothes	Common clean-room smocks
	Common synthetic personnel garments
	Nonconductive shoes
	Virgin cotton*
Chairs	Finished wood
	Vinyl
	Fiberglass
Packaging and handling	Common plastic bags, wraps, envelopes
	Common bubble-pack, foams
	Common plastic trays, plastic tote boxes, vials, parts bins
Assembly, cleaning, test and repair areas	Spray cleaners
	Common plastic solder suckers
	Solder irons with ungrounded tips
	Brushes (synthetic bristles)
	Cleaning or drying by fluid or evaporation
	Temperature chambers
	Cryogenic sprays
	Heat guns and blowers
	Sand blasting
	Electrostatic copiers
	Cathode ray tubes

*Virgin cotton can constitute a source of electrostatic charge at low relative humidity (30% or less).

Table 10.3. Typical electrostatic voltage levels*

	Electrostatic voltage level (volts)	
Means of electrostatic charge generation	Relative humidity 10% to 20%	Relative humidity 65% to 90%
Walking across carpet	35,000	1,500
Common poly bag picked up from bench	20,000	1,200
Work chair padded with polyurethane foam	18,000	1,500
Walking over vinyl floor	12,000	250
Vinyl envelopes for work instructions	7,000	600
Worker at bench	6,000	100

*Caution should be exercised when attempting to correlate these or actual measured voltages with the potential to damage ESD-sensitive items.

Figure 10.25. ESD susceptibility symbol. Source: ANSI ESD S8.1-1993, *ESD Awareness Symbols,* ESD Association, Rome, NY.

charged object. Figure 10.25 presents the ESD Susceptibility Symbol, broadly used to indicate ESDS integrated circuits, PCBs, and assemblies that are ESD sensitive. The symbol literally translates to "*ESD sensitive material, don't touch.*" It indicates that handling or use of this item may result in damage from ESD if proper precautions are not taken. If desired, the sensitivity level of the item may be added to the label.

Electrostatic discharges can introduce either intermittent (i.e., upset) or catastrophic (hard) failures, resulting in permanent damage to the ESDS equipment. Intermittent failures typically occur as a result of an ESD spark occurring near the equipment. The electrical noise associated with the discharge may couple into electronic equipment either as conducted or radiated interference. In such near field discharges, either capacitive or inductive coupling may be dominant, depending on the impedances of the ESD source and the receiver. In high-impedance circuits, capacitive coupling dominates and ESD-induced voltage will constitute the primary concern. In low-impedance circuits, inductive coupling will dominate and the ESD-induced currents will cause the problem. Upset may occur to equipment when it is in operation and is usually characterized by loss of data or temporary disturbance to or loss of its functions. No apparent hardware damage occurs and proper operation can resume automatically or by operator's intervention after the exposure to the ESD event.

Upset failures occur when the equipment is operating, but catastrophic (hard) failures can occur at any time. Catastrophic ESD failures can be the result of electrical overstress of electronic parts caused by an ESD such as a discharge from a person or object, an electrostatic field, or a high-voltage spark discharge. Some catastrophic failures may not occur until after exposure to multiple ESD events (also known as the latent failure process). Marginally damaged ESDS parts, which require operating stress and time to cause further degradation, may ultimately experience catastrophic failure.

Since the electrical stresses necessary to cause hard failures or damage are one to two orders of magnitude greater than those required to cause upset, damage is more likely to be the result of conductive coupling, that is, the discharge spark must directly couple to the circuit. Radiated coupling will normally cause upset only.

10.3.3 ESD Protected Areas (EPAs) in Facilities

The objective of an ESD protected area is to preclude or control electrostatic discharge effects by maintaining the lowest possible electrostatic field intensity and voltages in the protected area. The ESD protected area (EPA) concept requires careful considera-tion of two elements. The first of these is related to the primary purpose of the protect-ed area, namely the requirement to provide an adequate level of protection for ESDS items handled in the protected area. The second element is to maintain personnel elec-trical safety at all times. This element is directly related to the types of materials (con-ductive or dissipative) selected for use in the protected area and the specific grounding architecture and procedures to be implemented.

The sophistication of the design of an ESD protected area is directly related to the work processes to be carried out, its location, and environmental or physical con-straints.* For instance, during field maintenance, a protected area could consist of a temporary area free from static-generating materials and equipped with appropriate protective covering or packaging materials, a personnel wrist ground strap, portable protective work mat, and provisional grounding measures. A protective area in a man-ufacturing facility could include local or room humidity and air ionization controls, a comprehensive ESD protective workbench constructed of protective materials and grounded appropriately, and conductive flooring with the associated heel grounders or conductive footwear. Permanent grounding provisions for ensuring safety of personnel and equipment would also be included in such facilities.

The protected area is the focal point for effective ESD controls. It should be noted that the protected area concept is intended for use only when ESDS parts, assemblies, and equipment are handled outside of their protective covering or packaging. ESDS items that are properly protected by technically adequate protective covering or pack-aging require no unique handling or storage procedures as long as the protective cover-ing or packaging integrity is maintained. Protected area concepts consist of various complementary elements, which include:

1. Grounding considerations
2. Safety and grounding requirements
3. Tools, materials, and equipment
4. Operating procedures

Of these four items, the first two, directly associated with grounding, are discussed in detail in Section 10.3.5. The third, ESD protective tools, materials, and equipment, are discussed to the necessary extent in Section 10.3.4. Excellent resources exist covering the fourth item, namely operating procedures for ESD control, which are beyond the scope of this book. See, for instance, [30,27,28].

10.3.4   ESD Protective Tools, Materials, and Equipment

For controlling human-generated electrostatic discharge, ESD protected areas (EPAs) are commonly supplemented by tools, materials, and equipment used in the workplace.

---

*The design of the protected area is directly impacted by the susceptibility of the ESDS item and the com-plexity of the handling procedures used in the protected area. As the protected area becomes more compre-hensive, the handling procedures used in the protected area become less complex.

For guaranteeing their performance, adequate grounding provisions and procedures must be implemented and safety of personnel must not compromised. The grounding procedures for control of ESD are discussed in Section 10.3.5. Some examples of commonly used tools, materials, and equipment are now discussed.*

10.3.4.1 ESD Protective Workbenches and Work Surfaces. An ESD protective workbench refers to the work area of a single individual that is constructed and equipped with materials and equipment to limit damage to ESD-sensitive items. Stand-alone or multiple ESD protective workbenches are commonly found, for instance, in production or assembly lines, in inspection and test stations, and in controlled areas such as clean rooms. ESD protective workbenches may also be used in field locations such as in equipment bays on board ships or commercial aircraft or near weapon checkup and loading stations near military aircraft.

Key ESD control elements comprising typical workbenches include a static dissipative work surface, fixtures, handling equipment, means of grounding personnel (usually a wrist strap, discussed below), tools, and accessories used on the workbench. A typical workbench is shown in Figure 10.26 [28] and Figure 10.27 [29]. Figure 10.28 shows the application of a static dissipative work surface and surface resistivity measurement setup [28].

The static protective work surface is connected to the facility's common point ground (see Section 10.3.5). This ground connection should be accomplished through a resistance soft grounding, which should be located at or near the point of contact with the workbench top. The resistance to ground should be sufficiently high (i.e., between 10^6 Ω to 10^9 $\Omega^\dagger$ and 10^7 Ω/sq to 10^{10} Ω/sq, typically‡ [26,30]) in order to provide an electrical path for controlled bleeding of any static charge on materials on that surface to ground while precluding electrical shock hazard to personnel by limiting possible fault current to the perception level in accordance with applicable electrical safety codes and standards,** considering all parallel resistances to ground such as wrist ground straps (see Section 10.3.6).

10.3.4.2 Personnel Wrist Straps. Wrist straps (shown in Figure 10.29) are primary means of preclusion of static charge buildup on personnel at the workbench. When properly worn and connected to ground, a skin-contact wrist strap virtually keeps the person wearing it at ground potential. Since the person and other grounded

---

*Requirements of particular key ESD control elements can be found, for instance, in ANSI/ESD S20.20-2007, *ESD Association Standard for the Development of an Electrostatic Discharge Control Program for Protection of Electrical and Electronic Parts, Assemblies and Equipment (Excluding Electrically Initiated Explosive Devices)*, ESD Association, Rome, NY.

†Source: ANSI ESD S4.1, *Worksurfaces—Resistance Measurements,* 1997.

‡The factor of 10 between the two ranges is due to the geometry of concentric rings used in electrode assemblies; to measure surface resistivity in ohms per square.

**Examples of sources for electrical safety specifications are, for instance: ANSI/NFPA 70-1966, *National Electrical Code,* CENELEC EN 60950, *Safety of Information Technology Equipment Including Electrical Business Equipment;* and MIL-STD-454, *Standard General Requirements for Electronic Equipment (Requirement 1).* A discussion of grounding considerations for ensuring electrical safety of personnel is included in Chapter 6.

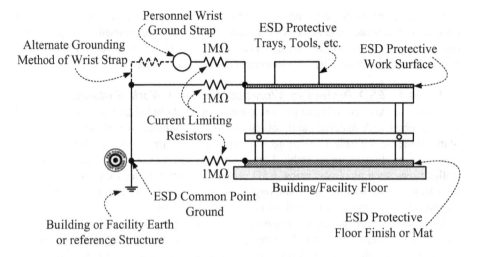

Figure 10.26. Components of a typical ESD protected workbench.

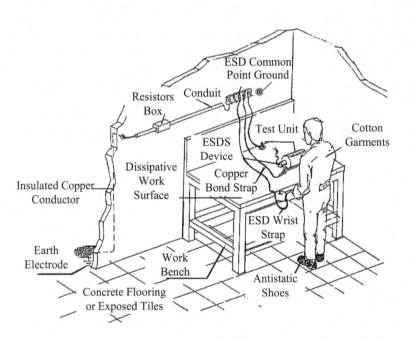

Figure 10.27. Construction details of a typical ESD protected workbench in a production, assembly, or test area. (Image courtesy of M. Netzer.)

<u>Figure 10.28.</u> Static dissipative work surface and surface resistivity measurement set-up. (Image courtesy of M. Netzer.)

objects in the work area are maintained at or near the same voltage potential, no hazardous discharge between them can occur. Furthermore, static charges do not accumulate on the person as they are safely dissipated to ground.

Wrist straps have two major components: the cuff that goes around the person's wrist and the ground cord that connects the cuff to the common point ground. As described above, personnel wrist straps should be connected to ground via a protective resistor to preclude safety hazards (see Section 10.3.6). Most wrist straps have an adequate current-limiting resistor molded into the ground cord head on the end that connects to the cuff. The most commonly used resistor is a 1 MΩ, 0.25 watt with a minimum working voltage rating of 250 V. For personnel safety, the resistor should be located near the point of contact with the individual's skin to reduce the likelihood of accidental cable shorting to ground, shunting the strap's resistance [29].

<u>Figure 10.29.</u> Application of personnel ESD ground straps (testing ground strap resistance). (Image courtesy of M. Netzer.)

10.3.4.3 Protective Floors, Floor Mats, and Floor Finishes. In lieu of a personnel wrist strap connections, an alternate method for controlling electrostatic charge on personnel is accomplished through the use of ESD protective flooring or floor materials (such as ESD protective floor mats) in conjunction with ESD control footwear or heel grounding straps, providing the necessary electrical ground path between the person and ground for dissipation of electrostatic charge (see Figure 10.30) [29].

In addition to dissipating charge, some floor materials (and floor finishes) also reduce triboelectric charging. The use of such floor materials is especially appropriate in those areas where increased personnel mobility is necessary.

ESD protective floor materials can similarly minimize charge accumulation on chairs, work stools, and other objects that move across the floor. People sitting at work benches in an ESD protected area often lift their feet from the floor to the work stool, counteracting the benefits of the protective flooring. Such items should be equipped, therefore, with dissipative or conductive casters or wheels to make electrical contact with the floor. When used as the primary personnel grounding system, the resistance to ground, including the person, footwear, and floor, must be the same as specified for wrist straps.

10.3.5 Essentials of Grounding for ESD Control

Effective ESD grounding is of utmost importance in any operation. It is essential in even the most basic of environments and critical in protected electronic manufacturing environments, and explosive, ordnance, and munitions environments. It serves as a primary means of protecting of ESD susceptible (ESDS) items and should be clearly defined and regularly evaluated. All conducting bodies in the environment, including flooring materials, work surfaces, tools, equipment, and personnel, must be bonded or electrically connected to a natural or contrived ground structure, creating an equipoten-

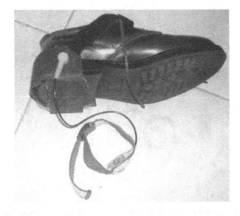

Figure 10.30. ESD control footwear and heel grounding straps. (Image courtesy of M. Netzer.)

tial balance between all items and personnel. By grounding charged conductive bodies or bonding them to inherently grounded objects such as metallic water pipes or large metallic storage tanks placed under or directly above the earth, a path is provided for neutralizing static charges present on such bodies. Positively charged bodies achieve equilibrium by drawing electrons from the earth onto the body, whereas negatively charged bodies deposit electrons into the earth. Electrostatic protection can also be maintained at a potential other than "zero" voltage ground reference as long as all items in the environment including ESD protective materials and personnel are kept at the same electrostatic potential.

Note: It is important to note that nonconductors in an electrostatic protected area (EPA) cannot dissipate their electrostatic charge through their attachment to ground. Static charges on such nonconductive bodies must be eliminated by means of ionizers or, in the case of nonconductive surfaces, by coating them with a conductive/dissipative compound and then grounding them.

Alternative grounding techniques may be suitable for providing ESD protection. The selected grounding scheme must ensure, however, that electrical safety of personnel is not compromised at any time. Electrical safety codes and standards* should thus be adhered to in the construction of ESD protected areas to reduce the likelihood of electrical shock hazards to personnel. Maximum current levels in ESD protected areas should be limited to the perception level dictated by the applicable electrical safety codes.

Design and construction of ESD protected areas is commonly achieved through grounding ESD protective equipment in a two-step procedure. The first step is to ensure that everything in the work space is maintained at a common electrical potential at all times during normal operation. This is achieved by grounding all components of the work area, including all ESD protective elements such as external parts, work surfaces, personnel, and electronic test equipment and power tools in an electronics production and test area, to the same electrical ground point called the common-point ground [31,32]. This ESD common-point ground should be properly identified. Use of the symbol in Figure 10.31 is recommended to identify the common-point ground. Figure 10.32 shows a practical implementation of ESD common-point grounding [29].

The second step for achieving ESD protective grounding is to connect the common-point ground to the local equipment electrical safety ground (ESG) or protective earth (PE) connection point. This is the preferred ESD ground connection because all electrical equipment on the site should already be connected to this point. Connecting the ESD control materials or equipment to the equipment electrical safety ground guarantees that all components in the work space are at the same electrical potential. If this were not the case, a possible electrical potential difference between the electrical safety ground (to which the ESDS item is connected) and an auxiliary ground (to which the work space surface containing the ESDS item was to be connected) could result in

*Examples of sources for electrical safety specifications are, for instance: ANSI/NFPA 70-1966, *National Electrical Code,* CENELEC EN 60950, *Safety of Information Technology Equipment Including Electrical Business Equipment;* and MIL-STD-454, *Standard General Requirements for Electronic Equipment (Requirement 1).* A discussion of grounding considerations for ensuring electrical safety of personnel is included in Chapter 6.

Figure 10.31. ESD Common point ground symbol. (Source: ANSI ESD S8.1-1993, *ESD Awareness Symbols,* ESD Association, Rome, NY.)

damage to the ESDS item, counteracting the purpose of the ESD ground and possibly even constituting a safety hazard for personnel if they come into electrical contact with both grounding systems simultaneously. Subsequently, any auxiliary ground structures (e.g., water pipe, building frame, or ground stake) present and used in the work space must be bonded to the equipment electrical safety ground in order to maintain the integrity of the connections from the common-point ground to this point and to minimize electrical potential differences between the two. The resistance between the two connections should be as low as possible, typically less than 1 Ω. Figure 10.26 illustrates the implementation of the procedure for grounding of ESD protective equipment in ESD protected areas.

Note: Some areas may preclude the use of protective flooring due to electrical safety requirements. This is particularly true of certain applications in military facilities and platforms that require the use of insulative floor mats for electrical safety. In these cases, personnel ground straps should provide the required degree of ESD protection.

10.3.6   Safety Considerations in ESD Grounding

Since ESD grounding requires that the ESD ground system be connected to the facility electrical safety ground connection, a potential electrical safety hazard may arise in sit-

Figure 10.32. Implementation of ESD common-point ground. (Image courtesy of M. Netzer.)

uations of electrical fault currents flowing through the ESD ground. This is of particular concern when personnel at the workplace come into electrical contact with tools and test equipment on grounded work benches with metal or other conductive coverings, which can shunt the protective resistance in the work bench ground cable if allowed to contact the work surface. The design of the ESD grounding system should, therefore, include as a primary consideration the preclusion of electrical safety hazards to personnel.

In order to reduce the hazard of severe electrical shock, conductive work surfaces and wrist ground straps should not be directly connected to a "hard ground," that is, a connection to ground through a path that has little resistance. Any conductive work surfaces and wrist straps must be properly "soft grounded." A soft ground constitutes a connection through a resistance sufficiently high such that it limits current flow through any parts of the human body to levels below perception (5 mA). The resistance required to achieve a "soft ground" is dependent upon voltage levels and AC frequencies that could be contacted by personnel at the work space. On the other hand, too high resistance to ground will affect the static charge decay rate of the work surfaces, and the work surfaces will not drain static electrical charges within the interval required for safe handling of sensitive electronic devices. For 50 or 60 Hz AC power, the minimum required resistance to ground to achieve an acceptable safe "soft ground" could be computed using

$$R_{min} = \frac{V_{AC\,max}}{5\,mA} \tag{10.1}$$

where:
R_{min} = Minimum resistance required to achieve a safe ground, ohms
V_{Acmax} = Highest AC voltage within a person's reach, volts (RMS)

Nevertheless, under normal circumstances, where only 110 VAC, 50 or 60 Hz, power sources are within reach of a person, a typical value of resistance is 250 kΩ. Based on Equation (10.1), a 1 MΩ resistor, for instance, provides an ample safety margin, even if 220 VAC, 50 or 60 Hz power sources are within reach.

The actual use of current-limiting resistors in grounding circuits is closely related to the material selected for ESD protective work surfaces (which may range from solid metallic surfaces such as stainless steel to dissipative material exhibiting surface resistivity of up to 10^{12} Ω/sq).

Caution must be observed when parallel paths to ground, even those beyond the immediate work space, from people, metal furniture, electrical equipment, floor mats, table mats, wrist straps, and so on, could exist (e.g., the work surface, the wrist ground strap, etc.). Such parallel paths could reduce the equivalent resistance of persons to ground, required for personal protection, consequently resulting in unsafe current levels.

As an added precaution for personnel safety, ground fault circuit interrupters (GFCI) can be used with electrical equipment. The GFCI senses leakage current from faulty equipment and interrupts the circuit almost instantaneously when these currents reach a potentially hazardous level (see Chapter 6).

Lessons Learned

- Electrostatic charges, and the personnel discharge process in particular, is the fundamental cause of damage to sensitive parts, assemblies, and equipment handled in production, maintenance, and operational use if ESD control measures are not properly implemented.
- ESD protected areas (EPAs) preclude or control electrostatic discharge effects by maintaining the lowest possible electrostatic field intensity and voltages in the environment.
- Grounding practices in EPAs are essential for ensuring performance of the protection measures, by providing a shunt for electrostatic charges to earth.
- For maintaining safety of personnel, only soft grounds should be used and checked regularly.
- Parallel paths that may bypass the soft-ground connection must be precluded, as they may compromise safety.
- Facility water pipes should not be depended on for providing electrical earth grounding. Water supply lines may contain a length of nonconductive pipe, or they may have insulative, sound-absorbing couplings.
- The entire ESD grounding system should be connected before handling any ESD-sensitive items.

10.4 GROUNDING PRINCIPLES IN MOBILE PLATFORMS AND VEHICLES

Grounding in mobile platforms and vehicles, such as mobile tactical shelters, aircraft, and space vehicles, as well as ships and marine vehicles, follows similar principles to those outlined throughout other sections of this book, as applicable. However, some aspects associated with grounding in these vehicles require some dedicated attention and discussion. This section addresses the special features associated with implementation of grounding in such applications.

10.4.1 Grounding in Transportable Tactical Shelters

Transportable tactical shelters are commonly used for commercial telecommunications or military command, control, communication, and Intelligence (C^3I) applications (Figure 10.33). Transportable shelters are typically constructed of metallic materials, offering some measure of shielding. The inner and outer skin of the shelters serves as an equipotential ground plane, and as such, as the reference structure for equipment contained in it. This requires that all seams be continuously welded so that the resistance of the seam does not exceed that of the conductive panel, creating a circumferential, low-impedance, conductive path to ground. Multiple ground terminals should be welded to the outer skin so that multiple ground conductors may be connected between the shelter structure and the earth electrode subsystem (EESS). When installing the

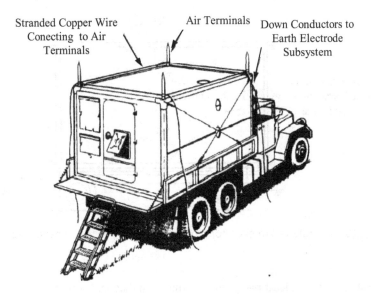

Stranded Copper Wire Conecting to Air Terminals

Air Terminals

Down Conductors to Earth Electrode Subsystem

Figure 10.33. A typical transportable shelter.

ground terminals, all paint or other nonconductive protective substances should be removed prior to installation for providing a mechanically strong and good electrically conductive bond. Welds should be circumferential to assure proper and effective bonding [20].

Whenever possible, the internal grounding schemes of transportable shelters follow the same principles and procedures applied for fixed facilities. As a minimum, internal grounds consist of threaded terminals of 25 mm (1 inch) by 6 mm (0.25 inch) copper ground bus, welded or brazed to the shelter's skin. The skin of the shelter should be bonded to the transporting frame (if the shelter is permanently affixed to the frame) by welding (typically No. 2 AWG stranded copper) ground conductors to the skin and frame at multiple points to assure that potential differences are minimized (Figure 10.34) [20].

Shelters constructed of nonconductive material should have a copper mesh screen installed between the inner and outer skins to form an equipotential ground plane. The screen should be installed in the floor, ceiling, and all walls of the shelter and should cover all surfaces except apertures of air conditioners and cable entrance plates. Air conditioner mounts and cases are subsequently bonded to the metallic screen. Cable entrance plates are likewise circumferentially welded to the screen as well as to the EESS through bonding conductors. Since the screen in the door must be detached from the remainder of the screen to permit the door to open and close, multiple flexible conductors should be bonded to the door and main screens [20].

Similar to fixed facilities (discussed in depth in Section 10.2), the ground systems of transportable shelters connect any metallic element of the associated subsystems to earth by way of an earth electrode subsystem (EESS) configuration. All transportable

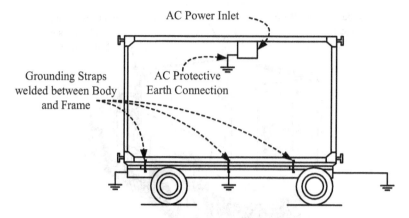

Figure 10.34. Preferred method for grounding shelters to the transporting frame.

shelters should have at least two ground terminals at diagonal corners of the shelter for providing versatile and optimal grounding of the shelter.

A good, basic earth electrode subsystem is the fundamental network for establishing a ground point for the three remaining ground subsystems, namely, lightning, signal reference, and fault protection. An ideal earth electrode subsystem will provide a common ground reference to any equipment or subsystem incorporated in the shelter. Tactical earth electrode subsystems are connected to existing buried low-resistance facilities, if available, or to driven ground rods or ground-rod configurations.

Contrary to fixed C^3I facilities (discussed in depth in Section 10.2), in which a high-quality earthing system can be achieved, the grounding scheme of transportable facilities and shelters must often be implemented in a manner that finds the middle ground between effectiveness of the grounding system and operational constraints. In particular, grounding of transportable facilities is dependent on the type of operation and on the terrain. Short-term operation and rapid deployment or frequent relocations may require the use of a single earth electrode, whereas operation in one area for longer periods (such as several hours) would permit the use of more extensive and effective grounding systems.

Three basic methods may be considered for improving the grounding performance of tactical equipment and systems, namely, utilization of (1) earth electrode subsystems of existing/permanent facilities; (2) recently configured earth electrode, earth rod configurations; and (3) chemical treatment of the soil around the earth rod and its interconnecting conductors (see Chapter 6 for a discussion of these design techniques).

Consequently, it is not possible to provide a fixed set of rules governing the grounding of all conceivable electrical or electronic equipment or system configurations. However, the principles outlined here may be adapted to the requirements of a particular tactical or transportable shelter installation.

This section contains grounding considerations for tactical deployments of transportable equipment. The tactical deployments of transportable equipment are considered to be of four types: stand-alone equipment, stand-alone shelters, collocated equipments, and collocated shelters.

10.4.1.1 *Stand-Alone Equipment.*
Stand-alone equipment of transportable systems is generally self-contained transportable (transportable or hand-held) field equipment, not contained in a shelter. Such equipment interfaces with other equipment over balanced or coaxial cables and its performance is, therefore, independent of any common reference connection. Stand-alone equipment is generally totally self-contained, with integral power supplies and grounding system. As a result, the primary grounding objective for stand-alone equipment is to assure personnel safety and lightning protection. Lightning protection is needed to protect operating personnel and preclude damage to equipment from the effects of lightning that may impinge upon interfacing cables or from direct strike on the shelter. These are achieved through low-impedance grounds and installation of surge arresters on interfacing cabling.

For small communications equipment (e.g., personal computers, ground radios, etc.) that is powered by small, single-phase generator sets, the third wire [electrical safety ground conductor, ESGC, or protective earth (PE) wire] contained within the power cord provides a sufficient connection to ground, provided that the generator set is properly grounded [33].

10.4.1.2 *A Stand-Alone Shelter.*
Stand-alone shelters are comprised of equipment housed in a transportable metallic shelter and are typically not situated sufficiently close to other equipment to merit construction of a common extensive earth electrode subsystem between their interfacing systems. Power to the shelter may be supplied from a generator or a commercial source. Interfacing with the shelter may be through the power cable. The need for grounding stand-alone shelters is to provide for: (1) fault protection, (2) "bleeding off" static charges or EMI from interfacing signal cables, (3) signal reference, and (4) lightning protection.

The signal reference and fault protection subsystems in the shelter are typically connected to the earth electrode subsystem given that (1) the skin of the shelter generally serves as the equipotential structure for the signal reference subsystem, (2) the electronic equipment in the shelter is typically bonded directly to its skin, and (3) the fault protection subsystem of the shelter is connected to the grounding bus in the power entry panel and, in turn, to the earth electrode subsystem.

Owing to the metallic construction of the shelter, its interior may be considered and treated as an isolated zone* from the standpoint of the electromagnetic environment and EMI control. No ground interfaces should penetrate the enclosure of the shelter, as that would constitute a severe violation of the shielding and zoning integrity of the shelter. The entire grounding system of the equipment in the shelter should, thus, be contained within and referenced to the internal structure of the shelter.

The earth electrode subsystem may consist of a single ground rod or interconnected ground rods in a multirod configuration. The most effective method of providing the

---

*The zoning technique, as applies to grounding in complex integrated systems, facilities, and platforms is discussed in Chapter 4.

earth ground is through the use of a circumferential EESS that consists of multiple copper-clad steel rods installed around the shelter. Recommended length of the rods is 1.5 m (5 feet) and 19 mm (0.75 inch) in diameter. At least two ground rods should be installed at diagonal corners of the shelter. Additional rods may be installed if the shelter size permits. For ensuring effectiveness of the EESS, it is important that a minimum separation between the rods of no less than one rod length and no more than two rod lengths be maintained (Figure 10.35)* [20]. For shelters that are too small to permit proper spacing when using rods at each corner, one rod will be installed at the front corner and another at the diagonal rear corner. The rods are connected to the outer skin or frame of the shelter and are bonded together by means of a No. 2 AWG stranded copper wire. The objective for the EESS resistance is 10 Ω or less [20].

An alternative way to provide effective earth grounding is to use three-rod delta or seven-rod star ground arrangements for normal soils or rocky or other problematic soils, respectively (Figure 10.36) [20,33,34]. One rod is installed as the center with the remaining six rods installed around it at a distance of 1.5 to 3 times the length of the rod from the center and from each adjacent rod in the star arrangement. All rods should be interconnected using a No. 1/0 to 2 AWG stranded copper conductor. The shelter is then connected to the EESS through multiple ground conductors between the structure of the shelter and the outer rods. If only one ground terminal is provided on the shelter, it should be connected to the center ground rod.

Adverse conditions may exist that prevent use of conventional grounding methods for transportable shelters and systems on granite, coral, rocky areas, desert areas, or coastal regions. To compensate for this, a copper mesh screen, not less than 3 m² (10 ft²) is laid on the earth surface (Figure 10.37) [20]. The screen should be constructed of No. 12 AWG stranded copper wire with apertures of no more than 100 mm² (4 in²). All crossover points should then be brazed, providing effective low-impedance bonds across the surface of the screen. Installation in granite, coral, rock, or hard-packed desert areas is accomplished by drilling holes 250 to 300 mm (10 to 12 in) deep, spaced 0.75 m (2.5 ft) apart around the perimeter of the screen. Fluted drive pins, 300 to 350 mm (12 to 14 in) long and 17.5 mm (0.688 in) in diameter should thereafter be driven into the wholes and bonded to the screen. After installation of the screen, the ground should be covered with at least 25 mm (1 in) of sand and treated with a thin layer of magnesium sulphate (epsom salts) and water. Note that once the drive pins are installed it may not be possible to remove them. When installing the screen in sandy soil or in coastal regions, drive pins should not be used; instead, the screen should be buried approximately 150 mm (6 in) below the earth surface and covered with sand. It should be periodically treated with magnesium sulphate to ensure effective performance of the grounding system.

When grounding for EMP/HEMP† vulnerability is required, special grounding measures must be applied. Power fault grounding is critical for lightning and EMP/HEMP

---

*The purpose of using multiple rods is to provide more than one conductive path through which ground currents may flow. Installation of vertical rods and considerations for mutual separation are discussed extensively in Chapter 6.
†Grounding considerations for EMP/HEMP in fixed C³I facilities are discussed in Section 10.2.5.

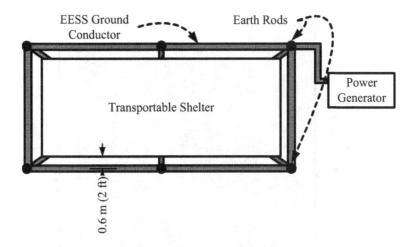

Figure 10.35. Preferred grounding scheme for transportable shelters.

protection; therefore, only circumferential EESS described above (see Figure 10.35) should be utilized. Multiple ground conductors should be installed between the shelter and the EESS, thus providing more than one path for HEMP-induced ground currents.

Grounds for equipment enclosed within HEMP-protected shelters (the inner surfaces of which are to be employed as the equipotential ground planes for this equipment) should be electrically bonded to the inside surface of an enclosure's shield by

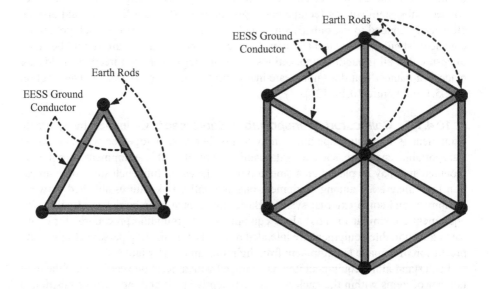

Figure 10.36. Multirod grounding arrangements for transportable shelters.

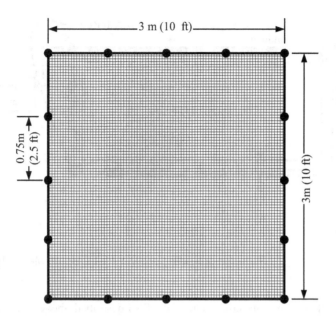

Figure 10.37. Mesh screen and drive pin positioning for grounding transportable shelters under adverse conditions.

the shortest practical path. Grounds for equipment and structures outside the electromagnetic barriers of the shelter should be electrically bonded to the outside surface of a shield or to the earth electrode subsystem (EESS). Grounding conductors used to connect subsystem shields (equipotential ground planes) to the EESS should also be electrically bonded to the outside surface of a shield, and at least one such grounding conductor should also be located at each penetration entry area. Care should be taken to ensure that all grounding connections to the shelter's HEMP-protective shields are made in a manner that does not create inadvertent points of entry (POEs) (see Section 10.2.5.1 and Figure 10.19) [35].

10.4.1.3 Collocated Transportable Equipment. Collocated transportable equipment is apparatus operating individually but hosted together within a single transportable enclosure such as a tarpaulin. Typically, this equipment is not rack mounted and may be placed on a table top or on the earth. Intraenclosure communication links may exist among equipment but, normally, links are established between equipment and some external system. Basic operational characteristics of collocated equipment are similar to stand-alone equipment and grounding procedures for collocated transportable equipment are intended primarily for ensuring personnel safety and preclusion of damage to equipment from lightning and power faults.

Each stand-alone equipment item is deployed with at least one earth rod. If the total number of items within the enclosure is sufficiently small and they can be positioned such that the earth rod for each can be used without compromising grounding integrity, then existing low-resistance grounding structures or a single driven earth rod per piece of equipment may be used to ground collocated items. Conversely, where a large number of items are housed within an enclosure for which the individual grounding procedure is not practical, a simple earth electrode subsystem should be deployed around the enclosure. The size of the earth subsystem and the number of attached rods needed to achieve the required ground resistance should be determined according to the procedures described in Chapter 6.

10.4.1.4 Collocated Shelters. Collocated transportable shelters constitute a system or network of interconnected shelters sharing common signal and/or power cables. Grounding for collocated shelters is primarily intended to provide personnel and equipment protection from the effects of lightning and power faults and to provide a common reference for signal grounds.

Collocated shelters' configurations are commonly classified as (1) shelters located within 8 m (26.5 ft)* of one another and (2) shelters located at distances greater than 8 m from one another. Shelters located within 8 m of each other should share a common earth electrode subsystem [Figure 10.38(a)] [1]. Grounding of collocated shelters located more than 8 m apart should be accomplished in accordance with Figure 10.38(b) [1], whereby each shelter has its dedicated EESS, which are thereafter typically interconnected using two bare 1/0 AWG copper cables. If distance between collocated shelters makes the interconnection between them impractical, based on the above criterion, each shelter should have its own independent EESS.

The actual need to establish an all-encompassing shelter grounding system for collocated shelters situated more than 8 m apart should be a function of ground resistance measurements taken at each shelter site compared to the tolerable value. Chemical treatment enhancement may also be used to reduce resistance at the location exhibiting higher resistance, as described in Chapter 6.

An alternative approach for site-level grounding is presented in Figure 10.39. This approach known as the "site central-ground" system is particularly advantageous where short-term operation and rapid deployment or frequent relocations are required, prohibiting the possibility of installing a circumferential earth electrode subsystem (EESS). In this method, a site central EESS is constructed and all shelters and subsystems collocated on-site are connected to it through grounding conductors. The EESS should have low impedance to earth.[†] The earth electrode can vary depending upon soil conditions, location of the site, and available materials.

The site central-ground scheme is constructed of three ground rods installed in a triangle, similar to the "delta arrangement" discussed above [see Figure 10.36(a)] and also shown in Figure 10.40 for this particular application. If the desired earth impedance goal is not reached, additional ground rods should be used to lower the resistance of the electrode system to earth, for instance, by applying the seven-rod "star arrangement" shown in Figure 10.36(b).

---

*When lightning and HEMP survivability are a concern, a separation criterion of 3.6 m (12 ft) is more appropriate and should be considered instead of 8 m (26.5 ft) [20].
†Criteria and objectives of earth resistance are discussed in Chapter 6.

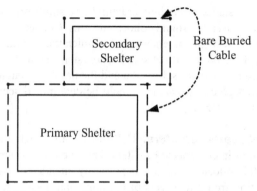

(a) Interconnecting the Grounding Subsystems of
Collocated Shelters Less than 8 Meters Apart

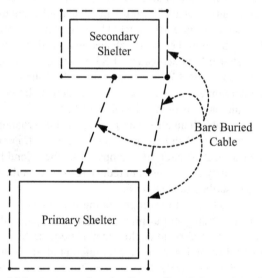

(b) Interconnecting the Grounding Subsystems of
Collocated Shelters More than 8 Meters Apart

Figure 10.38. Grounding of collocated shelters.

Vertical ground rods, at least 3 m (10 ft) long and 19 mm (3/4 in) in diameter should be used for the EESS. The ground rods should be spaced at least two rod lengths apart. When addition of rods has only a marginal effect in reducing the earth resistance, lowering the resistance in some cases may be aided by adding rock salt or magnesium sulfate and water to the holes around the ground rods.

Selection of the deployment site may also assist in improving the grounding system performance. Whenever feasible, it is recommended to select a site that is at the lowest

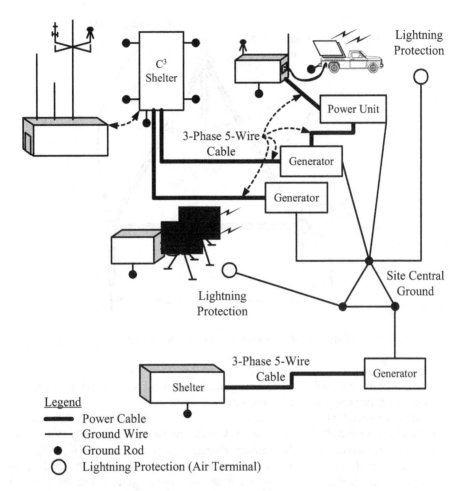

Figure 10.39. Collocated shelters with a site central ground.

point of the surrounding terrain. The water table should be closer to the surface of the earth at such a site. In particular, the site should be in near proximity to the location of the power generating equipment.

Particular consideration must be given when establishing grounding procedures for collocated shelters receiving power from a common power source or communicating over copper intershelter interconnects [1]. In the power distribution subsystem, in particular, voltage potential differences between return (neutral) leads of the different shelters must be minimized for preclusion of high circulating currents. Shelters powered from a single, common power source should, therefore, have all return (neutral) conductors bonded together and grounded at one common point near the generator. When several generators are connected in parallel, the return (neutral) terminals of all generators should also be interconnected and grounded at a single common point. For

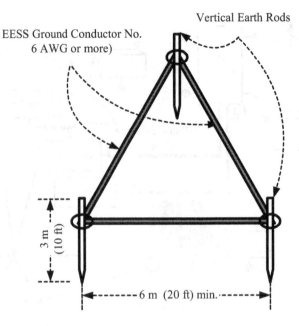

Figure 10.40. Site central ground installation.

collocated shelters supplied from separate power sources, all return (neutral) leads may be bonded together and grounded at one common point.

All power generators should have a separate electrical safety ground conductor (ESGC) wire run from their ground rod back to the central ground facility (see Figure 10.39). A copper conductor with minimum diameter equivalent to a No. 6 AWG or as appropriate for carrying the maximum current load should be used to connect the generator ground terminal and its ground rod to the site central ground. Diameters larger than No. 6 AWG are required for a current load exceeding 200 A, for instance [33,34]:

- No. 6 AWG is appropriate for current load 200 A or less
- No. 4 AWG is appropriate for current load 201 A to 400 A
- No. 2 AWG is appropriate for current load 401 A to 500 A
- No 1 AWG is appropriate for current load 501 A to 600 A
- No. 1/0 AWG is appropriate for current load 601 A to 800 A

On long ground conductor runs, additional ground rods should be located every 30 m (100 ft). Similarly, all communication–electronic (C–E) equipment using three-phase, four-wire power cables should also have an insulated electrical safety ground conductor (ESGC) as a fifth wire, with a minimum diameter of No. 6 AWG. This wire, if it is not part of the power cable, should be taped or strapped to the four-wire power cable every 30 cm (12 in) along its run. This wire is connected directly to the shelter earth terminal or the ground lug on the four- or five-wire cable adapter. The other end of the cable is bonded to the earth terminal for the power generator. This ESCG wire should be continuous, have no breaks in the runs of the cable length, and always be connected directly to the power source, never crossing through a junction box. On long runs, install additional ground rods each 100 feet. When connecting power cables of different length, a 3-ft ground rod section will be used to tie the fifth wire together at the power cable connector junction. Each fifth wire should be connected to the ground rod by its own ground lug; never combine the two.

Lessons Learned

- Grounding in transportable and fixed shelters follows the same rules as for terrestrial facilities.
- Internal grounding of equipment in the shelter and external interfaces of the shelter should follow the fundamental principles of "zoned grounding."

10.4.2 Grounding in Aircraft

One of the most important factors in the design and maintenance of aircraft and associated electrical and electronic systems is proper grounding and bonding. Inadequate bonding or grounding can lead to unreliable operation of systems, including EMI, electrostatic discharge (ESD) damage to sensitive equipment, personnel shock hazard, or damage from lightning strikes.

Aircraft are typically constructed as complete or partial metallic enclosures, designated as the reference ground structure for equipment installed on board. In certain cases, aircraft are constructed of predominantly nonconductive composite materials, such as Kevlar, which is totally nonconductive, or carbon fiber composite (CFC), which is conductive to some extent. On such nonmetallic vehicles, an equipotential ground system should be installed.

Grounding in aircraft is concerned, therefore, with two aspects: external earth connections (when parked) and on-board grounding [26,36,37,38].

10.4.2.1 *Earthing of Aircraft and Ground Services.* Contrary to most terrestrial facilities, aircraft, and especially military aircraft, are characterized by the increased potential for hazardous situations to occur, due to the presence of flammable materials, such as jet fuel, and explosive substances, such as weapon systems, on board. Hazards may further increase when the aircraft is parked on the ground, particularly when servicing and maintenance actions, and moreover, refueling/defueling and loading of weapons and ordnance systems, take place.

Of particular concern is the hazardous situation resulting from electrostatic discharge (ESD), particularly in the presence of potentially hazardous materials. Various mechanisms of static electrification of aircraft, such as triboelectric, frictional, and induced charging, may result in static charges deposited on the aircraft structure, yielding electric potentials exceeding 50 kV. In general, direct effects of static electricity on aircraft and associated systems are not serious, although such shocks can cause involuntary reflex movements resulting in injury to the affected person or others nearby. On the other hand, sparks generated due to voltage potential differences between aircraft and servicing equipment during fueling operation can be sufficient to ignite fuel vapor, bringing about disastrous results. The motion of fuel during refueling operations may also significantly contribute to static charging of the aircraft. Catastrophic results can occur during handling of ordnance due to discharge of electrostatic electricity into susceptible electrically initiated explosive devices (EIEDs), which may lead to their inadvertent ignition. Care in ensuring that aircraft weapon systems, fueling system, and personnel are maintained at ground potential will minimize the dangerous effects of static electricity. A resistance to ground of up to 10 kΩ is considered an adequate safety measure for EIEDs.*

Grounding of aircraft when operated on the ground is also required for preclusion of hazards associated with electrical ground power faults.[†] Ground power faults occur when the high-voltage side of an external power connection is brought into contact with the airframe.

When the aircraft is on the ground, AC and DC power is supplied to the aircraft by mobile or fixed generators external to the aircraft. Figure 10.41 shows a typical example of an aircraft supplied with external power from a ground power unit (GPU) cart in a flight-line configuration. The 115 V AC, 400 Hz, three-phase power is a four-wire system consisting of phases A, B, and C, and a neutral (N). The neutral wire is connected to the aircraft structure and the GPU cart chassis through the power connector. The 28 V DC supply is a two-wire system comprising positive and negative leads. The negative lead is connected to the aircraft structure, on the one hand, and to the GPU cart chassis through the power connector.

Since the predominant fault mode concerns the external power supply connector, one cannot rely on the neutral connection to the aircraft to provide protection against ground current faults. Bonding the aircraft to the external electrical power source (e.g., GPU) and to a dedicated and approved power ground point ("certified power ground") exhibiting a ground resistance of less than 10 Ω through a suitable grounding cable (Figure 10.41) will help avert hazardous situations. The dedicated grounding cable provides a controlled path for ground fault current while maintaining the aircraft at ground potential, concurrently ensuring that overload or ground fault protective devices (e.g., circuit breakers) are efficiently and rapidly triggered. If a ground fault occurs and the protective device operates slowly or not at all, electrocution of personnel, fire, or destruction of equipment may be the result.

Electrical fault conditions within ground support and maintenance equipment connected to the aircraft can also cause hazardous voltages to appear on the structure of the equipment. Both the aircraft and any ground support equipment must, therefore, be

---

*Note that the requirements for ESD grounding in aircraft and facilities differ significantly. Whereas in facilities highly resistive "soft" grounding is required for protection of personnel from power system fault currents flowing through the grounding system also utilized for ESD purposes, in aircraft grounding practices are driven by requirements for preclusion of hazards to ordnance and fuel systems.

[†]Power fault hazards and associated mitigation techniques are discussed in Chapter 6 with respect to terrestrial power distribution systems. The concepts described there apply to power faults in aircraft as well.

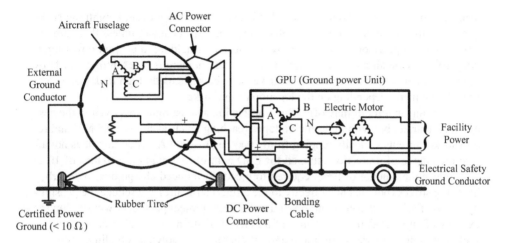

Figure 10.41. Aircraft supplied with external power.

properly bonded and referenced to a true earth point before any connection is made between the two.

When a ground fault occurs, the airframe is at power line potential, 115 V to 230 V, and may be capable of supplying currents as high as 200 A. The maximum allowable earth resistance is dictated by the requirement that sufficient current flow to trigger the power circuit protective devices. Personnel safety requires that the earth path resistance be sufficiently low to allow a 500% to 600% overload current with a trip time of approximately 200 msec. Thus a 120 V, 50 A service mandates an earth path resistance of 0.5 Ω for ensuring safety under power fault conditions. Unfortunately, rarely can this goal be achieved. A resistance of 10 Ω is recommended as a practical attainable value to provide some degree of safety and, at the same time, an adequate airframe system ground. Neither will it cause a grounding cable to burn [26,36,38,39]. In case of high earth path resistance, adequate shock protection has to be provided by ground fault circuit interrupter that employs the residual current principle.

Due to difference between the characteristics of electrical power and ESD energy sources, a ground point with an impedance to earth much lower than that allowed for ESD purposes is required. A distinction must be made between the power or maintenance ground (<10 Ω) and the ESD ground (<10 kΩ).

To accommodate for connections between the aircraft and earth, as well as to ground service and maintenance equipment, grounding jacks should be provided at strategic locations on the aircraft to provide ease of service and maintenance operations. Grounding jacks are provided at least at (1) fuel nozzle inlets, (2) servicing and maintenance equipment connection points, and (3) locations convenient for use in handling of weapons or other explosive devices.

Earthing aircraft also offers some degree of protection against lightning interaction with aircraft. The tail of an exposed aircraft constitutes a local high point that is at or near ground potential and, therefore, often provides the primary point for lightning attachment to the aircraft (Figure 10.42). Little can be done to prevent a lightning strike to an aircraft parked on a concrete/asphalt apron outside a hangar, unless there are tall objects such as poles or masks that provide adequate shielding to prevent direct lightning strikes to the aircraft. An aircraft may be struck numerous times during its life by lightning. Energy levels, even of near strikes, can be sufficiently high to damage electronic equipment, depending on the location of the strike.

Grounding an aircraft may offer some protection against lightning, even in the case of a direct strike, but with lightning voltages reaching levels as high as 0.5 MV and return strike currents typically ranging from 20 kA up to 200 kA, grounding does not afford the degree of protection attained for ESD control. A ground impedance of 10 Ω will reduce the amount of time required to bleed off induced charge due to "nearby" lightning strikes (not attached directly to the aircraft structure) and reduce the possibility of side flashes. Of course, no ground will totally protect personnel working on the exterior of an aircraft in the vicinity of an electrical storm. If weather conditions are such that lightning strikes are imminent, all operations involving refueling, ordnance loading, or maintenance must be suspended.

In summary, use of a proper ground connection ensures that the airframe is maintained at the same potential as the earth connection point for the hazard sources considered. In addition, since the airframe resistance to ground is greatly reduced, the duration of such effects as induced voltages is reduced to fractions of a millisecond. As an additional advantage, the use of proper grounding procedures ensures that any arcing or electrical discharge associated with the act of connecting grounds take place at the earth connection point rather than near the airframe. Illustration of the manner of aircraft grounding during different modes of operation ("evolutions") is provided in Figure 10.43 [36].

10.4.2.2 Internal Aircraft Grounding. With metallic aircraft, voltage drops through the structure are typically very low. The aircraft structure commonly serves as the reference structure for all electrical and electronic systems installed on board. Most

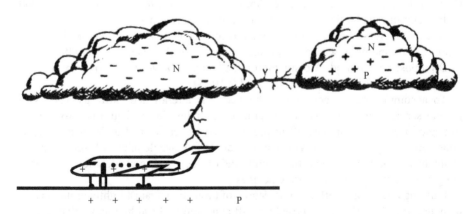

Figure 10.42. Lightning discharge to aircraft on ground.

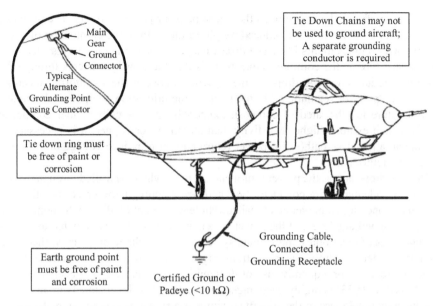

(a) Grounding during Parking, Maintenance and Weapons Loading/Unloading Evolutions

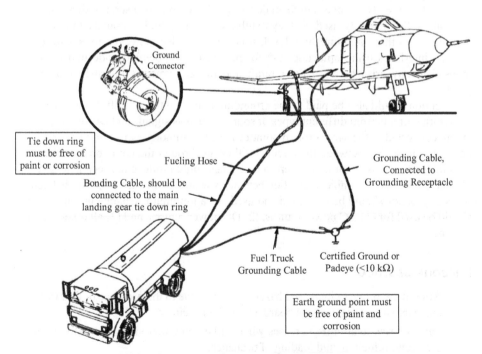

(b) Grounding during Fueling/Defueling Evolution

Figure 10.43. Illustration of the F-4 aircraft grounding.

metallic aircraft use the structure as the return path for power currents. The principle advantage of this system is the reduced weight resulting from the elimination of many heavy power return circuits [39]. The disadvantage of the common return system, using the structure as the common return path, is that the flow of currents through the structure produces voltage drops in the structure between the monitoring location, termed the "point of regulation," to the input of the utilization equipment. If the aircraft structure is fabricated of composite materials such as carbon fiber composite (CFC), which exhibits higher resistivity than aluminum or copper, it is necessary to provide an alternative path(s) for power return current to reduce voltage drops along the current return path.

Voltage drops across the power return current path, when controlled through the application of adequate wire conductor type and size, bonding between structural members, and proper signal/power electrical current levels, are normally small compared to the operating voltage levels of the power system but may be large compared to the operating levels of electronics and avionic systems. These voltage drops may, therefore, result in interference problems in electronic signal circuits using the structure as a path for signal return. Consequently, use of differential and balanced interfaces (e.g., RS-422, Mux-Bus 1553) is highly encouraged in sensitive, low-level signal circuits. Such circuits do not make use of the aircraft structure as a signal current return path and exhibit a high level of immunity to high-level interference currents that flow through the aircraft structure. High-level single-ended (e.g., 28 V/ground discrete) signal interfaces are still commonly used, as they exhibit sufficiently high immunity to interference thanks to their high threshold level. It is recommended, however, to isolate such discrete interfaces using optocouplers* for preclusion of potential undesired ground loops that may arise between the aircraft and the equipment's internal grounding points.

Attention should also be paid to the grounding points on the aircraft. If wires carrying return currents from different types of sources, such as signals and DC or AC power, are connected to the same ground point or have a common connection in the return paths, an interaction between the circuits will occur through the common impedance. This interaction may be of no concern but at times could produce unacceptable levels of interference. To minimize interaction between various return current paths, different types of grounds should be identified and used. As a minimum, separate ground types should be used for (1) AC power returns, (2) DC power returns, and (3) all other signal returns.

Lessons Learned

- Aircraft are prone to increased hazard potential due to the presence of flammable and explosive substances on board and in their vicinity.
- Potential risk to aircraft increases when parked on the ground, especially during servicing, refueling, and loading of ordnance.

---

*Optocouplers are discussed in Chapter 4 as a solution for mitigation of ground loops.

- External grounding of the aircraft is intended for protection against electrostatic discharge (ESD), electrical power fault conditions, and, to a lesser extent, against effects of lightning, and should be lower than 10 Ω for preclusion of power fault hazards and 10 kΩ for preclusion of ESD hazards.
- Due to the difference between the characteristics of electrical power and ESD energy sources, distinction must be made between power or maintenance ground (<10 Ω) and ESD ground (<10 kΩ).
- The aircraft structure typically serves as the return path for power current; thus, the on-board equipment should exhibit signal interface isolation for ensuring acceptable performance and preclusion of undesired ground loops or high immunity to interference produced by such ground loops.

10.4.3 Grounding in Spacecraft

Implementation of grounding on spacecraft, including launch* and space[†] vehicles is dictated by several considerations, unique to space systems, including (1) the diversity of systems on board the spacecraft, (2) the criticality of certain functions, (3) the multiplicity of power sources for the different systems, and (4) the presence of large masses of solid or liquid fuel as well as critical electrically initiated explosive systems on board. Unlike aircraft, which may take off and land without much ground support equipment, spacecraft rely until the very last moments before launch on complex ground support and launch facilities.[‡] Consequently, grounding in spacecraft is concerned both with the grounding interface to the launch facility** as well with as on-board grounding [3,26,40–43].

10.4.3.1 Earthing and External Grounding Connections between the Spacecraft and the Earth Facility.
Similar to aircraft, grounding and earthing of spacecraft at the launch site is governed by the objectives of lightning and ESD and ground fault current protection. The presence of a large mass of solid or liquid fuel and numerous electrically initiated explosive devices (EIEDs) used for various mission and safety-critical functions of the spacecraft (e.g., ignition, stage separation, etc.) necessitate that even more attention be paid to space systems. The spacecraft should, there-

---

*A launch vehicle is a composite of the initial stages, injection stages, space vehicle adapter, and fairing having the capability of launching and injecting a space vehicle or vehicles into orbit.

[†]A space vehicle is a complete, integrated set of subsystems and components capable of supporting an operational role in space. A space vehicle may be an orbiting vehicle, a major portion of an orbiting vehicle, or a payload that performs its mission while attached to a recoverable launch vehicle. The airborne support equipment that is peculiar to programs utilizing a recoverable launch vehicle is considered a part of the space vehicle being carried by the launch vehicle.

[‡]A space system facility is an earth-based facility that houses technical equipment for the operational support of a space system. The technical equipment may comprise of electrical, electronic, mechanical, optical, or any combination thereof.

**Grounding architecture of the space system facility itself follows the rules and principles of command, control and communication (C^3) facility grounding outlines earlier in this Chapter. See [1] and [4] for a discussion of grounding requirements in C^3 facilities.

fore, be earth grounded at the launch site through adequate bonding provisions. A suitable terminal must be provided for connection of the spacecraft to the launch facility ground network during assembly, test, and preparation for launch activities.

It is imperative that ground loops be controlled for electrical interfaces between the launch facility and the spacecraft in order to prevent any problems that may arise due to the interface between the systems. Correspondingly, it is required that the power output circuits of ground power sources intended for spacecraft supply be isolated from the spacecraft power input and control circuits and the equipment enclosure by an insulation resistance of at least 1 MΩ [40]. Similarly, all data and control interfaces between the spacecraft and the ground support equipment in the launch facility should maintain the grounding architecture for the spacecraft. In general, all interfaces between flight hardware and support equipment should be electrically isolated, for instance, by use of differential, balanced, or transformer coupling or by means of optical isolation.

10.4.3.2 Spacecraft Internal Grounding Considerations [40–43]. The typically metallic structure of the spacecraft constitutes a natural ground reference network. All spacecraft conductive and semiconductive structural members, segments, flight elements, subsystems, surfaces, and metallic enclosures of electrical and electronic assemblies will also be referenced to the spacecraft structure. Despite this fact, use of the spacecraft structure as a path for intentional electrical return current flow is prohibited, in general, except under fault conditions, as this may result in single-fault failure of the power system.

For achieving this objective, the spacecraft primary electrical power system typically utilizes a distributed single-point grounding* architecture. Each separately derived electrical power[†] source is connected electrically to the structure at no more than one point. Accordingly, when not terminated to the spacecraft chassis or structure, primary electrical power return is DC isolated[‡] from the chassis, structure, equipment conditioned power returns, and signal returns by a minimum of 1 MΩ.

Distributed single-point grounding architecture minimizes voltage potential differences between the circuits and metallic elements of the spacecraft and reduces common-mode noise at the user interfaces and radiated EMI emissions from wiring of the power distribution system. However, this could result in a single unfused power fault from the positive lead of the power system to the spacecraft chassis, which, in severe cases, may result in loss of the mission or vehicle.

In order to overcome this potential risk, it is strongly recommended that the power return leads be isolated from the spacecraft chassis by some modest impedance, sufficiently high to limit fault current but low enough to provide a stable reference.** For

---

*A distributed single-point grounding system is a system of multiple, isolated system ground points common to an isolated set of equipment (i.e., single-point ground) referenced to a common, large, conductive structure (e.g., equipotential "islands").

[†]A separately derived electrical power supply is power derived from a generator, a transformer, or a converter and with no direct electrical connection to conductors in another system.

[‡]Isolation means that little or no current flows through a structural path.

**This configuration is known as "soft grounding" and is discussed in Section 10.3.4.1 with respect to ESD control.

instance, isolation of 2 kΩ limits chassis currents to milliamps for a 28 V power system (28 V/2000 Ω = 14 mA and worst-case power loss of 390 mW). This maintains the power return close to chassis potential but prevents loss of mission. Figure 10.44 illustrates the above described principles.

If the power system is isolated from the chassis, all power inputs to the loads attached to the power bus must similarly be isolated from the chassis so that isolation is not disturbed anywhere in the system. This may result in higher power-bus common-mode noise so loads should have greater immunity to common-mode noise. Alternatively, the isolation impedance could be bypassed with capacitors to the chassis. Such modest measures are considered a tolerable side effect when balanced against the greater advantage of tolerance to high-side shorts to the chassis. If the power system is hard grounded to the chassis, it is necessary to ensure that unfused power shorts to the chassis are not credible failure modes by design (e.g., by means of double insulation). Common spacecraft power sources, such as solar arrays, batteries, and other power sources, are normally electrically DC isolated from the chassis as manufactured. In order to maintain a single-point grounding system, it is convenient to retain their isolation. Even in a multipoint grounded spacecraft, there is little need to deliberately ground the power sources.

Another consideration associated with spacecraft power grounding architecture is whether the primary power distribution system constitutes a primary power source providing a single voltage (typically 28 VDC) or all voltages (e.g., 3.3 VDC, 5 VDC, etc.) required by the user loads. A single voltage simplifies the power distribution system grounding architecture. Each load produces the secondary and tertiary* electrical power and voltages required by its internal circuits by means of embedded switch-mode power supplies (DC/DC converters). Such converters also provide isolation from the primary bus to the loads' secondary and tertiary power circuits. This system is preferred and highly recommended. The architecture shown in Figure 10.44 demonstrates this power system configuration.

The multiple voltage distribution system is more appropriate in a small spacecraft environment, for example, when all electronic subsystems are collocated in a single cage and interconnected through a common backplane. A corollary of this architecture is that secondary signal and power interfaces within the case may not be isolated from each other. This does not preclude them from being DC isolated from the chassis for fault isolation, however.

The final consideration in spacecraft grounding is associated with control and signal interface circuits. Design of these interfaces must follow the spacecraft-wide grounding architecture. If the spacecraft grounding system is totally isolated across all interfaces, then all signal circuits (control, signal and data, etc.) must be isolated as well. Note that only one end of an interface (sending or receiving) needs to be isolated. Recommended isolation impedance between signal interfaces (either a signal wire or its return) and chassis is 1 MΩ and 400 pF. For signals that share a single return wire, the requirement for the return may be relaxed on a case-by-case basis [41].

---

*Secondary electrical power is an electrical power that has been isolated from the primary electrical power before it is distributed to subsystems. Tertiary power is electrical power derived from secondary power.

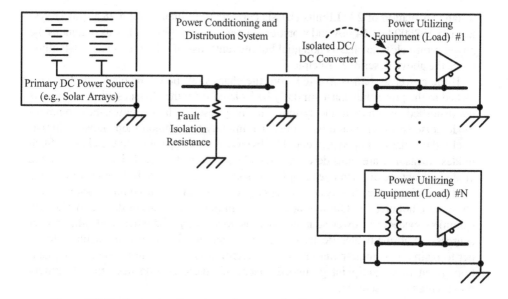

Figure 10.44. Example of spacecraft power distribution grounding architecture.

Digital and analog signal interfaces typically utilize standard differential interface driver/receiver pairs (e.g., the unidirectional RS-422 or the bidirectional MIL-STD-1553 bus circuits). Such devices are commonly designed so that zero-to-low currents flow through the spacecraft ground reference structure. For that reason, such circuits are preferred when interface isolation is required. When not energized, however, these devices may exhibit lower or no isolation and degraded common-mode rejection. Single-ended signal interfaces (such as TTL discrete signals) may utilize optocouplers or transformer coupling. Figure 10.45 illustrates an example application of a grounding scheme in spacecraft signal and control circuits.

RF signals normally process frequencies too high to permit application of single-point grounding, thus most RF circuits employ the multipoint grounding scheme. If DC isolation is essential in an RF system, frequency-selective grounding may be im-

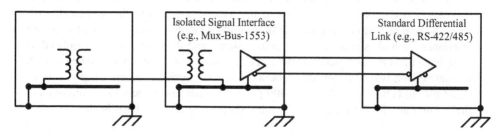

Figure 10.45. Example of spacecraft signal and control interface circuits grounding architecture.

plemented by numerous capacitors in the RF circuits. In order to maintain an isolated grounding architecture, non-RF interface circuits within RF subsystems must be isolated in a manner similar to general interface signal circuits. Figure 10.46 illustrates the application of a grounding scheme in a spacecraft RF circuit.

Pyrotechnic ("pyro") firing circuits typically require special attention, particularly in conservative designs. The greatest threat associated with grounding in such circuits is the phenomenon known as "pyro ground-fault currents." Chassis currents as high as 20 A may flow through the spacecraft chassis or structure during pyro firing events. This current is produced by a short circuit formed between the positive lead of the EIED* bridge wire through the ionized conductive path created when it is fired (this phenomenon occurs quite frequently, in approximately 25% of all firing events, and was shown to be the cause of several spacecraft anomalies and failures). In a direct-energy-transfer system (pyrotechnic devices switched directly from the main battery bus) that is not isolated by a deliberate turn-off switch, this ground fault current could continue indefinitely. The ground fault current could also result in momentary near total power loss as well as magnetic field interference, coupling into adjacent sensitive circuits. Figure 10.47 illustrates the application of a grounding scheme in a spacecraft pyrotechnic firing circuit.

In order to preclude adverse electromagnetic effects and prevent pyro ground-fault currents, pyro firing circuits should be electrically isolated from other electrical circuits as well as from each other by no less than 1 MΩ, typically. Return leads of each firing circuit should, therefore, be routed back to the firing power source and isolated anywhere elsewhere from the structural ground by a minimum impedance of 20 kΩ. One way to achieve the desired isolation (if the primary power bus is not already isolated) is by means of a switch-mode power supply, isolating the pyro firing circuits from both the primary DC power bus and from the spacecraft chassis. Alternatively, a dedicated battery for the pyro circuits could be utilized. Isolation of the pyro firing circuit may not be necessary if all critical interface circuits are guaranteed to incorporate sufficient intrinsic immunity to EMI, if some momentary power loss can be endured, or if the functional circuitry is software-tolerant to transient-type interference.

For safeguarding against adverse effects of ESD, the firing circuit should be balanced and referenced to the spacecraft structural ground by a resistance in the range of 20 kΩ and 1 MΩ ("soft grounding").[†] This grounding scheme precludes electrostatic charge buildup across the EIED bridge wire and provides a path for "bleeding" stray electrostatic charge. In addition, in a "safe" condition, EIED leads must be shorted together to prohibit pickup of stray RF fields by the firing circuit [42].

In addition to the standard schemes discussed above, a need may arise for special grounding applications for reasons associated with specific instrumentation and payloads. Grounding needs depend on the individual sensitivity and performance requirements of each such system and are often so unique that it is difficult to generalize the recommended grounding approach. All such special cases must be handled on an indi-

---

*Also known as a bridge-wire actuated device (BWAD).

[†]Although most standards and design guides recommend that soft grounding be accomplished through resistance levels on the order of 100 kΩ, experience has revealed that such high values preclude sufficiently fast response to 25 kV step charge buildup, for instance. Resistances in the order of 100 Ω to 1 kΩ were empirically found to be much more effective [29].

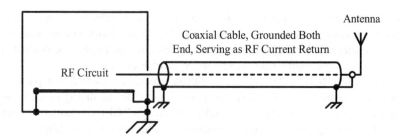

Figure 10.46. Example of spacecraft RF circuit grounding architecture.

vidual basis, and a well-thought-out grounding scheme of the spacecraft electrical and electronic systems must be devised at the outset of the particular program.

Grounding within individual spacecraft electrical or electronic equipment should be accomplished as appropriate for each circuit type and its electrical characteristics, as long as isolation of external power and signal return interfaces are maintained. In order to prevent violation of spacecraft-wide isolation, the topology of equipment power supply must be carefully laid out. Secondary and tertiary electrical-power-supplying circuits within electrical and electronic equipment should be DC isolated from equipment chassis and the spacecraft structure except at no more than one electrically conductive common point, thus creating a secondary single-point grounding scheme. Except at that single chassis connection, an isolation impedance of 1 MΩ minimum should be maintained in the secondary electrical power circuits when all grounds are not terminated to chassis or structure. The same applies to power distribution schemes in which the isolated secondary power from a "master" equipment unit is further used to supply an external "slave" load.

Lessons Learned

- Spacecraft constitute a unique type of airborne vehicle, due to the increased risk associated with space missions, the presence of a large mass of solid or liquid fuel, and numerous susceptible mission- and safety-critical explosive devices.

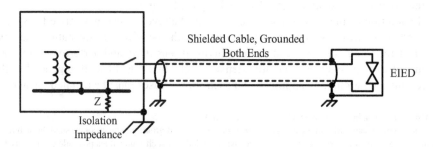

Figure 10.47. Example of spacecraft pyrotechnic firing circuit grounding architecture.

- Mitigation of hazards from ground power fault currents and ESD constitute the primary objectives of spacecraft grounding systems.
- Contrary to aircraft, use of the spacecraft structure as a path for intentional electrical return current flow is prohibited, except under fault conditions, suggesting that single-point grounding (SPG) or distributed SPG in large spacecraft vehicles and high levels of isolation elsewhere be employed.
- Special consideration should be paid to the primary power distribution system grounding and power bus fault isolation, signal and control interface isolation, and individual equipment grounding schemes, with particular attention paid to pyrotechnic (pyro) firing circuits.
- Unique grounding schemes may be necessary for specific instrumentation or payloads, to be applied on an individual basis.

10.4.4 Grounding in Ships

The increased use of electrical and electronic equipment aboard naval ships introduces electromagnetic interference (EMI) problems to ship operation and performance. In particular, the topside electromagnetic environment of a modern navy ship with its many transmitters and antennae is an area of concentrated RF fields (Figure 10.48). As systems are added, they all contribute and become susceptible to this intense electromagnetic environment (EME). Considering the corrosive saltwater environment in which ships must operate and the interaction of a ship's electrically conductive metallic superstructure, topside hardware, antenna systems, and so on, the potential for interoperability problems is significantly increased. Potential EMI and personnel safety problems related to electronic equipment operating in these environments are magni-

Figure 10.48. Topside antenna arrangement of a modern Navy ship. (Photograph courtesy of Richard T. Ford.)

fied as a result of (1) the need to establish and maintain a low-impedance, common-reference ground for all electrical/electronic equipment, and (2) the detrimental effects of (a) natural and manmade, (b) spurious and intentional, and (c) off-ship and on-ship electromagnetic (EM) energy.

In addition to the dense electromagnetic environment, one must consider another factor. A combat-ready ship is heavily loaded with fuel and ammunition, both of which are susceptible to ESD and EMI under proper conditions. The USS Forrestal incident serves as an excellent example of the potential devastating effects of a shipboard electromagnetic environment on safety (Figure 10.49).

On July 29, 1967, the U.S. aircraft carrier USS Forrestal cruised off the coast of North Vietnam, conducting combat operations. At 10:52 am, electromagnetic fields produced by radar penetrated an improperly mounted shielded connector, producing RF voltages in the arming mechanism of a Zuni rocket attached to the wing of an F-4 Phantom jet aircraft, causing it to be accidentally fired. The rocket streaked across the deck into a fully fueled A-4 Skyhawk aircraft, loaded with two 1000 lb bombs and air-to-ground and air to air missiles. The impact caused the belly fuel tank to break open, spilling JP5 jet fuel onto the flight deck and the two 1000 pound bombs on the Skyhawk to fall off and explode. The subsequent massive chain reaction of explosions engulfed half the aircraft on deck. Wingtip to wingtip, the aircraft burned and the bombs exploded, blowing huge holes in the steel flight deck. Fuel and bombs spilled into the holes in the flight deck, igniting fires further into the bowels of the ship. Before the fire was extinguished, 134 pilots and support personnel were trapped and burned alive [44].

The need to provide for shipboard EM compatibility (EMC) and safe operation dictates that adequate grounding and bonding practices be applied, whether on metallic or nonmetallic hull ships. In particular, six major items must be addressed: (1) shipboard ground reference structure, (2) hull-generated EMI, (3) grounding at ship hull penetrations, (4) hull (structure) power current return scheme, (5) signal grounding scheme, and (6) unique grounding architecture in nonmetallic hull ships. These are discussed in the next paragraphs [45,46].

10.4.4.1 *Ground Reference Structure.*

On both metallic and nonmetallic hull ships, an adequate ground potential must be established and maintained, as required for electrical and electronic equipment operation, EMC, and electrical safety grounding. On metallic hull ships, the metal hull, when in contact with the water, is utilized to establish and be designated as the ship's ground reference structure. Any metallic item, topside and below deck, such as metallic superstructure, shielded enclosures, equipment foundations and racks, stranded or solid conductor cable (e.g., a separate computer ground system), bonded to the ship's hull by welding or brazing* are, therefore, considered extensions of the ship's ground reference structure. On nonmetallic hull ships, such as minesweepers, an equipotential ground reference structure

---

*According to MIL-STD-1310G terminology, this is designated a class A bond. Hardware that is class B (through bolting or clamping) or class C (through bridging by a metallic bond strap) bonded to the ship's ground plane can be designated as grounded, but not as an element of the ground plane for grounding other items.

(a) USS Forrestal

(b) The Fire on USS Forrestal,
July 29, 1967

(c) USS Forrestal Antenna Topside
Arrangement

Figure 10.49. USS Forrestal incident. (Source: http:\\www.navysite.de\cvn\cv59.)

should be established. Ground plate(s) are installed to provide an earth ground connection via contact with seawater. The ground plate(s) are installed at the lowest point of the structural hull, as close as possible to the vertical of the mast. Any cable grounding system's branches and branch extensions connected to or terminating at the ground plate(s) of a nonmetallic hull ship's ground plane are also considered part of the ship's ground reference structure.

10.4.4.2 Hull-Generated EMI. One of the most significant considerations in shipboard grounding for the purpose of hull-generated EMI control is that associated with the marine corrosive atmosphere. For preclusion of harmful EMI produced by nonlinear or intermittent metal-to-metal contact junctions in the topside EME, the ship's topside areas must form, as much as possible, a single conducting surface, free of metallic discontinuities that might act as source(s) of intermodulation interference (IMI) and broadband noise (BBN). In particular, joining of dissimilar metals by bolting or riveting should be avoided in topside areas.* Hull-generated EMI may be precluded through (1) use of nonmetallic or insulation materials, (2) application of appropriate topside stowage practices, and (3) proper implementation of grounding and bonding.

Insulation of topside metal-to-metal contacts is encouraged (Figure 10.50), unless isolating the item from ground may result in its becoming an RF burn hazard. Insulating material should be placed on the contact surfaces of metallic hardware located in the topside EME whenever a suitable nonmetallic substitute is not available and insulation is preferable, or the item is not suited to bonding. Whenever metallic hardware in the topside EME is identified as a source of hull-generated EMI that cannot be resolved using any of the above described methods, metal-to-metal bonding of the hardware item to the ship's hull must be applied to eliminate IMI/BBN.

As an example, the lifelines illustrated in Figure 10.51 must be isolated or grounded using bonding straps in order to prevent IMI. Notice the isolators between the lifelines and the stanchion, and the bonding straps provided for assuring a solid connection to the ship (some of the bonding straps were not yet connected at the time of photography). If not connected in this manner, movement of the lifeline would constitute a source of IMI and BBN.

10.4.4.3 Grounding at Ship Hull Penetrations. Hull penetrations, such as cables, conduits, waveguides, and pipes traversing the hull (or shielded compartments) can result in transfer of harmful electromagnetic energy present above the ship's deck to below-deck systems (hull penetration EMI). Preclusion of topside EME energy intrusion through the skin of the ship into below-deck volume and equipment mandates that penetrations across boundaries established by the hull and shielded compartments be carefully controlled through application of grounding practices at the point of entry.

In the case of shielded cables and metallic conduits, waveguides, and pipes, their outer perimeter must be grounded to hull ground, providing a 360° peripheral bond at the penetration point. Unshielded cables routed from topside areas into the ship hull should be controlled by means of filters and terminal protection devices installed at the penetration point. Figure 10.52 illustrates, for example, the implementation of grounding and bonding practices and cable hull penetrations through wireways [45].

10.4.4.4 Hull (Structure) Power Current Return Scheme. Hull currents in ships can create problems in sensitive equipment. Very sensitive low-frequency ra-

---

*Metal electrochemical compatibility and consideration of bonding between dissimilar metals are discussed in Chapter 5.

Figure 10.50. Insulation of a topside lifeline where bonding is not mandatory. (Photograph courtesy of Richard T. Ford.)

dio and sonar receivers are commonly used on board naval surface ships and submarines. At low frequencies, currents flowing through the hull and across surfaces of electronic enclosures will penetrate into the ship's hull. The magnetic fields created by these currents can couple into critical circuits, degrading performance of electronic equipment, upsetting ground detectors, and counteracting degaussing. Extensive investigations were conducted to find solutions for this problem. Subsequently, the ship's electric power systems are ungrounded, so that power return currents are never al-

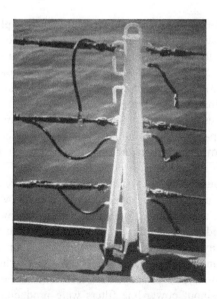

Figure 10.51. Isolators installed across lifelines, stanchion, and bonding straps to prevent IMI (some bonding straps not yet connected). (Photograph courtesy of Richard T. Ford.)

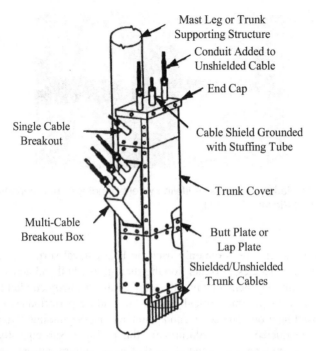

Figure 10.52. Cable penetrations at wireway trunk.

lowed to flow through the metallic hull of naval and marine platforms. Power mains circuits are always isolated from the ship hull. The rationale for this power return isolation is manifold, including (1) preventing damage to the ship's hull due to dissimilar metal anodic corrosive effects, (2) reduction of the ship's magnetic signature (particularly in submarines), (3) enhancement of personnel safety, and (4) prevention of hazards of power fault current flowing through the ship's hull.

At higher frequencies (i.e., greater than 100 kHz, approximately) the combination of power-line filter capacitance-to-ground limitation, skin effect of equipment enclosures, and reduced harmonic currents tend to minimize the problems associated with structure currents.

Even with an ungrounded power system, hull currents were found to be caused by leakage through large line-to-chassis capacitors included in EMI filters, feeding power line harmonics into the structure. Such filters establish low-impedance paths for structure (common-mode) currents through the ship's metallic hull. Leakage currents as large as 20 A may exist in an electric power system as a result of capacitive coupling between cables and equipment filters connected to the ship's hull. Tests carried out on the 400 Hz power line of the USS Guitarro (SSN-665) submarine demonstrated that capacitive-input power-line filters were producing large structural currents through the line-to-ground (common-mode) capacitors of approximately 135 μF per phase. The switch-mode power supplies constituted major offenders, not primarily from harmonics of their switching frequency but, rather, due to the rectification harmonics of the power frequency flowing through the capacitive input filters [47,48].

As a result, user equipment in navy ships and submarines is required to be ungrounded as related to the mains power distribution system ground. In order to minimize leakage current through the input filter capacitors, use of line-to-chassis (common-mode) capacitive filters for EMI control should be minimized. If such filters must be employed, the line-to-ground capacitance for each line is limited by applicable standards to sufficiently small values, such as 0.1 μF or 20 nF for surface ships utilizing 60 Hz and 400 Hz AC powered equipment, respectively, whereas for submarine DC-powered equipment, the filter capacitance from each line to ground at the user interface is limited to 75 nF/kW of connected load (for DC loads smaller than 0.5 kW, the filter capacitance is limited to 30 nF) [49]. If leakage current exceeds 30 mA, in cases where electrical grounding, solidly or capacitively, is required, isolation is mandated between the user equipment and the grounding system. In such cases, the neutral lead of the power system must not be grounded at the user equipment [47].

When floating and isolated AC/DC or DC/DC switch-mode power converters are used in user equipment circuits for producing secondary voltages (e.g., 5 V, 3.3 V, etc.), the return lead of the secondary circuits may be grounded at a single point to the ship's hull for the purpose of signal reference. The grounding architecture of circuits powered from isolated secondary power sources should follow the grounding scheme of the particular circuits.

10.4.4.5 Shipboard Signal Return Grounding Scheme [50]. Shipboard digital computers and their peripheral components utilize low signal level DC currents in interconnecting circuits. Differences in ground potential between equipment units directly affect a circuit's ability to discriminate between logical signals and may increase bit error rates in data transmissions to unacceptable levels. In equipment assemblies where the internal signal ground is the common-to-cabinet ground, the internal signal ground should be connected to the grounding point on the equipment or cabinet. The branch ground cable should connect to this point on the branch cable (see below). Figure 10.53 illustrates the basic elements of a typical digital equipment grounding system.

In order to achieve enhanced system operational reliability, a separate computer grounding system is typically allocated. In this fashion, isolation of major digital systems' power supplies from other electrically powered systems onboard the ship is achieved, ensuring compatibility between the ship's general grounding systems and digital computer systems/equipment [50]. Consequently, interface characteristics of the ships' digital computer systems and equipment impose certain constraints on the design and installation of the ship's grounding systems.

Digital computer systems and their peripherals are often installed on multiple equipment racks or cabinets. Two or more adjacent equipment cabinets that have interconnected circuits requiring common signal ground potential levels should be consid-

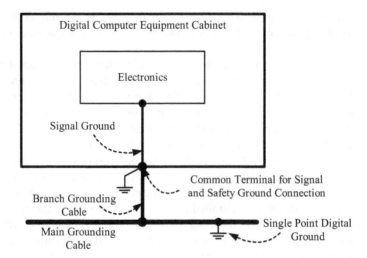

Figure 10.53. Basic elements of a typical shipboard digital equipment grounding system.

ered and treated as a single equipment assembly. The signal ground system of such equipment is, therefore, connected to the branch ground cable at one common point only (see Figure 10.54). Multiple equipment common ground requirements may apply, for instance, to central processors, data processing peripheral equipment, and similar devices.

A common signal reference for all digital equipment is established through the installation of a signal grounding system in all major digital equipment spaces. The grounding system consists of a main ground cable and branch ground cables interconnecting all units of a computer complex. The grounding system further extends to and includes the first digital-to-analog peripheral equipment, including remotely located equipment requiring a tight ground reference to properly interface with a computer complex. The grounding system should not normally extend, however, to single remotely located ancillary equipment such as signal monitors, indicators, displays, and similar devices (electrical, mechanical, or pneumatic), which do not require a tight reference ground for proper operation; motors, generators, and similar equipment should not connect to this grounding system. Figure 10.55 illustrates the manner of installation of the digital equipment ground system, cabinet grounding, and signal grounding.

A main ground-type SSGU-1000 cable is used to interconnect all major digital equipment spaces and is bonded to the ship's hull at one point only. Except for this single-point attachment, the main ground cable is otherwise electrically insulated from the ship's metallic hull or structure. The single-point ground connection to the main ground cable should be located approximately central to all equipment included in the digital system and should be accomplished by installing a short section of type SSGU-1000 cable between the main ground cable and hull structure (Figure 10.55). Care

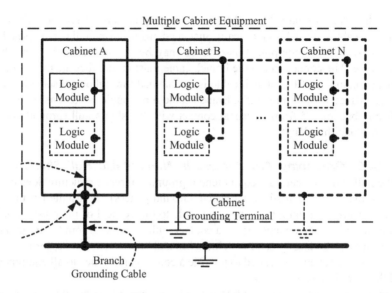

Figure 10.54. Signal grounding schemes in multiple cabinet equipment. (Note: For signal/cabinet grounding details see Figure 10.53.)

should be taken to ensure a low-impedance metal-to-metal bond at this point. Routing of the main ground cable should be such that the lengths of branch ground cables are minimized.

Connection between the signal grounds of the digital equipment to the main ground cable is accomplished through branch ground cables. Branch cables are made with a cross section area of 20 mm^2 (40,000 circular mils) or a cross-section area in square millimeters determined by the cable length in meters multiplied by 0.85 (in circular mils, cable length in feet multiplied by 500), whichever is the larger. Except for attach-

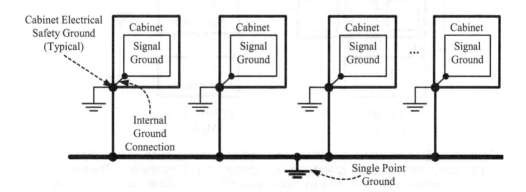

Figure 10.55. Digital equipment grounding scheme.

ment to the main ground cable and to a common signal/cabinet ground (when used), the branch cables must be electrically insulated from the metallic hull (see Figure 10.53). Two or more equipment assemblies may be connected to the same branch ground cable. The connection to the branch cable must assure that the mechanical and electrical continuity of the branch cable is maintained throughout. Installation should be in accordance with Figure 10.56. Branch cable runs should be kept as short as practicable and be isolated from electrical contact with the ship's hull (ground potential) anywhere along their length.

10.4.4.6 Grounding Architecture in Nonmetallic Hull Ships.

On nonmetallic hull ships, a "natural" equipotential ground reference structure does not exist and must, therefore, be artificially created. Ground plate(s) are installed at the lowest point of the structural hull, as close as possible to the vertical of the mast, in order to provide an earth ground connection via contact with seawater. From the ground plates stems a cable grounding system, with branches and branch extensions, forming a multitree grounding scheme, intended to provide a common reference to all equipment onboard the ship (Figure 10.57) [45].

Separate branches of the multitree ground reference plane are directly connected from the ground plate(s) to (1) all shielded spaces, (2) the metallic mast, (3) the ship's service switchboards, and (4) the ship's service and emergency generator sets.

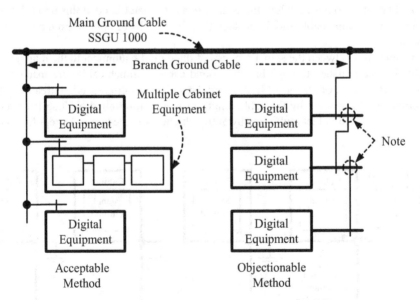

Figure 10.56. Grounding of digital equipment to branch ground cables. (Note: This connection method is not acceptable for shipboard digital installations. The mechanical connection is prone to long-term corrosion effects and deterioration, which could result in degraded performance. Each branch should be mechanically and electrically continuous along its entire run.)

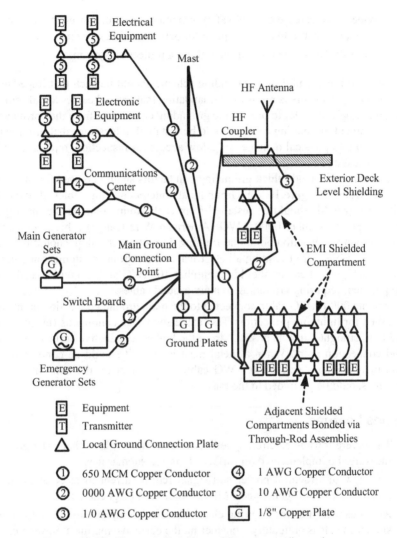

Figure 10.57. Nonmetallic ship multitree grounding scheme.

In particular, separate branches of the cable ground plane may also be required to:

- Provide common ground reference for operation of electrical and electronic equipment located outside shielded spaces as well as for bonding of metallic items located in the topside EME, when required for control of hull-generated EMI

- Directly connect equipment utilizing electrical power inside shielded spaces in the event that the short-circuit current capacity of the enclosure and the primary branch, trunks, and ground plate connection points thereto could be exceeded

- Provide a dedicated electrical safety ground connection for electrical and electronic equipment to the ground plate for personnel shock hazard protection
- Serve as down-conductors for the lightning protection system

Implementation of the multitree grounding scheme should be such that impedance of the grounding conductors is minimized, mandating that ground cables are of minimum achievable length. In order to protect the grounding conductors from the corrosive marine environment, grounding cables should be installed in locations that provide minimum exposure to physical damage and allow access for inspection, repair, or replacement, as necessary.

In addition, sufficiently thick grounding conductors should be used for minimizing impedance (refer to Figure 10.57). Ground plate interconnecting cables of 650 million circular mils (MCM) should be installed from each ground plate to the main ground connection point. Number 0000 AWG (American Wire Gauge) cables, connected to the main ground connection point and running throughout the ship, should be used as main ground cables for grounding all electrical and electronic equipment and metallic items. Branch ground cables should be number 10 AWG and should be used to connect equipment requiring grounding to main ground cables.

A separate 0000 AWG cable connected to the main ground connection point should be installed for lightning protection. This cable should be continuous (unspliced) and routed in as straight a line as possible from the highest conductive surface to the main ground connection point. When a metal mast is used, it should be connected to the ground plates using a size 0000 AWG cable. Equipment on the mast that requires grounding should be grounded to the mast.

Lessons Learned

- The severe topside electromagnetic environment mandates that strict grounding measures be implemented on surface ships and submarines.
- The flow of structural power and signal return current must be avoided, as it could result in hull-generated EMI and undesired magnetic fields.
- Rigorous measures for ensuring electrochemical compatibility when mating dissimilar metals is obligatory, considering the corrosive marine atmosphere.

BIBLIOGRAPHY

[1] MIL-HDBK-419A, *Military Handbook, Grounding, Bonding, and Shielding for Electronic Equipments and facilities,* December 1987.

[2] MIL-HDBK-1857, *Department of Defense Handbook, Grounding, Bonding and Shielding Design Practices,* March 1998.

[3] MIL-STD-1542B (USAF), *Military Standard, Electromagnetic Compatibility and Grounding Requirements for Space System Facilities,* November 1991.

[4] MIL-STD-188-124B, *Military Standard, Grounding, Bonding and Shielding for Common Long Haul/Tactical Communication Systems, Including Ground Based Communications-Electronics Facilities and Equipment,* February 1992.

[5] IEEE STD 1100, *IEEE Recommended Practice for Powering and Grounding Electronic Equipment (IEEE Emerald Book),* New York, IEEE Press, 2005.

[6] IEC 60364-5-548, *Electrical Installations of Buildings—Part 5: Selection and Erection of Electrical Equipment, Section 548: Earthing Arrangements and Equipotential Bonding for Information Technology Installations,* IEC, 1999.

[7] ITU-T Recommendation K.27, *Bonding Configurations and Earthing Inside a Telecommunication Building,* International Telecommunication Union, 1996.

[8] BS EN 50310:2000, *Application of Equipotential Bonding and Earthing in Buildings with Information Technology Equipment,* British Standards Institution, 2000.

[9] FAA-STD-019d, Department of Transport Federal Aviation Administration Standard, *Lightning and Surge Protection, Grounding, Bonding and Shielding Requirements for Facilities and Electronic Equipment,* 2002.

[10] IEC 60364-5-548, *Electrical Installations of Buildings—Part 5: Selection and Erection of Electrical Equipment, Section 548: Earthing Arrangements and Equipotential Bonding for Information Technology Installations,* IEC, 1999.

[11] EN 50174-2, *Information Technology—Cabling Installation—Part 2: Installation Planning and Practices Inside Buildings,* August 2000.

[12] ANSI T1.333-2001, *Grounding and Bonding of Telecommunications Equipment,* ANSI, 2001.

[13] ANSI T1.313-1997, *Electrical Protection for Telecommunications Central Offices and Similar Type Facilities,* ANSI, 1997.

[14] ANSI T1.334-2002, *Electrical Protection of Communications Towers and Associated Structures,* ANSI, 2002.

[15] IEEE STD 142-1991, *Grounding of Industrial and Commercial Power Systems,* IEEE, 1991.

[16] NFPA 70, *National Electrical Code 2005,* National Fire Protection Association, 2005.

[17] IEC 61000-5-2, *Electromagnetic Compatibility (EMC)—Part 5: Installation and Mitigation Guidelines—Section 2: Earthing and Cabling,* IEC, 1997.

[18] IEC 61312-2, *Protection against Lightning Electromagnetic Impulse—Part 2: Shielding of Structures, Bonding Inside Structure and Earthing,* IEC, 1999.

[19] Roger R. Block, *The "Grounds" for Lightning and EMP Protection,* 2nd ed., PolyPhaser Corporation, 1993.

[20] MIL-HDBK-232A, *Military Handbook, Red/Black Engineering, Installation Guidelines,* March 1987.

[21] MIL-HDBK-411B, *Military Handbook, Power and the Environment for Sensitive DoD Electronic Equipment (General),* May 1990.

[22] MIL-STD-188-125-1, *Department of Defense Interface Standard, High-Altitude Electromagnetic Pulse (HEMP) Protection for Ground-Based C4I Facilities Performing Critical Time-Urgent Missions, Part 1: Fixed Facilities,* July 1998.

[23] Air Force Qualification Training Package 2EXXX-202D, *EI TEMPEST Installation Handbook,* 1 October 1999.

[24] IEEE STD 1050:2004, *Guide for Instrumentation and Control Equipment Grounding in Generating Stations,* 2004.

[25] IEEE STD 80-2000, *Guide for Safety in AC Substation Grounding.*

[26] MIL-STD-464A, *Department of Defense Interface Standard, Electromagnetic Environmental Effects Requirements for Systems,* December 2002.

[27] Netzer, M., *Electrostatic Electricity* (in Hebrew), The Israel Institute for Occupational Safety and Hygiene (IIOSH), Israel, August, 1994.

[28] MIL-HDBK-263B, *Military Handbook, Electrostatic Discharge Control Handbook for Protection of Electrical and Electronic Parts, Assemblies and Equipment (Excluding Electrically Initiated Explosive Devices),* July 1994.

[29] Netzer, M., RAFAEL/ADA (Unclassified) Report 84-8750-910 (Hebrew), *A Design Guide for Hardening of Ordnance Initiation Circuits for Preclusion of Hazards from Electromagnetic Energy and Electrostatic Discharge,* Israel, February, 1993.

[30] ATIS Standard ATIS-0600321.2005, *Electrical Protection for Network Operator-Type Equipment Positions,* ATIS, December, 2005.

[31] ANSI/ESD Association Standard, ESD S6.1-1999, *Grounding—Recommended Practice,* Rome, NY, 1999.

[32] ESD Association Advisory Handbook 2.0-1994, *Grounding,* Rome, NY, 1994.

[33] Air Force Air Combat Command (ACC) Instruction 33-165, *Communications and Information Grounding Techniques,* September 1998.

[34] Pacific Air Forces PACAF Instruction 33-103, *Communications and Information Deployable Communications Standards—Safety,* April 1999.

[35] MIL-STD-188-125-2, *Department of Defense Interface Standard, High-Altitude Electromagnetic Pulse (HEMP) Protection for Ground-Based C⁴I Facilities Performing Critical Time-Urgent Missions, Part 2: Transportable System,* July 1998.

[36] MIL-HDBK-274, Notice 1, *Military Handbook, Electrical Grounding for Aircraft Safety,* June 1990.

[37] SAE Aerospace Recommended Practice, ARP 1870, *Aerospace Systems Electrical Bonding and Grounding for Electromagnetic Compatibility and Safety,* Warrendale, PA, 1987 (Reaffirmed 1999).

[38] SAE Aerospace Recommended Practice, ARP 4043, *Flight Line Grounding and Bonding of Aircraft,* Warrendale, PA, 1986 (Reaffirmed 1999).

[39] AFSC Design Handbook DH 1-4, *Electromagnetic Compatibility,* 4th ed., Rev. 1, January 1991.

[40] MIL-STD-1541A (USAF), *Military Standard, Electromagnetic Compatibility Requirements for Space Systems,* December 1987.

[41] NASA-HDBK-4001, *NASA Technical Handbook, Electrical Grounding Architecture for Unmanned Spacecraft,* February, 1998.

[42] International Space Station Standard SSP 30240, *Space Station Grounding Requirements,* Revision, C, December, 1998.

[43] International Space Station Standard SSP 30245, *Space Station Electrical Bonding Requirements,* Revision, E, October, 1991.

[44] *USS Forrestal (CV 59),* URL: http:\\www.navysite.de\cvn\cv59.

[45] MIL-STD-1310G (Navy), *Department of Defense, Standard Practice for Shipboard Bonding, Grounding, and Other Techniques for Electromagnetic Compatibility and Safety,* June 1996.

[46] Preston, E. L. Jr., *Shipboard Electromagnetics,* Norwood, MA: Artech House, 1987.

[47] MIL-STD-1399-300A (Navy), Notice 1, *Military Standard, Interface Standard for Shipboard Systems, Section 300A, Electric Power, Alternating Current (Metric),* March 1992.

[48] MIL-HDBK-241B, *Military Handbook, Design Guide for Electromagnetic Interference (EMI) Reduction in Power Supplies,* September 1983.

[49] MIL-STD-461E, *Department of Defense Interface Standard, Requirements for the Control of Electromagnetic Interference Characteristics of Subsystems and Equipment,* August 1999.

[50] MIL-STD-1399-406B (Navy), *Military Standard, Interface Standard for Shipboard Systems, Section 406B, Digital Computer Grounding,* July 1995.

GLOSSARY OF GROUNDING-RELATED TERMS AND DEFINITIONS

Balanced Line. A line or circuit using two conductors instead of one conductor and ground (common conductor). The two sides of the line are symmetrical with respect to ground. Line potentials to ground and line currents are equal but of opposite phase at corresponding points along the line. [MIL-HDBK-419A]*

Balanced Termination. A load presenting the same impedance to ground for each of the output terminals. [IEEE-100: 146-1980]

Balanced Three-Wire System. A three-wire system in which no current flows in the conductor connected to the neutral point of the supply. [IEEE-100: 1999]

Balanced Voltages. The voltages between corresponding points of a balanced circuit (voltages at a transverse plane) and the reference plane relative to which the circuit is balanced. [IEEE-100: 1999]

Balanced Wire Circuit. A circuit whose two sides are electrically alike and symmetrical with respect to ground and other conductors. The term is commonly used to indicate a circuit whose two sides differ only by chance. [IEEE-100: 599-1985]

Balun. (1) In networking, a passive device with distributed electrical constants, used to couple a balanced system or device to an unbalanced system or device. For example, a transformer used to connect balanced twisted-pair cables to unbalanced coaxial cables. *Note:* Derived from "balanced-to-unbalanced" transformer [IEEE-

*Appropriate standards are shown in brackets. See Appendix D.

100: 610.7-1995]. (2) A network for the transformation from an unbalanced transmission line or system to a balanced line or system, or vice versa [IEEE-100: 1999].

Black. As it pertains to C-E equipment, transmission lines, and associated wiring, "black" signifies both physical and electrical areas in which data/voice signals are encrypted or unclassified and, therefore, relatively safe from compromise. [EI TEMPEST Installation Handbook 2EXXX-202D]

Black Equipment. Equipment processing unclassified information or encrypted information. [MIL-HDBK-232A]

Black Equipment Area (BEA). An area in a limited exclusion are designated for the installation of equipment processing unclassified information or encrypted information. [MIL-HDBK-232A]

Bond. (1) The electrical connection between two metallic surfaces established to provide a low resistance path between them [MIL-HDBK-419A]. (2) Any fixed union existing between two objects that results in electrical conductivity between the objects. Such union occurs either from physical contact between conductive surfaces of the objects, or from the addition of a firm electrical connection between the objects. [SSP 30240 Revision C].

Bond, Direct. An electrical connection utilizing continuous metal-to-metal contact between the members being joined. [MIL-HDBK-419A]

Bond, Indirect. An electrical connection employing an intermediate electrical conductor or jumper between the bonded members. [MIL-HDBK-419A]

Bonding. (1) The electrical interconnecting of conductive parts, designed to maintain a common electrical potential. (2) The permanent joining of metallic parts to form an electrically conductive path that will assure electrical continuity and the capacity to conduct safely any current likely to be imposed [IEEE-100: 1100-1999]. (3) The process of establishing the required degree of electrical continuity between the conductive surfaces of members to be joined. [MIL-HDBK-419A].

Bonding, Equipotential. Electrical connection maintaining various exposed conductive parts and extraneous conductive parts at substantially the same potential. [BS 7176]

Bonding Network (BN). (1) Network interconnecting all conductive parts of the structure and of internal systems (live conductors excluded) to the earth-termination system [IEC 62305-4 (FDIS)]. (2) A set of interconnected conductive structures that provide an electromagnetic shield for electronic systems and personnel at frequencies from DC to low RF. The term "electromagnetic shield" denotes any structure used to divert, block, or impede electromagnetic energy. In general, a BN need not be connected to earth, but all BNs in this recommendation have an earth connection [ITU-T K.27].

Bond, Permanent. A bond not expected to require disassembly for operational or maintenance purposes. [MIL-HDBK-419A]

Bond, Semipermanent. Bonds expected to require periodic disassembly for maintenance or system modification, and that can be reassembled to continue to provide a low-resistance interconnection. [MIL-HDBK-419A]

Chassis or Structure Ground. Conducting connection, whether intentional or accidental, between an electrical circuit or equipment and the structure ground. [SSP 30240 Revision C]

Circuit Reference. The collection of bonded circuit and conductive structural elements. [SSP 30240 Revision C]

Circuit Reference/Common Connection (Power or Signal). A return path or physical tie point usually at a potential of zero volts. [SSP 30240 Revision C]

Common Bonding Network (CBN). The CBN is the primary way to create effective bonding and earthing inside a telecommunications building. It is the set of metallic components that are intentionally or unintentionally interconnected to form the principal BN in a building. These components include structural steel or reinforcing rods, metallic plumbing, AC power conduit, PE conductors, cable racks, and bonding conductors. The CBN always has a mesh topology and is connected to the earthing network. [ITU-T K.27]

Common DC Return (DC-C). A DC power system in which the return conductor is connected to the surrounding BN at many locations. This BN could be either a mesh-BN (resulting in a DC-C-MBN system) or an IBN (resulting in a DC-C-IBN system). More complex configurations are possible. [ITU-T K.27]

Common Mode. The instantaneous algebraic average of two signals applied to a balanced circuit, both signals referred to a common reference. [IEEE-100-1999]

Common-Mode Conversion. The process by which differential-mode interference is produced in a signal circuit by a common-mode interference applied to the circuit. [ANSI C37.100-1992]

Common-Mode Interference. (1) Interference that appears between both signal leads and a common reference plane (ground) and causes the potential of both sides of the transmission path to be changed simultaneously and by the same amount relative to the common reference plane (ground) [ANSI C37.13-1991]. (2) A form of interference that appears between any measuring circuit terminals and ground [ANSI C37.1-1994].

Common-Mode Noise (Longitudinal). The noise voltage that appears equally and in phase from each current-carrying conductor to ground. [IEEE-1100-1999]

Common-Mode Radio Noise. Conducted radio-noise voltage that appears between a common reference plane (ground) and all wires of a transmission line, causing their potentials to be changed simultaneously and by the same amount relative to the common reference plane (ground). [IEEE-1100: C63.4-1991]

Common-Mode Rejection (In-Phase Rejection). The ability of certain amplifiers to cancel a common-mode signal while responding to an out-of-phase signal. [IEEE-1100-1999]

Common-Mode Rejection Quotient (In-Phase Rejection Quotient). The quotient obtained by dividing the response to a signal applied differentially by the response to the same signal applied in common mode, or the relative magnitude of a common-mode signal that produces the same differential response as a standard differential input signal. [IEEE-1100-1999]

Common-Mode Rejection Ratio (CMRR). The ratio of the common-mode interference voltage at the input of a circuit to the corresponding interference voltage at the output. [ATIS Committee T1A1]

Common-Mode Signal. The average value of the signals at the positive and negative inputs of a differential input waveform recorder. If the signal at the positive input is designated V_+ and the signal at the negative input is designated V_-, then the common mode voltage (V_{CM}) is

$$V_{CM} = \frac{V_+ + V_-}{2}$$

[IEEE-1100: 1057-1999]

Common-Mode to Normal-Mode Conversion. In addition to the common-mode voltages that are developed in single conductors by the general environmental sources of electrostatic and electromagnetic fields, differences in voltages exist between different ground points in a facility due to the flow of ground currents. These voltage differences are considered common mode when connection is made to them either intentionally or accidentally, and the currents they produce are common mode. These common-mode currents can develop normal-mode noise voltages across unequal circuit impedances. [IEEE-1100: 422-1977, 525-1992]

Common-Mode Voltage. (1) The voltage that, at a given location, appears equally and in phase from each signal conductor to ground [IEEE-1100: C37.90.1-1989]. (2) The instantaneous algebraic average of two signals applied to a balanced circuit, with both signals referenced to a common reference [IEEE-1100: 802.3-1999].

Common Return. A return conductor common to several circuits. [IEEE-100: 1999]

Compromising Emanations (CE). Unintentional, intelligence-bearing signals which, if intercepted and analyzed, disclose classified information transmitted, received, handled, or otherwise processed by any information-processing equipment. [EI TEMPEST Installation Handbook 2EXXX-202D]

Conductive Material. For the purpose of ESD protection, material with the following characteristics. (1) Surface-conductive type: Materials with a surface resistivity less than 10^5 ohms per square. (2) Volume-conductive type: Materials with a volume resistivity less than 10^4 ohm-cm. [MIL-HDBK-263B]

Coupling. Circuit element or elements, or network, that may be considered common to the input and the output mesh and through which energy may be transferred from one to the other. [IEEE-1100-1999]

Critical Technical Load. Critical technical load is principally associated with systems having a requirement of 100% continuity in power service, such as command, control, communications, and information (C^3I) facilities. Such systems normally require on-line uninterruptible power systems (UPS). Systems that would require extensive reprocessing due to loss of data during a power failure might also qualify as critical technical electronic loads. These loads must not be shed intentionally if sufficient power is available to supply them. [MIL-HDBK-411B]

Dissimilar Metals. Two metal specimens that are in contact or otherwise electrically connected to each other in a conductive solution and generate an electric current. [MIL-STD-889B]

Dissipative Material. For the purpose of ESD protection, material with the following characteristics. (1) Surface-conductive type: Materials with a surface resistivity equal to or greater than 10^5 but less than 10^{12} ohms per square. (2) Volume-conductive type: Materials with a volume resistivity equal to or greater than 10^4 but less than 10^{11} ohm-cm. [MIL-HDBK-263B]

Distributed Single-Point Ground. A system of multiple, isolated system ground points common to an isolated set of equipment (i.e., single-point ground), referenced to a common, large, conductive structure (e.g., equipotential "islands"). [SSP 30240 Revision C]

Earth. The conductive mass of the earth, whose electric potential at any point is conventionally taken as zero. [BS 7671]

Earth Electrode/Ground Electrode. A conductor or group of conductors in intimate contact with, and providing an electrical connection to, earth. [BS 7671]

Earth Electrode Resistance. The resistance of an earth electrode to earth. [BS 7671]

Earth Fault Loop Impedance. The impedance of the earth fault current loop starting and ending at the point of the earth fault. [BS 7671]

Earth Ground. The portion of a circuit that is at zero potential with respect to Earth. [MIL-HDBK-263B]

Earthing Conductor. A protective conductor connecting the main earthing terminal of an installation to an earth electrode or to another means of earthing. [BS 7176]

Earth Mat. A solid metallic plate or a system of closely spaced bare conductors that are connected to and often placed at shallow depths above a ground grid or elsewhere at the earth's surface, in order to obtain an extra protective measure that minimizes the danger of the exposure to high step or touch voltage voltages in a critical operating area or places that are frequently used by people. Grounded metal gratings, placed on or above the soil surface, or wire mesh placed directly under the surface material, are common forms of ground mats. [IEEE 80-2000]

Eddy Current. An electric current caused by varying electromotive forces that are due to varying magnetic fields, and causing heating in motors, transformers, and so on. Also called Foucault current.

Electrically Small (Conductors). Conductors, the largest linear dimensions of which do not exceed one-tenth of a wavelength ($\lambda/10$). Some sources use $\lambda/2\pi$ rather than $\lambda/10$.

Electric Power System Ground (Ships). A ground is a plane or surface used by the electric power system as a common reference to establish zero potential. Usually, this surface is the metallic hull of the ship. On a nonmetallic hull ship, a special ground system is installed for this purpose. [MIL-STD-1399-300A]

Electromagnetic Compatibility (EMC). The capability of equipment or systems to be operated in their intended operational environment, within designed levels of efficiency, without causing or receiving degradation due to unintentional EMI/EMC is the result of an engineering planning process applied during the life cycle of the equipment. The process involves careful considerations of frequency allocation, design, procurement, production, site selection, installation, operation, and maintenance. [MIL-HDBK-419A]

Electromagnetic Interference (EMI). Any electrical or electromagnetic phenomenon, manmade or natural, either radiated or conducted, that results in unintentional and undesirable responses, performance degradation, or malfunction of electronic equipment. [MIL-HDBK-419A]

Electromagnetic Pulse (EMP). A large impulsive type electromagnetic wave generated by nuclear or chemical explosions. [MIL-HDBK-419A]

Electrostatic Discharge (ESD). A transfer of electrostatic charge between objects at different potentials caused by direct contact or induced by an electrostatic field. [MIL-HDBK-263B]

Electrostatic Discharge Sensitive (ESDS). The relative tendency of a device's performance to be affected or damaged by an ESD event. [MIL-HDBK-263B]

Emanation Security (EMSEC). EMSEC is a short unclassified name referring to investigation and studies of compromising emanations. It also refers to those measures used to control compromising emanations. EMSEC is interrelated with the "red/black" concept, which requires that electrical and electronic components, equipment, and systems processing classified plain text information be kept separate from those that process encrypted or unclassified information in order to prevent or reduce the effects of compromising emanations. [EI TEMPEST Installation Handbook 2EXXX-202D]

Equipment Grounding Conductor. The conductor used to connect the noncurrent-carrying parts of conduits, raceways, and equipment enclosures to the grounding electrode at the service equipment (main panel) or secondary of a separately derived system (e.g., isolation transformer). [IEEE-1100-1999]

Equipotential Plane. A grounding grid that provides a low-impedance path for signals and currents to return from the load, back to the generator. This system overcomes the limitations of the older, single-shunt grounding systems that are more affected by noise and increased use of filters. [EI TEMPEST Installation Handbook 2EXXX-202D]

ESD Protected Area. An area that is constructed and equipped with the necessary ESD protective materials, equipment, and procedures to limit ESD voltages below the sensitivity level of ESDS items handled therein. [MIL-HDBK-263B]

Exposed Conductive Part. A conductive part of equipment that can be touched and that is not a live part but which may become live under fault conditions. [BS 7176]

Extraneous Conductive Part. A conductive part liable to introduce a potential, generally an earth potential, and not forming part of the electrical installation. [BS 7176]

Facility Ground System. The electrically interconnected system of conductors and conductive elements that provides multiple current paths to earth. The facility ground system includes the earth electrode subsystem, lightning protection subsystem, signal reference subsystem, and fault protection subsystem, as well as the building structure, equipment racks, cabinets, conduits, junction boxes, raceways, duct work, pipes, and other normally noncurrent-carrying metal elements. [MIL-HDBK-419A]

Fault. A circuit condition in which current flows through an abnormal or unintended path. This may result from an insulation failure or a bridging of insulation. Conventionally, the impedance between live conductors or between live conductors and exposed or extraneous conductive parts at the fault position is considered negligible. [BS 7671]

Fault Current. The current that may flow in a circuit as a result of specified abnormal conditions. [MIL-HDBK-411B]

Functional Grounding (Earthing). Connection to earth necessary for proper functioning of electrical equipment [BS7671]. The earthing of a point in a system or in an installation or in equipment, which is necessary for a purpose other than protection against electric shock [IEC60364-5-548].

Galvanic Corrosion. Accelerated corrosion of the more active metal (anode) of a dissimilar metal couple in an electrolyte solution or medium, and decreased corrosive effects on the less active metal (cathode) as compared to the corrosion of the individual metals, when not connected, in the same electrolyte environment. [MIL-STD-889B]

Galvanic Series. A listing of metals and alloys based on their order and tendency to corrode independently, in a particular electrolyte solution or other environment. This tendency for dissolution or corrosion is related to the electrical potential of the metal in a conductive medium. Galvanic corrosion is inherently affected by the relative position of the galvanic series of the metals constituting the couple. Metals closely positioned in the series will have electrical potentials nearer one another, whereas with greater divergence in position, greater differences in potential will prevail. Galvanic effects such as corrosion of the anode will be minimal in the former condition, whereas the latter condition will exhibit more significant corrosive effects. [MIL-STD-889B]

Ground. (1) A direct conducting connection to the earth or body of water that is part thereof [IEEE-100: 315-1975]. (2) A conducting connection to a structure that serves a function similar to that of an earth ground (that is, a structure such as a frame of an air, space, or land vehicle that is not conductively connected to earth) [IEEE-100: 315-1975]. (3) A conducting connection, whether intentional or accidental, by which an electric circuit or equipment is connected to the earth, or to some conducting body of relatively large extent that serves in place of the earth. *Note:* Grounds are used for establishing and maintaining the potential of the earth (or of the conducting body), or approximately that potential, on conductors connected to it and for conducting ground currents to and from earth (or the conducting body) [IEEE-100: 1100-1999]. (4) The electrical connection to earth, primarily through an earth electrode subsystem. This connection is extended throughout the facility via the facility ground system consisting of the signal reference subsystem, the fault protection subsystem, the lightning protection subsystem, and the earth electrode subsystem. (5) A mass, such as the earth or a ship or vehicle hull, capable of supplying or accepting an electrical charge [MIL-HDBK-263B].

Ground Bar (Lightning). A conductor forming a common junction for a number of ground conductors.

Ground Bus. A bus to which the grounds from individual pieces of equipment are connected, and that, in turn, is connected to ground at one or more points. [ANSI C37.100-1992]

Ground Conductivity. A property of a ground, expressed as the ratio of electric current density to electric field strength. [IEEE-100: 1260-1996]

Ground Conductor (Lightning). A conductor providing an electric connection between part of the system, of the frame of a machine, or piece of apparatus, and a ground electrode or a ground bar

Ground Current. (1) Current flowing in the earth or in a grounding connection [IEEE-100: 81-1983]. (2) A current flowing into or out of the earth or its equivalent serving as a ground [IEEE-100: 80-2000].

Ground Electrode. (1) A conductor or group of conductors in intimate contact with the earth for the purpose of providing a connection with the ground [IEEE-100: 1100-1999]. (2) A conductor embedded in the earth and used for collecting ground current from or dissipating ground current into the earth.

Ground-Fault Circuit Interrupter. A device to protect personnel by deenergizing a circuit or part of a circuit, within an established period of time, when the current to ground exceeds a set value that is less than the supply protection value. [MIL-HDBK-411B]

Ground Grid. A system of interconnected bare conductors arranged in a pattern over a specified area and buried below the surface of the earth. The primary purpose of the ground grid is to provide safety for workmen by limiting potential differences within its perimeter to safe levels in case of high currents, which could flow if the circuit being worked became energized for any reason or if an adjacent energized circuit faulted. Metallic surface mats and gratings are sometimes utilized for the same purpose [IEEE-1100-1999]. A typical grid usually is supplemented by a number of ground rods and may be further connected to auxiliary ground electrodes to lower its resistance with respect to remote earth.

Ground Loop. (1) A potentially detrimental loop formed when two or more points in an electrical system that are nominally at ground potential are connected by a conducting path such that either or both points are not at the same ground potential [IEEE-1100-1999]. (2) More than one path to ground for equipment or a system such that the designated ground is not maintained at a common (zero voltage) potential; normally caused by multipoint grounding. Lack of a common reference point (ground) results in common mode (interference) currents in the ground loop. [MIL-STD-1310G]

Ground Mat. See Earth Mat.

Ground Overcurrent. (1) A conducting or reflecting plane functioning to image a radiating structure. Synonym: imaging plane [IEEE-145-1993, IEEE-145-1983s]. (2) (Audio Noise Emissions) A conducting surface or plane used as a common reference point for circuit returns and electrical or signal potentials [ANSI C63.4-1981]. (3) An assumed plane or true ground or zero potential [IEEE-100]. The net (phasor sum) current flowing in the phase and neutral conductors or the total current flowing in the normal neutral-to-ground connection that exceeds a predetermined value. [ANSI C37.100-1992]

Ground, Personnel Safety. Bonding of electrical and electronic equipment cabinets, cases, housings, and exposed metal (conductive) components to ground potential by any class bond described herein, or by the installation of a ground wire to establish a contact resistance of 0.1 ohm or less. [MIL-STD-1310G]

Ground Plane. (1) A conducting surface or plate used as a common reference point for circuit returns and electric or electric potentials [ANSI C63.5-1988, IEEE-1004-1987, ANSI C63.4-1988]. (2) A conducting or reflecting plane used to image a radiating structure [IEEE-145-1983s, IEEE-145-1993].

Ground Plane Field. The electromagnetic field in near proximity to a conducting surface, with the boundary conditions that the tangential electric field approaches zero and the normal magnetic field remains continuous. The total normal electric field is related to the surface charge density by Gauss's Law and the total tangential magnetic field to the surface current density by Ampere's Law. [IEEE-1309-1996]

Ground Plate (Grounding Plate). A plate of conducting material buried in the earth to serve as a grounding electrode. [IEEE-100]

Ground Potential. A point, plane, or surface designated as the zero potential (nominally) common reference point for electrical or electronic equipment. [MIL-STD-1310G]

Ground Potential Difference Voltage. The voltage that results from current flow through the finite resistance and inductance between the receiver and driver circuit ground voltages. [IEEE-1596.3-1996]

Ground Potential Rise (GPR). (1) The voltage that a station grounding grid may attain relative to a distant grounding point assumed to be at the potential of remote earth [ANSI C62.33-1995]. (2) The product of a ground electrode impedance, references to remote earth, and the current that flows through that electrode impedance [IEEE-367-1996]. (3) The difference in ground potential between a location in proximity to a point of large current injection into the ground and any remote ground point. GPR is usually caused by a sort circuit of an energized power conductor to ground and is a result of the injected current flowing through the impedance of the ground circuit [ANSI C37.100-1992]. (4) The maximum electrical potential that a substation grounded grid may attain relative to a distant grounding point assumed to be at the potential of remote earth [IEEE-1268-1997]. (5) The maximum electrical potential that a substation grounded grid may attain relative to a distant grounding point assumed to be at the potential of remote earth. This voltage, GPR, is equal to the maximum grid current times the grid resistance. *Note:* Under normal conditions, the grounded electrical equipment operates at near zero ground potential. That is, the potential of a grounded neutral conductor is nearly identical to the potential of remote earth. During a ground fault, the portion of fault current that is conducted by a substation grounding grid into the earth causes the rise of the grid potential with respect to remote earth [IEEE-80-2000].

Ground Potential Shift. The difference in voltage between grounding or grounded (earthed) structures such as the opposite corners of a metal building. Generally, ground potential shift increases with distance or separation of ground locations and with the frequency or wave front rise time of the resulting current flow. Ground potential shift problems are generally exacerbated by surge events from lightning and utility power sources. [IEEE-1100-1999]

Ground, Radial. A conductor connection by which separate electrical circuits or equipment are connected to earth at one point. Sometimes referred to as a star ground. [IEEE-1100-1999]

Ground Reference Plane (GRP). A flat conductive surface whose potential is used as a common reference. Where applicable, the operating voltage of the EUT and the operator ground should also be referenced to the ground plane. [C63.16-1993]

Ground Resistance (Grounding Electrode). The ohmic resistance between the grounding electrode and a remote grounding electrode of zero resistance. *Note:* By "remote" is meant "at a distance such that the mutual resistance of the two electrodes is essentially zero." [IEEE-81-1983]

Ground Return Circuit. (1) In ground systems, a circuit in which the earth is utilized to complete the circuit [IEEE-81-1983]. (2) In data transmission, a circuit that has a single conductor (or two or more in parallel) between two points and is completed through the ground or earth [IEEE-100: 599-1985].

Ground Return System. A system in which one of the conductors is replaced by the ground. [IEEE-100-1999]

Ground Rod. (1) A rod that is driven into the ground to serve as a ground terminal, such as copper-clad rod, solid copper rod, galvanized iron rod, or galvanized iron pipe. Copper-clad rods are commonly used during conductor stringing operations to provide a means of obtaining an electrical ground using portable grounding devices. Synonym: ground electrode [IEEE-1048-1990]. (2) A conducting rod serving as an electrical connection with the ground [IEEE-145-1993].

Ground Terminal. (1) In lightning protection systems, the portion extending into the ground, such as a ground rod, ground plate or the conductor itself, serving to bring the lightning protection system into electrical contact with the ground [IEEE-100-1999]. (2) In surge arresters, the conducting part provided for connecting the arrester to ground [ANSI C62.11-1999].

Ground, 360 Degree (Peripheral). A continuous metal-to-metal bond, existing or provided, around the outer perimeter of a metallic item or cable shield terminating at or penetrating through a metal surface that is at ground potential. [MIL-STD-1310G]

Grounded Circuit. A circuit in which one conductor or point (usually the neutral conductor or neutral point of transformer or generator windings) is intentionally grounded, either solidly or through a noninterrupting current-limiting grounding device. [IEEE-100: 32-1972r]

Grounded Conductor. (1) A system or circuit conductor that is intentionally grounded [NEC]. (2) A conductor in a power distribution system (usually designated the neutral) that is intentionally earth grounded, either solidly or through a grounding device. The outer jacket of the conductor, if insulated, is white in color [MIL-HDBK-411B].

Grounded Conductor (Neutral). The circuit conductor that is intentionally grounded (at first service disconnect or power source). [MIL-HDBK-419A]

Grounded Electric Power System. A grounded electric power system is a system in which at least one conductor or point (usually the middle wire or neutral point of the transformer or generator winding) is intentionally and effectively connected to system ground. [MIL-STD-1399-300A]

Grounded Parts. Parts that are intentionally connected to ground. [ANSI C37.40-1993, ANSI-C37.100-1993]

Grounded System. A system of conductors in which at least one conductor or point is intentionally grounded, either solidly or through a noninterrupting current-limiting device. [ANSI C2-1997, C2.2-1960]

Grounding Conductor. The conductor that is used to connect the equipment or the wiring system to a grounding electrode or electrodes [ANSI C2.2-1960]. (1) The conductor that is used to establish a ground and that connects an equipment, devices, wiring systems, or another conductor (usually the neutral conductor) with the grounding electrode or electrodes. Synonym: Ground System. (2) A metallic conductor used to connect the metal frame or enclosure of a device, equipment, or wiring system with a mine track or other effective grounding medium. (3) A conductor used to connect equipment or the grounded circuit of a wiring system to a grounding electrode or electrodes. (4) A conductor used to connect equipment or the grounded circuit of a wiring system to a grounding electrode or electrodes (i.e., ground grid) [IEEE-100]. (5) A conductor that carries no current under normal conditions. It serves to connect exposed metal surfaces to an earth ground to prevent hazards in case of breakdown between current-carrying parts of a power distribution system and the exposed surfaces. The outer jacket of the conductor, if insulated, is green in color, with or without a yellow conductor. [MIL-HDBK-411B]

Grounding Conductor (Equipment). The conductor used to connect the noncurrent-carrying metal parts of equipment, raceways, and other enclosures to the system-grounded conductor, the grounding electrode conductor, or both, at the service equipment or at the source of a separately derived system. [NEC]

Grounding Conductor (Green Wire). A conductor used to connect equipment or the grounded circuit of a power system to the earth electrode subsystem. [MIL-HDBK-419A]

Grounding Connection (Ground Systems). A connection used in establishing a ground that consists of a grounding conductor, a grounding electrode, and the earth (soil) that surrounds the electrode or some conductive body that serves instead of the earth. [IEEE-81-1983]

Grounding Electrode. A conductor used to establish a ground. Synonyms: Ground System, Ground Electrode. [IEEE-81-1983]

Grounding Electrode Conductor. A conductor used to connect the grounding electrode to the equipment grounding conductor and/or to the grounded conductor of the circuit at the service equipment or at the source of a separately derived system. [NEC]

Grounding Grid. A system of horizontal ground electrodes that consists of a number of interconnects and bare conductors buried in the earth, providing a common ground for electrical devices or metallic structures, usually in one specific location. *Note:* Grids buried horizontally near the earth's surface are also effective in controlling the surface potential gradients. A typical grid usually is supplemented by a number of ground rods and may be further connected to auxiliary ground electrodes to lower its resistance with respect to a remote earth. [IEEE-80-2000]

Grounding Jumper (Electric Appliances). A strap or wire to connect the frame of the range to the neutral conductor of the supply circuit. [IEEE-100]

Grounding Outlet. An outlet equipped with a receptacle of the polarity type having, in addition to the current-carrying contacts, one grounded contact that can be used for the connection of an equipment grounding conductor. *Note:* This type of outlet is used for connection of portable appliances. Synonym: safety outlet. [IEEE-100]

Grounding System. Comprises all interconnects grounding facilities in a specific area. [IEEE-80-2000]

Hard Ground. A connection directly to earth ground. [MIL-HDBK-263B]

Hazardous Voltage. A voltage exceeding 42.4 V peak, or 60 V DC, existing in a circuit that does not meet the requirements for either limited current circuit or a TNV circuit. [IEC 60950-1999]

Higher Frequency Ground. The interconnected metallic network intended to serve as a common reference for currents and voltages at frequencies above 30 kHz and in some cases above 300 kHz. Pulse and digital signals with rise and fall times of less than 1 microsecond are classified as higher frequency signals. [MIL-HDBK-419A]

Induced Current. Current in a conductor due to the application of a time-varying electromagnetic field. [IEEE-1100-1999]

Induced Voltage. A voltage produced around a closed path or circuit by a change in magnetic flux linking that path. [IEEE-1100-1999]

Inductance. The property of an electric circuit by virtue of which a varying current induces an electromotive force in that circuit or in a neighboring circuit. [IEEE-1100-1999]

Inductance Coupling. The type of coupling in which the mechanism is mutual inductance between the interference, induced in the signal system by a magnetic field produced by the interference source. [IEEE-1100-1999]

Induction. The process of generating time-varying voltages and/or currents in conductive objects or electric circuits by influence of the time-varying electric, magnetic, or electromagnetic fields. [IEEE-539-1990]

Isolated DC Return (DC-I). A DC power system in which the return conductor has a single point connection to a bonding network. More complex configurations are possible. [ITU-T K.27]

Isolated Electrical Power. Secondary electrical power isolated from primary electrical power. [SSP 30240 Revision C]

Isolated Equipment Ground. An insulated equipment grounding conductor run in the same conduit or raceway as the supply conductors. This conductor is insulated from the metallic raceway and all ground points throughout its length. It originates at an isolated ground-type receptacle or equipment input terminal block and terminates at the point where neutral and ground are bonded at the power source. [IEEE-1100-1999]

Isolation. Separation of one section of a system from undesired influences of other sections. [IEEE-1100-1999]

IT Earthing System. A system having no direct connection between live parts and earth, the exposed conductive parts of the electrical installation being earthed. [BS 7671]

Lightning Protection Subsystem. A complete subsystem consisting of air terminals, interconnecting conductors, ground terminals, arresters, and other connectors or fittings required to assure that a lightning discharge will be safely conducted to earth. [MIL-HDBK-419A]

Longitudinal Conversion Loss (LCL). The LCL of a one- or two-port network is a measure (a ratio expressed in decibels) of the degree of unwanted transverse signal produced at the terminals of the network due to the presence of a longitudinal signal on the connecting leads. [ITU-T O.9]

Longitudinal (Common-Mode) Signal. The longitudinal voltage is half the algebraic sum of the voltages to ground in the two conductors (tip and ring). The longitudinal current is the algebraic sum of the current in these conductors. [IEEE-100: 820-1984r]

Longitudinal Conversion Transfer Loss (LCTL). The LCTL is a measure (a ratio expressed in decibels) of an unwanted transverse signal produced at the output of a two-port network due to the presence of a longitudinal signal on the connecting leads of the input port. [ITU-T O.9]

Lower Frequency Ground. A dedicated, single-point network intended to serve as a reference for voltages and currents, whether signal, control, or power, from DC to 30 kHz and in some cases to 300 kHz. Pulse and digital signals with rise and fall times greater than 1 µs are considered to be lower frequency signals. [MIL-HDBK-419A]

Main Earthing Terminal/Main Ground Bus. The terminal or bar provided for the connection of protective conductors, including equipotential bonding conductors, and conductors for functional grounding, if any, to the earth electrode system. [BS 7671]

Mesh Isolated Bonding Network (M-IBN). A type of IBN in which the components of the IBN (equipment frames) are interconnected to form a mesh-like structure. This may, for example, be achieved by multiple interconnections between cabinet rows or by connecting all equipment frames to a metallic grid (bonding mat) that extends away from beneath the equipment. The bonding mat is, of course, insulated from the adjacent CBN. If necessary, the bonding mat could include vertical extensions that result in an approximation of a Faraday cage. The spacing of the grid depends upon the frequency range of the electromagnetic environment. [ITU-T K.27]

Metallic Signal (Differential). The metallic voltage is the algebraic difference between the voltages to ground in the two conductors (tip and ring). The metallic current is half the algebraic difference between the current in these conductors. [IEEE-100: 820-1984r]

Magnetomotive Force (MMF). Any physical cause that produces magnetic flux. The standard definition of magnetomotive force involves current passing through an electrical conductor, which accounts for the magnetic fields of electromagnets. The unit of magnetomotive force is the ampere turn (At), represented by a steady, direct electric current of one ampere flowing in a single-turn loop of electrically conducting material in a vacuum.

Mission-Critical System. A system in which failure of operation, or faulty operation, may have catastrophic outcomes, including loss of life, serious injury, loss or serious damage to plant or platform, large financial loss, serious inability to conduct business, or serious operational chaos.

National Electrical Code (NEC). A standard governing the use of electrical wire, cable, and fixtures installed in buildings. It is sponsored by the National Fire Protection Association (NFPA-70) under the auspices of the American National Standards Institute. [MIL-HDBK-419A]

Neutral. The AC power system conductor that is intentionally grounded on the supply side of the first service-disconnecting means. It is the low potential (white) side of a single phase AC circuit or the low potential fourth wire of a three-phase wye distribution system. The neutral (grounded conductor) provides a current return path for AC power currents, whereas the grounding (or green) conductor does not, except during fault conditions. [MIL-HDBK-419A]

Neutral Conductor (N). A conductor connected to the neutral point of a system and capable of contributing to the transmission of electrical energy. [ITU-T K.27]

Noncritical Technical Load. Noncritical technical load is that portion of the technical load that directly supports the routine accomplishment of facility missions. This load includes general lighting and power systems, HVAC, and similar equipment that can tolerate brief power outages without loss of data or without an adverse effect on mission accomplishment. Backup power for this load category would normally be an automatic-start, automatic-transfer engine generator. In some cases, an off-line uninterruptible power system (UPS) will be specified for power restoration. [MIL-HDBK-411B]

Nonlinear Junction. A contact area between two metallic surfaces that exhibits nonlinear voltage-current transfer characteristics when subjected to an RF voltage. This nonlinearity is usually caused by corrosion or other semiconducting materials in the contact area. [MIL-STD-1310G]

Phase Conductor. A conductor of an AC system for the transmission of electrical energy other than a neutral conductor, protective conductor, or EN conductor. [BS 7671]

Primary Electrical Power. Electrical power taken from power generation units without conditioning or isolation. [SSP 30240 Revision C, Generalized]

Protective Bonding Conductor. A conductor in the equipment, or a combination of conductive parts in the equipment, connecting a main protective earthing terminal to a part of the equipment that is required to be earthed for safety purposes in the building installation. [IEC 60950-1999]

Protective Conductor. A conductor used for some measures of protection against electric shock and intended for connecting together any of the following parts: exposed conductive parts, extraneous conductive parts, the main earth terminal, earth electrode(s), or the earthed point of the source. [BS 7176]

Protective Conductor Current. Current flowing through the protective earthing conductor under normal operating conditions. Protective conductor current was previously included in the term "leakage current." [IEC 60950-1999]

Protective Earthing Conductor. A conductor in the building installation wiring or in the power supply cord, connecting a main protective earthing terminal in the equipment to an earth point in the building installation. In some countries, the term "grounding conductor" is used instead of "protective earthing conductor." [IEC 60950-1999]

Red. As it pertains to C-E equipment, transmission lines, and associated wiring, "red" signifies both physical and electrical areas where classified data/voice signals are in plain text (unencrypted) and highly susceptible to compromise. [EI TEMPEST Installation Handbook 2EXXX-202D]

Red/Black Concept. The red/black concept requires electrical and electronic circuits, components, and systems that handle classified unencrypted information ("red") be separated from those that handle encrypted or unclassified information ("black"). Under this concept, "red" and "black" terminology is used to clarify and to differentiate between circuits, components, equipment, and systems. The terminology also differentiates between the physical areas in which they are contained. [EI TEMPEST Installation Handbook 2EXXX-202D]

Red Equipment. Red equipment is any equipment that processes classified information before encryption and after decryption, and should therefore be TEMPEST and physically protected. [MIL-HDBK-232A]

Red Equipment Area (REA). The space within a limited exclusion area that is designated for installation of red information processing equipment and associated power, signal, control, ground, and distribution facilities. [MIL-HDBK-232A]

Resistivity. A measure of the resistance of a material to electric current either through its volume or on its surface. Surface resistivity is the ratio of direct current (DC) voltage to the current that passes across the surface of a material. The unit measurement for surface resistivity (ρ_S) is ohms per square. Volume resistivity is the ratio of DC voltage per unit of thickness applied across two electrodes in contact with a specimen to the amount of current per unit area passing through the material. The unit of measurement for volume resistivity (ρ_V) is ohm-centimeter. [MIL-HDBK-263B]

Secondary Electrical Power. Electrical power a system that has been isolated from the primary electrical power before it is distributed to subsystems. [SSP 30240 Revision C, Generalized]

Separately Derived Power. Power derived from a generator, transformer, or converter, with no direct electrical connection to conductors in another system. [SSP 30240 Revision C]

Shield. (1) A metal barrier of solid, screen, or braid construction used to protect electronic components, wires, or cables from EM energy; or used to reduce the emission of EM energy from components, wires, or cabling [MIL-STD-1310G]. (2) As normally applied to instrumentation cables, a conductive sheath (usually metallic) applied over the insulation of a conductor or conductors, for the purpose of providing a means to reduce coupling between the conductors so shielded and other conductors that may be susceptible to, or that may be generating, unwanted electrostatic or electromagnetic fields (noise) [IEEE-1100-1999]. (3) A housing, screen, or cover that substantially reduces the coupling of electric and magnetic fields into or out of circuits, or prevents the accidental contact of objects or persons with parts or components operating at hazardous voltage levels. [MIL-HDBK-419A]

Signal Reference Subsystem. A conductive sheet or cable network/mesh providing an equipotential reference for C-E equipment to minimize interference and noise. [MIL-HDBK-419A]

Signal Return. A current-carrying path between a load and the signal source. It is the low side of the closed-loop energy transfer circuit between a source–load pair. [MIL-HDBK-419A]

Single-Point Connection (SPC). The unique location in an IBN where a connection is made to the CBN. In reality, the SPC is not a mere "point" but has sufficient size to accommodate the connection of multiple conductors. Usually, the SPC is a copper bus bar. If cable shields or coaxial outer conductors are to be connected to the SPC, the SPC could be a frame with a grid or sheet metal structure. [ITU-T K.27]

Single-Point Connection Window (SPCW). The interface or transition region between an IBN and the CBN. Its maximum dimension is typically 2 m. The SPC bus bar (SPCB) or frame lies within this region and provides the interface between IBN and CBN. Conductors (e.g., cable shields or DC return conductors) that enter a system block and connect to its IBN must enter via the SPCW and connect to the SPC bus bar or frame. [ITU-T K.27]

Single-Point Ground. A common electrical connection to ground, used as a common reference (typically, the common electrical connection is characterized by a minimum or zero potential difference between each physical connection). [SSP 30240 Revision C]

Skin Effect. The tendency of alternating current to flow near the surface of a conductor, thereby restricting the current to a small part of the total cross-sectional area and increasing the resistance to the flow of current. [Federal Standard 1037C]

Soft Ground. A connection to ground through a resistance sufficient to limit current flow to safe levels for personnel. [MIL-HDBK-263B]

Star IBN. A type of IBN comprising clustered or nested IBNs sharing a common SPC. [ITU-T K.27]

Step Voltage. The difference in surface potential experienced by a person whose feet bridge a distance of 1 m without contacting any grounded object. [IEEE 80-2000]

Surge Protective Device (SPD). Device intended to limit transient overvoltages and divert surge currents. It contains at least one nonlinear component. [IEC 62305-4 (FDIS)]

System Block. All the equipment whose frames and associated conductive parts form a defined BN. [ITU-T K.27]

TEMPEST. A term used to describe a methodology for controlling radiated and conducted emanations of classified information. [MIL-STD-188-125-1]

Terminal Protection Device (TPD). A quick-reaction switching device installed between the susceptible circuit and ground to protect electronic components from lightning and EMP damage. TPDs may also serve as transient protection devices or surge protection devices. [MIL-STD-1310G]

Tertiary Power. Electrical power derived from secondary power. [SSP 30240 Revision C]

Touch Current. Electric current through a human body when it touches one or more accessible parts. Touch current was previously included in the term "leakage current." [IEC 60950-1999]

Touch Voltage. The potential difference between the ground potential rise (GRP) and the surface potential at the point where a person is standing while at the same time having a hand in contact with a grounded structure. [IEEE 80-2000]

Transferred Potential. A special case of the touch voltage in which a potential is transferred into or out of the substation from or to a remote point external to the substation site. [IEEE 80-2000]

TN Earthing System. A system having one or more points of the source of energy directly earthed, the exposed conductive parts of the installation being connected to that point by protective conductors. [BS 7671]

TN-C Earthing System. A system in which neutral and protective functions are combined in a single conductor throughout the system. [BS 7671]

TN-S Earthing System. A system having separate neutral and protective conductors throughout the system. [BS 7671]

TN-C-S Earthing System. A system in which neutral and protective functions are combined in a single conductor in part of the system. [BS 7671]

Transient. A phenomenon or quantity that varies between two consecutive steady states during a time interval that is short compared with the time scale of interest. [IEC-50(161)-02-01]

Transverse Mode Noise. Noise signals measurable between or among active circuit conductors feeding the subject load, but not between the equipment grounding conductor or associated signal reference structure and the active circuit conductors. [IEEE-1100-1999]

TT Earthing System. A system having one point of the source of energy directly earthed, the exposed conductive parts of the installation being connected to earth electrodes electrically independent of the earth electrodes of the source. [BS 7671]

Ungrounded Electric Power System. An ungrounded electric power system is a system that is intentionally not connected to the metal structure or the grounding system of a ship, except for test purposes. [MIL-STD-1399-300A]

Voltage Surge. A transient voltage wave propagating along a line or a circuit and characterized by a rapid increase followed by a slower decrease of the voltage. [IEC-50(161)-08-11]

ACRONYMS

AC	Alternating Current
AWG	American Wire Gauge
BalUn	Balanced/Unbalanced (Transformer)
BEA	Black Equipment Area
BBN	Broadband Noise
BN	Bonding Network
CBN	Common Bonding Network
C^4	Command, Control, Communication, and Computer
C^3I	Command, Control, Communication, and Intelligence
C^4I	Command, Control, Communication, Computer, and Intelligence
CIIC, CI^2C	Common-Impedance Interference Coupling
CM	Common Mode
CMF-PDGS	Common-Mode Filter [using] Periodic Defected Ground Structure
CMR	Common-Mode Rejection
CMRR	Common-Mode Rejection Ratio
DC	Direct Current
DC-C	Common DC Return
DC-I	Isolated DC Return
DGS	Defected Ground Structure
DM	Differential Current

Grounds for Grounding. By Elya B. Joffe and Kai-Sang Lock
Copyright © 2010 The Institute of Electrical and Electronics Engineers, Inc.

EBG	Electromagnetic Band Gap
ECG	Electrical Grounding Conductor
EESS	Earth Electrode Subsystem
EIEDs	Electrically Initiated Explosive Devices
EMC	Electromagnetic Compatibility
EME	Electromagnetic Environment
EMF	Electromotive Force
EMI	Electromagnetic Interference
EMIC	Electromagnetic Interference Control
EMP	Electromagnetic Pulse
ESD	Electrostatic Discharge
ESDC	Electrostatic Discharge Control
ESGC	Electrical Safety Ground Conductor
FPSS	Fault Protection Subsystem
GBN	Ground Bounce Noise
GBS	Grounding, Bonding, Shielding
GFCI	Ground Fault Circuit Interrupter
GLIC	Ground Loop Interference Coupling
GPR	Ground Potential Rise
GRP	Ground Reference Point
HEMP	High-Altitude Electromagnetic Pulse
HF	High Frequency
IBN	Isolated Bonding Network
IEEE	Institute of Electrical and Electronic Engineers
IG	Isolated/Insulated Ground
IMI	Intermodulation Interference
IT	Isolation Transformer
IT	Information Technology
LCL	Longitudinal Conversion Loss
LCTL	Longitudinal Conversion Transfer Loss
LEMP	Lightning Electromagnetic Pulse
LPC-EBG	Low-Period Coplanar Electromagnetic Band Gap
LPF	Low-Pass Filter
LPSS	Lightning Protection Subsystem
MBN	Mesh Bonding Network
MCB	Miniature Circuit Breaker
MCBN	Mesh Common-Bonding Network
MET	Main Earthing Terminal
M-IBN	Mesh Isolated Bonding Network
MMF	Magnetomotive Force
MOV	Metal-Oxide Varistor
MPG	Multipoint Grounding
NEC	National Electrical Code
NEMP	Nuclear Electromagnetic Pulse
PCB	Printed Circuit Board

PDGS	Periodic Defected Ground Structure
PDN	Power Distribution Network
PE	Protective Earth
PEN	Protective Earth + Neutral
PME	Protective Multiple Earthing
PRF	Pulse Repetition Frequency
RCBO	Residual Current Circuit Breaker with Overcurrent Protection
RCD	Residual Current Device
REA	Red Equipment Area
RFI	Radio Frequency Interference
SG	Solid Grounding; Solidly Grounded
SMPS	Switch Mode Power Supply
SNR	Signal-to-Noise Ratio
SPC	Single Point of Connection
SPCW	SPC Window
SPD	Surge Protective Device
SPG	Single-Point Grounding
SRG	Signal Reference Grid
SRP	Signal Reference Plane
SRS	Signal Reference Structure
SRSS	Signal Reference Subsystem
SSN	Simultaneous Switching Noise
SSO	Simultaneous Switching Outputs
TEM	Transverse Electromagnetic
TPD	Terminal Protection Device
TPD	Transient Protection Devices
TSD	Transient Suppression Device
TSP	Twisted Shielded Pair
TVS	Transient Voltage Suppressor
TWP	Twisted Wire Pair
UTP	Unshielded Twisted Pair

SYMBOLS AND CONSTANTS

Symbol	Definition	Dimensions
I	Current	ampere (A)
$\vec{B}$	Magnetic flux density	tesla (T)
c	Phase velocity of EM waves ("speed of light"), $$c = \frac{1}{\sqrt{\mu_0 \varepsilon_0}} \approx 2.9979 \times 10^8 \, \text{m/sec}$$	meter/second (m/sec)
$\vec{D}$	Electric flux density	coulomb/meter^2 (C/m^2)
$\vec{E}$	Electric field strength	volt/meter (V/m)
$\vec{H}$	Magnetic field strength	ampere/meter (A/m)
$\vec{J}$	Conduction current density	ampere/meter^2 (A/m^2)
q	Charge	coulomb (C)
ε	Permittivity (dielectric constant): $\varepsilon = \varepsilon_0 \cdot \varepsilon_r$	farad/meter (F/m)
ε_0	Permittivity of free space: $\varepsilon_0 \cong 1/36\pi \times 10^{-9}$	farad/meter (F/m)
ε_r	Relative permittivity	farad/meter (F/m)
μ	Permeability	henry/meter (H/m)
μ_0	Permeability of free space: $\mu_0 = \pi \times 10^{-7}$	henry/meter (H/m)
μ_r	Relative permeability	
σ	Conductivity	siemens/meter (S/m)

σ_0	Conductivity of copper (Cu): $\sigma_0 \approx 5.82 \times 10^7$	siemens/meter (S/m)
σ_r	Relative conductivity	
ρ	Resistivity	ohm(meter ($\Omega{\cdot}$m)
ρ_S	Surface resistivity	ohm/square ($\Omega/\square$ or Ω/Sq.)
ρ_V	Volume resistivity	ohm(meter ($\Omega{\cdot}$m)
ρ_V	Charge volume density	columb/meter^3 (C/m^3)

∇ Del operator: $\nabla \equiv \dfrac{\partial}{\partial x}\hat{x} + \dfrac{\partial}{\partial y}\hat{y} + \dfrac{\partial}{\partial z}\hat{z}$

$\vec{s}$	Surface vector	meter^2 (m^2)
v	Volume	meter^3 (m^3)
$\vec{l}$	Linear length vector	meter (m)
V	Potential, electromotive force	volt (V)
t	Time	second (Sec)
Φ	Magnetic flux	webber (Wb)
$\eta_{\vec{B}}$	Energy density in magnetic field	joule/meter^3 (J/m^3)
L	Inductance	henry (H)
I_C	Conduction current	ampere (A)
I_d	Displacement current	ampere (A)

β Phase constant, $\beta = \dfrac{2\pi}{\lambda} = \omega\sqrt{\mu_0 \varepsilon_0}$ radians/meter (rad/m)

GROUNDING-RELATED STANDARDS, SPECIFICATIONS, AND HANDBOOKS

Important Notice and Disclaimer

The list of grounding-related standards, specifications, guides, and handbooks is included here for information only, and in no way is the list all-encompassing. Many other national and international codes, standards, and handbooks exist, and for practical purposes could not be included in this Appendix. Also, standards and handbooks are often revised or canceled, and new ones are produced. Users of this book should, therefore, obtain the latest and valid version of applicable national and international codes, standards, guidelines, and handbooks for their particular application.

D.1. ANSI/ESD STANDARDS

D.1.1. ANSI/ESD S20.20-2007. *ESD Association Standard for the Development of an Electrostatic Discharge Control Program for Protection of Electrical and Electronic Parts, Assemblies, and Equipment (Excluding Electrically Initiated Explosive Devices).* This document applies to activities that manufacture, process, assemble, install, package, label, service, test, inspect, transport, or otherwise handle electrical or electronic parts, assemblies, and equipment susceptible to damage by electrostatic discharge greater than or equal to 100 V HBM. Activities that handle items that are susceptible to less than 100 V HBM may require additional control elements or adjusted limits. Processes designed to handle items that have an ESD sensitivity less than 100 V HBM can still claim compliance to this standard. This document does not apply to electrically initiated explosive devices, flammable liquids, or powders.

D.2. ATIS STANDARDS

D.2.2. ATIS Standard ATIS-0600321.2005 (December 2005). *Electrical Protection for Network Operator-Type Equipment Positions.* The purpose of this standard is to provide ESD-mitigative measures that are intended to control ESDE in the operator–user environment, and to provide electrical protection measures that are intended to minimize potential differences at the network operator-type equipment position. It is no guarantee against the occurrence of ESD or other electrical disturbances caused by ESD.

D.2.2. ATIS Standard ATIS-0600332.2005 (December 2005). *Electrical Protection of Network-Powered Broadband Facilities.* The electrical protection, bonding, and grounding measures presented in this standard are intended to assist in protecting persons, equipment, and property from the effects of lightning, commercial power system faults, and electromagnetic interference (EMI) on the network-powered broadband facilities.

D.3. BRITISH STANDARDS

D.3.1. BS 7430:1998. *Code of Practice for Earthing.* This British standard gives guidance on the methods that may be adopted to earth an electrical system for the purpose of limiting the potential (with respect to the general mass of the earth) of current-carrying conductors forming part of the system, and noncurrent-carrying metalwork associated with equipment, apparatus, and appliances connected to the system. The standard applies only to land-based installations. It does not address electromagnetic compatibility requirements for earthing, nor does it give recommendations for functional earthing.

D.3.2. BS 6651:1992. *Protection of Structures against Lightning.* This British standard outlines the general technical aspects of lightning, illustrating its principal electrical, thermal, and mechanical effects. Guidance is given on how to assess the risk of being struck and it offers a method of compiling an index figure as an aid in deciding if a particular structure is in need of protection. It provides guidance on the design of systems for the protection of structures against lightning and on the selection of materials. Recommendations are made for special cases such as explosives stores and temporary structures such as cranes and spectator stands constructed of metal scaffolding. Off-shore oil and gas installations are not included.

D.4. CENELEC AND ETSI PUBLICATIONS

D.4.1. ETS 300 253 (January 1995). *European Telecommunication Standard, Equipment Engineering (EE); Earthing and Bonding of Telecommunication Equipment in Telecommunication Centres.* This European Telecommunication Standard

(ETS) applies in telecommunication centers and similar installations to the bonding network of the building, the bonding network of the equipment, and the interconnection between these two networks. It contributes to the standardization of telecommunication equipment and coordinates with the prerequisites of installation conditions to achieve safety from electrical hazards, reliable signal reference, and satisfactory electromagnetic compatibility (EMC) performance. A defined bonding configuration down to the equipment level facilitates the installation, operation, and maintenance of telecommunication centers in telecommunication buildings or similar installations, independent of the equipment supplier.

D.4.2. CENELEC Report R044-001:1999. *Safety of Machinery Guidance and Recommendations for the Avoidance of Hazards Due to Static Electricity.* This document gives guidance and recommendations for avoiding ignition and electric shock hazards arising from static electricity. It deals with the problems of static electricity that can give rise to ignition of flammable substances and to electric shock. Basic information about the generation of undesirable static electricity in solids, liquids, and gases, and also on persons, together with descriptions of how the charges generated cause ignitions or electric shocks, is given in the annexes. The processes that most commonly give rise to problems of static electricity are described in detail. The processes include the handling of different types of liquids, powders, gases, and sprays. In each case, the source and nature of the electrostatic hazard are identified and specific recommendations are given for dealing with them. This report is not applicable to the hazards of static electricity relating to lightning, damage to electronic components, medical hazards, or the handling and care of detonators and explosives.

D.5. IEC STANDARDS

D.5.1. IEC 60950-1:2005. *Information Technology Equipment—Safety—Part 1: General requirements.* IEC 60950-1 includes the basic requirements for the safety of information technology equipment. Additional parts of IEC 60950-1 cover specific safety requirements for information technology equipment having limited applications or having special features as follows: Part 21: Remote feeding; Part 22: Equipment installed outdoors; Part 23: Large data storage equipment.

D.5.2. IEC 61000-5-1:1996. *Electromagnetic Compatibility (EMC)—Part 5: Installation and Mitigation Guidelines—Section 1: General Consideration.* This technical report covers general considerations and guidelines on mitigation methods aimed at ensuring electromagnetic compatibility (EMC) among electrical and electronic apparatus or systems used in industrial, commercial, and residential installations. This technical report is intended for use by installers, users, and, to some extent, manufacturers of sensitive electrical or electronic installations and systems, and equipment with high emission levels that could degrade the overall electromagnetic (EM) environment. It applies primarily to new installations but, where economically feasible, it may be applied to extensions or modifications to existing facilities. Specific topics, such as recommendations on the design and implementation of the earthing system, including the earth electrode and the earth network, the design and implementation of bonding apparatus or systems to earth or to the earth network, the selection and installation of appropriate cables, and the design and implementation mitigation means involving shielded enclosures, high-frequency filters, isolating transformers, surge-protective devices, and so on are addressed in other sections of Part 5. The recommendations presented in this report address the EMC concerns of the installation, not the safety aspects of the installation nor the efficient transportation of power within the installation. Nevertheless, these two prime objectives are taken into consideration in the recommendations concerning EMC. These two primary objectives can be implemented concurrently for enhanced EMC of the installed sensitive apparatus or systems without conflict by applying the recommended practices presented in this report and the relevant safety requirements such as those of IEC 364. As each installation is unique, it is the responsibility of the designer and the installer to select the relevant recommendations most appropriate to a particular installation.

D.5.3. IEC 61000-5-2. *Electromagnetic Compatibility (EMC)—Part 5: Installation and Mitigation Guidelines, Section 2: Earthing and Cabling.* This technical report covers guidelines for the earthing and cabling of electrical and electronic systems and installations aimed at ensuring electromagnetic compatibility (EMC) among electrical and electronic apparatus or systems. It deals with earthing practices and with cables used in industrial, commercial, and residential installations. The guide is intended for use by installers and users, and, to some extent, manufacturers of sensitive electrical or electronic installations and systems, and equipment with high emission levels that could degrade the overall electromagnetic environment. It applies primarily to new installations but, where economically feasible, it may be applied to extensions or modifications to existing facilities.

D.5.4. IEC 60364-1:2001. *Electrical Installations of Buildings—Part 1: Fundamental Principles, Assessment of General Characteristics, Definitions.* This part contains the rules for the design and erection of electrical installations so as to provide safety and proper functioning for the use intended. It states the guiding principles without specifying detailed technical requirements. The various types of distribution system earthing are covered.

D.5.5. IEC60364-4-41:2005. *Low-Voltage Electrical Installations—Part 4-41: Protection for Safety—Protection against Electric Shock.* This part specifies essential requirements regarding protection against electric shock, including basic protection (protection against direct contact) and fault protection (protection against indirect contact) of persons and livestock. It also deals with the application and coordination of these requirements in relation to external influences. Requirements are also given for the application of additional protection in certain cases. Has the status of a group safety publication in accordance with IEC Guide 104.

D.5.6. IEC 60364-4-44: 2007-08. *Low-Voltage Electrical Installations—Part 4-44: Protection for Safety—Protection against Voltage Disturbances and Electromagnetic Disturbances.* Part 4-44 of IEC 60364 covers the protection of electrical installations and measures against voltage disturbances and electromagnetic disturbances. The rules of this part are not intended to apply to systems for distribution of energy to the public, or power generation and transmission for such systems (see the scope of IEC 60364-1), although such disturbances may be conducted into or between electrical installations via these supply systems. The requirements are arranged into four clauses as follows: Clause 442 Protection of low-voltage installations against temporary overvoltages due to earth faults in the high-voltage system and due to faults in the low-voltage system; Clause 443 Protection against overvoltages of atmospheric origin or due to switching; Clause 444 Measures against electromagnetic influences; and Clause 445 Protection against undervoltage.

D.5.7. IEC 60364-5-54:2002. *Electrical installations of buildings—Part 5-54: Selection and Erection of Electrical Equipment—Earthing Arrangements, Protective Conductors, and Protective Bonding Conductors.* This part addresses earthing arrangements, protective conductors, and protective bonding conductors in order to satisfy the safety of an electrical installation. It stipulates the minimum sizes for earth electrodes of commonly used materials from the point of view of corrosion and mechanical strength where embedded in soil. Minimum cross-sectional areas of protective conductors and protective bonding conductors are also indicated.

D.5.8. IEC 61024-1:1993. *Protection of Structures against Lightning—Part 1: General Principles.* This part establishes the fundamental definitions and general principles of lighting protection and also provides the necessary information concerning design, construction, and materials to facilitate the basic installation of external and internal lighting protection systems of common structures. It also stipulates the basic requirements for good maintenance and inspection practices. It provides information on the classification of structures according to the consequential effects of a lightning strike and on procedures for selection of a lightning protection system (LPS) giving an adequate level of protection.

D.5.9. IEC 61312-1:1995. *Protection against Lightning Electromagnetic Impulse—Part 1: General Principles.* This standard provides information for the design, installation, inspection, maintenance, and testing of an effective lightning protection system for information systems in or on a structure. The system equipment itself is not considered in this document. However, the content provides guidelines for the cooperation between the designer of the information system and the designer of the protection system against lightning electromagnetic impulses in order to achieve optimum protection effectiveness.

D.5.10. IEC 61312-2:1999. *Protection against Lightning Electromagnetic Impulse—Part 2: Shielding of Structures, Bonding inside Structures, and Earthing.* This document provides methods for the evaluation of the effectiveness of shielding measures against lightning electromagnetic impulses (LEMP) for structures with information equipment such as electronic systems in case of direct and nearby lightning strikes. In addition it provides rules for bonding measures inside structures and for earthing methods relating to LEMP.

D.5.11. IEC 61312-4:1998. *Protection against Lightning Electromagnetic Impulse—Part 4: Protection of Equipment in Existing Structures.* The document gives guidelines for protection of information technology equipment against lightning electromagnetic impulses. It provides a checklist to determine whether protective measures for the electronic system are needed and, if so, to identify the most cost-effective protection measures for that equipment.

D.5.12. IEC 61643-12. *Low-Voltage Surge Protective Devices—Part 12: Surge Protective Devices Connected to Low-Voltage Power Distribution Systems—Selection and Application Principles.* This IEC document provides information to the user about characteristics useful for the selection of a surge protection device (SPD). It gives guidance to perform a risk analysis and how to evaluate the need for using SPDs in low-voltage systems. Information is given on selection and coordination of SPDs, while taking into account the entire environment in which they are applied. Details are provided on the temporary overvoltages in the low-voltage system due to faults between high-voltage systems and earth.

D.5.13. IEC 62305-1:2006. *Protection against Lightning—Part 1: General Principles.* This part of IEC 62305 provides the general principles to be followed in the protection against lightning of structures, including their installations and contents as well as persons, and services connected to a structure.

D.5.14. IEC 62305-3:2006. *Protection against Lightning—Part 3: Physical Damages to Structure and Life Hazards.* This part of IEC 62305 deals with the protection, in and around a structure, against physical damage and injury to living beings due to touch and step voltages. The main and most effective measure for protection of structures against physical damage is considered to be the lightning protection system (LPS). It usually consists of both external and internal lightning protection systems. This standard is applicable to (a) design, installation, inspection, and maintenance of an LPS for structures without limitation of their height, and (b) establishment of measures for protection against injury to living beings due to touch and step voltages.

D.5.15. IEC 62305-4:2006. *Protection against Lightning—Part 4: Electrical and Electronic Systems within Structures.* This part of IEC 62305 provides information for the design, installation, inspection, maintenance, and testing of a lightning electromagnetic impulse protection measures system (LPMS) for electrical and electronic systems within a structure, able to reduce the risk of permanent failures due to lightning electromagnetic impulses. This standard does not cover protection against electromagnetic interference due to lightning, which may cause malfunctioning of electronic systems. However, the annexed information can also be used to evaluate such disturbances. This standard provides guidelines for cooperation between the designer of the electrical or electronic system and the designer of the protection measures, in an attempt to achieve optimum protection effectiveness.

D.6. IEEE STANDARDS

D.6.1. IEEE-STD-80:2000. *Guide for Safety in AC Substation Grounding.* This guide provides guidance and information pertinent to safe grounding practices in AC substation design. It gives a review of substation grounding practices with special reference to safety, and develops criteria for a safe design. Procedures for the design of practical grounding systems are discussed. As a basis for design, it establishes the safe limits of potential differences that can exist in a substation under fault conditions between points that can be contacted by the human body. Analytical methods are developed as an aid in the understanding and solution of typical gradient problems. This guide is primarily concerned with safe grounding practices for power frequencies in the range of 50–60 Hz.

D.6.2. IEEE-STD-81:1983. *Guide for Measuring Earth Resistivity, Ground Impedance, and Earth Surface Potentials of a Ground System.* The guide describes and discusses the techniques of measuring ground resistance and impedance, earth resistivity, potential gradients from currents in the earth, and the prediction of the magnitudes of ground resistance and potential gradients from scale model tests. Factors influencing the choice of instruments and the techniques for various types of measurements are covered. These include the purpose of the measurement, the accuracy required, the type of instruments available, possible sources of error, and the nature of the ground or grounding system under test. The guide is intended to provide assistance in obtaining and interpreting accurate, reliable data. It describes test procedures that promote the safety of personnel and property, and prevent interference with the operation of neighboring facilities.

D.6.3. IEEE-STD-141:1993. *Recommended Practice for Electric Power Distribution for Industrial Plants.* This comprehensive publication provides a recommended practice for the electrical design of industrial facilities and power distribution in industrial plants. It presents a thorough analysis and detailed procedures of the electrical power distribution system. Complete information on electrical design criteria is provided to ensure safety and preservation of property. The guide is of greatest value to power-oriented engineers with limited industrial plant experience.

D.6.4. IEEE-STD-142:1991. *IEEE Recommended Practice for Grounding of Industrial and Commercial Power Systems.* The problem of system grounding, that is, connection to ground of the neutral, corner of the delta, or midtap of one phase, are covered. The advantages and disadvantages of grounded versus ungrounded systems are discussed. Information is given on how to ground the system, where the system should be grounded, and how to select equipment for the grounding of neutral circuits. Connecting the frames and enclosures of electric apparatus, such as motors, switchgear, transformers, buses, cables conduits, building frames, and portable equipment, to a ground system is addressed. The fundamentals of making the interconnection of ground-conductor system between electric equipment and the ground rods, water pipes, and so on are outlined. The problems of static electricity—how it is generated, what processes may produce it, how it is measured, and what should be done to prevent its generation or to drain the static charges to earth to prevent sparking—are treated. Methods of protecting structures against the effects of lightning are also covered. Obtaining a low-resistance connection to the earth, use of ground rods, connections to water pipes, and so on are discussed. A separate chapter on sensitive electronic equipment is included.

D.6.5. IEEE-STD-242:1986. *Recommended Practice for Protection and Coordination of Industrial and Commercial Power Systems.* This guide deals with the proper selection, application, and coordination of the components that constitute system protection for industrial plants and commercial buildings. System protection and coordination serve to minimize damage to a system and its components in order to limit the extent and duration of any service interruption occurring on any portion of the system. It presents complete information on protection and coordination principles designed to protect industrial and commercial systems against any abnormalities that could reasonably be expected to occur in the course of system operation.

D.6.6. IEEE-STD-1050:1996. *Guide for Instrumentation Control Equipment Grounding in Generating Stations.* This application guide provides information on instrumentation and control (I&C) equipment-grounding methods to achieve both a suitable level of protection for personnel and equipment, and to provide suitable electric noise immunity for signal ground references in generating stations. Details are given on grounding of instrument chassis, racks, cable sheaths, or cable shields and signal pairs. Basic theory and guidelines are provided for designing I&C grounding.

D.6.7. IEEE-STD-1100:1992. *IEEE Recommended Practice for Powering and Grounding Sensitive Electronic Equipment.* Recommended design, installation, and maintenance practices for electrical power and grounding (including both power-related and signal-related noise control) of sensitive electronic processing equipment used in commercial and industrial applications are presented. The main objective is to provide a consensus of recommended practices in an area where conflicting information and confusion, stemming primarily from different view points of the same problem, have dominated. Practices herein address electronic equipment performance issues while maintaining a safe installation. A brief description is given of the nature of power quality problems, possible solutions, and the resources available for assistance in dealing with problems. Fundamental concepts are reviewed. Instrumentation and procedures for conducting a survey of the power distribution system are described. Site surveys and site power analysis are considered. Case histories are given to illustrate typical problems.

D.7. INTERNATIONAL SPACE STATION (ISS) PROGRAM STANDARDS

D.7.1. SSP 30240, Revision C (22 December 1998). *International Space Station Program, Space Station Grounding Requirements.* This document defines the requirements for Space Station grounding, including primary power grounds, secondary power grounds, signal reference grounds, signal return grounds, and grounds for user-conditioned power return/references. These requirements are in accordance with the specifications in SSP 30243.

D.7.2. SSP 30245, Revision E (15 October 1999). *International Space Station Program, Space Station Electrical Bonding Requirements.* This requirements document defines the Space Station electrical bonding requirements in accordance with SSP 30243, Space Station Systems Requirements for Electromagnetic Compatibility (EMC), and SSP 30240, Space Station Grounding Requirements. These requirements address the characteristics, application, and testing of electrical bonding. Joining of surfaces, objects, equipment, and structures must satisfy the requirements of this document, regardless of application or conductivity.

D.8. ITU-T RECOMMENDATIONS

D.8.1. ITU-T Recommendation K.27 (05/96). *Protection against Interference, Bonding Configurations, and Earthing Inside a Telecommunication Building.* This recommendation provides guidance on bonding and earthing of telecommunication equipment in telephone exchanges and similar telecommunication switching centers and is intended to achieve safety requirements on AC power installations. It is applicable to installation of new telecommunication centers, and for expansion and replacement of systems in existing centers. The document treats coordination with external lightning protection, but does not provide details of protective measures specific to telecommunication buildings. It covers the shielding contribution of the effective elements of the building and addresses shielding provided by cabinets, cable trays, and cable shields.

D.8.2. ITU-T Recommendation K.31 (03/93). *Protection against Interference, Bonding Configurations, and Earthing of Telecommunication Installation inside a Subscriber's Building.* This recommendation is a guide to bonding and earthing of telecommunication equipment in residential and commercial subscribers' premises. It is intended to comply with IEC or national standardizing bodies on AC power installations, and is intended for use with new installations as well as for expansion and replacement of existing installations. It is intended to encourage planning for electromagnetic compatibility, which should include bonding and earthing arrangements that accommodate installation tests and diagnostics, but does not necessarily provide protection for the installation in the case of a direct lightning stroke to the building, nor is it intended to replace national regulations on bonding configurations and earthing.

D.8.3. ITU-T Recommendation K.35 (05/96). *Protection Against Interference, Bonding Configurations and Earthing at Remote Electronic Sites.* This recommendation covers bonding configurations and earthing for equipment located at remote electronic sites such as switching or transmission huts, cabinets, or controlled environmental vaults with only one level, a need for AC mains power service, and a floor space of about 100 m^2 without an antenna tower on the roof of the building or nearby, but which are more substantial than small electronic housings, such as carrier repeaters or distribution terminals. Experience in the operation of electronic equipment enclosures shows that the use of a bonding configuration and earthing that are coordinated with equipment capability and with electrical protection devices has the following attributes: promotes personnel safety and reduces fire hazards; enables signaling with earth return (functional earthing); minimizes service interruptions and equipment damage caused by lightning, exposures to power lines, and faults in internal DC power supplies; minimizes radiated and conducted emissions and susceptibility; and improves system tolerance to discharge of electrostatic energy.

D.8.4. ITU-T Recommendation O.9 (03/99). *Specifications of Measuring Equipment, General, Measuring Arrangements to Assess the Degree of Imbalance about Earth.* This recommendation describes arrangements for measuring the following parameters: longitudinal conversion loss, transverse conversion loss, longitudinal conversion transfer loss, transverse conversion transfer loss, input longitudinal interference loss, common-mode rejection, and output signal balance. In practice, the above parameters are the seven most significant imbalance parameters. Limits for these parameters, special considerations for terminating impedances, and the measurement frequencies to be used are given in the relevant recommendation for the item under test.

D.9. MILITARY STANDARDS AND HANDBOOKS

D.9.1. MIL-STD-171E (23JUNE 1989). *Department of Defense Manufacturing Process Standards, Finishing of Metal and Wood Surfaces.* This standard establishes and updates general finish codes and serves as a general guide to the selection of suitable materials, procedures, and systems for cleaning, plating, painting, and otherwise finishing metal and wood surfaces.

D.9.2. MIL-STD-188-114A (30 September 1985). *Military Standard, Electrical Characteristics of Digital Interface Circuits.* This document specifies the electrical characteristics of unbalanced voltage and balanced voltage digital interface circuits, normally implemented in integrated circuit (IC) technology. These circuits are employed for the interchange of serial digital binary signals between and among data terminal equipment (DTE) and data circuit terminating equipment (DCE), or in any interconnection of binary signals between physically separated equipment, regardless of the type of information, such as digitized voice or data, that is represented by the binary signals. The technical parameters promulgated by this document represent, in general, minimum interoperability and performance characteristics, which may be exceeded in order to satisfy specific requirements. Additional nonstandard interface characteristics may also be implemented to satisfy specific interoperability requirements, provided that a basic capability exists to interoperate with a standard interface as stated in this document.

D.9.3. MIL-STD-188-124B (1 May 1998). *Military Standard, Grounding, Bonding and Shielding for Common Long Haul/Tactical Communication Systems, Including Ground-Based Communication–Electronics Facilities and Equipment.* This standard establishes the minimum basic requirements and goals for grounding, bonding, and shielding of ground-based communications–electonics (C–E) equipment installations, subsystems, and facilities, including buildings and structures supporting tactical and long-haul military communication systems. This standard addresses the facilities ground systems, as well as grounding, bonding, and shielding and lightning protection for telecommunications C–E facilities and equipment. Grounding for building and structures is listed under the headings of Earth Electrode Subsystem, Fault Protection Subsystem, Lightning Protection Subsystem, and Signal Reference Subsystem.

D.9.4. MIL-STD-188-125-1 (17 July 1998). *Department of Defense Interface Standard, High-Altitude Electromagnetic-Pulse (HEMP) Protection for Ground-Based C⁴I Facilities Performing Critical Time-Urgent Missions, Part 1: Fixed Facilities.* This standard establishes minimum requirements and design objectives for high-altitude electromagnetic-pulse (HEMP) hardening of fixed, ground-based facilities that perform critical, time-urgent command, control, communications, computer, and intelligence (C^4I) missions. This standard prescribes minimum performance requirements for low-risk protection from mission-aborting damage or upset due to HEMP threat environments defined in MIL-STD-2169. This standard also addresses minimum testing requirements for demonstrating that prescribed performance has been achieved and for verifying that the installed protection subsystem provides the operationally required hardness for the completed facility.

D.9.5. MIL-STD-188-125-2 (3 March 1999). *Department of Defense Interface Standard, High-Altitude Electromagnetic-Pulse (HEMP) Protection for Ground-Based C^4I Facilities Performing Critical Time-Urgent Missions, Part 2: Transportable Systems.* This standard establishes minimum requirements and design objectives for high-altitude electromagnetic-pulse (HEMP) hardening of transportable ground-based systems that perform critical, time-urgent command, control, communications, computer, and intelligence (C^4I) missions. This standard prescribes minimum performance requirements for low-risk protection from mission-aborting damage or upset due to HEMP threat environments defined in MIL-STD-2169. The standard also addresses minimum testing requirements for demonstrating that prescribed performance has been achieved and for verifying that the installed protection measures provide the operationally required HEMP hardness for the completed system.

D.9.6. MIL-HDBK-232A (20 March 1987). *Military Handbook, Red/Black Engineering—Installation Guidelines.* This handbook provides guidance within the red/black concept for the engineering and installation of systems and facilities processing classified information. The engineering installation concepts contained herein should be selectively applied for control of TEMPEST at all Department of Defense (DoD) facilities where classified information is processed. This handbook addresses and applies to the following general areas: (a) power distribution, installation, and protection; (b) equipment installation and protection; (c) signal distribution, installation, and protection; (d) filters and isolators; (e) grounding, bonding, and shielding (GBS); (f) physical security; and (g) administrative telephones.

D.9.7. MIL-HDBK-241B (30 September 1983). *Military Handbook, Design Guide for Electromagnetic Interference (EMI) Reduction in Power Supplies.* This handbook offers guidance to power supply designers in techniques that have been found effective in reducing conducted and radiated interference generated by power supplies. It is a compilation of information from library sources; pertinent military laboratory programs, including contracts to universities and industry; and practical fixes derived from the experience of engineers.

D.9.8. MIL-HDBK-263B (31 July 1994). *Military Handbook, Electrostatic Discharge Control Handbook for Protection of Electrical and Electronic Parts, Assemblies, and Equipment (Excluding Electrically Initiated Explosive Devices).* This handbook provides guidance for developing, implementing, and monitoring an ESD control program in accordance with the requirements of MIL-STD-1686. This handbook is not applicable to electrically initiated explosive devices. The specific guidance provided is supplemented by the technical data contained in the appendices.

D.9.9. MIL-HDBK-274 (AS), Notice 1 (29 June 1990). *Military Handbook, Electrical Grounding for Aircraft Safety.* The purpose of this handbook is to provide aircraft maintenance personnel with the information required for electrical safety grounding of each type of operational aircraft in the U.S. Navy inventory [Navy aircraft use power equipment not equipped with ground fault interrupters (GFIs)]. In addition, this handbook provides background information pertaining to the operational concerns for aircraft grounding, static electricity theory and how it affects aircraft, and techniques used for measurement of grounding points.

D.9.10. MIL-HDBK-419A (29 December 1987). *Military Handbook, Grounding, Bonding and Shielding for Electronic Equipment and Facilities.* This handbook addresses the practical considerations for engineering of grounding systems, subsystems, and other components of ground networks. Electrical noise reduction is discussed as it relates to the proper installation of ground systems. Power distribution systems are covered to the degree necessary to understand the interrelationships between grounding, power distribution, and electrical noise reduction. The information provided in this handbook primarily concerns grounding, bonding, and shielding of fixed-plant telecommunications–electronics facilities; however, it also provides basic guidance in the grounding of deployed transportable communications/electronics equipment. Grounding, bonding, and shielding are approached from a total system concept, which comprises four basic subsystems in accordance with current Department of Defense (DOD) guidance. These subsystems are (1) an earth electrode subsystem, (2) a lightning protection subsystem, (3) a fault protection subsystem, and (4) a signal reference subsystem.

D.9.11. MIL-HDBK-411B (15 MAY 1990). *Military Handbook, Power and the Environment for Sensitive DoD Electronic Equipment (General).* This three-volume handbook is a reference for the planning and engineering of power and environmental control systems for fixed Department of Defense (DoD) communications, data processing, and information systems facilities. The engineering concepts contained herein should be selectively applied to the power and environmental elements of DoD fixed facilities. DoD communications and data processing installations include equipment rooms and spaces needing more precisely controlled environments than comfort spaces. The more limiting parameters within these rooms and spaces are established specifically for the equipment being used. Outside these specially designed areas, where PCs and other electronic office equipment are used, the environment is the responsibility of the user. Power protection or conditioning for this equipment should follow the guidance provided in Volume II. Environmental control of this space should follow the guidance contained in Volume III. Volume I addresses these subjects in general terms for the planner, manager, or executive. Volume II addresses power system engineering considerations. Volume III addresses environmental control system engineering considerations.

D.9.12. MIL-STD-461E (20 August 1999). *Department of Defense Interface Standard, Requirements for the Control of Electromagnetic Interference Characteristics of Subsystems and Equipment.* This standard establishes interface and associated verification requirements for the control of the electromagnetic interference (emission and susceptibility) characteristics of electronic, electrical, and electromechanical equipment and subsystems designed or procured for use by activities and agencies of the Department of Defense. Such equipment and subsystems may be used independently or as an integral part of other subsystems or systems. This standard is best suited for items that have the following features: electronic enclosures that are no larger than an equipment rack, electrical interconnections that are discrete wiring harnesses between enclosures, and electrical power input derived from prime power sources. This standard should not be directly applied to items such as modules located inside electronic enclosures or entire platforms. The principles in this standard may be useful as a basis for developing suitable requirements for those applications. Data item requirements are also included.

D.9.13. MIL-STD-464A (19 December 2002). *Department of Defense Interface Standard, Electromagnetic Environmental Effects Requirements for Systems.* This standard establishes electromagnetic environmental effects (E^3) interface requirements and verification criteria for airborne, sea, space, and ground systems, including associated ordnance.

D.9.14. MIL-STD-889B, Notice 3 (USAF) (17 May, 1993). *Military Standard, Dissimilar Metals, May 1993.* This standard defines and classifies dissimilar metals, and establishes requirements for protecting coupled dissimilar metals, with attention directed to the anodic member of the couple, against corrosion.

D.9.15. MIL-STD-1310G (NAVY) (28 June 1996). *Department of Defense Standard, Practice for Shipboard Bonding, Grounding, and Other Techniques for Electromagnetic Compatibility and Safety.* This document specifies the performance requirements for shipboard bonding, grounding and shielding; identifies requirements for EMI control in the areas of hull-generated and equipment-generated EMI, hull and cable penetration EMI, and superstructure blockage/reflections; specifies requirements for protection of personnel from electrical shock, and identifies the applicable requirements for measuring the effectiveness of EMI control and safety measures implemented when requirements herein are invoked.

D.9.16. MIL-STD-1377 (NAVY) (20 August 1971). *Military Standard, Effectiveness of Cable, Connector, and Weapon Enclosure Shielding and Filters in Precluding Hazards of Electromagnetic Radiation to Ordnance, Measurement of.* This standard is intended to provide a weapon developer or designer with shielding and filter effectiveness test methods for determining whether the particular weapon design requirements of MIL-P-24014 have been properly implemented. It is not intended to be a substitute for full-scale electromagnetic hazard evaluation tests of the weapon system but, rather, an aid in developing a weapon system with a high probability of successfully passing such environmental tests.

D.9.17. MIL-STD-1399 (Navy), Section 300A, Notice 1 (11 March 1992). *Military Standard, Interface Standard for Shipboard Systems, Section 300A, Electric Power, Alternating Current (Metric).* This section establishes electrical interface characteristics for shipboard equipment utilizing AC electric power to ensure compatibility between user equipment and the electrical power system. Characteristics of the electric power system are defined and tolerances are established. User equipment must operate from a power system having these characteristics and be designed within these constraints in order to reduce adverse effects of the user equipment on the electric power system. Test methods are included for verification of compatibility.

D.9.18. MIL-STD-1399 (SH), Section 406B (1 July 1995). *Military Standard, Interface Standard for Shipboard Systems, Section 406, Digital Computer Grounding (Metric).* This section establishes a standard ground system and the requirements for grounding shipboard digital computer equipment, to ensure compatibility between such grounding systems and digital computer systems/equipment, which will enhance system operational reliability.

D.9.19. MIL-STD-1541A (USAF) (30 DEC 1987). *Military Standard, Electromagnetic Compatibility Requirements for Space Systems.* This standard establishes the electromagnetic compatibility requirements for space systems, including frequency management and the related requirements for the electrical and electronic equipment used in space systems. It also includes requirements designed to establish an effective ground reference for the installed equipment and inhibit adverse electrostatic effects. The purposes of this standard are (a) to define minimum performance requirements for electromagnetic compatibility; (b) to identify the system relationships pertinent to electromagnetic compatibility; (c) to identify requirements for system and equipment engineering designed to enable achieving compatibility in a timely, predictable, and economical manner; and (d) to define requirements for equipment and system tests and analyses to demonstrate compliance with this standard.

D.9.20. MIL-STD-1542B (USAF) (15 November 91). *Military Standard, Electromagnetic Compatibility and Grounding Requirements for Space System Facilities.* The purpose of this standard is to specify the design, performance, and verification requirements for electrical subsystems for space system facilities, including electromagnetic compatibility (EMC), electrical power, grounding, bonding, shielding, lightning protection, and TEMPEST security. These requirements are interrelated and interdependent, and, therefore, require an integrated approach in the design.

D.9.21. MIL-STD-1568B (USAF), Notice 1 (12 October 1994). *Military Standard, Materials and Processes for Corrosion Prevention and Control in Aerospace Weapons Systems.* This standard establishes the requirements for materials, processes and techniques, and identifies the tasks required to implement an effective corrosion prevention and control program during the conceptual, validation, development and production phases of aerospace weapon systems. The intent is to minimize life cycle cost due to corrosion and to obtain improved reliability.

D.9.22. MIL-STD-1576 (31 July, 1984). *Military Standard, Electroexplosive Subsystem, Safety Requirements and Test Methods for Space Systems.* This standard establishes the general requirements and test methods for the design and development of electroexplosive subsystems to preclude hazards from unintentional initiation and from failure to fire. These requirements apply to all subsystems utilizing electrically initiated explosive or pyrotechnic components. This standard applies to all space vehicle systems (e.g., launch vehicles, upper stages, boosters, payloads, and related systems).

D.9.23. MIL-HDBK-1857 (27 March 1998). *Department of Defense Handbook, Grounding, Bonding, and Shielding Design Practices.* This standard covers the characteristics of grounding, bonding and shielding design practices to be applied in the construction and installation of marine, fixed station, transportable, and ground mobile electronic equipment subsystems and systems.

D.9.24. MIL-B-5087B Interim Amendment 3, USAF (Cancelled) (24 December 1984). *Military Specification, Bonding, Electrical, and Lightning Protection, for Aerospace Systems.* This specification covers the characteristics, application, and testing of electrical bonding (see Section 6.2.2) for aerospace systems, as well as bonding for the installation and interconnection of electrical and electronic equipment therein, and lightning protection

D.9.25. AFSC Design Handbook DH 1-4 (27 March 1998). *Electromagnetic Compatibility.* This handbook provides system designers with electromagnetic compatibility design principles, information, guidance, and criteria, and establishes a central source of electromagnetic-compatibility design data (any type of factual information that can be used as a basis for design decisions). This handbook partially implements AFSCR 8-4.

D.9.26. Air Force Qualification Training Package 2EXXX-202D (1 October 1999). *EI TEMPEST Installation Handbook.* This handbook is intended for use by Air Force Communications Electronic Engineering Installation (EI) personnel. It provides guidance on the installation of C–E equipment in an environment where emission security (EMSEC) is a consideration IAW red/black installation criteria. It is primarily designed for the inexperienced EI team members and team chiefs who are unfamiliar with TEMPEST-related installations. Team chiefs who are familiar with EMSEC and red/black installation criteria can use the handbook as a reference guide and as a tool to train team members. This handbook does not deal with standard installation practices; rather, it explains the fundamentals and concepts behind EMSEC and red/black installation criteria. Formal installation training is provided by each EI unit. EI trainers use AFJQS 2EXXX-202B, *Standard Installation Practices—Electronics/Inside Plant,* to plan, conduct, and document qualification training. Contents of the handbook are *not* to be used as a basis for inspection or evaluation. This handbook is a specialized publication for familiarization and training purposes only; it is *not* a technical reference.

D.9.27. Pacific Air Forces PACAF Instruction 33-103 (26 April 1999). *Communications and Information Deployable Communications Standards—Safety.* This instruction implements policy found in Air Force Policy Directive 33-1, Command, Control, Communications, and Computer (C^4) Systems. This publication provides policy and guidance for Ground Theater Air Control System (GTACS) units, combat communications units, Pacific Initial Communications Package (PICP), and communications squadrons supporting the Expeditionary Air Forces (EAF). By standardizing safety guidance and procedures, responsiveness of units in meeting worldwide contingencies is maximized.

D.9.28. Air Force Air Combat Command (ACC) Instruction 33-165 (18 September 1998). *Communications and Information Grounding Techniques.* This publication provides guidance for grounding and lightning protection for deployable communications–electronics (C–E) equipment associated with theater air control system units, combat communications units, and communications squadrons supporting wing-deployable communications packages. It implements policy found in AFPD 33-1, Command, Control, Communications, and Computer (C^4) Systems.

D.10. NASA STANDARDS AND HANDBOOKS

D.10.1. NASA-HDBK-4001 (February 17, 1998). *NASA Technical Handbook, Electrical Grounding Architecture for Unmanned Spacecraft.* This handbook describes spacecraft grounding architecture options at the system level. Implementation of good electrical grounding architecture is an important part of overall mission success for spacecraft. The primary objective of proper grounding architecture is to aid in the minimization of electromagnetic interference (EMI) and unwanted interaction between various spacecraft electronic components and/or subsystems. Success results in electromagnetic compatibility (EMC). This handbook emphasizes that spacecraft grounding architecture is a system design issue, and all hardware elements must comply with the architecture established by the overall system design. A further major emphasis is that grounding architecture must be established during the early conceptual design stages (before subsystem hardware decisions are made). The preliminary design review (PDR) stage is too late. The purpose of this handbook is to provide a ready reference for spacecraft systems designers and others who need information about system grounding architecture design and rationale. The primary goal of this handbook is to show design choices that apply to a grounding system for a given size and mission of spacecraft and to provide a basis for understanding those choices and trade-offs.

D.10.2. NASA-STD-P023 (Draft) (August 17, 2001). *NASA Technical Standard, Electrical Bonding for NASA Launch Vehicles, Spacecraft Payloads, and Flight Equipment.* This standard defines the basic electrical bonding requirements for NASA launch vehicles, spacecraft, payloads, and equipment.

D.10.3. NASA Reference Publication 1368 (June 1995). *Marshall Space Flight Center Electromagnetic Compatibility Design and Interference Control (MEDIC) Handbook, CDDF Final Report, Project No. 93-15.* This handbook is intended to be used primarily by those organizations involved in the electrical design of payload equipment and subsystems. The purpose of this handbook is to provide practical and helpful information in the design of electrical equipment for electromagnetic compatibility (EMC).

D.11. NFPA CODES AND STANDARDS

D.11.1. NFPA 70 (2005). *National Electrical Code 2005, National Fire Protection Association.* The National Electrical Code (NEC) contains provisions that are considered necessary for practical safeguarding of persons and property from hazards arising from the use of electricity. The Code covers (a) installations of electric conductors and equipment within or on public and private buildings or other structures, (b) installations of conductors and equipment that connect to the supply of electricity, (c) installations of other outside conductors and equipment on the premises, (d) installations of optical fiber cables and raceways, and (e) installations in buildings used by the electric utility that are not an integral part of a generating plant, substations, or control center.

D.11.2. NFPA 780 (2004). *Standard for the Installation of Lightning Protection Systems, National Fire Protection Association.* This standard developed by the National Fire Protection Association is intended to provide for the safeguarding of persons and property from hazards arising from exposure to lightning. The document covers traditional lightning protection system installation requirements for (a) ordinary structures, (b) miscellaneous structures and special occupancies, (c) heavy-duty stacks, (d) watercraft, and (e) structures containing flammable vapors, flammable gases, or liquids that give off flammable vapors.

D.12. SAE RECOMMENDED PRACTICES

D.12.1. ARP1870 (April 1999). *Aerospace Recommended Practice, Aerospace Systems Electrical Bonding and Grounding for Electromagnetic Compatibility and Safety.* This document establishes the minimum requirements for the electrical bonding and grounding of electric, avionic, armament, communication, and electronic equipment installations for aeronautical and aerospace applications. The bonding and grounding requirements specified herein ensure that an adequate low-resistance return path for electric, avionic, armament, communication, and electronic equipment is achieved that can withstand operating conditions and corrosion. This is essential for the reduction of coupling of electromagnetic fields into or out of the equipment as well as for providing electrical stability to control the currents and/or voltages caused by static charges and discharges, and for suppressing the hazardous effects thereof.

D.12.2. ARP4043, REV. A (January 1999). *Aerospace Recommended Practice, Flight Line Grounding and Bonding of Aircraft.* This ARP provides the rationale and theory of charges being present on aircraft while on the ground. The necessary implementation of safety practices are explained and defined.

D.13. TIA/EIA STANDARDS

D.13.1. TIA/EIA-607: 1994. *Commercial Building Grounding and Bonding Requirements for Telecommunications.* The purpose of this standard is to enable the planning, design, and installation of telecommunications grounding systems within a building with or without prior knowledge of the telecommunication systems that will subsequently be installed. This telecommunications grounding and bonding infrastructure supports a multivendor, multiproduct environment as well as the grounding practices for various systems that may be installed on customer premises. This standard specifies the requirements for a uniform telecommunications grounding and bonding infrastructure that must be followed within commercial buildings where telecommunications equipment is intended to be installed. This telecommunications grounding and bonding infrastructure, in conjunction with other grounding and bonding systems (e.g., electrical power grounding system, water piping, lightning protection system), make up the building grounding system.

D.14. UL STANDARDS

D.14.1. UL 96A:1998. *Standard for Installation Requirements for Lightning Protection Systems.* This UL standard covers the installation requirements of lightning protection systems on all types of structures other than structures used for the production, handling, or storage of ammunition, explosives, flammable liquids or gasses, and explosive ingredients. These requirements apply to lightning protection systems that are complete and cover all parts of a structure.

D.15. OTHER (MISCELLANEOUS) STANDARDS

D.15.1. API (American Petroleum Institute) Recommended Practice 2003 (6th Edition, 1998). *Protection against Ignitions Arising Out of Static, Lightning, and Stray Currents.* This recommended practice presents the current state of knowledge and technology in the fields of static electricity, lightning, and stray currents applicable to the prevention of hydrocarbon ignition in the petroleum industry.

D.15.2. FAA-STD-019D (August 2002). *Department of Transportation, Federal Aviation Administration Standard, Lightning and Surge Protection, Grounding, Bonding and Shielding Requirements for Facilities and Electronic Equipment.* This document mandates standard lightning protection, transient protection, electrostatic discharge (ESD), grounding, bonding, and shielding configurations and procedures for new facilities, facility modifications, facility upgrades, new equipment installations, and new electronic equipment used in the National Airspace System (NAS). It provides requirements for the design, construction, modification, or evaluation of facilities and equipment. This document does not apply to existing facilities unless the facility is undergoing upgrade or receiving new electronic equipment. This version of the document applies to equipment and facilities and procurement initiated after the effective date of this document. However, if the procurement contract is not in accordance with the version of FAA-STD-019 in effect on the initiation date of the procurement, then the procurement shall be required to meet the version of this standard in effect when the noncompliance is noted. The requirements of this standard provide a systematic approach to minimize electrical hazards to personnel, electromagnetic interference and damage to facilities and electronic equipment from lightning, transients, ESD, and power faults.

ON THE CORRESPONDENCE BETWEEN OHM'S LAW AND FERMAT'S LEAST TIME PRINCIPLE

In Chapter 2, the principle of path of least impedance with respect to return current propagation path was discussed. It was concluded that "current, if not altered by significant amount of impedance, always assumes a distribution that minimizes the impedance of the loop formed by the signal and return path." At lower frequencies, the "path of least impedance" was shown to correspond to the "path of least resistance," whereas, at higher frequencies, it corresponds to the "path of least reactance" or "path of least inductance."

It was shown that this principle is traceable to the principle of conservation of energy, whereby it was stated that "current will flow in the path such that the energy stored in the consequent magnetic field is minimized."

It was further shown that conservation of energy is yet but a special case of the greatest generalization in all physical science, that of "least action," which, simply stated, asserts that nature always finds the most efficient course from one point to another. Fermat's premise of "least time" is one of the manifestations of the "least action" principle.

The objective of this appendix is to demonstrate the equivalence between Ohm's Law,

$$\vec{E} = \frac{1}{\sigma}\,\vec{J}$$

(Equation F of Maxwell's eight original equations) to Fermat's premise of "least time." With this demonstration established, the notion that "current follows the path of least impedance because it constitutes the path of least time and, thus, the path of least action" is proven.

This appendix begins with a fundamental scientific formulation and proceeds to introduce the least time/maximum probability (LT/MP) principle. The LT/MP principle is Fermat's principle of least time and classical probability theory combined [1].

E.1   ORIGIN OF THE LT/MP PRINCIPLE

The origin of the least time/maximum probability (LT/MP) principle can be traced back to a fundamental scientific hypothesis. This hypothesis is phrased as follows: In any frame of reference where the total number of possible outcomes of an observation is N, the number of similar outcomes n is inversely proportional to the average time τ required to complete the process that generates the observed outcome.

The scientific hypothesis is founded on the basis of two classical and elementary concepts:

- The proportionate number of similar outcomes (n/N) represents the classical definition of probability of an observed outcome
- The average time τ needed to complete the process that generates the observed outcome. In a process involving a group of outcome, the inverse of the mean time (τ^{-1}) measures the frequency of occurrence of the observed outcome.

Thus, the scientific hypothesis is a simple statement of a commonly known fact: The probability of the outcome of an observation P is directly proportional to the frequency of occurrence of the observed outcome.

E.2   STATEMENT OF THE LT/MP PRINCIPLE

The parametric equation for the LT/MP scientific premise can be written as follows:

$$P(p) = \frac{1}{N} \cdot \left(\frac{1}{\tau(p)} \right) = \frac{\left(\dfrac{1}{\tau(p)} \right)}{\displaystyle\sum_{all\ \tau} \left(\dfrac{1}{\tau(p)} \right)} \tag{E.1}$$

In Equation (E.1), N represents the total number of possible outcomes and, thus, it clearly demonstrates that the maximum probability is inversely proportional to the least time. This is the least time/maximum probability (LT/MP) principle, given by

$$P_{max}(p) = \frac{1}{N} \cdot \left(\frac{1}{\tau_{min}(p)} \right) \tag{E.2}$$

Electrical current constitutes the transport of an ensemble of charged particles between two points in a circuit. Given that a number of alternative paths (or channels) are available, the total current for which the total number of charged particles is N will be distributed over all available paths. The manner in which the current is distributed over the different paths or channels is in direct proportion to the probability that charged particles fall into a specific channel. The channel that carries that larger current is, by definition, the channel of greater probability, since for that one channel

$$\left(\frac{n_1}{N} \right) = P(n_1) > \left(\frac{n_2}{N} \right) = P(n_2) \tag{E.3}$$

Where n_1 and n_2 represent the number of charged particles in (the alternate) channels 1 and 2, respectively. Therefore, the channel with the largest current is the channel corresponding to the largest probability, also called the "main channel."

The LT/MP principle is, therefore, a selection rule for determining the primary current propagation channel, and states that current follows the path of least time because it is the path of maximum probability.

E.3 DERIVATION OF THE EQUIVALENCE BETWEEN OHM'S LAW AND FERMAT'S LEAST TIME PRINCIPLE

The derivation commences from Ohm's Law of Materials:

$$\vec{J} \triangleq \sigma \vec{E} \tag{E.4}$$

where σ is the conductivity of the medium, $\sigma = \rho^{-1}$, and the current density, $\vec{J}$ is expressed as

$$\vec{J} = ne < \vec{u} > \tag{E.5}$$

In Equation (E.5), ne represents the electron volume (Vol) charge density and is given by

$$ne = \frac{dQ}{dVol} = \frac{dQ}{A \cdot d\ell} \tag{E.6}$$

where:
A = cross section of the current channel passing an average charge particles velocity $<\vec{u}>$
$d\ell$ = distance traveled along the electric field, $\vec{E} <u> \cdot dt$ (refer to Figure E.1).

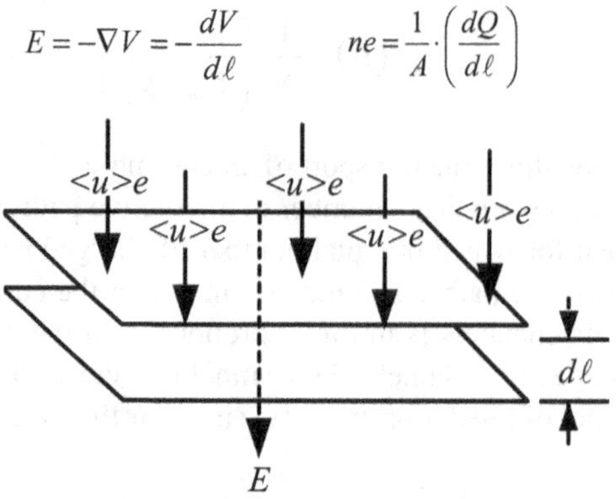

$$E = -\nabla V = -\frac{dV}{d\ell} \qquad ne = \frac{1}{A}\left(\frac{dQ}{d\ell}\right)$$

Figure E.1. Geometry for demonstrating the parameters of the charge propagation channel.

The electric field $\vec{E}$ is responsible for driving the charges from a source region to a drain region through a voltage drop, V, and is given by

$$\vec{E} = -\nabla V = -\frac{dV}{d\ell} \tag{E.7}$$

where ∇ represents the "del" or "grad" operator.

Ohm's Law can therefore be rearranged and is rewritten in the form

$$\vec{J} = \frac{dQ}{A \cdot d\ell} \cdot <\vec{u}> = \left(\frac{dQ}{A}\right)\left(\frac{<\vec{u}>}{d\ell}\right) = \frac{1}{\rho}\vec{E} = -\frac{1}{\rho}\frac{dV}{d\ell} \tag{E.8}$$

This equation can again be rearranged to obtain the equation for the incremental time to transport the charges through the path $d\ell$:

$$\frac{d\ell}{<\vec{u}>} = \left(\rho \cdot \frac{1}{A}\right)\left(\frac{dQ}{\vec{E}}\right) = -\left(\frac{dQ}{dV}\right)\left(\rho \cdot \frac{1}{A} \cdot d\ell\right) \tag{E.9}$$

Equation (E.9) can now be integrated to obtain the equation for the incremental time required to transport the charges through the distance $d\ell$. Note that since the electron charge is negative, the incremental time in Equation (E.9) is positive. Integrating both sides of Equation (E.9) with respect to distance, ℓ, we obtain

$$\tau = \frac{\ell}{<u>} = \left(\rho \cdot \frac{1}{A}\right)\left(\frac{Q}{E}\right) = -\left(\frac{dQ}{dV}\right)\left(\rho \cdot \frac{\ell}{A}\right) \tag{E.10}$$

Also, observe that the right-hand term in Equation (E.10) is the classical expression for the electrical resistance, R, of the current path:

$$R = \rho \cdot \frac{\ell}{A} \tag{E.11}$$

Hence, Equation (E.10) reduces to

$$\tau = \frac{\ell}{<u>} = \left(\frac{-dQ}{dV}\right) \cdot R \tag{E.12}$$

In any given medium, the time to transport a charged particle is the distance traveled divided by its velocity in that medium. Any ensemble of particles is characterized by a velocity stochastic distribution of its constituents and in a given medium the majority of the particles travel at an average velocity, $<\vec{u}>$. Therefore, for the majority of charged particles in an ensemble, the average transport time $\tau(p)$ is

$$\tau(p) = \frac{\ell}{<u>} \tag{E.13}$$

Note that Equation (E.13) is equivalent to the left-hand term of Equation (E.10), considering the transport time $\tau(p)$ as a stochastic process. Equation (E.13) thus corresponds to the key equation in the LT/MP theory.

E.4   EQUIVALENCE OF OHM'S LAW AND THE LT/MP THEORY

In a typical circuit, the factor Q/E appearing in Equation (E.10) is constant. Equation (E.10) can be minimized from the equivalence of Ohm's Law (minimum resistance) and Fermat's Principle (least time). Recalling that capacitance, C, is defined as

$$C \triangleq \frac{dQ}{dV} \tag{E.14}$$

It follows therefore that

$$\min\{\tau\} = \min\left\{\frac{\ell}{<u>}\right\} = \min\left\{\left(\frac{-dQ}{dV}\right)\left(\rho \cdot \frac{\ell}{A}\right)\right\} = \min\{C \cdot R\} \tag{E.15}$$

Consequently, τ (Fermat's Minimum Time) = $C \times R$ (Ohm's Minimum Resistance)

Equation (E.15) illustrates an unequivocal correspondence between minimum time and minimum resistance. The derivation to present a direct correspondence with the least impedance is practically the same, and follows the following reasoning, casting the transmission line equation into the framework of statistics can easily demonstrate this equivalence:

$$V' = \frac{dV}{dx} = I \cdot Z \tag{E.16}$$

Solving for the current, I, yields

$$I = \frac{V'}{Z}$$ (E.17)

In a given electric field, $\vec{E} = -\nabla V = -(dv/dx)$, the path of maximum probability where current, I, is at its maximum, the impedance of the channel, Z, must, therefore, be at its minimum. Therefore, in conclusion:

1. Least Time $\rightarrow$ Maximum Probability

   (Theory of Fourier transform, noting that probability $\leftrightarrow$ frequency of occurrence)
2. Maximum Probability $\rightarrow$ Maximum Current

   (Fourier Transforms $\rightarrow$ Maxwell's Equations through Ohm's Law in Materials)
3. Maximum Current $\rightarrow$ Least Impedance

   (Maxwell's Equations $\rightarrow$ Transmission Line Theory)

BIBLIOGRAPHY

[1] Briët, R., "Application of the LT-MP Principle to the Theory of Lightning Propagation," *Interference Technology Engineering Master (ITEM)*, West Conshohocken, PA: ITEM Publications,, 1997.

APPENDIX **F**

OVERVIEW OF *S* PARAMETERS

In many places throughout this book, particularly in Chapter 9, *S* parameters are used to describe performance of circuits and systems. This appendix provides a short overview of *S* parameters and their applications to the extent necessary for the understanding of the material included in this book. For a more detailed understanding of *S* parameters and the theoretical background of their derivation, the reader is encouraged to refer to the many resources available on the topic.

F.1 BACKGROUND

Historically, an electrical network comprises a "black box" containing various interconnected basic electrical circuit components or lumped elements such as resistors, capacitors, inductors, and transistors. Elementary circuit theory provides many methods for describing complex electronic networks. Those methods, however, best describe DC and low-frequency circuits. They fall short when wavelengths of the signals of interest shrink to become comparable to the physical dimensions of the circuit of interest.

To better characterize high-frequency circuits, scattering parameters or *S* parameters are commonly used. *S* parameters constitute properties of electrical, electronic, and communication systems engineering for describing the electrical behavior of typically high-frequency (electrically large) linear electrical networks when subjected to

Grounds for Grounding. By Elya B. Joffe and Kai-Sang Lock
Copyright © 2010 The Institute of Electrical and Electronics Engineers, Inc.

various steady-state stimuli by small signals.

Many electrical properties of networks or components may be expressed using *S* parameters, such as gain, return loss, voltage standing wave ratio (VSWR), network stability, and reflection coefficient. The term "scattering," as related to electromagnetic wave propagation, commonly refers to the effect observed when a plane electromagnetic wave is incident on an obstruction or passes across dissimilar dielectric media. In the context of *S* parameters, scattering refers to the way in which traveling current and voltage waveforms in a transmission line are affected as they interact with discontinuities along the transmission line, for instance, due to the presence of a lumped electrical network. Such networks may include many typical communication system components or "blocks," such as amplifiers, attenuators, filters, and couplers, provided that they operate under linear and defined conditions.

S parameters change with the frequency, thus, frequency dependence must be included for any *S* parameter characterization of a network in addition to its characteristic impedance or system impedance. *S* parameters are readily represented in matrix form and obey the rules of matrix algebra.

F.2 PORTS AND INTERACTION MATRICES

The external behavior of any "black box" can be predicted without regard for its contents. The black box could contain anything, ranging from a resistor, a transmission line, or an integrated circuit (Figure F.1). Network analysis can be significantly simplified when circuit nodes are grouped into appropriate pairs, yielding the concept of ports. In fact, knowledge of the internal circuit topology is not required for utilizing the port concept.

An electrical network or "black box" to be characterized using *S* parameters may have any number, *N*, of ports. Ports are the points at which electrical currents either enter or exit the network. Sometimes these are referred to as pairs of "terminals"; thus a two-port network is equivalent to a four-terminal network. Figure F.1 shows a simple network with just two ports.

Any linear two-port (and multiport) network ("black box") can thus be characterized by a number of equivalent circuit parameters associated with interactions occurring at each of its ports so as to produce a simple matrix representation of the internal circuitry, such as their transfer matrix, impedance matrix, admittance matrix, and scattering matrix. A network may have any number of ports. A two-port network can be

Figure F.1. Concept of a "black box" and its associated ports.

represented by a 2×2 matrix. In general, an N-port network having N pairs of nodes can similarly be represented by an $N \times N$ matrix. Figure F.2 shows a typical two-port network.

All interaction matrices relate the signal incident upon a port of a network or "black box" to that leaving from each port. With an RF signal incident on one port, a fraction of the signal bounces back out of that port, whereas another portion of it scatters and exits other ports (possibly even amplified), while some of it is dissipated as heat or is emitted as electromagnetic radiation. Power, voltage, and current can be considered to be in the form of waves traveling in both directions (Figure F.3).

The transfer matrix, also known as the $ABCD$ matrix, relates the voltage and current at Port 1 to those at Port 2, whereas the impedance matrix relates the two voltages V_1 and V_2 to the two currents I_1 and I_2:*

$$\text{Transfer Matrix} \quad \begin{bmatrix} V_1 \\ I_1 \end{bmatrix} = \begin{bmatrix} A & B \\ C & D \end{bmatrix} \begin{bmatrix} V_2 \\ I_2 \end{bmatrix} \tag{F.1}$$

$$\text{Impedance Matrix} \quad \begin{bmatrix} V_1 \\ V_2 \end{bmatrix} = \begin{bmatrix} Z_{11} & Z_{12} \\ Z_{21} & Z_{22} \end{bmatrix} \begin{bmatrix} I_1 \\ -I_2 \end{bmatrix} \tag{F.2}$$

Thus, the transfer (T) and impedance (Z) matrices for two-port networks are the 2×2 matrices:

$$T = \begin{bmatrix} A & B \\ C & D \end{bmatrix} \quad Z = \begin{bmatrix} Z_{11} & Z_{12} \\ Z_{21} & Z_{22} \end{bmatrix} \tag{F.3}$$

The admittance matrix is simply the inverse of the impedance matrix, $Y = Z^{-1}$.

F.3   THE SCATTERING MATRIX AND S PARAMETERS

S parameters are members of a family of similar impedance or admittance two-port parameters used in two-port theory, such as $ABCD$, Z, and Y parameters. Analogous to

*In Figure F.X, I_2 flows out of port 2 and, hence, $-I_2$ flows into it. In the usual convention, both currents I_1 and I_2 are taken to flow into their respective ports.

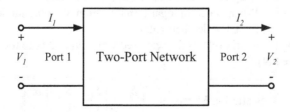

Figure F.2. Representation of a two-port network.

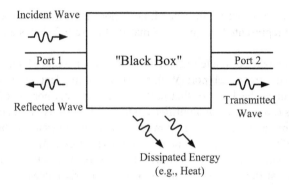

<u>Figure F.3.</u> A fraction of the signal incident at Port 1 reflects back out of this port and another portion of the signal exits through other ports (e.g., Port 2).

the Y or Z parameter, they describe the performance of a two-port network completely. They differ from these, however, in the sense that S parameters do not make use of open-circuit (O.C.) or short-circuit (S.C.) conditions to characterize a linear electrical network but, rather, relate to the traveling waves that are incident, scattered, or reflected when a two-port network is imbedded into a transmission line of a certain characteristic impedance, Z_0. Furthermore, the waves' quantities are expressed in terms of power and can be interpreted in terms of normalized voltage or current amplitudes.

S parameters are conceptually simple, analytically convenient, and capable of providing detailed insight into a measurement and modeling problem. Still, it must kept in mind that similar to all other two-port parameters, S parameters are linear by default, that is, they represent the linear behavior of the two-port network.

F.3.1   The Scattering (S) Matrix

The scattering (S) matrix describing an N-port network contains N^2 elements or S parameter coefficients, each representing a possible input–output interaction. The number of rows and columns in an S matrix is equal to the number of ports. For the S-parameter subscripts ij, j stands for the excited port (the input port) and i denotes the output port.

At each frequency, each element or S parameter is expressed as a unitless complex number, representing magnitude and angle, or amplitude and phase, commonly presented in polar form. When stated in logarithmic fashion, the magnitude of the S parameter is typically expressed in decibels (dB).

The scattering (S) matrix relates the scattered waves b_1 and b_2 to the incident waves a_1 and a_2 (Figure F.4):

$$\text{Scattering Matrix} \qquad \begin{bmatrix} b_1 \\ b_2 \end{bmatrix} = \begin{bmatrix} S_{11} & S_{12} \\ S_{21} & S_{22} \end{bmatrix} \begin{bmatrix} a_1 \\ a_2 \end{bmatrix} \tag{F.4}$$

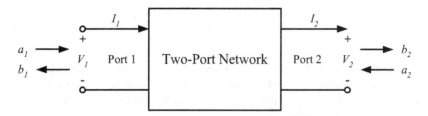

<u>Figure F.4.</u> Representation of incident and reflected waves in a two-port network.

Thus, the scattering (S) matrix for a two-port network is the 2×2 matrix

$$S = \begin{bmatrix} S_{11} & S_{12} \\ S_{21} & S_{22} \end{bmatrix}$$

(F.5)

The S parameter matrix for the two-port network is probably the most common and it serves as the basic building block for generating the higher order matrices for larger networks.

Parameters along the leading diagonal of the S-matrix, S_{22} and S_{11}, are referred to as reflection coefficients because they refer to the reflection occurring at one port only. Off-diagonal S parameters, S_{12} and S_{21}, are referred to as transmission coefficients because they refer to what happens from one port to another.

The normalized, traveling incident (applied) power waves for each port are designated by the letter an, whereas the normalized reflected power wave is designed by b_n, where n is the port number of the network. These can be expressed in general form as

$$a_n = \frac{1}{2\sqrt{Z_0}} \cdot (V_n + Z_0 I_n)$$

$$b_n = \frac{1}{2\sqrt{Z_0}} \cdot (V_n - Z_0 I_n)$$

(F.6)

In a two-port system the traveling wave variables a_1 and b_1 at Port 1 and a_2 and b_2 at Port 2 are defined in terms of the port voltages and currents, V_1 and I_1 and V_2 and I_2 and a real-valued positive reference impedance Z_0 as follows:

$$a_1 = \frac{V_1 + Z_0 I_1}{2\sqrt{Z_0}} \quad a_2 = \frac{V_2 + Z_0 I_2}{2\sqrt{Z_0}}$$

$$b_1 = \frac{V_1 - Z_0 I_1}{2\sqrt{Z_0}} \quad b_2 = \frac{V_2 + Z_0 I_2}{2\sqrt{Z_0}}$$

(F.7)

The definitions at Port 2 appear to be different from those at Port 1, but they are actually the same if expressed in terms of the incoming current, $-I_2$:

$$a_2 = \frac{V_2 - Z_0 I_2}{2\sqrt{Z_0}} = \frac{V_2 + Z_0\left(-I_2\right)}{2\sqrt{Z_0}} \tag{F.8}$$

Similar to the T and Z parameters, S parameters are frequency dependent, so when looking at the formulae for S parameters it is important to note that frequency dependence is implied (even if not explicitly written). For this reason, S parameters are often called complex scattering parameters.

F.3.2 S_{21}, or "Forward Transmission Gain/Loss"

S_{21} refers to the signal exiting at Port 2 for the signal incident at Port 1 expressed as the ratio of the two waves, b_2 and a_2 (Figure F.5):

$$S_{21} = \left.\frac{b_2}{a_1}\right|_{a_2=0} = \frac{\text{Transmitted power and Port 2}}{\text{Incident power and Port 1}} \tag{F.9}$$

The greatest value of S_{21} to a circuit or system designer is to indicate how much frequency-dependent forward transmission gain or loss may be expected at a given frequency. The plot of the magnitude of S_{21} versus frequency allows a straightforward comparison of lossy structures.

For a two-port network, S_{21} represents the complex linear gain, G:

$$G = S_{21} \tag{F.10}$$

S_{21}, therefore, corresponds to linear ratio of the output voltage divided by the input voltage, all values expressed as complex quantities. The magnitude is given in linear form by

$$|G| = |S_{21}| \tag{F.11}$$

The scalar logarithmic (decibel or dB) expression for gain (g) is more commonly used than the scalar linear gain, and is expressed as

$$g = 20\log_{10}|S_{21}|, \text{dB} \tag{F.12}$$

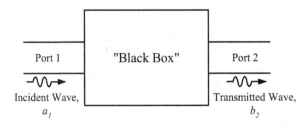

Figure F.5. Illustration of S_{21}, or "forward transmission gain/loss."

$g > 0$ is normally understood to represent "gain," whereas $g < 0$ is interpreted as "negative gain" or, more commonly, as "loss" equivalent to its magnitude in decibels.

F.3.3 S_{11}, or "Input Return Loss"

S_{11} refers to the signal reflected at Port 1 for the signal incident at Port 1 expressed as the ratio of the two waves, b_1 and a_1 (Figure F.6):

$$S_{11} = \frac{b_1}{a_1}\bigg|_{a_2=0} = \frac{\text{Reflected power and Port 1}}{\text{Incident power and Port 1}} \tag{F.13}$$

Engineers most often use S_{11} to compare the quality of electrically short structures that constitute impedance discontinuities, such as connectors and vias. When designing a transmission path, a lower return loss is preferable, because it indicates that less energy is reflected due to impedance mismatches and their resultant discontinuities to the energy flow.

Input return loss (RL_{in}) is a scalar measure of how close the actual input impedance of the network is to the nominal system impedance value and, when expressed in logarithmic magnitude, is given by

$$RL_{in} = \left|20\log_{10}\left|S_{11}\right|\right|, \text{dB} \tag{F.14}$$

By definition, return loss is a positive scalar quantity implying the two pairs of magnitude (|) symbols.

F.3.4 S_{22}, or "Output Return Loss"

The output return loss (RL_{out}) has a similar definition to the input return loss but applies to the output port (Port 2) instead of the input port and is expressed as the ratio of the two waves, b_2 and a_2 (Figure F.7):

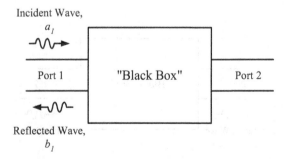

Incident Wave,
a_1

Port 1   "Black Box"   Port 2

Reflected Wave,
b_1

Figure F.6. Illustration of S_{11}, or "input return loss."

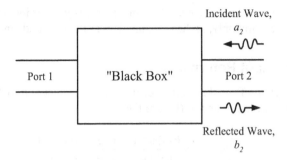

Incident Wave,
a_2

Reflected Wave,
b_2

Figure F.7. Ilustration of S_{22}, or "output return loss."

$$S_{22} = \frac{b_2}{a_2}\bigg|_{a_1=0} = \frac{\text{Reflected power and Port 2}}{\text{Incident power and Port 2}} \qquad \text{(F.15)}$$

When expressed in logarithmic magnitude, the output return loss is given by

$$RL_{out} = \left|20\log_{10}\left|S_{22}\right|\right|, dB \qquad \text{(F.16)}$$

F.3.5 S_{12}, or "Reverse Gain and Reverse Isolation"

S_{12} refers to the signal exiting at Port 1 for the signal incident at Port 2 expressed as the ratio of the two waves, b_1 and a_2 (Figure F.8):

$$S_{12} = \frac{b_1}{a_2}\bigg|_{a_1=0} = \frac{\text{Transmitted power and Port 1}}{\text{Incident power and Port 2}} \qquad \text{(F.17)}$$

The greatest value of S_{21} to a circuit or system designer is to indicate how much frequency-dependent forward transmission gain or loss may be expected at a given frequency. The plot of the magnitude of S_{21} versus frequency allows a straightforward comparison of lossy structures.

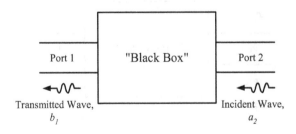

Transmitted Wave,
b_1

Incident Wave,
a_2

Figure F.8. Illustration of S_{12} or "reverse gain/isolation."

When expressed in logarithmic magnitude, the reverse gain is given by

$$g_{rev} = 20\log_{10}|S_{12}|, \text{dB} \tag{F.18}$$

Often, this expression will be used to represent reverse isolation (I_{rev}), in which case it becomes a positive quantity equal to the magnitude of g_{rev} and the expression becomes

$$I_{rev} = |g_{rev}| = |20\log_{10}|S_{12}||, \text{dB} \tag{F.19}$$

F.4   CHARACTERISTIC VALUES OF *S* PARAMETERS

The magnitude of S_{11} and S_{22} is always less than 1. Otherwise, it would represent a negative ohmic value. On the other hand, the magnitude of S_{21} and S_{12} can exceed the value of 1 in the case of active amplification. Also, S_{21} and S_{12} can be positive or negative. If they are negative, this implies the existence of a phase shift. Table F.1 is a summary of the interpretation of the different values of the *S* parameters.

F.5   *S* PARAMETERS IN LOSS-FREE AND LOSSY NETWORKS

F.5.1   The Loss-Free Network

A loss-free network is one that does not dissipate any power; therefore, the sum of the incident power at all ports is equal to the sum of the reflected power at all ports:

$$\sum_{n}|a_n|^2 = \sum_{n}|b_n|^2 \tag{F.20}$$

Table F.1. Interpretation of *S* parameters

S_{11} and S_{22}	Interpretation
−1	Incident voltage wave is reflected and inverted (load = short circuit, $Z_L = 0\ \Omega$)
0	Impedance matching; no reflections (matched load, $Z_L = Z_0$)
+1	Incident voltage wave is reflected (load = open circuit, $Z_L = \infty\Omega$)
S_{21} and S_{12}	Interpretation
0	No signal transmission
0... +1	Input signal is damped in the Z_0 environment
+1	Unity gain signal transmission in the Z_0 environment
> +1	Input signal is amplified in the Z_0 environment

This implies that the S-matrix is unitary, or

$$(S)^H (S) = (I)$$

(F.21)

where $(S)^H$ is the conjugate transpose of (S) and (I) is the identity matrix.

F.5.2   Lossy Networks

A lossy passive network is one in which the sum of the incident power at all ports is greater than the sum of the reflected power at all ports. It, therefore, dissipates power, thus:

$$\sum_n |a_n|^2 \neq \sum_n |b_n|^2$$

(F.22)

In this case, $\sum_n |a_n|^2 > \sum_n |b_n|^2$ and $(S)^H(S) \leq (I)$.

F.5.3   Insertion Loss

Insertion loss (IL), usually also represented in decibels, is given by

$$IL = -10\log_{10} \frac{|S_{21}|^2}{1 - |S_{11}|^2}, \text{dB}$$

(F.23)

Since insertion loss is, by definition, a loss (or negative gain), the leading negative sign is often neglected. Insertion loss is commonly confused with g (above). The difference between the two is that g penalizes the network for mismatch at the input, whereas insertion loss, IL, is not a function of the input or source impedance. This point can be further made clear using the following expressions:

$$g = 20\log_{10}\left(\frac{P_{out}}{P_{av}}\right)$$

$$IL = 20\log_{10}\left(\frac{P_{out}}{P_{in}}\right)$$

(F.24)

P_{av} represents the power available from the source, whereas P_{in} stands for the power incident into Port 1 of the network.

F.5.4   Radiation Loss

In networks in which power is "lost" in the form of radiated emissions, the radiation loss, LR, can be expressed using S_{11} and S_{12}, respectively (note that S_{11} and S_{12} are expressed in their linear, nondecibel, values):

$$L_R = 1 - |S_{11}|^2 - |S_{21}|^2 \qquad \text{(F.25)}$$

This expression implies that the RF energy "lost" to the environment is the fraction of the incident that is not transmitted forward into the network (S_{12}) or reflected at the input port (S_{11}). This represents a situation different from a closed lossless system, in which the total combination of reflection coefficient and transmission coefficient equals unity.

BIBLIOGRAPHY

[1]   Paul, C. R., *Introduction to Electromagnetic Compatibility*, 2nd ed., New York: Wiley, 2006.

INDEX